MODERN CONTROL SYSTEM THEORY AND DESIGN

MODERN CONTROL SYSTEM THEORY AND DESIGN

SECOND EDITION

Stanley M. Shinners

Program Manager
Lockheed Martin Federal Systems, Inc.
Great Neck, New York

and

Adjunct Professor of Electrical Engineering
The Cooper Union for the Advancement of Science and Art
New York, New York

A WILEY-INTERSCIENCE PUBLICATION

JOHN WILEY & SONS, INC.

New York / Chichester / Weinheim / Brisbane / Toronto / Singapore

This book is printed on acid-free paper. ♾

Published simultaneously in Canada.

Library of Congress Cataloging-in-Publication Data:

Shinners, Stanley M.
Modern control system theory and design / Stanley M. Shinners – 2nd ed.
p. cm.
Includes index.
ISBN 0-471-24906-8 (cloth : alk. paper)
1. Automatic control. 2. Control theory. I. Title.
TJ213.S455443 1998 97-34538
629.4–dc21 CIP

Printed in the United States of America.

10 9 8 7 6 5 4 3 2

To my wife, Doris,
and to
Sharon, Stuart, Jessica, Jonathan,
Walter, Laurie, Jason, Rebecca
and
Daniel

CONTENTS

PREFACE

The goal of the Second Edition of *Modern Control System Theory and Design* is to present a unified treatment of conventional and modern continuous control systems, and to show how to apply the theory presented to realistic design problems. My main objective has been to produce an easily understandable book from the reader's viewpoint. It is written in an integrated manner so one can clearly proceed from the first to the last chapter. Having been a visitor for many years for the Accrediting Board for Engineering and Technology (ABET), I can appreciate the problems faced by professors who are trying to meet the design requirements of control-system courses. This book addresses this problem by including many realistic design examples.

This book is intended for an introductory course in control systems and for the practicing control-system engineer who wishes to master control systems through self-study. Its contents correspond to courses that I have taught at universities and in industry. It reflects my industrial experience with these corporations, which is coupled with extensive teaching experience in both university and industrial programs.

New Enhancements to the Second Edition

This Second Edition contains many of the new developments and applications in the control-system field that have occurred since the 1992 publication date of the first edition. The following are the enhancements and new features which have been incorporated into the Second Edition of *Modern Control System Theory and Design*:

1. Addition of a new chapter (Chapter 8) entitled "Modern Control-System Design Using State-Space, Pole Placement, Ackermann's Formula, Estimation, Robust Control, and H^{∞} Techniques".

No other control-system textbook contains all of this recent material presented in a cohesive, easy to understand, manner for undergraduate students and practicing engineers.

2. Addition of new sections at the end of each chapter containing illustrative problems and their solutions which cover the material in each chapter.

These sections include completely worked out illustrative problems which, as well as covering the salient features of the material presented in the individual chapters, emphasize recent applications. *These problems constitute an integral part of the text.*

3. MATLAB™ (MATLAB is a trademark of The MathWorks, Inc.) is integrated into the text in the Second Edition, in addition to the free software contained in the *Modern Control System Theory and Design Toolbox* which complements the

book and contains MATLAB solutions to computational problems, and can be retrieved free from The MathWorks anonymous File Transfer Protocol (FTP) server at **ftp://ftp.mathworks.com/pub/books/shinners**. The Modern Control System Theory and Design Toolbox available with the Second Edition of this book is designed to run with MATLAB for Windows (and predecessor versions), and the Student Edition of MATLAB for Windows (and predecessor versions).

No prior knowledge of MATLAB™ is assumed in this book. MATLAB is now considered a control-system industry standard. All computational problems are solved with MATLAB.

In addition to the solutions to computational problems, the *Modern Control System Theory and Design* Toolbox software contains the following: (a) a Tutorial File explaining the MATLAB fundamentals, notations, and the use of the software; (b) a Demonstration M-file which gives an overview of the various utilities; (c) a synopsis file that reviews and highlights features of each chapter. Together, the integrated learning package and the MATLAB software are self-contained, so it is not necessary to purchase additional books/materials to learn how to use MATLAB, the *Modern Control System Theory and Design* Toolbox, or the various applications within this book. In this manner, the user of this book has available a toolbox of features/utilites created to enhance The MathWorks' Control System Toolbox, optionally utilize The MathWorks' Simulink, Nonlinear and Symbolic Toolboxes, as well as supply the computer-generated solutions of the problems in the book. The *Modern Control System Theory and Design* Toolbox runs equally well with The Student Edition of MATLAB as it does with the professional version of MATLAB (with or without The MathWorks' Control System Toolbox, Simulink, the Nonlinear Toolbox, or the Symbolic Toolbox). Accordingly, the resultant presentation of this book is an integration of computer-aided software engineering (CASE) and computer-aided design (CAD) techniques with the control-system analysis and design methods illustrated.

For those who do not have MATLAB, a feature of this book is a complete set of working digital-computer programs containing their logic flow diagrams, listings, and representative output based on the programs' application to practical design problems.

4. Many new problems have been added at the ends of the chapters. Coupled with the additional worked out illustrative problems added to the new sections in each chapter, this greatly increases the total number of problems in the Second Edition.

5. The Solutions Manual has been updated to contain solutions to new problems at the ends of the chapters. Answers contained in the Answers to Selected Problems in the back of the book have been increased to contain answers to the remaining problems at ends of chapters.

Primary Features of Modern Control System Theory and Design

1. *Unifies and blends the conventional and modern approaches.* It uses the conventional block-diagram-transfer function and state-variable methods in parallel throughout the book. Modern control-system design techniques using the state-space approach, pole placement, estimation, robust control, and H^∞ are presented in addition to the classical Nyquist, Bode, Nichols, and root-locus methods. In

addition, the modern state-space, pole placement, Ackermann's formula for pole placement, estimation, robust control, and the H^∞ method present ideas for design projects at the undergraduate level and ideas for specialization and future research at the graduate level.

2. *Emphasizes design.* The development of theoretical topics is coupled with clear applications of the theory in engineering design. Recognizing that control theory is interdisciplinary and cuts across all specialized engineering fields, I have presented modern illustrations and practical problems from the fields of robotics, space-vehicle systems, aircraft, submarines, hydrofoils, servomechanisms, economics, management, biomedical engineering, and nuclear reactor systems. This should prove to be of interest to electrical, mechanical, aerospace, system, chemical, nuclear, biochemical, and industrial engineers.

3. *Presents a complete set of working digital computer programs and summarizes commercially available software packages available for control system analysis in addition to the many solutions presented in the text using MATLAB.* Although computer techniques are extremely valuable to the student and the practicing engineer, I believe one must first understand the details of the techniques being applied so that computer results can be properly interpreted and judged for their analytical reasonableness. Therefore, I present basic theory while applying it to design problems with and without computer solutions.

4. *Covers a variety of topics of recent importance.* In addition to covering modern state-space design techniques including pole placement, Ackermann's formula for pole placement, estimation, robust control, and the H^∞ method for design, the presentation also includes controllability, observability, and linear-state-variable feedback (including the design of controllers, estimators, and compensators).

5. *Adapts to the student's training, ability, and to various curricula.*

6. *Contains problems (with answers to approximately one-third).* This is helpful to the student and also to the practicing control-system engineer using the book for self-study. The available MATLAB computer-generated solutions will guide the reader in understanding the approach used for solving the problems.

7. *Offers an accompanying Solutions Manual (available to professors) containing detailed solutions to the remaining problems whose answers are not found in this book. As an aide to the professors using this book, methods for enchancing the students' learning environment, a listing of control system journals, and design chart/graphical templets are also contained in the Solutions Manual.*

By means of these features, *Modern Control Systems Theory and Design* is a computer-oriented, state-of-the-art book that is comprehensive and unique and fills a rather large gap in the existing literature on feedback control-system design.

Chapter Organization

Chapter 1 introduces the concept of open-loop and closed-loop control systems, and presents the state of the developments of the control-system field today, with a preview of the future. Chapter 2 reviews Fourier methods, the Laplace transform, and the transfer-function method. In addition, it introduces the signal-flow graph, state-variable concepts, and the state transition matrix. A tutorial on MATLAB is introduced into the text in Chapter 2 for the reader who has had no prior knowledge of MATLAB. MATLAB programs for obtaining partial-fraction expansion and transforming between the state-space form and the transfer function are developed. The transfer-function and state-variable representation of several common devices found in control systems are derived in Chapter 3. The concepts of conservation and analogy are also introduced in this chapter. Chapter 4 focuses attention on second-order systems because they occur so frequently, most basic transient response definitions are based on second-order systems, and many higher-order systems can be approximated as second-order systems. It also provides guidelines for selecting the optimum damping ratio of a specific system. A method for modelling the transfer functions of control systems from available data is presented.

Chapter 5 addresses control-system sensitivity and accuracy, and presents various performance criteria such as the ITAE criterion. A comprehensive treatment of techniques for determining stability, together with several useful and practical examples, is given in Chapter 6. The viewpoints presented include the state-variable, the Routh–Hurwitz, Nyquist and Bode-diagrams, Nichols-chart, and root-locus methods. MATLAB programs for obtaining the Nyquist and Bode diagrams, Nichols chart, and root locus are developed and applied. In addition, non-MATLAB working digital-computer programs are also developed and applied. A review of commercially available software packages is also presented. A comparison is presented of the Nyquist-diagram, Bode-diagram, Nichols-chart, and root-locus methods for 12 commonly used transfer functions; guidelines are presented for using these stability methods. The concepts of stability, presented in Chapter 6, are applied in great detail to the design of linear control systems in Chapter 7. Single-degree and two-degrees of-freedom compensation techniques using cascade-compensation and minor-loop feedback compensation techniques are developed and applied. Phase-lag, phase-lead, and phase-lag–lead networks are illustrated, in addition to proportional-plus-integral-plus-derivative (PID) compensators.

This volume concludes with Chapter 8, a new chapter for the Second Edition, which is dedicated to presenting modern control-system design using state-space methods, including pole-placement design using linear-state-variable-feedback, Ackermann's formula for design using pole placement, and estimator design in conjunction with the pole placement approach using linear-state-variable feedback. The concepts of controllability and observability are introduced. Robust control is developed and applied to a variety of problems. The modern H^{∞} control concepts are introduced, developed, and applied. The chapter concludes with a presentation of linear algebraic aspects of control-system design computations, and several illustrative problems and solutions.

The flexibility of *Modern Control System Theory and Design* in adapting to the course level, the various curricula, and the student's training and ability is quite evident from this content summary. Its design orientation will also make it an excellent choice for the practicing engineer.

The Learning Package

The following educational materials comprise the learning package for the Second Edition of *Modern Control System Theory and Design*, and supplement and enhance this textbook:

1. *Modern Control System Theory and Design Toolbox*, available free from The MathWorks Inc. anonymous FTP server at ftp://ftp.mathworks.com/pub/books/shinners. This software contains MATLAB solutions to the computational problems in this book. In addition, the *Modern Control System Theory and Design* Toolbox software contains the following features:

    - Tutorial File explaining the MATLAB fundamentals, notations, and the use of the software.
    - Demonstration M-File which gives an overview of the various utilities.
    - Synopsis file that reviews and highlights features of each chapter.

    The *Modern Control System Theory and Design* Toolbox runs equally well with the professional and Student Edition of MATLAB. By virtue of the *Modern Control System Theory and Design* Toolbox, and the fact that MATLAB is also integrated into this textbook throughout, it is not necessary for the reader to buy a supplementary book to learn how to use MATLAB.
2. *Solutions Manual for Modern Control System Theory and Design*, second edition, by Stanley M. Shinners contains the detailed solutions to approximately 66% of the problems at the ends of each chapter. The *Solutions Manual* is available only to qualifying faculty members.

ACCOMPANYING COMPANION VOLUME

An accompanying companion volume has been written to *Modern Control System Theory and Design*, titled *Advanced Modern Control System Theory and Design*. It serves as an excellent book for a follow-up advanced college course to the introductory course which this book serves. In addition, it can serve as an excellent text for practicing control system engineers who need to learn more advanced control system subjects which are required to perform their tasks. *Advanced Modern Control System Theory and Design* begins with a review of introductory control system analysis concepts and is then followed by Chapters 7 and 8 of the first volume and includes the following chapters:

- *Digital Control-System Analysis and Design* which extends the continuous concepts presented to discrete systems.
- *Nonlinear Control-System Design* which extends the linear concepts presented in this volume to nonlinear systems.
- *Introduction to Optimal Control Theory and its Applications* presents such important topics as dynamic programming and the maximum principle, and applies it to the space attitude control problem and the lunar soft-landing problem.
- *Control-System Design Examples: Complete Case Studies* presents the complete case studies of five control system design examples which illustrate practical design projects.

Advanced Modern Control System Theory and Design contains the important subjects that a professor requires in a follow-up course. It is also very important and useful to the practicing control system engineer who had only an introductory control system course as an undergraduate, and now must be concerned with digital control systems, nonlinearities and the other advanced techniques for a design project. This volume also contains these same features as *Modern Control System Theory and Design*:

- Sections at the end of each chapter containing illustrative problems and their solutions, separate from a problems section at ends of each chapter.
- MATLAB is also integrated into this textbook. In addition, free software which complements the book and contains MATLAB solutions to computational problems can be retrieved from The MathWorks anonymous FTP server.
- A Solutions Manual is available to professors and contains solutions to more than one-half of the problems at the end of each chapter. Answers to the remaining problems are contained at the end of the book.

ACKNOWLEDGMENTS

I am most grateful to my wife Doris for her encouragement, understanding, patience, and word-processing assistance throughout this project. In addition, I express thanks and appreciation to my parents for their efforts, encouragement, and inspiration. I thank the editor of my book at John Wiley & Sons, Mr. George Telecki, who managed the smooth production of the book, and to Lisa Van Horn of John Wiley & Sons for her professional production of this book.

STANLEY M. SHINNERS

Jericho, New York
March 1998

MODERN CONTROL SYSTEM THEORY AND DESIGN

1

GENERAL CONCEPT OF CONTROL-SYSTEM DESIGN

1.1. INTRODUCTION

The desire of people to control nature's forces successfully has been the catalyst for progress throughout history. Our goal has been to control these forces in order to help perform physical tasks which were beyond our own capabilities. During the dynamic and highly motivated 20th century, the control-system engineer has transformed many of our hopes and dreams into reality.

Control-system engineers have made very important contributions to our advancements in the 20th century, and they are building the foundation for greater advancements as we approach the 21st century. As we look back, control-system engineers have made contributions to robotics; space-vehicle systems, including the successful accomplishment of the lunar soft landing; aircraft autopilots and controls; control systems for ships and submarines; guidance systems for intercontinental missiles; and automatic control systems for hydrofoils, surface-effect ships, high-speed rail systems; and, most recently, control systems for the magnetic-levitation rail systems. The future lies in our imagination, creativity, and ability to transform ideas into working automatic control systems that we can build to work reliably and accurately, and which can be manufactured at a profit and on schedule. It is hoped that this book contributes to the attainments of our future goals and dreams in the advancement of automatic control systems.

The control of systems is an interdisciplinary subject and cuts across all specialized engineering fields. This book recognizes this fact and presents illustrations of control systems from the fields of electrical, mechanical, aeronautical, chemical, nuclear, economics, management, bioengineering, and other related fields. The versatile subject of automatic control ranks today as one of the most promising fields, and its growth potential appears unlimited.

Control systems can be defined as devices which regulate the flow of energy, matter, or other resources. Their arrangement, complexity, and appearance vary with their purpose and function. In general, control systems can be categorized as being either open loop or closed loop. The distinguishing feature between these two types of control systems is the use of feedback comparison for closed-loop operation.

Properties characteristic of open-loop and closed-loop control systems are discussed in this chapter. We shall give several examples of each type so that the reader may gain a thorough understanding and a good foundation for further studies in this book. Included will be a qualitative, philosophical comparison between the behavior of closed-loop control systems and that of living creatures. The feedback concept can also be applied to model our economic system. A discussion of modern control-system applications will conclude the presentation of this chapter.

1.2. OPEN-LOOP CONTROL SYSTEMS

Open-loop control systems represent the simplest form of controlling devices. Their concept and functioning are illustrated by several simple examples in this section.

Figure 1.1 illustrates a simple tank-level control system. We wish to hold the tank level h within reasonable acceptable limites even though the outlet flow through valve V_1 is varied. This can be achieved by irregular manual adjustment of the inlet flow rate by valve V_2. This system is not a precision system, as it does not have the capability of accurately measuring the output flow rate through valve V_1, the input flow rate through valve V_2, or the tank level. Figure 1.2 shows the simple relationship that exists in this system between the input (the desired tank level) and the output (the actual tank level). This signal-flow representation of the physical system is called a *block diagram*. Arrows are used to show the input entering and the output leaving the control system. This control system does not have any feedback comparison, and the term *open loop* is used to describe this absence.

The angular position of a missile launcher being controlled from a remote source is illustrated in Figure 1.3. Commands from a potentiometer located at a remote location activate the positioning of the missile launcher. The control signal is amplified and drives a motor which is geared to the launcher. The block diagram corre-

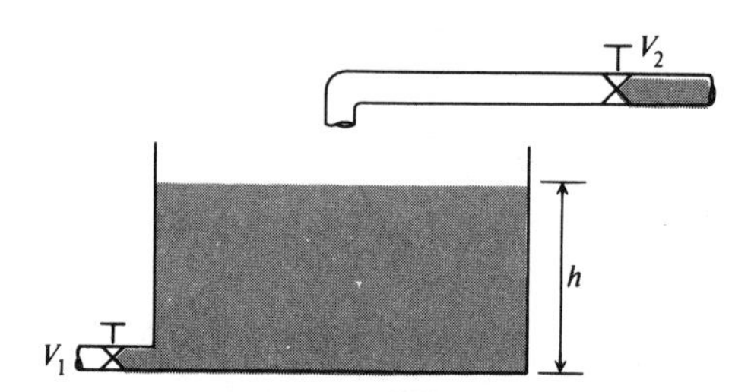

Figure 1.1 Tank-level control system.

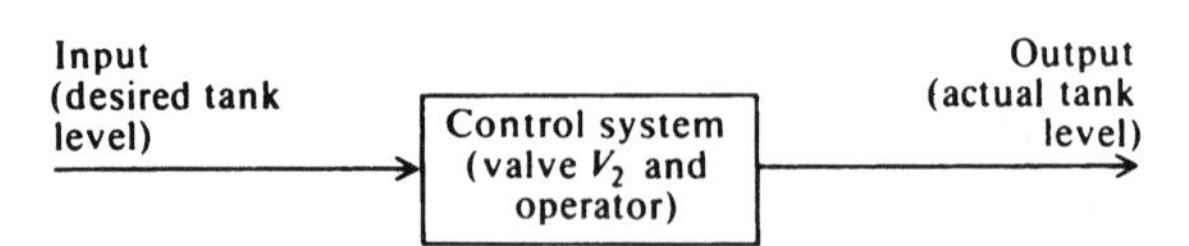

Figure 1.2 Tank-level control-system block diagram.

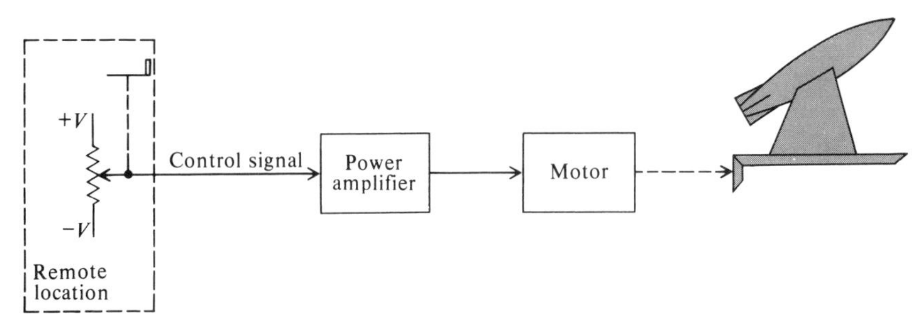

Figure 1.3 Controlling the position of a missile launcher from a remote location.

sponding to Figure 1.3 is illustrated in Figure 1.4. The input is the desired angular position of the missile launcher, the output is the actual angular position of the missle launcher, and the control system consists of the potentiometer, power amplifier, motor, gearing between the motor and missile launcher, and the missile launcher. For accurate positioning, the missile launcher should be precisely calibrated with reference to the angular position of the potentiometer, and the characteristics of the potentiometer, amplifier, and motor should remain constant. Except for the potentiometer, the components that comprise this open-loop control system are not precision devices. Their characteristics can easily change and result in false calibration and poor accuracies. In practice, simple open-loop control systems are never used for the accurate positioning of fire-control systems because of the inherent possibility of inaccuracies and the risks involved.

Figure 1.5 illustrates a field-controlled dc motor turning a cutting wheel at a constant speed. When a piece of wood is applied to the surface of the cutting wheel, it acts as a disturbing torque to the driving torque of the motor and results in a reduction of the speed of the cutting wheel, assuming that the control signal remains constant. This situation can be represented as shown in Figure 1.6. The symbol appearing between the motor and the load represents a subtractor.

The effect of disturbance torques, or other secondary inputs, is detrimental to the accurate functioning of an open-loop control system. It has no way of automatically correcting its output, because there is no feedback comparison. We must resort to changing the input manually in order to compensate for secondary inputs.

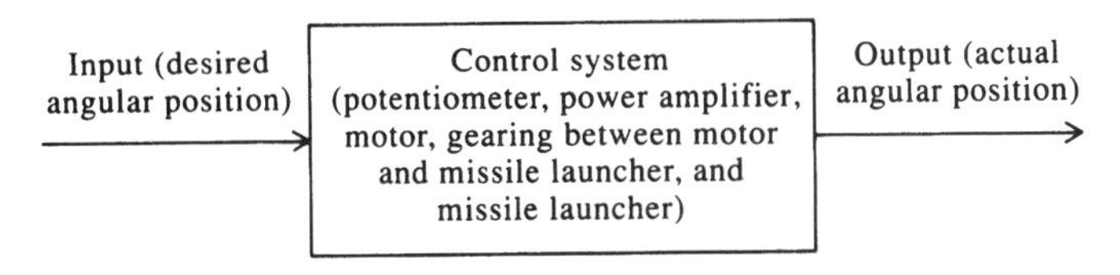

Figure 1.4 Missile launcher control-system block diagram.

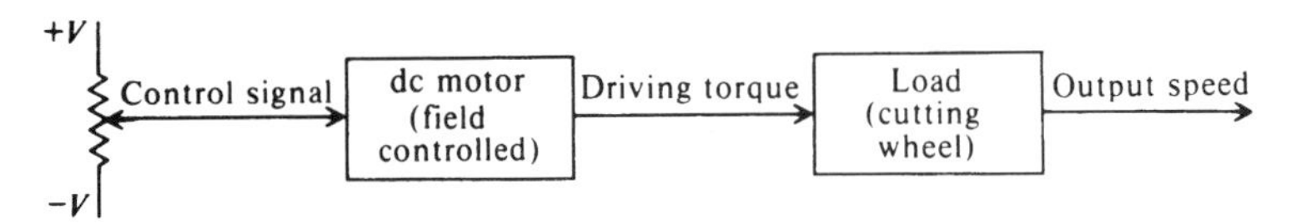

Figure 1.5 Field-controlled dc motor.

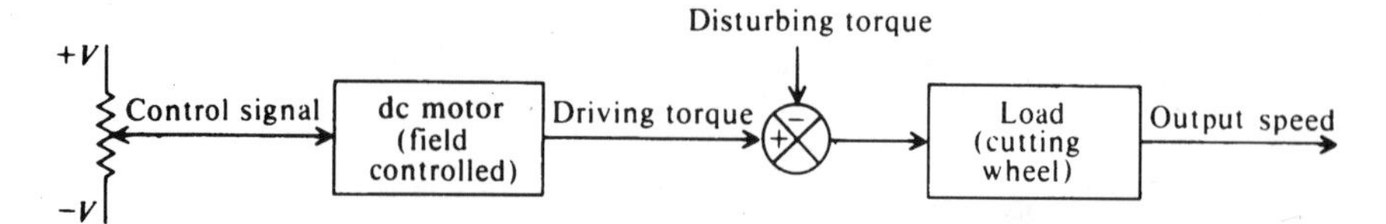

Figure 1.6 Field-controlled dc motor having a disturbance torque.

1.3. CLOSED-LOOP CONTROL SYSTEMS

Closed-loop control systems derive their valuable accurate reproduction of the input from feedback comparison. An error detector derives a signal proportional to the differences between the input and output. The closed-loop control system drives the output until it equals the input and the error is zero. Any differences between the actual and desired output will be automatically corrected in a closed-loop control system. Through proper design, the system can be made relatively independent of secondary inputs and changes in component characteristics. This section illustrates the closed-loop control system versions of the open-loop control systems considered in Section 1.2.

Figure 1.7 illustrates an automatic tank-level control version of the system shown in Figure 1.1. It can maintain the desired tank level h within quite accurate tolerances even though the output flow rate through valve V_1 is varied. If the tank level is not correct, an error voltage is developed. This is amplified and applied to a motor drive which adjusts valve V_2 in order to restore the desired tank level by adjusting the inlet flow rate. A block diagram analogous to this system is shown in Figure 1.8. Because feedback comparison is present, the term *closed loop* is used to describe the system's operation.

Figure 1.9 illustrates an automatic missile launcher position control version of the system shown in Figure 1.3. This feedback system can be designed to position the launcher quite accurately on commands from potentiometer R_1. Potentiometer R_2 feeds a signal back to the difference amplifier, which functions as an error detector. Should an error exist, it is amplified and applied to a motor drive which adjusts the output-shaft position until it agrees with the input-shaft position, and the error is zero. The block diagram shown in Figure 1.8 is also applicable to this system. The input would be the desired angular position, the output would be the actual angular

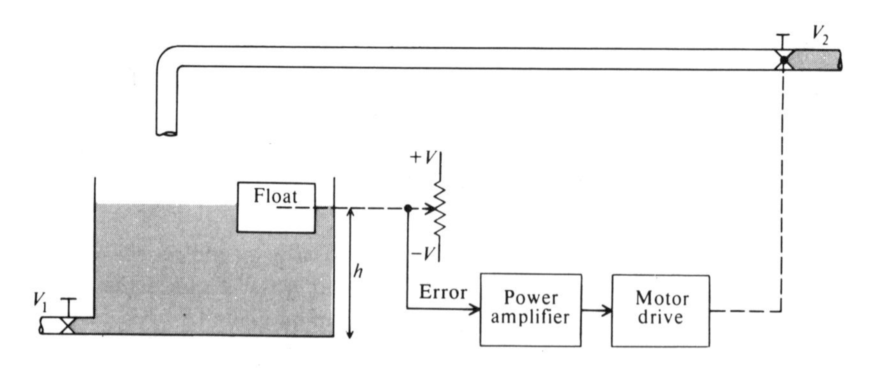

Figure 1.7 Automatic tank-level control system.

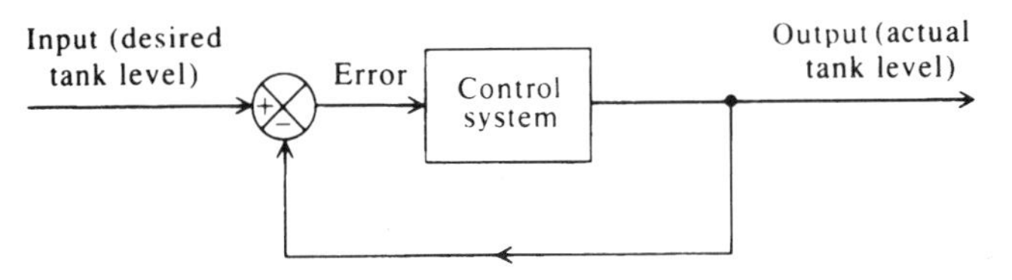

Figure 1.8 Block diagram of a closed-loop system.

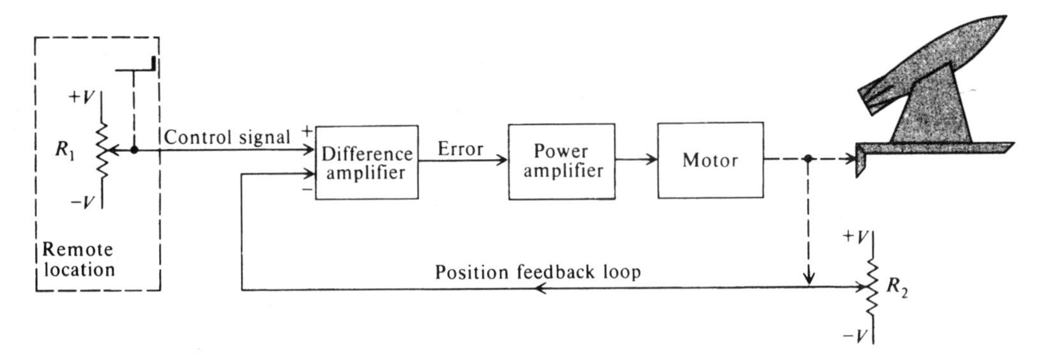

Figure 1.9 An automatic positioning system for a missile launcher.

position, and the control system would consist of the potentiometer power, amplifier, motor gearing between the motor and missile launcher, and the missile launcher.

An automatic speed-control version of a field-controlled dc motor, which was shown in Figure 1.6, is illustrated in Figure 1.10. This feedback sysem has the capability of maintaining the output speed relatively constant even though disturbing torques may occur. A tachometer, which functions as a transducer that transforms speed to voltage, is the feedback element for this control system. Should the output speed differ from the desired speed, the difference amplifier develops an error signal which adjusts the field current of the motor in order to restore the desired output speed.

Feedback control systems used to control position, velocity, and acceleration are very common in industrial and military applications. They have been given the special name of *servomechanisms*. With all their advantages, feedback systems have a very serious disadvantage, because the closed-loop system may inadvertently act as an oscillator. Through proper design, however all the advantages of feedback

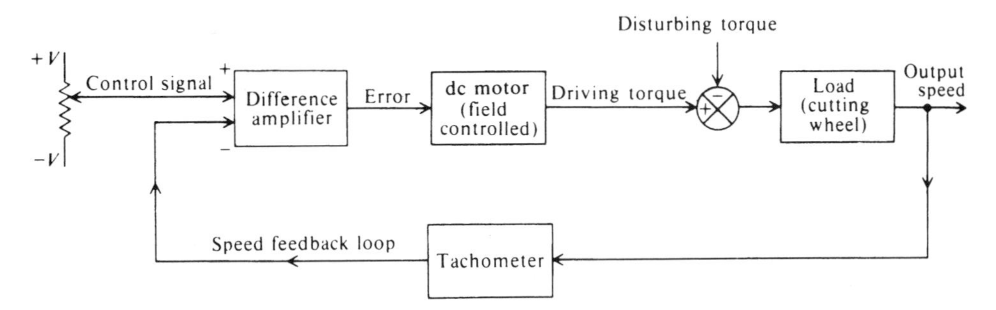

Figure 1.10 Automatic speed control for a field-controlled dc motor.

can be utilized without having an unstable system. A major task of this book is to determine how this may be accomplished for several kinds of systems.

1.4. HUMAN CONTROL SYSTEMS

The relation between the behavior of living creatures and the functioning of feedback control systems has recently gained wide attention. Wiener [1] implied that all systems, living and mechanical, are both information and feedback control systems. He suggested that the most promising techniques for studying both systems are information theory and feedback control theory.

Several characteristics of feedback control systems can be linked to human behavior. Feedback control systems can "think" in the sense that they can replace, to some extent, human operations. These devices do not have the privilege of freedom in their thinking process and are constrained by the designer to some predetermined function. Adaptive feedback control systems which are capable of modifying their functioning in order to achieve optimum performance in a varying environment, have gained wide attention. These systems are a step closer to the adaptive capability of human behavior [2].

The human body is, indeed, a very complex and highly perfected adaptive feedback control system. Consider, for example, the human actions required to steer an automobile. The driver's objective is to keep the automobile traveling in the center of a chosen lane on the road. Changes in the direction of the road are compensated for by the driver turning the steering wheel. His objective is to keep the differences between the output (the car's actual position on the road) and the input (the car's desired position on the road) as close to zero as possible.

Figure 1.11 illustrates the block diagram of the feedback control system involved in steering an automobile. The error detector in this case is the brain. This in turn activates the driver's muscles, which control the steering wheel. Power amplification is provided by the automobile's steering mechanism, which controls the position of the front wheels. The feedback element represents the human's sensors (visual and tactile). Of course, this description is very crude—any attempt to construct a mathematical model of the process should somehow account for the adaptability of the human being and the effects of learning, fatigue, motivation, and familiarity with the road.

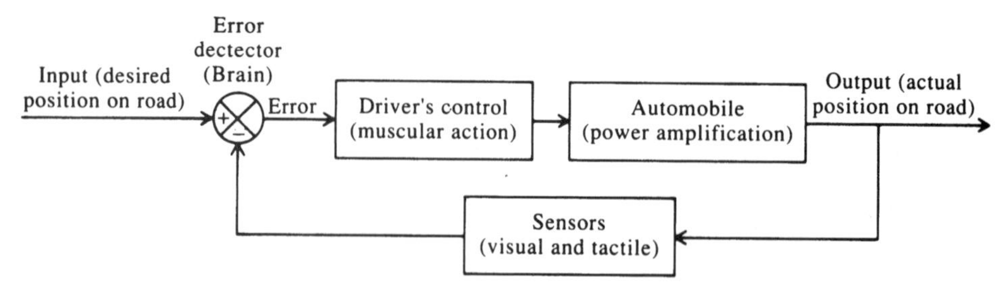

Figure 1.11 Steering of an automobile—a feedback control system involving a human.

1.5. MODERN CONTROL-SYSTEM APPLICATIONS WITH A PREVIEW OF THE FUTURE

Feedback control systems are to be found in almost every aspect of our daily environment. In the home, the refrigerator utilizes a temperature-control system. The desired temperature is set and a thermostat measures the actual temperature and the error. A compressor motor is utilized for power amplification. Other applications of control in the home are the hot-water heater (see Problem 1.3), the central heating system, and the oven, which all work on a similar principle. We also encounter feedback control systems when driving our automobile. Automatic control systems are used for maintaining constant speed (cruise control), constant temperature (climate control) steering, suspension, engine control, and to control skidding (anti-skid sysem) [3–5].

In industry the term *automation* is very common. Modern industrial plants utilize robots for obtaining temperature controls, pressure controls, speed controls, position controls, etc. The chemical process control field is an area where automation has played an important role [6]. Here, the control engineer is interested in controlling temperature, pressure, humidity, thickness, volume, quality, and many other variables. Areas of additional interest include automatic warehousing [7] and inventory control [8] and automation of farming [9].

In this section, I present the state of the control-system field by illustrating its application in the following important aspects of engineering: robotics, space travel, commercial rail and air transportation, military systems, surface effect ships, hydrofoils, biomedical control systems and photographic automatic focus control systems. In addition, the use of control theory to model systems unrelated to control is illustrated. Illustrations will be given of current applications and future plans of control-system design. Because a feature of this book is the application of the theory presented to practical design examples, these illustrations will also serve as models for the problems used throughout the book.

A. Robotics

Robotics has been a very important field for the application of control sytems. In the 1960s robots began to be recognized as important devices to aid manufacturing; their application to a wide variety of manufacturing systems has mushroomed since then. Control theory is needed for obtaining the desired motion or force needed; the field of robotics also depends on the use of sensors for vision, and computers for programming these devices to accomplish their desired tasks.

Robots have been created to perform a wide variety of tasks spanning the fields of manufacturing, space vehicle missions, and hospital care. A major application of robotics has been in automating the manufacturing process. Figure 1.12 illustrates a line of automobiles and robots at the Ford Motor Company's Chicago assembly plant. This is the same plant which produced the legendary Model T. Following the plant's modernization program, it now produces the Ford Taurus and Mercury Sable car lines. Some of the new high-technology additions at the plant include robotic sealing of side windows and vehicle underbodies, automatic feeding of car parts to the assembly process, and new overhead conveyors.

Robots have also been incorporated into NASA's space shuttle. Figure 1.13 is an artist's concept of the Canadian-built remote manipulator system (RMS), which is a

Figure 1.12 A line of automobiles and robots at Ford Motor Company's Chicago Assembly Plant which produced the legendary Model T. (Courtesy of the Ford Motor Company)

robot arm-like device. The robot arm is shown lifting the West German MBB shuttle pallet satellite (SPAS-01) from the cargo bay of the Earth-orbiting space shuttle *Challenger*.

Robots have also been applied to helping people. Figure 1.14 illustrates HelpMate^{TM}, made by Transitions Research Corporation (TRC), which is a mobile robotic materials transport system designed to perform mundane fetch-and-carry tasks that would ordinarily be performed by nurses and aides in hospitals and long-term-care facilities [10]. HelpMate^{TM} is a self-guided robot which relieves highly trained personnel to carry out more important duties. This robot uses vision, ultrasonic and infrared proximity sensors, and dead reckoning to navigate through hallways, ride elevators, and avoid obstacles encountered along its route. It can transport special-request meal trays, laboratory samples and specimens, pharmacy medications, patient medical records, administrative reports, etc. [11].

Analysis and design of robots similar to these illustrations are contained in several examples in this book.

B. NASA Space Programs and Associated Trainers

At the forefront of the space program is the International Space Station. Figure 1.15 shows an artist's concept of the baseline configuration of the International Space Station. It is an international space complex comprised of the permanently manned

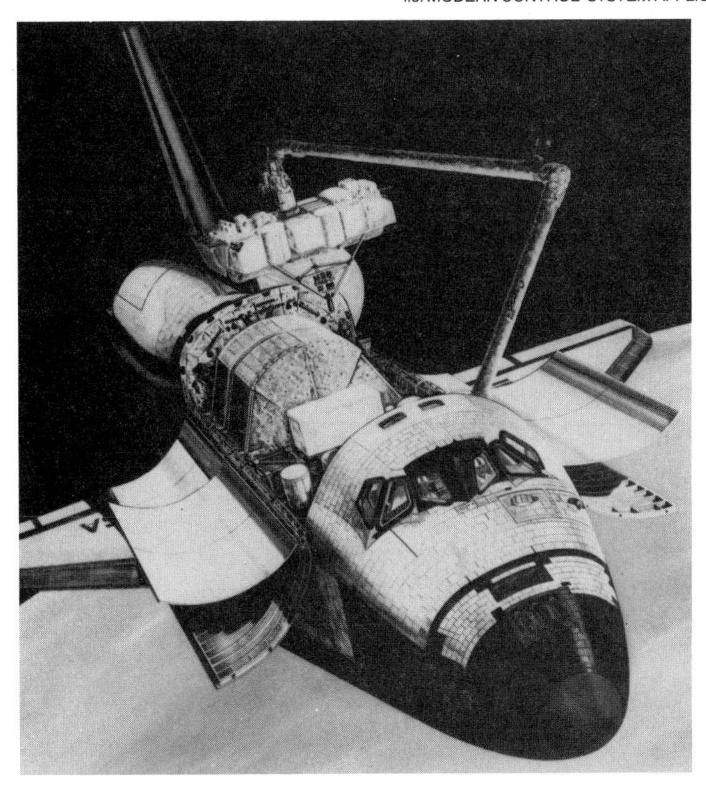

Figure 1.13 Artist's concept of the Canadian-built remote manipulator system (RMS), a robot arm-like device. The robot arm is shown lifting the West Germany MBB shuttle pallet satellite from the cargo bay of the Earth-orbiting space shuttle *Challenger*. (Official NASA photo)

orbiting and unmanned scientific platforms to be placed into orbit. Overall configuration is about 500 ft across, 200 ft tall, and 145 ft deep. The illustration shows a space shuttle orbiter being used to bring up sections of the station for assembly in orbit by astronauts. An eight-person crew will live and work aboard the manned base's three laboratory modules—one each from the United States, Japan, and the European Space Agency—and the United States-provided habitation module, which was designed for use for up to six months at a time. Canada is providing the Mobile Servicing System, which features a robot arm three times stronger than the shuttle's remote manipulator discussed previously under robotics. Ultimately, the International Space Station will serve as a gateway to the stars, providing an in-orbit assembly, test, and departure point for piloted missions back to the Moon, to Mars, and beyond. This book analyzes several applications of space vehicles. Figure 1.16 illustrates the block diagram of the attitude control system of a typical space vehicle. Chapter 11 is dedicated to optimal control theory, which is applied to the space-attitude-control problems concerned with minimizing response time, fuel consumption, and energy consumption of space vehicles. The lunar soft-landing probe, is synthesized from the viewpoint of optimal control theory in Chapter 11, as well.

An important aspect needed to support the success of the NASA Shuttle and the International Space Station is training the astronauts to successfully perform their missions. Figure 1.17 is a photograph of the NASA Shuttle Mission Simulator

Figure 1.14 Robot HelpMate^{TM}, which is a mobile robotic materials transport system used to perform mundane fetch-and-carry tasks in hospitals and nursing homes. (Courtesy of TRC)

Motion-based trainer which simulates the shuttle cockpit with full six-degree-of-freedom motion cues. This simulator enables astronauts to train for the operation of the mission. Additionally, it must train for assembling the space station using the shuttle robotic manipulator arm (see Figure 1.13), controlling the embryonic space station through station-compatible communications, docking with the space station, and transferring equipment and personnel. This simulator provides the capability to train astronauts and for mission-control ground crews to control the space shuttle through launch, ascent, orbit, de-orbit, and landing as well as for in-orbit activities such as experiment activation/deactivation, and satellite deployment and maintenance.

C. Rail Transportation

Great advances have been made in rail and air transportation, and more are planned. Magnetic-levitation (MAGLEV) train systems are in various stages of planning, design, test, and limited operation in Japan, Berlin, and the United

Figure 1.15 Artist's concept of the International Space Station. (Official NASA photo)

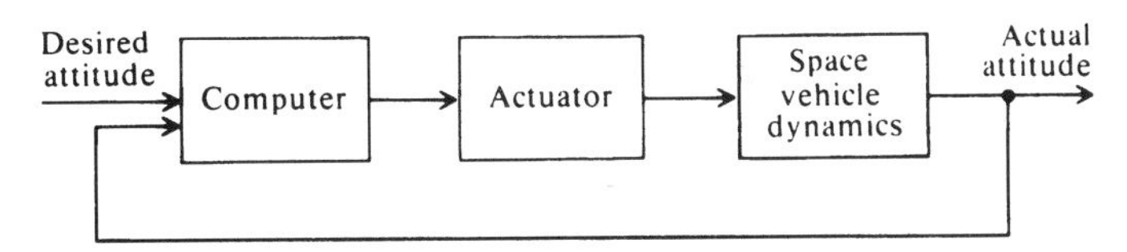

Figure 1.16 The attitude-control system of a typical space vehicle.

States [12,13]. MAGLEV's main advantages over conventional rail systems are its relative lack of noise and vibration, lower power consumption, and lower maintenance costs, because there is no need for components such as a rotating motor [13]. In Japan, a MAGLEV system is being constructed for a 40-km-long test track in Yamanashi prefecture, 70 percent of which will be a tunnel. The 500-kph, or 310-mph, electrodynamically suspended (repulsion mode) technology used by Japan's MAGLEV, named the Linear Express, will be tested under operational conditions [14]. The test track could form part of a new Chuo line, supplying service between Tokyo and Osaka early in the next century. It is estimated that it would consume half the energy of an airplane in moving the same amount of people [15].

The Central Japan Railway Company's Linear Express MAGLEV technology makes use of superconductivity magnets which function on the principle of magnetic repulsion [16]. The superconducting magnets are attached to the bottom of the train, and the U-shaped guideway has coils which have a current induced in them as the

Figure 1.17 NASA Shuttle Mission Motion-based trainer simulates the shuttle cockpit with full six-degree-of-freedom motion cues. (Courtesy of: Hughes Training, Inc./Arlington, TX)

train gains speed. The coils then become magnetic and repel the magnets on the bottom of the train. The Linear Express uses these principles to levitate 10 cm above the ground. (Regular magnets could only elevate the train 1 cm, which would mean that high-precision tracks and control systems would be necessary.) It is equipped with rubber wheels which function until the vehicle reaches 100 kph, when the repulsion is sufficient to elevate the train (in a manner similar to the takeoff of an airplane.) The Linear Express obtains forward propulsion along the tracks using magnets having south and north poles arranged in an alternating manner. The attraction and repulsion of the magnets, which are changed continuously by computers so that repulsion forces act from behind and attraction forces act from the front, propel the train forward. Figure 1.18 is a photograph of the Linear Express (serial 002) which was used on a test track on the island of Kyushu.

Figure 1.19 is a photograph of Japan's new Series 100 "bullet" high-speed trains, or Shinkansen, which have greatly facilitated intercity travel in Japan [17–20]. The

Figure 1.18 Photograph of the magnetic levitation train, the Linear Express, test vehicle (serial 002) which was used on a test track on the island of Kyushu, Japan. (Courtesy of Japan Railways Group)

Figure 1.19 New Series 100 "bullet" high-speed trains, or Shinkansen, have greatly facilitated intercity travel in Japan. (Courtesy of Japan Railways Group)

Shinkansen long-distance high-speed railways include the Tokaido and Sanyo Shinkansen, which runs southwest from Tokyo, and the Tohoku and Joetsu Shinkansen, which serves the regions to the northeast. The maximum speed for the Tohoku and Joetsu Shinkansen is now 240 kph (148.8 mph). (It is interesting to note that the MAGLEV Linear Express can attain speeds twice as fast as the Shinkansen bullet trains. Energy consumption of the Shinkansen is only 21 percent that of an airplane, and 19 percent that of an automobile [19,20].

The Shinkansen uses automatic verification of signals and brake control. In addition, an automated traffic control system regulates the trains and their routing, permitting safe, high-speed operation.

Figure 1.20 illustrates the general position-control concepts of such a high-speed automated train system [17]. Observe from the block diagram that it contains a position-measuring loop and a velocity-measuring loop. Position can be measured from the rotation of the train's wheels. Speed can be measured by using velocity-sensing devices such as tachometers. The control computer system monitors the positions and speeds of all trains in the system and issues control signals via a high-speed communication system.

D. Air Transportation

In November 1990, McDonnell Douglas delivered the first of its MD-11 jets to the Finnish airline, Finnair Oy [21]. Figure 1.21 is a photograph of the MD-11, and Figure 1.22 is a photograph of its advanced flight deck [22,23]. The MD-11 has a two-seat cockpit with automatic system controllers in the overhead panel that monitor and configure fuel, hydraulic, electrical, and environmental systems. Each system is controlled by two redundant computers. The advanced flight deck features six cathode-ray-tube displays, digital instrumentation, wind-shear detectors and guidance devices, a dual flight management system that helps conserve fuel, and a dual digital automatic flight control system (autopilot) with fail-operational capability. Computerized system controllers perform automated normal, abnormal, and emergency checklist duties for the MD-11's major systems. The flight control com-

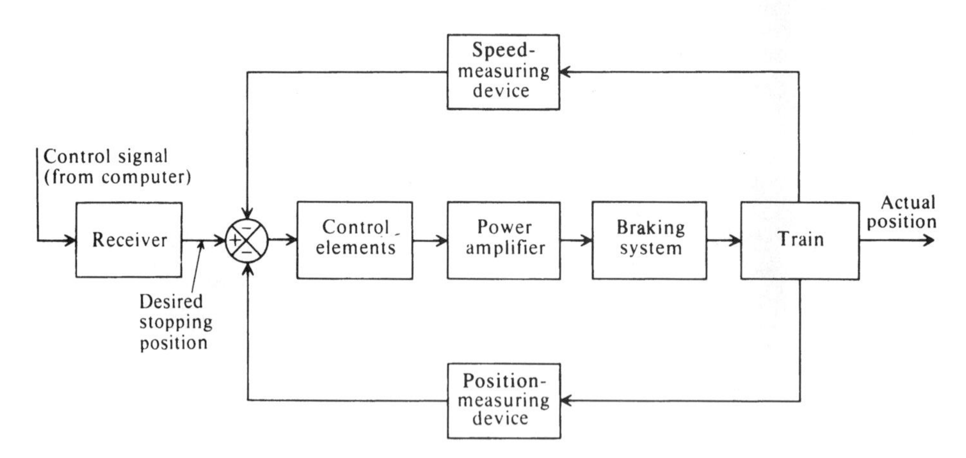

Figure 1.20 Automatic position-control system of a high-speed automated train system.

Figure 1.21 Photograph of the McDonnell Douglas MD-11. (Courtesy of the McDonnell Douglas Corporation)

Figure 1.22 Photograph of the advanced flight deck of the McDonnell Douglas MD-11. (Courtesy of the McDonnell Douglas Corporation)

puter, formerly called the automatic pilot and flight director, includes an automatic throttle and a longitudinal stability augmentation system, which enhances the pitch stability of the aircraft.

Several problems are included in this book that analyze and design the attitude-control systems of aircraft. Figure 1.23 illustrates the block diagram of a typical autopilot system for an aircraft.

E. Tilt Rotor Aircraft

Tilt rotor aircraft which can take off and land vertically, and also fly as a fixed-wing aircraft, have found great use in military applications. They provide the speed and range of a turboprop airplane and the vertical takeoff and landing capability of a helicopter. Figure 1.24 is a photograph of the Bell Boeing V-22 Osprey military tilt rotor which is a twin-turbine, vertical-lift, transport that is used by the U.S. Marine

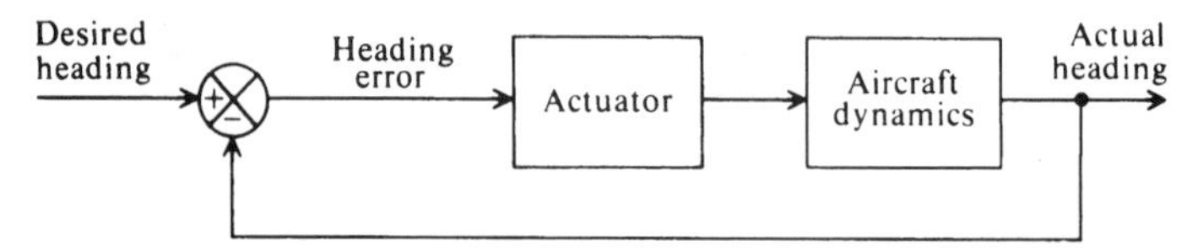

Figure 1.23 A typical autopilot system for an aircraft.

Figure 1.24 Photograph of the Bell Boeing V-22 Osprey military tilt rotor which is a twin-turbine, vertical-lift, transport used by the U.S. Marine Corps, Navy, Air Force, and Army. (Photo by Boeing)

Corps, Navy, Air Force, and Army [24]. Bell Helicopter Textron and the Boeing Company have initiated a program to develop and manufacture the Bell Boeing 609 illustrated in Figure 1.25. It will be a six-to-nine passenger civil tilt rotor aircraft, which builds on technology developed for the V-22 Osprey military tilt rotor, [25, 26]. Anticipated uses for the Bell Boeing 609 are for executive transportation, search-and-rescue, offshore oil operations, emergency medical evacuation, disaster relief, and government support for drug enforcement, and border patrol. It is anticipated that the Bell Boeing 609 will be the first of a range of tilt rotor aircraft for civil and commercial markets which will eventually include large tilt rotors that could be used by airlines to fly passengers between city centers.

F. Helicopters and Associated Trainers

Helicopters are a very important transportation link for the military and for civilian use. As an example, Figure 1.26 is a photograph of the UH-60 Black Hawk combat assault helicopter airlifting M-102 mm howitzers during Operation Desert Shield. How do you train a pilot to fly a complex helicopter like the UH-60 Black Hawk helicopter? Figure 1.27 is a photograph of the UH-60 Black Hawk Flight Simulator which provides Black Hawk pilots and co-pilots with basic, transition, refresher, and advanced training in flight operations, emergency procedures, and combat tactics

Figure 1.25 Photograph of the planned Bell Boeing 609 civil tilt rotor which can be used in such applications as offshore oil exploration. It is estimated that it could reduce the time it takes to ferry personnel and equipment to an oil rig to half the current time it takes helicopters. (Photo by Bell Boeing)

Figure 1.26 A UH-60 Black Hawk Helicopter airlifitng M-102mm howitzers during Operation Desert Shield. (Official U.S. Navy Photo (released))

and advanced combat skills. Figure 1.27 illustrates the trainer station, which moves on a dynamic six-degree-of-freedom motion system, and provides a realistic cockpit environment for the pilot and the co-pilot. The simulated cockpit is an authentic replica of the interior of the production helicopter. All items look, feel, move, and operate as they do in a UH-60 Black Hawk helicopter.

Basic training tasks which can be accomplished in the UH-60 trainer include ground operations, takeoff, hovering, basic/advanced flight, formation flight maneuvers, emergency procedures, and navigation. All UH-60 simulators have an instructor operator station which enables the instructor to initialize the training mission, select and edit initial conditions, position and reset the trainer, insert malfunctions, and program demonstration maneuvers [27].

G. Military Systems

Great advances have been made in the design of military aircraft, ships, and submarines. Figure 1.28*a* is a photograph of the U.S. Air Force's F-15 eagle, an all-weather, extremely maneuverable tactical fighter designed to gain and maintain air superiority in aerial combat [28]. It contains electronic systems and weaponry to detect, acquire, track, and attack enemy aircraft. The avionic system includes an inertial navigation system, a tactical navigation system, an instrument landing system, and a central digital computer. The inertial navigation system permits the Eagle to navigate anywhere in the world.

Figure 1.28*b* is a photograph of the new F-22 which is regarded as the world's most advanced fighter [29]. Plans are to have the F-22 fighter replace the F-15C

Figure 1.27 U.S. Army's Black Hawk (UH-60) Flight Simulator simulates the helicopter and accomplishes all of the functions related to helicopter operations. The trainer station pictured moves on a six-degree-of-freedom motion system and provides a realistic cockpit environment for the pilot and co-pilot. (Courtesy of Hughes Training, Inc., Arlington, TX)

Eagle fighter after the turn of the century. The primary objective of the F-22 is to establish absolute control of the skies through the conduct of counter-air operations. This fighter also has an inherent precision ground attack capability. In order to make the F-22 twice as effective as the F-15, extensive use has been made of computers. The exponential explosion of computer technology in the last 10 years has allowed the F-22 to exploit this capability. For example, the computer used in the Lunar Module (which is analyzed in this book in Chapter 2, Problem 2.46 and illustrated in

Figure 1.28(*a*) Photograph of the U.S. Air Force's F-15 Eagle Aircraft. (Official U.S. Air Force photo)

Figure 1.28(*b*) The F-22 Raptor is unveiled at the Lockheed Martin Aeronautical Systems in Marietta, Georgia. (F-22 Team Photo courtesy of Lockheed Martin Aeronautical Systems)

Figure P2.46, and in Section 11.7) operated at 100,000 operations per second and had 37 kilobytes of memory. Today, the F-22's main mission computers, which are called Common Integrated Procesors or CIPs, operate at 10.5 billion instructions per second and have 300 megabytes of memory. These numbers represent 100 000 times the computing speed and 8000 times the memory of the Apollo moon lander.

Another illustration of the application of advanced automatic control theory to the military is in the area of strategic missile submarines. Figure 1.29 is a bow view of the nuclear-powered strategic submarine USS Tennessee (SSBN-734) underway. The

Figure 1.29 Photograph of the nuclear-powered strategic missile submarine USS Tennessee (SSBN-734). (Official U.S. Navy Photograph)

Trident fitted Strategic Missile Submarine (SSBN) force provides the principle USA strategic deterrent. These huge submarines are 560 ft × 42 ft × 36.4 ft in size and they contain 24 Lockheed Trident II (D5) stellar inertial guidance missiles . Trident missile accuracy is as good as ground-base systems [45]. Control systems are used in almost every aspect for controlling this modern submarine. For example, suppose the submarine is required to maintain a fixed depth automatically. This operation is called "hovering." It is accomplished by an automatic control system which compares the desired depth (derived as a voltage) with the voltage output of a depth detector which operates on the principle that the water pressure increases with depth. This difference is used to control the angular position of the submarine's planes which then controls the submarine's depth through the submarine's dynamics. The block diagram for this control system is illustrated in Figure I1.3ii, which is presented as part of Illustrative Problem I1.3, and in Figure P2.57 which is presented as part of Problem 2.57.

Several examples of analyzing and designing control systems associated with these kinds of military weapon systems are contained in this book (e.g., automatic depth-control systems for submarines).

H. Surface-Effect Ships and Hydrofoils

Modern control has made a major contribution in the control of two new classes of ships: surface-effect ships and hydrofoils. Figure 1.30*a* is a photograoh of U.S. Coast Guard surface-effect ships (WSES). These new patrol craft are used by the U.S. Coast Guard, primarily for enforcement of laws and treaties. Using the surface-effect principle, it can attain a maximum speed which is greater than 30 knots [30]. Lift engines drive fans that create a pressurized air cushion under the vessel; this lifts the vessel and, therefore, reduces drag and draft. The two solid side walls pierce the water surface and form a catamaran hull; the air cushion is sealed by flexible rubberized skirts at the bow and stern.

Hydrofoils are another important illustration of the advancement in the design of ships. Figure 1.30*b* is a photograph of the U.S. Navy hydrofoil USS Pegasus (PHM-1) which is a modern, missile-equipped patrol boat [31]. The PHM class of vessels can be adapted to such diverse roles as antisubmarine warfare, fisheries law enforcement, and the protection of offshore resources. The bow foil provides approximately 31.8 percent of the dynamic lift, and the aft foil provides approximately 68.2 percent. The bow foil uses a strut that rotates in order to provide directional control and reliable turning rates in heavy seas. Control and lift are also provided at takeoff and in flight by flaps connected to the trailing edges. An automatic control system, the helm, and the throttle provide continuous dynamic control during takeoff, foil-borne operation, and landing. Control of the PHM is achieved by sensing its attitude, boat-motion rates, and acceleration. These quantities are then compared to the desired values in a closed-loop feedback system such as those described in Section 1.3. Errors between the desired and actual quantities are then used by hydraulic actuators to reposition the control surfaces (trailing edge flaps on each of the foils and the rotating bow foil strut) in order to achieve the desired quantities.

The control systems used on these kinds of ships are analyzed and designed in several examples in this book.

(a)

(b)

Figure 1.30 (*a*) Photograph of the U.S. Coast Guard Surface-Effect Ships (WSES) (Courtesy of the U.S. Department of Transportation). (*b*) Photograph of the U.S. Navy Hydrofoil, USS Pegasus (PHM-1). (Official U.S. Navy photo)

I. Biomedical Control Systems

There are many applications of control-system concepts in the biomedical field, including prostheses [32–38], and controlling the human heart using a pacemaker. In this section we illustrate the former; the latter is illustrated in an application problem in Chapter 5 (Problem 5.14).

As an illustration of the application of control system concepts to aid above-elbow amputees, let us consider the working of the Boston Elbow [36–40] which is illustrated in Figure 1.31. Use is made of minute electrical signals on the skin in the stump, which are generated by tension in the underlying muscles. These minute signals, which range from 5 to 1000 μV, are denoted as myoelectric, and great care must be given to providing very good instrumentation for their amplification because these minute signals must operate in the presence of much larger external noise caused by interference due to such devices as fluorescent lamps and radio transmitters. The amplified signal is used to drive a battery-powered miniature high-speed electric motor containing a gear reducer to reduce its speed to that compatible with human hand functioning.

The Boston Elbow can be operated in various control-system configurations. For example, it can be controlled in an on-off mode using a logical on-off muscular signal, and it can be operated in a mode in which the signals are proportional to muscle tension. Variation in muscle tension, which results in a proportional myoelectric sygnal, is used as part of a proportional control system. Because an amputee uses the muscles in much the same way as before the amputation, a small amount of tension will cause the Boston Elbow to move slowly, and increased muscle activity will make it move faster. Because the prosthesis has no feeling, both elbow position and grasp must be contrtolled by visual feedback. Therefore, the feedback loop is closed by the humans visual feedback in a manner similar to that discussed in Section 1.4 for driving an automobile.

The Boston Elbow has also been operated in a closed-loop positioning control system mode on an experimental basis [37,38]. In the application of such a configuration to a prosthetic hand, the actual output signal is the finger position. For generation of the feedback signal, the finger drives a second transducer. Differences in signals between the finger's feedback transducer and the input transducer are amplified to drive the motor. For example, if the transducer is moved one-third of the way, the hand will settle at a position one-third open. This closed-loop configuration has the advantage that the user can feel the position of the body part

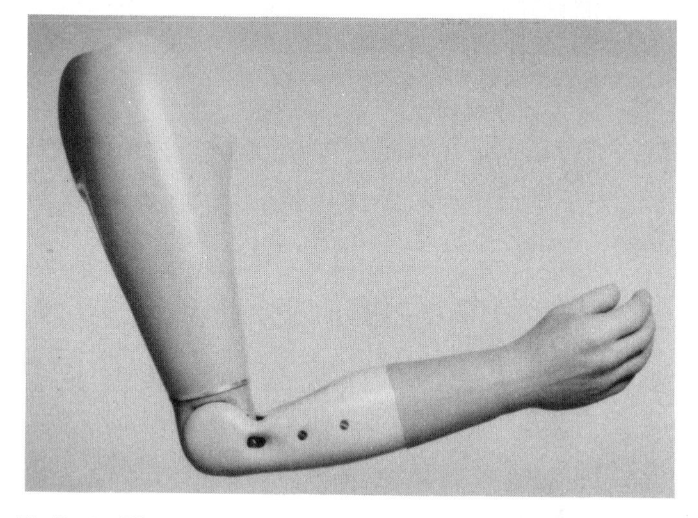

Figure 1.31 The Boston Elbow uses myoelectric signals on the skin in the stump, which are generated by tension in the underlying muscles, for controlling purposes. (Courtesy of the Liberty Mutual Insurance Group, Boston)

operating the control-system transducer. Therefore, the user's hand will open to a given position without having to use visual feedback.

J. Photographic Automatic Focus Control Systems [41, 42]

It is not necessary to have to consider robotics, aircraft, space, and military systms to illustrate examples of automatic control systems. Automatic control systems are also used in a very wide array of everyday applications such as your camera. Figure 1.32*a* is a cutaway view of the Nikon N6006 35-mm camera which uses a charge-coupled device (CCD) that converts images into electrical signals as part of its automatic

Figure 1.32(*a*) Cutaway view of the Nikon N6006 35-mm camera which shows portions of its automatic focusing system. (Courtesy of Nikon Inc.)

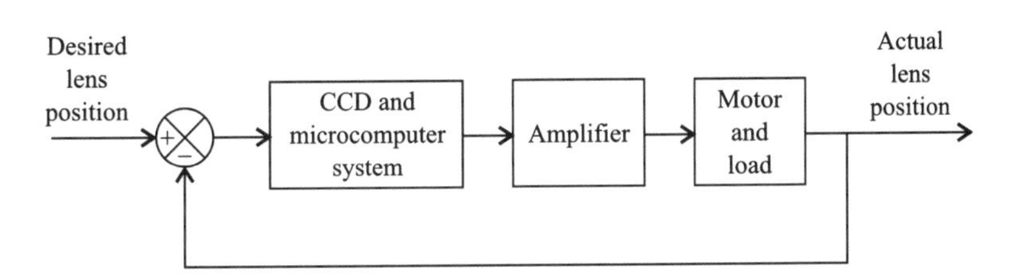

Figure 1.32(*b*) Functional block diagram representation of an automatic focusing system.

focusing system. At the core of the N6006's automatic focus is Nikon's Advanced AM200 Autofocus Sensor Module, which is an autofocus detection system, containing a high-density 200 CCD sensor which can detect detail even in dim light conditions. Automatic focusing systems are implemented by focusing the center of the image on a CCD array through two lenses. Focusing, which is related to the separation of the two images on the CCD, is sensed by the camera and a microcomputer drives the lens and focuses the image. Automatic focusing systems compare the actual position of the lens with the desired position of the lens which is obtained by pointing the camera at the object. Figure 1.32*b* shows a block diagram representation of automatic focusing control systems which compare the desired and actual lens positions. The control system contains a microcomputer, a CCD, an amplifier, and a motor.

K. Modeling Systems Unrelated to Control

Automatic control theory has also been applied for modeling the feedback processes of many systems unrelated to control. An example of this is the modeling of our economic system in order to understand it better. Our economic system contains many feedback systems and regulatory agencies. Figure 1.33 illustrates a crude model of the economics concerned with national income, government policy on spending, private business investment, business production, taxes, and consumer spending. This type of feedback model assists the analyst in understanding the overall effects of government policy and private business investment on the national income.

Several other examples of using control theory to model systems unrelated to control are contained in the book (e.g., model of a fish pond in Problem, 2.59).

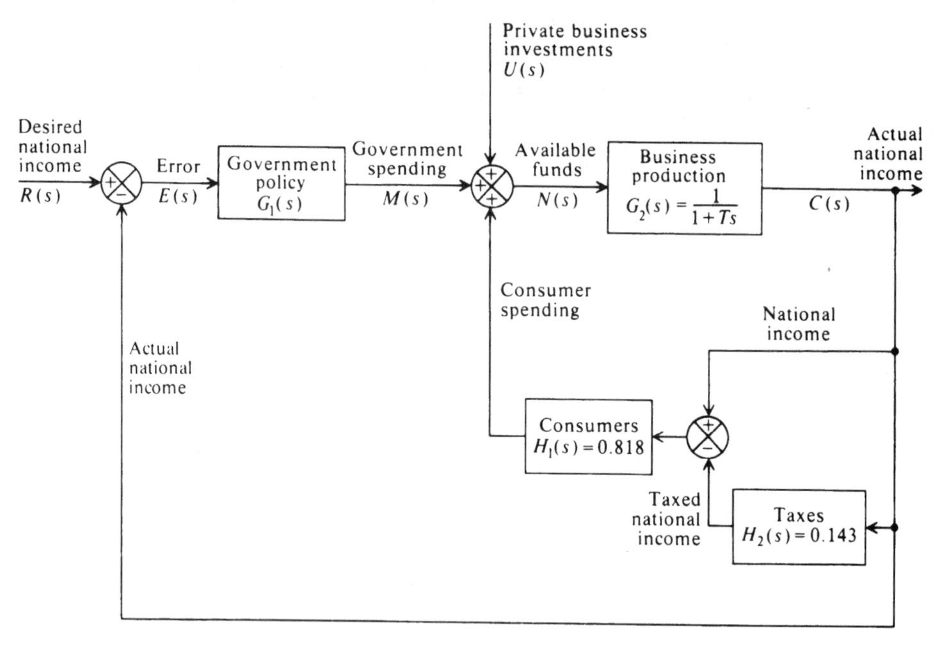

Figure 1.33 Economic feedback relationship model concerning national income, government policy on spending, private business investment, business production, taxes, and consumer spending.

1.6. ILLUSTRATIVE PROBLEMS AND SOLUTIONS

This section provides a set of illustrative problems and their solutions to supplement the material presented in Chapter 1.

I1.1. Draw a block diagram representation of a thermostatically controlled electric oven in the kitchen of a home.

SOLUTION:

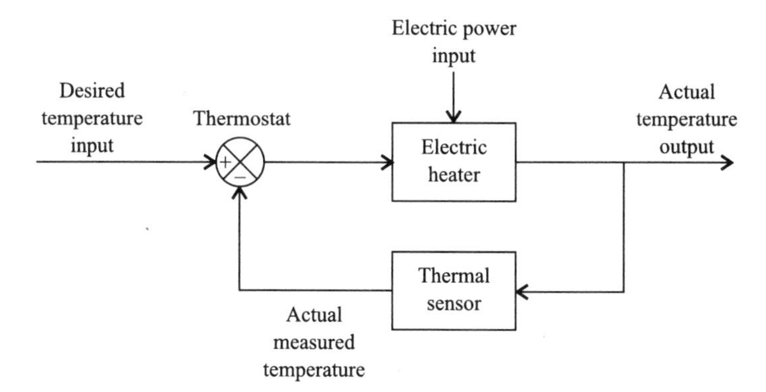

Figure I1.1

I1.2. The control of a nuclear reactor is a very interesting control-system problem. The rate of fission is controlled by rods inserted into the radioactive core. The position of the rods inserted into the core determines the flow of neutrons. The automatic control of the rod position determines the fission process and its resulting heat which is used to generate steam in the turbine. If the rods are pulled out completely, an uncontrolled fission occurs; if the rods are fully inserted into the core, the fission process stops. Draw a block diagram representation for the control of the radiation of the nuclear reactor system shown.

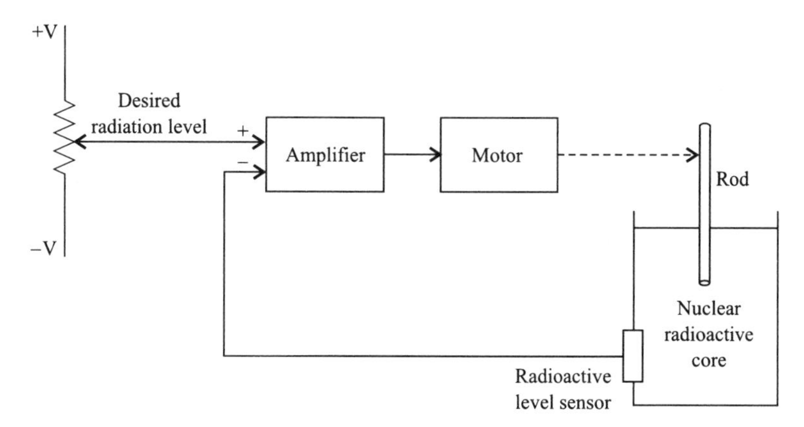

Figure I1.2i

SOLUTION:

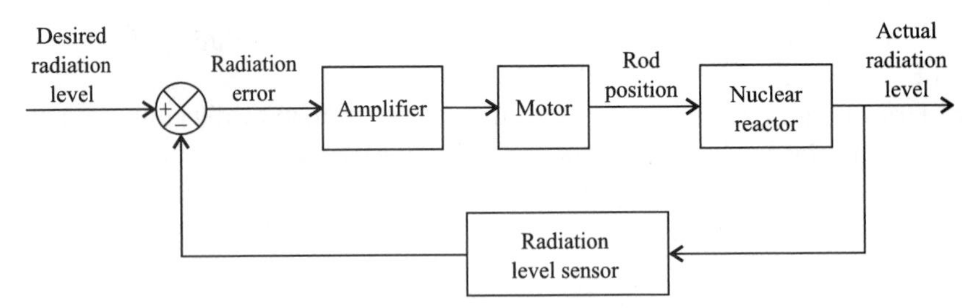

Figure I1.2ii

I1.3. The automatic depth control of a submarine is an interesting control system problem. Suppose the captain of the submarine wants the submarine to "hover" at a desired depth, and sets the desired depth as a voltage from a calibrated potentiometer. The actual depth is measured by a pressure transducer which produces a voltage proportional to depth. The following figure illustrates the problem, where the actual depth of the submarine is denoted as C. Any differences are amplified which then drives a motor that rotates the stern plane actuator angle θ in order that the stern plane rotation reduces the depth error of the submarine to zero. Draw the block diagram representation of the automatic depth control system of the submarine.

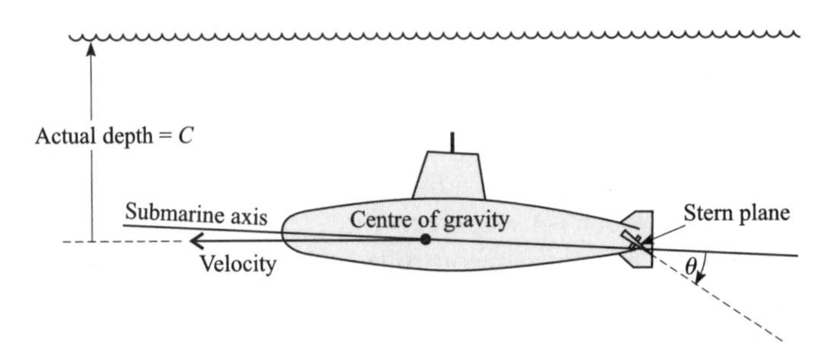

Figure I1.3i

SOLUTION:

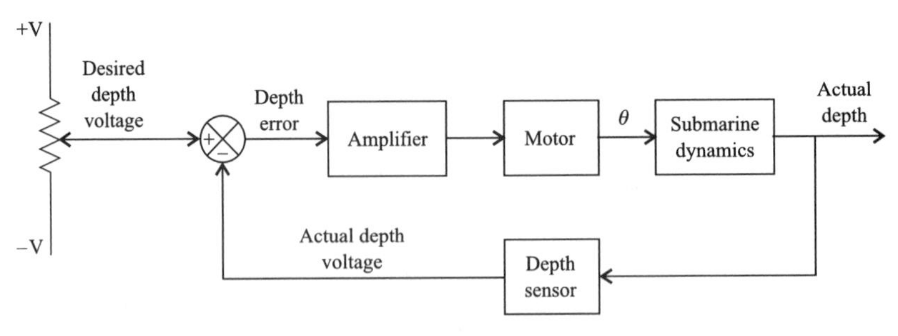

Figure I1.3ii

I1.4. An elevator-position control system used in an apartment building is a very interesting control-system problem. Draw the block diagram representation of an elevator control system in a three-floor building which obtains the desired floor reference position as a voltage from the elevator passenger pressing a button on the elevator, and compares this voltage with a voltage from a position sensor that represents the actual floor position the elevator is at. The difference is an error voltage which is amplified and connected to an electric motor that positions the elevator car to the desired floor selected.

SOLUTION:

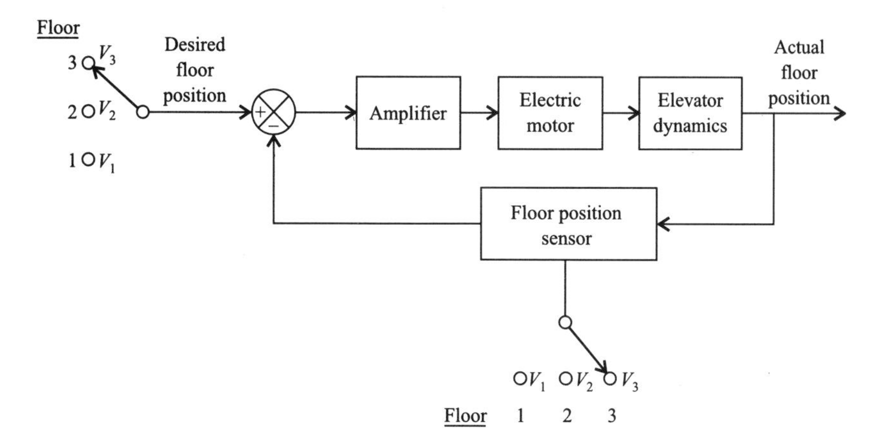

Figure I1.4

PROBLEMS

1.1. Figure P1.1 illustrates the control system for controlling the angular position of a ship's heading. The desired heading, which is determined by the gyroscope setting, is the reference. An electrical signal proportional to the reference is obtained from a resistor that is fixed to the ship's frame. Qualitatively explain how this control system operates in following a command for a southbound direction.

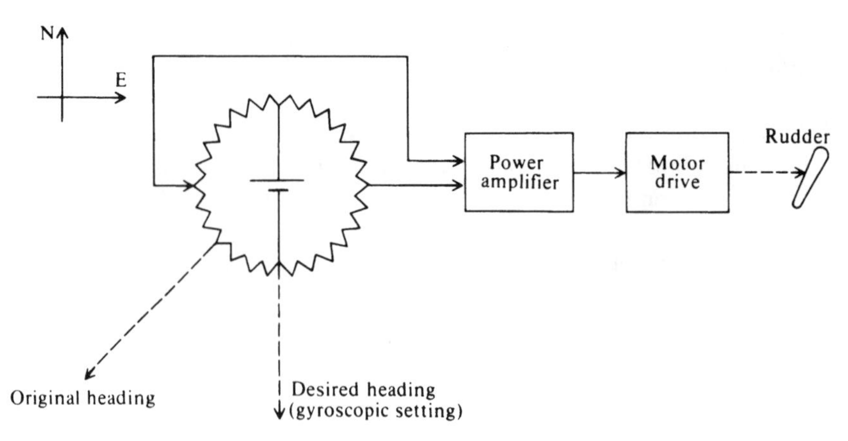

Figure P1.1

1.2. Explain how the rudder positioning system discussed in Problem 1.1 can be modified to become a closed-loop control system.

1.3. An electric hot-water heater is illustrated in Figure P1.3. The heating element is turned on and off by a thermostatic switch in order to maintain a desired temperature. Any demand for hot water results in hot water leaving the tank and cold water entering. Draw the simple functional block diagram for this closed-loop control system and qualitatively explain how it operates if the desired temperature of the thermostat is changed.

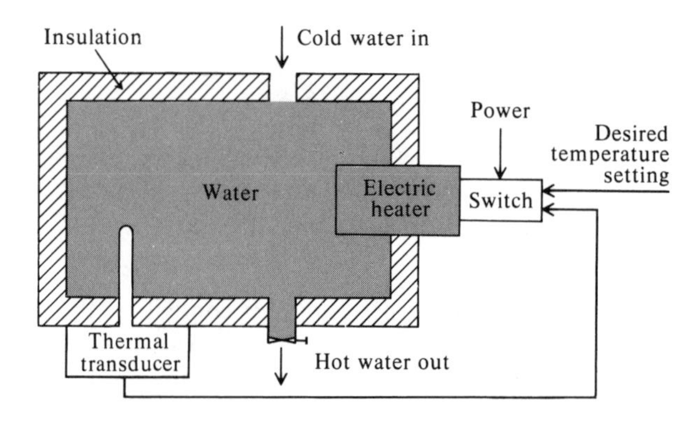

1.4. Explain how the system discussed in Problem 1.3 operates if the ambient temperature surrounding the tank suddenly changes. Refer to the block diagram.

1.5. Figure P1.5 illustrates a control system using a human operator as part of the closed-loop control system. Draw the block diagram of this liquid volume rate control system.

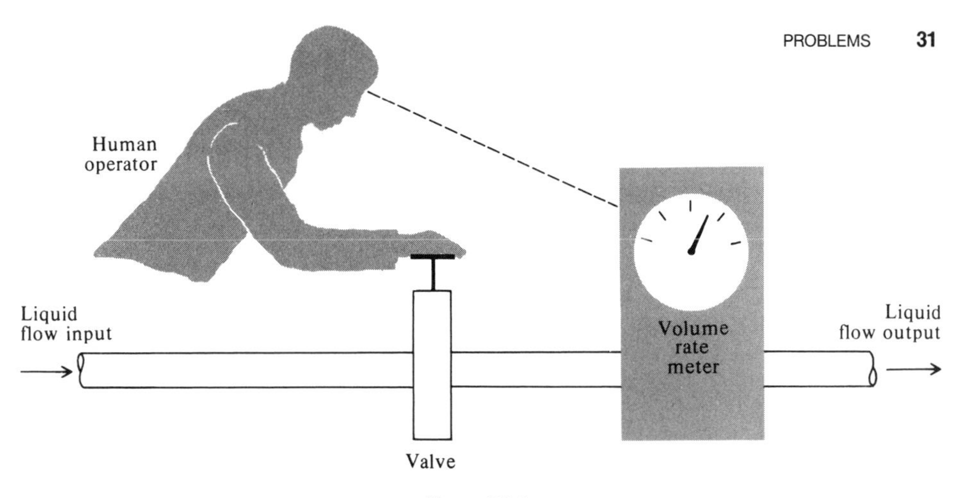

Figure P1.5

1.6. Devise a system that can control the speed of an internal-combustion engine in accordance with a command in the form of a voltage. Explain the operation of your system.

1.7. Devise a system that can control the position, rate, and acceleration of an elevator used in an apartment house. What specifications or limits would you place on the position,velocity, and acceleration capabilities of the system?

1.8. For the control system devised in Problem 1.7, describe what happens when a man weighing 200 lb enters the elevator that has stopped at one of the floors of the apartment building. Utilize a functional block diagram.

1.9. The economic model illustrated in Figure P1.9 illustrates the relationship between wages, prices, and cost of living. Note that an automatic cost of living increase results in a positive feedback loop. Indicate how additional feedback loops in the form of legislative control can stabilize the economic system.

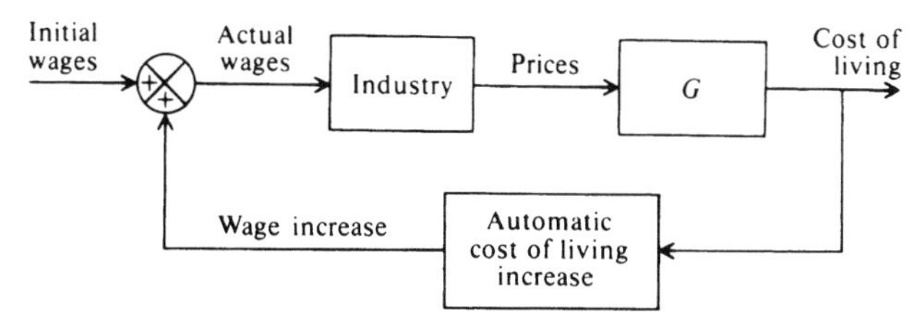

Figure P1.9

1.10. Explain what happens to the automatic position control system illustrated in Figure 1.20 if the speed-measuring device fails. Can the system still operate?

1.11. Determine what happens in the autopilot system illustrated in Figure 1.23 if the aircraft suddenly enters a turbulent atmosphere.

1.12. Modify the block diagram of the attitude-control system of the space vehicle illustrated in Figure 1.16, to allow for sudden failure of the computer and manual control of the vehicle.

1.13. The pH factor of a liquid is a very important factor to be controlled in chemical process control systems. Figure P1.13 illustrates an example of a system which controls the pH factor of a liquid flowing through a tank. The pH probe measures the actual pH level and compares it with the desired pH level. If the liquid in the tank does not have the required acidic level, then the motor adjusts the valve permitting additional acid to enter the tank. The limitation of the system shown is that it does not have any provisions for adding a base solution in the event that the liquid in the tank is too acidic. Modify this control system so that it could add an acid or a base to the liquid in the tank, and achieve the desired pH factor for all kinds of liquid pH factors.

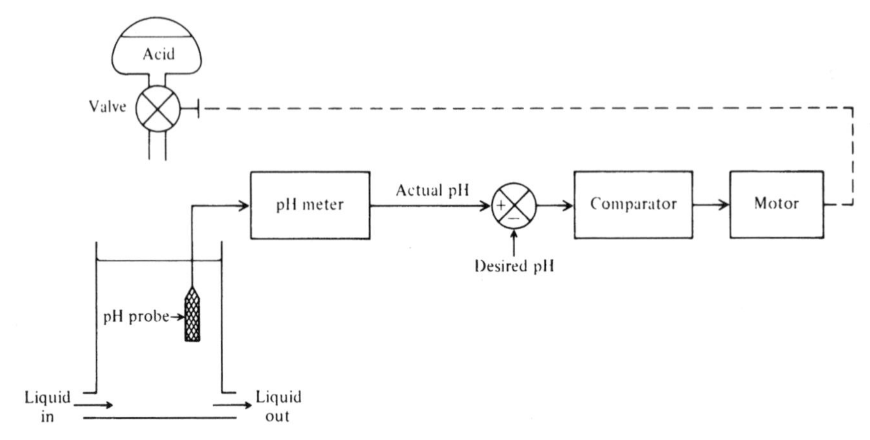

Figure P1.13

1.14. The actual sales price of a commodity in a free economic market is governed by the law of supply and demand that states that the market demand for the item increases as its price decreases. In addition, this concept states that an equilibrium sales price is achieved when the supply is equal to the demand. The control engineer can represent this concept as a feedback system with the actual sales price being represented as the output, and the desired market sales price change as the command input. Draw the block diagram for this economic system which includes the supplier, demander, prices, and market. Under which conditions is an equilibrium operation of supply and demand achieved?

1.15. The level of a liquid in a chemical process control system must be maintained constant within a narrow range. In order to achieve this, a float-operated device is used to sense the level of liquid and control the valve opening which determines the flow of liquid into the tank to replace that which is consumed in the process. Show how this float-valve control system can be operated in a closed-loop control system.

1.16. Repeat Problem 1.15 if the level of the liquid is controlled by a float switch which opens the valve whenever the level of the liquid reaches a certain predetermined level. The valve is automatically closed by a timer switch four minutes after it is opened by the float switch, even though the liquid may be more or les than the desired level. Is this an open-loop or closed-loop control system?

1.17. Figure P1.17 represents the general concept of a command, control, and communication system which is essentially a closed-loop system having the capability of changing an environment based on information presented to a central human controller [43]. The decisions are based on information obtained by sensors, communicated by a data link to a data processor, and appropriately displayed to a commander who then closes the loop by communicating commands to the personnel and equipment at his disposal in order to change the overall environment. A complete command, control, and communication system includes all subsystems, related facilites, equipment, material, services, and personnel required to operate the self-sufficient system. From a functional systems engineering viewpoint, a command, control, and communication system includes sensors, comunications, data processing, displays, and the man-machine interface. Synthesize command and control systems for the following applications:

(a) An airline reservations system for a trunk carrier that has one central terminal, five major terminals, and six minor terminals.

(b) A worldwide satellite communications system containing two major land stations and one major shipborne station.

(c) A strategic command, control, and communication system responsible for the defense of the northeastern section of the United States.

(d) An antisubmarine warfare system aboard a destroyer.

(e) A ground traffic control system for a city of 4 million population.

(f) A police command system controlling the operation of 20 stationary posts and 40 mobile units from one central headquarters station.

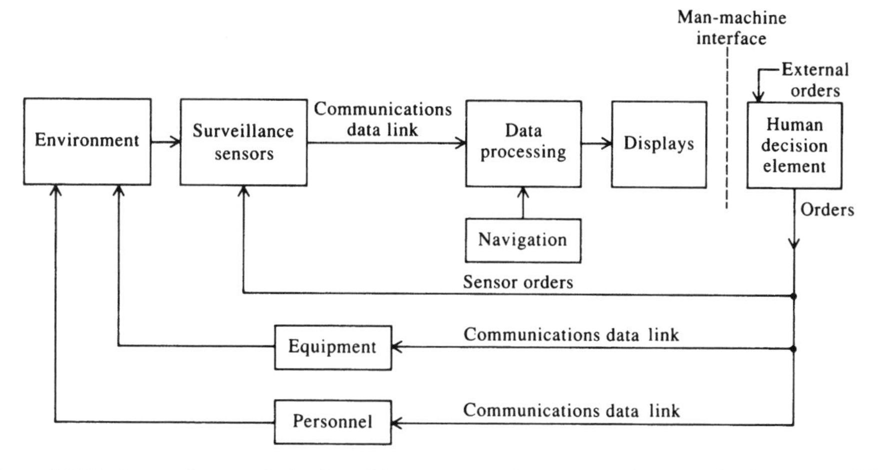

Figure P1.17 General concept of a closed-loop command, control, and communication system [43].

1.18. Robots are being used to a greater degree in almost all aspects of industry. Figure P1.18 shows a photograph of RoboKent™, an automatic floor scrubber for commercial cleaning of supermarkets, warehouses, airports, etc. It has been developed by Transitions Research Corporation (TRC) [44] under a contract from Electrolux AB of Sweden, through its U.S. subsidiary, Kent Company, manufacturer of floor cleaning equipment. The machine is a Kent KA-201 BST fitted with Transitions Research Corporation's drive, scanning, and sensing system. It is designed to automatically clean 100 percent of the floor area in simple unobstructed environments, and 70 percent in more complicated areas. The user need only walk the machine through the middle of the area to be cleaned one time, and then RoboKent will execute its preprogrammed cleaning patterns automatically. This robot contains 16 ultrasonic range sensors, bumpers, a light curtain, dead-reckoning, and a rate gyro for navigation. Draw a block-diagram representation of this robot with the input being represented as the desired position of the robot, and the output being the actual position of the robot. Is this an open-loop or closed-loop control system?

Figure P1.18 Photograph of RoboKent™, an automatic floor scrubber for commercial cleaning of supermarkets, warehouses, airports, and so on. (Courtesy of TRC)

1.19. An automobile leasing agency desires to develop a control-system model which represents the actual number of automobiles being leased as an output and compares it with the desired automobiles being leased as an input. In this control system model, the management of this automobile leasing agency determines the rate of automobiles being leased based on their maintenance facilities, which service the returned automobiles, and the availability of automobiles in their agency. The Maintenance Department then uses the rate established by the management to hire a sufficient staff and buy the necessary support equipment. Draw a block-diagram representation which illustrates the input, the output, management, and the Maintenance Department as blocks of this control system. In addition, show the return rate of automobiles, the rate of leased automobiles, and the net rate of automobiles being leased.

1.20. The organization of an engineering corporation is a feedback process in which the desired product represents the referrence input, and the actual resulting product represents the output. The groups involved in the process include the management of the engineering organization, which monitors continually the process and compares the desired product and the actual resulting product, and the following departments:

- Research and Development
- Laboratory Development
- Laboratory Test
- Prototype Development
- Prototype Test
- Product Design
- Production Department
- Quality Assurance and Test

Draw a feedback control system illustrating the representation of this control system. In addition, indicate any inner feedback loops which occur in this process.

REFERENCES

1. N. Wiener, *Cybernetics or Control and Communication in the Animal and the Machine*, 2nd ed. MIT Press and Wiley, New York, 1961.
2. S. M. Shinners, *Techniques of System Engineering*. McGraw-Hill, New York, 1967.
3. R. Surgen, "Detroit '88 driver friendly innovation." *IEEE Spectrum*, Dec. 1987, pp. 53–56.
4. T. Tale et al., "Automatic climate control." *IEEE Control Syst.*, Oct. 1986, pp. 20–23.
5. J. G. Rivaid, "The automobile in 1997," *IEEE Spectrum*, Oct. 1987, pp. 67–71.
6. S. C. Lyman, "Computer directs process control at textile plant." *Control Eng.* **13**, 20–21 (1966).
7. J. C. Keebler, "Warehouse automation." *Automation* **13**, 64 (1966).

8. S. Hearne, "In-plant data collection tracks material.' Control Eng. **14**, 100 (1967).
9. S. W. R. Cox, "Automation in agriculture." Control **9**, 247–252 (1965).
10. *Helpmate*™ *Specification Sheet*. Transitions Research Corporation, Danbury, CT.
11. J. Evans, B. Krishnamurthy, W. Pong, R. Croston, C. Weiman, and G. Engelberger, "Helpmate™: A robotic materials transport system." *Rob. Auton. Syst.* **5**, 251–256 (1989).
12. A. Daniels, "Computer rail service expands." *IEEE Spectrum*, Jan. 1990, p. 48.
13. E. E. Murphy, "Transportation." *IEEE Spectrum*, Jan. 1989, pp. 62–63.
14. A. R. Eastham, "A year of evolutionary changes. *IEEE Spectrum*, Jan. 1991, p. 70.
15. J. Teresko, "Japan's new idea." Ind. Week, Sept. 3, 1990, pp. 39–69.
16. *The Linear Express: Tomorrow's Transportation*, Central Japan Railway Company, Nagoya 450, Japan.
17. I. Nakamura and S. Yamazaki, "On the centralized system for train operation and traffic control—Including signaling and routing information." *Railway Tech. Res. Inst.*, **5**, (1) (1964).
18. Technical Aspects on the New Tokaido Line. Japanese National Railways, Tokyo, 1966.
19. *Shinkansen*. Japan Railways Group, New York.
20. *New Shinkansen*, Ser. 100. Central Japan Railway Company (JR Tokal), Nagoya 450, Japan, 1987.
21. K. T. Chen, "Transportation: MD-11 debuts." *IEEE Spectrum*, Jan. 1991, p. 69.
22. K. T. Chen, "Transportation (technology 1991 report)." *IEEE Spectrum*, Jan. 1991, pp. 69–70.
23. *NEWS from McDonnell Douglas*, MD-11. McDonnell Douglas, June 1990 and February 1991, Douglas Aircraft Company, Long Beach, CA.
24. *Fact Sheet of Bell Boeing V-22 Osprey*.
25. First Civil Tiltrotor, Bell Boeing 609, *Aviation Week & Space Technology*, Nov. 25, 1996. p. 75.
26. *Fact Sheet of the Bell Boeing 609*.
27. *Fact Sheet of the UH-60 Black Hawk Flight Simulator*, Hughes Training, Inc., Arlington, TX.
28. *United States Air Force Fact Sheet*, F-15 Eagle, Secretary of the Air Force, Office of Public Affairs, Washington, DC, May 1989.
29. *F-22 Raptor Media Guide*, Air Dominance for the 21st Century, edited by the Lockheed Martin Aeronautical Systems Communications Office,
30. United States Department of Transportation Fact Sheet, The Surface Effect Ships (WSES). U.S. Coast Guard, Public Affairs Office, Washington, DC.
31. R. L. Trillo, ed., *Jane's High-Speed Marine Craft*, Jane's Information Group, Alexandria, VA, 1990.
32. N. A. Coulter, Jr. and O. L. Updike, Jr., "Biomedical control developments." In *Proceedings of the 1965 Joint Automatic Control Conference*, p. 258–272.
33. R. W. Mann, "Efferent and afferent control of an electromyographic proportional-rate, force sensing artificial elbow with cutaneous display of joint angle," *Proc. Inst. Mech. Eng.* **183**, 86–91 (1968–1969).
34. *The Boston Arm*. Liberty Mutual Insurance Company, Research Center, Boston, MA.
35. "Designers still grope for efficient 'limbs'," *Electron. Des.* **18**, U92–U93 (1970).
36. *The Newsletter for Professionals in Prosthetics and Rehabilitation*, Vol. 1 : 2. Liberty Mutual Research Center, Hopkinton, MA, 1986/1987.

37. J. A. Doubler, "An analysis of extended physiological proprioception as a control technique for upper extremity prostheses." PhD. thesis, Northwestern University, Evanston, IL, 1982.
38. L. E. Carlson and G. Scott, *Extended Physiological Proprioception for the Control of Arm Prostheses, ICAART88*, Montreal, 1988.
39. T. Walley Williams III, "Practical methods for controlling powered upper-extremity Prostheses." *Assist. Technol.*, **2**(1), 3–18 (1990).
40. T. Walley Williams III, "The Boston elbow." *SOMA*, July 1986, pp. 29–33.
41. *Nikon N6006*, Daring Photography, Nikon Inc., 1300 Walt Whitman Road, Melville, NY, 1996.
42. Auto Focus SLR Update. *Popular Photography*, Dec. 1987, pp. 72–75.
43. S. M. Shinners, *A Guide to Systems Engineering and Management*, Lexington Books, Heath, Lexington, MA. 1976.
44. *Specification Sheet of RoboKent*. Transitions Research Corporation, Danbury, CT.
45. *Jane's Fighting Ships*, Ninety-ninth Edition, 1996–97, Captain Richard Sharpe, OBE RN, ed. Jane's Information Group Limited, Sentinel House, 163 Brighton Road, Coulsdon, Surrey CR5 2NH, UK, p. 794.

2

MATHEMATICAL TECHNIQUES FOR CONTROL-SYSTEM ANALYSIS

2.1 INTRODUCTION

The design of linear, continuous, feedback control systems is dependent on mathematical techniques such as the Laplace transformation, the signal-flow graph, and the state-variable concept. In addition to these techniques, the design of linear, discrete, feedback control systems requires a knowledge of the z and w transforms, the Fourier transform, and some aspects of information theory. The design of nonlinear, continuous, feedback control systems is dependent on mathematical techniques such as the Fourier transform, and the state-variable concept. The scope of this book does not permit a detailed discussion of all these mathematical devices. The philosophy followed here is to review the theory of those techniques necessary for understanding the design of linear continuous and discrete control systems, and nonlinear continuous control systems, and to focus attention on the specific application of these mathematical tools to these classes of control systems.

This chapter logically develops the many mathematical tools used by the control-system engineer. Starting with a review of the complex variable, complex functions, and the s plane, the presentation follows with the trigonometric and complex forms of the Fourier series. The Fourier integral is next presented, from which the Fourier transform is developed. The limitation of the Fourier transform to control systems is illustrated, and the presentation then develops the Laplace transform. Besides being a logical road to the development of the Laplace transform, the purpose of presenting all of these concepts is that they will be used in the ensuing discussion. For example, the trigonometric form of the Fourier series is used for the describing function analysis of nonlinear control systems. The Laplace transform is the fundamental tool used in the classical transfer function, signal-flow graph, and block-diagram approach for analyzing linear systems which is used extensively throughout the book.

The presentation in this chapter then proceeds to modern control theory aspects which are based on the foundation of matrix theory. After a brief review of those aspects of matrix theory used by the control-system engineer, the state variable is then defined, and the generation of the state and output equations (phase-variable canonical form) of a control system is illustrated. The chapter then concludes with the definition of the state transition matrix and its application. The z and w transforms needed to analyze digital control systems are presented in Chapter 9, where digital control systems are analyzed and designed.

The duality of using the classical Laplace transform/transfer function/block diagram and the modern state-variable approaches is pursued throughout the book. The message of this book is that they enhance each other, and the control-system engineer can benefit by looking at many problems from both viewpoints.

2.2 REVIEW OF COMPLEX VARIABLES, COMPLEX FUNCTIONS, AND THE *s* PLANE

The design of control systems depends greatly on the application of complex-variable theory. In what follows, the complex variable s is composed of a real part σ and an imaginary part ω, where

$$s = \sigma + j\omega \tag{2.1}$$

In the complex s plane, σ is plotted horizontally and $j\omega$ vertically. A complex function $F(s)$ is considered to be a function of the complex variable s if there is at least one value of $F(s)$ for every value of s. The function $F(s)$ will have real and imaginary components, because s has real and imaginary components, and it has the following form:

$$F(s) = \text{Real } F(s) + j \text{ Imaginary } F(s). \tag{2.2}$$

If there is only one value of $F(s)$ for every value of s, the function $F(s)$ is called a *single-valued function*. However, if there is more than one point in the $F(s)$ plane for every value of s, then $F(s)$ is a *multivalued function*. Most complex functions used in linear control systems are single-valued functions of s.

Figure 2.1 illustrates the mapping of a single-valued function from the s plane to the $F(s)$ plane. Figure 2.2 illustrates the corresponding mapping for a multivalued function.

Five notions of complex-variable theory that are important to the control-systems engineer are those of analytic functions, ordinary points, singularities, poles, and zeros of a function.

A. Analytic Functions and Ordinary Points

A complex-variable function $F(s)$ is *analytic* in a region if the function and all of its derivatives exist at every point in that region. As an example, the function

$$F(s) = \frac{1}{s(s+4)(s+6)} \tag{2.3}$$

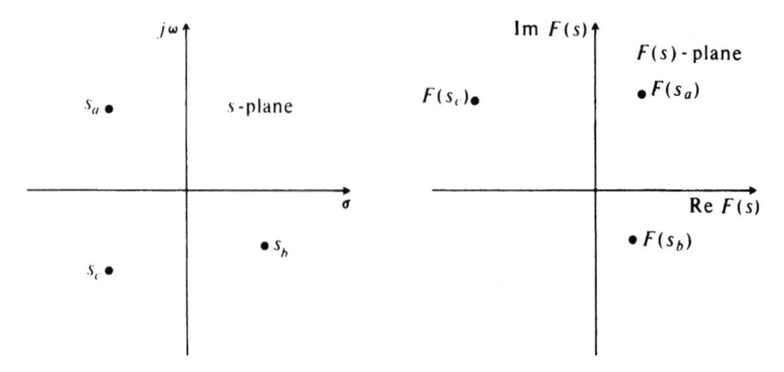

Figure 2.1 Mapping of a single-valued function from the s plane to the $F(s)$ plane.

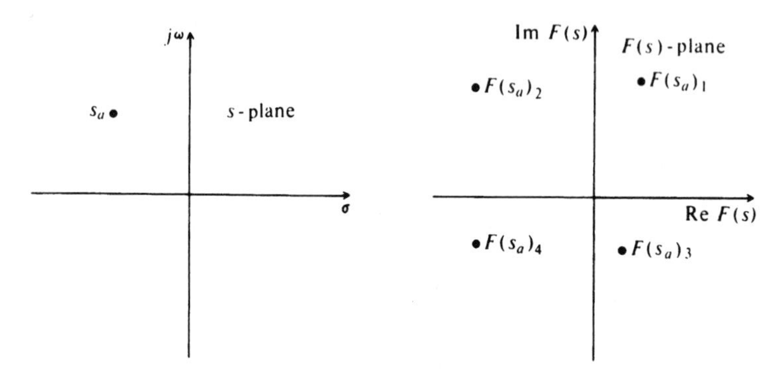

Figure 2.2 Mapping of a multivalued function from the s plane to the $F(s)$ plane.

is analytic at every point in the s plane except at the points $s = 0, -4$ and -6. At these three points, the value of the function $F(s)$ is infinity. However, the function

$$F(s) = s(s+4)(s+6) \tag{2.4}$$

is analytic at every point in the s plane. Points in the s plane where the function $F(s)$ is analytic are defined as *ordinary* points.

The derivative of an *analytic function* $F(s)$ is given by

$$\frac{d}{ds}F(s) = \lim_{\Delta s \to 0} \frac{F(s+\Delta s) - F(s)}{\Delta s} = \lim_{\Delta s \to 0} \frac{\Delta F(s)}{\Delta s}.$$

Because $\Delta s = \Delta\sigma + j\Delta\omega$, Δs can approach zero along an infinite number of different paths. If the derivative is taken along two specific paths, $\Delta s = \Delta\sigma$ (parallel to the real axis) and $\Delta s = j\Delta\omega$ (parallel to the imaginary axis) are equal, then the derivative is unique for any other path $\Delta s = \Delta\sigma + j\Delta\omega$ and, therefore, the derivative exists.

To illustrate this, first let us consider the case where $\Delta s = \Delta\sigma$. Then,

$$\frac{d}{ds}F(s) = \lim_{\Delta\sigma \to 0}\left(\frac{\Delta F_x}{\Delta\sigma} + j\frac{\Delta F_y}{\Delta\sigma}\right) = \frac{\partial F_x}{\partial\sigma} + j\frac{\partial F_y}{\partial\sigma}. \tag{2.5}$$

Secondly, let us consider the path $\Delta s = j\Delta\omega$. For this case,

$$\frac{d}{ds}F(s) = \lim_{j\Delta\omega \to 0}\left(\frac{\Delta F_x}{j\Delta\omega} + j\frac{\Delta F_y}{j\Delta\omega}\right) = -j\frac{\partial F_x}{\partial\omega} + \frac{\partial F_y}{\partial\omega}. \tag{2.6}$$

Setting the values of these derivatives from Eqs. (2.5) and (2.6) equal to each other, we obtain the following:

$$\frac{\partial F_x}{\partial\sigma} + j\frac{\partial F_y}{\partial\sigma} = \frac{\partial F_y}{\partial\omega} - j\frac{\partial F_x}{\partial\omega}. \tag{2.7}$$

Therefore, if the equations of

$$\frac{\partial F_x}{\partial\sigma} = \frac{\partial F_y}{\partial\omega} \tag{2.8}$$

and

$$\frac{\partial F_y}{\partial\sigma} = -\frac{\partial F_x}{\partial\omega} \tag{2.9}$$

hold, then the derivative

$$\frac{dF(s)}{ds}$$

is uniquely determined. Equations (2.8) and (2.9) are referred to as the Cauchy-Riemann conditions and, if these conditions are satisfied, then the function $F(s)$ is analytic.

As an example, let us analyze the following function:

$$F(s) = \frac{1}{s+4} \tag{2.10}$$

Therefore,

$$F(\sigma + j\omega) = \frac{1}{(\sigma + j\omega + 4)} \tag{2.11}$$

where

$$F(\sigma + j\omega) = \frac{1}{(\sigma + 4 + j\omega)}\frac{(\sigma + 4 - j\omega)}{(\sigma + 4 - j\omega)} = \frac{\sigma + 4 - j\omega}{(\sigma + 4)^2 + \omega^2}. \tag{2.12}$$

Therefore,

$$F_x = \frac{\sigma + 4}{(\sigma + 4)^2 + \omega^2} \tag{2.13}$$

and

$$F_y = \frac{-\omega}{(\sigma+4)^2+\omega^2}. \tag{2.14}$$

Except for the condition of $s = -4$ (e.g., $\sigma = -4$ and $\omega = 0$), $F(s)$ satisfies the Cauchy–Riemann conditions:

$$\frac{\partial F_x}{\partial \sigma} = \frac{\partial F_y}{\partial \omega} : \frac{(\sigma+4)^2+\omega^2(1)-(\sigma+4)(2)(\sigma+4)}{[(\sigma+4)^2+\omega^2]^2} = \frac{\omega^2-(\sigma+4)^2}{[(\sigma+4)^2+\omega^2]^2} \tag{2.15}$$

$$\frac{\partial F_y}{\partial \sigma} = -\frac{\partial F_x}{\partial \omega} : \frac{\omega(2)(\sigma+4)}{[(\sigma+4)^2+\omega^2]^2} = \frac{2\omega(\sigma+4)}{[(\sigma+4)^2+\omega^2]^2} \tag{2.16}$$

Therefore, the function $F(s)$ of Eq. (2.10) is analytic in the entire s plane except at $s = -4 (\sigma = -4, \omega = 0)$.

We can obtain the derivative of the function $F(s)$ from Eq. (2.10) at all points except at $s = -4$ from:

$$\frac{d}{ds}F(s) = \frac{\partial F_x}{\partial \sigma} + j\frac{\partial F_y}{\partial \sigma} = \frac{\partial F_y}{\partial \omega} - j\frac{\partial F_x}{\partial \omega} = -\frac{1}{(\sigma + j\omega + 4)^2} = -\frac{1}{(s+4)^2}. \tag{2.17}$$

Derivatives of an analytic function such as $F(s)$ of Eq. (2.10) can be determined simply by differentiating $F(s)$ with respect to s:

$$\frac{d}{ds}\left(\frac{1}{s+4}\right) = -\frac{1}{(s+4)^2} \tag{2.18}$$

B. Singularities, Poles, and Zeros

Singularities are defined as points in the s plane where the function, or its derivatives, do not exist. An important example of a singularity is a *pole*. If a function $F(s)$ is analytic in the region of s_j, except at the poles of s_j, then $F(s)$ has a pole of order q (where q is finite) at $s = s_j$ if

$$\lim_{s \to s_j}\left[F(s)(s-s_j)^q\right], \quad q = 1, 2, 3... \tag{2.19}$$

is finite. Therefore, the denominator of $F(s)$ must contain the factor $(s - s_j)^q$, and the function becomes infinite when $s = s_j$. In Eq. (2.3), the function has a simple pole (i.e., poles of order one) at $s = 0, -4$, and -6. As a more complex example consider the function

$$F(s) = \frac{100(s+1)(s+8)^2}{s(s+4)(s+10)(s+20)^2}. \tag{2.20}$$

This function has simple poles at $s = 0, -4$, and -10. In addition, it has a pole of order 2 at $s = -20$.

Finally, we consider the concept of *zeros*. If a function $F(s)$ is analytic at $s = s_j$, then $F(s)$ has a zero of order q (where q is finite) at $s = s_j$, if

$$\lim_{s \to s_j} \left[F(s)(s - s_j)^{-q} \right], \quad q = 1, 2, 3... \tag{2.21}$$

is finite. Therefore, the numerator of $F(s)$ must contain the factor $(s - s_j)^q$, and the function becomes zero where $s = s_j$. For example, in Eq. (2.20), there is a simple zero (i.e., a zero of order 1) at $s = -1$ and a zero of order 2 at $s = -8$.

Another important point to emphasize before leaving this discussion of poles and zeros is that the total number of poles has to equal the total number of zeros (including zeros at infinity) for functions that are defined to be the quotient of two polynomials in s. For example, in Eq. (2.20), we have five finite poles at $s = 0, -4, -10$, and a double pole at -20. We have three finite zeros: one at -1 and a double one at -8. However, there have to be two zeros at infinity because

$$\lim_{s \to \infty} F(s) = \lim_{s \to \infty} \frac{100}{s^2} = 0. \tag{2.22}$$

We conclude from this that $F(s)$ has a total of five zeros (three finite zeros and two infinite zeros) and five poles in the overall s plane.

2.3. REVIEW OF FOURIER SERIES AND FOURIER TRANSFORM

In the analysis of the behaviour of systems evolving in time, it is often convenient to introduce mathematical transformations that take us from the time domain to a new domain called the frequency domain. Such transformations are called transforms. Here we will focus on the Fourier series, which is used to analyze periodic functions of time, and the Fourier integral which is used to examine aperiodic time functions of a restricted class. Section 2.4 introduces the Laplace transform, which is of great value in the analysis of time functions that vanish for negative time.

A. Fourier Series

Given a single-valued periodic function $f(t)$ whose period is T (and fundamental frequency is $\omega = 2\pi/T$), then

$$f(t) = f(t + T). \tag{2.23}$$

Functions that satisfy Eq. (2.23) can be represented by a Fourier series provided the function is bounded and contains only a finite number of discontinuities in a finite interval. The classical trigonometric form of the Fourier series is given by

$$f(t) = \frac{A_0}{2} + \sum_{K=1}^{K=\infty} A_K \cos K\omega t + \sum_{K=1}^{K=\infty} B_K \sin K\omega t, \tag{2.24}$$

where ω represents the fundamental frequency, and $K\omega$ represents the Kth harmonic:

$$A_K = \frac{2}{T}\int_{-T/2}^{T/2} f(t)\cos K\omega t\,dt, \qquad K = 0, 1, 2, 3, \ldots, \tag{2.25}$$

$$B_K = \frac{2}{T}\int_{-T/2}^{T/2} f(t)\sin K\omega t\,dt, \qquad K = 1, 2, 3, \ldots. \tag{2.26}$$

If $f(t) = -f(-t)$, the the function is odd and $A_K = 0$; Figure 2.3*a* illustrates an odd function. If $f(t) = f(-t)$, then the function is even and $B_K = 0$; Figure 2.3*b* illustrates an even function. The two terms in the series whose frequency is $K\omega$ when added comprise the Kth *harmonic* of $f(t)$. The amplitude of the harmonic is given by

$$\sqrt{A_K^2 + B_K^2}, \tag{2.27}$$

and the phase is given by

$$tan^{-1}(-B_K/A_K). \tag{2.28}$$

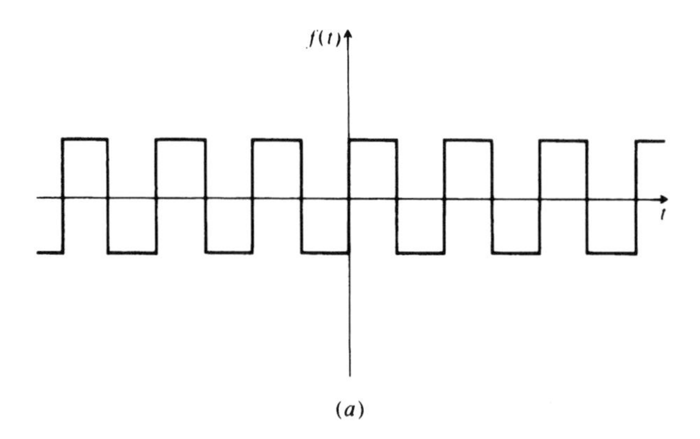

(*a*)

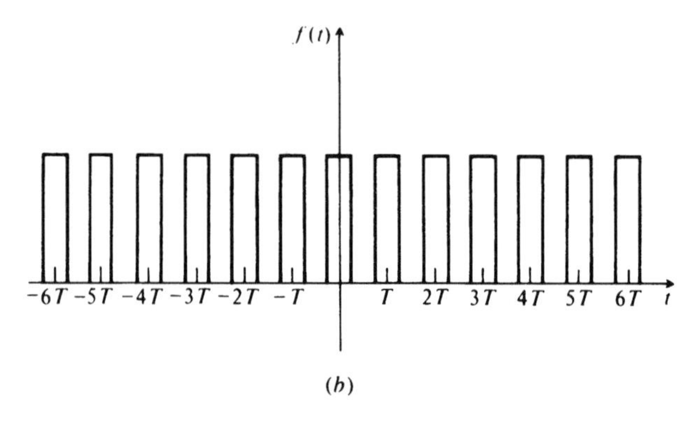

(*b*)

Figure 2.3 (*a*) An odd function. (*b*) An even function.

As an example of the application of the Fourier series, consider the periodic waveshape illustrated in Figure 2.4*a*. This function represents a periodic switching function whose period is $2\pi/\omega_c$. In addition, it is an even function, because $f(t) = f(-t)$, so that the B_K terms are zero. The evaluation of A_0 and A_K follows from Eq. (2.25):

$$A_0 = \frac{1}{\pi/\omega_c}\left[\int_{-\pi/\omega_c}^{-\pi/2\omega_c} (-1)\,dt + \int_{-\pi/2\omega_c}^{\pi/2\omega_c} (1)\,dt + \int_{\pi/2\omega_c}^{\pi/\omega_c} (-1)\,dt\right] = 0, \qquad (2.29)$$

$$A_K = \frac{1}{\pi/\omega_c}\left[\int_{-\pi/\omega_c}^{-\pi/2\omega_c} (-1)\cos K\omega_c t\,dt \right.$$
$$\left. + \int_{-\pi/2\omega_c}^{\pi/2\omega_c} (1)\cos K\omega_c t\,dt + \int_{\pi/2\omega_c}^{\pi/\omega_c} (-1)\cos K\omega_c t\,dt\right], \qquad (2.30)$$

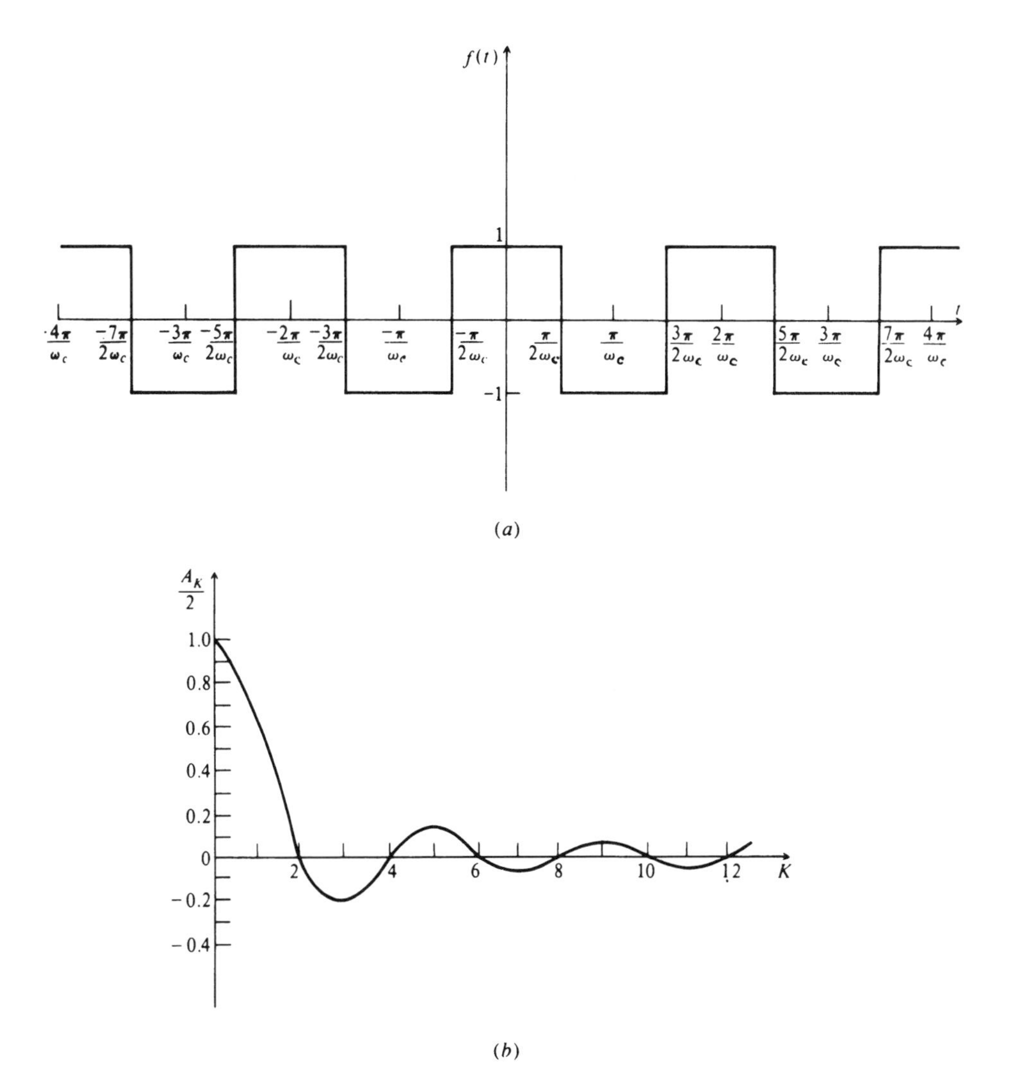

Figure 2.4 (*a*) A periodic function. (*b*) Envelope of the function $A_K/2 = (\sin K\pi/2)/(K\pi/2)$ showing the $(\sin x)/x$ pattern.

$$A_K = \frac{2\sin(K\pi/2)}{K\pi/2}, \qquad K \neq 0. \tag{2.31}$$

Therefore,

$$f(t) = 2\sum_{K=1}^{K=\infty} \frac{\sin(K\pi/2)}{K\pi/2}\cos K\omega t. \tag{2.32}$$

A plot of the envelope of the function $A_K/2$, which follows a $(\sin x)/x$ wave pattern, is illustrated in Figure 2.4*b*.

B. Complex Form of the Fourier Series

The Fourier series given by Eqs. (2.24)–(2.26) can be converted to a complex form by means of the following substitutions:

$$\sin K\omega t = \frac{1}{2j}(e^{jK\omega t} - e^{-K\omega t}), \tag{2.33}$$

$$\cos K\omega t = \frac{1}{2}(e^{jK\omega t} + e^{-jK\omega t}). \tag{2.34}$$

Therefore,

$$f(t) = \frac{A_0}{2} + \sum_{K=1}^{\infty} \frac{A_K}{2}(e^{jK\omega t} + e^{-jK\omega t}) + \sum_{K=1}^{\infty} \frac{B_K}{2j}(e^{jK\omega t} - e^{-jK\omega t}). \tag{2.35}$$

This can also be written as

$$f(t) = \frac{A_0}{2} + \frac{1}{2}\sum_{K=1}^{\infty}(A_K - jB_K)e^{jK\omega t} + \frac{1}{2}\sum_{K=1}^{\infty}(A_K + jB_K)e^{-jK\omega t}. \tag{2.36}$$

Note that from Eqs. (2.25) and (2.26),

$$(A_K - jB_K) = \frac{2}{T}\int_{-T/2}^{T/2} f(t)e^{-jK\omega t}\,dt$$

and

$$(A_K + jB_K) = \frac{2}{T}\int_{-T/2}^{T/2} f(t)e^{jK\omega t}\,dt.$$

Therefore, Eq. (2.36) becomes

$$f(t) = \frac{A_0}{2} + \sum_{K=1}^{\infty} \frac{e^{jK\omega t}}{T}\int_{-T/2}^{T/2} f(t)e^{-jK\omega t}\,dt + \sum_{K=1}^{\infty} \frac{e^{-jK\omega t}}{T}\int_{-T/2}^{T/2} f(t)e^{jK\omega t}\,dt. \tag{2.37}$$

Because the second and third terms on the right side of Eq. (2.37) differ by a sign, the complex form of the Fourier series can be written as

$$f(t) = \sum_{K=-\infty}^{K=\infty} \frac{e^{jK\omega t}}{T} \int_{-T/2}^{T/2} f(t)e^{-jK\omega t}\, dt. \tag{2.38}$$

This equation is usually written in the following form:

$$f(t) = \frac{1}{T} \sum_{K=-\infty}^{K=\infty} C_K e^{jK\omega t}, \tag{2.39}$$

where

$$C_K = \int_{-T/2}^{T/2} f(t)e^{-jK\omega t}\, dt. \tag{2.40}$$

Equation (2.39) gives the complex Fourier series of $f(t)$, and Eq. (2.40) defines the complex Fourier coefficient C_K.

As an example of the application of the complex form of the Fourier series, consider the periodic pulse train illustrated in Figure 2.5*a*. The complex Fourier coefficient of this periodic pulse train is given by

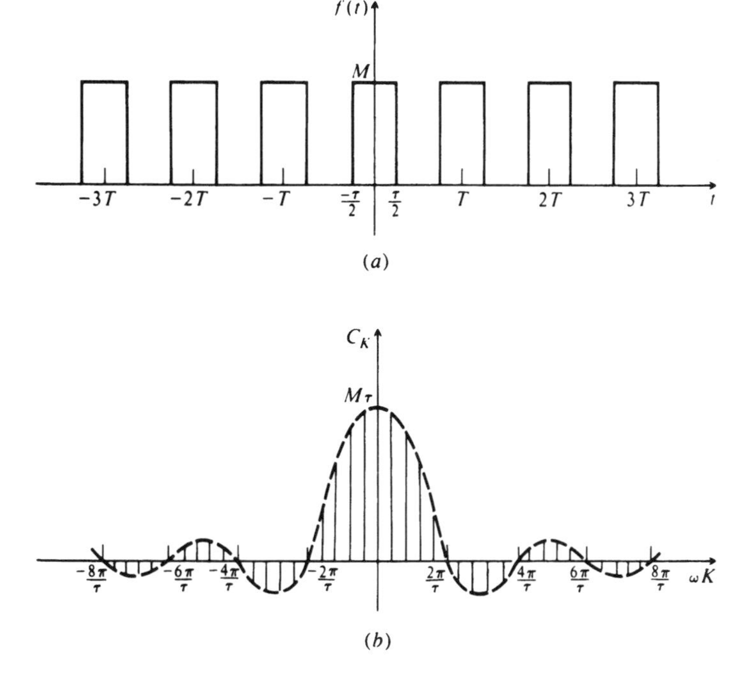

Figure 2.5 (*a*) Periodic pulse train. (*b*) Complex Fourier coefficient C_K for the pulse train of (*a*).

$$C_K = \int_{-\tau/2}^{\tau/2} Me^{-jK\omega t}\,dt = \left[-\frac{Me^{-jK\omega t}}{jK\omega}\right]_{-\tau/2}^{\tau/2} = \frac{2M}{K\omega}\sin\omega K\tau/2. \tag{2.41}$$

Equation (2.41) can also be written as

$$C_K = M\tau\frac{\sin\frac{1}{2}\omega K\tau}{\frac{1}{2}\omega K\tau}.$$

A plot of C_K for all values of frequency is illustrated in Figure 2.5*b*. The envelope of C_K follows a $(\sin x)/x$ function and the spacing between lines is given by $\Delta\omega K = 2\pi/T$.

C. Fourier Integral

The control-system engineer finds many functions of interest that are not periodic. For these functions, the Fourier series cannot be applied. However, the Fourier integral can be used to analyze a wide class of aperiodic functions.

Let us assume that instead of the period pulse train in Figure 2.5*a*, we have only one nonperiodic pulse as illustrated in Figure 2.6*a*. By considering this single pulse as the fundamental function during a period and assuming that its period is infinity, this function can be represented by the Fourier integral. Basically, the approach is to

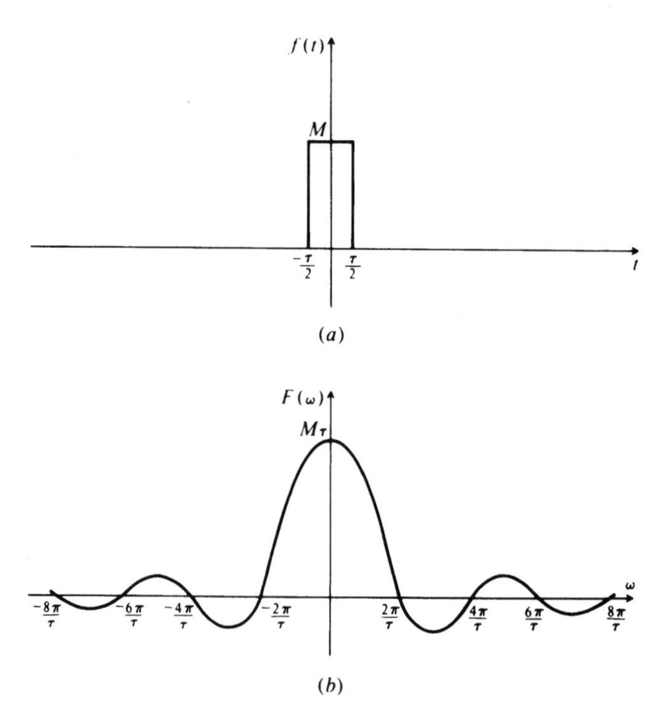

Figure 2.6 (*a*) A nonperiodic pulse train. (*b*) Fourier transform for the pulse illustrated in (*a*).

obtain the Fourier series of the function of Figure 2.6*a* assuming that it is periodic with a period of infinity. Let us write Eq. (2.39) as

$$f(t) = \frac{1}{2\pi} \sum_{K=-\infty}^{K=\infty} C_K e^{jK\omega t} \Delta K\omega$$

by letting $T = 2\pi/\Delta K\omega$. As T approaches infinity, $\Delta K\omega$ approaches zero and we have the following relationships:

$$\lim_{\substack{T\to\infty \\ \Delta K\omega\to 0}} f(t) = \lim_{\substack{T\to\infty \\ \Delta K\omega\to 0}} \frac{1}{2\pi} \sum_{K=-\infty}^{K=\infty} C_K e^{jK\omega t} \Delta K\omega, \tag{2.42}$$

$$\lim_{\substack{T\to\infty \\ \Delta K\omega\to 0}} f(t) = \frac{1}{2\pi} \int_{-\infty}^{\infty} C_K e^{j\omega t}\, d\omega. \tag{2.43}$$

Therefore, the discrete lines of C_K in Figure 2.5*b* merge into a continuous frequency spectrum as illustrated in Figure 2.6*b*. As the period T approaches infinity, Eq. (2.40) can be written as

$$C_K = F(\omega) = \int_{-\infty}^{\infty} f(t) e^{-j\omega t}\, dt. \tag{2.44}$$

This is the definition of the *Fourier transform* of a nonperiodic function $f(t)$, and is commonly denoted as $F(\omega)$. It is important to note here that the function must satisfy the following condition of absolute convergence before the Fourier transform of a function is taken:

$$\int_{-\infty}^{\infty} |f(t)| dt < \infty. \tag{2.45}$$

As an example of the calculation of a Fourier transform, consider the following exponential time function, where $\theta < 0$.

$$f(t) = e^{\theta t}, \qquad \text{for} \quad t \geqslant 0, \tag{2.46}$$

$$f(t) = 0, \qquad \text{for} \quad t < 0. \tag{2.47}$$

Let us first check on the convergence of the function. From Eq. (2.45),

$$\int_{-\infty}^{\infty} |e^{\theta t}| dt < \infty, \qquad \textit{since} \quad \theta < 0. \tag{2.48}$$

Therefore, this exponential function is absolutely integrable and its Fourier transform can be obtained from Eq. (2.44), with the lower limit replaced by zero, as follows:

$$F(\omega) = \int_{0}^{\infty} e^{\theta t} e^{-j\omega t} dt, \theta < 0. \tag{2.49}$$

Integrating, we obtain

$$F(\omega) = \left[\frac{e^{(\theta - j\omega)t}}{\theta - j\omega} \right]_0^\infty, \tag{2.50}$$

$$F(\omega) = \frac{1}{j\omega - \theta}. \tag{2.51}$$

The Fourier transform is a very powerful mathematical tool that is used to a great extent in engineering. However, the limitation defined by Eq. (2.45) restricts its use in important situations. For example, the control engineer is usually interested in the response of a control system to unit step, ramp, and parabolic time functions denoted by $U(t)$, $tU(t)$, and $t^2U(t)$, respectively. Unfortunately, the Fourier transforms of these functions do not exist. For these types of functions, the engineer modifies the Fourier transform by adding a convergence factor $e^{-\sigma t}$ where σ is a real number that is large enough to maintain absolute convergence. Therefore, the new transform is given by

$$F(\sigma, \omega) = \int_0^\infty f(t) e^{-\sigma t} e^{-j\omega t}\, dt. \tag{2.52}$$

Notice that the lower limit is defined as zero rather than minus infinity, so that the new transform only applies to time functions that vanish for negative time. However, this is not a serious limitation in control problems, because the time reference is usually chosen to be $t = 0$. By defining

$$s = \sigma + j\omega, \tag{2.53}$$

Eq. (2.52) can be written as

$$F(s) = \int_0^\infty f(t) e^{-st}\, dt, \tag{2.54}$$

where $F(s)$ is defined as the Laplace transform of the function $f(t)$.

2.4. REVIEW OF THE LAPLACE TRANSFORM

The Laplace transform [1–5] is helpful in the solution of ordinary differential equations describing the behavior of systems. When the transform operates on a differential equation, a transformed equation results. It is expressed in terms of an arbitrary complex variable s. The resulting transformed equation is in purely algebraic terms, which can be easily manipulated to obtain a solution for the desired quantity as an explicit function of the complex variable. In order to obtain a solution in terms of the original variable, it is necessary to carry out an inversion process to determine the desired time function. The inverse Laplace transform is given by

$$f(t) = \frac{1}{2\pi j} \int_{c-j\infty}^{c+j\infty} F(s) e^{st}\, ds, \tag{2.55}$$

where c is a real constant greater than the real part of any singularity of $F(s)$. The evaluation of this integral is usually difficult, and the inverse transformations are usually obtained by expanding $F(s)$ into simpler components using a partial fraction expansion, and then obtaining the equivalent time-domain representation of the simpler components utilizing a table of Laplace transforms. The Laplace transform $F(s)$ of a certain function of time $f(t)$ is conventionally written as

$$F(s) \triangleq \mathscr{L}[f(t)]. \tag{2.56}$$

From the definition of the Laplace transform given by Eq. (2.54), the integral exists if

$$\int_0^\infty |f(t)e^{-\sigma t}|\, dt < \infty. \tag{2.57}$$

2.5. USEFUL LAPLACE TRANSFORMS

The Laplace transforms for various time functions will now be considered. These are readily obtainable through a direct application of Eq. (2.54).

A. Laplace Transform of a Unit Step

For the unit step function defined by

$$f(t) = U(t), \qquad f(t) = \begin{cases} 0, & \text{for} \quad t < 0, \\ 1, & \text{for} \quad t \geqslant 0, \end{cases}$$

the Laplace transform is

$$\mathscr{L}[f(t)] = \int_0^\infty (1)e^{-st}\, dt = \left[-\frac{1}{s}e^{-st}\right]_0^\infty.$$

Therefore,

$$\mathscr{L}[U(t)] = 1/s, \qquad \text{Re}[s] > 0. \tag{2.58}$$

From here on we assume that $f(t) = 0$ for $t < 0$.

B. Laplace Transform of an Exponential Decay

For the function

$$f(t) = \begin{cases} e^{-\alpha t}, & t \geqslant 0, \\ 0, & t < 0, \end{cases}$$

we have the Laplace transform

$$\mathscr{L}[f(t)] = \int_0^\infty e^{-(s+\alpha)t}\,dt = \left[-\frac{1}{s+\alpha}e^{-(s+\alpha)t}\right]_0^\infty.$$

Therefore,

$$\mathscr{L}(e^{-\alpha t}) = \frac{1}{s+\alpha}, \qquad \mathrm{Re}[s] > -\alpha. \tag{2.59}$$

C. Laplace Transform of a Unit Ramp

For the function

$$f(t) = \begin{cases} t, & t \geqslant 0, \\ 0, & t < 0, \end{cases}$$

the Laplace transform is

$$\mathscr{L}[f(t)] = \int_0^\infty te^{-st}\,dt. \tag{2.60}$$

Integrating by parts,

$$\int u\,dv = uv - \int v\,du, \tag{2.61}$$

with $u = t$, $dv = e^{-st}\,dt$, the following is obtained:

$$\int_0^\infty te^{-st}\,dt = \left[t\frac{e^{-st}}{-s}\right]_0^\infty - \int_0^\infty \frac{e^{-st}}{-s}\,dt.$$

Therefore,

$$\mathscr{L}(t) = \frac{1}{s^2}, \qquad \mathrm{Re}[s] > 0. \tag{2.62}$$

D. Laplace Transform of a Sinusoidal Function

For the function

$$f(t) = \begin{cases} \sin\omega t, & t \geqslant 0, \\ 0, & t < 0, \end{cases}$$

The Laplace transform is

$$\mathscr{L}[f(t)] = \mathscr{L}(\sin\omega t) = \int_0^\infty \sin\,\omega t\; e^{-st}\,dt. \tag{2.63}$$

The solution to Eq. (2.63) is simplified by using the exponential form of sin ωt,

$$\sin \omega t = \frac{e^{j\omega t} - e^{-j\omega t}}{2j}.$$

Therefore,

$$\begin{aligned}\int_0^\infty \sin \omega t \, e^{-st} \, dt &= \int_0^\infty \frac{e^{j\omega t} - e^{-j\omega t}}{2j} e^{-st} \, dt \\ &= \frac{1}{2j} \int_0^\infty (e^{-(s-j\omega)t} - e^{-(s+j\omega)t}) \, dt \\ &= \frac{1}{2j} \left(\frac{1}{s - j\omega} - \frac{1}{s + j\omega} \right).\end{aligned} \tag{2.64}$$

Therefore,

$$\mathscr{L}(\sin \omega t) = \frac{\omega}{s^2 + \omega^2}, \qquad \text{Re}[s] > 0. \tag{2.65}$$

Once the Laplace transform for any function $f(t)$ is obtained and tabulated, it need not be derived again. The foregoing results and other important transform pairs useful to the control engineer appear in Table 2.1. An extended table is shown in Appendix A. In addition, the location of the poles of the transformed function in the s-plane is listed in Table 2.1.

Table 2.1. Important Laplace Transform Pairs

Name of Function	Time Function, $f(t)$	Laplace Transform, $F(s)$	Location of Poles in s-plane
1. Unit impulse at $t = 0$	$\delta(t)$	1	None
2. Unit step	$U(t)$	$\frac{1}{s}$	One pole at the origin
3. Unit ramp	$tU(t)$	$\frac{1}{s^2}$	Double pole at the origin
4. Parabolic	$t^2 U(t)$	$\frac{2}{s^3}$	Triple pole at the origin
5. nth order ramp	$t^n U(t)$	$\frac{n!}{s^{n+1}}$	Pole of multiplicity $(n+1)$ at the origin
6. Exponential decay	$e^{-\alpha t} U(t)$	$\frac{1}{s+\alpha}$	One pole on the real axis at $-\alpha$
7. Sine wave	$\sin \omega t U(t)$	$\frac{\omega}{s^2+\omega^2}$	Two poles on the imaginary axis at $\pm j\omega$
8. Cosine wave	$\cos \omega t U(t)$	$\frac{s}{s^2+\omega^2}$	Two poles on the imaginary axis at $\pm j\omega$
9. Exponentially decaying sine wave	$e^{-\alpha t} \sin \omega t U(t)$	$\frac{\omega}{(s+\alpha)^2+\omega^2}$	Two complex poles located at $-\alpha \pm j\omega$

2.6. IMPORTANT PROPERTIES OF THE LAPLACE TRANSFORM

The Laplace transform has been introduced in order to simplify several mathematical operations. These operations center upon the solution of linear differential equations. Several basic properties of the Laplace transform are given here.

A. Addition and Subtraction

If the Laplace transforms of $f_1(t)$ and $f_2(t)$ are $F_1(s)$ and $F_2(s)$, respectively, then

$$\mathcal{L}[f_1(t) \pm f_2(t)] = F_1(s) \pm F_2(s).$$

B. Multiplication by a Constant

If the Laplace transform of $f(t)$ is $F(s)$, the multiplication of the function $f(t)$ by a constant K results in a Laplace transform $KF(s)$.

C. Direct Transforms of Derivatives

If the Laplace transform of $f(t)$ is $F(s)$, the transform of the first time derivative $\dot{f}(t)$ of $f(t)$ is given by

$$\mathcal{L}\left[\dot{f}(t)\right] = sF(s) - f(0^+), \tag{2.66}$$

where $f(0^+)$ is the initial value of $f(t)$, evaluated as $t \to 0$ from the positive region. The transform of the second time derivative $\ddot{f}(t)$ of $f(t)$ is given by

$$\mathcal{L}\left[\ddot{f}(t)\right] = s^2F(s) - sf(0^+) - \dot{f}(0^+) \tag{2.67}$$

where $\dot{f}(0^+)$ is the first derivative of $f(t)$ evaluated at $t = 0^+$. The Laplace transform of the nth derivative of a function is given by

$$\mathcal{L}\left\{\frac{d^n f}{dt^n}\right\} = s^n F(s) - s^{n-1} f(0^+) - s^{n-2}\dot{f}(0^+) - \cdots - f^{(n-1)}(0^+), \tag{2.68}$$

The notation $f^{(n-1)}(0^+)$ represents the $(n-1)$th derivative of $f(t)$ with respect to time evaluated at $t = 0^+$.

D. Direct Transforms of Integrals

If the Laplace transform of $f(t)$ is $F(s)$, the transform of the time integral of $f(t)$ is given by

$$\mathcal{L}\left[\int f(t)\,dt\right] = \frac{F(s)}{s} + \frac{1}{s}\left[\int f(t)\,dt\right]_{t=0^+}, \tag{2.69}$$

where $[\int f(t)\,dt]_{t=0^+}$ signifies that the integral is evaluated as $t \to 0$ from the positive region. In general, for nth-order integration,

$$\mathscr{L}\left[\int\int\cdots\int f(t)\,dt^n\right] = \frac{F(s)}{s^n} + \frac{1}{s^n}\left[\int f(t)\,dt\right]_{t=0^+} + \frac{1}{s^{n-1}}\left[\int\int f(t)\,dt^2\right]_{t=0^+} + \cdots + \frac{1}{s}\left[\int\int\cdots\int f(t)\,dt^n\right]_{t=0^+}. \tag{2.70}$$

E. Time-Shifting Theorem

The Laplace transform of a time function $f(t)$ delayed in time by T equals the Laplace transform of $f(t)$ multiplied by e^{-sT}:

$$\mathscr{L}[f(t-T)U(t-T)] = e^{-sT}F(s), \qquad t \geqslant T. \tag{2.71}$$

F. Frequency-Shifting Theorem

If the Laplace transform of $f(t)$ is $F(s)$, then the Laplace transform of

$$e^{-at}f(t)$$

is obtained as follows:

$$\mathscr{L}[e^{-at}f(t)] = \int_0^\infty e^{-at}f(t)e^{-st}\,dt = \int_0^\infty f(t)e^{-(s+a)t}\,dt = F(s+a). \tag{2.72}$$

Therefore, multiplying $f(t)$ by e^{-at} is equivalent to replacing s by $(s+a)$ in the Laplace transform. In addition, changing s to $(s+a)$ is equivalent to multiplying $f(t)$ by e^{-at}.

G. Initial-Value Theorem

If the Laplace transform of $f(t)$ is $F(s)$, and if $\lim_{s\to\infty} sF(s)$ exists, then the initial value of the time function is given by

$$\lim_{t\to 0} f(t) = \lim_{s\to\infty} sF(s). \tag{2.73}$$

H. Final-Value Theorem

If the Laplace transform of $f(t)$ is $F(s)$, and if $sF(s)$ is analytic on the imaginary axis and in the right half-plane, then the final value of the time function is given by

$$\lim_{t\to\infty} f(t) = \lim_{s\to 0} sF(s). \tag{2.74}$$

2.7. INVERSION BY PARTIAL FRACTION EXPANSION

The time response is the quantity of ultimate interest to the control-system designer. The process of inversion of a function $F(s)$ to find the corresponding time function $f(t)$ is denoted symbolically by

$$\mathscr{L}^{-1}F(s) = f(t). \tag{2.75}$$

In applications, $F(s)$ is usually a rational function of the form

$$F(s) = \frac{A_X s^X + A_{X-1}s^{x-1} + \cdots + A_1 s + A_0}{s^Y + B_{Y-1}s^{Y-1} + \cdots + B_1 s + B_0} \tag{2.76}$$

In practical systems, the order of the polynomial in the denominator is equal to, or greater than, that of the numerator. For the cases where $Y > X$, partial fraction expansion is directly applicable. When $Y \leqslant X$, it is necessary to reduce $F(s)$ to a polynomial in s plus a remainder (ratio of polynomials in s).

The simplest method for obtaining inverse transformations is to use a table of transforms. Unfortunately, many forms of $F(s)$ are not found in the usual table of Laplace-transform pairs. When the form of the solution cannot be readily reduced to a form available in a table, we must use the technique known as partial fraction expansion. This method permits the expansion of the algebraic equation into a series of simpler terms whose transforms are available from a table. It is then possible to obtain the inverse transformation of the original algebraic expression by adding together the inverse transformations of the terms in the expansion. Equation (2.75) expresses this operation symbolically. The function $F(s)$ represents the original algebraic expression and $F_1(s), F_2(s), F_3(s), \ldots, F_n(s)$ are terms of the partial fraction expansion:

$$\mathscr{L}^{-1}[F(s)] = \mathscr{L}^{-1}[F_1(s)] + \mathscr{L}^{-1}[F_2(s)] + \cdots + \mathscr{L}^{-1}[F_n(s)]. \tag{2.77}$$

As an example of the method, consider the transform

$$F(s) = \frac{As + B}{(s + C)(s + D)}. \tag{2.78}$$

In this equation A, B, C, and D are constants. This function can be expanded into partial fractions:

$$\frac{As + B}{(s + C)(s + D)} = \frac{K_1}{s + C} + \frac{K_2}{s + D}, \tag{2.79}$$

where K_1 and K_2 are the coefficients of the expansion, and the coefficients K_i are called the *residues* at the pole $s = -p_i$. The residues are determined by multiplying both sides of Eq. (2.79) by the denominator factor, $(s + p_i)$, corresponding to K_i and setting s equal to its root, $s = -p_i$.

To determine K_1, both sides of Eq. (2.79) are multiplied by $s + C$, yielding

$$\frac{As + B}{s + D} = K_1 + \frac{K_2(s + C)}{s + D}.$$

By substituting $s = -C$, the last term vanishes and a numerical value for K_1 can be obtained. An analogous procedure leads to the value for K_2. One finds

$$K_1 = \frac{B - AC}{D - C}, \quad K_2 = \frac{B - AD}{C - D}.$$

An expression for the expanded form of $F(s)$ can now be obtained by substituting this result into Eq. (2.79):

$$F(s) = \frac{B - AC}{D - C} \frac{1}{s + C} + \frac{B - AD}{C - D} \frac{1}{s + D}. \tag{2.80}$$

It is now a simple process to obtain the inverse Laplace transform from this last equation. The corresponding time function can be obtained by inspecting the terms and comparing them with transform pairs listed in Table 2.1 or Appendix A:

$$f(t) = \frac{B - AC}{D - C} e^{-Ct} + \frac{B - AD}{C - D} e^{-Dt}, \text{ for } t \geqslant 0. \tag{2.81}$$

Transforms with multiple poles are also encountered. For example, consider the transform

$$F(s) = \frac{As + B}{(s + C)^2 (s + D)}. \tag{2.82}$$

In this equation A, B, C, and D are constants. The partial fraction expansion is written as

$$\frac{As + B}{(s + C)^2 (s + D)} = \frac{K_1}{(s + C)^2} + \frac{K_2}{s + C} + \frac{K_3}{s + D}. \tag{2.83}$$

To find K_1, both sides of Eq. (2.83) are multiplied by $(s + C)^2$:

$$\frac{As + B}{s + D} = K_1 + K_2 (s + C) + \frac{K_3 (s + C)^2}{s + D}. \tag{2.84}$$

The constant K_1 can now be evaluated by simply substituting $s = -C$:

$$K_1 = \frac{B - AC}{D - C}. \tag{2.85}$$

In order to determine the constant K_2, both sides of Eq. (2.84) must be differentiated with respect to s, and s is then set equal to $-C$:

$$\left[\frac{d}{ds} \frac{As + B}{s + D}\right]_{s=-C} = K_2 + K_3 \left[\frac{d}{ds} \frac{(s + C)^2}{s + D}\right]_{s=-C} \tag{2.86}$$

The resulting numerical value for K_2 is given by

$$K_2 = \left[\frac{d}{ds} \frac{As + B}{s + D}\right]_{s=-C} = \frac{AD - B}{(D - C)^2}. \tag{2.87}$$

The constant K_3 can be obtained by the same procedure used in evaluating the expression given by Eq. (2.78). Its value is

$$K_3 = \frac{B - AD}{(C - D)^2}. \tag{2.88}$$

The corresponding time function can be obtained by inspecting the terms in Eq. (2.83) and comparing them with the transform pairs listed in Table 2.1 or Appendix A:

$$f(t) = \frac{B - AC}{D - C} te^{-Ct} + \frac{AD - B}{(D - C)^2} e^{-Ct} + \frac{B - AD}{(C - D)^2} e^{-Dt}, \quad \text{for} \quad t \geqslant 0.$$

2.8. APPLICATION OF MATLAB TO CONTROL SYSTEMS [6]

At this juncture of the presentation, we can show that partial-fraction expansion can also be accomplished with a simple command from MATLAB. An important feature of this book is that most of the solutions shown in this book were generated using the commercially available software package called MATLAB which is available from the MathWorks, Inc. The *Modern Control System Theory and Design (MCSTD) Toolbox*, which complements this book, and the M-files that were used to develop these solutions can be retrieved free from The Mathworks, Inc. anonymous FTP server at ftp://ftp.mathworks.com/pub/books/shinners. These M-files are the *MCSTD Toolbox* and were used to develop the graphical figures and problem solutions in this book. In this manner, the user of this book has available a toolbox of features/utilities created to enhance The Mathworks' Control System Toolbox, and the computer-generated solutions to the problems in this book. The *Modern Control System Theory and Design Toolbox* runs equally well with *The Student Edition of MATLAB* (authored by The Mathworks, Inc., and published by Prentice-Hall, Englewood Cliffs, NJ), as well as with or without the professional versions of MATLAB (with or without The MathWorks' Control System, Simulink, Nonlinear, and the Symbolic Toolboxes).

- MATLAB, an abbreviation for *MAT*rix *LAB*oratory, is a matrix-based system used for engineering calculations which has progressed to become the language used by most control-system engineers worldwide. This section focuses attention on the required background needed for designing control systems with MATLAB. MATLAB is a very useful language which operates interactively with the user, and will respond to all of the user's attempts at conversation. It offers a variety of graphic output displays useful to control-system engineers such as system block diagram development (Simulink required), linear, log, semilog, polar, and contour plots. It is envisioned that this section will serve as a supplement to Reference 6, and enable students, professors, and practicing engineers to quickly learn the basics needed to use MATLAB with its associated toolboxes and the *Modern Control System Theory and Design Toolbox*.

This integrated textbook and software learning package is self-contained, and is designed for undergraduate courses on control systems and for the practicing engineer. This toolbox (software) contains extensive new features/utilities created to enhance MATLAB and several of The MathWorks' toolboxes. *Since this integrated learning package and the MATLAB software is self-contained, it is not necessary to purchase additional books/material to learn how to use MATLAB with this textbook.*

The software contained in the *Modern Control System Theory and Design Toolbox* contains the following features that makes this book self-contained for use with MATLAB:

- A *Tutorial File* has been created that contains the essentials necessary to understand and effectively utilize the MATLAB interface. This tutorial file aids the user in understanding the MATLAB interface where most other books require additional books for full comprehension. Features of this file are as follows:
  - MATLAB installation assistance
  - MATLAB performance improvement suggestions
  - MATLAB fundamentals
  - Understanding notations used by MATLAB
  - Control analysis using MATLAB
  - *Modern Control System Theory and Design Toolbox* use with MATLAB

- A *Demonstration m-file* gives the users a feel for the various utilities included in the *Modern Control System Theory and Design Toolbox*. Included are the following:
  - General purpose utilities
  - Linear, frequency domain, Bode diagrams (used starting in Chapter 6)
  - Linear, frequency domain, Nichols charts (used starting in Chapter 6)
  - Linear, frequency domain, Root locus (used starting in Chapter 6)
  - Nonlinear, frequency domain, Describing function (used in Chapter 5 of the accompanying volume)
  - Linear, time domain
  - Conversions between discrete time domain and continuous time systems (used in Chapter 4 of the accompanying volume)

  This demonstration helps the user learn how to use the MATLAB package easily with the *Modern Control System Theory and Design Toolbox* and, with the tutorial, makes this integrated package self-contained.

- Online HELP is available for all *Modern Control System Theory and Design Toolbox* utilities. Additionally, the "lookfor" command is fully supported in the online help for each *Modern Control System Theory and Design Toolbox* utility.

- A *Synopsis File* reviews and highlights the features of each chapter in *Modern Control System Theory and Design* in a concise manner which is helpful in guiding the professor on subjects to emphasize, and to the student and practicing engineer for reviewing and remembering important aspects of the coverage.

- The software is compatible with all editions the *The Student Edition of MATLAB* [31] and the Professional Versions of MATLAB [6], and it is compatible with The MathWorks' following software packages:
  - Control System Toolbox
  - Simulink Toolbox
  - Nonlinear Toolbox
  - Symbolic Toolbox

  I have attempted to make the presentation of MATLAB easy reading for the reader about to learn and use MATLAB. Most of this material can be read with little or no hands-on practice while getting a full grasp of most of the MATLAB capabilities.

A. First Time Usage—Software For Engineering*

This section deals with the software. It does assume that you already have an understanding of the theoretical principals of the topics about to be presented. Any software that you may use now, or any time in the future, is only as good as your comprehension of the topics, and your ability to interpret the results presented to you by the computer. This cannot be stressed enough, because misinterpreted results due to whatever cause (incorrect inputs, computer round-offs, misinterpreted printouts, faulty programming code, etc.) can lead to any number of problems, depending on what you are doing. Computer literate people are all too aware of this, and have even given this a name: GIGO (garbage-in/garbage out).

With the advent of modern computers, engineers (as well as many other technical professionals) have had available a tool that can help them do their jobs quicker, more accurately, presentably, etc. The main question is what software are you going to use? Programming it all yourself is cumbersome, time-consuming, must be customized for each application, requires extensive understanding of all the topics encompassed, and on top of it all, is possibly ridden with errors. However, existing software packages have most of the features you require at a simple-to-use level ("user-friendly").

The major problem that most people have is which software package to buy! The correct answer has to depend on your current needs, and your expected near-term future needs. Far future needs probably should not be addressed due to the continually evolving computer software and hardware world, and should be re-evaluated when you get there. Some topics to consider are:

1. Ease of learning/use (user-friendly)
2. Analytic capability
3. Graphic capability
4. Expandability/modularity
5. Technical support/support groups

*It is suggested that those readers who are unfamiliar with MATLAB read Sections A through E before writing MATLAB programs.

6. Hardware requirements
7. Price

Considering that this section is primarily prepared for a course in which it is beneficial for the student to own their own software (maximizing-capability/minimizing-costs), versus having a classroom licence (which usually ends up being a one-time experience), I suggest using "*The Student Edition of MATLAB*" by the MathWorks, Inc. It includes Software, Manual, User Groups, etc., and has most of the essentials that assist in learning a wide variety of topics at a very modest price.*

The Student Edition of MATLAB package gives you extensive mathematical/analytical/graphical capabilities, as well as a "taste" of two of the different professional-level toolboxes (Signal Toolbox and Control Toolbox) that they combine to call the "Signals and Systems Toolbox" (quite adequate for classroom training purposes of both topics). This is adequate for you to realize many control-theory applications, the programming potential of MATLAB, and the ease of dealing with toolbox add-ons (conceptually and mathematically). Another item of interest to all, is that there are available "FREE" toolboxes to all who desire. In their newsletters (that they willingly send to registered users) they list textbooks that use MATLAB and/or have toolboxes associated with the books. To top it off, they maintain a MATLAB User Group software library (nicknamed MUG) that contains all past newsletters, and FREE User software, as well as examples, and other useful information, available at no charge to you.

B. First Time Usage—MATLAB Installation

Installation is quite straightforward, just a little time-consuming. This step-by-step installation procedure focuses on the IBM (or compatible) PC. However, there are additional instructions in the MCSTD Toolbox which apply equally as well to Macintosh and UNIX users. The user will need some basic knowledge of DOS and the system that you are working on. Certain typos in the first printing of *The Student Edition of MATLAB* for the IBM PC have been identified (subsequently corrected in later printings), and it is suggested that you correct them prior to reading and installing MATLAB:

1. p. 16: "three or more 360K" should read "two 360K"
2. p. 26: "\MATLABGEO" should read "\MATLAB\GEO"

Now simply follow the directions in Chapter 4. Once completed, you are ready to start using the student version of MATLAB. You may want to read the section on MATLAB fundamentals before trying to use MATLAB, but definitely read the following first.

If you have followed the installation section of MATLAB, all you have to do to start using MATLAB, is to type "MATLAB" and hit enter at the DOS prompt.

*All references to pages or manuals herein is for *The Student Edition*, unless otherwise annotated.

After a moment to load, you should see the MATLAB copyright message on the top of the screen, followed by the line "HELP, DEMO, and INFO are available" (if proper installation has occurred). If this much does not come up whenever you activate MATLAB, something has gone wrong.

If you encounter any difficulties with the package:

1. Re-read the appropriate chapters in the textbook.
2. Check for obvious typographical errors.
3. For the *Student Edition*, consult the instructor (who has the available technical support). For the professional version, consult The MathWorks.

The first time (and only the first time) that you activate MATLAB, it will prompt you for registration information (serial number and your name) following its normal messages. The MATLAB book does not tell you anything about this step because it is a leftover feature from the professional version not described in any (student version or professional version) of their manuals (it is explained with a separate letter attached to the professional version). It is a formality, so that you can identify your copy of MATLAB without having to pull out your original diskettes when calling for technical support. Fill it in by answering the prompts that come on the screen. I do suggest entering your name, if nothing else but to personalize this copy as yours.

To add additional toolboxes to MATLAB, merely follow the directions on the bottom of page 26 (as if you are adding a library of your own). For installing the FREE *Modern Control System Theory and Design Toolbox* (henceforth called the MCSTD toolbox), proceed as follows:

1. "MD\MATLAB\MCSTD," I chose the name MCSTD as an abbreviation of the name.
2. "COPY A:\TOOLBOX\MATLAB\MCSTD," copy the toolbox part of the MCSTD diskette.
3. "EDLIN\MATLAB\BIN\MATLAB.BAT," update the file to reflect the new toolbox. In MATLAB version 4.0, the file "MATLABRC.M" must be updated. the "SET MATLABPATH=" has now got to have ";\MATLAB\MCSTD" appended to it.
4. The toolbox is fully installed and ready to be used in MATLAB.

The problems, figures, Fortran and demo directories that come with that toolbox are non-essentials (but give a nice demo of the MCSTD features and a variety of worked examples from this book). They can be installed in a similar manner to the above. The readme.m, contents.m, tutorial, and synopsis files should be printed as reference material.

C. First Time Usage—Performance Tuning MATLAB

I wish to provide to the reader some of the things that I have learned in the proper use of MATLAB. For those of you who will spend hours at a time using MATLAB (student or professional version), this material will greatly benefit you. These are the suggestions that I make to my fellow colleagues using MATLAB, for significantly improving performance at an insignificant cost!

The first thing you notice when starting MATLAB, is how long it takes to start before you can type. Those of you with sharp eyes and/or minds might even notice a lot of hard disk activity occurring during this time. This is because MATLAB is an interpreter. That means for every command that you give, it must read the file with that command in it (and similarly the same for every command called inside the original command). With that in mind (and the fact that you can't change this), the faster the drive that has all these commands on them, the faster your performance.

The limiting factors for your choice, as you will see, is available computer memory and/or your desire to try something different. Buying a new and better drive is overkill and may not give you peak performance. Software that improves your current drive's speed is a step in the right direction; however, creating a drive out of computer memory (which is the fastest possible drive for your machine) is the ideal choice. The super-intelligent among you will even use a combination of these techniques (as I am about to explain). I have seen speed factor improvements of 3 to 15 times using these techniques.

With the advantages come the disadvantages (and I feel I should point them out before going any farther). When creating memory drives, as the computer is turned off, all information on that drive is lost. My solution is to copy the standard toolboxes and software to that drive when I am going to use it (requires a short additional length of time), and keep my personal routines and data on my hard disk (the best of both worlds). Also, most PCs have limited memory which you use sparingly. My solution is: if you don't mind the speed, stick with what you have. Otherwise, merely buy and install more memory (the appropriate type: expanded or extended) for your computer.

The amount of memory that you desire for your memory drive (henceforth called ramdrive) depends on how much software you want to put there, the more you put the faster you go. Demo routines with their data do not have to be kept on the ramdrive after you have become familiar with them. But for now, you can add up the sizes of the files you want to move (add 10 percent for blank space) to see how big this ramdrive should be.

The commands that add this ramdrive to your system (provided that the memory is already there) are either vdisk, ramdrive (in DOS), or whatever appropriate software came with the memory that you purchased. An example of my computer setup (using DOS version 5.00 with high memory; your machine may different) for *The Student Edition of MATLAB* is:

1. Adding "DEVICE=C:\DOS\RAMDRIVE.SYS 2000/E" to the file "C:\CONFIG.SYS" makes a 2000KB (2MB) ramdrive. Most ramdrives make you add one or more lines to the file "C:\CONFIG.SYS." When your computer is rebooted, the ramdrive should be available for use.
2. Adding "DEVICE=C:\DOS\SMARTDRV.SYS" to the file "C:\CONFIG.SYS," adds disk caching upon reboot.
3. Adding "FASTOPEN C:" to the file "C:\AUTOEXEC.BAT," improves file access performance for that drive.

In step one, you may choose an appropriate size for your specific needs. I chose 2MB because it is adequate to hold all of *Student Edition of MATLAB* and the complete MCSTD *Toolbox*, serving my purpose quite well. You may choose not to put all that on your ramdrive, but only the parts that you are going to use (such as no demo files, compiler linking files, example files, etc.). On the other hand, you may have other toolboxes that you want to put on this ramdrive. This is a choice that you must make.

Now that you have your ramdrive set up (all changes have been made and you have rebooted), you want to take advantage of these changes. I will use "F:" as the letter of the ramdrive created—change according to your individual system. Your file that starts MATLAB up ("C:\MATLAB\BIN\MATLAB.BAT") has to reflect these changes. This is how I suggest altering that file:

1. Add to the top of the file a line for each directory that you want to copy which copies that directory from your hard drive to your ramdrive. One example is "XCOPY C:\MATLAB\MATLAB F:\MATLAB\MATLAB\". This would copy MATLAB from your hard disk C: to your ramdrive F:.
2. Alter the "SET MATLABPATH = " line by including the drive letter where the directory exists for each directory listed (you may mix between hard disk and ramdrive as you choose). In MATLAB version 4.0, the MATLABPATH is set in the file "MATLABRC.M".

The next performance suggestion concerns improving the speed of shelling to DOS with the MATLAB "!" command. If the computer uses the file "COMMAND.COM" from a ramdrive, it can load and execute the "!" command faster. Not only MATLAB will benefit from this tip, but also any program that shells to DOS. Further DOS performance tips I leave for the experts to suggest (but I do like this one in particular). This is my suggestion for altering the file "C:\AUTOEXEC.BAT":

1. Add "COPY C:\COMMAND.COM F:\" to the bottom of this file.
2. Add "SET COMSPEC = F:\COMMAND.COM" following the copy line. This tells DOS to use that file when shelling to DOS.

The last performance suggestion, as mentioned earlier, concerns demo files which take room but also make the computer search for each MATLAB command a little wider (since it checks all file names in the MATLABPATH until the file is found). I do not like losing anything, so what I do is move the demos/examples of each toolbox to separate directories (as I have already done in the MCSTD toolbox). This way I can always add the demos/examples to the MATLABPATH in the MATLAB.BAT file. Personally, I like to do as little as possible, and just create a copy of the MATLAB.BAT file (I call MATLABD.BAT) with those changes made to it. Also, I do not copy the demos/examples to the ramdrive (which would reduce the room on the ramdrive as well) unless I want them (as in MATLABD.BAT).

D. First Time Usage—MATLAB Fundamental Concepts

MATLAB is a very simple, but efficient, interpreter (versus compiler). This allows you to type in a line and have it execute immediately. The tremendous advantage is that you do not have to re-compile your program (as you would in Fortran, Pascal, or C) for every change you make before re-executing the code, as well as the ability to view the values of the variables without adding debugging write statements to your program. MATLAB does include an interface so that you can execute your Fortran or C code from inside MATLAB (see MATLAB manual section on MEX files). Like any other language, it has a small subset of commands (statements), from which other more sophisticated commands are developed by the users. There are four fundamental items in MATLAB: variables, functions, programming code and algebraic operators.

Variables, in MATLAB, are all treated as matrices. Text strings (which are stored as matrices), however, are automatically given a special attribute, so that the ASCII text value is printed (not the numeric value).

Functions in MATLAB operate slightly differently than most programming languages (with the possible exception of C++). Most MATLAB functions take none, one or more variables as parameters (within the parenthesis), and return none, one, or more variables as a result. An unusual aspect of this is that a single function can be used many different ways depending on how many parameters are being passed in, the type of data being passed in, and how many parameters are being retrieved.

Programming code, in MATLAB, is mainly for control flow, directory manipulation, data storage, and debugging purposes. Like most languages, this includes: if, else, end, for, while, ... etc. These will be discussed in more detail later in this text.

Algebraic operators, in MATLAB, are very similar to the math operators that you are used to, with only a few minor exceptions. The main consideration to be taken when using them is that they are (almost all) matrix math operators. If you want to use scalar mathematics, you must take care to use the scalar versions of the algebraic operators.

Another topic that should be mentioned in this section, is the concept of toolboxes. A toolbox, in MATLAB, is a collection of functions/utilities that work together on a specific topic (i.e., signal processing, control processing, system identification toolbox, ...) so that it (hopefully) meets all your needs on this topic, and you do not need to program any functions for your own needs. As your skills expand into other areas of expertise, there are other toolboxes (the MATLAB toolbox list is always growing) available for (in the professional world) a reasonable cost (versus you programming–debugging–testing your own code for each specific need). The MathWorks' desire to express this highly important fact has lead them to include the "Signal and Systems" toolbox with their student version. This toolbox gives the user a taste of the professional versions of the Signal and Systems Toolbox and the Control-System toolbox. A more detailed discussion on these and other toolboxes will follow later.

E. First Time Usage—Matrix Representations

Every variable of any sort in MATLAB is a matrix. The approach is to represent a wide variety of items as matrices. Once certain conventions (of which MATLAB has many predefined) are agreed upon, all that remains is to learn how to use them. All the types about to be described here have values which are either real or complex.

Scalar values start very simply, represented as a 1×1 matrix. They can take on any double precision value that the computer can represent, including some unusual ones: Infinity (Inf) and Not-A-Number (NaN).

Inf represents a number the representation of which is beyond the computer's ability. Almost any mathematical operation dealing with such a number yields the value Inf, as should be expected.

The use of NaN is difficult to illustrate; consider a set of data with one value in it that is totally absurd. If you use that value in your data set, your analysis results will be totally corrupted. Instead of removing that data from your set (which sometimes cannot be done) it would be nice to say that value is *N*ot really *a* valid *N*umber, hence NaN. Now we can agree to check for this value in our analysis routines, which is already done in many MATLAB routines.

Vector values become a little more sophisticated, represented as a $1 \times N$ matrix or a $N \times 1$ matrix (N being the length of the vector). The distinction between the two representations is used by MATLAB, and will be examined later. The typical uses of a vector are as a set of data, polynomial representations, etc. These uses will become more apparent to you as you practice with the topics presented later.

Matrix values are straightforward, as they represent a matrix. In MATLAB version 4.0, matrices have an added feature of being "sparse." Sparse matrix technology is a method for storing a large matrix filled with mostly zeros, in a much smaller storage space. This can aid in speed when dealing with these matrices. This feature does not change the features of a matrix, just the physical storage and speed of execution when used. Some typical matrix representation examples are sets of data samples, transition matrices, covariance matrices, etc. Once again, these uses will become more apparent to you as you practice with the topics presented later.

Strings are a little more confusing to understand. They are stored, like vectors, as a $1 \times N$ matrix. The difference between strings and vectors is that strings are given a special attribute (automatically set by MATLAB) so that they will be displayed properly on the screen. A non-vital detail (trivial and for reference purposes only) is that each letter in a string is stored as the numeric value of its ASCII representation.

F. First Time Usage—MATLAB Fundamentals

The MATLAB prompt that is listed/described in the manual is the "≫". At this prompt you will begin to learn how to use MATLAB. This practice session is created to give you the idea of how exit, find help, and inquire about variables in MATLAB.

MATLAB, like most modern software packages, tries to make the commands it uses very simple and logical. Most commands are like their English counterparts, making (hopefully) the commands simple for everyone to understand. One important fact that should be stated now is that MATLAB is case sensitive (this means that entering things in upper case is different than entering things in lower case). All commands that are to be entered in the following practice session should be in lower case unless specifically stated. After each command has been entered, to start the execution of the command press Enter or Return (this shall be assumed with each of the following commands, unless otherwise stated).

The first thing that any software should teach is how to exit the software and get back to your operating system (DOS). There are two wasy to do this: "exit" or "quit". Both are good examples of English-like commands. A good idea before exiting any software program is to save your work, but for now we shall not do this (especially since we have done no work, and you have not been shown how to save as of yet). After executing either of these commands, if you choose to continue using MATLAB, you must start MATLAB up because you have been returned to DOS.

The next thing that we are going to do is to "demo" some of the MATLAB features. Hopefully, you will have guessed that the command to do this is "demo", and you would be right. An extensive demo has been prepared by MATLAB, showing you a wide variety of its capabilities. You should take the time to try a few (or all) of these demos at your leisure.

The "help" feature of MATLAB is very useful to the user. This was designed so that you do not have to pull out the MATLAB manual (or toolbox manual(s)) in order to be able to understand the various commands currently available to you. There are several different ways to use the help command, depending on the version of MATLAB that you possess.

By typing "help" by itself, you will get either a list of all installed toolboxes (version 4.0 or later) or a list of all the currently available commands organized on different screen according to the toolboxes that you have installed (prior to version 4.0). As you add additional toolboxes to MATLAB, each toolbox will have its own screen listing the commands available in that toolbox. This is a nice feature which helps the individual think in a modular fashion when going to the professional world.

In version 4.0 or later, typing "help" followed by a space and the installed toolbox directory name, will list the commands in the toolbox with a oneline explanation of their functions (e.g., "help matlab").

Additionally, by typing "help" followed by a space and any listed command (in the manual or on the help screens), you will get an explanation of the command and how to use it (e.g., "help quit").

It is highly recommended that you review *The Student Edition of MATLAB.* I suggest getting some hands on practice while reading it. Functions worth reading about (or at least glancing at the help screens) are:

1. "!" allows you to execute DOS commands (from inside the MATLAB), without having to quit your MATLAB session (called shelling to DOS). When the DOS command that issues following the "!" has completed, you are returned to MATLAB without losing any of your MATLAB session.

   TSR warning: Do not use the "!" to run any TSR program. Start all TSR programs prior to entering MATLAB. TSRs take memory to add themselves to your environment (DOS). MATLAB is not aware of this memory usage, and will consequently use the memory, at some point in time, clobbering software and probably corrupting MATLAB and/or DOS (in memory). Do not be fooled if you do not see this happen, you may have been lucky in one of several ways this time. This time MATLAB may not have used the memory. This time what it clobbered may not have had any effect. Or you simply did not realize that the corrupted system is now returning corrupted results (MATLAB and/or DOS)!

2. "load" allows you to get data that has been stored. Two forms of data currently can be loaded: their own special form (called MAT files) and ASCII files that contain one numeric matrix (any size allowed) in it. This is useful to get data into MATLAB from other software packages.
3. "save" is the counterpart to load. This allows you to save data for future use. Like load, it saves as a MAT file or as an ASCII file (upon request). This is useful if you intend to export data to other programs after processing them in MATLAB.
4. "plot" is your first step into MATLAB's graphic world. There are too many different graphic commands to list, each with accompanying support functions. But as you may have guessed, "plot" plots data on the screen.

   Getting printed copies of the graphs is done in several different ways (see Apendix C in *The Student Edition* manual): using a graphic screen dump to the printer utility (printed quality is only as good as the screen resolution), using the "meta" command with their "GPP" software (professional MATLAB prior to version 4.0 or later), and using the print command or print menu (professional MATLAB version 4.0 or later). Since this is for the student version, I shall discuss the screen dump utility.

   DOS and MATLAB both come with screen dump TSRs (see prior TSR warning). If you intend to print any plots, you should load either one of these utilities prior to starting up MATLAB! The DOS utility is called "GRAPHICS." The MATLAB utility is "EGAEPSON" for Epson compatible printers, or "EGALASER" for laserjet compatible printers. If you have problems printing and have a VGA monitor, a suggestion is to restart MATLAB in the EGA mode.

Most functions are recognizable by their name and one-line descriptions. In *The Student Edition* manual (the only book that I reference by pages), you will find this section on pp. 185–194.

G. Control Systems—Data Representation

I will now show how to represent real systems as mathematical models. Any type of control system is going to be represented as matrices in MATLAB. The trick is to become fluent with their usage (which only really comes with practice). Each of the following subsections will make sense as you have been properly introduced to their topics by your instructor.

First and foremost, in most control-theory classes, is the *transfer function* representation of a system. This is where a system is represented by a series of polynomials, usually organized (as in most control-theory textbooks) as:

$$G(s) = \frac{C(s)}{R(s)} = \frac{A_m s^m + \ldots + A_1 s^1 + A_0 s^0}{B_n s^n + \ldots + B_1 s^1 + B_0 s^0}.$$

Here, a single-input/single-output (SISO) transfer function can be represented as two polynomials of decreasing powers (numerator polynomial divided by a demoninator polynomial). In MATLAB, a polynomial (numerator, denominator, or other) is represented as a row vector ($1 \times N$ matrix) containing the scale factors for each of the sequentially decreasing powers of s. This is a very important point to comprehend before reading the following examples:

1. "$17s^5 + 23s^3 + 8s^2 + 6$" converts to "[17 0 23 8 0 6]". The highest power of s is five, so there will be six numbers in this vector (s-powers 5 to 0). Powers of s that are not there have scale factors of zero (they hold the s-power sequence and are NOT to be forgotten).
2. "$5 + 12s + 4s^3 + 3$" converts to "[4 0 12 8]". If you cannot visually see this, then the best thing you can do for yourself is to rewrite the polynomial in "decreasing powers" of s as: "$4s^3 + 0s^2 + 12s^1 + (5 + 3)s^0$" or more simply "$4s^3 + 12s + 8$." You should now be able to see how to do this.

Next is the *zero-pole-gain method* that is similar to the transfer function in that the two methods represent the same equations, the equations have merely been manipulated into a different form. The difference is though that the polynomials have been factored into their roots, and "k" is the gain that makes the equations equal (namely the highest s-power numerator scale-factor divided by the highest s-power denominator scale-factor). This representation of the system is:

$$G(s) = \frac{Z(s)}{P(s)} = k\frac{(s - Z_1)(s - Z_2)\ldots(s - Z_N)}{(s - P_1)(s - P_2)\ldots(s - P_M)}$$

Here, a single-input/single-output (SISO) system (called the zero-pole-gain function) can be represented as two sets of roots (numerator and denominator) and a gain factor. When any of the numerator roots values goes to zero, the overall function value goes to zero; hence the numerator roots are called "Zeros." When any of the denominator root values goes to zero, the overall function value goes to infinity; hence, the donominator roots are called "Poles."

In MATLAB, a set of polynomial roots (numerator, denominator or other) is represented as a column vector ($N \times 1$ matrix) with each element representing one of these roots. *The Student Edition* attempts to show this "simple" transformation with one example followed by one example of the zero-pole-gain method. A few examples follow to challenge your understanding of "polynomial root" representation:

1. "$(s - 3)(s - 4.5)(s + 100)$" converts to "[3; 4.5; −100]". By now you should recognize the semicolon from MATLAB as meaning (among other things) a new line (new row). This makes the matrix listed above become a 3 rows by 1 column matrix (column matrix, three in length).
2. "$(s - 2)(s - 500)(s - 30)$" converts to "[500 ; 30 ; 2]". Your first guess would probably be different ([2 ; 500 ; 30] is also correct), because the order of the elements in this column vector (roots) is irrelevant, since in multiplication the order in which you multiply is irrelevant (at least in scalar math). Therefore, any order, as long as all values are accounted for, is correct.

3. "$(s-40)^3(s-80)$" converts to "[40 ; 40 ; 40 ; 80]". Each root has to be accounted for. Multiple roots must be accounted for each time they are multiplied.
4. "$(s^3-11s^2+38s-40)(s-3)$" converts to "[2 ; 4 ; 5 ; 3]". You must always remember to factor to a single power of s (even if this makes complex values). $s^3-11s^2+38s-40$ factors into $(s-2)(s-4)(s-5)$.

Continuing, the easiest of the system representations to explain arises. State-variable has become popular in the professional industry, due largely to the ease of representing multi-input/multi-output systems. Several signals can be evaluated simultaneously in very sophisticated systems without having to trace through the transfer functions for each input to output. Several other subjects that may cross your path in your progression of control-theory knowledge (e.g. Kalman filter, sensitivity analysis), use this base structure to build upon. State-variable concepts are discussed in great detail starting in Section 2.21. For the present, let us consider the MATLAB representation of the following state and output equations, respectively:

$$\dot{\mathbf{x}}(t) = \mathbf{A}\mathbf{x}(t) + \mathbf{B}\mathbf{u}(t)$$
$$\dot{\mathbf{y}}(t) = \mathbf{C}\mathbf{x}(t) + \mathbf{D}\mathbf{u}(t).$$

The "**A**" matrix represents the system dynamics, namely how the system is connected and where the integrators are located. "**B**" represents how inputs couple into the system. "**C**" represents how the outputs couple out of the system. "**D**" represents what portion of the inputs couples directly into the output. This representation may abstract the original design a little, but the advantages are quite significant. One such advantage is the improved precision of the representation, it suffers from fewer computer round-off-type problems.

Each of these three (transfer function, zero-pole-gain and state variable) representations have their advantages for visual inspection, but the conversion of them from one form to another can be cumbersome. MATLAB has realized this, and created several routines that allow you to convert from one representation to the next. A list of them, as well as their relations to each other, continuous and discrete versions, has been prepared and is stated under Model Conversions. A good exercise would be to pick a system, convert it to one of these representations, and then convert this representation to others with these routines. Practice in becoming familiar with these representations and converting between them is left to the readers' discretion.

Another aspect to consider is whether your system is a continous-time or discrete-time system. In reality, modern systems (digitally controlled) read the continous input (analog data) at discrete intervals making hybrid (sampled-data) systems. Breaking these systems apart, we can separately look at the continuous parts and discrete parts, or we can create "equivalent" models of these systems by using conversion routines. The concepts of hybrid systems are mind-boggling, and it is often preferable to convert their representations to one or the other (continuous or discrete) "equivalent" system for analysis.

A utility to convert from continuous to discrete, called "C2D" comes with the "Signal and Systems" toolbox. (The professional Control System Toolbox has many utilities that go back and forth with a variety of methods for converting.) The MCSTD toolbox comes with the conceptual essentials (further elaborated in the demo m-file) to do almost all the conversions that the professional toolbox does. The MCSTD toolbox even gives and uses examples of these conversions, since they are mostly polynomial substitutions (see polysbst in the MCSTD toolbox). Once again, this is an excellent teaching tool for control-systems theory, while not giving you all of the professional Control System Toolbox features.

A visual implementation of the system modeling is also available. It utilizes the above data representations of the basic building blocks, and the capability to interconnect the blocks. The Simulink Toolbox, used for system simulations, allows building of the system models in the visual sense. The Simulink developed model can be "queried" for the states contained in the system model.

H. Summary of MATLAB and Modern Control System Theory and Design Toolbox Commands

An abbreviated list of the commonly predefined functions available from MATLAB, which are of prime interest to the control-system engineer, is listed in Table 2.2. By entering these commands, MATLAB processes them immediately and displays the results determined.

A list of the additional commands available from the *Modern Control System Theory and Design Toolbox*, to supplement the MATLAB commands, which are also of prime interest to the control-system engineer, is given in Table 2.3. By entering these commands, the data are processed immediately and the results are displayed. The control-system engineer will find these commands very useful in the analysis and design of control systems. For example, some practical control-system examples of the polynomial utilities are as follows (Bode plot and root locus are discussed starting in Chapter 6.):

- Polysbst { go from s to $j\omega$ domain,
  { transform (scale/transport/rotate) axis in root locus plot
  { continuous to discrete transformations (substitutions)
- Polymag { Bode plot: pick any gain frequency (Odb = Phase margin (P.M.), 3db = Band Width (B.W.)
- Polyangl { Bode plot: pick any phase frequency (−180 = Gain Margin (G.M.)
- Rootmag { Root locus: pick a particular s/z magnitude (Unit Circle)
- Rootangl { Root locus: pick a particular damping angle

2.9. INVERSION WITH PARTIAL FRACTION EXPANSION USING MATLAB [6]

Having presented the fundamentals of MATLAB in Section 2.8, let us now return to the problem we addressed in Section 2.7 on obtaining the inverse Laplace transform

Table 2.2. Commands and Functions used by MATLAB [6]

Commands and Matrix Functions	Description
angle	Phase angle
ans	Answer when expression is not assigned
asin	Arcsine
atan	Arctangent
axis	Manual axis scaling on plots
bode	Plot Bode diagram (available only with Control System Toolbox, and the Student Edition of MATLAB)
clear	Remove items from memory and clear workspace
clg	Clear graph's screen
conj	Complex conjugate
conv	Multiplication; convolution
cos	Cosine
cosh	Hyperbolic cosine
cov	Covariance
deconv	Division; deconvolution
det	Determinant
disp	Display text or matrix
end	Terminates scope or "for," "while," and "if" statements
exit	Terminates program
exp	Exponential base *e*
eye	Identity matrix
figure	Opens new graphic window by creating figure object
grid	Draws grid lines for 2-D and 3-D plots
help	Online help for MATLAB functions and m-files
hold	Hold the current graph
home	Sends the cursor "home" to the upper left of the screen
imag	Imaginary part
inf	Infinity
inv	Matrix inverse
length	Length of vector
linspace	Generate linearly spaced vectors
log	Natural logarithm
loglog	Log–log scale plot
log10	Common logarithm (base 10)
logspace	Generate logarithmically spaced vectors
max	Maximum elements of a matrix
mean	Average or mean value of vectors and matrices
menu	Generate a menu of choices for user input
meshgrid	Generate X and Y arrays for 3-D plots
min	Minimum elements of a matrix
nyquist	Plot Nyquist frequency response (available only with the Control System Toolbox and, the Student Edition of MATLAB)
pi	Provide pi (π)
plot	Linear 2-D plot
plot3	Plots lines and point in 3-space
polar	Polar coordinate plot
poly	Characteristic polynomial

Table 2.2. (*Continued*)

Commands and Matrix Functions	Description
polyfit	Polynomial curve fitting
prod	Product of the elements
quit	Terminate MATLAB
real	Real part
residue	Partial-fraction expansion
rlocus	Plot root locus (available only with the MCSTD Toolbox and with the Student Edition of MATLAB)
roots	Polynomial roots
save	Save workspace variables on disk
semilogx; semilogy	Semi-logarithmic 2-D plot (x-axis or y-axis logarithmic)
sin	Sine
sinh	Hyperbolic sine
size	Matrix dimensions
ss2tf	State-space to transfer function conversion
step	Plot unit step response (available only with the Control System Toolbox, and the Student Edition of MATLAB)
sum	Sum of the elements
tan	Tangent
tanh	Hyperbolic tangent
text	Add text to plot by creating text object
tf2ss	Transfer function to state-space conversion
title	Graph title
trace	Trace of a matrix
type	List file
what	Directory listing of m-files
who	List directory of variables in memory
xlabel	x-axis label
ylabel	y-axis label

using partial fraction expansion. However, in this section, we will obtain the partial fraction expansion using MATLAB.

Let us consider the following transfer function:

$$Y(s) = \frac{4s^2 + 24s + 12}{s^3 + 5s^2 + 6s}.$$

We wish to first find its partial fraction expansion, prior to obtaining its Laplace transform using Table 2.1.

In order to use MATLAB's command "*residue*" (num,den) for obtaining the partial fraction expansion, we must first describe this problem in MATLAB format where row vectors "num" and "den" specify the coefficients of the numerator and denominator of the transfer function as follows:

$$\text{num} = [4 \quad 24 \quad 12]$$
$$\text{den} = [1 \quad 5 \quad 6 \quad 0].$$

Table 2.3. Commands used by the Modern Control System Theory and Design Toolbox

Category	Command	Description
Polynomial utilities	poly_add	— add two polynomials
	polysbst	— substitution of a polynomial variable with a polynomial
	polyder	— derivative of a polynomial (supplied with MATLAB)
	polyintg	— integral of a polynomial
	polymag	— locate roots of a polynomial that generate a given magnitude
	polyangl	— locate roots of a polynomial that generate a given angle
	rootmag	— locate roots of a polynomial that are at given magnitude
	rootangl	— locate roots of a polynomial that are at given angle
Interpolation utilities	crossing	— interpolates the index of specified values from a data set
	crosses	— interpolates the value of a data set at specified indices
	crosser	— iterates the solution of a function to a specified value
Control utilities	margins	— analytic calculations of all the phase and gain margins
	rlpoba	— root locus point of break-away/break-in
	rlaxis	— real axis portion of the root locus
	wpmp	— maximum feedback frequency location
	nichgrid	— Nichols grid at user requested inputs
Nonlinear Functions (using the describing function implementation)	back_lsh	— backlash response
	dead_zn	— deadzone response
	relays	— hysteresis response
Compatibility Functions	sbplot	— subplot replacement
	frz_axis	— axis replacement

The command residue (num, den) provides the residue r, the poles p, and the direct term k:

$$[r, p, k] = \text{residue (num,den)}.$$

Applying this command for this problem, we obtain the following result:

$$r = \begin{matrix} 2.0000 \\ -8.0000 \\ 10.0000 \end{matrix}$$

$$p = \begin{matrix} 0 \\ -3.0000 \\ -2.0000 \end{matrix}$$

$$k = 0.$$

Therefore, the partial fraction expansion of the transfer function $Y(s)$ is given by:

$$Y(s) = \frac{2}{s} - \frac{8}{s+3} + \frac{10}{s+2} + 0 \text{ (direct term)}.$$

Therefore, the inverse Laplace transform is obtained as follows using Table 2.1:

$$y(t) = (2 - 8e^{-3t} + 10e^{-2t})U(t).$$

If we use the following command,

$$[\text{num,den}] = \text{residue } (r, p, k)$$

where r, p, and k are given from the previous MATLAB output, then MATLAB converts the partial-fraction expansion back to the original polynomial ratio of $Y(s)$ as follows:

num = 4.0000 24.0000 12.0000
den = 1.0000 5.0000 6.0000 0.0000.

Therefore,

$$Y(s) = \frac{4s^2 + 24s + 12}{s^3 + 5s^2 + 6s}$$

2.10. LAPLACE-TRANSFORM SOLUTION OF DIFFERENTIAL EQUATIONS

After the physical relationships of a linear system have been described by means of its integrodifferential equation, the analysis of the system's dynamic behavior can be carried out by solving the equations and incorporating the initial conditions into the solution. Two examples are given in this section to illustrate the application of the Laplace transform to solve a linear differential equation. In general, we take the Laplace transform of each term in the differential equation. This step eliminates time and all of the time derivatives from the original equation and results in an algebraic equation in s. The resulting equation is then solved for the transform of the desired time function. The final step inolves obtaining the inverse Laplace transform, which yields the solution directly.

Example 1. Consider the following linear differential equation:

$$\frac{d^2y(t)}{dt^2} + 5\frac{dy(t)}{dt} + 6y(t) = 6 \tag{2.89}$$

Assume the initial conditions are

$$\dot{y}(0^+) = 2, \quad y(0^+) = 2.$$

By taking the Laplace transform of both sides of Eq. (2.89), the following equation is obtained [using Eqs. (2.66) and (2.67)]:

$$s^2 Y(s) - sy(0^+) - \dot{y}(0^+) + 5sY(s) - 5y(0^+) + 6Y(s) = 6/s. \tag{2.90}$$

Substituting the values of the initial conditions and solving for $Y(s)$ yields the following equation:

$$Y(s) = \frac{2s^2 + 12s + 6}{s(s^2 + 5s + 6)} = \frac{2s^2 + 12s + 6}{s(s+3)(s+2)}. \tag{2.91}$$

If Eq. (2.91) is expanded by means of partial fractions as discussed previously, the following expansion is obtained:

$$Y(s) = \frac{1}{s} - \frac{4}{s+3} + \frac{5}{s+2}. \tag{2.92}$$

The inverse Laplace transform of Eq. (2.92) is given, using Table 2.1 or Appendix A, by

$$y(t) = 1 - 4e^{-3t} + 5e^{-2t}, \quad t \geqslant 0. \tag{2.93}$$

This solution is composed of two portions: the steady-state solution given by 1, and the transient solution given by $-4e^{3t} + 5e^{-2t}$. As a check of the steady-state solution, we can apply the final-value theorem given by Eq. (2.74):

$$\lim_{t\to\infty} y(t) = \lim_{s\to 0} sY(s) = \lim_{s\to 0} \frac{2s^2 + 12s + 6}{(s+3)(s+2)} = 1. \tag{2.94}$$

Example 2. As a second example, the following differential equation is considered and the inapplicability of the final-value theorem when the function is not analytic in the right half-plane is illustrated:

$$\frac{d^2 y(t)}{dt^2} + \frac{dy(t)}{dt} = e^{4t}. \tag{2.95}$$

The initial conditions are assumed to be

$$y(0^+) = 2, \quad \dot{y}(0^+) = 0. \tag{2.96}$$

Taking the Laplace transform of both sides of Eq. (2.95), the following equation is obtained:

$$[s^2 Y(s) - sy(0^+) - \dot{y}(0^+)] + sY(s) - y(0^+) = \frac{1}{s-4}. \tag{2.97}$$

By substituting the values of the initial conditions, and solving for $Y(s)$, the following equation is obtained:

$$Y(s) = \frac{2s^2 - 6s - 7}{s(s+1)(s-4)}. \tag{2.98}$$

Expansion of Eq. (2.98) by means of partial fractions, as discussed previously, gives

$$Y(s) = \frac{\frac{7}{4}}{s} + \frac{\frac{1}{5}}{s+1} + \frac{\frac{1}{20}}{(s-4)}. \tag{2.99}$$

From Table 2.1 or Appendix A, the inverse Laplace transform of Eq. (2.99) is given by

$$y(t) = 1.75 + \frac{1}{5}e^{-t} + \frac{1}{20}e^{4t}, \quad t \geqslant 0. \tag{2.100}$$

It is obvious from Eq. (2.100) that the final value of this function is infinite. However, if one were to apply the final-value theorem to Eq. (2.98), the incorrect final value of 1.75 would be obtained. This example, therefore, illustrates very clearly that Eq. (2.74) cannot be applied when the function $F(s)$ is not analytic in the right half-plane.

2.11. TRANSFER-FUNCTION CONCEPT

For analysis and design, control systems are usually described by a set of differential equations. A block diagram is a device for displaying the interrelationships of the equations pictorially. Each component is described by its *transfer function*. This is defined as the ratio of the transform of the output of the component to the transform of the input. The component is assumed to be at rest prior to excitation, and all initial values are assumed to be zero when determining the transfer function.

Consider the block diagram of the simple system shown in Figure 2.7. The only assumption made concerning this system is that the input and output are related by a linear differential equation whose coefficients are constant and can be written in the form

$$A_n \frac{d^n c(t)}{dt^n} + \cdots + A_1 \frac{dc(t)}{dt} + A_0 c(t) = B_m \frac{d^m r(t)}{dt^m} + \cdots + B_1 \frac{dr(t)}{dt} + B_0 r(t). \tag{2.101}$$

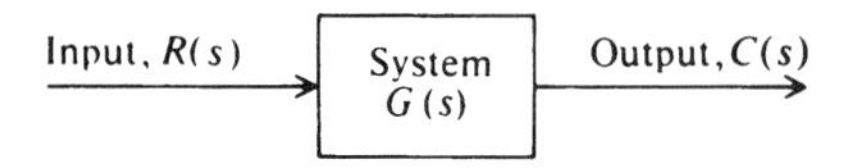

Figure 2.7 Block diagram of a simple linear system.

The Laplace transform of Eq. (2.101), assuming zero initial conditions, can be written as

$$(A_n s^n + \cdots + A_1 s + A_0)C(s) = (B_m s^m + \cdots + B_1 s + B_0)R(s). \tag{2.102}$$

The ratio $C(s)/R(s)$ is called the transfer function of the system and completely characterizes its performance. Designating the transfer function of the element as $G(s)$, we obtain

$$G(s) = \frac{C(s)}{R(s)} = \frac{B_m s^m + \cdots + B_1 s + B_0}{A_n s^n + \cdots + A_1 s + A_0}. \tag{2.103}$$

Therefore, assuming that the initial conditions are zero, the Laplace transform of the output is

$$C(s) = G(s)R(s). \tag{2.104}$$

In general, the function $G(s)$ is the ratio of two polynomials in s:

$$G(s) = P(s)/Q(s). \tag{2.105}$$

The transfer function $G(s)$ is a property of the system elements only, and is not dependent on the excitation and initial conditions. In addition, transfer functions can be used to represent closed-loop as well as open-loop systems

2.12. TRANSFER FUNCTIONS OF COMMON NETWORKS

The control engineer depends heavily on passive networks to modify the transfer function of the feedback control system in order to promote stability, improve closed-loop performance, and minimize the effects of noise and other undesirable signals.

Figure 2.8*a* represents an electrical network which is used for integration or to provide a phase lag. This circuit obtains its integrating property from the fact that the voltage across the capacitor is proportional to the integral of the current through it. To determine the behavior of this network we must determine the transfer function relating input and output signals. The integrodifferential equations describing the behavior of this network are given by

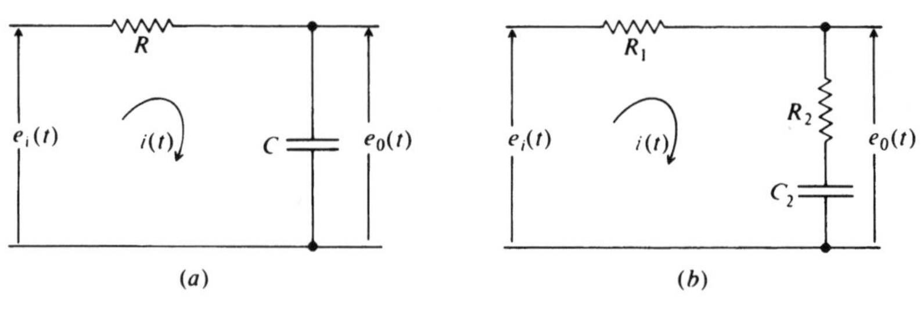

Figure 2.8 (*a*) An integrating or phase-lag network. (*b*) An integrating network with fixed high-frequency attenuation.

$$e_i(t) = Ri(t) + \frac{1}{C}\int i(t)\,dt, \tag{2.106}$$

$$e_o(t) = \frac{1}{C}\int i(t)\,dt. \tag{2.107}$$

These differential equations can be solved for the relation between $e_o(t)$ and $e_i(t)$ by means of the Laplace transform; $e_o(0^+)$ is assumed equal to zero. Thus one finds

$$E_i(s) = RI(s) + I(s)/Cs, \tag{2.108}$$

$$E_o(s) = I(s)/Cs. \tag{2.109}$$

Eliminating $I(s)$ results in the transfer function

$$\frac{E_o(s)}{E_i(s)} = \frac{1}{RCs + 1}. \tag{2.110}$$

Notice from Eq. (2.110) that integration is possible only for those frequencies where $|RCs| \gg 1$. It is important to note that this same transfer function can be obtained quite simply by determining the impedance of each circuit element and using the voltage divider rule. Thus

$$\frac{E_o(s)}{E_i(s)} = \frac{1/(Cs)}{1/(Cs) + R} = \frac{1}{RCs + 1}. \tag{2.111}$$

The impedance of the capacitance approaches a short circuit at very high frequencies, and ultimately it will have no output. This circuit is basically a low-pass filter whose high-frequency response is attenuated to a very large degree. Very often it is undesirable to have such a large attenuation at high frequencies. The circuit of Figure 2.8*b* limits this attenuation to a value of $R_2/(R_1 + R_2)$. The transfer function of the network is obtained directly by using the voltage divider rule. Thus,

$$\frac{E_o(s)}{E_i(s)} = \frac{R_2 + (1/C_2 s)}{R_1 + R_2 + (1/C_2 s)} = \frac{R_2 C_2 s + 1}{(R_1 + R_2)C_2 s + 1}. \tag{2.112}$$

A tabulation of the foregoing results, together with other useful transfer functions which the control engineer will most likley encounter, is shown in Table 2.4. Network 1 is known as a differentiating or phase-lead network. Notice that it is basically a high-pass filter possessing very large attenuation at low frequencies. This attenuation can be limited to a finite value of $R_2/(R_1 + R_2)$ by network 3. A lag-lead network, which provides a phase lag at low frequencies and a phase lead at high frequencies, is shown as network 5. Networks 6 and 7 are slight modifications of network 5. Network 8 is used for eliminating unwanted frequency bands. Network 9 is used for passing signals in a narrow band of frequencies.

Table 2.4. Transfer Functions of Common Networks

	Network	Transfer function
1	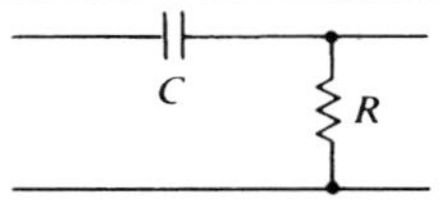 C, R	$\dfrac{RCs}{RCs+1}$
2	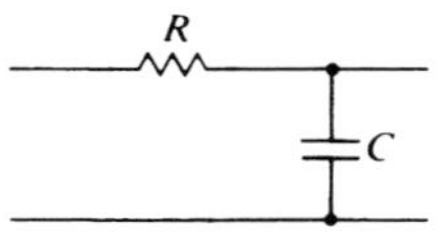 R, C	$\dfrac{1}{RCs+1}$
3	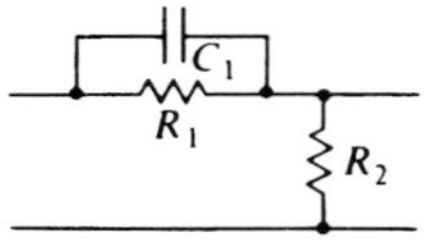 C_1, R_1, R_2	$\dfrac{R_2}{R_1+R_2}\,\dfrac{1+R_1C_1s}{1+\dfrac{R_2}{R_1+R_2}R_1C_1s}$
4	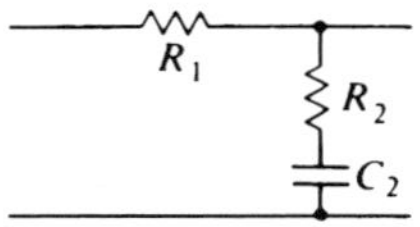 R_1, R_2, C_2	$\dfrac{R_2C_2s+1}{(R_1+R_2)C_2s+1}$
5	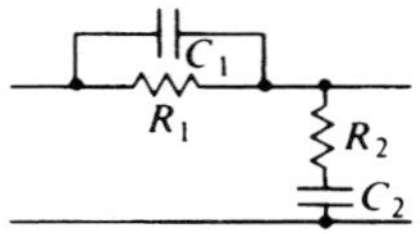 C_1, R_1, R_2, C_2	$\dfrac{(1+R_1C_1s)(1+R_2C_2s)}{R_1R_2C_1C_2s^2+(R_1C_1+R_2C_2+R_1C_2)s+1}$

Table 2.4. *(Continued)*

	Network	Transfer function
6	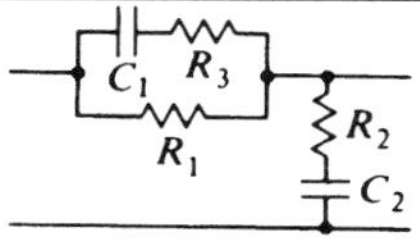 C_1, R_3, R_1, R_2, C_2	$\dfrac{R_2(R_1+R_3)C_1C_2s^2+(R_1C_1+R_2C_2+R_3C_1)s+1}{(R_1R_2+R_2R_3+R_1R_3)C_1C_2s^2+(R_1C_1+R_2C_2+R_1C_2+R_3C_1)s+1}$
7	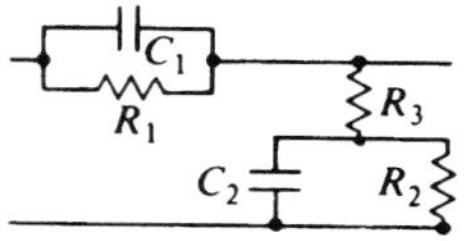 C_1, R_1, R_3, C_2, R_2	$\dfrac{\dfrac{R_1^2R_3}{R_2+R_3}C_1C_2s^2+\left[R_2C_1+\dfrac{R_2^2R_3C_2}{R_1(R_2+R_3)}\right]s+\dfrac{R_2+R_3}{R_1}}{R_2R_3C_1C_2s^2+\left(R_3C_1+R_2C_1+\dfrac{R_2R_3}{R_1}C_2\right)s+\dfrac{R_2+R_3}{R_1}+1}$
8	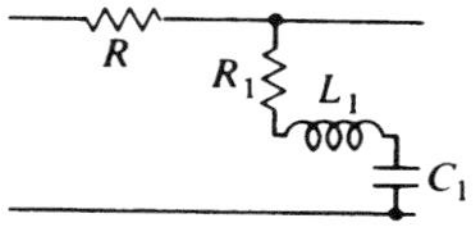 R, R_1, L_1, C_1	$\dfrac{s(L_1C_1s+R_1C_1)+1}{L_1C_1s^2+(R_1+R_2)C_1s+1}$
9	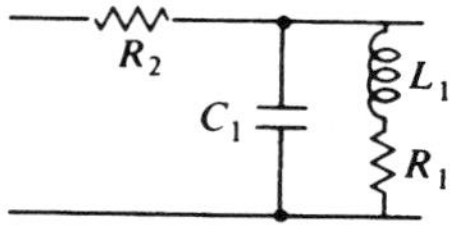 R_2, C_1, L_1, R_1	$\dfrac{(L_1/R_1)s+1}{\dfrac{L_1}{R_1}R_2C_1s^2+\left(\dfrac{L_1}{R_1}+R_2C_1\right)s+\dfrac{R_1+R_2}{R_1}}$

2.13. TRANSFER FUNCTIONS OF SYSTEMS

In order to determine the transfer function of complex systems, it is necessary to eliminate intermediate variables of the elements that comprise the system. This will enable the designer to obtain a relation between the input and output of the overall system. This section will consider the transfer functions of cascaded elements, single-loop feedback systems, and multiple-loop feedback systems. A nonloading cascaded system is shown in Figure 2.9. We assume that the input impedance of each system is infinite, and the outputs of the preceding systems are not affected by connecting to the succeeding element. The transfer function of the overall system can be obtained by solving the following set of equations:

$$E_2(s) = G_1(s)E_1(s), \tag{2.113}$$

$$E_3(s) = G_2(s)E_2(s), \tag{2.114}$$

$$E_4(s) = G_3(s)E_3(s), \tag{2.115}$$

$$E_5(s) = G_4(s)E_4(s). \tag{2.116}$$

By inspection, it can be seen that the transfer function of the cascaded system, $E_5(s)/E_1(s)$ is the product of the transfer functions of the individual elements:

$$E_5(s)/E_1(s) = G_1(s)G_2(s)G_3(s)G_4(s). \tag{2.117}$$

Consider the elementary linear feedback system in Figure 2.10. $G(s)$ and $H(s)$ represent the transfer functions of the direct-transmission and feedback portions of the loop, respectively. They may be individually composed of cascaded elements and minor feedback loops.

The following three equations are required in order to compute the overall system transfer function, $C(s)/R(s)$:

$$B(s) = H(s)C(s), \tag{2.118}$$

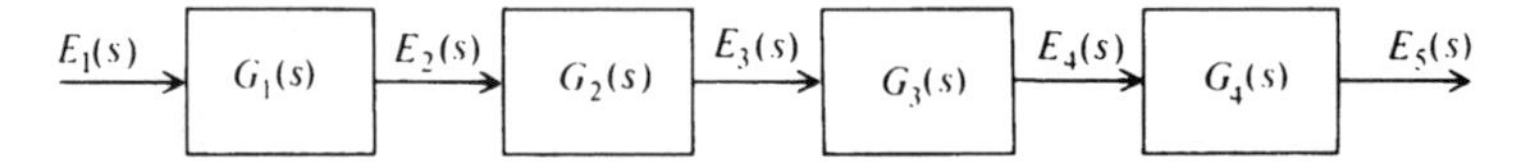

Figure 2.9 A nonloading cascaded system.

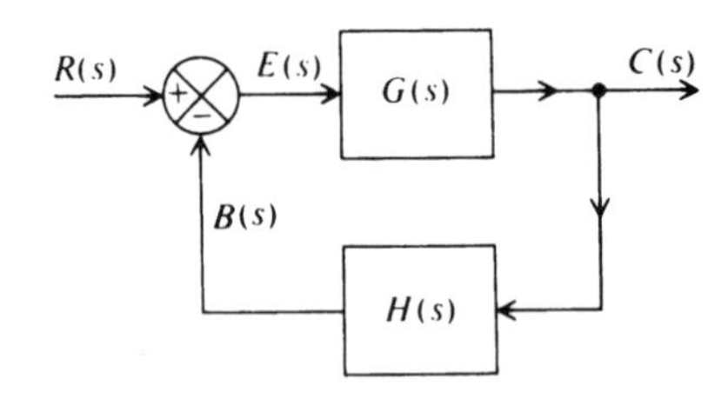

Figure 2.10 General block diagram of a single-loop feedback system.

$$E(s) = R(s) - B(s), \tag{2.119}$$

$$C(s) = G(s)E(s). \tag{2.120}$$

Solution of these three equations results in the following transfer function relating $R(s)$ and $C(s)$:

$$\frac{C(s)}{R(s)} = \frac{G(s)}{1 + G(s)H(s)}. \tag{2.121}$$

For cases where $|G(s)H(s)| \gg 1$, the closed-loop transfer function can be approximated by

$$\frac{C(s)}{R(s)} \simeq \frac{1}{H(s)}. \tag{2.122}$$

This implies that the closed-loop transfer function is independent of the direct transmission transfer function $G(s)$, and only depends on the feedback transfer function $H(s)$. Use is made of this characteristic in feedback amplifier design in order that the overall amplifier gain may be insensitive to circuit parameter variations. Note also that the approximate transfer function is the inverse of the feedback transfer function. This property is used for producing behavior which may be difficult to achieve directly.

The characteristic equation for the system can be obtained by setting the denominator of the system transfer function equal to zero:

$$1 + G(s)H(s) = 0. \tag{2.123}$$

This equation is known as the characteristic equation and it determines system stability. It will receive much attention in later chapters.

Another useful relationship is the transfer function $E(s)/R(s)$ when $H(s) = 1$. Then $E(s)$ represents the error, $R(s) - C(s)$. One finds

$$\left.\frac{E(s)}{R(s)}\right|_{H(s)=1} = \frac{1}{1 + G(s)}. \tag{2.124}$$

When $|G(s)| \gg 1$, this can be approximated by

$$\left.\frac{E(s)}{R(s)}\right|_{H(s)=1} \simeq \frac{1}{G(s)}. \tag{2.125}$$

Thus the error is small when the magnitude of the open-loop transfer function is large.

Practical feedback systems usually contain multiple feedback loops and several inputs. All multiple-loop systems can be reduced to the basic form shown in Figure 2.10 by means of step-by-step feedback loop reduction, or by means of *signal-flow graphs* and Mason's theorem which are considered in Sections 2.14–2.16 of this chapter. Multiple inputs, which are present in all control systems because unwanted inputs (such as noise and drift) are present, can occur anywhere in the feedback system. Successive block-diagram reduction techniques permit the designer to determine their effect on overall performance. Table 2.5 illustrates several transformations that can be used to simplify the reduction of a multiple feedback system. Note that the criterion in transformations 1–4 is to maintain the loop gains constant. The technique of multiple feedback loop reduction can best be understood by means of an example.

Figure 2.11 illustrates a multiple-loop feedback system containing three feedback loops. The original feedback system is illustrated in Figure 2.11*a*. Figure 2.11*b–e* shows successive steps in reducing this system using the transformations illustrated in Table 2.5. Figure 2.11*f* shows the resulting closed-loop transfer function of the system.

Table 2.5. Block Diagram Transformations

Transformation	Original block diagram	Equivalent block diagram
1. Moving a pickoff point behind a block	e_1 → G → e_2; pickoff e_1	e_1 → G → e_2; pickoff → $\frac{1}{G}$ → e_1
2. Moving a pickoff point ahead of a block	e_1 → G → e_2; pickoff e_2	e_1 → G → e_2; pickoff → G → e_2
3. Moving a summing point behind a block	e_1 → (+ −) → G → e_2; e_3	e_1 → G → (+ −) → e_2; e_3 → G
4. Moving a summing point ahead of a block	e_1 → G → (+ −) → e_2; e_3	e_1 → (+ −) → G → e_2; e_3 → $\frac{1}{G}$
5. Eliminating a feedback loop	R → (+ −) → G → C; feedback H	R → $\frac{G}{1+GH}$ → C

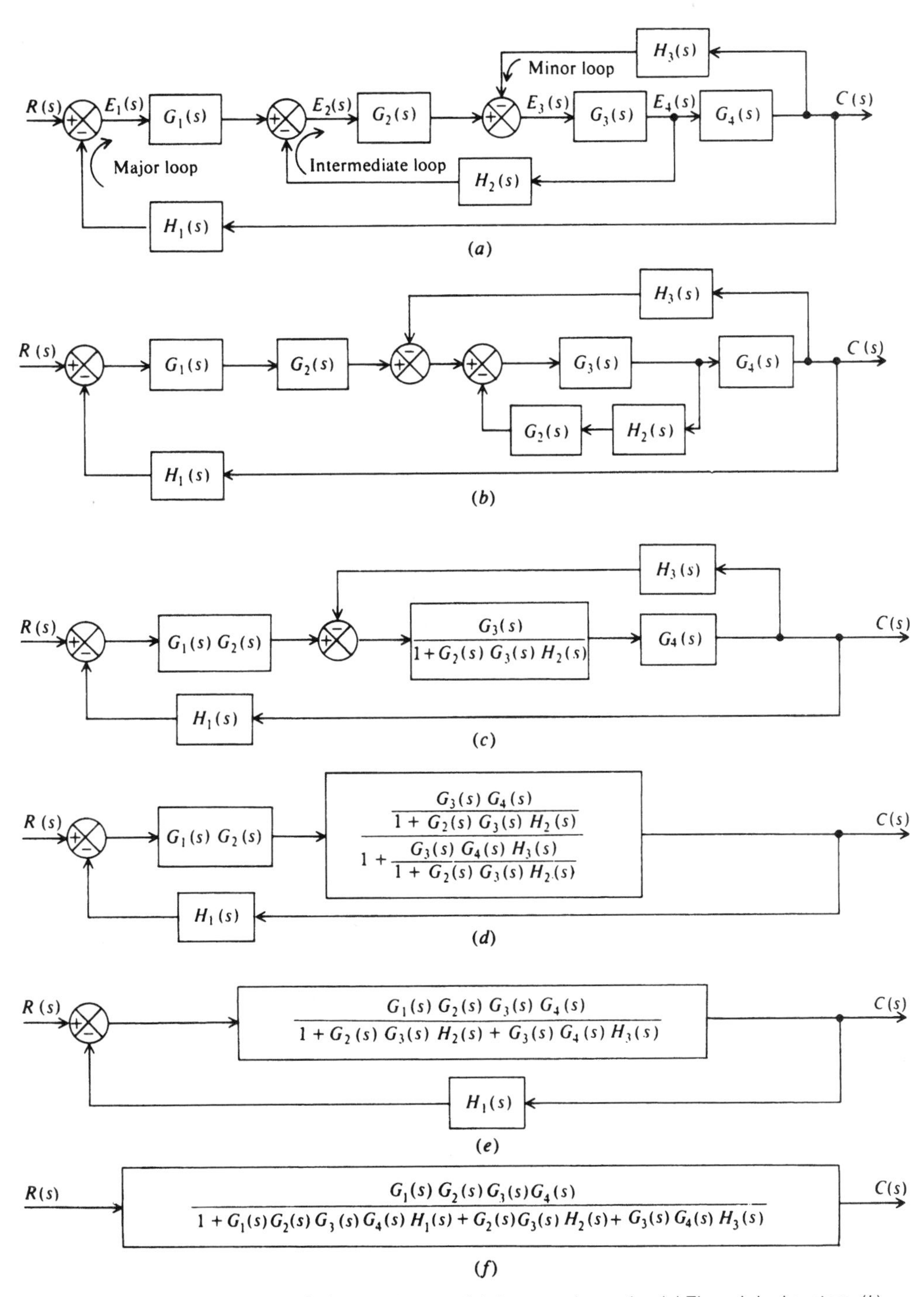

Figure 2.11 Reducing a multiple-loop system containing complex paths. (*a*) The original system. (*b*) Rearrangement of the summing points of the intermediate and minor loops. (*c*) Reduction of the equivalent intermediate loop. (*d*) Reduction of the equivalent minor loop. (*e*) The equivalent feedback system. (*f*) The system transfer function.

Block-diagram reduction techniques get very tedious and time-consuming as the number of feedback paths increases, as is illustrated in Figure 2.11. In order to solve complex problems, it is much simpler to make use of Mason's theorems and properties of signal-flow graphs, which permit a solution almost by inspection.

2.14. SIGNAL-FLOW GRAPHS AND MASON'S THEOREM

Signal-flow graphs and Mason's theorem [7, 8] enable the control engineer to determine the response of a complicated linear, multiloop system to any input much more rapidly than do block-diagram reduction techniques.

A signal-flow graph is a topological representation of a set of linear equations having the form

$$y_i = \sum_{j=1}^{n} a_{ij} y_j, \quad i = 1, 2, \ldots, n. \tag{2.126}$$

The equation expresses each of the n variables in terms of the others and themselves. A signal-flow graph represents a set of equations of this type by means of branches and nodes. A node is assigned to each variable of interest in the system. For example, node i represents variable y_i. Branch gains are used to relate the different variables. For example, branch gain a_{ij} relates variable y_i to y_j, where the branch originates at node i and terminates at node j. Consider the following set of linear equations

$$y_2 = ay_1 + by_2 + cy_4, \tag{2.127}$$

$$y_3 = dy_2, \tag{2.128}$$

$$y_4 = ey_1 + fy_3, \tag{2.129}$$

$$y_5 = gy_3 + hy_4. \tag{2.130}$$

The signal-flow graph which represents this set of equations is shown in Figure 2.12. Here y_1 can be interpreted as the input to the system and y_5 as its output. Usually we would be interested in obtaining the ratio of y_5/y_1 although we might also be interested in determining the ratio of y_2/y_1.

Before proceeding further, several terms used in signal-flow diagrams must be defined.

(a) A *source* is a node having only outgoing branches, such as y_1 in the preceding illustration.

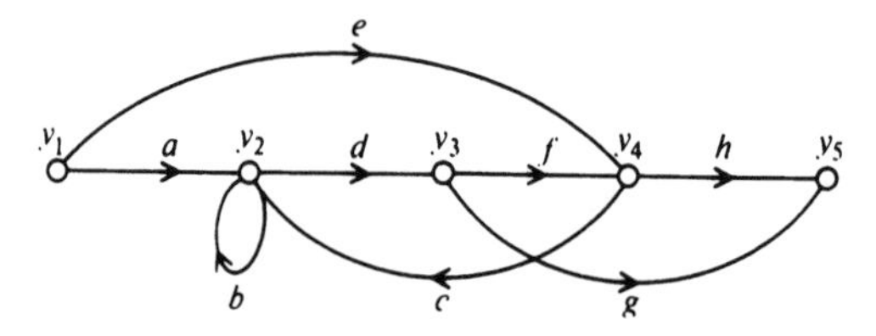

Figure 2.12 Signal-flow graph.

(b) A *sink* is a node having only incoming branches, such as y_5.

(c) A *path* is a group of connected branches having the same sense of direction. In Figure 2.12, *eh*, *adfh*, and *b* are paths.

(d) *Forward paths* are paths which originate from a source and terminate at a sink and along which no node is encountered more than once, as *eh*, *ecdg*, *adg*, and *adfh*.

(e) *Path gain* is the product of the coefficients associated with the branches along the path.

(f) *Feedback loop* is a path originating from a node and terminating at the same node. In addition, a node cannot be encountered more than once. In the preceding example *b* and *dfc* are feedback loops.

(g) *Loop gain* is the product of the coefficients associated with the branches forming a feedback loop.

2.15 REDUCTION OF THE SIGNAL-FLOW GRAPH

Several preliminary simplifications can be made to the complex signal-flow graphs of a system by means of the following signal-flow graph algebra.

(a) *Addition*

1. The signal-flow graph in Figure 2.13*a* represents the linear equation

$$y_3 = ay_1 + by_2. \tag{2.131}$$

2. The signal-flow graph in Figure 2.13*b* represents the linear equation

$$y_2 = (a + b)y_1. \tag{2.132}$$

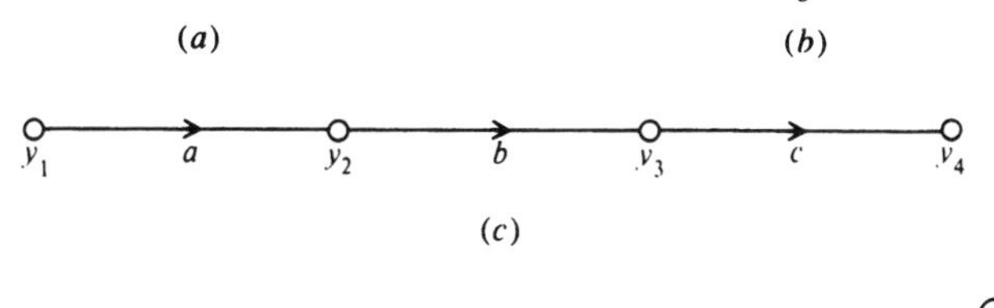

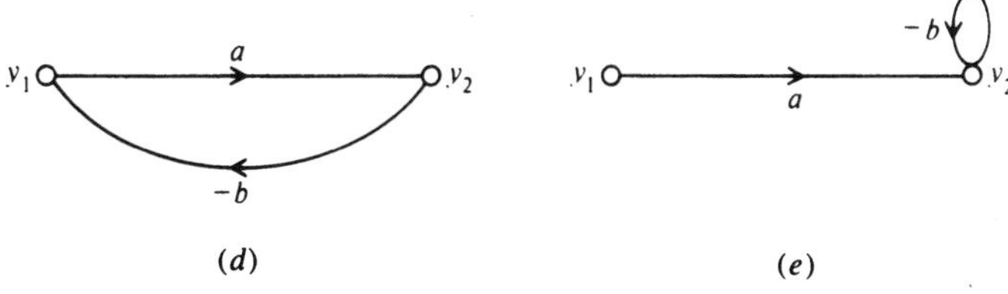

Figure 2.13 Signal-flow graph algebra.

(b) *Multiplication.* The signal-flow graph in Figure 2.13*c* represents the linear equation

$$y_4 = abcy_1. \tag{2.133}$$

(c) *Feedback loops*

1. The signal-flow graph in Figure 2.13*d* represents the linear equation

$$y_2 = \frac{a}{1+ab}y_1. \tag{2.134a}$$

2. The signal-flow graph in Figure 2.13*e* represents the linear equation

$$y_2 = \frac{a}{1+b}y_1. \tag{2.134b}$$

It is possible to apply the preceding signal-flow graph algebra to a complicated graph and reduce it to one containing only a source and a sink. This process requires repeated applications until the final desired form is obtainable. An interesting property of network and system topology, based on Mason's theorem [7, 8] permits the writing of the desired answer almost by inspection. The general expression for signal-flow graph gain G is given by

$$G = \frac{\sum_k G_K \Delta_K}{\Delta}, \tag{2.135}$$

where

$\Delta = 1 - \sum \mathrm{L}_1 + \sum \mathrm{L}_2 - \sum \mathrm{L}_3 + \cdots + (-1)^{\mathrm{m}} \sum \mathrm{L}_{\mathrm{m}}$

L_1 = gain of each closed loop in the graph

L_2 = product of the loop gains of any two nontouching closed loops (loops are considered nontouching if they have no node in common)

. . . .

L_{m} = product of the loop gains of any m nontouching loops

G_K = gain of the Kth forward path

Δ_K = the value of Δ for that part of the graph not touching the Kth forward path (value of Δ remaining when the path producing G_K is removed).

Δ is known as the determinant of the graph and Δ_K is the cofactor of the forward path K. Basically, Δ consists of the sum of the products of loop gains taken none at a time (1), one at a time (with a minus sign), two at a time (with a plus sign), etc.; Δ_K contains the portion of Δ remaining when the path producing G_K is removed. The proof of this general gain expression is contained in Reference [8]. A few examples follow in order to show how this expression may be used.

Example 1. For Figure 2.14*a*,

$$\Delta = 1 - bd,$$
$$G_1 = abc,$$
$$\Delta_1 = 1.$$

Therefore,

$$G = \frac{y_2}{y_1} = \frac{abc}{1 - bd}.$$

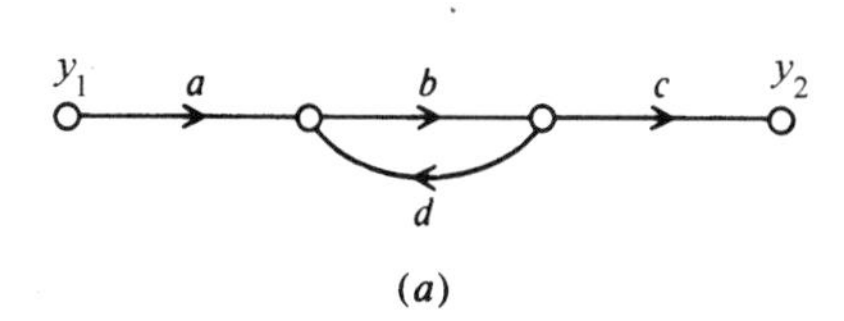

(*a*)

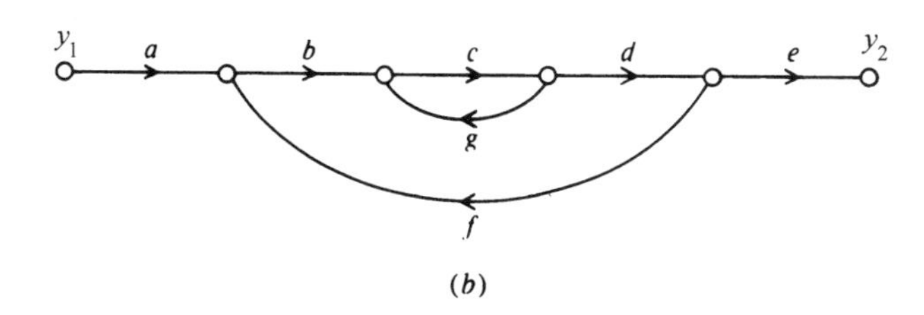

(*b*)

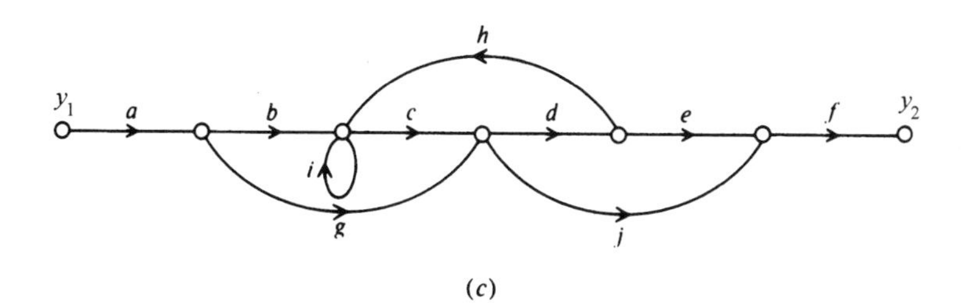

(*c*)

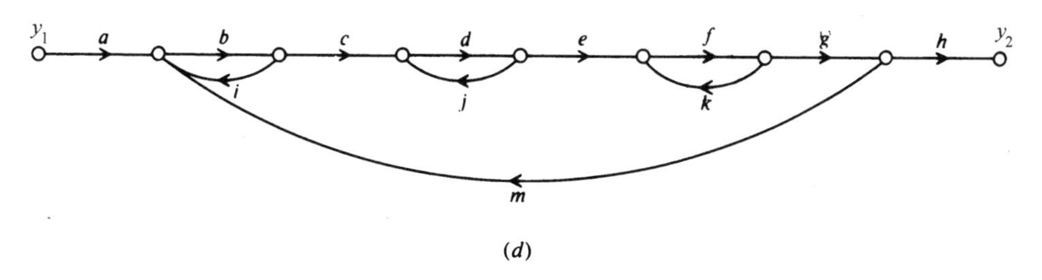

(*d*)

Figure 2.14 Signal-flow graph examples: (*a*) Example 1, (*b*) Example 2, (*c*) Example 3, (*d*) Example 4.

Example 2. For Figure 2.14*b*,

$$\Delta = 1 - cg - bcdf$$
$$G_1 = abcde.$$

Therefore,

$$\Delta_1 = 1,$$
$$G = \frac{y_2}{y_1} = \frac{abcde}{1 - cg - bcdf}.$$

Example 3. For Figure 2.14*c*,

$$\Delta = 1 - (i + cdh),$$
$$G_1 = abcdef,$$
$$G_2 = agdef,$$
$$G_3 = agjf,$$
$$G_4 = abcjf,$$
$$\Delta_1 = 1, \qquad \Delta_3 = 1 - i,$$
$$\Delta_2 = 1 - i, \quad \Delta_4 = 1.$$

Therefore,

$$G = \frac{y_2}{y_1} = \frac{abcdef + agdef(1 - i) + agjf(1 - i) + abcjf}{1 - (i + cdh)}.$$

Example 4. For Figure 2.14*d*,

$$\Delta = 1 - (bi + dj + fk + bcdefgm) + (bidj + bifk + djfk) - bidjfk,$$
$$G_1 = abcdefgh,$$
$$\Delta_1 = 1.$$

Therefore,

$$G = \frac{y_2}{y_1} = \frac{abcdefgh}{1 - (bi + dj + fk + bcdefgm) + (bidj + bifk + djfk) - bidjfk}.$$

2.16. APPLICATION OF MASON'S THEOREM AND THE SIGNAL-FLOW GRAPH TO MULTIPLE-FEEBACK SYSTEMS

It is important at this point to differentiate between signal-flow graphs and block diagrams. Basically, the signal-flow graph represents a detailed picture of a system's topological structure, whereas the block diagram focuses on the transfer functions that comprise the various elements of the system. The signal-flow graph is useful in analyzing multiple-loop feedback systems and in determining the effect of a particular element or parameter in an overall feedback system. Essentially both present the same information in different ways, and Mason's theorem can be applied to both. However, Mason's theorem is conventionally used with the signal-flow graph, because the topology is more clearly depicted by the signal-flow graph.

Let us next analyze various control-system configurations, depicted by both their block diagrams and signal-flow graphs, using Mason's theorem.

Example 1. The signal-flow graph corresponding to the block diagram shown in Figure 2.10 is illustrated in Figure 2.15:

Applying Mason's theorem to this feedback control system results in the following:

$$\begin{aligned} \Delta &= 1 - [-G(s)H(s)] = 1 + G(s)H(s) \\ G_A &= G(s) \\ \Delta_A &= 1 \end{aligned}$$

Therefore,

$$\frac{C(s)}{R(s)} = \frac{G(s)}{1 + G(s)H(s)}$$

which agrees with Eq. (2.121) that was derived from block-diagram relationships.

Example 2. Let us consider the feedback control system illustrated in Figure 2.16*a* which contains two feedback paths. Its corresponding signal-flow graph representation is illustrated in Figure 2.16*b*.

Applying Mason's theorem to this control system results in the following:

$$\begin{aligned} \Delta &= 1 - [-G_1(s)G_2(s)H_1(s) - G_2(s)H_2(s)] \\ G_A(s) &= G_1(s)G_2(s) \\ \Delta_A &= 1 \end{aligned}$$

Therefore,

$$\frac{C(s)}{R(s)} = \frac{G_1(s)G_2(s)}{1 + G_1(s)G_2(s)H_1(s) + G_2(s)H_2(s)}$$

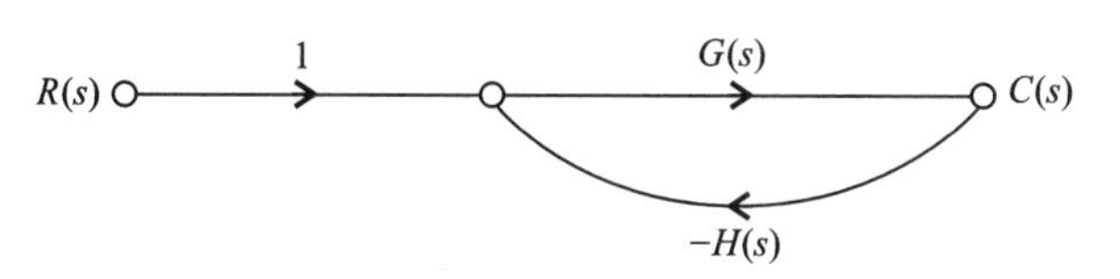

Figure 2.15 Signal-flow graph corresponding to the block diagram shown in Figure 2.10.

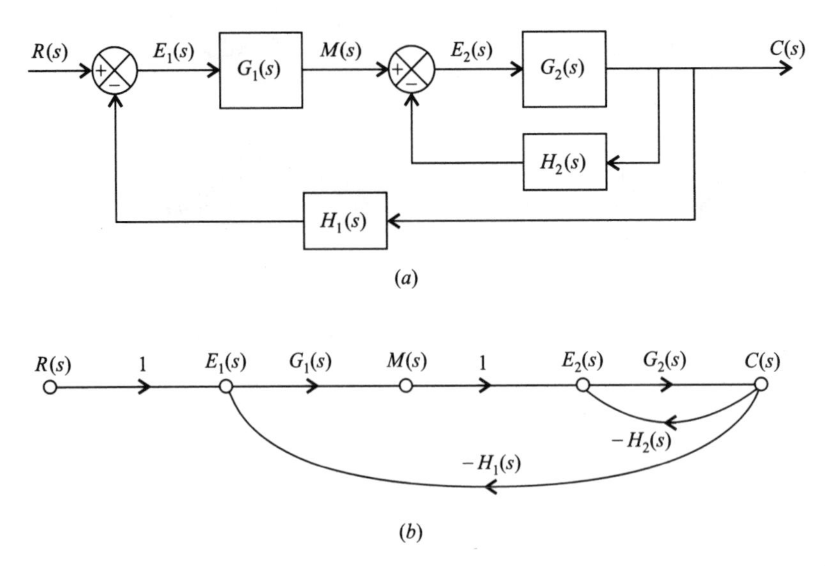

Figure 2.16 Control system containing two feedback paths. (*a*) Block-diagram representation. (*b*) Signal-flow graph representation.

Example 3. Let us next consider the feedback control system illustrated in Figure 2.17*a* which contains four feedback paths. Its corresponding signal-flow graph representation is illustrated in Figure 2.17*b*.

Applying Mason's theorem to this control system results in the following:

$$\Delta = 1 - [-G_1G_2G_3G_4G_5G_6G_7G_8H_0 - G_2H_2 - G_4H_4 - G_6H_6].$$

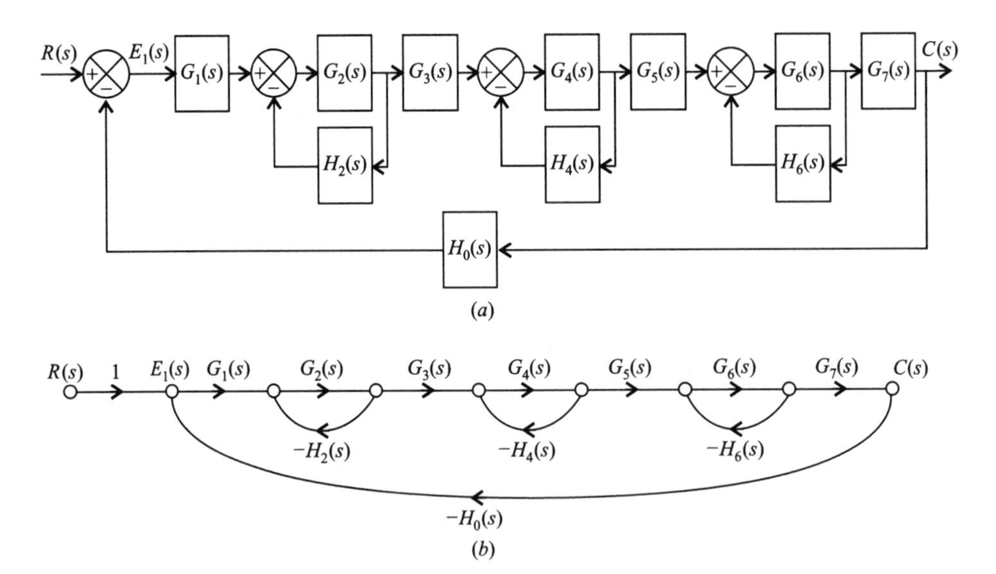

Figure 2.17 Control system containing four feedback paths. (*a*) Block diagram representation. (*b*) Signal-flow graph representation.

(All parameters in Δ are functions of s.)

$$G_A = G_1(s)G_2(s)G_3(s)G_4(s)G_5(s)G_6(s)G_7(s)G_8(s)$$

$$\Delta_A = 1$$

Therefore,

$$\frac{C(s)}{R(s)} = \frac{G_1G_2G_3G_4G_5G_6G_7G_8}{\begin{array}{l}1-[-G_2H_2-G_4H_4-G_6H_6-G_1G_2G_3G_4G_5G_6G_7G_8H_0]+[G_2H_2G_4H_4\\ +G_2H_2G_6H_6+G_4H_4G_6H_6]-[G_2H_2G_4H_4G_6H_6]\end{array}}$$

(All parameters in $C(s)/R(s)$ are functions of s.)

Example 4. For our fourth example, we will analyze the control system whose block diagram is illustrated in Figure 2.11*a*. Previously, it took us five steps and an entire page of block-diagram reductions to determine $C(s)/R(s)$ using the tedious block-diagram reduction methodology. We will now show how simple the process is using Mason's theorem which will only require one step for obtaining $C(s)/R(s)$.

The signal-flow graph for Figure 2.11*a* is shown in Figure 2.18. By inspection, the overall system transfer function is

$$\begin{aligned}\Delta &= 1+G_1(s)G_2(s)G_3(s)G_4(s)H_1(s)+G_2(s)G_3(s)H_2(s)\\ &\quad +G_3(s)G_4(s)H_3(s),\\ G_A &= G_1(s)G_2(s)G_3(s)G_4(s),\\ \Delta_A &= 1\end{aligned}$$

Therefore,

$$\frac{C(s)}{R(s)} = \frac{G_1(s)G_2(s)G_3(s)G_4(s)}{\begin{array}{l}1+G_1(s)G_2(s)G_3(s)G_4(s)H_1(s)+G_2(s)G_3(s)H_2(s)\\ +G_3(s)G_4(s)H_3(s)\end{array}}.$$

This result agrees with the transfer function shown in Figure 2.11*f*.

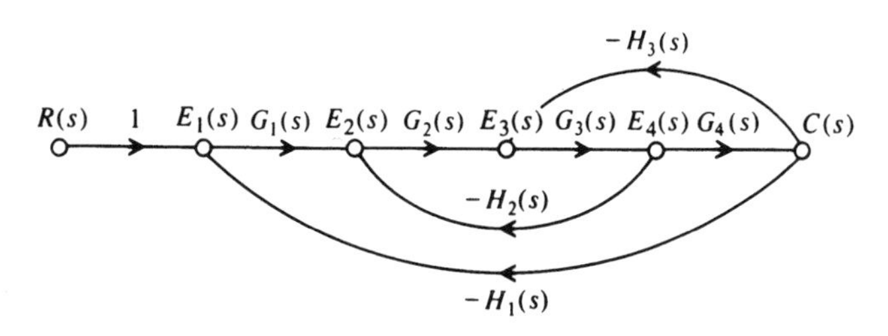

Figure 2.18

2.17. DISTURBANCE SIGNALS IN FEEDBACK CONTROL SYSTEMS

A disturbance signal is commonly found in control systems. For example, wind gusts hitting the antenna dish of a tracking radar create large unwanted torques which affect the position of the antenna. Another example, are sea waves hitting a hydrofoil's foil which create very large unwanted torques which affect the foil's position. (See Figure 1.30*b*). Maximum lift for the hydrofoil is achieved when the bow foil is at an optimum angle of approximately 40°. Disturbing this angle by torques created from sea waves crashing into this foil can seriously affect the lift capability of the hydrofoil.

Disturbance signals represent unwanted inputs which affect the control-system's output, and result in an increase of the system error. It is the job of the control-system engineer to properly design the control system to partially eliminate the affects of disturbances on the output and system error.

Consider the feedback control system shown in Figure 2.19*a*, which contains four feedback paths and a disturbance input $U(s)$. Its corresponding signal-flow graph is shown in Figure 2.19*b*. We wish to find the transfer functions $C(s)/R(s)$ and the transfer function of the error due to the disturbance. $E_1(s)/U(s)$. By inspection, these transfer functions are given by

$$\frac{C(s)}{R(s)} = \frac{G_1(s)G_2(s)G_3(s)G_4(s)G_5(s)G_6(s)G_7(s)G_8(s)}{\begin{array}{l}1 - [G_3(s)H_3(s) - G_6(s)H_6(s) - G_5(s)G_6(s)G_7(s)G_8(s)H_5(s) \\ \quad - G_1(s)G_2(s)G_3(s)G_4(s)G_5(s)G_6(s)G_7(s)G_8(s)H_0(s)] \\ \quad + [(G_3(s)H_3(s))(G_5(s)G_6(s)G_7(s)G_8(s)H_5(s)) \\ \quad + (G_3(s)H_3(s))(G_6(s)H_6(s))]\end{array}}$$

$$\frac{E_1(s)}{U(s)} = \frac{-G_8(s)H_0(s)\left\{\begin{array}{l}[1 - (-G_3(s)H_3(s) - G_6(s)H_6(s)) \\ \quad +(G_3(s)H_3(s))(G_6(s)H_6(s))]\end{array}\right\}}{\begin{array}{l}1 - [-G_3(s)H_3(s) - G_6(s)H_6(s) - G_5(s)G_6(s)G_7(s)G_8(s)H_5(s) \\ \quad -G_1(s)G_2(s)G_3(s)G_4(s)G_5(s)G_6(s)G_7(s)G_8(s)H_0(s)] \\ \quad +[(G_3(s)H_3(s))(G_5(s)G_6(s)G_7(s)G_8(s)H_5(s)) \\ \quad +(G_3(s)H_3(s))(G_6(s)H_6(s))]\end{array}}$$

The foregoing examples illustrate the simplifications made possible by use of Mason's theorem in conjunction with the signal-flow graphs. It will be extended to the state-variable concept in this chapter and, in the remaining chapters, the signal-flow graph approach will be used to simplify the sollutions of problems. In addition, further properties and applications of this powerful tool will be demonstrated.

2.18. OPERATIONAL AMPLIFIERS [9]

Operational amplifiers, usually referred to as "op amps," are very commonly used in control systems for performing differencing of signals, integration, amplifying sig-

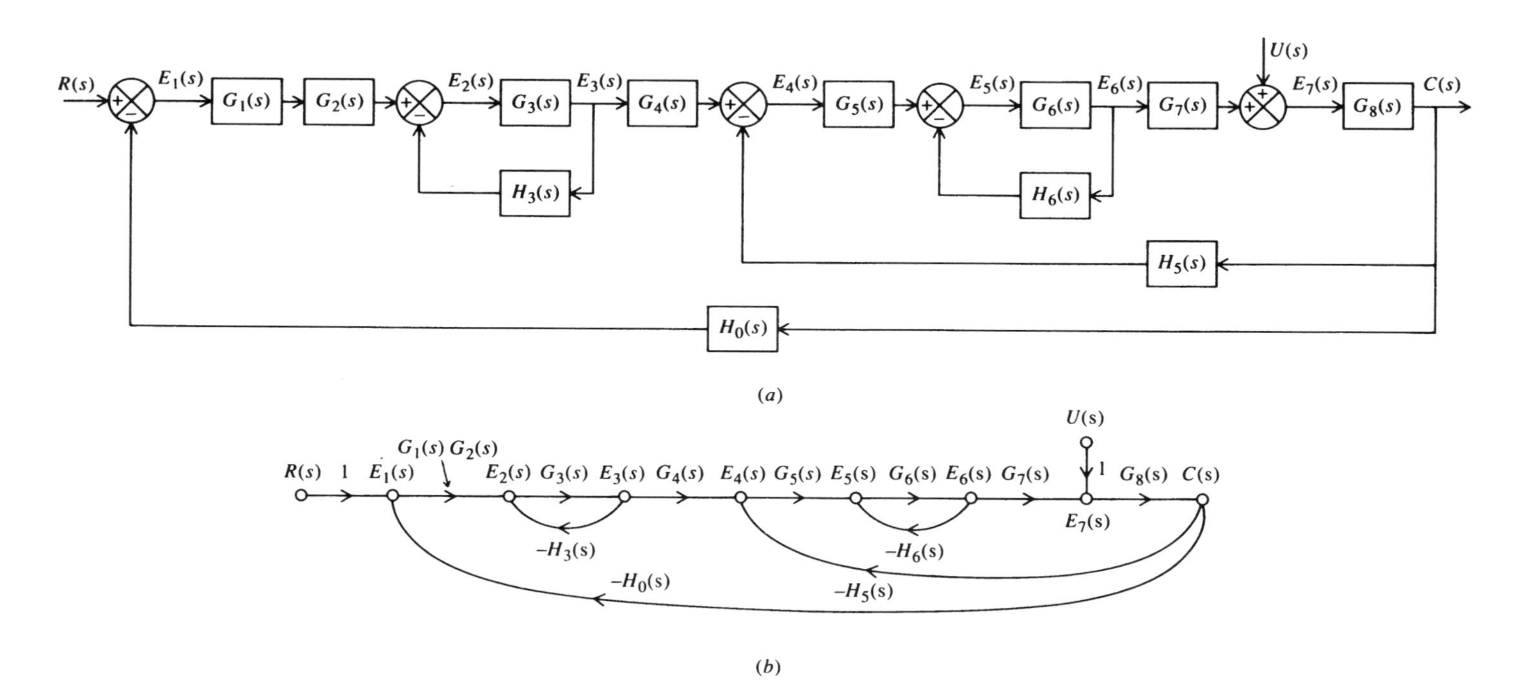

Figure 2.19. A feedback control system containing four feedback paths, an input, $R(s)$, and a disturbance input, $U(s)$. (*a*) The block diagram. (*b*) The signal-flow graph.

nals in sensor circuits, and in filters used for compensation. The fundamental component of an operational amplifier is a high gain dc voltage amplifier. By appropriately choosing the input and feedback impedances, any desired analog computer characteristic can be obtained easily. These devices are of relatively high gain in the range of 10^5 to 10^7. Due to these high gains, the drift problems associated with dc amplifiers are greatly magnified. To minimize the drift, highly regulated power supplies, temperature-compensated precision resistors, and feedback techniques are designed into each dc amplifier.

Let us consider the basic operational amplifier circuit illustrated in Figure 2.20. It consists of an input circuit impedance, Z_i, a feedback impedance, Z_f, and the input impedance of the dc amplifier, Z_g. The open-loop gain of the amplifier is assumed to be K. The gain of the feedback circuit, e_o/e_i, can be obtained from the following set of equations:

$$e_i - e_g = i_i Z_i \tag{2.136}$$

$$e_o - e_g = i_f Z_f \tag{2.137}$$

$$e_g = (i_i + i_f) Z_g \tag{2.138}$$

$$e_o = -K e_g. \tag{2.139}$$

Solving Eq. (2.136) and (2.137) for i_i and i_f, respectively, and then substituting these values into Eq. (2.138), the following equation is obtained:

$$e_g = \left(\frac{e_i - e_g}{Z_i} + \frac{e_o - e_g}{Z_f} \right) Z_g. \tag{2.140}$$

Substituting e_g into Eq. (2.139), the overall gain of the feedback operational amplifier is obtained.

$$\frac{e_o}{e_i} = -\frac{Z_f}{Z_i} \frac{1}{1 + \frac{1}{K}\left[1 + \frac{Z_f(Z_i + Z_g)}{Z_i Z_g}\right]}. \tag{2.141}$$

In practice, Z_g is usually very much larger than Z_i and Z_f. In addition, as mentioned previously, K is a very large number, greater than 10^5. Therefore, Eq. (2.141) can be approximated by the following relationship:

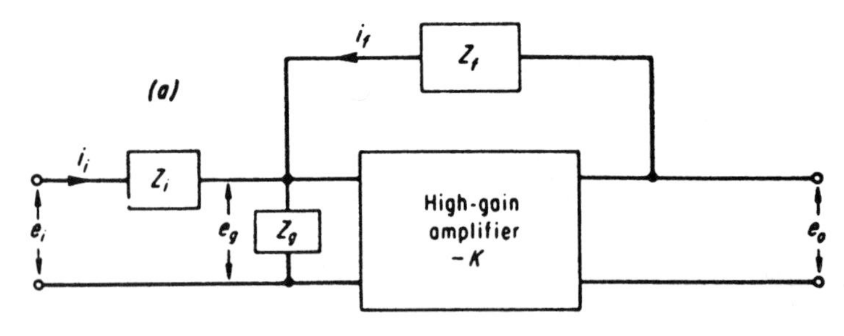

Figure 2.20 Basic operational amplifier circuit.

$$\frac{e_o}{e_i} \approx -\frac{Z_f}{Z_i}. \tag{2.142}$$

The corresponding simplified representation of the operational amplifier is illustrated in Figure 2.21.

Observe from Eq. (2.142) that the ratio of Z_f to Z_i determines the characteristics of the operational amplifier. For example, if Z_i and Z_f are resistances of equal value, we obtain a sign changer. However, if Z_f is greater than Z_i, we obtain an amplifier. In addition, if the feedback is a capacitor and the input element is a resistor, the operational amplifier behaves as an integrator where

$$e_o(t) = e_o(0) - \int_o^t \frac{1}{RC} e_i(t)dt. \tag{2.143}$$

By reversing the capacitor and resistor, a differentiator results where

$$e_o(t) = RC\frac{d}{dt}e_i(t). \tag{2.144}$$

The circuit of Figure 2.20 can be easily modified to produce an analog adder, as indicated in Figure 2.22. This circuit can be analyzed easily if it is recognized that e_g is essentially zero for operational amplifiers. This can be proved for the circuit of

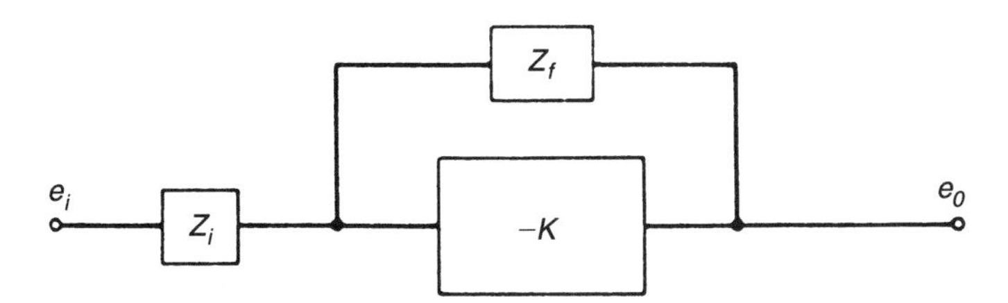

Figure 2.21 Simplified representation of the operational amplifier.

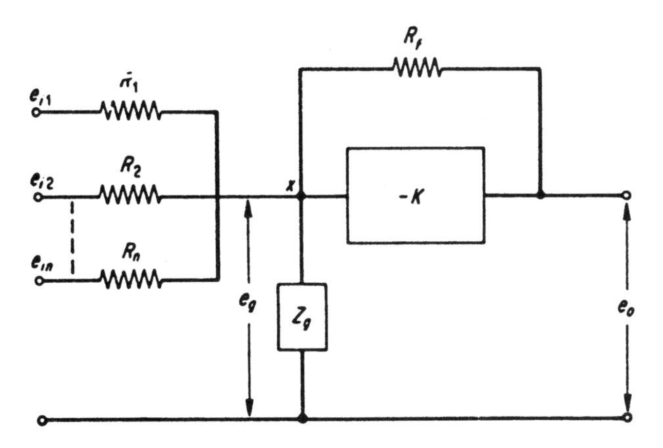

Figure 2.22 An adder operational amplifier.

Figure 2.20 from Eqs. (2.136) through (2.139). Solving for the ratio of e_g to e_i, and simplifying, the following equation is obtained:

$$\frac{e_g}{e_i} = \frac{1}{1 + \dfrac{Z_i}{Z_g} + \dfrac{Z_i}{Z_f} + K\left(\dfrac{Z_i}{Z_f}\right)}. \tag{2.145}$$

Since K is greater than 10^5, this equation clearly illustrates that e_g is maintained practically at zero volts. This is also true for the operational amplifier circuit of Figure 2.22. Therefore, by summing all of the currents flowing into node X of Figure 2.22 the following equation is obtained.

$$\frac{e_{i_1} - e_g}{R_1} + \frac{e_{i_2} - e_g}{R_2} + \ldots + \frac{e_{i_n} - e_g}{R_n} + \frac{e_o - e_g}{R_f} = \frac{e_g}{Z_g}. \tag{2.146}$$

Since we proved that e_g is zero, this equation reduces to the following form:

$$e_o = -\frac{R_f}{R_1}e_{i_1} - \frac{R_f}{R_2}e_{i_2} - \ldots - \frac{R_f}{R_n}e_{i_n}. \tag{2.147}$$

Equation (2.147) can be put in the following simpler format:

$$e_o = -\sum_{n=1}^{N} A_n e_{i_n} \tag{2.148}$$

where

$$A_n = R_f/R_n.$$

Equation (2.148) clearly illustrates the additive characteristic of this circuit. The gains associated with each input are conventionally marked on the block diagram of the operational amplifier.

In the case of operational amplifiers, one or more components of the device are purposely left out in order to make it more flexible. These components can then be externally selected in order to vary the closed-loop gain and bandwidth, and have the operational amplifier perform as an integrator, multiplier, sign changer, or adder.

Table 2.6 summarizes the operation, circuit, and symbology of several common operational amplifier computing devices. The circuits illustrated depend on the high gain of a dc amplifier, K. The higher the gain of this amplifier, the more exact is the operation performed.

2.19. SIMULATION DIAGRAMS

From the development of signal-flow graphs, Mason's theorem, and operational amplifiers, we can now develop simulation diagrams. The simulation diagram can be either a block diagram or a signal-flow graph which is constructed to have a

Table 2.6. Summary of Basic Operational Amplifier Computing Elements

Operation	Circuit	Symbol
Multiplication and sign change: $e_o = -Ae_i$ where $A = R_f/R_i$	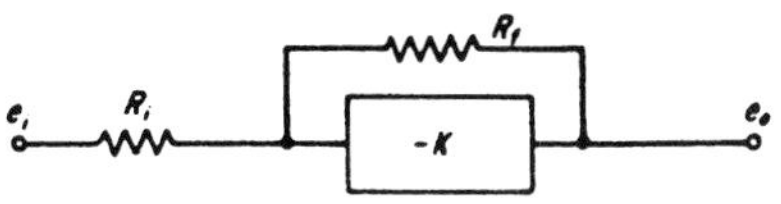	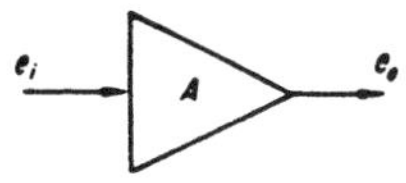
Addition (case 1): $e_o = -\sum_{n=1}^{N} A_n e_{in}$ where $A_n = R_f/R_n$	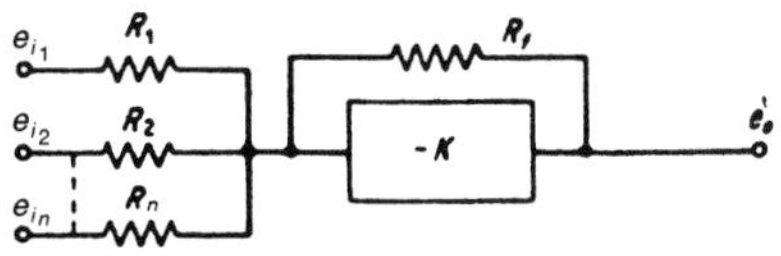	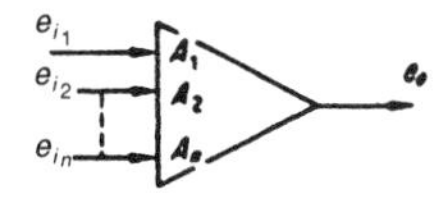
Addition (case 2): $e_o = -\sum_{n=1}^{N} A\ e_{in}$ where $A = R_f/R$	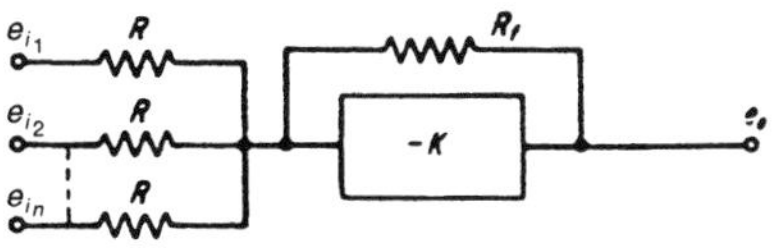	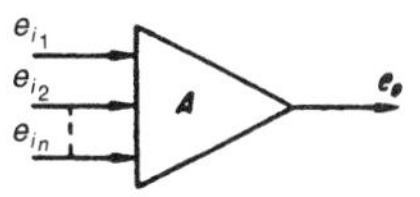
Integration: $e_o(t) = e_o(0) - \dfrac{1}{RC}\int_0^t e_i(t)\mathrm{d}t$	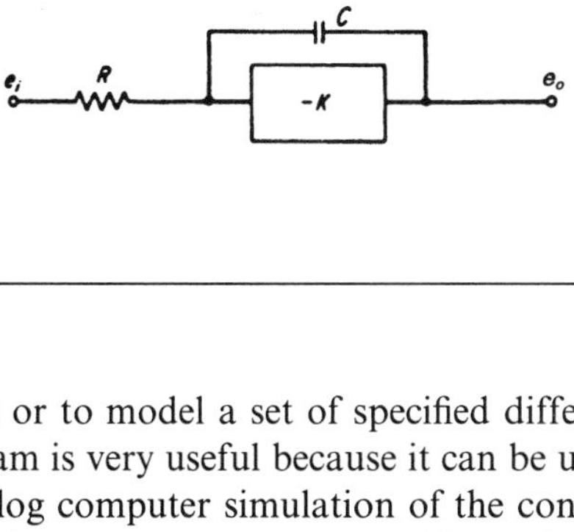	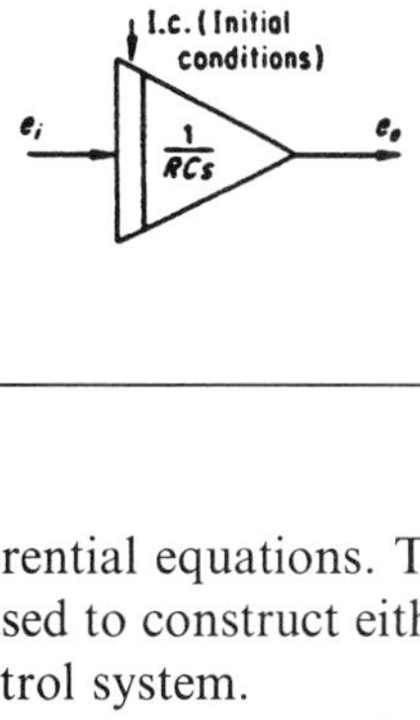

specified transfer function or to model a set of specified differential equations. The resulting simulation diagram is very useful because it can be used to construct either a digital computer or analog computer simulation of the control system.

From the previous section on operational amplifiers, the basic element of the simulation diagram is an integrator (see Eq. (2.143) and Table 2.6). In using the integrator in a simulation diagram, therefore, it is important to recognize that if the output of the integrtor is $x(t)$, then the input to the integrator is $dx(t)/dt$. Similarly, if two integrators are connected in cascade and the output of the last integrator is $x(t)$, then the input to the last integrator is $dx(t)/dt$, and the input to the first integrator is $d^2x(t)/dt^2$. We can use integrators, amplifiers, and summers, which were all described in Section 2.18 on operational amplifiers and summarized in Table 2.6, to represent a given differential equation by a simulation diagram.

In addition to representing a differential equation by a simulation diagram, we can reverse this procedure and start with a given transfer function to construct the simulation diagram. A simulation diagram constructed from a differential equation representing the system is usually unique. If the simulation diagram is constructed from the transfer function, however, the simulation diagram is not unique. This is illustrated next.

Let us consider representing the closed-loop transfer function

$$\frac{C(s)}{R(s)} = \frac{4s^2 + 3s + 2}{s^3 + 6s^2 + 7s + 8} \tag{2.149}$$

by a simulation diagram. It can be accomplished by Figures 2.23 or 2.24. Both are correct. Figure 2.23 is denoted as the *observer-canonical form*, and Figure 2.24 is denoted as the *control-canonical form*. We use the term observer-canonical form to denote the configuration in Figure 2.23 because all the feedback loops come from the output or the "observed" signal. We use the term control-canonical form to denote

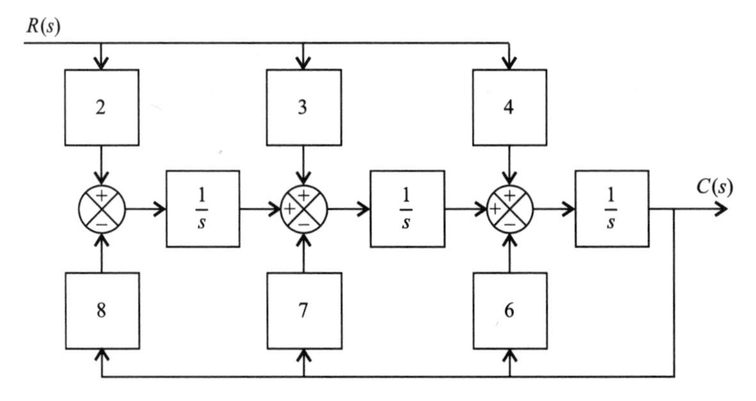

Figure 2.23 Observer-canonical form representation of Eq. (2.149).

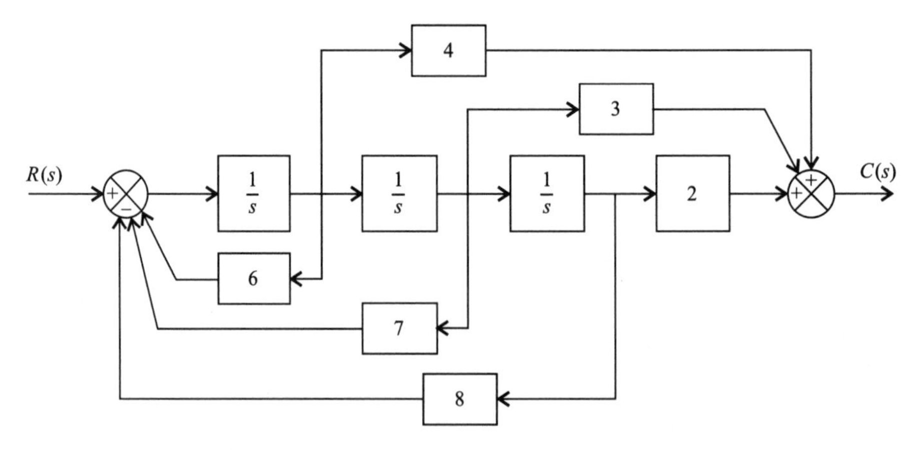

Figure 2.24 Control-canonical form representation of Eq. (2.149).

the configuration in Figure 2.24 because all the feedback loops return to the input or the "control" variable. Applying Mason's theorem to these two simulation diagrams, we find that they both have the same transfer function. For example, from the observer-canonical form of Figure 2.23, we find that

$$\frac{C(s)}{R(s)} = \frac{4s^{-1} + 3s^{-2} + 2s^{-3}}{1 + 6s^{-1} + 7s^{-2} + 8s^{-3}} = \frac{4s^2 + 3s + 2}{s^3 + 6s^2 + 7s + 8}. \tag{2.150}$$

For the control-canonical form of Figure 2.24, we find that

$$\frac{C(s)}{R(s)} = \frac{4s^{-1} + 3s^{-2} + 2s^{-3}}{1 + 6s^{-1} + 7s^{-2} + 8s^{-3}} = \frac{4s^2 + 3s + 2}{s^3 + 6s^2 + 7s + 8}. \tag{2.151}$$

Therefore, we conclude that simulation diagrams constructed from transfer functions are not unique, and can have different forms.

2.20. REVIEW OF MATRIX ALGEBRA

The classical methods of describing a linear system by means of transfer functions, block diagrams, and signal-flow graphs have thus far been presented in this chapter. An inherent characteristic of this type of representation is that the system dynamics are described by definable input-output relationships. Disadvantages of these techniques, however, are that the initial conditions have been neglected and intermediate variables lost. The method cannot be used for nonlinear, or time-varying systems. Furthermore, working in the frequency domain is not convenient for applying modern optimal control theory, discussed in Chapter 6 of the accompanying volume, which is based on the time domain. The use of digital computers also serves to focus on time-domain methods. Therefore, a different set of tools for describing the system in the time domain is needed and is provided by state-variable methods. As a necessary preliminary, matrix algebra is reviewed in this section [5, 10].

A matrix $\mathbf{A}$ is a collection of elements arranged in a rectangular or square array defined by

$$\mathbf{A} = \begin{bmatrix} a_{11} & a_{12} & \cdots & a_{1n} \\ a_{21} & a_{22} & \cdots & a_{2n} \\ \vdots & \vdots & & \vdots \\ a_{m1} & a_{m2} & \cdots & a_{mn} \end{bmatrix}. \tag{2.152}$$

The order of a matrix is defined as the total number of rows and columns of the matrix. For example, a matrix having m rows and n columns is referred to as an $m \times n$ matrix. A square matrix is one for which $m = n$. A column matrix, or vector, is one for which $n = 1$, and is represented in the following manner:

$$\mathbf{a} = \begin{bmatrix} a_1 \\ a_2 \\ \vdots \\ a_m \end{bmatrix}. \tag{2.153}$$

A row matrix or row vector is one that has one row and more than one column (e.g., a $1 \times n$ matrix). Capital letters are used to denote matrices and lower-case letters to denote vectors. Matrix **A** equals matrix **B** if element a_{ij} equals element b_{ij} for each i and each j, where the subscripts refer to elements in row i and column j of the respective matrices.

Before the basic types of properties of matrices are presented, let us note the differences between a matrix and a determinant. First, a matrix is an array of numbers or elements having m rows and n columns, while a determinant is an array of numbers or elements with m rows and n columns and is always square ($m = n$). The second major difference is that a matrix does not have a value while a determinant has a value. In addition, note that a square matrix ($m = n$) has a determinant. A matrix is said to be *singular* if the value of its determinant is zero.

In order to present the basic types and properties of matrices, matrix **A** will be considered in the following discussion where the representative element is a_{ij}.

A. Identity Matrix

The identity matrix, or unit matrix, is a square matrix whose principal diagonal elements are unity, all other elements being zero. This matrix, which is denoted by **I**, is given by

$$\mathbf{I} = \begin{bmatrix} 1 & 0 & \cdots & 0 \\ 0 & 1 & \cdots & 0 \\ \vdots & \vdots & & \vdots \\ 0 & 0 & \cdots & 1 \end{bmatrix}. \tag{2.154}$$

An interesting property of the identity matrix is that multiplication of any matrix **A** by an identity matrix **I** results in the original matrix **A**:

$$\mathbf{AI} = \mathbf{A}, \tag{2.155}$$

where matrix multiplication is defined later on.

B. Diagonal Matrix

A diagonal matrix, which is denoted by diag x_i, is given by

$$\text{diag } \mathbf{x}_i = \begin{bmatrix} x_1 & 0 & \cdots & 0 \\ 0 & x_2 & \cdots & 0 \\ \vdots & \vdots & & \vdots \\ 0 & 0 & \cdots & x_n \end{bmatrix} \tag{2.156}$$

C. Transpose of a Matrix

A matrix is transposed by interchanging its rows and columns. To form the transpose of a matrix, element a_{ij}, which is the element of row i and column j, is interchanged with element a_{ji}, which is the element of row j and column i, for all i and j. For example, the transpose $\mathbf{A}$, where

$$\mathbf{A} = \begin{bmatrix} a_{11} & a_{12} & a_{13} \\ a_{21} & a_{22} & a_{23} \\ a_{31} & a_{32} & a_{33} \end{bmatrix} \tag{2.157}$$

is written as $\mathbf{A}^T$, where

$$\mathbf{A}^T = \begin{bmatrix} a_{11} & a_{21} & a_{31} \\ a_{12} & a_{22} & a_{32} \\ a_{13} & a_{23} & a_{33} \end{bmatrix}. \tag{2.158}$$

Note that the transpose of $\mathbf{A}^T$ is $(\mathbf{A}^T)^T = \mathbf{A}$.

D. Symmetric Matrix

A symmetric matrix is defined by the condition

$$a_{ij} = a_{ji}. \tag{2.159}$$

This matrix, which is denoted by $\mathbf{A}_s$, is represented by

$$\mathbf{A}_s = \begin{bmatrix} a_{11} & a_{12} & \cdots & a_{1n} \\ a_{12} & a_{22} & \cdots & a_{2n} \\ \vdots & \vdots & & \vdots \\ a_{1n} & a_{2n} & \cdots & a_{nn} \end{bmatrix}. \tag{2.160}$$

Notice that it is symmetrical about the principal diagonal, and that a symmetric matrix and its transpose are identical.

E. Skew-Symmetric Matrix

A skew-symmetric matrix is defined by the condition

$$a_{ij} = -a_{ji}, \quad a_{ii} = 0. \tag{2.161}$$

This matrix, which is denoted by $\mathbf{A}_a$, is represented by

$$\mathbf{A}_a = \begin{bmatrix} 0 & a_{12} & \cdots & a_{1n} \\ -a_{12} & 0 & \cdots & a_{2n} \\ \vdots & \vdots & & \vdots \\ -a_{1n} & -a_{2n} & \cdots & 0 \end{bmatrix}. \tag{2.162}$$

Note that a skew-symmetric matrix is equal to the negative of its transpose.

F. Zero Matrix

The zero matrix or null matrix, denoted by $\mathbf{0}$, is defined as the matrix whose elements are all zero. It has the property that

$$\mathbf{A} + \mathbf{0} = \mathbf{A}, \tag{2.163}$$

where addition of matrices is defined later on.

G. Adjoint Matrix

The adjoint of a square matrix is formed by replacing each element of the matrix by its corresponding cofactor, and then transposing the result. For example, if a matrix $\mathbf{A}$ is given by

$$\mathbf{A} = \begin{bmatrix} a_{11} & a_{12} & a_{13} \\ a_{21} & a_{22} & a_{23} \\ a_{31} & a_{32} & a_{33} \end{bmatrix}, \tag{2.164}$$

then the cofactors of the elements a_{11}, a_{21} are given, respectively, by the determinants

$$A_{11} = \begin{vmatrix} a_{22} & a_{23} \\ a_{32} & a_{33} \end{vmatrix} = a_{22}a_{33} - a_{23}a_{32}, \tag{2.165}$$

$$A_{21} = -\begin{vmatrix} a_{12} & a_{13} \\ a_{32} & a_{33} \end{vmatrix} = a_{32}a_{13} - a_{12}a_{33}. \tag{2.166}$$

The element a_{ij} has for its cofactor A_{ij} with the proper algebraic sign, $(-1)^{i+j}$, prefixed. The new matrix formed by replacing the original elements in the matrix with their corresponding cofactors is given by

$$\begin{bmatrix} A_{11} & A_{12} & A_{13} \\ A_{21} & A_{22} & A_{23} \\ A_{31} & A_{32} & A_{33} \end{bmatrix}. \tag{2.167}$$

The transpose of this matrix results in the expression for the adjoint (adj) of matrix $\mathbf{A}$:

$$\operatorname{adj} \mathbf{A} = \begin{bmatrix} A_{11} & A_{21} & A_{31} \\ A_{12} & A_{22} & A_{32} \\ A_{13} & A_{23} & A_{33} \end{bmatrix}. \tag{2.168}$$

In order to present the basic operations of matrix analysis, three matrices $\mathbf{A}$, $\mathbf{B}$, and $\mathbf{C}$ will be considered in the following discussion where the representative elements are denoted by a_{ij}, b_{ij}, c_{ij}.

H. Addition or Subtraction

The sum (or difference) of two matrices $\mathbf{A}$ and $\mathbf{B}$ with the same numbers of rows and columns is obtained by adding (or subtracting) corresponding elements. The result is a new matrix $\mathbf{C}$, where

$$\mathbf{C} = \mathbf{A} + \mathbf{B}, \tag{2.169}$$

and

$$c_{ij} = a_{ij} + b_{ij}, \tag{2.170}$$

For example, if matrices **A** and **B** are given by

$$\mathbf{A} = \begin{bmatrix} a_{11} & a_{12} & \cdots & a_{1n} \\ a_{21} & a_{22} & \cdots & a_{2n} \\ \vdots & \vdots & & \vdots \\ a_{n1} & a_{n2} & \cdots & a_{nn} \end{bmatrix}, \quad \mathbf{B} = \begin{bmatrix} b_{11} & b_{12} & \cdots & b_{1n} \\ b_{21} & b_{22} & \cdots & b_{2n} \\ \vdots & \vdots & & \vdots \\ b_{n1} & b_{n2} & \cdots & b_{nn} \end{bmatrix}, \tag{2.171}$$

then the sum of matrix **A** and matrix **B** is given by

$$\mathbf{C} = \mathbf{A} + \mathbf{B} = \begin{bmatrix} a_{11} + b_{11}) & (a_{12} + b_{12}) & \cdots & (a_{1n} + b_{1n}) \\ (a_{21} + b_{21}) & (a_{22} + b_{22}) & \cdots & (a_{2n} + b_{2n}) \\ \vdots & \vdots & & \vdots \\ (a_{n1} + b_{n1}) & (a_{n2} + b_{n2}) & \cdots & (a_{nn} + (b_{nn}) \end{bmatrix}. \tag{2.172}$$

I. Multiplication of a Matrix by a Scalar

Multiplication of a matrix **A** by a scalar d is equivalent to multiplying each element of the matrix by d. The result is a new matrix **C**, where

$$\mathbf{C} = d\mathbf{A} \tag{2.173}$$

and

$$c_{ij} = da_{ij}. \tag{2.174}$$

For example, if matrix **A** is given by

$$\mathbf{A} = \begin{bmatrix} a_{11} & a_{12} & \cdots & a_{1n} \\ a_{21} & a_{22} & \cdots & a_{2n} \\ \vdots & \vdots & & \vdots \\ a_{n1} & a_{n2} & \cdots & a_{nn} \end{bmatrix}, \tag{2.175}$$

then the multiplication of matrix **A** by scalar d is given by

$$\mathbf{C} = d\mathbf{A} = \begin{bmatrix} da_{11} & da_{12} & \cdots & da_{1n} \\ da_{21} & da_{22} & \cdots & da_{2n} \\ \vdots & \vdots & & \vdots \\ da_{n1} & da_{n2} & \cdots & da_{nn} \end{bmatrix}. \tag{2.176}$$

J. Multiplication of Two Matrices

Postmultiplication* of matrix **A** by matrix **B** results in a new matrix **C**, where

$$\mathbf{C} = \mathbf{AB} \tag{2.177}$$

and

$$c_{ij} = \sum_{k=1}^{n} a_{ik} b_{kj}. \tag{2.178}$$

These equations state that the result of postmultiplication of matrix **A** by matrix **B** is a matrix **C** whose element located in row i and column j is obtained by multiplying each element in row i of matrix **A** by the corresponding element in column j of matrix **B** and then adding the results. It is important to note that the number of columns in matrix **A** must equal the number of rows in matrix **B** so that matrices **A** and **B** may be multiplied together. For example, if matrices **A** and **B** are given by

$$\mathbf{A} = \begin{bmatrix} a_{11} & a_{12} & a_{13} \\ a_{21} & a_{22} & a_{23} \\ a_{31} & a_{32} & a_{33} \end{bmatrix}, \quad \mathbf{B} = \begin{bmatrix} b_{11} & b_{12} \\ b_{21} & b_{22} \\ b_{31} & b_{32} \end{bmatrix} \tag{2.179}$$

then the product of matrices **A** and **B** is given by

$$\mathbf{C} = \mathbf{AB} = \begin{bmatrix} (a_{11}b_{11} + a_{12}b_{21} + a_{13}b_{31}) & (a_{11}b_{12} + a_{12}b_{22} + a_{13}b_{32}) \\ (a_{21}b_{11} + a_{22}b_{21} + a_{23}b_{31}) & (a_{21}b_{12} + a_{22}b_{22} + a_{23}b_{32}) \\ (a_{31}b_{11} + a_{32}b_{21} + a_{33}b_{31}) & (a_{31}b_{12} + a_{32}b_{22} + a_{33}b_{32}) \end{bmatrix}. \tag{2.180}$$

Matrix multiplication is associative and distributive with respect to addition, but in general not commutative:

$$\mathbf{A}(\mathbf{BC}) = (\mathbf{AB})\mathbf{C} \quad \text{(associative)}, \tag{2.181}$$

$$\mathbf{A}(\mathbf{B} + \mathbf{C}) = \mathbf{AB} + \mathbf{AC} \quad \text{(distributive)}, \tag{2.182}$$

$$\mathbf{AB} \neq \mathbf{BA} \quad \text{(commutative)}. \tag{2.183}$$

K. Inverse of a Square Matrix

The inverse of a square matrix **B** is denoted by $\mathbf{B}^{-1}$ and has the property

$$\mathbf{BB}^{-1} = \mathbf{B}^{-1}\mathbf{B} = \mathbf{I}. \tag{2.184}$$

It can be shown that the product of an adjoint matrix with the matrix itself has the property that it is equal to the product of an indentity matrix and the determinant of the matrix:

$$\mathbf{B}\,\mathrm{adj}\,\mathbf{B} = \mathbf{I}|\mathbf{B}|. \tag{2.185}$$

*The terms post- or premultiplication are used to indicate whether the matrix is multiplied from the right or the left, respectively.

Using these two relationships, we can derive the expression for the inverse matrix $\mathbf{B}^{-1}$. By solving for $\mathbf{I}$ from Eq. (2.185), the following relationship is obtained:

$$\mathbf{I} = \frac{\mathbf{B}\,\text{adj}\,\mathbf{B}}{|\mathbf{B}|}, \quad |\mathbf{B}| \neq 0. \tag{2.186}$$

Using Eqs. (2.184) and (2.186), we obtain

$$\mathbf{B}^{-1}\mathbf{B} = \mathbf{B}\mathbf{B}^{-1} = \frac{\mathbf{B}\,\text{adj}\,\mathbf{B}}{|\mathbf{B}|}, \quad |\mathbf{B}| \neq 0. \tag{2.187}$$

Solving for the inverse matrix $\mathbf{B}^{-1}$ in terms of the adjoint matrix and the determinant of the matrix, we obtain the following expression:

$$\mathbf{B}^{-1} = \frac{\text{adj}\,\mathbf{B}}{|\mathbf{B}|}, \quad |\mathbf{B}| \neq 0, \tag{2.188}$$

or

$$\mathbf{B}^{-1} = \frac{1}{|\mathbf{B}|}\begin{bmatrix} B_{11} & B_{21} & \cdots & B_{n1} \\ B_{12} & B_{22} & \cdots & B_{n2} \\ \vdots & \vdots & & \vdots \\ B_{1n} & B_{2n} & \cdots & B_{nn} \end{bmatrix}, \quad |\mathbf{B}| \neq 0. \tag{2.189}$$

Note that $\mathbf{B}^{-1}$ does not exist if $|\mathbf{B}| = 0$, although Eq. (2.185) still holds.

L. Differentiation of a Matrix

The usual concepts regarding the differentiation of scalar variables carry over to the differentiation of a matrix. Let $\mathbf{A}$ be an $m \times n$ matrix whose elements $a_{ij}(t)$ are differentiable functions of the scalar variable t. The derivative of $\mathbf{A}$ with respect to the variable t is given by

$$\frac{d}{dt}[\mathbf{A}] = \dot{\mathbf{A}} = \begin{bmatrix} \frac{da_{11}(t)}{dt} & \frac{da_{12}(t)}{dt} & \cdots & \frac{da_{1n}(t)}{dt} \\ \frac{da_{21}(t)}{dt} & \frac{da_{22}(t)}{dt} & \cdots & \frac{da_{2n}(t)}{dt} \\ \vdots & \vdots & & \vdots \\ \frac{da_{m1}(t)}{dt} & \frac{da_{m2}(t)}{dt} & \cdots & \frac{da_{mn}(t)}{dt} \end{bmatrix}. \tag{2.190}$$

In addition, the derivative of the sum of two matrices is the sum of the derivatives of the matrices:

$$\frac{d}{dt}[\mathbf{A} + \mathbf{B}] = \dot{\mathbf{A}} + \dot{\mathbf{B}}. \tag{2.191}$$

M. Integration of a Matrix

The usual concepts regarding the integration of scalar variables also carry over to the integration of a matrix. Again, let $\mathbf{A}$ be an $m \times n$ matrix whose elements $a_{ij}(t)$ are integrable functions of the scalar variable t. The integration of $\mathbf{A}$ with respect to the variable t is given by

$$\int \mathbf{A} dt = \begin{bmatrix} \int a_{11}(t)\, dt & \int a_{12}(t)\, dt & \cdots & \int a_{1n}(t)\, dt \\ \int a_{21}(t)\, dt & \int a_{22}(t)\, dt & \cdots & \int a_{2n}(t)\, dt \\ \vdots & & & \vdots \\ \int a_{ml}(t)\, dt & \int a_{m2}(t)\, dt & \cdots & \int a_{mn}(t)\, dt \end{bmatrix}. \tag{2.192}$$

2.21. STATE-VARIABLE CONCEPTS

In the analysis of a system via the state-variable approach, the system is characterized by a set of first-order differential or difference [3, 5] equations that describe its "state" variables. System analysis and design can be accomplished by solving a set of first-order equations rather than a single, higher-order equation. This approach simplifies the problem and has several advantages when utilizing a digital computer for solution. It is also the basis of optimal-control theory.

What is meant by the *state* of a system? Qualitatively, a system's state refers to the initial, current, and future behavior of a system. Quantitatively, it is defined as the minimum set of variables, denoted by $x_1(t_0), x_2(t_0), \ldots, x_n(t_0)$ that are specified at an initial time $t = t_0$, which together with the given inputs $u_1(t), u_2(t), \ldots, u_m(t)$ for $t \geqslant t_0$ determine the state at any future time $t \geqslant t_0$ [4, 11–14]. We can view the state of a system, therefore, as describing the past, present, and future behavior of the system.

What is meant by the term *state variables*? These are the variables which define the smallest set of variables which determine the state of a system. Physically, this means that a set of state variables $x_1(t_0), x_2(t_0), \ldots, x_n(t_0)$ define the initial state of the system based on past history. In addition, the set of state variables $x_1(t), x_2(t), \ldots, x_n(t)$ characterizes the future behavior of the system once the inputs $u_1(t), u_2(t), \ldots, u_m(t)$ for $t \geq t_0$ are specified, together with the knowledge of the initial state. It is important to emphasize that state variables are not necessarily system outputs and may not always be accessible, measurable, observable, or controllable. (Controllability and observability are discussed in Chapter 8, Sections 8.4 and 8.5, respectively.)

What is meant by the term *state vector*? This is a vector which completely describes a system's dynamics in terms of its n-state variables. Therefore, given the initial state and the input $\mathbf{u}(t)$, a system's state vector $\mathbf{x}(t)$ completely defines the system's state for $t \geqslant t_0$.

What is meant by the term *state space*? This refers to an n-dimensional space consisting of the $x_1, x_2, \ldots, x_n$ axes. A specific state is a point in the state space.

The motivation for using the state-variable model is to use a representation of the system dynamics that contain the system's input-output relationship (similar to that of a transfer function) but in terms of n first-order differential equations to represent the nth order system. This approach has the following advantages:

1. The solution to a set of n first-order differential or difference equations is much easier to determine on a digital computer than the solution of the equivalent higher-order differential or difference equation. This facilitates computer-aided analysis and design on digital computers for higher-order systems. It is important to note that the Laplace transform/transfer function/block-diagram approach is inadequate for this purpose.
2. The state-variable concept greatly simplifies the mathematical notation by utilizing vector matrix notation for the set of first-order equations.
3. The inclusion of the initial conditions of a system in the analysis of control systems, which is difficult using conventional techniques, can be accounted for readily in the state-variable approach.
4. The state-variable representation lends itself to system synthesis using optimal control techniques (discussed in Chapter 6 of the accompanying volume).

Let us consider the simple system illustrated in Figure 2.25 in order to define the nomenclature used in the state-variable approach. The system's inputs are represented by inputs $u_1(t), u_2(t), \ldots, u_m(t)$, and the system's outputs by $c_1(t), c_2(t), \ldots, c_q(t)$. Systems with more than one input and/or more than one output are denoted as *multiple-input/multiple-output* or *multivariable* systems. It is desirable to represent the inputs by an *input vector* $\mathbf{u}(t)$, where

$$\mathbf{u}(t) = \begin{bmatrix} u_1(t) \\ u_2(t) \\ \vdots \\ u_m(t) \end{bmatrix}; \tag{2.193}$$

the outputs are represented by an *output vector* $\mathbf{c}(t)$, where

$$\mathbf{c}(t) = \begin{bmatrix} c_1(t) \\ c_2(t) \\ \vdots \\ c_q(t) \end{bmatrix}. \tag{2.194}$$

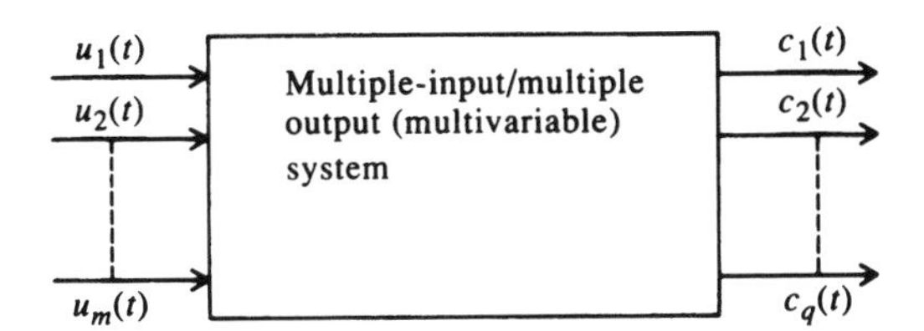

Figure 2.25 A multivariable system.

Let us next develop the general form of the state-variable equations. The system represented by Figure 2.25 has an output vector $\mathbf{c}(t)$ that depends on the initial value $\mathbf{c}(t_0)$ and the input vector $\mathbf{u}(t)$ for $t \geqslant t_0$. Because the system must contain devices which remember values of the input for $t \geqslant t_0$, we will assume that this memorization is performed by integrators, and their outputs represent the variables defining the system's states. Therefore, we will represent the outputs of these integrators as the state variables.

For our general representation, we will assume a multiple-input and multiple-output system as shown in Figure 2.25, which has m inputs $u_1(t), u_2(t), u_3(t), \ldots, u_m(t)$, q outputs $c_1(t), c_2(t), c_3(t), \ldots, c_q(t)$, and contains n integrators. Because an nth-order system requires n states to represent it, we will represent the n-state variables to the outputs of the integrators as $x_1(t), x_2(t), \ldots, x_n(t)$. We will assume that we are considering initially linear, time-invariant, systems. Nonlinear and time-varying systems will be considered afterwards. Therefore, the system can be represented by the following linear n first-order differential equations, where the dot notation is used to indicate derivatives:

$$
\begin{aligned}
\dot{x}_1(t) &= p_{11}x_1(t) + \cdots + p_{1n}x_n(t) + b_{11}u_1(t) + \cdots + b_{1m}u_m(t), \\
\dot{x}_2(t) &= p_{21}x_1(t) + \ldots + p_{2n}x_n(t) + b_{21}u_1(t) + \cdots + b_{2m}u_m(t), \\
&\vdots \\
\dot{x}_n(t) &= p_{n1}x_1(t) + \cdots + p_{nn}x_n(t) + b_{n1}u_1(t) + \cdots + b_{nm}u_m(t).
\end{aligned} \tag{2.195}
$$

We can simplify this set of n first-order differential equations by using matrix equations:

$$
\begin{bmatrix} \dot{x}_1(t) \\ \dot{x}_2(t) \\ \vdots \\ \dot{x}_n(t) \end{bmatrix} = \begin{bmatrix} p_{11} & p_{12} & \cdots & p_{1n} \\ p_{21} & p_{22} & \cdots & p_{2n} \\ \vdots & & & \\ p_{n1} & p_{n2} & \cdots & p_{nn} \end{bmatrix} \begin{bmatrix} x_1(t) \\ x_2(t) \\ \vdots \\ x_n(t) \end{bmatrix} + \begin{bmatrix} b_{11} & b_{12} & \cdots & b_{1m} \\ b_{21} & b_{22} & \cdots & b_{2m} \\ \vdots & & & \\ b_{n1} & b_{n2} & \cdots & b_{nm} \end{bmatrix} \begin{bmatrix} u_1(t) \\ u_2(t) \\ \vdots \\ u_m(t) \end{bmatrix}. \tag{2.196}
$$

Using the state-vector differential equation representation of Eq. (2.196), we obtain the following simpler representation of the system:

$$
\dot{\mathbf{x}}(t) = \mathbf{P}\mathbf{x}(t) + \mathbf{B}\mathbf{u}(t), \tag{2.197}
$$

where $\mathbf{P}$ and $\mathbf{B}$ are $n \times n$ and $n \times m$ coefficient matrices, respectively, $\mathbf{x}(t)$ is the state vector defined by

$$
\mathbf{x}(t) = \begin{bmatrix} x_1(t) \\ x_2(t) \\ \vdots \\ x_n(t) \end{bmatrix},
$$

$\dot{\mathbf{x}}(t)$ is its derivative, and $\mathbf{u}(t)$ is the input vector defined by Eq. (2.193). Matrix **P** is defined as the state or *companion matrix*, and **B** is defined as the input matrix. The solution to this vector matrix equation is uniquely determined by $\mathbf{x}(t_0)$ and $\mathbf{u}(t_0, T)$ where $\mathbf{u}(t)$ is defined for the interval $t_0 \leqslant t \leqslant T$. The system's output is related to the inputs and the state variables of the system by the following vector matrix equation:

$$\mathbf{c}(t) = \mathbf{Lx}(t) + \mathbf{Du}(t), \tag{2.198}$$

where $\mathbf{c}(t)$ is the output vector $(q \times 1)$ defined by Eq. (2.194), **L** is defined as the output matrix $(q \times n)$, and **D** is a coefficient matrix $(q \times m)$ which represents the direct transmission between input and output. Equation (2.197) is defined as the state equation of the system, and Eq. (2.198) is defined as the output equation of the system. Together, they are referred to as the *phase-variable canonical form*.

It is interesting to analyze the block-diagram representation of the phase-variable canonical form (Eqs. (2.197) and (2.198) which is shown in Figure 2.26. In most practical problems, there is no direct transmission between input and output. Therefore, **D** is zero in most cases, and the phase variable canonical form reduces to

$$\dot{\mathbf{x}}(t) = \mathbf{Px}(t) + \mathbf{Bu}(t), \tag{2.199}$$

$$\mathbf{c}(t) = \mathbf{Lx}(t). \tag{2.200}$$

For this case, the block diagram of Figure 2.26 reduces to that illustrated in Figure 2.27. It is this system and the phase-variable canonical form representation of Eqs. (2.199) and (2.200) that will be used to represent most of the systems in this book.

For simpler cases where there is only one input, the matrix **B** becomes a column vector, and $u(t)$ becomes a scalar. For systems containing one input and one output, the output $c(t)$ becomes a scalar, and the matrix **L** becomes a row vector.

Equations (2.199) and (2.201) are the state and output equations for linear time-invariant systems only. The state and output equations of general nonlinear and/or time-varying systems are given by

$$\dot{\mathbf{x}}(t) = \mathbf{f}(\mathbf{x}, \mathbf{u}, t), \tag{2.201}$$

$$\mathbf{c}(t) = \mathbf{g}(\mathbf{x}, \mathbf{u}, t). \tag{2.202}$$

In these equations, **u** is an m vector, **x** is an n vector, and **c** is a q-vector.

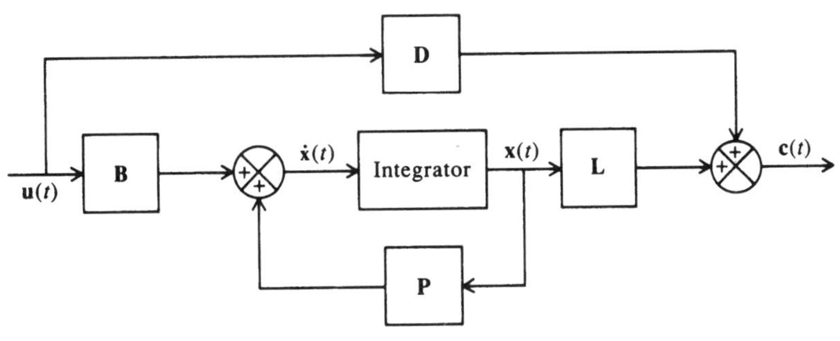

Figure 2.26 Block-diagram representation of the phase-variable canonical form.

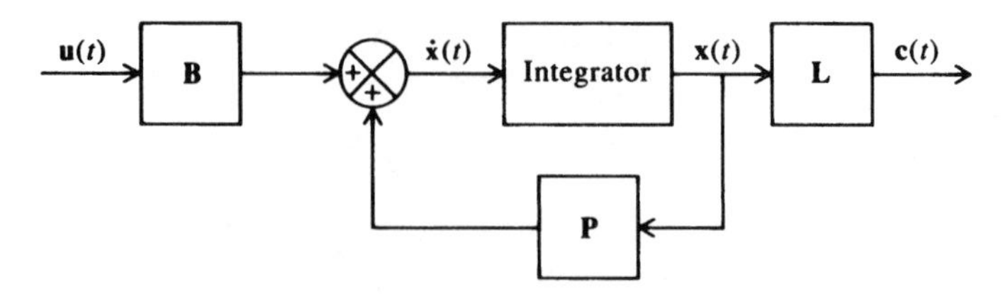

Figure 2.27 Block-diagram representation of the phase-variable canonical form without direct transmission between input and output ($\mathbf{D} = 0$).

A time description of the controlled process can be obtained by solving the differential equations (2.199) or (2.201). The solution is represented as

$$\mathbf{x}(t) = \phi_{\mu}(T, \mathbf{x}_0, t_0). \tag{2.203}$$

The above equation is interpreted as the value of $\mathbf{x}(t)$ at time t after starting at time t_0, in state $\mathbf{x}_0$, and governed by the control input $\mathbf{u}(t)$, defined for the interval $t_0 \leqslant t \leqslant T$.

In the remainder of this section, several examples are given for converting the dynamics of a system (given in any of several forms) into the phase-variable canonical form.

Let us next apply this concept to several examples:

Example 1. For the first example, let us consider an open-loop system, the transfer function of which is given by

$$P(s) = \frac{C(s)}{U(s)} = \frac{5}{s^3 + 8s^2 + 9s + 2}. \tag{2.204}$$

The differential equation corresponding to this system is given by

$$\frac{d^3c(t)}{dt^3} + 8\frac{d^2c(t)}{dt^2} + 9\frac{dc(t)}{dt} + 2c(t) = 5u(t). \tag{2.205}$$

Defining the state variables as

$$x_1(t) = c(t), \quad x_2(t) = \dot{c}(t), \quad x_3(t) = \ddot{c}(t), \tag{2.206}$$

the system can now be described by the following three first-order differential equations:

$$\dot{x}_1(t) = x_2(t) = \dot{c}(t), \tag{2.207}$$
$$\dot{x}_2(t) = x_3(t) = \ddot{c}(t), \tag{2.208}$$
$$\dot{x}_3(t) = -2x_1(t) - 9x_2(t) - 8x_3(t) + 5u(t). \tag{2.209}$$

Therefore, the system can be described in the phase-variable canonical form by

$$\dot{\mathbf{x}}(t) = \mathbf{P}\mathbf{x}(t) + \mathbf{B}u(t), \tag{2.210}$$
$$c(t) = \mathbf{L}\mathbf{x}(t), \tag{2.211}$$

where

$$\mathbf{P} = \begin{bmatrix} 0 & 1 & 0 \\ 0 & 0 & 1 \\ -2 & -9 & -8 \end{bmatrix}, \quad \dot{\mathbf{x}}(t) = \begin{bmatrix} \dot{x}_1(t) \\ \dot{x}_2(t) \\ \dot{x}_3(t) \end{bmatrix},$$

$$\mathbf{B} = \begin{bmatrix} 0 \\ 0 \\ 5 \end{bmatrix},$$

$$\mathbf{x}(t) = \begin{bmatrix} x_1(t) \\ x_2(t) \\ x_3(t) \end{bmatrix}, \quad \mathbf{L} = [1 \quad 0 \quad 0].$$

The set of state variables selected for this problem in Eq. (2.206) is not unique for this system. Another set of state variables may be defined as long as the state vector describes the same information about the system's behavior. Figure 2.28 shows the block diagram representation of the state and output equations from Eqs. (2.210) and (2.111), respectively. Observe that the transfer function of the feedback elements are identical to the negatives of the coefficients of the differential Eq. (2.205).

Example 2. As a second example concerned with obtaining the phase-variable canonical form of a system, consider the closed-loop system shown in Figure 2.29. The closed-loop transfer function of this system is given by

$$\frac{C(s)}{R(s)} = \frac{2}{s^2 + s + 2}. \tag{2.212}$$

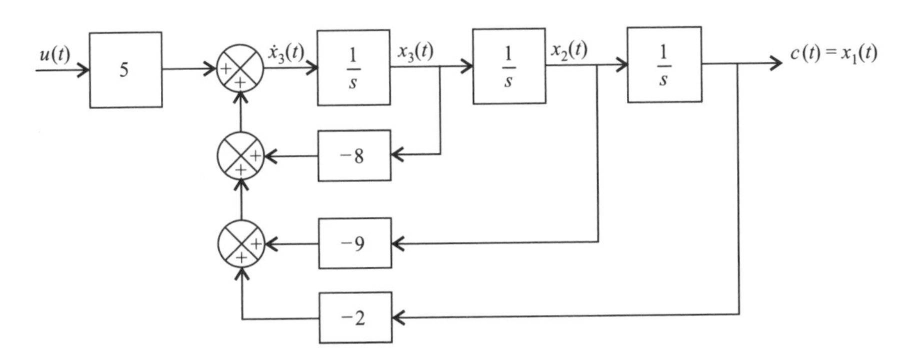

Figure 2.28 Block diagram representation of the system described by Eqs. (2.210) and (2.211).

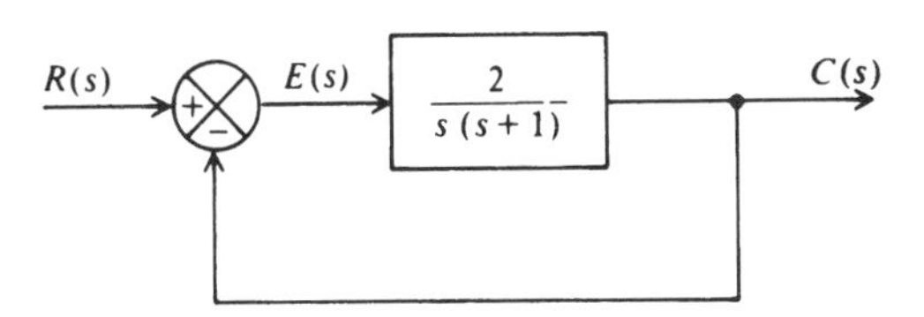

Figure 2.29 A feedback control system.

The corresponding differential equation is given by

$$\frac{d^2c(t)}{dt^2} + \frac{dc(t)}{dt} + 2c(t) = 2r(t). \tag{2.213}$$

By defining the state variables as

$$x_1(t) = c(t), \quad x_2(t) = \dot{c}(t), \tag{2.214}$$

the system can be described by the following two first-order differential equations:

$$\begin{aligned} \dot{x}_1(t) &= x_2(t) = \dot{c}(t) \\ \dot{x}_2(t) &= -2x_1(t) - x_2(t) + 2r(t). \end{aligned} \tag{2.215}$$

Therefore, the entire system can be described in the phase-variable canonical form by

$$\dot{\mathbf{x}}(t) = \mathbf{P}\mathbf{x}(t) + \mathbf{B}r(t), \tag{2.216}$$

$$c(t) = \mathbf{L}\mathbf{x}(t), \tag{2.217}$$

where

$$\mathbf{P} = \begin{bmatrix} 0 & 1 \\ -2 & -1 \end{bmatrix}, \quad \mathbf{B} = \begin{bmatrix} 0 \\ 2 \end{bmatrix}, \quad \mathbf{x}(t) = \begin{bmatrix} x_1(t) \\ x_2(t) \end{bmatrix},$$

$$\dot{\mathbf{x}}(t) = \begin{bmatrix} \dot{x}_1(t) \\ \dot{x}_2(t) \end{bmatrix}, \quad \mathbf{L} = [1 \quad 0].$$

Example 3. In the third example used to illustrate the representation of the dynamics of a system in state-variable form, consider the problem of rocket flight in two dimensions. Representing the vertical and horizontal axes by $v(t)$ and $r(t)$, respectively, the describing equations are given by

$$\ddot{r}(t) = F(t)\cos\theta(t), \tag{2.218}$$

$$\ddot{v}(t) = F(t)\sin\theta(t) - g, \tag{2.219}$$

where F is thrust force per unit mass, θ is thrust direction relative to the r axis, and g is the gravitational force. The control inputs are considered to be $F(t)$ and $\theta(t)$. Defining

$$\begin{aligned} x_1(t) &= r(t), & x_2(t) &= \dot{r}(t), \\ x_3(t) &= v(t), & x_4(t) &= \dot{v}(t), \\ u_1(t) &= F(t), & u_2(t) &= \theta(t), \end{aligned}$$

we find that the dynamics are described by

$$\begin{aligned} \dot{x}_1(t) &= x_2(t), \\ \dot{x}_2(t) &= u_1(t)\cos u_2(t), \\ \dot{x}_3(t) &= x_4(t), \\ \dot{x}_4(t) &= u_1(t)\sin u_2(t) - g. \end{aligned}$$

This system can also be described in phase-variable canonical form by

$$\dot{\mathbf{x}}(t) = \mathbf{Px}(t) + \mathbf{Bu}(t), \tag{2.220}$$

$$\mathbf{c}(t) = \mathbf{Lx}(t), \tag{2.221}$$

where

$$\mathbf{x}(t) = \begin{bmatrix} x_1(t) \\ x_2(t) \\ x_3(t) \\ x_4(t) \end{bmatrix}, \quad \dot{\mathbf{x}}(t) = \begin{bmatrix} \dot{x}_1(t) \\ \dot{x}_2(t) \\ \dot{x}_3(t) \\ \dot{x}_4(t) \end{bmatrix}, \quad \mathbf{P} = \begin{bmatrix} 0 & 1 & 0 & 0 \\ 0 & 0 & 0 & 0 \\ 0 & 0 & 0 & 1 \\ 0 & 0 & 0 & 0 \end{bmatrix}, \quad \mathbf{c}(t) = \begin{bmatrix} r(t) \\ v(t) \end{bmatrix}$$

$$\mathbf{B} = \begin{bmatrix} 0 & 0 & 0 & 0 \\ 0 & 1 & 0 & 0 \\ 0 & 0 & 0 & 0 \\ 0 & 0 & 0 & 1 \end{bmatrix}, \quad \mathbf{u}(t) = \begin{bmatrix} 0 \\ u_1(t)\cos u_2(t) \\ 0 \\ u_1(t)\sin u_2(t) - g \end{bmatrix}, \quad \mathbf{L} = \begin{bmatrix} 1 & 0 & 0 & 0 \\ 0 & 0 & 1 & 0 \end{bmatrix}.$$

2.22. STATE-VARIABLE DIAGRAM

The state-variable diagram provides a physical picture that is useful in understanding the state-variable concept. In addition, the differential equations relating the state variables are easily obtained by inspection of the diagram. A state-variable diagram consists of integrators, summing devices, and amplifiers. Outputs from the integrators denote the state variables. It should be noted that the state-variable diagram is the same as an analog computer simulation diagram [15].

Example 1. As an example of determining the state-variable diagram, consider a system whose transfer function is given by

$$P(s) = \frac{C(s)}{U(s)} = \frac{s^2 + 4s + 1}{s^3 + 9s^2 + 8s}. \tag{2.222}$$

Dividing numerator and denominator by s^3, to obtain the result in terms of integrators:

$$P(s) = \frac{C(s)}{U(s)} = \frac{s^{-1} + 4s^{-2} + s^{-3}}{1 + 9s^{-1} + 8s^{-2}}. \tag{2.223}$$

Defining the error node of the control system to be $E(s)$,

$$E(s) = \frac{U(s)}{1 + 9s^{-1} + 8s^{-2}}, \tag{2.224}$$

Eq. (2.223) may be rewritten as follows:

$$C(s) = (s^{-1} + 4s^{-2} + s^{-3})E(s). \tag{2.225}$$

From Eq. (2.225) and the relation

$$E(s) = U(s) - 9s^{-1}E(s) - 8s^{-2}E(s), \tag{2.226}$$

the state-variable diagram can easily be obtained as indicated in Figure 2.30.* The state variables are indicated in the diagrams as $x_1(t), x_2(t)$, and $x_3(t)$. Also, the differential equations relating the state variables are easily obtained from Figure 2.30 by inspection. From the state-variable diagram, the differential equations relating the state variables are as follows:

$$\begin{aligned} \dot{x}_1(t) &= x_2(t), \\ \dot{x}_2(t) &= x_3(t), \\ \dot{x}_3(t) &= u(t) - 8x_2(t) - 9x_3(t). \end{aligned} \tag{2.227}$$

Therefore, the entire system can be described in phase-variable canonical form by

$$\dot{\mathbf{x}}(t) = \mathbf{P}\mathbf{x}(t) + \mathbf{B}u(t), \tag{2.228}$$

$$c(t) = \mathbf{L}\mathbf{x}(t) \tag{2.229}$$

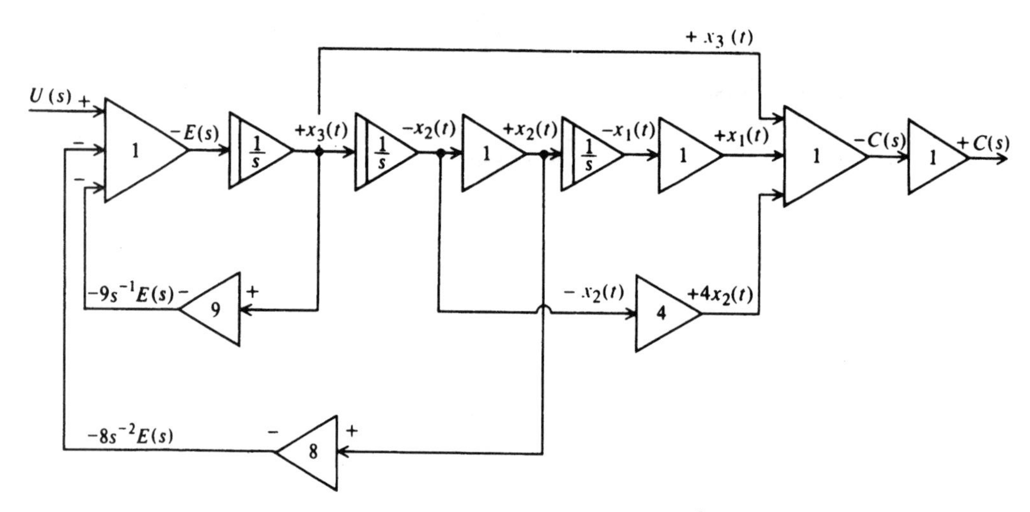

Figure 2.30 State-variable diagram for system where $P(s) = (s^2 + 4s + 1)/(s^3 + 9s^2 + 8s)$.

*This analog computer simulation diagram contains more inverters than are actually required to implement Eqs. (2.225) and (2.226). It has been presented in this manner to indicate the proper phase relationships existing among the various states, as indicated in Eq. (2.227). From now on, all state-variable diagrams will also contain a sufficient number of inverters to indicate the proper phase relationships among the various states. It is left as an exercise to show how an analog computer simulation of Eqs. (2.225) and (2.226) can be performed using the same number of integrators and summers as shown in Figure 2.30, but with only one inverter.

where

$$\mathbf{P} = \begin{bmatrix} 0 & 1 & 0 \\ 0 & 0 & 1 \\ 0 & -8 & -9 \end{bmatrix}, \quad \mathbf{B} = \begin{bmatrix} 0 \\ 0 \\ 1 \end{bmatrix},$$

$$\mathbf{x}(t) = \begin{bmatrix} x_1(t) \\ x_2(t) \\ x_3(t) \end{bmatrix}, \quad \dot{\mathbf{x}}(t) = \begin{bmatrix} \dot{x}_1(t) \\ \dot{x}_2(t) \\ \dot{x}_3(t) \end{bmatrix}, \quad \mathbf{L} = [1 \quad 4 \quad 1].$$

The output $c(t)$ can be obtained by a linear combination of the three state variables as follows:

$$c(t) = x_1(t) + 4x_2(t) + x_3(t). \tag{2.230}$$

Example 2. As a second example for determining the state-variable diagram, consider a system whose transfer function is given by

$$P(s) = \frac{C(s)}{U(s)} = \frac{2}{s^2(s^2 + s + 1)}. \tag{2.231}$$

Dividing through by s^4 we obtain

$$P(s) = \frac{C(s)}{U(s)} = \frac{2s^{-4}}{1 + s^{-1} + s^{-2}}. \tag{2.232}$$

Defining

$$E(s) = \frac{2U(s)}{1 + s^{-1} + s^{-2}},$$

Eq. may be rewritten as

$$C(s) = s^{-4}E(s). \tag{2.233}$$

From Eq. (2.233) and the relation

$$E(s) = 2U(s) - s^{-1}E(s) - s^{-2}E(s), \tag{2.234}$$

the state-variable diagram for this system can easily be obtained (see Figure 2.31). The state variables are referred to as $x_1(t)$, $x_2(t)$, $x_3(t)$, and $x_4(t)$. They are defined as

$$x_1(t) = c(t), \quad x_2(t) = \dot{c}(t), \quad x_3(t) = \ddot{c}(t), \quad x_4(t) = \dddot{c}(t). \tag{2.235}$$

The differential equations relating the state variables are as follows:

$$\begin{aligned} \dot{x}(t) &= x_2(t) = \dot{c}(t), \\ \dot{x}_2(t) &= x_3(t) = \ddot{c}(t), \\ \dot{x}_3(t) &= x_4(t) = \dddot{c}(t), \\ \dot{x}_4(t) &= 2u(t) - x_3(t) - x_4(t). \end{aligned} \tag{2.236}$$

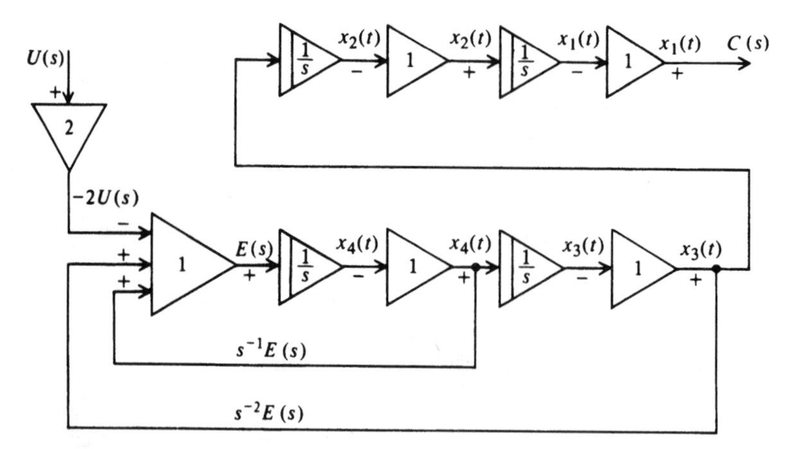

Figure 2.31 State-variable diagram for system where $P(s) = [2/s^2(s^2 + s + 1)]$.

The corresponding phase-variable canonical form is given by

$$\dot{\mathbf{x}}(t) = \mathbf{P}\mathbf{x}(t) + \mathbf{B}u(t), \tag{2.237}$$

$$c(t) = \mathbf{L}\mathbf{x}(t), \tag{2.238}$$

where

$$\mathbf{P} = \begin{bmatrix} 0 & 1 & 0 & 0 \\ 0 & 0 & 1 & 0 \\ 0 & 0 & 0 & 1 \\ 0 & 0 & -1 & -1 \end{bmatrix}, \quad \mathbf{B} = \begin{bmatrix} 0 \\ 0 \\ 0 \\ 2 \end{bmatrix},$$

$$\mathbf{x}(t) = \begin{bmatrix} x_1(t) \\ x_2(t) \\ x_3(t) \\ x_4(t) \end{bmatrix}, \quad \dot{\mathbf{x}}(t) = \begin{bmatrix} \dot{x}_1(t) \\ \dot{x}_2(t) \\ \dot{x}_3(t) \\ \dot{x}_4(t) \end{bmatrix}, \quad \mathbf{L} = [1 \quad 0 \quad 0 \quad 0].$$

From this discussion it can be seen that it is possible to apply the signal-flow graph technique to state-variable analysis. Mason's theorem can be applied to the signal-flow graph which is obtained directly by inspection of the state-variable diagram. The signal-flow graph also provides a physical interpretation of the state-variable concept since its nodes actually represent the different states of the system.

For example, the signal-flow graphs corresponding to the state-variable diagrams of Figures 2.30 and 2.31 are given in Figures 2.32 and 2.33, respectively. The physical meaning of system state is quite clear from these diagrams.

2.23. TRANSFORMATION BETWEEN THE STATE-SPACE FORM AND THE TRANSFER FUNCTION FORM USING MATLAB [36]

The presentation in this chapter so far has shown that the state-space and transfer-function forms are both very useful for analyzing control systems and complement

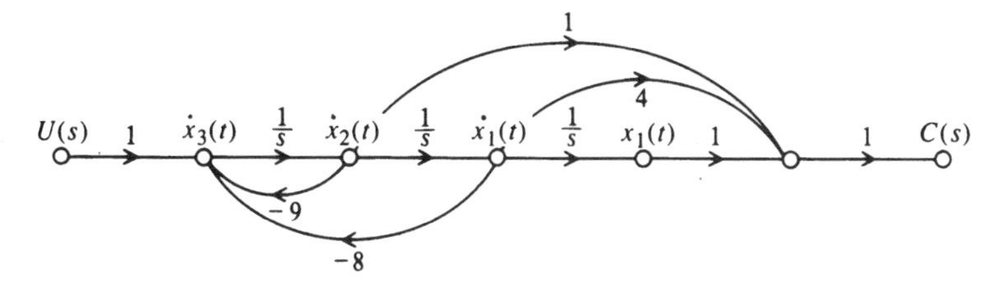

Figure 2.32 Signal-flow graph corresponding to the state-variale diagram of Figure 2.30.

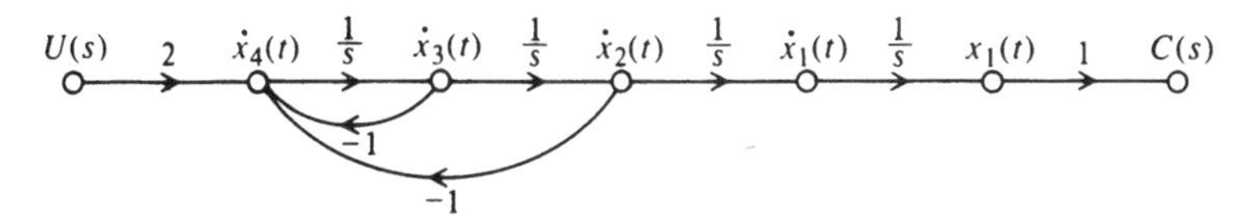

Figure 2.33 Signal-flow graph corresponding to the state-variable diagram of Figure 2.31.

each other. The state-space formulation has the major advantage of representing control systems containing multiple inputs and multiple outputs. The transfer function formulation is limited primarily to control systems containing a single input and having a single output, although it has limited use in control systems containing more than one input and output (see Problem 2.21). In the previous Sections 2.21 and 2.22, it was shown how transfer functions can be converted to the state-space formulation. In this section, the transformation of control systems from transfer function to the state-space form, and vice versa will be demonstrated using MATLAB.

A. Transformation from Transfer Function to State-Space Form

Let us first consider the transformation from the transfer function representation of a control system to the state-space formulation. We will reconsider the transfer function of the control system represented by Eq. (2.222) in the previous section, and is repeated here:

$$\frac{C(s)}{U(s)} = \frac{s^2 + 4s + 1}{s^3 + 9s^2 + 8s} \tag{2.239}$$

The command found in the *Control System Toolbax* and *The Student Edition* of MATLAB to perform this conversion is given by

$$[\mathbf{A}, \mathbf{B}, \mathbf{C}, \mathbf{D}] = \text{tf2ss (num,den)}$$

The matrix **A** used in MATLAB is synonymous with the companion matrix **P** used in this book. Therefore, the MATLAB program to transform from the transfer function given by Eq. (2.222) [and (2.239)] is given by the MATLAB Program shown in Table 2.7.

It is interesting to observe that MATLAB provided a different state-space formulation than was obtained in the previous section as given by Eqs. (2.228) and

Table 2.7. MATLAB Program to Transform from the Transfer Function to State-Space Form

```
num=[0  1  4  1]
den=[1  9  8  0]
[A,B,C,D]=tf2ss(num,den)
A=
    -9  -8   0
     1   0   0
     0   1   0
B=
     1
     0
     0
C=
     1  4  1
D=
     0
```

(2.229), but both are correct. The state-space formulation provided by Eqs. (2.228) and (2.229) are one of a great many possible state-space representations for this control system. There is no unique solution in transforming a transfer function into a state-space formulation. The **P**, **B**, **L**, and **D** matrices provided in Eqs. (2.228) and (2.229) are correct, and the **A** (same as **P** in Eq. (2.228)), **B**, **C** (same as L in Eq. (2.229), and **D** matrices in the results shown in Table 2.7 are also correct.

B. Transformation from State-Space to Transfer Function Form

We will rework the previous problem analyzed in this section, but reverse the process. Now, we will take the state-space formulation, as provided by MATLAB in Table 2.7, and obtain the transfer function representation. The command found in the *Control System Toolbox* and *The Student Edition* of MATLAB to perform this conversion is given by

$$[\text{num,den}] = \text{ss2tf}(\mathbf{A},\mathbf{B},\mathbf{C},\mathbf{D},\text{iu})$$

where "iu" is specified for control systems containing more than one input. If the control system has only one input, as in this example, then we can use

$$[\text{num,den}] = \text{ss2tf}(\mathbf{A},\mathbf{B},\mathbf{C},\mathbf{D})$$

or

$$[\text{num,den}] = \text{ss2tf}(\mathbf{A},\mathbf{B},\mathbf{C},\mathbf{D},1)$$

Therefore, the MATLAB program to transform from the state-space formulation provided in Table 2.7 to a transfer function form is given by the MATLAB program in Table 2.8.

Therefore, the MATLAB Program in Table 2.8 will result in the following transfer function which is identical to the transfer function of Eq. (2.222) (and Eq. (2.239)) which we used to obtain the state-space formulation which was used in the MATLAB Program in Table 2.8:

$$\frac{C(s)}{U(s)} = \frac{s^2 + 4s + 1}{s^3 + 9s^2 + 8s}. \tag{2.240}$$

It is very interesting that the MATLAB state-space formulation as provided in the MATLAB Program of Table 2.7, which is different from that obtained in the previous section and given by Eqs. (2.228) and (2.229), also results in the same transfer function of Eq. (2.222) (and Eq. (2.239)). This result reinforces the concept that there is no unique state-space formulation in going from a transfer function form to a state-space form.

2.24. DIGITAL COMPUTER EVALUATION OF THE TIME RESPONSE

The state-variable representation of a system's dynamics easily lends itself to analysis by means of a digital computer. The technique involves the division of the time axis into sufficiently small increments $t = 0, T, 2T, 3T, 4T, \ldots$, where T is the incremental time of evaluation $\Delta\tau$. This time increment must be made small enough for accurate results. Round-off errors in the computer, however, limit how small the time increment can be.

To illustrate the procedure, let us consider the equation

$$\dot{\mathbf{x}}(t) = \mathbf{P}\mathbf{x}(t) + \mathbf{B}\mathbf{u}(t). \tag{2.241}$$

By definition of a derivative,

$$\dot{\mathbf{x}}(t) = \lim_{\Delta\tau \to 0} \frac{\mathbf{x}(t + \Delta\tau) - \mathbf{x}(t)}{\Delta\tau}. \tag{2.242}$$

Table 2.8. MATLAB Program to Transform from State-Space to Transfer Function Form

```
A=[-9  -8  0;1  0  0;0  1  0];
B=[1;0;0];
C=[1  4  1];
D=[0];
[num,den]=ss2tf(A,B,C,D)
num=
        0      1.0000    4.0000    1.0000
den=
        1      9         8         0
```

Utilizing this definition, the value of $\mathbf{x}(t)$ when t is subdivided into the increments $\Delta\tau$ can be determined. Because $\Delta\tau = T$, we can say (approximately) that

$$\dot{\mathbf{x}}(t) = \frac{\mathbf{x}(t+T) - \mathbf{x}(t)}{T}. \tag{2.243}$$

Substituting Eq. (2.243) into Eq. (2.241), we obtain

$$\frac{\mathbf{x}(t+T) - \mathbf{x}(t)}{T} = \mathbf{P}\mathbf{x}(t) + \mathbf{B}\mathbf{u}(t). \tag{2.244}$$

Equation (2.244) may be solved for $\mathbf{x}(t+T)$ as follows

$$\mathbf{x}(t+T) = T\mathbf{P}\mathbf{x}(t) + \mathbf{x}(t) + T\mathbf{B}\mathbf{u}(t). \tag{2.245}$$

This equation can be written as

$$\mathbf{x}(t+T) = (T\mathbf{P} + \mathbf{I})\mathbf{x}(t) + T\mathbf{B}\mathbf{u}(t). \tag{2.246}$$

To generalize this expression for the intervals mT, let

$$t = mT, \tag{2.247}$$

where $m = 0, 1, 2, 3, 4, \ldots$. Therefore, Eq. (2.246) can be written as the recurrence relation

$$\mathbf{x}[(m+1)T] = (T\mathbf{P} + \mathbf{I})\mathbf{x}(mT) + T\mathbf{B}\mathbf{u}(mT). \tag{2.248}$$

Equation (2.248) states that the value of the state vector at time $(m+1)T$ is based on the values of $\mathbf{x}$ and $\mathbf{u}$ at time mT. This resulting recurrence relation is a sequential series of calculations that is very suitable for digital-computer operation. Note that this is a very crude scheme—more sophisticated schemes involve more refined approximations to $\dot{\mathbf{x}}$[16–18].

To illustrate the capability of using the recurrence equation (2.248), let us evaluate the unit step response of the feedback control system illustrated in Figure 2.29 whose closed-loop transfer function is given by Eq. (2.212). This is the first illustration in this book of developing working digital computer programs to solve problems other than with MATLAB. This program was written using Fortran (FORmula TRANslation) [19]. Several additional problems are solved throughout the book using Fortran and Basic and their corresponding logic flow diagrams, program listings, and program outputs are all illustrated for teaching purposes. In addition, the majority of the solutions in this book are obtained using MATLAB™ (see Section 2.8). The *Modern Control System Theory and Design Toolbox*, which complements this book, and contains the M-files used for these solutions, is available free from The MathWorks, Inc. anonymous FTP server at ftp://ftp.mathworks.com/pub/books/Shinners. (See the Preface for additional information.) MATLAB was used to obtain the graphical results of this problem.

The logic flow diagram for developing the program that can be used to obtain the state responses of second- or higher-order systems as illustrated in Figure 2.34. Table

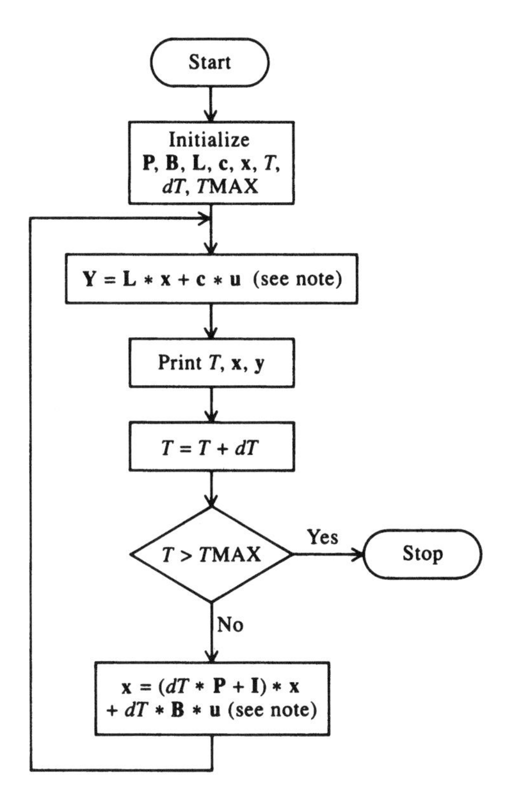

Figure 2.34 Logic flow diagram for obtaining the response time of a control system. Note: For unit step input, **u** = 1 for all positive time.

2.9 illustrates the actual program listing. Figure 2.34 and Table 2.9 should be compared in order to obtain a thorough understanding of the method. This program was then applied to obtain the unit step response of the system illustrated in Figure 2.29 whose phase-variable canonical equations are defined in Eqs. (2.216) and (2.217). The computer output is illustrated in Table 2.10. The results are plotted in Figure 2.35*a* and *b* using MATLAB for sampling times T of 0.05 sec (where the response is as expected and very close to the theoretical response, as shown in Figure 2.35*b*) to 0.5 sec (where an unstable, and obviously erroneous response is indicated). The procedure for obtaining transient responses using MATLAB is described in Section 2.25.

Based on the sensitivity of the responses to the sampling time T, a few words of guidance are in order here. In practice, a good guideline is to sample at a rate at least $\frac{1}{10}$ of the smallest time constant of the system. We will show in Chapter 6 when we discuss the Bode frequency response diagram that this system being analyzed has an approximate bandwidth of 2 rad/sec. Therefore, its time constant is approximately 0.5 sec. As shown in Figure 2.35*a*, a sampling time of 0.5 sec indicates that it yields an unstable response (although we know that the system response is an exponentially damped sinusoid, and is stable as will be proven in Chapter 4). However, as we increase the sampling rate T, the digital computer result gets better, as illustrated in Figure 2.35*a*. Using the guideline of sampling at $\frac{1}{10}$ the system's time constant, or

Table 2.9. Computer Program Listing for Obtaining the State and Output Response of a Control System from Knowledge of its P and B Matrices, Input and State Vectors, and the Computer Implemented Algorithm of Eq. (2.248)

```
      PARAMETER (NSTATE=2,NIN=1,NOUT=1)
      PARAMETER (DT=0.25,TMAX=10.0)
C
      DIMENSION P(NSTATE,NSTATE),B(NSTATE,NIN)
      DIMENSION RL(NOUT,NSTATE),C(NOUT,NIN)
      DIMENSION X(NSTATE),Y(NOUT),TEMP(NSTATE)
      DATA P /0.0,-2.0,1.0,-1.0/, B /0.0,2.0/ , RL /1.0,0.0/ , C /0.0/
      DATA X /NSTATE*0.0/ , T /0.0/
C
C *** UPDATE THE OUTPUT (Y)
100 DO 110 I=1,NOUT
       Y(I)=0
       DO 120 J=1,NSTATE
120       Y(I)=Y(I)+RL(I,J)*X(J)
C *** APPLY A UNIT STEP TO ALL INPUTS (U=1)
       DO 130 J=1,NIN
130       Y(I)=Y(I)+C(I,J)*1
110   CONTINUE
      WRITE(6,*)T,X,Y
      T=T+DT
C     ***IS THE INTERVAL COMPLETED (T>TMAX)
      IF(T.GT.TMAX)GO to 999
C     ***UPDATE THE STATE VARIABLES (X)
      DO 200 I=1,NSTATE
           TEMP(I)=X(I)
           DO 210 J=1,NSTATE
210          TEMP(I)=TEMP(I)+DT*P(I,J)*X(J)
C *** APPLY A UNIT STEP INPUT TO ALL THE INPUTS (U=1)
           DO 220 J=1,NIN
220          TEMP(I)=TEMP(I)+DT*B(I,J)*1
200    CONTINUE
      DO 250 I=1,NSTATE
250   X(I)=TEMP(I)
      GO TO 100
C
999   STOP
      END
```

$\frac{1}{10} \times 0.5 = 0.05$ sec, we obtain a quite accurate solution. Figure 2.35*a* illustrates the improvement as T is decreased from 0.5 to 0.05 sec in incremental steps of $DT = 0.5/n$, where $1 \leqslant n \leqslant 10$. In Figure 2.35*b*, the time response obtained using the fast sampling time of 0.05 sec is compared with the theoretical response of the system. The result is very good.

Notice the accuracy and simplicity of the digital computer's solution. In addition, the speed of the computer's solution is of great benefit. Its usefulness as an aid to the control system engineer is quite evident.

2.25. OBTAINING THE TRANSIENT RESPONSE OF SYSTEMS USING MATLAB [6]

In the previous section, an algorithm (Eq. 2.248) was developed for obtaining the time response of a control system based on a knowledge of the **P** and **B** matrices, and

Table 2.10. Results of Computer Analysis for the Unit Step Response of the System Illustrated in Figure 2.29

Time	State $x_1(t)$	State $x_2(t)$	Output $y(t)$
0.00000000	0.00000000	0.00000000	0.00000000
0.25000000	0.00000000	0.50000000	0.00000000
0.50000000	0.12500000	0.87500000	0.12500000
0.75000000	0.34375000	1.0937500	0.34375000
1.0000000	0.61718750	1.1484375	0.61718750
1.2500000	0.90429688	1.0527344	0.90429688
1.5000000	1.1674805	0.83740234	1.1674805
1.7500000	1.3768311	0.54431152	1.3768311
2.0000000	1.5129089	0.21981812	1.5129089
2.2500000	1.5678635	−0.91590881–001	1.5678635
2.5000000	1.5449657	−0.35262489	1.5449657
2.7500000	1.4568095	−0.53695154	1.4568095
3.0000000	1.3225716	−0.63111842	1.3225716
3.2500000	1.1647920	−0.63462463	1.1647920
3.5000000	1.0061359	−0.55836450	1.0061359
3.7500000	0.86654477	−0.42184132	0.86654477
4.0000000	0.76108444	−0.24965338	0.76108444
4.2500000	0.69867110	−0.67782253–001	0.69867110
4.5000000	0.68172553	0.99827766–001	0.68172553
4.7500000	0.70668247	0.23400806	0.70668247
5.0000000	0.76518448	0.32216481	0.76518448
5.2500000	0.84572569	0.35903137	0.84572569
5.5000000	0.93548352	0.34641069	0.93548352
5.7500000	1.0220862	0.29206626	1.0220862
6.0000000	1.0951027	0.20800661	1.0951027
6.2500000	1.1471044	0.10845359	1.1471044
6.5000000	1.1742178	0.77880025–002	1.1742178
6.7500000	1.1761648	−0.81267886–001	1.1761648
7.0000000	1.1558478	−0.14903331	1.1558478
7.2500000	1.1185895	−0.18969889	1.1185895
7.5000000	1.0711648	−0.20156892	1.0711648
7.7500000	1.0207725	−0.18675908	1.0207725
8.0000000	0.97408278	−0.15045559	0.97408278
8.2500000	0.93646889	−0.99883087–001	0.93646889
8.5000000	0.91149812	−0.43146759–001	0.91149812
8.7500000	0.90071143	0.11890873–001	0.90071143
9.0000000	0.90368415	0.58562443–001	0.90368415
9.2500000	0.91832475	0.9207959–001	0.91832475
9.5000000	0.94134469	0.1098744	0.94134469
9.7500000	0.96881905	0.11175074	0.96881905
10.0000000	0.99675673	0.99403530–001	0.99675673

a digital computer program was written in Fortran and applied to obtain the unit response for the control system illustrated in Figure 2.29. In this section, the relatively simple procedure for obtaining transient responses of control systems using MATLAB will be provided.

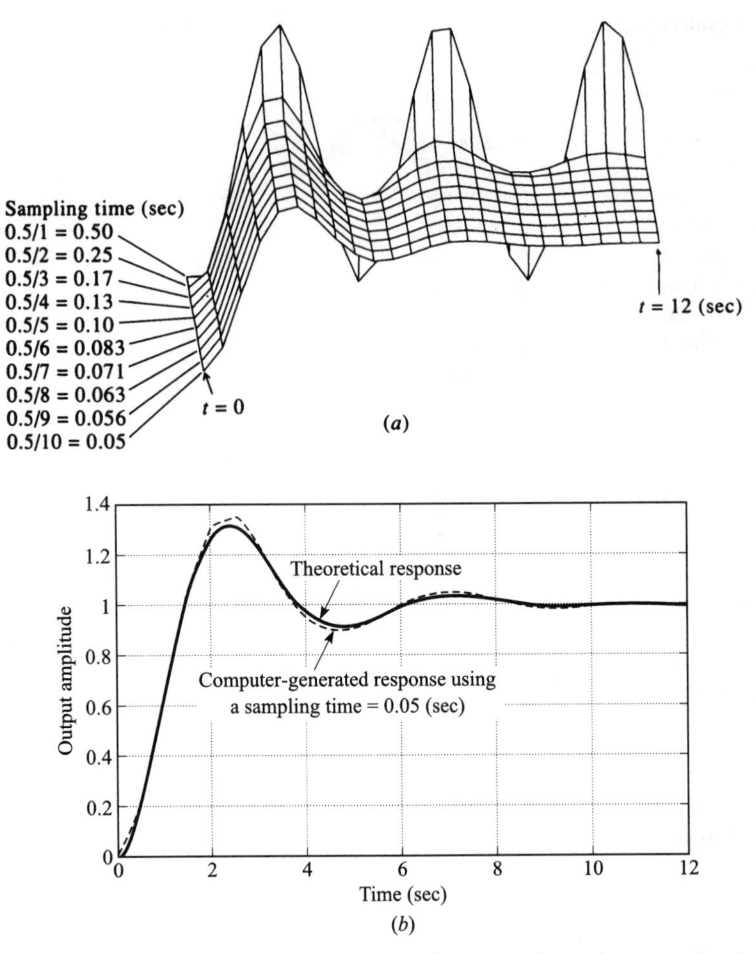

Figure 2.35 Response of the system shown in Figure 2.29 to a unit step input for sampling times of 0.05–0.5 sec (a), and a comparison of the case of a 0.05-sec sampling time with the theoretical response (b).

The MATLAB commands, found in the *Control System Toolbox* and *The Student Edition* of MATLAB, for obtaining the unit step response of a control system are as follows:

(a) If the numerator (num) and denominator (den) of a closed-loop system are known in transfer function form:

step(num,den)

If the user wishes to supply the time t at which the step response will be computed at regularly spaced times, the following command is used:

step(num,den,t).

The time vector is automatically determined when t is excluded from the command.

As an illustration for obtaining the transient response to a unit setp input, let us reconsider the problem analyzed in the previous section where we obtained the unit step response to the control system illustrated in Figure 2.29. The closed-loop transfer function of the control system illustrated in Figure 2.29 is given by

$$\frac{C(s)}{R(s)} = \frac{\frac{2}{s(s+1)}}{1 + \frac{2}{s(s+1)}} = \frac{2}{s^2 + s + 2}. \tag{2.249}$$

Therefore, the MATLAB program for obtaining the unit step response for this control system is given by the MATLAB Program in Table 2.11.

The resultant transient response is illustrated in Figure 2.36. Observe that it is the same as the theoretical curve illustrated in Figure 2.35*b*.

(b) If the state-space form (including **A** (or **P**), **B**, **C** (or **L**), and **D**) are known:

$$\text{step}(\mathbf{A},\mathbf{B},\mathbf{C},\mathbf{D}).$$

Table 2.11. MATLAB Program for Obtaining the Unit Step Response of System Shown in Figure 2.29 given the Transfer Function in Eq. (2.249)

```
num=[0    0    2]
den=[1    1    2]
step(num,den)
grid
title('Unit-Step Response of System Shown in Figure 2.29')
```

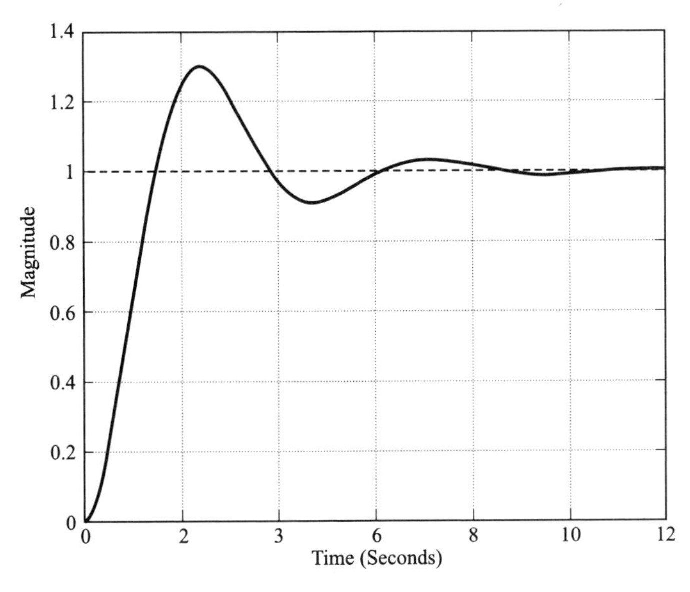

Figure 2.36 Unit step response of system shown in Figure 2.29 obtained using MATLAB Program in Table 2.11.

If the user wishes to supply the time t at which the response will be computed at regularly spaced times, use the following command:

step(**A**,**B**,**C**,**D**,t).

If the control system has multiple inputs, the command

step(**A**,**B**,**C**,**D**,iu)

or

step(**A**,**B**,**C**,**D**,iu,t)

produces a step response from the single input "iu" specified to all the outputs of the system. The scalar, iu, is an index into the inputs of the system and specifies which input is to be used for the response.

These "step" commands do not result in a plot on the screen. Therefore, it is necessary to use the "plot" command in order to obtain the graphical transient response.

As an illustration for obtaining the transient response to a unit step input from knowledge of the state-space form, we will reconsider the same problem analyzed previously in this section.

The corresponding phase-variable canonical equations for the control system shown in Figure 2.29 are as follows:

$$\dot{x}_1(t) = x_2(t) \quad (2.250)$$

$$\dot{x}_2(t) = -2x_1(t) - x_2(t) + 2r(t). \quad (2.251)$$

Therefore, the matrices **A** (or **P**), **B**, **C** (or **L**), and **D** are given by:

$$\mathbf{A} = \begin{bmatrix} 0 & 1 \\ -2 & -1 \end{bmatrix}; \quad \mathbf{B} = \begin{bmatrix} 0 \\ 2 \end{bmatrix}; \quad \mathbf{L} = [1 \quad 0]; \quad \mathbf{D} = 0.$$

The resulting MATLAB program used to obtain the unit step response from knowledge of the phase variables canonical form is given in Table 2.12.

Table 2.12. MATLAB Program for Obtaining the Unit Step Response of System Shown in Figure 2.29 given the state equations of Eqs. (2.250) and (2.251)

```
A=[0    -1;-2    -1];
B=[-2    -1];
C=[1    0;0    0]
D=[0    0;0    0];
step(A,B,C,D)
grid
title ('Unit Step Response of System Shown in Figure 2.29')
```

The resultant transient response as illustrated in Figure 2.37. Observe that it is the same as that shown in Figure 2.36, and that shown as the theoretical curve in Figure 2.35*b*.

2.26. STATE TRANSITION MATRIX

The state transition matrix relates the state of a system at $t = t_0$ to its state at a subsequent time t, when the input $\mathbf{u}(t) = 0$. In order to define the state transition matrix of a system, let us consider the general form of the state equation [see Eq. 2.197]:

$$\dot{\mathbf{x}}(t) = \mathbf{P}\mathbf{x}(t) + \mathbf{B}\mathbf{u}(t). \tag{2.252}$$

The Laplace transform of Eq. (2.252) is given by

$$s\mathbf{X}(s) - \mathbf{x}(0^+) = \mathbf{P}\mathbf{X}(s) + \mathbf{B}\mathbf{U}(s), \tag{2.253}$$

where $\mathbf{X}(s)$ is the Laplace transform of $\mathbf{x}(t)$ and $\mathbf{U}(s)$ is the Laplace transform of $\mathbf{u}(t)$. Solving for $\mathbf{X}(s)$, we obtain

$$\mathbf{X}(s) = [s\mathbf{I} - \mathbf{P}]^{-1}\mathbf{x}(0^+) + [s\mathbf{I} - \mathbf{P}]^{-1}\mathbf{B}\mathbf{U}(s). \tag{2.254}$$

The inverse Laplace transform of Eq. (2.254) gives the state transition equation

$$\mathbf{x}(t) = \mathbf{\Phi}(t)\mathbf{x}(0^+) + \int_0^t \mathbf{\Phi}(t - \tau)\mathbf{B}\mathbf{u}(\tau)\, d\tau, \tag{2.255}$$

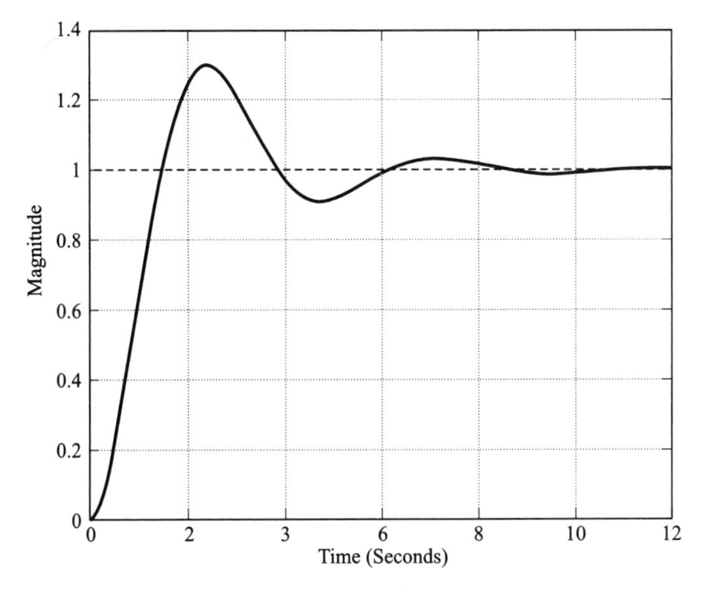

Figure 2.37 Unit step response of system shown in Figure 2.29 obtained using the MATLAB Program in Table 2.12.

where the state transition matrix is defined by

$$\mathbf{\Phi}(t) = \mathscr{L}^{-1}\{[s\mathbf{I} - \mathbf{P}]^{-1}\}, \text{ for } t \geq 0. \tag{2.256}$$

The first term on the right-hand side of Eq. (2.255) is known as the homogeneous solution and is due only to the initial conditions; the second term on the right-hand side of Eq. (2.255), the convolution integral, is known as the particular solution and is due to the external forcing function. Equation (2.256) is defined as the state-transition matrix of the system for $t \geqslant 0$. When the input $\mathbf{u} = 0$, Eq. (2.256), reduces to

$$\mathbf{x}(t) = \mathbf{\Phi}(t)\mathbf{x}(0^+). \tag{2.257}$$

The matrix $\mathbf{\Phi}(t)$ is defined as the state transition matrix, because it relates the transition of the system state at time $t_0 = 0^+$ to the state at some subsequent time t. It has the following properties:

$$\mathbf{\Phi}(0) = \mathbf{I}, \tag{2.258}$$

$$\mathbf{\Phi}(t_2 - t_0) = \mathbf{\Phi}(t_2 - t_1)\mathbf{\Phi}(t_1 - t_0), \tag{2.259}$$

$$\mathbf{\Phi}(t + \tau) = \mathbf{\Phi}(t)\mathbf{\Phi}(\tau), \tag{2.260}$$

$$\mathbf{\Phi}^{-1}(t) = \mathbf{\Phi}(-t). \tag{2.261}$$

Very often it is desired to use a more general initial time t_0. Equation (2.255) can be modified by letting $t = t_0$. Solving for $\mathbf{x}(0^+)$, we obtain the following expression:

$$\mathbf{x}(0^+) = \mathbf{\Phi}^{-1}(t_0)\mathbf{x}(t_0) - \mathbf{\Phi}^{-1}(t_0)\int_0^{t_0} \mathbf{\Phi}(t_0 - \tau)\mathbf{B}\mathbf{u}(\tau)\,d\tau. \tag{2.262}$$

Using Eq. (2.261), Eq. 2.262 can be rewritten as

$$\mathbf{x}(0^+) = \mathbf{\Phi}(-t_0)\mathbf{x}(t_0) - \mathbf{\Phi}(-t_0)\int_0^{t_0} \mathbf{\Phi}(t_0 - \tau)\mathbf{B}\mathbf{u}(\tau)\,d\tau \tag{2.263}$$

Substituting Eq. (2.263) into Eq. (2.255), the following expression is obtained:

$$\begin{aligned}\mathbf{x}(t) =& \mathbf{\Phi}(t)\mathbf{\Phi}(-t_0)\mathbf{x}(t_0) - \mathbf{\Phi}(t)\mathbf{\Phi}(-t_0)\int_0^{t_0} \mathbf{\Phi}(t_0 - \tau)\mathbf{B}\mathbf{u}(\tau)\,d\tau \\ &+ \int_0^{t} \mathbf{\Phi}(t - \tau)\mathbf{B}\mathbf{u}(\tau)\,d\tau.\end{aligned} \tag{2.264}$$

Using Eq. (2.259), Eq. (2.264) can be reduced to

$$\mathbf{x}(t) = \mathbf{\Phi}(t - t_0)\mathbf{x}(t_0) + \int_{t_0}^{t} \mathbf{\Phi}(t - \tau)\mathbf{B}\mathbf{u}(\tau)\,d\tau. \tag{2.265}$$

Equation (2.265) is the solution of Eq. (2.252) for $t \geqslant t_0$.

As an example of determining the state transition matrix, consider the open-loop system where the transfer function of the controlled process is given by

$$P(s) = \frac{C(s)}{U(s)} = \frac{1}{s^2}. \tag{2.266}$$

Its corresponding differential equation is given by $\ddot{c}(t) = u(t)$. Defining the state variables as

$$x_1(t) = c(t), \quad x_2(t) = \dot{c}(t), \tag{2.267}$$

the system can be described by the following two first-order differential equations:

$$\dot{x}_1(t) = x_2(t) = \dot{c}(t), \quad \dot{x}_2(t) = u(t). \tag{2.268}$$

Therefore, the entire system can be described by the state equation

$$\dot{\mathbf{x}}(t) = \mathbf{P}\mathbf{x}(t) + \mathbf{B}u(t), \tag{2.269}$$

where

$$\mathbf{P} = \begin{bmatrix} 0 & 1 \\ 0 & 0 \end{bmatrix}, \quad \mathbf{B} = \begin{bmatrix} 0 \\ 1 \end{bmatrix}, \quad \mathbf{x}(t) = \begin{bmatrix} x_1(t) \\ x_2(t) \end{bmatrix}, \quad \dot{\mathbf{x}}(t) = \begin{bmatrix} \dot{x}_1(t) \\ \dot{x}_2(t) \end{bmatrix}. \tag{2.270}$$

The state transition matrix, which is defined by

$$\boldsymbol{\Phi}(t) = \mathscr{L}^{-1}\{[s\mathbf{I} - \mathbf{P}]^{-1}\}, \tag{2.271}$$

can be obtained from Eq. (2.270). We find

$$[s\mathbf{I} - \mathbf{P}] = \begin{bmatrix} s & 0 \\ 0 & s \end{bmatrix} - \begin{bmatrix} 0 & 1 \\ 0 & 0 \end{bmatrix} = \begin{bmatrix} s & -1 \\ 0 & s \end{bmatrix}. \tag{2.272}$$

From Eq. (2.188), we know that

$$\mathbf{B}^{-1} = \frac{\operatorname{adj} \mathbf{B}}{|\mathbf{B}|}. \tag{2.273}$$

Therefore

$$[s\mathbf{I} = \mathbf{P}]^{-1} = \frac{\operatorname{adj}[s\mathbf{I} - \mathbf{P}]}{|s\mathbf{I} - \mathbf{P}|} = \frac{\begin{bmatrix} s & 1 \\ 0 & s \end{bmatrix}}{\begin{vmatrix} s & -1 \\ 0 & s \end{vmatrix}} = \frac{\begin{bmatrix} s & 1 \\ 0 & s \end{bmatrix}}{s^2} = \begin{bmatrix} \frac{1}{s} & \frac{1}{s^2} \\ 0 & \frac{1}{s} \end{bmatrix}. \tag{2.274}$$

The state transition matrix defined by Eq. (2.256) is the inverse transform of this matrix. It is given by

$$\boldsymbol{\Phi}(t) = \mathscr{L}^{-1}\{[s\mathbf{I} - \mathbf{P}]^{-1}\} = \begin{bmatrix} U(t) & t \\ 0 & U(t) \end{bmatrix}, \quad t \geqslant 0 \tag{2.275}$$

With knowledge of the state transition matrix, we can easily determine the values of states $\mathbf{x}_1(t)$ and $\mathbf{x}_2(t)$ as a function of time. Let us assume that the initial values of the states are given by the following initial-state vector:

$$\mathbf{x}(0^+) = \begin{bmatrix} x_1(0) \\ x_2(0) \end{bmatrix} = \begin{bmatrix} 1 \\ 2 \end{bmatrix}. \tag{2.276}$$

We can find the state vector $\mathbf{x}(t)$ as a function of time from Eq. (2.257) as follows:

$$\mathbf{x}(t) = \mathbf{\Phi}(t)\mathbf{x}(0^+). \tag{2.277}$$

Therefore, substituting Eqs. (2.275) and (2.276) into (2.277), we obtain

$$\begin{bmatrix} x_1(t) \\ x_2(t) \end{bmatrix} = \begin{bmatrix} U(t) & t \\ 0 & U(t) \end{bmatrix}, \quad t \geqslant 0. \tag{2.278}$$

Solving for $x_1(t)$ and $x_2(t)$, we obtain the following:

$$x_1(t) = U(t) + 2t, \quad t \geqslant 0, \tag{2.279}$$

$$x_2(t) = 2U(t), \quad t \geqslant 0. \tag{2.280}$$

These expressions are plotted in Figure 2.38.

The state transition matrix may also be derived from the state-variable diagram. As an example of the technique, consider a system described by the differential equation

$$\ddot{c}(t) + 4\dot{c}(t) + 3c(t) = r(t). \tag{2.281}$$

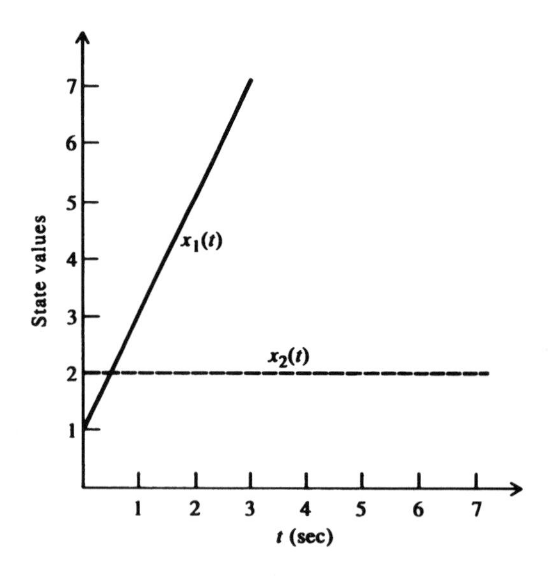

Figure 2.38 Plot of states $x_1(t)$ and $x_2(t)$ as defined in Eqs. (2.279) and (2.280), respectively.

The Laplace transform of Eq. (2.281) yields

$$\frac{C(s)}{R(s)} = \frac{1}{s^2 + 4s + 3}, \tag{2.282}$$

and dividing top and bottom by s^2, we obtain

$$\frac{C(s)}{R(s)} = \frac{s^{-2}}{1 + 4s^{-1} + 3s^{-2}}. \tag{2.283}$$

Defining

$$E(s) = \frac{R(s)}{1 + 4s^{-1} + 3s^{-2}}, \tag{2.284}$$

Eq. (2.283) may be rewritten as

$$C(s) = s^{-2}E(s). \tag{2.285}$$

From Eq. (2.285) and the relation

$$E(s) = R(s) - 4s^{-1}E(s) - 3s^{-2}E(s), \tag{2.286}$$

the state-variable diagram for this system is obtained as illustrated in Figure 2.39. In addition, for generality, it is assumed that the states of the system, $x_1(t)$ and $x_2(t)$, have the initial conditions $x_1(0)$ and $x_2(0)$, respectively, at the inputs to each integrator. The corresponding state-variable signal-flow graph is given by Figure 2.40. The resulting transformed state equations of the system are obtained from this state-variable signal-flow graph using Mason's formula [Eq. 2.135]:

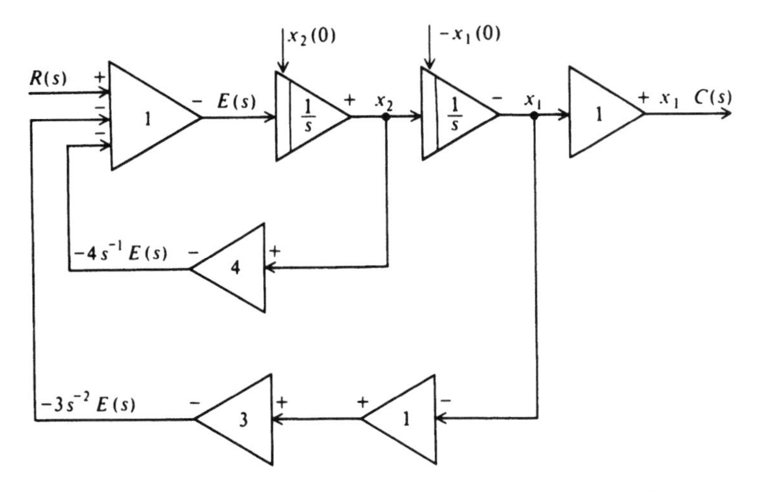

Figure 2.39 State-variable diagram for system where $C(s)/R(s) = 1/(s^2 + 4s + 3)$.

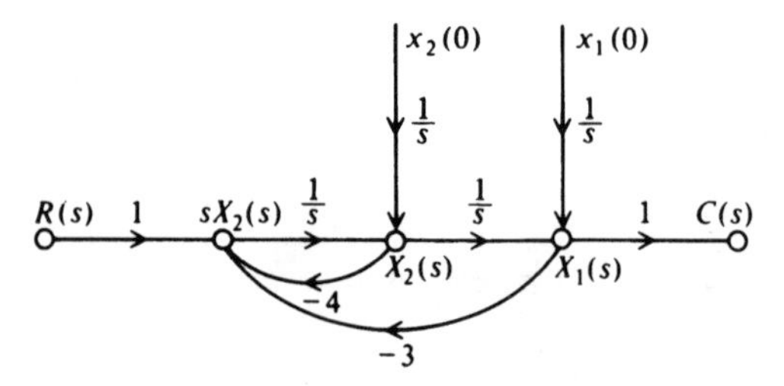

Figure 2.40 State-variable signal-flow graph corresponding to the state-variable diagram of Figure 2.39.

$$X_1(s) = \frac{s^{-1}(1+4s^{-1})x_1(0)}{\Delta} + \frac{s^{-2}x_2(0)}{\Delta} + \frac{s^{-2}R(s)}{\Delta}, \tag{2.287}$$

$$X_2(s) = \frac{-3s^{-2}x_1(0)}{\Delta} + \frac{s^{-1}x_2(0)}{\Delta} + \frac{s^{-1}R(s)}{\Delta}, \tag{2.288}$$

where

$$\Delta = 1 - (-4s^{-1} - 3s^{-2}) = 1 + 4s^{-1} + 3s^{-2}. \tag{2.289}$$

Simplifying Eqs. (2.287)–(2.289), we obtain the following pair of equations:

$$X_1(s) = \frac{s+4}{s^2+4s+3}x_1(0) + \frac{1}{s^2+4s+3}x_2(0) + \frac{R(s)}{s^2+4s+3}, \tag{2.290}$$

$$X_2(s) = \frac{-3}{s^2+4s+3}x_1(0) + \frac{s}{s^2+4s+3}x_2(0) + \frac{sR(s)}{s^2+4s+3}. \tag{2.291}$$

These two equations can be put into the following form:

$$\begin{bmatrix} X_1(s) \\ X_2(s) \end{bmatrix} = \frac{1}{(s+1)(s+3)}\begin{bmatrix} s+4 & 1 \\ -3 & s \end{bmatrix}\begin{bmatrix} x_1(0) \\ x_2(0) \end{bmatrix} + \begin{bmatrix} \dfrac{1}{(s+1)(s+3)} \\ \dfrac{s}{(s+1)s+3)} \end{bmatrix} R(s). \tag{2.292}$$

From Eq. (2.292), we can obtain the state transition matrix by taking the inverse Laplace transform. It is assumed in the following solution that $r(t) = U(t)$ and $R(s) = 1/s$:

$$\begin{bmatrix} x_1(t) \\ x_2(t) \end{bmatrix} = \begin{bmatrix} 1.5e^{-t} - 0.5e^{-3t} & 0.5e^{-t} - 0.5e^{-3t} \\ -1.5e^{-t} + 1.5e^{-3t} & -0.5e^{-t} + 1.5e^{-3t} \end{bmatrix}\begin{bmatrix} x_1(0) \\ x_2(0) \end{bmatrix} + \begin{bmatrix} 0.33U(t) - 0.5e^{-t} + 0.167e^{-3t} \\ 0.5e^{-t} - 0.5e^{-3t} \end{bmatrix}, \text{ for } t \geq 0. \tag{2.293}$$

Therefore, the state transition matrix is given by

$$\mathbf{\Phi}(t) = \begin{bmatrix} 1.5e^{-t} - 0.5e^{-3t} & 0.5e^{-t} - 0.5e^{-3t} \\ -1.5e^{-t} + 1.5e^{-3t} & -0.5e^{-t} + 1.5e^{-3t} \end{bmatrix}, \quad t \geqslant 0. \tag{2.294}$$

Once the state transition matrix is obtained, the evaluation of the time response is easily obtained.

This technique unfortunately shows that the use of the state-variable diagram and the state-variable signal-flow graph method for determining the state transition matrix is extremely inefficient compared with calculating it directly from Eq. (2.256). However, the method utilizing the state-variable diagram in conjunction with the signal-flow graph does have advantages in certain situations, because it offers other choices of state variables and permits application of Mason's theorems for determining the relationships among the various states required in the state transition matrix.

2.27. TOTAL SOLUTION OF THE STATE EQUATION

The purpose of this section is to illustrate how one may obtain the complete solution for the output in the time domain of a control system utilizing the state-variable method. In this example, we will want to determine the complete solution by evaluating Eq. (2.255), the state transition equation.

Consider a system described by the following differential equation:

$$\ddot{c}(t) + 2\dot{c}(t) + c(t) = \dot{r}(t) + r(t). \tag{2.295}$$

It is desired to determine the output $c(t)$, given that the input $r(t)$ is given by

$$r(t) = \sin t \tag{2.296}$$

and the initial conditions are $c(0) = 1$ and $\dot{c}(0) = 0$. The technique employed is to determine the state transition matrix from Eq. (2.256) and then evaluate Eq. (2.255) for $\mathbf{x}(t)$. The output $\mathbf{c}(t)$ is then evaluated from

$$\mathbf{c}(t) = \mathbf{L}\mathbf{x}(t). \tag{2.297}$$

If the state variables are defined by

$$x_1(t) = c(t), \quad x_2(t) = \dot{c}(t), \tag{2.298}$$

and $u(t)$ by

$$u(t) = r(t),$$

then the system can be described by the following two first-order differential equations:

$$\begin{aligned} \dot{x}_1(t) &= x_2(t), \\ \dot{x}_2(t) &= -2x_2(t) - x_1(t) + u(t) + \dot{u}(t). \end{aligned} \tag{2.299}$$

Therefore, the system can be described by

$$\dot{\mathbf{x}}(t) = \mathbf{P}\mathbf{x}(t) + \mathbf{B}(u(t) + \dot{u}(t)), \tag{2.300}$$

where

$$\mathbf{P} = \begin{bmatrix} 0 & 1 \\ -1 & -2 \end{bmatrix}, \quad \mathbf{B} = \begin{bmatrix} 0 \\ 1 \end{bmatrix}, \quad \mathbf{x}(t) = \begin{bmatrix} x_1(t) \\ x_2(t) \end{bmatrix}, \quad \dot{x}(t) = \begin{bmatrix} \dot{x}_1(t) \\ \dot{x}_2(t) \end{bmatrix}. \tag{2.301}$$

The state transition matrix, which is defined by Eq. (2.256), can be obtained from Eq. (2.301). We find

$$[s\mathbf{I} - \mathbf{P}] = \begin{bmatrix} s & 0 \\ 0 & s \end{bmatrix} - \begin{bmatrix} 0 & 1 \\ -1 & -2 \end{bmatrix} = \begin{bmatrix} s & -1 \\ 1 & s+2 \end{bmatrix}. \tag{2.302}$$

From Eq. (2.188), we know that

$$\mathbf{B}^{-1} = \frac{\operatorname{adj} \mathbf{B}}{|\mathbf{B}|}. \tag{2.303}$$

Therefore,

$$[s\mathbf{I} - \mathbf{P}]^{-1} = \frac{\operatorname{adj}[s\mathbf{I} - \mathbf{P}]}{|s\mathbf{I} - \mathbf{P}|} = \frac{\begin{bmatrix} s+2 & 1 \\ -1 & s \end{bmatrix}}{(s+1)^2} = \begin{bmatrix} \dfrac{s+2}{(s+1)^2} & \dfrac{1}{(s+1)^2} \\ -\dfrac{1}{(s+1)^2} & \dfrac{s}{(s+1)^2} \end{bmatrix} \tag{2.304}$$

The state transition matrix defined by Eq. (2.256) is the inverse transform of this matrix. It is given by

$$\mathbf{\Phi}(t) = \mathscr{L}^{-1}\{[s\mathbf{I} - \mathbf{P}]^{-1}\} = \begin{bmatrix} e^{-t}(t+1) & te^{-t} \\ -te^{-t} & e^{-t}(1-t) \end{bmatrix}, \quad t \geqslant 0. \tag{2.305}$$

The full solution for the output can be obtained from Eqs. (2.255) and (2.297) as follows:

$$\mathbf{x}(t) = \mathbf{\Phi}(t)\mathbf{x}(0^+) + \int_0^t \mathbf{\Phi}(t-\tau)\mathbf{B}\mathbf{u}(\tau)\, d\tau, \tag{2.306}$$

$$\mathbf{c}(t) = \mathbf{L}\mathbf{x}(t). \tag{2.307}$$

Substituting Eq. (2.306) into Eq. (2.307), we obtain the following relationship for the output in terms of the state transition matrix:

$$\mathbf{c}(t) = \mathbf{L}\mathbf{\Phi}(t)\mathbf{x}(0^+) + \int_0^t \mathbf{L}\mathbf{\Phi}(t-\tau)\mathbf{B}\mathbf{u}(\tau)\, d\tau. \tag{2.308}$$

We know $\mathbf{\Phi}$(t) from Eq. (2.305). We have looked at many similar systems in this chapter, and should know by inspection now that

$$\mathbf{L} = [1 \quad 0], \quad \mathbf{x}(0^+) = \begin{bmatrix} x_1(0) \\ x_2(0) \end{bmatrix} = \begin{bmatrix} 1 \\ 0 \end{bmatrix} \tag{2.309}$$

For this system, the input function $u(\tau) + \dot{u}(\tau)$ is obtained as follows:

$$u(\tau) + \dot{u}(\tau) = r(\tau) + \dot{r}(\tau) = \sin\tau + \cos\tau. \tag{2.310}$$

Substituting all of these values into Eq. (2.308), we obtain the following expression:

$$\begin{aligned} c(t) =& [1 \quad 0]\begin{bmatrix} e^{-t}(t+1) & te^{-t} \\ -te^{-t} & e^{-t}(1-t) \end{bmatrix}\begin{bmatrix} 1 \\ 0 \end{bmatrix} \\ &+ \int_0^t [1 \quad 0]\begin{bmatrix} e^{-(t-\tau)}(t-\tau+1) & (t-\tau)e^{-(t-\tau)} \\ -(t-\tau)e^{-(t-\tau)} & e^{-(t-\tau)}(1-t+\tau) \end{bmatrix}\begin{bmatrix} 0 \\ 1 \end{bmatrix} \\ &\times (\sin\tau + \cos\tau)\, d\tau. \end{aligned} \tag{2.311}$$

On simplifying, the result becomes

$$c(t) = e^{-t}(t+1) + \int_0^t \left[(t-\tau)e^{-(t-\tau)}\right][\sin\tau + \cos\tau]\, d\tau. \tag{2.312}$$

Integrating and simplifying, we finally obtain the output as

$$c(t) = \frac{3}{2}e^{-t} + te^{-t} + \frac{1}{2}\sin t - \frac{1}{2}\cos t, \quad t \geqslant 0. \tag{2.313}$$

We can check the reasonableness of this result by determining the initial value, $c(0)$. Substituting $t = 0$ into Eq. (2.313), we obtain

$$c(0) = \frac{3}{2} + 0 + 0 - \frac{1}{2}(1) = 1 \tag{2.314}$$

which agrees with the value of $c(0)$ specified. It is left as an exercise to the reader to also check that $\dot{c}(0) = 0$ which was also specified.

2.28. EVALUATION OF THE STATE TRANSITION MATRIX FROM AN EXPONENTIAL SERIES

The state transition matrix may be evaluated from an exponential series. Several methods have been proposed for its numerical evaluation. References [20] and [21] discuss one type of computational algorithm developed by Faddeev and Faddeeva for accomplishing this. However, this approach requires the Laplace-transform inversion of $\mathbf{\Phi}(s)$. Unfortunately, this approach is very tedious for matrices of any size. This section presents a straightforward method that evaluates the state transition matrix based on its infinite matrix series definition [22]. Direct application of the series definition gives a very efficient and fast method that depends only on matrix multiplication. It is based on assuming a solution to the homogeneous state equation, as is commonly done in the classical method of solving linear differential equations.

In order to derive the exponential series definition of the state transition matrix [23], let us assume that the solution to the homogeneous state equation

$$\mathbf{\Phi}(t) = \mathbf{P}\mathbf{\Phi}(t), \tag{2.315}$$

is given by

$$\mathbf{x}(t) = e^{\mathbf{P}t}\mathbf{x}(0), \quad t \geq 0 \tag{2.316}$$

where

$$\mathbf{\Phi}(t) = e^{\mathbf{P}t}, \tag{2.317}$$

and

$$e^{\mathbf{P}t} = \mathbf{I} + \mathbf{P}t + \frac{\mathbf{P}^2 t^2}{2!} + \cdots + \frac{\mathbf{P}^k t^k}{k!} + \cdots. \tag{2.318}$$

We shall now work backwards (as in the classical method of solving differential equations) and prove that Eq. (2.316) is indeed the correct solution to Eq. (2.315). Following this procedure, the value of $\dot{\mathbf{\Phi}}(t)$ is given by

$$\frac{d}{dt}[e^{\mathbf{P}t}] = \mathbf{P} + \mathbf{P}^2 t + \frac{\mathbf{P}^3 t^2}{2!} + \cdots + \frac{\mathbf{P}^{k+1} t^k}{k!} + \cdots. \tag{2.319}$$

A comparison of Eqs. (2.318) and (2.319) indicates that

$$\frac{d}{dt}\left[e^{\mathbf{P}t}\right] = \mathbf{P}e^{\mathbf{P}t}. \tag{2.320}$$

Therefore, from the definition of Eq. (2.317), we find that

$$\dot{\mathbf{\Phi}}(t) = \mathbf{P}\mathbf{\Phi}(t), \tag{2.321}$$

so that

$$\mathbf{\Phi}(t) = e^{\mathbf{P}t} = \sum_{k=0}^{\infty} \frac{\mathbf{P}^k t^k}{k!}, \quad t \geqslant 0 \tag{2.322}$$

or

$$\mathbf{\Phi}(t) = \mathbf{I} + \sum_{k=1}^{\infty} \frac{\mathbf{P}^k t^k}{k!}, \quad t \geq 0 \tag{2.323}$$

is indeed a correct solution to Eq. (2.315).

From this derivation, we can now extend our original definition of the state transition matrix (see Eq. 2.256) to the following:

$$\mathbf{\Phi}(t) = \mathscr{L}^{-1}\left\{[s\mathbf{I} - \mathbf{P}]^{-1}\right\} = e^{\mathbf{P}t} = \sum_{k=0}^{\infty} \frac{\mathbf{P}^k t^k}{k!}, \quad t \geqslant 0. \tag{2.324}$$

Because the matrix series is uniformly convergent for any finite interval, the state transition matrix can be determined within prescribed accuracy using only a finite number of terms [5].

Let us apply this series solution approach for obtaining the state transition matrix to the same problem we solved in Section 2.26 from the definition of the state transition matrix provided in Eq. (2.256). The problem solved in Section 2.26 had the companion $\mathbf{P}$ matrix given in Eq. (2.270) as:

$$\mathbf{P} = \begin{bmatrix} 0 & 1 \\ 0 & 0 \end{bmatrix}. \tag{2.325}$$

Substituting this value of $\mathbf{P}$ into Eq. (2.323), where $k = 2$ in this problem, we obtain the following:

$$\mathbf{\Phi}(t) = \mathbf{I} + \sum_{k=1}^{2} \frac{\mathbf{P}^2 t^2}{2!} = \begin{bmatrix} 1 & 0 \\ 0 & 1 \end{bmatrix} + \begin{bmatrix} 0 & 1 \\ 0 & 0 \end{bmatrix} t + \begin{bmatrix} 0 & 1 \\ 0 & 0 \end{bmatrix} \begin{bmatrix} 0 & 1 \\ 0 & 0 \end{bmatrix} \frac{t^2}{2!}. \tag{2.326}$$

This reduces to the following:

$$\mathbf{\Phi}(t) = \begin{bmatrix} U(t) & t \\ 0 & U(t) \end{bmatrix}, \quad t \geq 0 \tag{2.327}$$

which is the same result we had obtained in Eq. (2.275).

An iterative procedure for evaluating $e^{\mathbf{P}t}$, based on the definition of Eq. (2.322), is now presented, and is readily adapted for digital computer computation [22]. Let $e^{\mathbf{P}t}$ be represented as

$$e^{\mathbf{P}t} = \mathbf{M} + \mathbf{R} \tag{2.328}$$

where $\mathbf{M}$ is the approximating matrix for $e^{\mathbf{P}t}$,

$$\mathbf{M} = \sum_{k=0}^{k} \frac{\mathbf{P}^k T^k}{k!}, \quad T = t, \tag{2.329}$$

and $\mathbf{R}$ is the remainder matrix

$$\mathbf{R} = \sum_{k=K+1}^{\infty} \frac{\mathbf{P}^k T^k}{k!}. \tag{2.330}$$

Assuming that each element in the matrix $e^{\mathbf{P}t}$ is required to within an accuracy of at least b significant digits, then

$$|r_{ij}| \leqslant 10^{-b} |m_{ij}|, \tag{2.331}$$

where r_{ij} and m_{ij} represent elements of the matrices $\mathbf{R}$ and $\mathbf{M}$.

Let the norm of matrix $\mathbf{P}$ be given by

$$\|\mathbf{P}\| = \sum_{i,j=1}^{m} |k_{ij}|. \tag{2.332}$$

Then, it can be shown that

$$\|\mathbf{P}^k\| \leqslant \|\mathbf{P}\|^k, \quad k = 1, 2, 3, \ldots \tag{2.333}$$

Therefore, each element of the matrix $\mathbf{P}^k$ is less than or equal to $\|\mathbf{P}\|^k$. It follows that

$$|r_{ij}| \leqslant \sum_{k=K+1}^{\infty} \frac{\|\mathbf{P}\|^k T^k}{k!}. \tag{2.334}$$

Let us define the ratio of the second term to the first term of the previous series to be ϵ as follows:

$$\epsilon = \frac{\|\mathbf{P}\| T}{K+2}. \tag{2.335}$$

Therefore,

$$\frac{\|\mathbf{P}\| T}{k} \leqslant \epsilon, \quad k \geqslant K+2. \tag{2.336}$$

Substituting Eq. (2.336) into Eq. (2.334), we obtain the following expression:

$$|r_{ij}| \leqslant \frac{\|\mathbf{P}\|^{K+1} T^{K+1}}{(K+1)!} (1 + \epsilon + \epsilon^2 + \cdots). \tag{2.337}$$

Equation (2.337) can be rewritten in closed form as

$$|r_{ij}| \leqslant \frac{(\|\mathbf{P}\| T)^{K+1}}{(K+1)!} \frac{1}{1-\epsilon}. \tag{2.338}$$

Let us summarize the steps of this iterative procedure for evaluating $e^{\mathbf{P}t}$ before applying it to a problem.

(a) An initial value of K is chosen arbitrarily.
(b) The value of m_{ij} is evaluated by means of Eq. (2.329).
(c) The value of ϵ is determined by means of Eq. (2.335).
(d) The upper bound of $|r_{ij}|$ is calculated from Eq. (2.338).
(e) Each element of $\mathbf{M}$, obtained from Eq. (2.329), is compared with the upper bound of $|r_{ij}|$ obtained from Eq. (2.338).
(f) If the inequality of Eq. (2.331) is not satisfied, the value of K is increased, and steps (a)–(e) are repeated; otherwise, the procedure is ended.

As an example of applying this procedure, let us evaluate $\mathbf{\Phi}(t)$ numerically for the following example [22]:

$$\mathbf{P} = \begin{bmatrix} 0 & 1 & 0 \\ 0 & 0 & 1 \\ -0.75 & -2.75 & -3 \end{bmatrix},$$
$$T = 0.1.$$

Let us assume that each element in the matrix $e^{\mathbf{P}t}$ is required to within an accuracy of at least four significant digits and each number carries six significant digits. The state transition matrix is obtained approximately, using the procedure indicated:

$$e^{\mathbf{P}T} \cong \mathbf{M} = \begin{bmatrix} 0.999884 & 0.995717 \times 10^{-1} & 0.452513 \times 10^{-3} \\ -0.339385 \times 10^{-2} & 0.987440 & 0.859963 \times 10^{-1} \\ -0.644972 \times 10^{-1} & -0.239884 & 0.729451 \end{bmatrix}.$$

In this example, $b = 4$. When $K = 9$, the upper bound of $|r_{ij}|$ from Eq. (2.338) is 0.587945×10^{-7}. Therefore, $10^b|r_{ij}| = 0.587945 \times 10^{-3} < |m_{ij}|$, $(i, j = 1, 2, \ldots)$, where m_{ij} are the elements of $\mathbf{M}$ given in the example. This illustrative example indicates the simplicity and accuracy of the procedure for obtaining the state transition matrix utilizing its series definition.

2.29. SUMMARY

Many mathematical techniques have been presented in this chapter for use by the control-system engineer. Starting with complex-variable theory, we then developed the Fourier transform and Laplace transform. The transfer function, block diagram, and signal-flow graphs were then presented. It was pointed out that these concepts were not applicable to the more general multivariable inputs and outputs, nonlinear, time-varying systems. For this class of systems, the state-variable concept was then presented. Matrix algebra was reviewed, and the state-variable signal-flow graph and the state transition matrix were presented. It is reasonable for the reader at this point to ask which methods he or she should use.

There are no hard and fast guidelines, but reasonable rules of thumb can be outlined. In general, if the problem is one of analysis, if the system has one input and one output, and if its differential equation can be described by a linear differential equation having constant coefficients, then the engineer can use the simple Laplace-transform/transfer-function/block-diagram approach or the state-variable method, each technique complementing the other. On the other hand, if the analysis problem involves nonlinearities, time-varying characteristics and/or multivariable inputs and outputs, then the state-variable approach should be used. If the problem is one of synthesis involving optimal control theory which is discussed in Chapter 11, then again the state-variable approach has to be used.

Because the main purpose of this book is pedagogical, both the Laplace transform/transfer-function/block-diagram and state-variable concepts will be used wherever possible. For example, in the following chapter, where it is desired to represent mathematically various linear physical components, both approaches are used. The reader should also develop this dual capability and be able to handle a problem from either point of view whenever possible.

2.30. ILLUSTRATIVE PROBLEMS AND SOLUTIONS

This section provides a set of illustrative problems and their solutions to supplement the material presented in Chapter 2.

I2.1. Determine the poles and zeros of

$$F(s) = \frac{10(s+1)(s+4)}{s(s+2)(s+6)(s+8)^2}.$$

SOLUTION: Simple poles are located at $s = 0, -2, -6$
Pole of order two located at -8
Simple zeros located at $-1, -4$

I2.2. Determine the Laplace transform of f(t) which is given by

$$f(t) = te^{4t}, \quad t \geq 0$$
$$f(t) = 0, \quad t < 0.$$

SOLUTION: From Appendix A, eighth item: For $n = 2$ and $a = -4$, we obtain

$$F(s) = \mathscr{L}(te^{-4t}) = \frac{1}{(s+4)^2}.$$

I2.3. Determine the Laplace transform $F(s)$ for the function $f(t)$ illustrated:

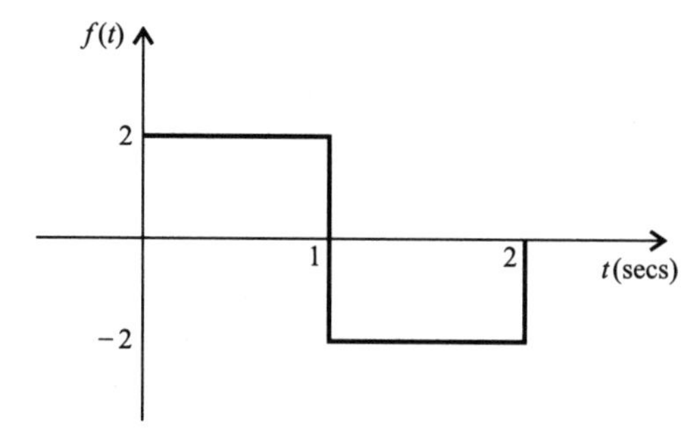

Figure I2.3

SOLUTION:

$$f(t) = 2U(t) - 4U(t-1) + 2U(t-2)$$

From Table 2.1, item 2, and the time-shifting theorem, we obtain

$$F(s) = \frac{2}{s} - \frac{4}{s}e^{-s} + \frac{2}{s}e^{-2s}$$

$$F(s) = \frac{2}{s}(1 - 2e^{-s} + e^{-2s}).$$

I2.4. Determine the initial value of $c(t)$ where the Laplace transform of $C(s)$ is given by:

$$C(s) = \frac{(4s+1)}{s(s^2+2s+1)}$$

SOLUTION: From the initial-value theorem

$$\lim_{t\to 0} c(t) = \lim_{s\to\infty} sF(s)$$

we obtain,

$$\lim_{t\to 0} c(t) = \lim_{s\to\infty} s\frac{(4s+1)}{s(s^2+2s+1)} = 0.$$

I2.5. Determine the final value of $c(t)$ when the Laplace transform of $C(s)$ is given by

$$C(s) = \frac{8(2s+3)}{s(s+4)(s^2+3s+2)}.$$

SOLUTION: From the final-value theorem

$$\lim_{t\to\infty} c(t) = \lim_{s\to 0} sF(s)$$

we obtain,

$$\lim_{t\to\infty} c(t) = \lim_{s\to 0} s\frac{8(2s+3)}{s(s+4)(s^2+3s+2)} = 3.$$

I2.6. Determine the final value of $c(t)$ when the Laplace transform of $C(s)$ is given by

$$C(s) = \frac{26.8(s+1)(s+10)}{s(s+2.8)(s+3)(s-2)(s+26.8)}.$$

SOLUTION: The final-value theorem cannot be applied in this problem because the function $sC(s)$ is not analytic in the right half-plane.

I2.7. Determine the Laplace transform of

$$c(t) = \sin \omega t \; U(t-2)$$

from knowledge of the Laplace transform of a sine wave (see Table 2.1, item 7) and the time-shifting theorem.

SOLUTION:

$$C(s) = \frac{\omega}{s^2 + \omega^2} e^{-2s}.$$

I2.8. Determine the Laplace transform of

$$c(t) = e^{-3t} \sin \omega t \; U(t)$$

from knowledge of the Laplace transform of a sine wave (see Table 2.1, item 7) and the frequency-shifting theorem.

SOLUTION:

$$C(s) = \frac{\omega}{(s+a)^2 + \omega^2.}$$

I2.9. Determine the Laplace transform of the function $f(t)$ where

$$f(t) = e^{-t(t-1)} U(t-1).$$

SOLUTION:

$$F(s) = \int_0^{\infty} e^{-st} e^{-t(t-1)} U(t-1)\, dt.$$

To integrate this equation, it is necessary to change the variable as follows:

Let $\tau = t - 1$.

Therefore,

$$F(s) = \int_0^{\infty} e^{-s(\tau+1)} e^{-(\tau+1)(\tau)} d\tau$$
$$F(s) = \int_0^{\infty} e^{-\tau^2-(s+1)\tau-s} d\tau$$
$$F(s) = e^{-s} \int_0^{\infty} e^{-\tau^2} e^{-(s+1)\tau} d\tau.$$

Adding the term

$$e^{0.5^2} e^{(s+1)^2}$$

inside and outside the integral to complete the square, we obtain the following:

$$F(s) = e^{-s} e^{0.5^2(s+1)^2} \int_0^{\infty} e^{-(\tau+0.5(s+1))^2} d\tau.$$

Let us now change the variable. By defining the new variable T as follows:

$$T = \tau + 0.5(s+1).$$

Therefore,

$$F(s) = e^{0.5^2(s-1)^2} \int_{0.5(s+1)}^{\infty} e^{-T^2} dT.$$

The integral term is equal to:

$$0.5\pi^{0.5}(1 - \text{error function})$$

Abbreviating the error function as "erf," the Laplace transform is given by the following:

$$F(s) = 0.5\pi^{0.5} e^{0.5^2(s-1)^2}(1 - \text{erf}(0.5)(s+1)).$$

12.10. Determine the residues of $F(s)$ where

$$F(s) = \frac{1}{(s+2)(s+4)}$$

which has single poles at $s = -2$ and $s = -4$.

SOLUTION: The residue at $s = -2$ is given by

$$\lim_{s \to -2} \frac{1}{(s+4)} = \frac{1}{2} \text{ for } s = -2.$$

The residue at $s = -4$ is given by

$$\lim_{s \to -4} \frac{1}{(s+2)} = -\frac{1}{2} \text{ for } s = -4.$$

I2.11. Determine the inverse Laplace transform of $C(s)$ where

$$C(s) = \frac{1}{s(s+2)(s+4)}$$

SOLUTION: Using partial-fraction expansion, we obtain the following:

$$C(s) = \frac{K_1}{s} + \frac{K_2}{s+2} + \frac{K_3}{s+4}$$

where

$$K_1 = \frac{1}{(s+2)(s+4)} = \frac{1}{8} \quad \text{for } s = 0$$

$$K_2 = \frac{1}{s(s+4)} = -\frac{1}{4} \quad \text{for } s = -2$$

$$K_3 = \frac{1}{s(s+2)} = \frac{1}{8} \quad \text{for } s = -4.$$

From Table 2.1, items 2 and 6 we obtain the following:

$$c(t) = \left(\frac{1}{8} - \frac{1}{4}e^{-2t} + \frac{1}{8}e^{-4t}\right)U(t).$$

I2.12. Determine the inverse Laplace transform of $C(s)$ where

$$C(s) = \frac{10(s+2)}{s^2(s+1)(s+4)}.$$

SOLUTION: Using partial-fraction expansion, we obtain the following:

$$C(s) = \frac{K_1}{s^2} + \frac{K_2}{s} + \frac{K_3}{s+1} + \frac{K_4}{s+4}$$

where

$$K_1 = \frac{10(s+2)}{(s+1)(s+4)} = \frac{10(2)}{1(4)} = 5 \quad \text{for } s = 0$$
$$K_2 = \frac{d}{ds}\left[\frac{10(s+2)}{(s+1)(s+4)}\right] = \frac{-30}{8} \quad \text{for } s = 0$$
$$K_3 = \frac{10(s+2)}{s^2(s+4)} = \frac{10}{3} \quad \text{for } s = -1$$
$$K_4 = \frac{10(s+2)}{s^2(s+1)} = \frac{5}{12} \quad \text{for } s = -4.$$

From Table 2.1, items 2, 3, and 6, we obtain the following:

$$c(t) = \left(5t - \frac{30}{8} + \frac{10}{3}e^{-t} + \frac{5}{12}e^{-3t)}\right)U(t).$$

I2.13. Determine the value of $c(t)$ for the following differential equation using the Laplace transform. Assume that the initial conditions are zero, and $U(t)$ represents a unit step.

$$\frac{d^2c(t)}{dt^2} + 7\frac{dc(t)}{dt} + 6c(t) = e^{-3t}U(t).$$

SOLUTION:

$$(s^2 + 7s + 6)C(s) = \frac{1}{(s+3)}$$
$$C(s) = \frac{1}{(s+1)(s+6)(s+3)}.$$

Using partial-fraction expansion, we obtain the following:

$$C(s) = \frac{K_1}{(s+1)} + \frac{K_2}{(s+6)} + \frac{K_3}{(s+3)}$$

where

$$K_1 = \frac{1}{(s+6)(s+3)} = 0.1 \quad \text{for } s = -1$$
$$K_2 = \frac{1}{(s+1)(s+3)} = 0.067 \quad \text{for } s = -6$$
$$K_3 = \frac{1}{(s+1)(s+6)} = -0.167 \quad \text{for } s = -3.$$

Therefore,

$$c(t) = (0.1e^{-t} + 0.067e^{-6t} - 0.167e^{-3t})U(t).$$

I2.14. Solve the following two differential equations for $x_1(t)$ and $x_2(t)$ by means of the Laplace transform.

$$\frac{dx_1(t)}{dt} = x_2(t)$$
$$\frac{dx_2(t)}{dt} = (-3x_1(t) - 4x_2(t) + 1)U(t).$$

The initial conditions are: $x_1(0) = 1$; $x_2(0) = 0$.

SOLUTION:

$$sX_1(s) - x_1(0) = X_2(s) \tag{I2.14-1}$$
$$\text{with } x_1(0) = 1$$
$$sX_2(s) - x_2(0) = -3X_1(s) - 4X_2(s) + \frac{1}{s} \tag{I2.14-2}$$
$$\text{with } x_2(0) = 0.$$

Therefore,

$$sX_1(s) - 1 = X_2(s).$$

Solving for $X_1(s)$:

$$X_1(s) = \frac{1}{s} + \frac{1}{s}X_2(s). \tag{I2.14-3}$$

Substituting $X_1(s)$ from Eq. (I2.14-3) into Eq. (I2.14-2), we obtain:

$$X_2(s) = \frac{-2}{(s+3)(s+1)}. \tag{I2.14-4}$$

Using partial-fraction expansion, we obtain:

$$X_2(s) = \frac{A}{(s+3)} + \frac{B}{(s+1)}$$

where

$$A = \frac{-2}{(s+1)} = 1 \quad \text{for } s = -3$$
$$B = \frac{-2}{(s+3)} = -1 \quad \text{for } s = -1.$$

Therefore,

$$X_2(s) = \frac{1}{(s+3)} - \frac{1}{(s+1)}.$$

Taking the inverse Laplace transform of $X_2(s)$, we obtain

$$x_2(t) = (e^{-3t} - e^{-t})U(t).$$

Substituting Eq. (I2.14-4) into Eq. (I2.14-1), we obtain

$$sX_1(s) - 1 = \frac{-2}{(s+3)(s+1)}.$$

Solving for $X_1(s)$, we obtain the following:

$$X_1(s) = \frac{s^2 + 4s + 1}{s(s+3)(s+1)}.$$

Using partial-fraction expansion, we obtain:

$$X_1(s) = \frac{A}{s} + \frac{B}{(s+3)} + \frac{C}{(s+1)}$$

where

$$A = \frac{s^2 + 4s + 1}{(s+3)(s+1)} = 0.33 \text{ for } s = 0$$
$$B = \frac{s^2 + 4s + 1}{s(s+1)} = -0.33 \text{ for } s = -3$$
$$C = \frac{s^2 + 4s + 1}{s(s+3)} = 1 \text{ for } s = -1.$$

Therefore,

$$X_1(s) = \frac{0.33}{s} - \frac{0.33}{(s+3)} + \frac{1}{(s+4)}.$$

Taking the inverse Laplace transform of $X_1(s)$, we obtain

$$x_1(t) = (0.33 - 0.33e^{-3t} + e^{-t})U(t)$$

I2.15. Determine the value of $c(t)$ for the following differential equation using the Laplace transform.

$$\frac{d^2c(t)}{dt^2} + 7\frac{dc(t)}{dt} + 6c(t) = \frac{dr(t)}{dt} + 4r(t)$$

where

$$r(t) = e^{-3t}U(t).$$

SOLUTION: Since

$$\frac{dr(t)}{dt} = -3e^{-3t}$$

then

$$\frac{dr(t)}{dt} + r(t) = -3e^{-3t} + 4e^{-3t} = e^{-3t}$$

$$\frac{d^2c(t)}{dt^2} + \frac{7dc(t)}{dt} + 6c(t) = e^{-3t}.$$

Taking the Laplace transform of this equation, we obtain the following:

$$s^2C(s) + 7sC(s) + 6C(s) = \frac{1}{s+3}.$$

Solving for $C(s)$, we obtain the following:

$$C(s) = \frac{1}{(s+1)(s+6)(s+3)}$$

Using partial-fraction expansion, we obtain the following:

$$C(s) = \frac{K_1}{s+1} + \frac{K_2}{s+6} + \frac{K_3}{s+3}$$

where

$$K_1 = \frac{1}{(s+6)(s+3)} = \frac{1}{10} \quad \text{for } s = -1$$

$$K_2 = \frac{1}{(s+1)(s+3)} = \frac{1}{15} \quad \text{for } s = -6.$$

$$K_3 = \frac{1}{(s+1)(s+6)} = \frac{-1}{6} \quad \text{for } s = -3$$

Therefore,

$$C(s) = \frac{\frac{1}{10}}{s+1} + \frac{\frac{1}{15}}{s+6} - \frac{\frac{1}{6}}{s+3}.$$

Taking the inverse Laplace transform of this equation, we obtain

$$c(t) = \left(\frac{1}{10}e^{-t} + \frac{1}{15}e^{-6t} - \frac{1}{6}e^{-3t}\right)U(t).$$

I2.16. A unit impulse is applied to an element whose transfer function is unknown. If the resulting output is given by

$$c(t) = 4e^{-6t}U(t)$$

where U(t) represents a unit step, determine the transfer function of the element.

SOLUTION: Since

$$C(s) = \frac{4}{(s+6)}$$

and

$$R(s) = 1$$

Therefore,

$$G(s) = \frac{C(s)}{R(s)} = \frac{\frac{4}{s+6}}{1} = \frac{4}{s+6}.$$

I2.17. A system is represented by the following differential equation:

$$4\frac{d^2c(t)}{dt^2} + 3\frac{dc(t)}{dt} + 6c(t) = r(t) + 4r(t-1).$$

In this differential equation, $r(t)$ represents the input, and $c(t)$ represents the output. Determine the transfer function $C(s)/R(s)$.

SOLUTION: Taking the Laplace transform of this differential equation, we obtain the following:

$$C(s)(4s^2 + 3s + 6) = R(s) + 4R(s)e^{-s}.$$

Therefore, the transfer function $C(s)/R(s)$ is given by

$$\frac{C(s)}{R(s)} = \frac{1 + 4e^{-s}}{4s^2 + 3s + 6}.$$

12.18. A system is represented by the following differential equation:

$$\frac{d^3c(t)}{dt^3} + 10\frac{d^2c(t)}{dt^2} + 6\frac{dc(t)}{dt} + c(t) + 7\int_0^t c(\tau)d\tau = \frac{dr(t)}{dt} + 3r(t).$$

In this differential equation, $r(t)$ represents the input, and $c(t)$ represents the output. Determine the transfer function $C(s)/R(s)$.

SOLUTION:

Taking the Laplace transform of this differential equation, we obtain the following:

$$C(s)\left(s^3 + 10s^2 + 6s + 1 + \frac{7}{s}\right) = sR(s) + 3R(s).$$

Simplifying this equation, we obtain the following:

$$C(s)(s^4 + 10s^3 + 6s^2 + s + 7) = s^2R(s) + 3sR(s).$$

Therefore, the transfer function $C(s)/R(s)$ is given by

$$\frac{C(s)}{R(s)} = \frac{s^2 + 3s}{s^4 + 10s^3 + 6s^2 + s + 7}.$$

12.19. We wish to determine the transfer function of an element in a control system. To determine its transfer function, a unit ramp is applied to the input $r(t)$. The output of this element is recorded, and is modelled according to the following equation:

$$c(t) = (4t - 2e^{-4t} - 2e^{-6t})U(t).$$

Determine the transfer function, $C(s)/R(s)$, of this element.

SOLUTION: Since

$$R(s) = \frac{1}{s^2}$$

and

$$C(s) = \frac{4}{s^2} - \frac{2}{s+4} - \frac{2}{s+6},$$

simplifying the expression for $C(s)$ results in the following:

$$C(s)=\frac{-4s^3-16s^2+40s+96}{s^2(s+4)(s+6)}$$

$$\frac{C(s)}{R(s)}=\frac{\dfrac{-4s^3-16s^2+40s+96}{s^2(s+4)(s+6)}}{\dfrac{1}{s^2}}=\frac{-4s^3-16s^2+40s+96}{(s+4)(s+6)}$$

$$\frac{C(s)}{R(s)}=-4\frac{(s+5.039)(s+1.72)(s-2.763)}{(s+4)(s+6)}.$$

I2.20. We wish to determine the transfer function of an element in a control system. Its input, $m(t)$, represents speed, and its output is represented by $y(t)$. To determine its transfer function, a unit step voltage of 1 volt, corresponding to a speed of 1 ft/sec, is applied to the input, $m(t)$. The output of the element is recorded and is modeled according to the following equation:

$$y(t)=(40-8e^{-5(t-1)})U(t-1)$$

Determine the transfer function, $Y(s)/M(s)$, of this element. Reduce your answer to its simplest form.

SOLUTION: The Laplace transform of the output $y(t)$ is given by

$$Y(s)=\left(\frac{40}{s}-\frac{8}{s+5}\right)e^{-s}.$$

Simplifying the expression for $Y(s)$ results in the following:

$$Y(s)=\frac{(32s+200)e^{-s}}{s(s+5)}.$$

Therefore, the transfer function $Y(s)/M(s)$ is given by

$$\frac{Y(s)}{M(s)}=\frac{\frac{(32s+200)e^{-s}}{s(s+5)}}{\frac{1}{s}}=\frac{32(s+6.25)e^{-s}}{(s+5)}.$$

I2.21. The block diagram of a control system is illustrated where $R(s)$ represents the reference input and $D(s)$ represents the disturbance input.

(a) Determine the transfer function $C(s)/R(s)$.

(b) Determine the transfer function $C(s)/D(s)$.

(c) Determine the value of the feedback element $H_1(s)$ in order that the disturbance input $D(s)$ has no effect on the output $C(s)$. Assume that $R(s) = 0$.

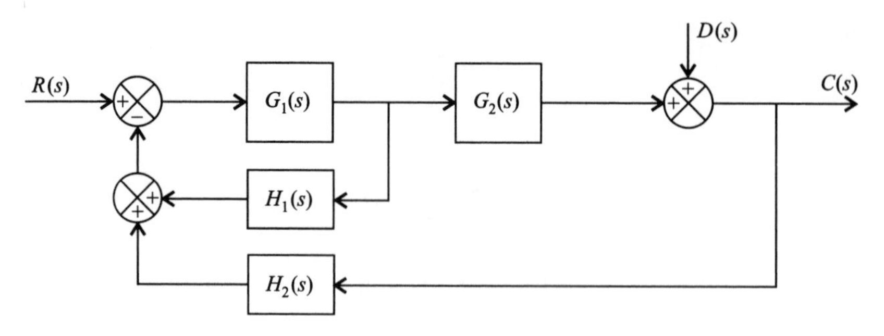

Figure I2.21

SOLUTION: (a)

$$\frac{C(s)}{R(s)} = \frac{G_1(s)G_2(s)}{1 + G_1(s)H_1(s) + G_1(s)G_2(s)H_2(s)}.$$

(b)

$$\frac{C(s)}{D(s)} = \frac{1 + G_1(s)H_1(s)}{1 + G_1(s)H_1(s) + G_1(s)G_2(s)H_2(s)}.$$

(c) Setting the numerator of the transfer function $C(s)/D(s) = 0$, we obtain the following:

$$1 + G_1(s)H_1(s) = 0.$$

Therefore,

$$H_1(s) = -\frac{1}{G_1(s)}.$$

I2.22. The signal-flow graph of a control system is represented by the following, where $D(s)$ represents an external disturbance input into the control system:

(a) Determine the transfer function $C(s)/R(s)$.

(b) Determine the transfer function $E(s)/D(s)$.

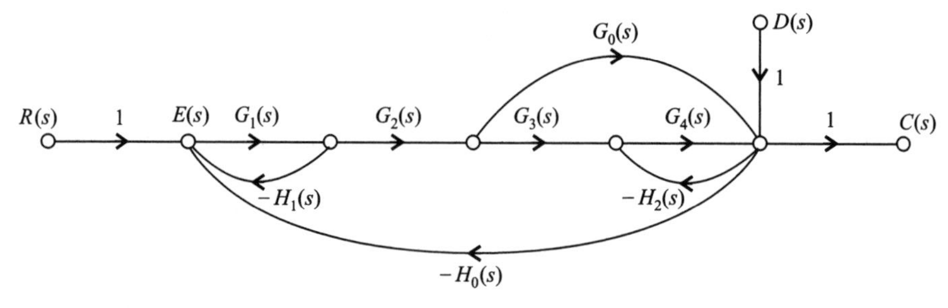

Figure I2.22

SOLUTION: (a)

$$\frac{C}{R} = \frac{G_1G_2G_3G_4 + G_1G_2G_0}{(1 + G_1H_1 + G_4H_2 + G_1G_2G_3G_4H_0 + G_1G_2G_0H_0) + (G_1H_1G_4H_2)}.$$

All terms in this transfer function are functions of s.

(b)

$$\frac{E(s)}{D(s)} = \frac{-H_0(s)}{\textit{Same denominator as in Part (a)}}.$$

I2.23. Consider the LC ladder network illustrated:

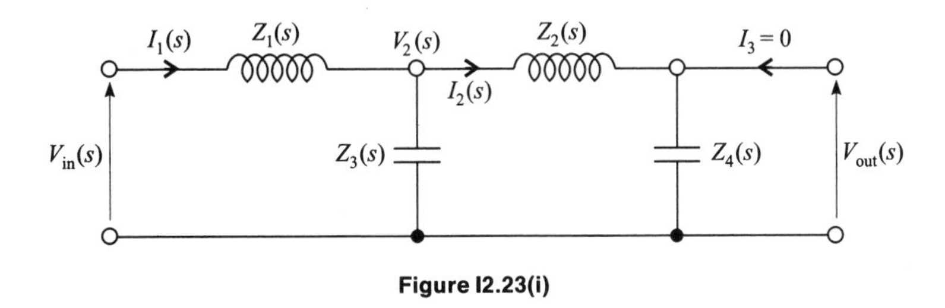

Figure I2.23(i)

The circuit equations representing this LC network are as follows:

$$I_1(s) = \frac{(V_{in}(s) - V_2(s))}{Z_1(s)}$$

$$I_2(s) = \frac{(V_2(s) - V_{out}(s))}{Z_2(s)}$$

$$V_2(s) = (I_1(s) - I_2(s))Z_3(s)$$

$$V_{out}(s) = I_2(s)Z_4(s).$$

(a) Draw the signal-flow graph representing these circuit equations.

(b) Using Mason's theorem, determine the transfer function, $V_{out}(s)/V_{in}(s)$.

SOLUTION: (a)

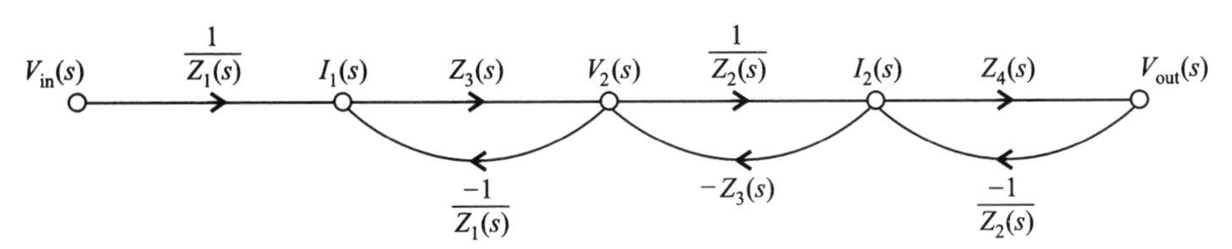

Figure I2.23(ii)

(b) $$\frac{V_{out}(s)}{V_{in}(s)} = \frac{\frac{1}{Z_1(s)}Z_3(s)\frac{1}{Z_2(s)}Z_4(s)}{1 - \left(\frac{-Z_3(s)}{Z_1(s)} - \frac{Z_3(s)}{Z_2(s)} - \frac{Z_4(s)}{Z_3(s)}\right) + \left(\frac{Z_3(s)\,Z_4(s)}{Z_1(s)\,Z_2(s)}\right)}.$$

I2.24. A feedback control system is represented by the following signal-flow graph:

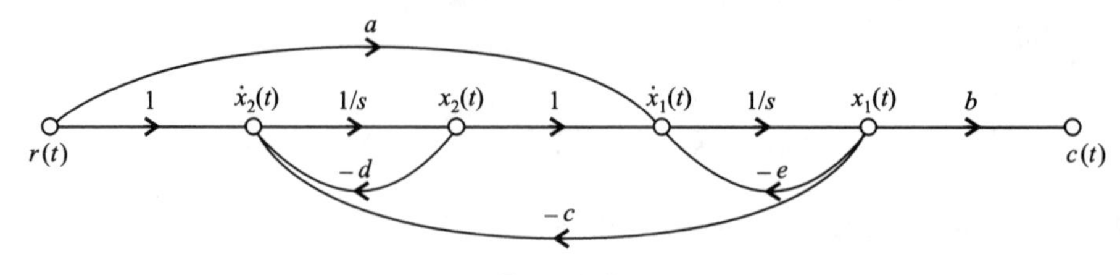

Figure I2.24

Determine the phase-variable canonical form equations for this feedback control system.

SOLUTION: The differential equations representing this signal-flow graph are given by the following:

$$\dot{x}_1(t) = -ex_1(t) + x_2(t) + ar(t)$$
$$\dot{x}_2(t) = -cx_1(t) - dx_2(t) + r(t).$$

The phase-variable canonical equations are given by:

$$\begin{bmatrix} \dot{x}_1(t) \\ \dot{x}_2(t) \end{bmatrix} = \begin{bmatrix} -e & 1 \\ -c & -d \end{bmatrix}\begin{bmatrix} x_1(t) \\ x_2(t) \end{bmatrix} + \begin{bmatrix} a \\ 1 \end{bmatrix} r(t)$$
$$c(t) = \begin{bmatrix} b & 0 \end{bmatrix}\begin{bmatrix} x_1(t) \\ x_2(t) \end{bmatrix}$$

I2.25. Determine the state transition matrix as a sum of an infinite series using Eq. (2.323) for a system whose **P** matrix is given by:

$$\mathbf{P} = \begin{bmatrix} 2 & 2 \\ 0 & 0 \end{bmatrix}.$$

Carry out the computation in Eq. (2.323) from $k = 0$ through $k = 2$.

SOLUTION:

$$\mathbf{\Phi}(t) = \begin{bmatrix} 1 & 0 \\ 0 & 1 \end{bmatrix} + \begin{bmatrix} 2t & 2t \\ 0 & 0 \end{bmatrix} + \begin{bmatrix} 2 & 2 \\ 0 & 0 \end{bmatrix}\begin{bmatrix} 2 & 2 \\ 0 & 0 \end{bmatrix}\frac{t^2}{2} = \begin{bmatrix} 1 + 2t + 2t^2 & 2t + 2t^2 \\ 0 & 1 \end{bmatrix}, \quad t \geqslant 0$$

I2.26. A system has a companion matrix, **P**, and an input vector, **B**, given by the following:

$$\mathbf{P} = \begin{bmatrix} 2 & 0 \\ 0 & -2 \end{bmatrix}; \ \mathbf{B} = \begin{bmatrix} 0 \\ 1 \end{bmatrix}$$

(a) Determine its state transition matrix using Eq. (2.256).

(b) Solve the state transition equation, Eq. (2.306), for $t \geq 0$. Assume that the input is a unit step function, and that the initial state vector is represented by $\mathbf{x}(0)$ where

$$\mathbf{x}(0) = \begin{bmatrix} 0 \\ 0 \end{bmatrix}.$$

SOLUTION: (a) From Eq. (2.256):

$$\mathbf{\Phi}(t) = \mathscr{L}^{-1}\{[sI - P]^{-1}\}, \quad \text{for } t \geqslant 0$$

where

$$[s\mathbf{I} - \mathbf{P}] = \begin{bmatrix} s & 0 \\ 0 & s \end{bmatrix} - \begin{bmatrix} 2 & 0 \\ 0 & -2 \end{bmatrix} = \begin{bmatrix} s-2 & 0 \\ 0 & s+2 \end{bmatrix}$$

and

$$\begin{bmatrix} s-2 & 0 \\ 0 & s+2 \end{bmatrix}^{-1} = \frac{\operatorname{adj}\begin{bmatrix} s-2 & 0 \\ 0 & s+2 \end{bmatrix}}{\begin{vmatrix} s-2 & 0 \\ 0 & s+2 \end{vmatrix}} = \frac{\begin{bmatrix} s+2 & 0 \\ 0 & s-2 \end{bmatrix}^T}{(s-2)(s+2)} = \frac{\begin{bmatrix} s+2 & 0 \\ 0 & s-2 \end{bmatrix}}{(s-2)(s+2)}.$$

Therefore,

$$\mathbf{\Phi}(t) = \mathscr{L}^{-I}\begin{bmatrix} \dfrac{1}{s-2} & 0 \\ 0 & \dfrac{1}{s+2} \end{bmatrix} = \begin{bmatrix} e^{2t} & 0 \\ 0 & e^{-2t} \end{bmatrix}, \quad t \geqslant 0.$$

(b) From Eq. (2.306):

$$\boldsymbol{x}(t) = \mathbf{\Phi}(t)\boldsymbol{x}(0^+) + \int_0^t \mathbf{\Phi}(t-\tau)\boldsymbol{Bu}(\tau)d\tau$$

where

$$x(t) = \int_0^t \begin{bmatrix} e^{2(t-\tau)} & 0 \\ 0 & e^{-2(t-\tau)} \end{bmatrix} \begin{bmatrix} 0 \\ 1 \end{bmatrix} \begin{bmatrix} 1 \end{bmatrix} d\tau.$$

This reduces to:

$$x(t) = \int_0^t \begin{bmatrix} 0 \\ e^{-2(t-\tau)} \end{bmatrix} (1) d\tau; \quad \int_0^t e^{-2t} e^{2\tau} d\tau = \frac{1}{2} e^{-2t} \left[e^{2\tau} \right]_0^t = \frac{1}{2}(1 - e^{-2t}).$$

Therefore,

$$\mathbf{x}(t) = \begin{bmatrix} 0 \\ \frac{1}{2}(1 - e^{-2t}) \end{bmatrix}, \quad t \geqslant 0.$$

I2.27. Using the elements presented in Section 2.18 on Operational Amplifiers, develop a flow diagram to simulate the following set of differential equations which represent the model of a system. The outputs from the system are the variables $x(t)$ and $y(t)$.

$$\ddot{x}(t) + 1.59\dot{y}(t) + \dot{x}(t) + 7.31y(t) + 0.5x(t) = 0.02$$
$$\ddot{y}(t) + 0.74\dot{y}(t) + 0.49\dot{x}(t) + 0.035y(t) + x(t) = 0.$$

Assume the initial conditions are zero.

SOLUTION:

$$\ddot{x}(t) = -\dot{x}(t) - 0.5x(t) + 0.02 - 1.59\dot{y}(t) - 7.31y(t)$$
$$\ddot{y}(t) = -0.74\dot{y}(t) - 0.03y(t) - 0.49\dot{x}(t) - x(t)$$

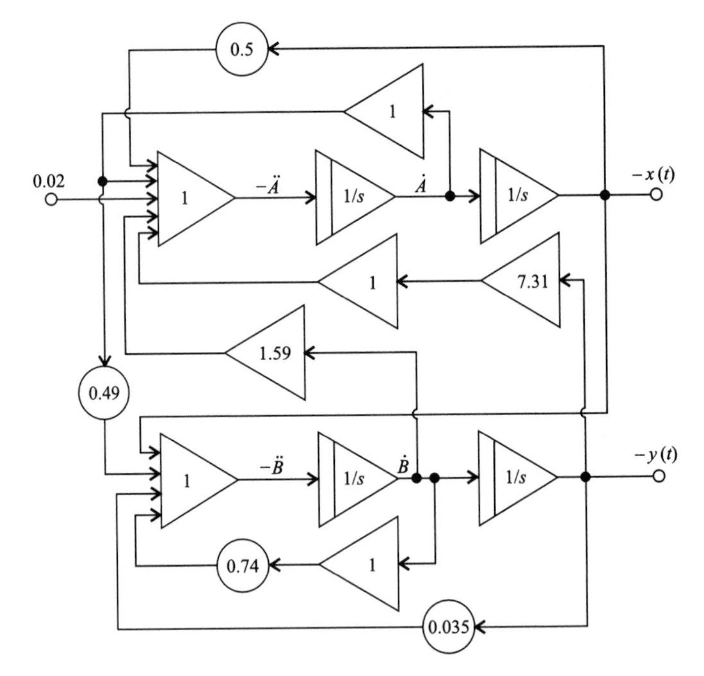

Figure I2.27

PROBLEMS

2.1 The periodic function $f(t)$ has the line spectrum shown in Figure P2.1. Assuming that $f(t)$ is an even function, determine the following:

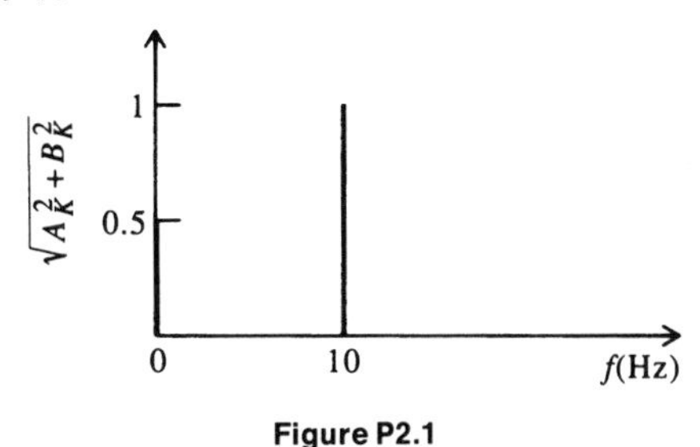

Figure P2.1

(a) The trigonometric Fourier series for $f(t)$.

(b) The complex Fourier series for $f(t)$.

2.2. Find $v(t)$ in the circuit of Figure P2.2 using the Laplace transform.

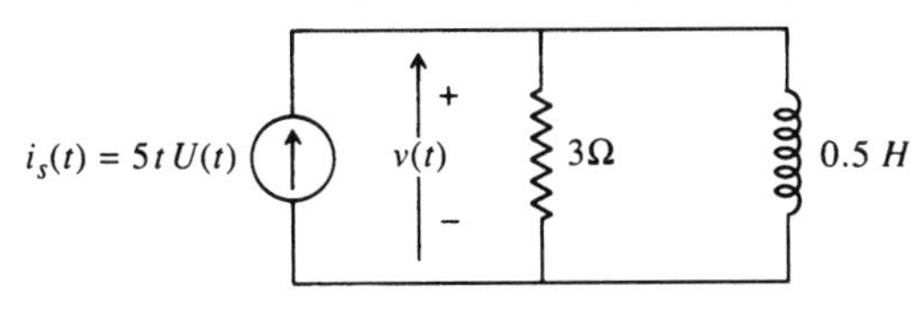

Figure P2.2

2.3. Find $x(t)$ for $t \geqslant 0$ for

$$d^2x(t)/dt^2 + x(t) = 0$$

where $x(0) = 1$ and $dx(0)/dt = -1$.

2.4. Consider the Laplace transform

$$F(s) = \frac{b}{s(s+b)}.$$

Using the final-value theorem, determine $f(\infty)$. Check your answer by finding the inverse Laplace transform $f(t)$ and letting $t \to \infty$.

2.5. The initial conditions for the following differential equation

$$2\frac{d^2y(t)}{dt^2} + 5\frac{dy(t)}{dt} + 3y(t) = 4$$

are given by

$$y(0) = 8,$$
$$\frac{dy(0)}{dt} = 10.$$

(a) Write in its simplest form the Laplace transform of the function $Y(s)$, by taking the Laplace transform of this differential equation.

(b) Expand $Y(s)$ by means of the partial fraction expansion method. Determine all unknown constants.

(c) Determine the function $y(t)$ by taking the inverse Laplace transform.

2.6. Repeat Problem 2.5 for the following differential equation and initial conditions:

$$6\frac{d^2y(t)}{dt^2} + 10\frac{dy(t)}{dt} + 4y(t) = 2,$$
$$y(0) = 3,$$
$$\frac{dy(0)}{dt} = 5.$$

2.7. Determine the initial and final values of the unit step response of the system of Figure P2.7 using the initial- and final-value theorems, respectively:

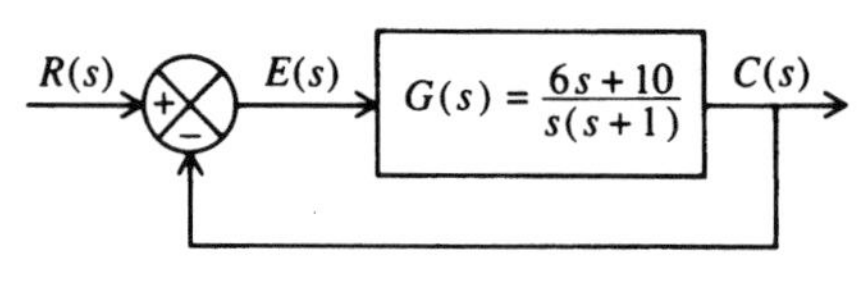

Figure P2.7

2.8. Using the final-value theorem, determine $f(\infty)$ for the following function:

$$F(s) = \frac{25}{s(0.5s - 1)(0.08s + 1)}.$$

2.9 Obtain the inverse Laplace transform for the following expressions:

(a) $$F_A(s) = \frac{20}{(s+2)^2(s^2 + 12s + 16)},$$

(b) $$F_B(s) = \frac{10(s+2)}{(s^2 - 16)(s+1)},$$

(c) $$F_C(s) = \frac{2(s+1)}{s(s^2 + 8s + 4)}.$$

2.10. Prove that the transfer function for the network illustrated in Table 2.4, item 3, is given by the expression shown.

2.11. Prove that the transfer function for the network illustrated in Table 2.4, item 6, is given by the expression shown.

2.12. Prove that the transfer function for the network illustrated in Table. 2.4, item 9, is given by the expression shown.

2.13. A step input $U(t)$ is applied to a linear system $G(s)$, and the resulting output is given by

$$(e^{-2t} - 1)U(t).$$

Two of these systems are now to be connected in cascade so that the output of the first forms the input to the second. A unit step $U(t)$ is applied to the input of the first system. Determine the output of each system.

2.14. A unit step input $U(t)$ is applied to a linear system $G(s)$ and the resulting output is given by

$$(1 - e^{-4t})U(t).$$

Determine the transfer function $G(s)$.

2.15. An input $tU(t)$ is applied to a linear system $G(s)$ and the resultant output is given by $(1 - 0.5e^{-2t} - 0.5e^{-4t})U(t)$. Determine the transfer function $G(s)$.

2.16. We wish to determine the transfer function of an element in a control system. Its input, $m(t)$, represents speed, and its output is repreented by $y(t)$. To determine its transfer function, a unit step voltage of 1 volt, corresponding to a speed of 1 ft/sec, is applied to the input, $m(t)$. The output of this element is recorded, and is modeled according to the following equation:

$$y(t) = (20 - 16e^{-2t} - 4e^{-4t})U(t).$$

Determine the transfer function, $Y(s)/M(s)$, of this element. Reduce your answer to its simplest form.

2.17. A feedback control system containing four feedback paths is shown in Figure P2.17.

(a) Draw its signal-flow graph.

(b) Using Mason's theorem, determine the transfer function $C(s)/R(s)$.

2.18. Determine the closed-loop transfer function, $C(s)/R(s)$, of the following control system whose signal flow graph is shown in Figure P2.18.

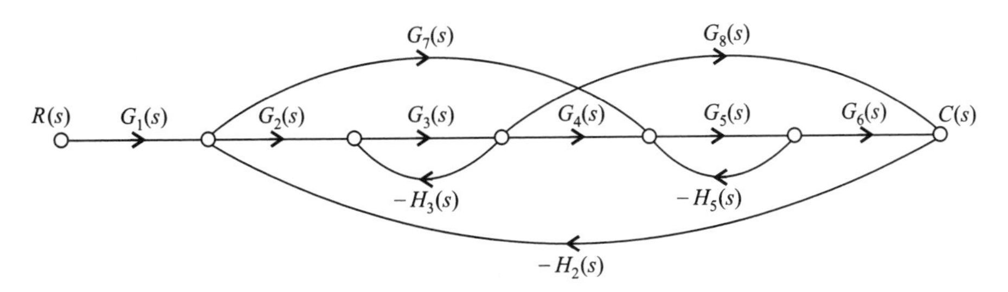

Figure P2.18

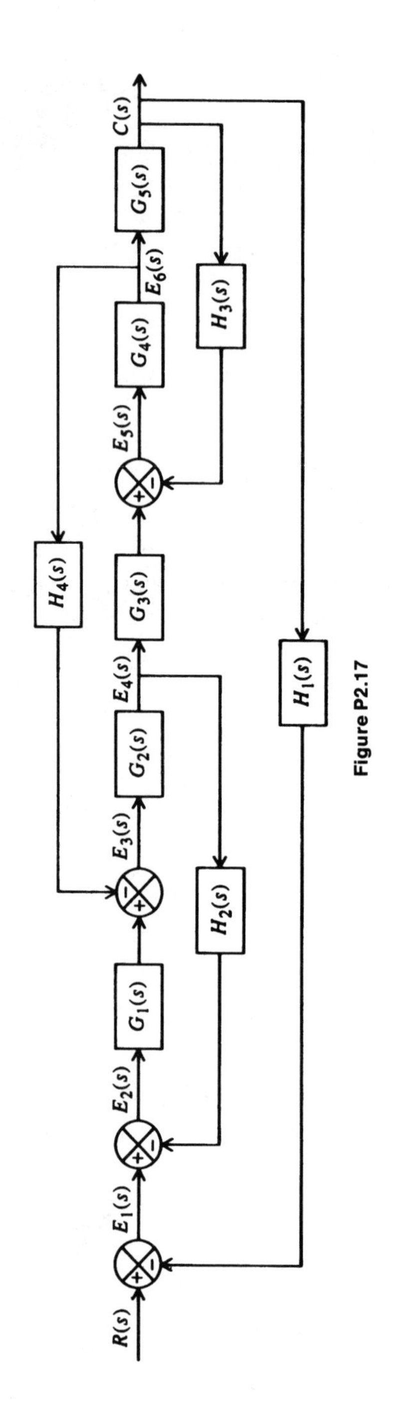

Figure P2.17

2.19. The signal-flow graph of a control system is illustrated in Figure P2.19.

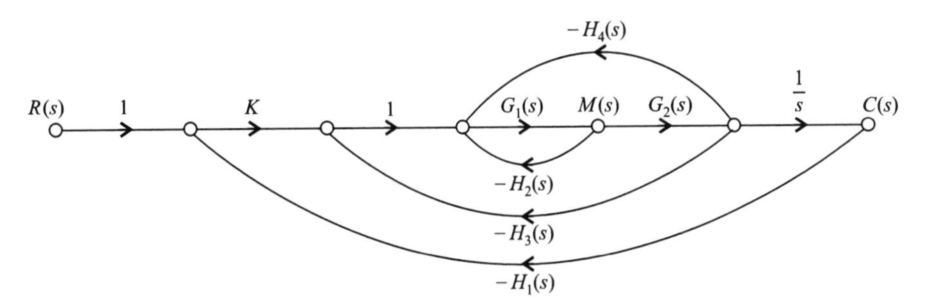

Figure P2.19

Determine the following:

(a) $C(s)/R(s)$

(b) $M(s)/R(s)$

(c) $C(s)/M(s)$.

2.20. The signal-flow graph of a control system is represented in Figure P2.20.

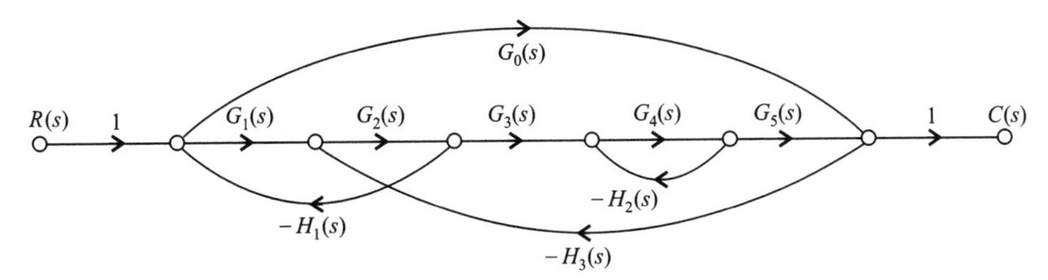

Figure P2.20

(a) Determine $C(s)/R(s)$.

(b) Determine $E(s)/R(s)$.

2.21. The signal-flow graph of a feedback control system containing two inputs and two outputs is illustrated in Figure P2.21.

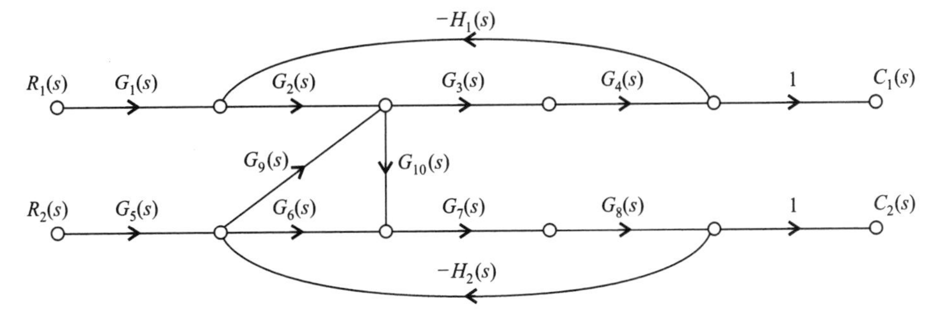

Figure P2.21

(a) Assuming that $R_1(s) = 0$, determine $C_1(s)/R_2(s)$ and $C_2(s)/R_2(s)$.

(b) Assuming that $R_2(s) = 0$, determine $C_1(s)/R_1(s)$ and $C_2(s)/R_1(s)$.

2.22. The block diagram for the system shown in Figure P2.22 represents the block diagram of one axis of a tracking radar which tries to maintain the actual radar pointing angle $C(s)$ identical to the desired pointing angle $R(s)$. In this block diagram, $U(s)$ represents a wind disturbance which tries to change the actual pointing angle.

(a) Using Mason's signal-flow graph method, determine $C(s)$ as a function of $R(s)$ and $U(s)$.

(b) Repeat part (a) for the transfer function $E_1(s)/U(s)$.

2.23. In the block diagram of the multiple feedback system shown in Figure P2.23, the error function $E_1(s)$ is added directly (through K_1) to the output in order to correct the control system's output for its error.

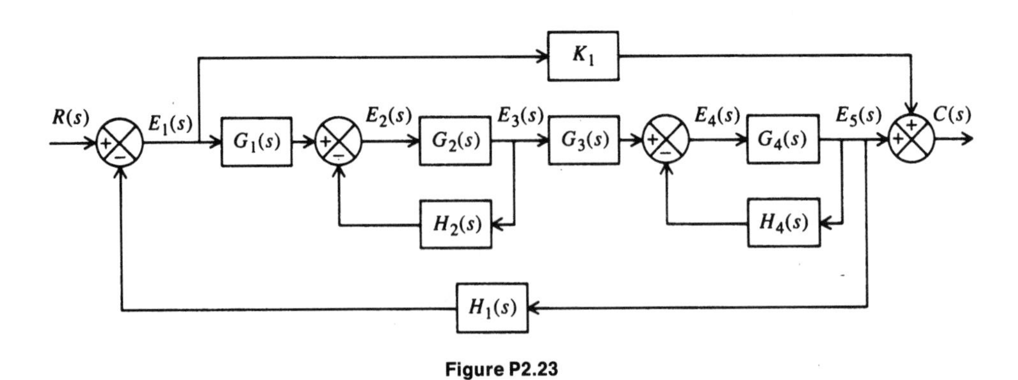

Figure P2.23

(a) Using Mason's signal-flow graph method, determine the transfer function $C(s)/R(s)$ for this system.

(b) Repeat part (a) for the transfer function $E_1(s)/R(s)$.

(c) Assuming that it is desired to have the output $C(s)$ follow the input $R(s)$, the value of $H_1(s)$ is made equal to 1. If the output $C(s)$ lags behind the input $R(s)$, what should K_1 be made to order to correct the output $C(s)$?

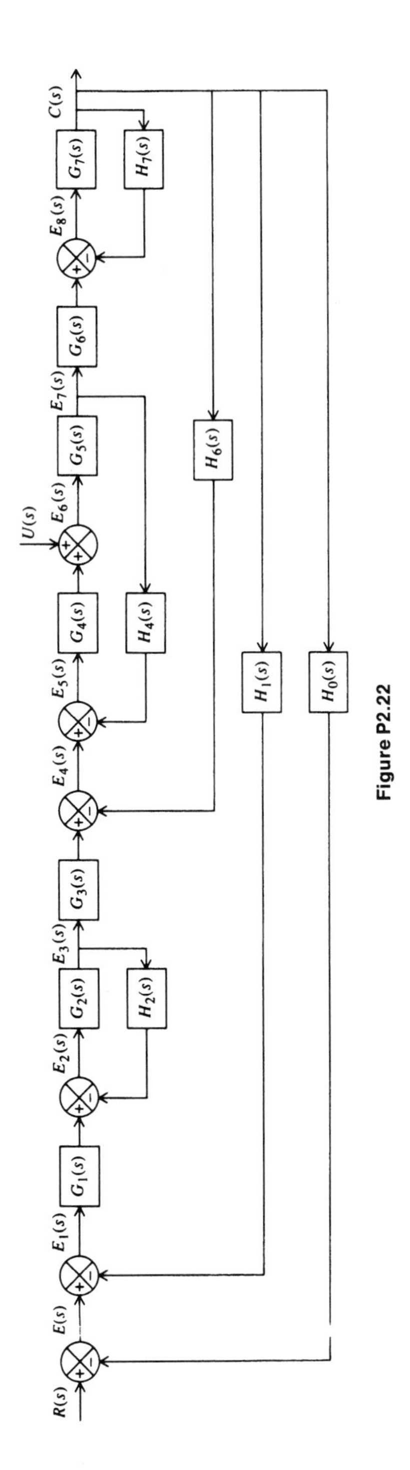
R(s)
+
-
E(s)
+
-
E1(s)
G1(s)
+
-
E2(s)
G2(s)
H2(s)
E3(s)
G3(s)
+
-
E4(s)
+
-
E5(s)
G4(s)
H4(s)
+
+
U(s)
E6(s)
G5(s)
E7(s)
G6(s)
+
-
E8(s)
G7(s)
H7(s)
C(s)
H6(s)
H1(s)
H0(s)

Figure P2.22

2.24. Using signal-flow graphs and Mason's theorem, find the following for the block diagram shown in Figure P2.24.

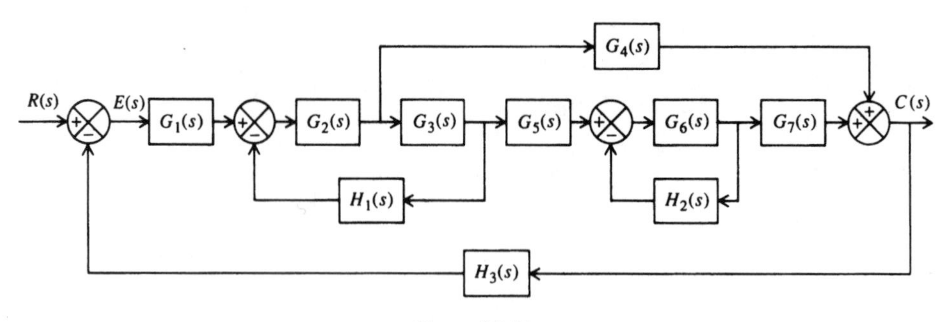

Figure P2.24

(a) $C(s)/R(s)$.

(b) E(s)/R(s).

How do the denominators in parts (a) and (b) compare?

2.25. By means of the signal-flow graph and Mason's theorem, find the transfer function of the closed-loop system, $C(s)/R(s)$, for the block diagram illustrated in Figure P2.25.

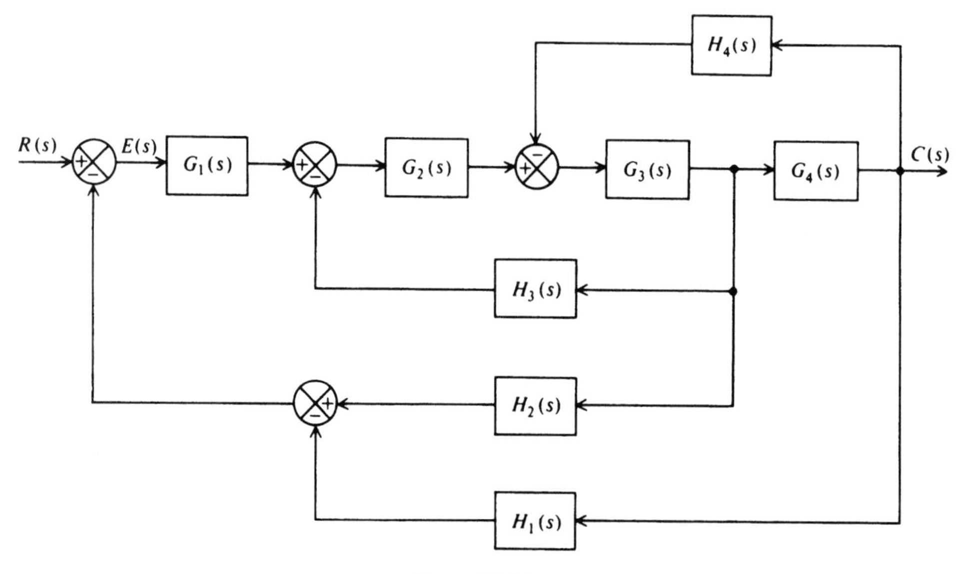

Figure P2.25

2.26. Determine the transfer function of the system shown in Figure P2.25 relating error $E(s)$ and input $R(s)$.

2.27. Determine the transfer function of the overall system shown in Figure P2.27, $C(s)/R(s)$, using the signal-flow graph and Mason's theorem.

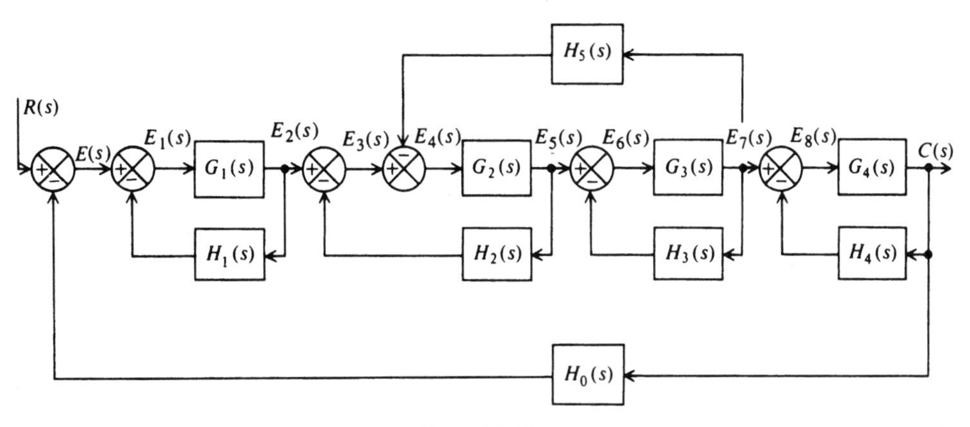

Figure P2.27

2.28. For the control system illustrated in Problem 2.27, determine $E_5(s)/E_2(s)$.

2.29. Repeat Problem 2.27 with the feedback path containing element $H_5(s)$ removed.

2.30. Determine the transfer function of the system shown in Figure P2.27 relating error $E(s)$ and input $R(s)$.

2.31. Repeat Problem 2.30 with the feedback path containing element $H_5(s)$ removed.

2.32. Determine the transfer function of the overall system shown in Figure P2.32, $C(s)/R(s)$, using the signal-flow graph and Mason's Theorem.

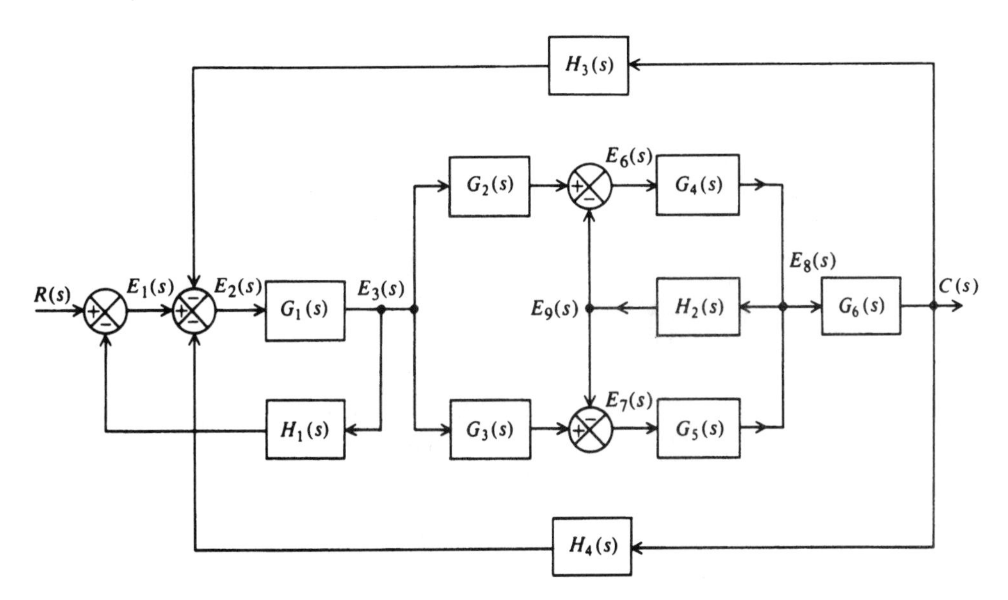

Figure P2.32

2.33. Determine the transfer function of the system shown in Figure P2.32 relating error $E_1(s)$ and input $R(s)$.

2.34. Repeat Problem 2.32, with the feedback path containing element $H_1(s)$ removed.

2.35. The block diagram of an antenna positioning control system is represented by the block diagram in Figure P2.35. The disturbance, $D(s)$, represents an external disturbance in the form of wind gusts that hit the antenna. The feedforward transfer function, $G_d(s)$, is used to eliminate the effect of the disturbance $D(s)$ on the output, $C(s)$. Assume that $R(s) = 0$.

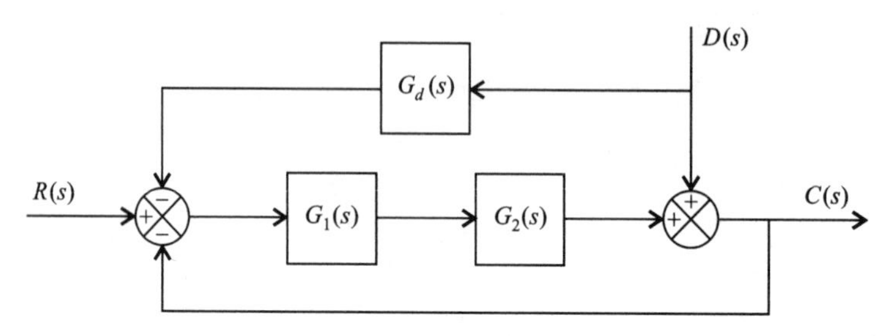

Figure P2.35

(a) Determine the transfer function $C(s)/D(s)$.

(b) Determine the explicit transfer function $G_4(s)$ which will eliminate the effect of $D(s)$ on $C(s)$.

2.36. The two-feedback loop control system illustrated in Figure P2.36 is to be designed so that the inner feedback loop has zero effect (the inner feedback loop's transfer function, $G(s)H(s)$, approaches zero) at very low and very high frequencies.

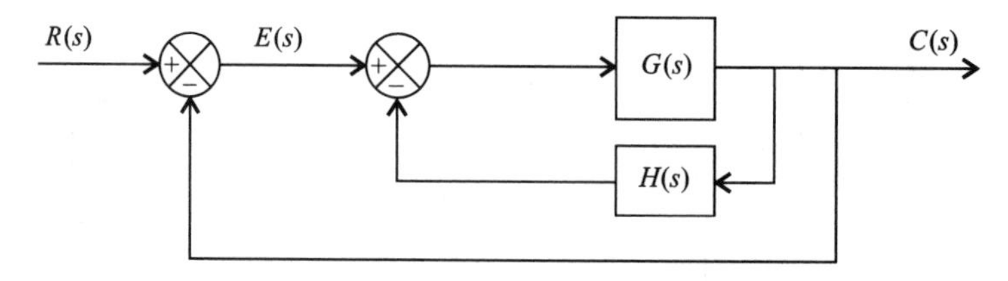

Figure P2.36

When the open-loop transfer function, $G(s)H(s)$, approaches zero at very low and very high frequencies, $C(s)/E(s)$ becomes approximately equal to $G(s)$. Assume that the transfer function of $G(s)$ is given by:

$$G(s) = \frac{K}{(T_1 s + 1)(T_2 s + 1)(T_3 s + 1)}.$$

Determine the transfer function of $H(s)$ in the inner feedback loop so that the inner feedback loop has zero effect at very low and very high frequencies.

2.37. Repeat Problem 2.33, with the feedback path containing element $H_1(s)$ removed.

2.38. Determine the system transmission $T(s) = C(s)/R(s)$ for the signal-flow graph shown in Figure P2.38.

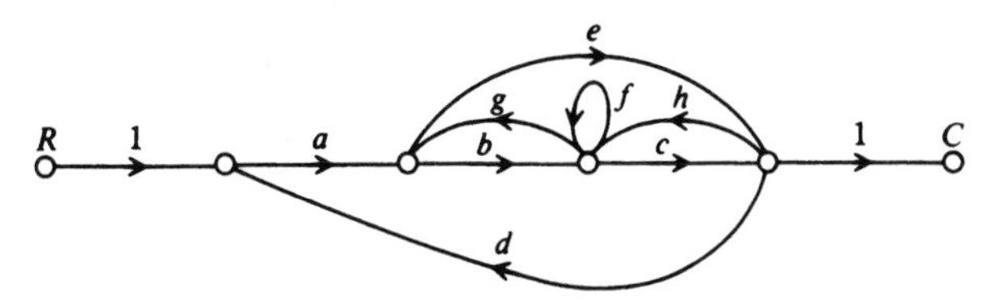

Figure P2.38

2.39. Synthesize a signal-flow graph for the following source-to-sink transmission $T(s)$, such that each branch transmission in the graph is a different letter (each branch transmission is a function of s):

$$H(s) = \frac{C(s)}{R(s)} = \frac{ah(1 - cf - dg)}{(1 - be)(1 - dg) - cf}.$$

2.40. Synthesize a signal-flow graph for the following source-to-sink transmission expression, such that each branch transmission in the graph is a different letter (each branch transmission is a function of s):

$$H(s) = \frac{C(s)}{R(s)} = \frac{(ag + adi + e)(f + bh + bcj) + ij(1 - abcd)}{1 - abcd}.$$

2.41. The signal-flow graph of a control system is illustrated in Figure P2.41.

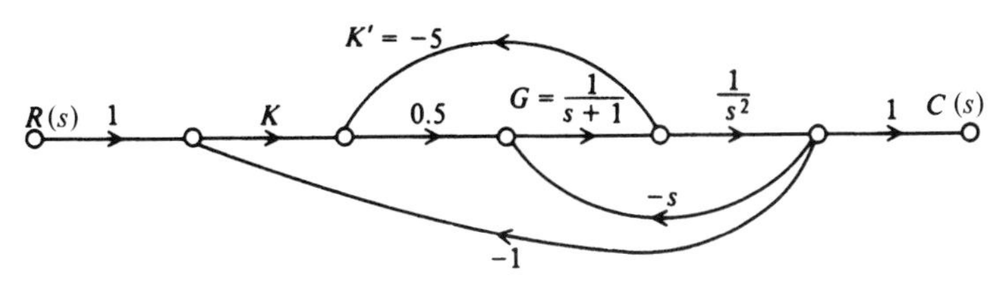

Figure P2.41

(a) Determine the overall tranmission $T(s) = C(s)/R(s)$.

(b) If the K' branch were made zero, the same overall transmission could be obtained by appropriately modifying the $G(s)$ branch. Determine the required modification.

2.42. Determine the phase-variable canonical form for the systems characterized by the following differential equations:

(a) $$\frac{d^2c(t)}{dt^2} + 2\frac{dc(t)}{dt} + c(t) = 0,$$

(b) $$\frac{d^2c(t)}{dt^2} + 2\frac{dc(t)}{dt} + c(t) = A,$$

(c) $$\frac{d^3c(t)}{dt^3} + \frac{3d^2c(t)}{dt^2} + 2\frac{dc(t)}{dt} + 2c(t) = 0,$$

(d) $$\frac{d^3c(t)}{dt^3} + 3\frac{d^2c(t)}{dt^2} + 2\frac{dc(t)}{dt} + 2c(t) = A.$$

2.43. The approximate linear equations for a spherical satellite are given by

$$I\ddot{\theta}_1(t) + \omega_0 I\dot{\theta}_3(t) = L_1,$$
$$I\ddot{\theta}_2(t) = L_2,$$
$$I\ddot{\theta}_3(t) - \omega_0 I\dot{\theta}_1(t) = L_3,$$

where $\theta_1(t), \theta_2(t), \theta_3(t)$ represent angular deviaitons of the satellite from a set of axes with fixed orientation, L_1, L_2, L_3 represent applied torques, I represents the moment of inertia, and ω_0 represents the angular frequency of the oriented axis. Determine the phase-variable canonical form of the system's dynamics.

2.44. The signal-flow graph of Figure P2.44 illustrates the process of interest accrual in a savings account. The initial deposit is represented as $r(t)$ and the total savings is represented as $c(t)$. The interest is assumed to be a constant of K% per year. It is interesting to note from this representation that it represents a positive feedback process.

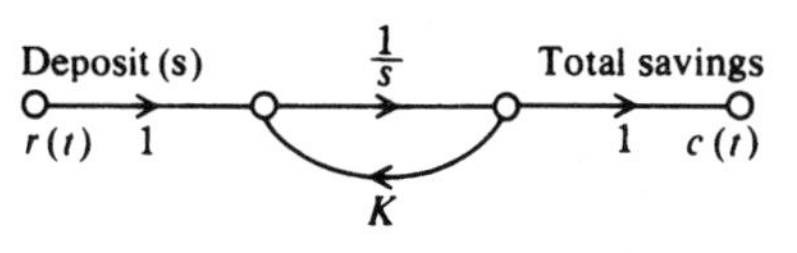

Figure P2.44

(a) Write the state equation of this system and determine the total savings as a function of deposit(s) and the interest rate.

(b) Compare the total savings at the end of a 10-year period for the following two conditions:

1. An initial deposit of \$10,000 held in a savings account over a 10-year period.

2. A yearly deposit of \$1000, totalling \$10,000 over a 10-year period, in a savings account.

In each case assume that the interest rate K is 5% per year.

(c) How do each of these cases compare with the case of savings without interest?

2.45. The voltage buildup of a simple electronic oscillator is given by the Van der Pol equation as follows:

$$\ddot{v}(t) - u(1 - v^2(t))\dot{v}(t) + v(t) = 0.$$

Determine the state equations.

2.46. The Apollo 11 mission, in which astronauts Neil Armstrong and Edwin Aldrin successfully soft landed the Lunar Excursion Module (LEM) on the lunar surface, was a historic event. Figure P2.46*a* is a photograph of the LEM vehicle taken by astronaut Michael Collins from the Apollo Command Module window after they separated. In order to obtain the state-variable vector equations of the LEM during the terminal soft-landing phase, let us consider the basic physics involved [24]. Figure P2.46*b* illustrates the forces acting on the LEM, assuming that the vehicle is vertical and subject to the following conditions:

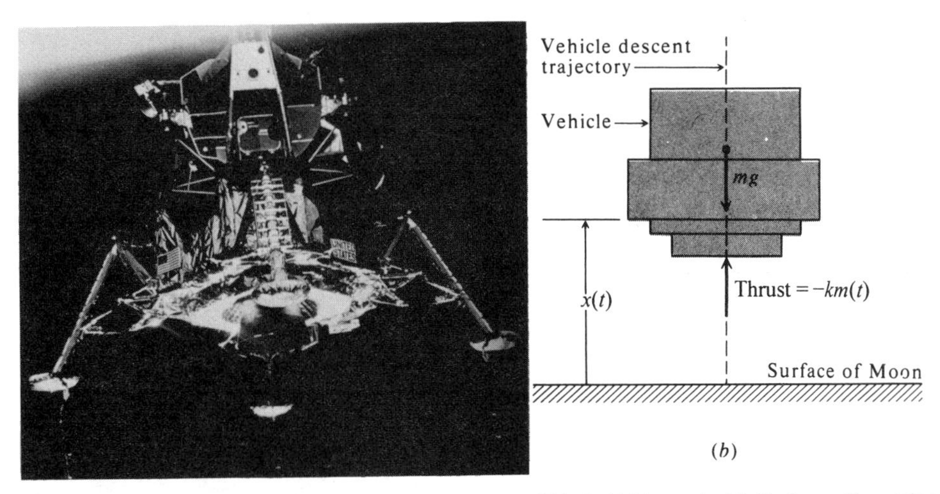

Figure P2.46 (*a*) Apollo 11 astronauts Neil Armstrong and Edwin Aldrin are inside the Lunar Excursion Module (LEM) separated from the Apollo command module. (Official NASA photo) (*b*) Forces acting on LEM.

(a) The only forces acting on the vehicle are its own weight and the thrust which acts as a braking force.

(b) The moon is flat in the vicinity of the desired landing point.

(c) The propulsion system is capable of producing a mass flow rate $\dot{m}(t)$.

Based on these assumptions, the motion of the vehicle is governed by the following relation:

$$\ddot{x}(t) = -\frac{K\dot{m}(t)}{m(t)} + g,$$

where

$$\begin{aligned} x(t) &= \text{altitude}, \\ m(t) &= \text{total mass} \\ \dot{m}(t) &= \text{mass flow rate} \leqslant 0, \\ g &= \text{acceleration of gravity at the surface}, \\ K &= \text{velocity of exhaust gases} = \text{constant} > 0. \end{aligned}$$

Defining the state variables as

$$\begin{aligned} x(t) &= x_1(t), & x_2(t) &= \dot{x}_1(t), \\ x_3(t) &= m(t), & u_1(t) &= \dot{m}(t), \end{aligned}$$

determine the phase-variable canonical equation of the system during the terminal soft-landing phase of the mission.

2.47. Determine the phase-variable canonical equation for the system shown in Figure P2.47.

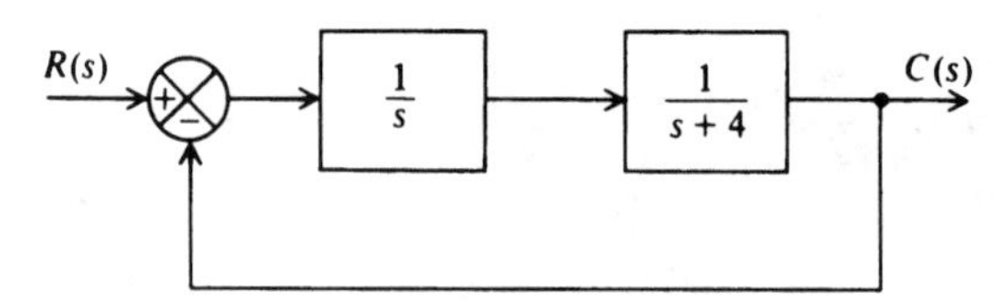

Figure P2.47

2.48. The landing of an aircraft consists of several phases [25]. First, the aircraft is guided towards the airport with approximately the correct heading by radio direction-finding equipment. Within a few miles of the airport, radio contact is made with the radio beam of the instrument landing system (ILS). In following this beam, the pilot guides the aircraft along a glide path angle of approximately 3° towards the runway. Finally, at an altitude of approximately 100 ft, the flare-out phase of the landing begins. During this final phase of the landing, the ILS radio beam is no longer effective, nor is the −3° glide path angle desirable from the viewpoint of safety and comfort. Therefore, the pilot must guide the aircraft along the desired flare path by making visual contact with the ground. Let us consider, in this problem, the states of the aircraft during the final phase of the landing (the last 100 ft of the aircraft's decent). Figure P2.48*a* defines the aircraft's coordinates and angles. Figure P2.48*b* illustrates the aircraft's block diagram in terms of measurable state signals from the elevator deflection angle δ_e to the altitude $h(t)$. The state signals being fed back are all measurable. For example, the altitude $h(t)$ can be measured with a radar altimeter, and the rate of ascent $\dot{h}(t)$ can be measured with a barometric ratemeter. In addition, the pitch angle $\theta(t)$ and the pitch rate $\dot{\theta}(t)$ can be easily measured using gyros. Defining the states of the aircraft-landing system as

$$x_1(t) = \dot{\theta}(t), \quad x_2(t) = \theta(t),$$
$$x_3(t) = \dot{h}(t), \quad x_4(t) = h(t), \quad u_1(t) = \delta_e(t),$$

determine the phase-variable canonical equation of this system.

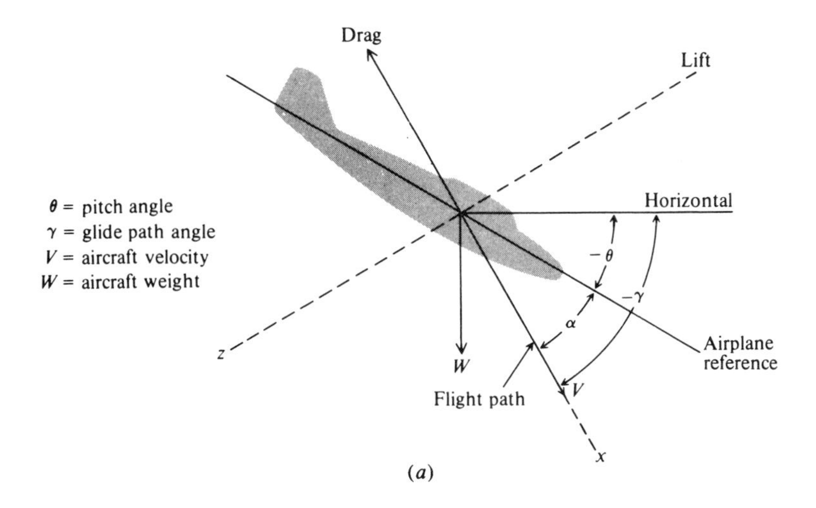

(a)

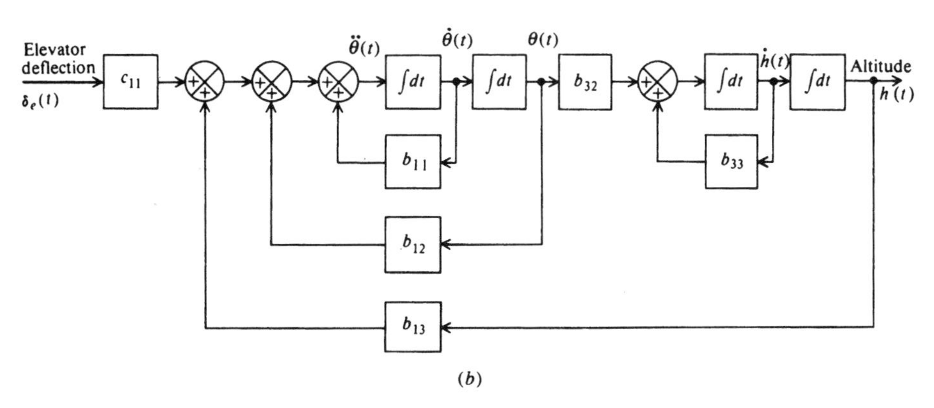

(b)

$$b_{11} = \frac{1}{T_s} - 2\zeta\omega_s, \quad b_{13} = \frac{1}{VT_s^2} - \frac{2\zeta\omega_s}{VT_s} + \frac{\omega_s^2}{V}, \quad b_{33} = -\frac{1}{T_s}$$
$$b_{12} = \frac{2\zeta\omega_s}{T_s} - \omega_s^2 - \frac{1}{T_s^2}, \quad b_{32} = \frac{V}{T_s}, \quad c_{11} = \omega_s^2 K_s T_s$$

where

K_s = short-period gain of the aircraft,
T_s = path time constant,
ω_s = short-period resonant frequency of the aircraft,
ζ = short-period damping ratio of the aircraft.

Figure P2.48

2.49. Determine the phase-variable canonical equation for the system shown in Figure P2.49.

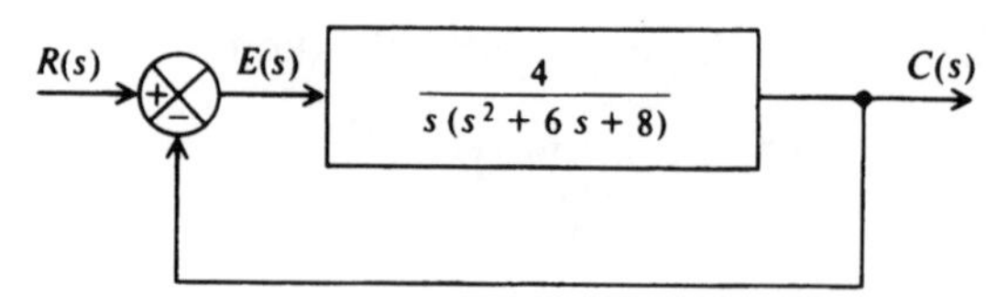

Figure P2.49

2.50. A finless torpedo is directionally unstable without control. However, its controlled performance is usually made better than that of a torpedo with fins by utilizing optimal control theory. It is possible to synthesize an intercept trajectory that overcomes the problems associated with random disturbances, measurement noise, and drift [26]. Figure P2.50 illustrates the motion of such a torpedo in the yaw axis. If it is assumed that the optimal intercept trajectory is a straight line, then the usual small-angle assumptions are reasonably accurate and the problem can be treated by linear techniques. The following equations describe yaw acceleration $\ddot{\theta}(t)$, sideslip effects $\phi(t)$, and the command input $\delta_c(t)$, for the finless torpedo:

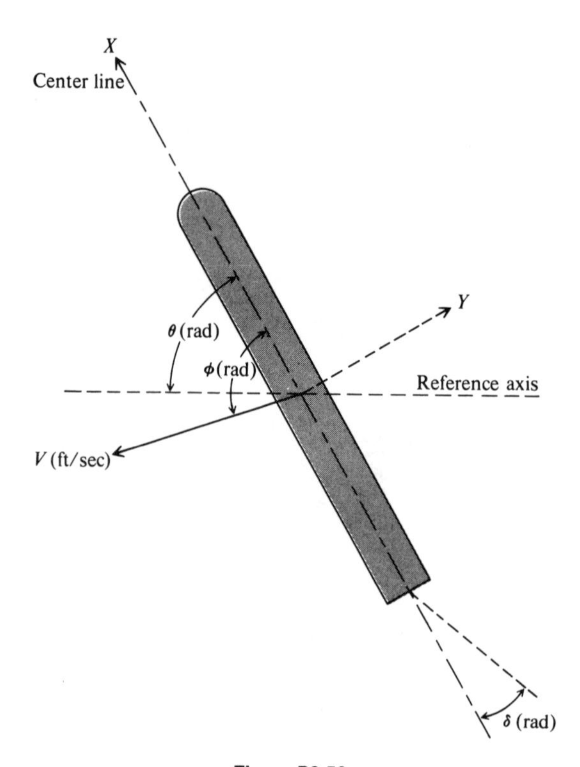

Figure P2.50

$$M_2\dot{\phi}(t) + Y_\phi \phi(t) + (Y_r - M_1)\dot{\theta}(t) + Y_\delta \delta(t) = 0,$$
$$I_2\ddot{\theta}(t) - N_\phi \phi(t) - N_r\dot{\theta}(t) - N_\delta \delta(t) = 0,$$
$$t'\dot{\delta}(t) = -\delta(t) + \delta_c(t).$$

The value t' is a normalized time parameter. Because the problem has been presented on the assumption that only small perturbations are being considered, no bounds need to be placed on $\delta_c(t)$. Determine the state equation in matrix/vector form for this system with the coefficients, normalized with respect to ship length, defined as follows:

M_1 = longitudinal virtual mass coefficient = 1.56,
M_2 = lateral virtual mass coefficient = 2.92,
I_2 = virtual moment of inertia cofficient = 0.138,
Y_ϕ = static side force rate coefficient = 0.60,
N_ϕ = static yaw moment rate coefficient = 0.99,
Y_r = damping force rate coefficient = 0.20,
N_r = damping moment rate coefficient = −0.08,
Y_δ = rudder force rate coefficient = 0.10,
N_δ = rudder moment rate coefficient = $Y_\delta e_R$ = −0.50
e_R = dimensionless coordinate of rudder side force = −0.50.

2.51. Given

$$\mathbf{A} = \begin{bmatrix} 1 & 2 & 4 \\ 2 & 1 & 1 \\ 2 & 2 & 2 \end{bmatrix} \text{ and } \mathbf{B} = \begin{bmatrix} 2 & 4 & 2 \\ 2 & 1 & 1 \\ 4 & 2 & 4 \end{bmatrix},$$

determine $\mathbf{AB}^{-1}$. How can you check your results?

2.52. Determine the state and output equations for the system shown in Figure P2.52.

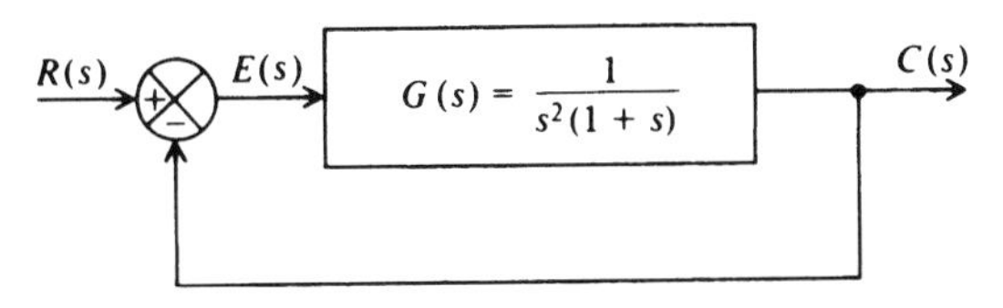

Figure P2.52

2.53. Determine the state and output equations for an open-loop system where

$$\frac{C(s)}{R(s)} = \frac{1}{s^2(s+10)}.$$

2.54. Repeat Problem 2.53 with

$$\frac{C(s)}{R(s)} = \frac{1}{s(s+1)(s+8)}.$$

2.55. Repeat Problem 2.53 with

$$\frac{C(s)}{R(s)} = \frac{5}{s(s^2+4s+2)}.$$

2.56. Determine the state-variable signal-flow graph and the state and output equations for the following open-loop system:

$$P(s) = \frac{C(s)}{U(s)} = \frac{s+1}{s(s^2+7s+1)}.$$

2.57. The automatic depth control of a submarine is an interesting control problem. Figure P2.57*a* illustrates a representative problem, where the actual depth of the submarine is denoted as $C(s)$. In practice, the actual depth of a submarine is measured by a pressure transducer. This measurement is then compared with the desired depth $R(s)$. Any differences are amplified in a control system which appropriately adjusts the stern plane actuator angle θ. An equivalent block diagram of such a system is illustrated in Figure P2.57*b*. Determine the state-variable signal-flow graph and the state and output equations for this system.

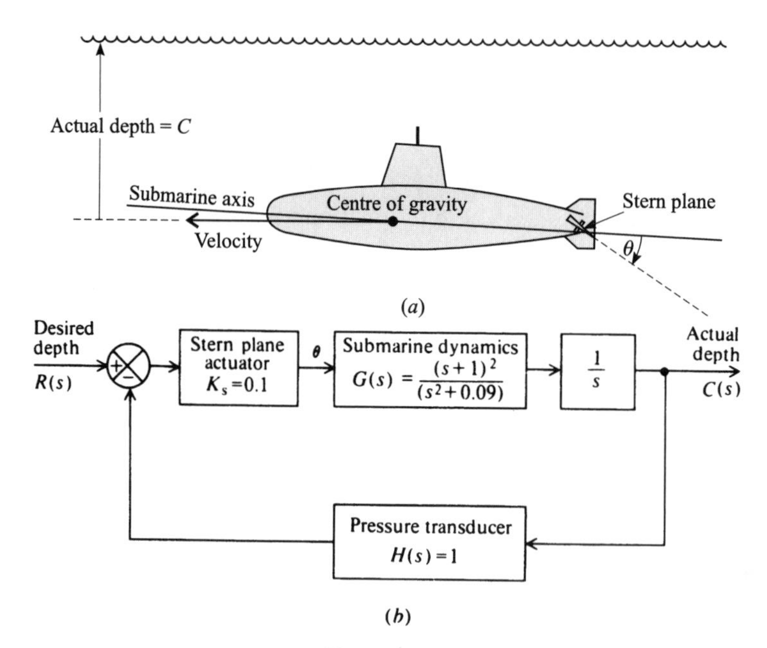

Figure P2.57

2.58. Determine the vector-matrix form of the following set of first-order differential equations:

$$\frac{dx_1(t)}{dt} = -x_1(t) + 3x_2(t)$$

$$\frac{dx_2(t)}{dt} = -2x_2(t) + 6x_3(t) + u_1(t)$$

$$\frac{dx_3(t)}{dt} = -x_1(t) - 2x_2(t) - x_3(t) + u_2(t)$$

2.59. Control techniques are being utilized to solve problems in the aquatic systems field. These ecological problems require a mathematical model to provide an understanding of the characteristics within the system. An aquatic ecosystem model [27] for Lago Pond, a Georgia farm pond, has been developed. This pond is a man-made pond created by an earth-fill dam and is used for sport fishing. The pond is fertilized in the spring and periodically throughout the summer in order to cause algal blooms; this increase in primary productivity travels up the food chain and causes an increase in the fish population. A constant coefficient model of Lago Pond, which assumes a constant temperature environment, is given by the following set of equations:

$$\begin{aligned}
\dot{x}_1(t) &= -\,30.8x_1(t) + 616 + 200\sin 0.524t, \\
\dot{x}_2(t) &= 2.13x_1(t) - 17.4x_2(t) + 0.0458x_9(t), \\
\dot{x}_3(t) &= 1.67x_1(t) - 8.66x_3(t) + 0.0553x_9(t), \\
\dot{x}_4(t) &= 0.457x_2(t) + 0.553x_3(t) - 0.941x_4(t) + 0.148x_6(t), \\
\dot{x}_5(t) &= 0.0925x_3(t) + 0.0814x_4(t) - 0.349x_5(t) + 0.00224x_6(t), \\
\dot{x}_6(t) &= 5.97x_2(t) - 1.94x_6(t), \\
\dot{x}_7(t) &= 0.346x_2(t) + 2.73x_3(t) + 0.0193x_6(t) - 0.314x_7(t) + 0.23, \\
\dot{x}_8(t) &= 0.0898x_4(t) + 0.0166x_5(t) + 0.0166x_7(t) - 0.104x_8(t), \\
\dot{x}_9(t) &= 13.5x_1(t) + 5.45x_2(t) + 4.23x_3(t) + 0.213x_4(t) + 0.0703x_5(t) \\
&\quad + 0.382x_6(t) + 0.0628x_7(t) + 0.0207x_8(t) - 0.816x_9(t).
\end{aligned}$$

The states $x_4(t), x_5(t), x_6(t)$, and $x_7(t)$ represent the levels of redbreast, warmouth, chaoborus, and bluegill sunfish in the lake; the state $x_8(t)$ represents the level of bass in the lake. States $x_1(t), x_2(t)$, and $x_3(t)$ represent nutrient factors for the fish as a result of pond fertilization, and state $x_9(t)$ represents detritus (rocks and other small material worn or broken away from a mass due to the action of water).

(a) Determine the companion matrix **P**, and $\mathbf{Bu}(t)$, for the state equation of the system:

$$\dot{\mathbf{x}}(t) = \mathbf{Px}(t) + \mathbf{Bu}(t).$$

(b) Simulate this constant coefficient model of Lago Pond on a digital computer, and determine the time responses of the states for 12 months using the algorithm given in Section 2.24 (see Eq. 2.248). Assume the following initial conditions for the nine states (units are in Kcal/m^2): $x_1 = 20$; $x_2 = 3.5$; $x_3 = 6.4$; $x_4 = 7.16$; $x_5 = 3.44$; $x_6 = 10.8$; $x_7 = 61$; $x_8 = 16.5$; $x_9 = 400$. What conclusions can you reach from your results?

2.60. Determine the phase-variable canonical form of the state and output vector equations for the following system, where $r(t)$ represents the reference input and $n(t)$ represents the noise entering the system:

$$\frac{d^3c(t)}{dt^3} + 3\frac{d^2c(t)}{dt^2} + 2\frac{dc(t)}{dt^2} + 4c(t) = 4r(t) + 2n(t).$$

2.61. Determine the phase-variable canonical state and output equations for the control system of Figure P2.61 from knowledge of the system's transfer function.

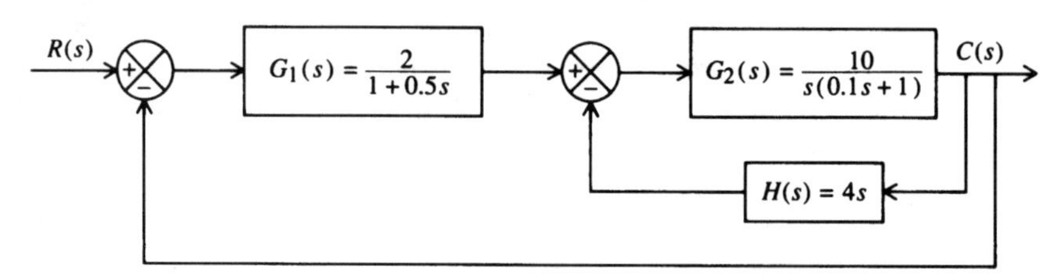

Figure P2.61

2.62. A control system has a transfer function given by

$$\frac{C(s)}{U(s)} = \frac{20s + 1}{s^4 + 4s^3 + 3s^2 + 2s + 1}$$

(a) Derive the state-variable signal-flow graph for this control system. Label the states on this diagram.

(b) Determine the phase-variable canonical state and output equations from the state-variable signal-flow graph of part (a).

2.63 The closed-loop transfer function of a control system is given by the following:

$$\frac{C(s)}{R(s)} = \frac{(s + 1)}{s(s^2 + 6s + 8)}.$$

(a) Determine and illustrate the state-variable signal-flow graph of this system.

(b) Find the state and output equations of this system from the state-variable signal-flow graph.

(c) Write the phase-variable canonical form of the system's equations.

2.64. Determine analytically the state transition matrix of the system considered in Problem 2.56.

2.65. Determine the state transition matrix of the following open-loop control system analytically:

$$P(s) = \frac{C(s)}{U(s)} = \frac{1}{s(s+2)(s+5)}.$$

2.66. Repeat Problem 2.65 with

$$P(s) = \frac{C(s)}{U(s)} = \frac{1}{s^2(s+6)(s+10)}.$$

2.67. A nuclear reactor has been operating in equilibrium for a long period of time at a high thermal-neutron flux level, and is suddenly shut down. At shutdown, the density of iodine 135 (I) is 5×10^{16} atoms per unit volume and that of xenon 135 (X) is 2×10^{15} atoms per unit volume. The equations of decay are given by the following state equations:

$$\dot{\mathrm{I}}(t) = -0.1\mathrm{I}(t), \quad \dot{X}(t) = -\mathrm{I}(t) - 0.05X(t).$$

(a) Determine the state transition matrix using Eq. (2.256).

(b) Determine the system response equations, $\mathrm{I}(t)$ and $\mathrm{X}(t)$.

(c) Determine the half-life time of I 135. The unit of time for $\dot{\mathrm{I}}$ and $\dot{\mathrm{X}}$ is hours.

2.68. Determine whether the following matrices can represent state transition matrices by checking whether they satisfy all of the necessary pròperties to be a state transition matrix:

(a)

$$\begin{bmatrix} 1 - e^{-3t} & 0 \\ 0 & -e^{-3t} \end{bmatrix}$$

(b)

$$\begin{bmatrix} 1 & 0 \\ 1 - e^{-2t} & e^{-2t} \end{bmatrix}$$

2.69. Control-system concepts have recently been applied in attempts to solve some problems of sociology and of economics. Let us consider one such example concerned with the problem of underdeveloped countries [28, 29]. Representing the number of underdeveloped countries by C_u, and developed countries by C_d, the state variable x_1 is defined as

$$x_1 = C_u/C_d.$$

The state variable x_1 is an indication of the development of the countries in the world. A second state variable x_2 is used to represent the tendency towards underdevelopment. In writing a set of dyamical equations for these state variables, it is important to recognize that the tendency towards under-development can be reduced by means of technical-assistance programs, education, etc. However, it is interesting to note that studies indicate that the gap between the developed and underdeveloped nations is growing, because the underdeveloped nations tend to remain in their present state relative to the developed area [29]. It has been proposed that the following set of state equations may be used to represent this process [28]:

$$\dot{x}_1(t) = -Ax_1(t) - Bx_2(t),$$
$$\dot{x}_2(t) = Cx_1(t) - Dx_2(t).$$

(a) Determine the state transition matrix of this system from the state-variable signal flow graph when $A = 4$, $B = C = 1$, and $D = 2$.

(b) Determine the response of this system when $x_1(0) = 2$ and $x_2(0) = 1$.

2.70. Determine the state transition matrix of the following open-loop control system analytically:

$$P(s) = \frac{C(s)}{U(s)} = \frac{1}{s^2(s+1)^2}.$$

2.71. The effectiveness of a mass-marketing campaign to sell a product can be predicted by analyzing its state-equation representation using control techniques. Consider the promotion of a new automobile in a city by the mass media including television, radio, newspapers, and periodicals. Consider the population under study to be made up of the following three groups:

$x_1(t)$ represents the group who might by receptive to the automobile;

$x_2(t)$ represents the group who buys the automobile;

$x_3(t)$ represents the group removed from groups $x_1(t)$ and $x_2(t)$ due to death or other factors that cause them to be isolated or removed from the groups $x_1(t)$ and $x_2(t)$.

The rate at which new receptives are added to the population is equal to $u_1(t)$, and the rate at which new buyers are added to the population is equal to $u_2(t)$. The model of this system can be represented by the following set of equations:

$$\dot{x}_1(t) = -mx_1(t) - nx_2(t) + u_1(t),$$
$$\dot{x}_2(t) = nx_1(t) - px_2(t) + u_2(t),$$
$$\dot{x}_3(t) = mx_1(t) + px_2(t).$$

Assume a closed population model ($u_1(t) = u_2(t) = 0$).

(a) Determine the state transition matrix of this system when $m = 4, n = 2$, and $p = 1$.

(b) Forecast how well the automobile will sell and whether the marketing campaign will be successful by determining the response of $x_2(t)$ of this system for the following initial conditions: $x_1(0) = 100{,}000$; $x_2(0) = 20{,}000$; $x_3(0) = 2000$. What conclusions can you reach from your results? The unit of time is in weeks.

2.72. A very interesting ecological problem is that of rabbits and foxes in a controlled environment. If the number of rabbits were left alone, they would grow indefinitely until the food supply was exhausted. Representing the number of rabbits by $x_1(t)$, their growth rate is given by

$$\dot{x}_1(t) = Ax_1(t).$$

However, rabbit-eating foxes in the environment change this relationship to the following:

$$\dot{x}_1(t) = Ax_1(t) - Bx_2(t),$$

where $x_2(t)$ represents the fox population. In addition, if foxes must have rabbits to exist, then their growth rate is given by

$$\dot{x}_2(t) = -Cx_1(t) + Dx_2(t).$$

(a) Assume that $A = 1, B = 2, C = 2$, and $D = 4$. Determine the state transitions matrix for this ecological model.

(b) From the state transition matrix, determine the response of this ecological model when $x_1(0) = 100$ and $x_2(0) = 50$. Explain your results.

2.73. [30] A very important ecological control problem is the treatment of wastes in water. Control engineers are attempting to solve this problem by adding bacteria, which grow by using the pollutant as a nutrient, to the waste flow. In this manner, wastes in sewage treatment plants can be eliminated by this natural biological process. Microbial growth models currently being used to analyze this kind of problem are based on differential equations. Defining $x(t)$ as the bacteria concentration required for pollutant removal, $s(t)$ as the pollutant serving as the nutrient substance for the bacteria, and D as the rate of flow/volume (dilution rate), the normalized material balance equations can be approximated by the following two state equations:

$$\dot{x}(t) = ux(t) - Dx(t)$$
$$\dot{s}(t) = -ux(t) - Ds(t).$$

(a) Assuming that the specific growth rate u is a constant, determine the state transition matrix of this ecological control system.

(b) Analyze the meaning of the state transition matrix obtained for this problem. Under what conditions does a balance exist in this ecological control system?

2.74. Substances $x_1(t)$ and $x_2(t)$ are involved in the reaction of a chemical process. The state equations representing this reaction are as follows:

$$\dot{x}_1(t) = -8x_1(t) + 4x_2(t),$$
$$\dot{x}_2(t) = 4x_1(t) - 2x_2(t).$$

(a) Determine the state transition matrix of this chemical process.

(b) Determine the response equations of this system, $x_1(t)$ and $x_2(t)$ when

$$x_1(0) = 2,$$
$$x_2(0) = 1.$$

(c) At what value of time will the amount of substances $x_1(t)$ and $x_2(t)$, be equal? The unit of time for $x_1(t)$ and $x_2(t)$ is hours.

2.75. Substances $x_1(t)$ and $x_2(t)$ are involved in the reaction of a chemical process. The state equations representing this reaction are as follows:

$$\dot{x}_1(t) = -4x_1(t) + 2x_2(t),$$
$$\dot{x}_2(t) = 2x_1(t) - x_2(t).$$

(a) Determine the state transition matrix of this chemical process.

(b) Determine the response of this system when:

$$x_1(0) = 200{,}000 \text{ units},$$
$$x_2(0) = 10{,}000 \text{ units}.$$

(c) At what value of time will the amount of substances $x_1(t)$ and $x_2(t)$ be equal?

2.76. The state equations of a chemical process which mixes chemicals $x(t)$ and $y(t)$ are given by the following:

$$\dot{x}(t) = -100x(t),$$
$$\dot{y}(t) = -10x(t) + 2y(t).$$

The initial conditions of states $x(t)$ and $y(t)$ are as follows:

$$x(0) = 1000,$$
$$y(0) = 4000.$$

(a) Determine the state transition matrix.

(b) Determine the system's response equations $x(t)$ and $y(t)$.

(c) Find the time it takes in minutes for element $x(t)$ to decay to 200 units. (The unit of time for $x(t)$ and $y(t)$ is hours.)

2.77. **(a)** Determine the state transition matrix of the control system of Figure P2.77 analytically from the definition of the state transition matrix.

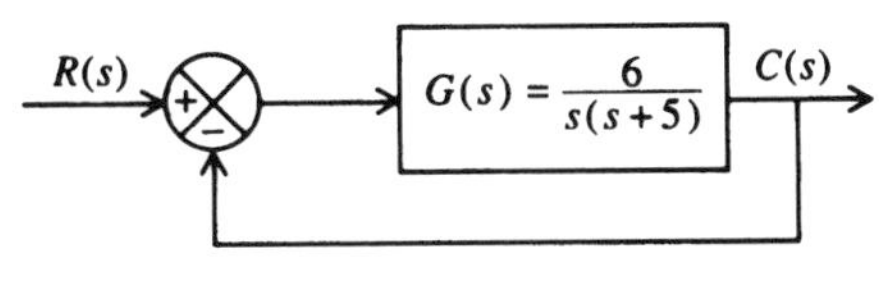

Figure P2.77

(b) If the initial position and velocity of the system are 10 units and 20 units/sec, respectively, determine the system response equations for position and velocity: $c(t)$ and $\dot{c}(t)$.

2.78. **(a)** Determine the state transition matrix for a system whose companion matrix **P** is given by

$$\mathbf{P} = \begin{bmatrix} 1 & 2 \\ 0 & 0 \end{bmatrix}$$

Use the definition of the state transition matrix as given by Eq. (2.256).

(b) Solve the state transition equation, Eq. (2.255), for the system whose **P** matrix is given in part (a). Assume that the input is a unit step function, its input vector, **B**, is given by

$$\mathbf{B} = \begin{bmatrix} 1 \\ 1 \end{bmatrix},$$

and the initial state vector, $\mathbf{x}(0)$, is given by

$$\mathbf{x}(0) = \begin{bmatrix} 0 \\ 0 \end{bmatrix}.$$

2.79. Repeat Problem 2.78 if the **P** matrix is changed to the following.

$$\mathbf{P} = \begin{bmatrix} 3 & 4 \\ 0 & 0 \end{bmatrix}$$

REFERENCES

1. J. W. Nillson, *Electric Circuits*, 2nd ed. Addison-Wesley, Reading, MA, 1986.
2. R. K. Livesley, *Mathematical Methods for Engineering*. Wiley, New York, 1989.

3. E. Kamen, *Introduction to Signals and Systems.* Macmillan, New York, 1987.
4. B. C. Kuo, *Linear Circuits and Systems.* McGraw-Hill, New York, 1987.
5. M. F. O'Flynn and G. Moriarti, *Linear Systems.* Harper and Row, New York, 1987.
6. MATLAB for MS-DOS Personal Computers, User's Guide; *The Control System Toolbox*; The MathWorks, Inc., 1997.
7. S. J. Mason, "Feedback theory: Some properties of signal flow graphs." *Proc. IRE* **41**, 1144 (1953).
8. S. J. Mason, "Feedback theory: Further properties of signal flow graphs." *Proc. IRE* **44**, 920 (1956).
9. S. M. Shinners, *A Guide to Systems Engineering and Management*, Lexington Books, Lexington, MA, 1976.
10. S. Barnet, "Matrices, polynomials, and linear time-invariant systems." *IEEE Trans. Autom. Control* **AC-18**, 1–10 (1973).
11. C. A. Desoer, "An introduction to state space techniques in linear systems." In *Proceedings of the 1962 Joint Automatic Control Conference*, New York, pp. 10-2-1–10-2-5.
12. L. A. Zadeh, "An introduction to state-space techniques." In *Proceedings of the 1962 Joint Automatic Control Conference*, New York, pp. 10-1-1–10-1-5.
13. D. F. Delchamps, *State Space and Input-Output Linear Systems.* Springer-Verlag, New York, 1988.
14. R. C. Rosenberg and D. C. Karnopp, *Introduction to Physical System Design.* McGraw-Hill, New York, 1987.
15. G. A. Korn and T. M. Korn, *Electronic Analog and Hybrid Computers.* McGraw-Hill, New York, 1964.
16. R. W. Hamming, *Numerical Methods for Scientists and Engineers.* McGraw-Hill, New York, 1962.
17. D. D. McCracken and W. S. Dorn, *Numerical Methods and FORTRAN Programming.* Wiley, New York, 1964.
18. K. E. Atkinson, *An Introduction to Numerical Analysis*, 2nd ed. Wiley, New York, 1989.
19. T. Ward and E. Bromhead, *FORTRAN and the Art of PC Programming.* Wiley, New York, 1989.
20. D. K. Faddeev and V. N. Faddeeva, *Computational Methods of Linear Algebra.* Freeman, San Francisco, CA, 1963.
21. B. S. Morgan, Jr., "Sensitivity analysis and synthesis of multivariable systems." *IEEE Trans. Automat. Control* **AC-11**, 506–512 (1966).
22. M. I. Liou, "A novel method of evaluating transient response." *Proc. IEEE* **54**, 20–23 (1966).
23. B. O. Watkins, *Introduction to Control Systems*, pp. 258–259. Macmillan, New York, 1969.
24. J. S. Meditch, "On the problem of optimal thrust programming for a lunar soft landing." *IEEE Trans. Autom. Control,* **AC-9**, 477–484 (1964).
25. F. J. Ellert and C. W. Merriam, III, "Synthesis of feedback controls using optimization theory—An example." In *Proceedings of the 1962 Joint Automatic Control Conference*, New York, p. 19-1.
26. N. W. Rees, "An application of optimal control theory to the guided torpedo problem." In *Proceedings of the 1968 Joint Automatic Control Conference*, pp. 820–825.
27. W. R. Emanuel and R. J. Mulholland, "Energy based dynamic model for Lago Pond, Georgia." In *Proceedings of the 1974 Joint Automatic Control Conference*, pp. 354–362.

28. R. C. Dorf, *Modern Control Systems*, 7th ed. Addison-Wesley, Reading, MA, 1995.
29. D. Rockefeller, "The population problem and economic progress." *Vital Speeches* **32**, 366–370 (1966).
30. G. D'Ans, P. W. Kotovic, and D. Gottlieb, "A nonlinear regulator problem for a model of biological water treatment." *IEEE Trans. Autom. Control* **AC-16**, 341–347 (1971).
31. *The Student Edition of MATLAB*, The MathWorks, Inc., Prentice Hall, Englewood Cliffs, NJ, 1997.

3

STATE EQUATIONS AND TRANSFER-FUNCTION REPRESENTATION OF PHYSICAL LINEAR CONTROL-SYSTEM ELEMENTS

3.1. INTRODUCTION

The purpose of this chapter is to illustrate the procedures used for deriving the state equations and transfer-function representation for several common linear control-system elements. This is a very important step that must be mastered before considering the determination of performance and stability of various kinds of control systems.

In general, the devices that are encountered can be classified as being electrical, mechanical, electromechanical, hydraulic, thermal, etc. The emphasis here will be to describe the state equations and transfer functions for linear control-system elements that are representative of these classes. Nonlinear devices are discussed in Chapter 5 on Nonlinear Control-System Design in the accompanying volume.

3.2. STATE EQUATIONS OF ELECTRICAL NETWORKS

Before introducing the techniques for formulating the state equations and transfer-function representation of some common linear control-system devices, the subject of electrical networks is considered [1]. Specifically, it is the purpose of this section to illustrate how the engineer can represent the state equations of electrical networks [2]. The very same method is applicable to the analysis of control-system devices. Network equations of electrical circuits can be formulated from the basic laws of Kirchhoff. It is assumed that the reader is familiar with the corresponding loop and nodal techniques for analyzing electrical networks [1].

The loop equation of the basic *RLC* circuit illustrated in Figure 3.1 is given by

$$e(t) = Ri(t) + L\frac{di(t)}{dt} + \frac{1}{C}\int i(t)dt. \tag{3.1}$$

The current in the inductor and the voltage across the capacitor are usually considered to be the state variables of a network. Then the resulting state equations for this circuit can be formulated by relating the voltage across the inductor L and the current in the capacitor C as follows:

$$L\frac{di(t)}{dt} = e(t) - e_c(t) - Ri(t), \tag{3.2}$$

$$i(t) = C\frac{de_c(t)}{dt}. \tag{3.3}$$

Defining the state variables as

$$x_1(t) = e_c(t), \tag{3.4}$$

$$x_2(t) = i(t), \tag{3.5}$$

the following state equations of the circuit are obtained:

$$\dot{x}_1(t) = \frac{1}{C}x_2(t), \tag{3.6}$$

$$\dot{x}_2(t) = -\frac{1}{L}x_1(t) - \frac{R}{L}x_2(t) + \frac{e(t)}{L}. \tag{3.7}$$

These equations can be written

$$\dot{\mathbf{x}}(t) = \mathbf{P}\mathbf{x}(t) + \mathbf{B}r(t), \tag{3.8}$$

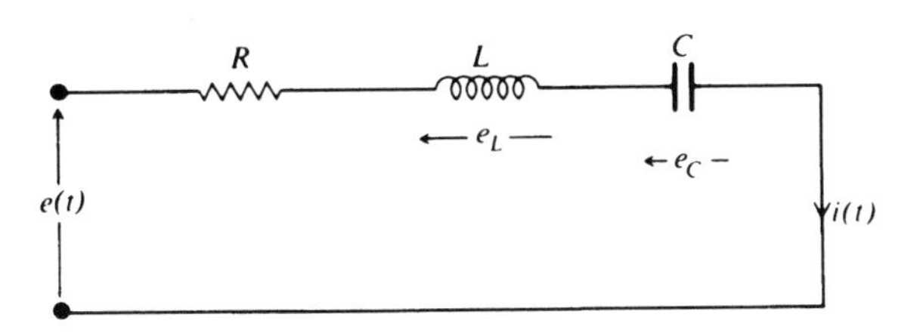

Figure 3.1 An *RLC* circuit.

where

$$\mathbf{x}(t) = \begin{bmatrix} x_1(t) \\ x_2(t) \end{bmatrix}, \quad \dot{\mathbf{x}}(t) = \begin{bmatrix} \dot{x}_1(t) \\ \dot{x}_2(t) \end{bmatrix}, \quad \mathbf{P} = \begin{bmatrix} 0 & \frac{1}{C} \\ -\frac{1}{L} & -\frac{R}{L} \end{bmatrix}$$

$$\mathbf{B} = \begin{bmatrix} 0 \\ \frac{1}{L} \end{bmatrix}, \quad r(t) = e(t).$$

As a second example, consider the electrical network of Figure 3.2. The state equations are obtained by writing the equation for the currents in the capacitors and the voltage across the inductors as follows:

$$i_1(t) - i_2(t) = C\frac{de_c(t)}{dt}, \tag{3.9}$$

$$L_1\frac{di_1(t)}{dt} = e(t) - R_1 i_1(t) - e_c(t), \tag{3.10}$$

$$L_2\frac{di_2(t)}{dt} = e_c(t). \tag{3.11}$$

If the state variables are defined as

$$x_1(t) = e_c(t), \tag{3.12}$$
$$x_2(t) = i_1(t), \tag{3.13}$$
$$x_3(t) = i_2(t), \tag{3.14}$$

the following state equations for this electrical network are obtained:

$$\dot{x}_1(t) = \frac{1}{C}x_2(t) - \frac{1}{C}x_3(t), \tag{3.15}$$

$$\dot{x}_2(t) = -\frac{1}{L_1}x_1(t) - \frac{R_1}{L_1}x_2(t) + \frac{1}{L_1}e(t), \tag{3.16}$$

$$\dot{x}_3(t) = \frac{1}{L_2}x_1(t). \tag{3.17}$$

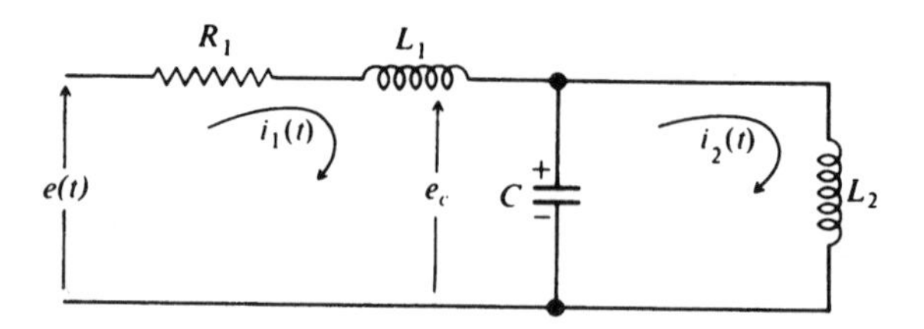

Figure 3.2 An electrical network.

These equations can be written as

$$\dot{x}(t) = \mathbf{P}\mathbf{x}(t) + \mathbf{B}r(t), \tag{3.18}$$

where

$$\mathbf{x}(t) = \begin{bmatrix} x_1(t) \\ x_2(t) \\ x_3(t) \end{bmatrix}, \quad \dot{\mathbf{x}}(t) = \begin{bmatrix} \dot{x}_1(t) \\ \dot{x}_2(t) \\ \dot{x}_3(t) \end{bmatrix}, \quad \mathbf{P} = \begin{bmatrix} 0 & \dfrac{1}{C} & -\dfrac{1}{C} \\ -\dfrac{1}{L_1} & -\dfrac{R_1}{L_1} & 0 \\ \dfrac{1}{L_2} & 0 & 0 \end{bmatrix},$$

$$\mathbf{B} = \begin{bmatrix} 0 \\ \dfrac{1}{L_1} \\ 0 \end{bmatrix}, \quad r(t) = e(t)$$

These simple examples illustrate the general method for analyzing electrical networks utilizing the state-variable technique. The reader is referred to Reference [2] for a more detailed discussion of this particular subject. The same general method is illustrated next for the analysis of commonly used linear mechanical control-system devices.

3.3. TRANSFER-FUNCTION AND STATE-VARIABLE REPRESENTATION OF TYPICAL MECHANICAL CONTROL-SYSTEM DEVICES

Mechanical control-system devices can generally be classified as being either translational or rotational. The major difference between the two is that we talk of forces and translational units in the former, and torque and angular units in the latter. Newton's three laws of motion [3] govern the action of both types of mechanical systems. Basically, these laws state that the sum of the applied forces, or torques, must equal the sum of the reactive forces, or torques, for a body whose acceleration is zero. Another way of stating this is that the sum of the forces must equal zero for a body at rest or moving at a constant velocity. We shall consider some representative translational systems and then some rotational systems. The basic concepts illustrated and developed here should be sufficient to enable the reader to handle more complex systems.

A. Mechanical Translation Systems

The three basic characteristics of a mechanical translational system are mass, stiffness, and damping. Mass represents an element having inertia. Stiffness represents the restoring force action such as that of a spring. Damping, or viscous friction, represents a characteristic of an element that absorbs energy. The symbols for the various quantities used in our analyses are shown in Table 3.1.

1. Force Applied to a Mechanical System Containing a Mass, Spring, and Damper. The case of a force $f(t)$ applied to a mass, spring, and damper is shown in Figure 3.3. The system produces a displacement of the mass $y(t)$ measured from a reference terminal y_0, which is assumed to be stationary. The application of Newton's law to this system yields

$$M\frac{d^2y(t)}{dt^2}+B\frac{dy(t)}{dt}+Ky(t)=f(t). \tag{3.19}$$

This second-order differential equation can be written as two first-order differential equations by defining

$$x_1(t)=y(t) \tag{3.20}$$

Table 3.1. Mechanical Translation Symbols

Quantity	Symbol
Distance	$y(t)$
Velocity	$v(t)$
Acceleration	$a(t)$
Force	$f(t)$
Mass	M
Damping factor	B
Stiffness factor	K

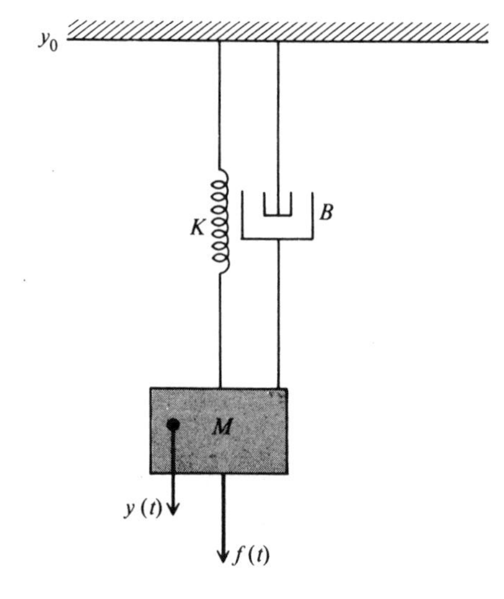

Figure 3.3 Force applied to a system containing a spring, a mass, and a damper.

and

$$x_2(t) = \dot{y}(t). \tag{3.21}$$

The resulting state equations are given by

$$\dot{x}_1(t = x_2(t), \tag{3.22}$$

$$\dot{x}_2(t) = -\frac{K}{M}x_1(t) - \frac{B}{M}x_2(t) + \frac{1}{M}f(t). \tag{3.23}$$

In phase-variable canonical form, these equations become

$$\dot{\mathbf{x}}(t) = \mathbf{P}\mathbf{x}(t) + \mathbf{B}r(t), \tag{3.24}$$

$$y(t) = \mathbf{L}\mathbf{x}(t), \tag{3.25}$$

where

$$\mathbf{x}(t) = \begin{bmatrix} x_1(t) \\ x_2(t) \end{bmatrix}, \quad \dot{\mathbf{x}}(t) = \begin{bmatrix} \dot{x}_1(t) \\ \dot{x}_2(t) \end{bmatrix}, \quad \mathbf{P} = \begin{bmatrix} 0 & 1 \\ -\frac{K}{M} & -\frac{B}{M} \end{bmatrix}, \quad \mathbf{B} = \begin{bmatrix} 0 \\ \frac{1}{M} \end{bmatrix},$$

$$r(t) = f(t), \quad \mathbf{L} = [1 \quad 0].$$

Assuming the system is initially at rest, the transfer function of this mechanical translation system, defined as the ratio of the output $Y(s)$ divided by the input $F(s)$, is readily found to be

$$\frac{Y(s)}{F(s)} = \frac{1/M}{s^2 + (B/M)s + K/M}, \tag{3.26}$$

and its simple block diagram appears in Figure 3.4.

2. *Force Applied to a Complex Mechanical System.* Consider next the complex mechanical system illustrated in Figure 3.5. A force $f(t)$ is applied to a mass, spring, and damper, which in turn applies a force to another mass, spring, and damper. Mass M_2 is displaced a distance $y_2(t)$, and mass M_1 is displaced a distance $y_1(t)$ with reference to y_0. In order to apply Newton's laws to this system, it is advantageous to first separate the mechanical system into a set of free-body diagrams as illustrated in Figure 3.6. Newton's equations for this resulting system are

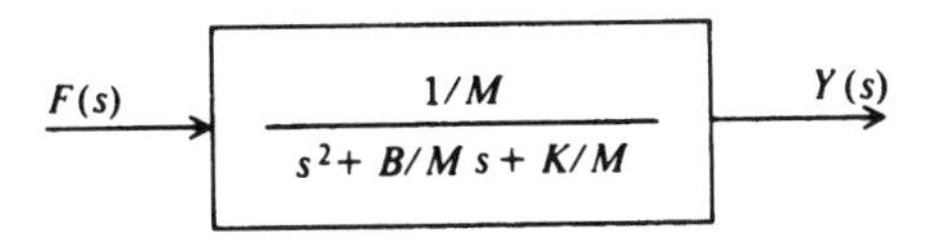

Figure 3.4 Block diagram for the spring-mass-damper system.

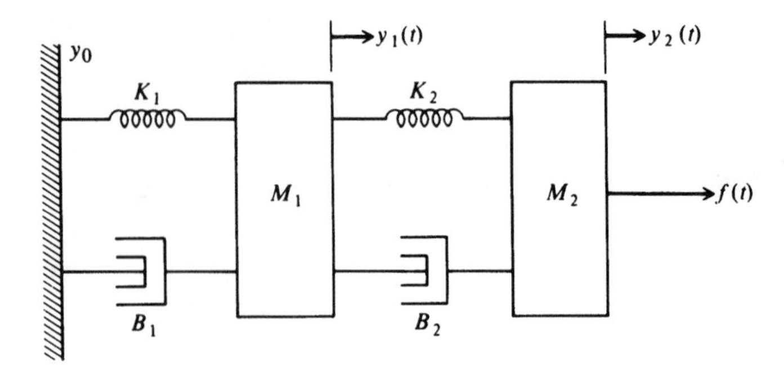

Figure 3.5 A mechanical system.

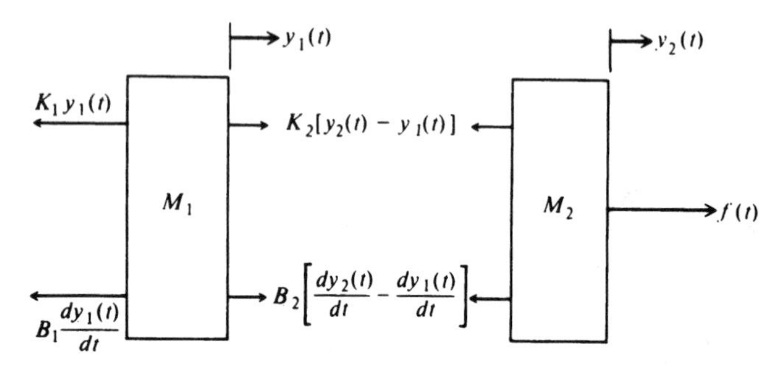

Figure 3.6 Free-body diagram for the system illustrated in Figure 3.5.

$$f(t) = M_2 \frac{d^2 y_2(t)}{dt^2} + B_2\left[\frac{dy_2(t)}{dt} - \frac{dy_1(t)}{dt}\right] + K_2[y_2(t) - y_1(t)], \tag{3.27}$$

$$\begin{aligned} 0 = B_2&\left[\frac{dy_2(t)}{dt} - \frac{dy_1(t)}{dt}\right] + K_2[y_2(t) - y_1(t)] \\ &- M_1 \frac{d^2 y_1(t)}{dt^2} - B_1 \frac{dy_1(t)}{dt} - K_1 y_1(t). \end{aligned} \tag{3.28}$$

These two second-order differential equations can be transformed into four first-order differential equations by defining

$$x_1(t) = y_1(t), \tag{3.29}$$

$$x_2(t) = y_2(t) \tag{3.30}$$

$$x_3(t) = \frac{dy_1(t)}{dt}, \tag{3.31}$$

$$x_4(t) = \frac{dy_2(t)}{dt}. \tag{3.32}$$

The resulting state equations are given by

$$\dot{x}_1(t) = x_3(t), \tag{3.33}$$

$$\dot{x}_2(t) = x_4(t), \tag{3.34}$$

$$\begin{aligned}\dot{x}_3(t) = &-\frac{1}{M_1}(K_1 + K_2)x_1(t) + \frac{K_2}{M_1}x_2(t) \\ &-\frac{1}{M_1}(B_1 + B_2)x_3(t) + \frac{B_2}{M_1}x_4(t),\end{aligned} \tag{3.35}$$

$$\dot{x}_4(t) = \frac{K_2}{M_2}x_1(t) - \frac{K_2}{M_2}x_2(t) + \frac{B_2}{M_2}x_3(t) - \frac{B_2}{M_2}x_4(t) + \frac{f(t)}{M_2}. \tag{3.36}$$

In phase-variable canonical form, these equations become

$$\dot{\mathbf{x}}(t) = \mathbf{Px}(\mathbf{t}) + \mathbf{B}r(\mathbf{t}), \tag{3.37}$$

$$\mathbf{y}(t) = \mathbf{Lx}(\mathbf{t}), \tag{3.38}$$

where

$$\mathbf{x}(t) = \begin{bmatrix} x_1(t) \\ x_2(t) \\ x_3(t) \\ x_4(t) \end{bmatrix}, \quad \dot{\mathbf{x}}(t) = \begin{bmatrix} \dot{x}_1(t) \\ \dot{x}_2(t) \\ \dot{x}_3(t) \\ \dot{x}_4(t) \end{bmatrix},$$

$$\mathbf{P} = \begin{bmatrix} 0 & 0 & 1 & 0 \\ 0 & 0 & 0 & 1 \\ -\frac{1}{M_1}(K_1 + K_2) & \frac{K_2}{M_1} & -\frac{1}{M_1}(B_1 + B_2) & \frac{B_2}{M_1} \\ \frac{K_2}{M_2} & -\frac{K_2}{M_2} & \frac{B_2}{M_2} & -\frac{B_2}{M_2} \end{bmatrix},$$

$$\mathbf{B} = \begin{bmatrix} 0 \\ 0 \\ 0 \\ \frac{1}{M_2} \end{bmatrix}, \quad \mathbf{L} = \begin{bmatrix} 1 & 0 & 0 & 0 \\ 0 & 1 & 0 & 0 \end{bmatrix}, \quad r(t) = f(t), \quad \mathbf{y}(t) = \begin{bmatrix} y_1(t) \\ y_2(t) \end{bmatrix}.$$

B. Mechanical Rotational Systems

The three basic characteristics of a mechanical rotational system are moment of inertia, stiffness, and damping. Rotational systems are quite similar to translational systems, except that torque equations are used to describe system equilibrium instead of force equations, and we use angular displacement, velocity, and acceleration quantities. The symbols for the various quantities used in our analysis are shown in Table 3.2.

Either the English set of units or the International System of units can be used for solution to the resulting equations. The primary thing to remember is to be consistent. Table 3.3 provides the conversion factors in going from the International System to the English set of units.

1. Torque Applied to a Body Having a Moment of Inertia, a Twisting Shaft, and a Damping Device. The configuration of a torque $T(t)$ applied to a body having a moment of inertia J, a twisting shaft having stiffness factor K, and a damper having a damping factor B, is shown in Figure 3.7. The system produces a displacement, $\theta(t)$, measured from an equilibrium position θ_0, which is assumed to be zero. The reference end of the damping device is assumed to be stationary. Applying Newton's law to this system yields

$$J\frac{d^2\theta(t)}{dt^2} + B\frac{d\theta(t)}{dt} + K\theta(t) = T(t). \tag{3.39}$$

Table 3.2. Mechanical Rotation Symbols

Quantity	Symbol
Angle	$\theta(t)$
Angular velocity	$\omega(t)$
Angular acceleration	$\alpha(t)$
Torque	$T(t)$
Moment of inertia	J
Damping factor	B
Stiffness factor	K

Table 3.3. Factors for Converting from the International System to the English Set of Units

Parameter	International System	Multiply By	English System
Speed	meters/second	2.2371	miles/hour
Force	newton	0.2248	pounds
Torque	kilogram meters	7.2307	foot pounds
Mass	kilograms	2.2046	pounds
Length	centimeter	0.0328	feet
Power	watts	0.0013	horsepower

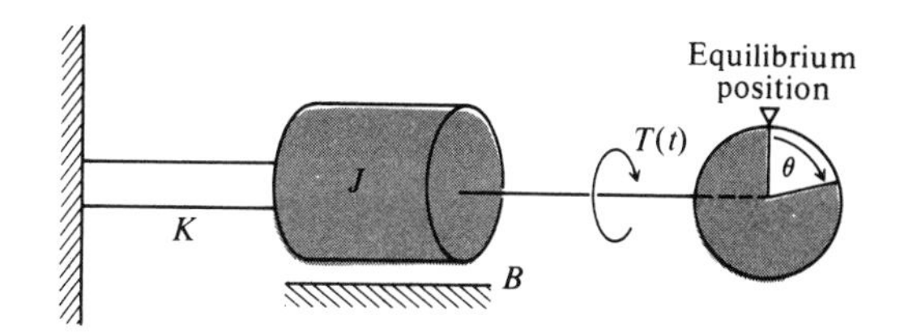

Figure 3.7 Torque applied to a moment of inertia, a twisting shaft, and a damping device.

The units of moment of inertia in the International System is kg m^2, and in the English system it is slug ft^2 or lb ft sec^2. The units of the spring constant are newton meters/radian in the International System, and lb ft/radian in the English system.

This second-order differential equation can be written as two first-order differential equations by defining

$$x_1(t) = \theta(t), \tag{3.40}$$

$$x_2(t) = \dot{\theta}(t). \tag{3.41}$$

The resulting state equations are given by

$$\dot{x}_1(t) = x_2(t), \tag{3.42}$$

$$\dot{x}_2(t) = -\frac{K}{J}x_1(t) - \frac{B}{J}x_2(t) + \frac{1}{J}T(t). \tag{3.43}$$

In vector form, these equations become

$$\dot{\mathbf{x}}(t) = \mathbf{Px(t)} + \mathbf{B}r(t), \tag{3.44}$$

$$\theta(t) = \mathbf{Lx(t)}, \tag{3.45}$$

where

$$\mathbf{x}(t) = \begin{bmatrix} x_1(t) \\ x_2(t) \end{bmatrix}, \quad \dot{\mathbf{x}}(t) = \begin{bmatrix} \dot{x}_1(t) \\ \dot{x}_2(t) \end{bmatrix}, \quad \mathbf{P} = \begin{bmatrix} 0 & 1 \\ -\frac{K}{J} & -\frac{B}{J} \end{bmatrix}, \quad \mathbf{B} = \begin{bmatrix} 0 \\ \frac{1}{J} \end{bmatrix},$$

$$\mathbf{L} = [1 \quad 0], \quad r(t) = T(t).$$

3.4. TRANSFER-FUNCTION AND STATE-VARIABLE REPRESENTATION OF TYPICAL ELECTROMECHANICAL CONTROL-SYSTEM DEVICES

This section illustrates the procedure used for deriving the transfer function and state-variable representation for commonly used devices from their basic differential equations. Examples of control systems which use electromechanical components are in robots, positioning systems such as tracking radars, guidance control systems, printer controls for computer tape and disk drive systems, and ship and aircraft control systems. An illustration of a control system which uses electromechanical components is illustrated in Figure 3.8 which represents the Operational Flight Trainer to simulate the operation of the MV-22A tilt-rotor, twin engine, vertical short takeoff and landing aircraft.

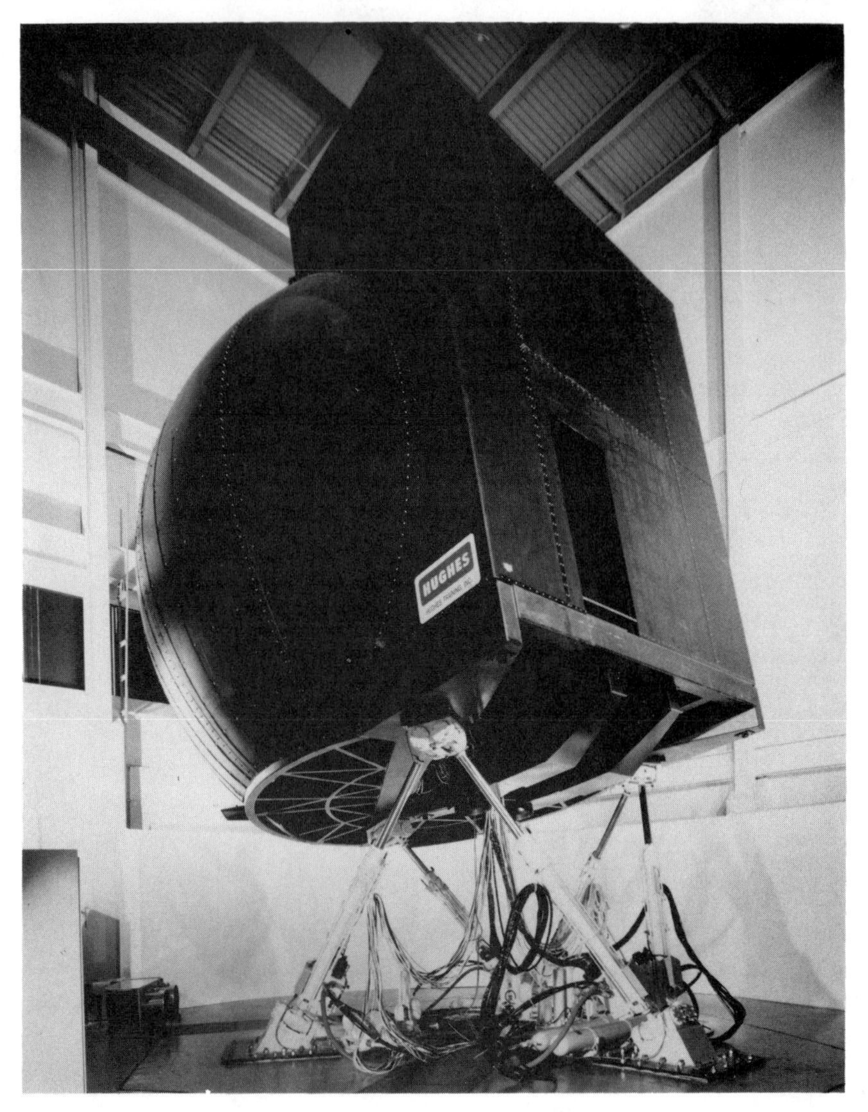

Figure 3.8 The Hughes Operational Flight Trainer simulates the MV-22A tilt-rotor, twin engine, vertical short takeoff and landing aircraft designed for assault support, troop lift, medical evacuation, and combat search and rescue. This simulator provides training for normal and emergency procedures, visual and instrument flight, and aerial refueling. Software-controlled hydraulic actuators provide six-degree-of-freedom motion to simulate rotary wings, fixed wings, and transition motion. A seat vibrator system simulates aerodynamic and proprotor vibration. (Courtesy of Hughes Training, Inc./ Arlington, TX).

We specifically analyze a dc generator, an armature-controlled dc servomotor, a Ward-Leonard system, a field-controlled dc servomotor, and a two-phase ac induction servomotor. The analysis is limited to the linear operating range of these devices.

A. dc Generator

A dc generator [4, 5] is commonly used in control systems for power amplification. The armature, which is driven at a constant speed n, is capable of producing a relatively large controllable current $i_a(t)$ as the field current $i_f(t)$ is varied.

The exact value of $i_a(t)$ is dependent on the load circuit impedance Z_L. A schematic diagram of the configuration is shown in Figure 3.9. The symbols R_f, L_f, R_g, and L_g represent the resistive and inductive components of the field and armature circuits, respectively.

The voltage induced in the armature, $e_g(t)$, is a function of the speed of rotation $n(t)$ and the flux developed by the field $\phi(t)$. It can be expressed as

$$e_g(t) = K_1 n(t)\phi(t). \tag{3.46}$$

The flux depends on the field current and the characteristics of the iron used in the field. This is a linear relationship up to a certain saturation point and can be expressed as

$$\phi(t) = K_2 i_f(t). \tag{3.47}$$

By substituting Eq. (3.47) into Eq. (3.46) and assuming that the armature rotation speed is constant, the relation between the induced armature voltage $e_g(t)$ and the field current $i_f(t)$ can be expressed as

$$e_g(t) = K_g i_f(t), \tag{3.48}$$

where $K_g = K_1 K_2 n$ = generator constant having units of V/A. The equation relating the field voltage $e_f(t)$ and field current $i_f(t)$ is

$$e_f(t) = R_f i_f(t) + L_f \frac{di_f(t)}{dt}. \tag{3.49}$$

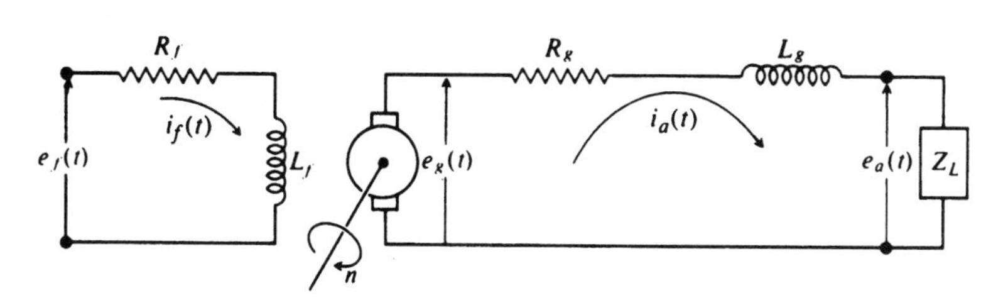

Figure 3.9 dc generator schematic diagram.

The field current can be eliminated between Eqs. (3.48) and (3.49) and an expression relating $e_f(t)$ and the induced armature voltage $e_g(t)$ can be obtained:

$$e_f(t) = \frac{R_f}{K_g} e_g(t) + \frac{L_f}{K_g} \frac{de_g(t)}{dt}. \tag{3.50}$$

By transforming, there results

$$E_f(s) = \frac{1}{K_g}(R_f + L_f s)E_g(s). \tag{3.51}$$

The transfer function of this device, defined as the ratio of the output $E_g(s)$ to the input $E_f(s)$, is given by

$$\frac{E_g(s)}{E_f(s)} = \frac{K_g/L_f}{s + R_f/L_f} \tag{3.52}$$

and its simple block diagram is shown in Figure 3.10.

If it is desired to obtain the transfer function $E_a(s)/E_f(s)$, we must first determine the nature of the actual load connected to the armature. For example, assume that the load impedance is $Z_L(s)$. Then the transfer function $E_a(s)/E_g(s)$ becomes

$$\frac{E_a(s)}{E_g(s)} = \frac{Z_L(s)}{R_g + L_g s + Z_L(s)}, \tag{3.53}$$

and the overall transfer function of the dc generator, $E_a(s)/E_f(s)$, is

$$\frac{E_a(s)}{E_f(s)} = \frac{E_g(s)}{E_f(s)} \times \frac{E_a(s)}{E_g(s)} = \frac{K_g/L_f}{s + R_f/L_f} \times \frac{Z_L(s)}{R_g + L_g s + Z_L(s)}. \tag{3.54}$$

B. Armature-Controlled dc Servomotor

Armature-controlled dc servomotors [6] are quite commonly used in control systems. As a matter of fact, a dc generator driving an armature-controlled dc servomotor is known as the Ward-Leonard system. We will study this configuration next drawing on the relations derived for both the dc generator and armature-controlled dc servomotor.

A schematic diagram of the armature-controlled dc servomotor is shown in Figure 3.11a. The symbols R_m and L_m represent the resistive and inductive compo-

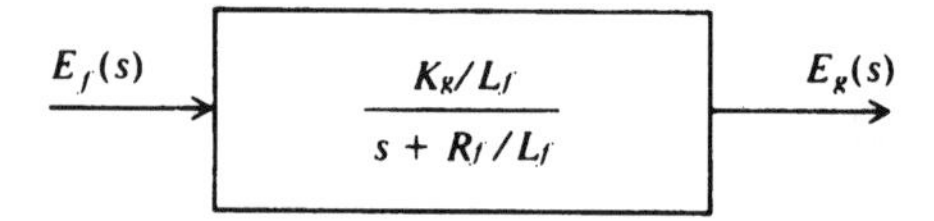

Figure 3.10 Block diagram of a dc generator.

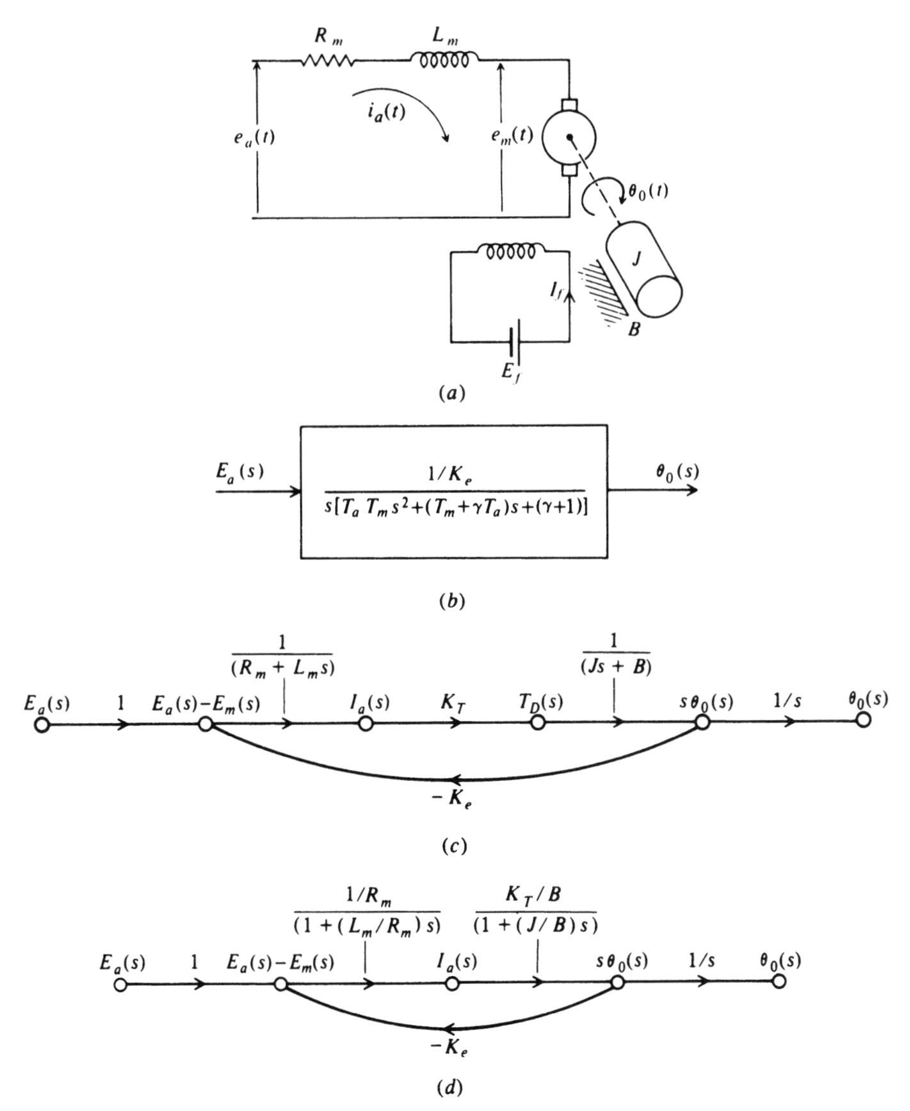

Figure 3.11 (*a*) Armature-controlled dc servomotor schematic diagram. (*b*) Block diagram of an armature-controlled dc motor. (*c*) Signal-flow graph representation for the armature-controlled dc servomotor. (*d*) Modification of the signal-flow graph to a more compact form.

nents of the armature circuit. The field excitation is constant, being supplied from a dc source. The motor is shown driving a load having an inertia J and damping B.

As the armature rotates, it develops an induced voltage $e_m(t)$ which is in a direction opposite to $e_a(t)$. The induced voltage is proportional to the speed of rotation $n(t)$ and the flux created by the field current. Because we are assuming that the field current is held constant, the flux must be constant. Therefore, the induced armature voltage is only dependent on the speed of rotation and can be expressed as

$$e_m(t) = K_e n(t) = K_e \frac{d\theta_0(t)}{dt}, \tag{3.55}$$

where K_e = voltage constant of the motor having units of V/(rad/sec). The voltage equation of the armature circuit is

$$e_a(t) = R_m i_a(t) + L_m \frac{di_a(t)}{dt} + e_m(t). \tag{3.56}$$

Substituting Eq. (3.55) into Eq. (3.56) and taking the Laplace transform, we obtain

$$E_a(s) = (R_m + L_m s)I_a(s) + K_e s\,\theta_0(s). \tag{3.57}$$

The developed torque of the motor, $T_D(t)$, is a function of the flux developed by the field current, the armature current, and the length and number of the conductors. Because we are assuming that the field current is held constant, the developed torque $T_D(t)$ can be expressed as

$$T_D(t) = K_T i_a(t), \tag{3.58}$$

where K_T = torque constant of the motor. The units of K_T in the International System are newton meter/ampere, and in the English system are lb ft/ampere.

The developed torque drives the mechanical load and the torque equation is

$$T_D(t) = J\frac{d^2\theta_0(t)}{dt^2} + B\frac{d\theta_0(t)}{dt}. \tag{3.59}$$

Substituting Eq. (3.58) into Eq. (3.59) and taking the Laplace transform, we obtain

$$K_T I_a(s) = (Js^2 + Bs)\theta_0(s). \tag{3.60}$$

The overall system transfer function $\theta_0(s)/E_a(s)$ obtained by eliminating $I_a(s)$ between Eqs. (3.57) and (3.60) is

$$\frac{\theta_0(s)}{E_a(s)} = \frac{K_T}{JL_m s^3 + (R_m J + L_m B)s^2 + (R_m B + K_e K_T)s}. \tag{3.61}$$

Equation (3.61) can be defined in terms of an armature time constant T_a, a motor time constant T_m, and a damping factor γ, as follows:

$$\frac{\theta_0(s)}{E_a(s)} = \frac{1/K_e}{s[T_a T_m s^2 + (T_m + \gamma T_a)s + (\gamma + 1)]}, \tag{3.62}$$

where

$$T_a = \frac{L_m}{R_m}, \quad T_m = \frac{JR_m}{K_e K_T}, \quad \gamma = \frac{R_m B}{K_e K_T}. \tag{3.63}$$

The simple block diagram of this system is illustrated in Figure 3.11*b*.

The state-variable model is derived directly for the armature-controlled dc servomotor from the original set of defining scalar equations, (3.55), (3.56), (3.58), and (3.59). Defining the state variables as

$$x_1(t) = \theta_0(t), \tag{3.64}$$

$$x_2(t) = \dot{\theta}_0(t), \tag{3.65}$$

$$x_3(t) = i_a(t), \tag{3.66}$$

the resulting state equations are given by

$$\dot{x}_1(t) = x_2(t), \tag{3.67}$$

$$\dot{x}_2(t) = -\frac{B}{J}x_2(t) + \frac{K_T}{J}x_3(t), \tag{3.68}$$

$$\dot{x}_3(t) = -\frac{K_e}{L_m}x_2(t) - \frac{R_m}{L_m}x_3(t) + \frac{1}{L_m}e_a(t). \tag{3.69}$$

In phase-variable canonical form, these equations become

$$\dot{\mathbf{x}}(t) = \mathbf{P}\mathbf{x}(t) + \mathbf{B}r(t), \tag{3.70}$$

$$\mathbf{c}(t) = \mathbf{L}\mathbf{x}(t), \tag{3.71}$$

where

$$\mathbf{x} = \begin{bmatrix} x_1 \\ x_2 \\ x_3 \end{bmatrix}, \quad \dot{\mathbf{x}} = \begin{bmatrix} \dot{x}_1 \\ \dot{x}_2 \\ \dot{x}_3 \end{bmatrix}, \quad \mathbf{B} = \begin{bmatrix} 0 \\ 0 \\ \dfrac{1}{L_m} \end{bmatrix}, \quad \mathbf{P} = \begin{bmatrix} 0 & 1 & 0 \\ 0 & -\dfrac{B}{J} & \dfrac{K_T}{J} \\ 0 & -\dfrac{K_e}{L_m} & -\dfrac{R_m}{L_m} \end{bmatrix},$$

$$\mathbf{c}(t) = \begin{bmatrix} \theta_0(t) \\ i_a(t) \end{bmatrix}, \quad \mathbf{L} = \begin{bmatrix} 1 & 0 & 0 \\ 0 & 0 & 1 \end{bmatrix}, \quad r(t) = e_a(t).$$

It is usually quite interesting and revealing to study the signal-flow graph for such a device, especially for purposes of analog computer simulation. The governing equations from which the signal-flow graph can be drawn for the armature-controlled dc servomotor are given by

$$E_m(s) = K_e s\theta_0(s) \qquad \text{[from Eq. (3.55)]}, \tag{3.72}$$

$$I_a(s) = \frac{E_a(s) - E_m(s)}{R_m + L_m s} \qquad \text{[from Eq. (3.56)]}, \tag{3.73}$$

$$T_D(s) = K_T I_a(s) \qquad \text{[from Eq. (3.58)]}, \tag{3.74}$$

$$\theta_0(s) = \frac{1}{s}\frac{T_D(s)}{Js + B} \qquad \text{[from Eq. (3.59)]}. \tag{3.75}$$

The signal-flow graph is illustrated in Figure 3.11*c*. It clearly illustrates the inherent feedback (back electromotive force) of this device. This property is sometimes used to stabilize a feedback control system (see Problem 3.8). It is left as an exercise to the reader to prove that the transfer function of this system, as derived from the signal-flow graph, agrees with Eqs. (3.62) and (3.63) and that the state equations derived from the state-variable diagram agree with Eqs. (3.70) and (3.71) (see Problem 3.28). For the case of the armature-controlled dc servomotor, the signal-flow graph of Figure 3.11*c* is modified to a more compact form in Figure 3.11*d*.

C. The Ward-Leonard System

A configuration having a dc generator driving an armature-controlled dc motor is known as a Ward-Leonard system [7, 8]. The dc generator acts as a rotating power amplifier that supplies the power which, in turn, drives the servomotor. Variations of the conventional Ward-Leonard system are known as the Amplidyne, the Metadyne, the Rotorol, and the Regulex. The reader is referred to more authoritative books on dc machinery [7, 8] for a description of these devices.

A schematic diagram of the basic Ward-Leonard system is shown in Figure 3.12. The notations used are the same as those of Figures 3.9 and 3.11. Many sophisticated variations of this basic configuration, using compensating windings, exist [8].

To enable us to combine the transfer-function relationships derived previously for the dc generator and armature-controlled dc motor, we assume that the generator voltage $e_g(t)$ is applied to the armature of the motor. Therefore, we are interested in applying the generator equation (3.52) rather than (3.54). In order to apply the motor transfer function (3.62), we must first combine the resistive and inductive components of the generator's and motor's armatures. It is assumed that R_g is much smaller than R_m, and L_g is much smaller than L_m. This will result in a set of new, modified motor constants as follows:

$$\gamma' = \frac{(R_g + R_m)B}{K_e K_t}, \quad T'_a = \frac{L_g + L_m}{R_g + R_m}, \quad T'_m = \frac{J(R_g + R_m)}{K_e K_T}. \tag{3.76}$$

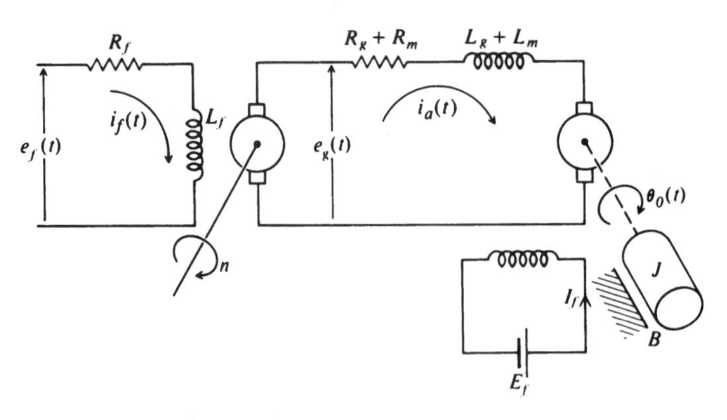

Figure 3.12 Ward-Leonard system schematic diagram.

Therefore, Eq. (3.62) may be written as follows in this case:

$$\frac{\theta_0(s)}{E_g(s)} = \frac{1/K_e}{s[T_a' T_m' s^2 + (T_m' + \gamma' T_a')s + (\gamma' + 1)]}. \tag{3.77}$$

It is now relatively simple to obtain the transfer-function representation of the configuration shown in Figure 3.12. Defining $e_f(t)$ as the input and $\theta_0(t)$ as the output of the system, we need merely to combine the transfer functions given by Eqs. (3.52) and (3.77). Therefore, the system transfer function is given by

$$\frac{\theta_0(s)}{E_f(s)} = \frac{K_g/L_f}{s + R_f/L_f} \frac{1/K_e}{s[T_a' T_m' s^2 + (T_m' + \gamma' T_a')s + (\gamma' + 1)]}, \tag{3.78}$$

and its simple block diagram is illustrated in Figure 3.13.

D. Field-Controlled dc Servomotor

A dc servomotor [9] can also be controlled by varying the field current and maintaining a constant armature current. The schematic diagram of such a configuration, known as the field-controlled dc servomotor, is shown in Figure 3.14. The notations are the same as those of Figure 3.11.

The developed torque of the motor, $T_D(t)$, is a function of the flux developed by the armature current, the field current, and the length of the conductors. Because we are assuming that the armature current is held constant, the developed torque $T_D(t)$ can be expressed as

$$T_D(t) = K_T i_f(t), \tag{3.79}$$

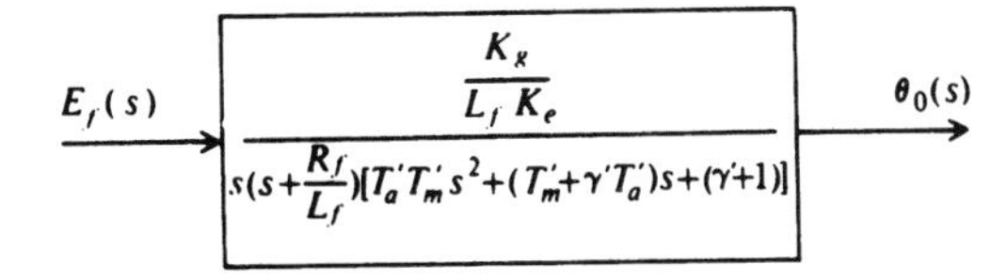

Figure 3.13 Block diagram of the Ward-Leonard system.

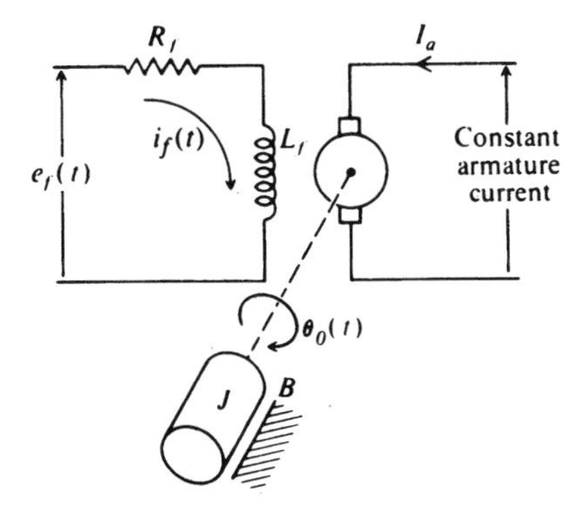

Figure 3.14 Field-controlled dc servomotor schematic diagram.

where K_T = torque constant of the motor having units of lb ft/A. The developed torque is used to drive the mechanical load with inertia J and damping B. The torque equation is

$$T_D(t) = J\frac{d^2\theta_0(t)}{dt^2} + B\frac{d\theta_0(t)}{dt}. \tag{3.80}$$

Substituting Eq. (3.79) into Eq. (3.80), we obtain a differential equation relating the field current and the output shaft position:

$$i_f(t) = \frac{J}{K_T}\frac{d^2\theta_0(t)}{dt^2} + \frac{B}{K_T}\frac{d\theta_0(t)}{dt}. \tag{3.81}$$

An expression for the value of the field current $i_f(t)$ can be obtained from the voltage equation of the field circuit:

$$e_f(t) = R_f i_f(t) + L_f\frac{di_f(t)}{dt}. \tag{3.82}$$

Substituting Eq. (3.81) into Eq. (3.82) and taking the Laplace transform, we obtain

$$E_f(s) = (R_f + L_f s)\left(\frac{J}{K_T}s^2 + \frac{B}{K_T}s\right)\theta_0(s). \tag{3.83}$$

The transfer function of the device, defined as the output $\theta_0(s)$ divided by the input $E_f(s)$, is given by

$$\frac{\theta_0(s)}{E_f(s)} = \frac{K_T/R_f B}{s(1 + T_f s)(1 + T_m s)}, \tag{3.84}$$

where

$$T_f = \frac{L_f}{R_f} = \text{ field time constant,}$$

$$T_m = \frac{J}{B} = \text{ motor time constant,}$$

and its simple block diagram is illustrated in Figure 3.15*a*.

The governing equations from which the signal-flow graph can be drawn for the field-controlled dc servomotor are given by

$$I_f(s) = \frac{E_f(s)}{R_f + L_f s} \quad \text{[from Eq. (3.82)],} \tag{3.85}$$

$$T_D(s) = K_T I_f(s) \quad \text{[from Eq. (3.79)],} \tag{3.86}$$

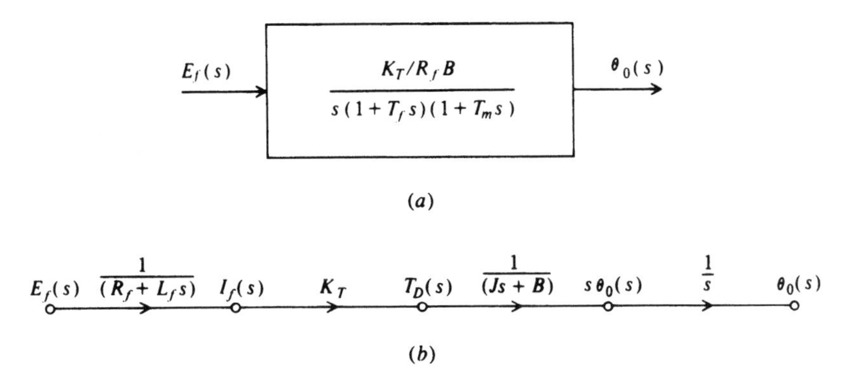

Figure 3.15 (*a*) Block diagram of a field-controlled dc servomotor. (*b*) Signal-flow graph representation for the field-controlled dc servomotor.

$$\theta_0(s) = \frac{1}{s}\frac{T_D(s)}{Js+B} \quad \text{[from Eq. (3.80)]}. \tag{3.87}$$

The simple signal-flow graph of this device is illustrated in Figure 3.15*b*.

The state-variable equations of the field-controlled dc servomotor can be obtained either by starting from the defining scalar equations, (3.81) and (3.82), or by utilizing the resultant transfer function derived in Eq. (3.84). The first method is usually preferable. However, because the first method was utilized in the case of the armature-controlled dc servomotor, the second method is utilized here for the field-controlled dc servomotor to permit comparison of the two approaches.

Equation (3.84) can be written in differential-equation form as follows:

$$T_fT_m\frac{d^3\theta_0(t)}{dt^3} + (T_f+T_m)\frac{d^2\theta_0(t)}{dt^2} + \frac{d\theta_0(t)}{dt} = \frac{K_T}{R_fB}e_f(t). \tag{3.88}$$

This third-order differential equation can be written as three first-order differential equations by defining

$$x_1(t) = \theta_0(t), \tag{3.89}$$

$$x_2(t) = \dot{\theta}_0(t), \tag{3.90}$$

$$x_3(t) = \ddot{\theta}_0(t). \tag{3.91}$$

The resulting state equations are given by

$$\dot{x}_1(t) = x_2(t), \tag{3.92}$$

$$\dot{x}_2(t) = x_3(t), \tag{3.93}$$

$$\dot{x}_3(t) = -\frac{1}{T_fT_m}x_2(t) - \frac{(T_f+T_m)}{T_fT_m}x_3(t) + \frac{K_T}{R_fBT_fT_m}e_f(t). \tag{3.94}$$

In phase-variable canonical form, these equations become

$$\dot{\mathbf{x}}(t) = \mathbf{P}\mathbf{x}(t) + \mathbf{B}r(t), \tag{3.95}$$

$$\theta_0(t) = \mathbf{L}\mathbf{x}(t), \tag{3.96}$$

where

$$\mathbf{x}(t) = \begin{bmatrix} x_1(t) \\ x_2(t) \\ x_3(t) \end{bmatrix}, \quad \dot{\mathbf{x}} = \begin{bmatrix} \dot{x}_1(t) \\ \dot{x}_2(t) \\ \dot{x}_3(t) \end{bmatrix}, \quad \mathbf{P} = \begin{bmatrix} 0 & 1 & 0 \\ 0 & 0 & 1 \\ 0 & -\dfrac{1}{T_f T_m} & -\dfrac{(T_f + T_m)}{T_f T_m} \end{bmatrix},$$

$$\mathbf{B} = \begin{bmatrix} 0 \\ 0 \\ \dfrac{K_T}{R_f B T_f T_m} \end{bmatrix},$$

$$\mathbf{L} = [1 \quad 0 \quad 0], \quad r(t) = e_f(t).$$

Notice that the resultant vector equation has eliminated the field current. This is a measurable state and it is preferable to retain it in the state-variable model. If the state-variable representation had proceeded from the defining scalar equations, (3.81) and (3.82), then the field current would have been retained—just as the armature current was retained as a state in the case of the armature-controlled dc servomotor (see Eq. (3.66)].

E. Two-Phase ac Servomotor

The two-phase ac servomotor [10–12] is probably the most commonly used type of servomotor. Its popularity stems from the fact that many error-sensing devices are carrier-frequency (ac) devices. By using two-phase ac servomotors, demodulation need not be performed and ac amplification can be used throughout the electrical portion of the system.

The ac servomotor is a two-phase induction motor having its two stator coils separated by 90 electrical degrees with a high-resistance rotor. A control signal is applied to one phase (the control winding) while the other phase (the reference winding) is supplied with a fixed signal that is phase-shifted by 90° relative to the control signal. The motor is used primarily for relatively low-power applications. A schematic diagram of an ac servomotor driving a load of inertia J and damping B is shown in Figure 3.16. The reference field voltage is denoted as $e_r(t)$ and the control field voltage is denoted as $e_c(t)$. The developed torque of this motor is proportional to $e_r(t)$, $e_c(t)$, and the sine of the angle between $e_r(t)$ and $e_c(t)$. In practice, the reference winding voltage is kept constant at its rated value, the angle between the reference winding and control winding is kept close to 90° by means of a large capacitor placed in series with the reference winding, and the input to the control winding is varied for control purposes.

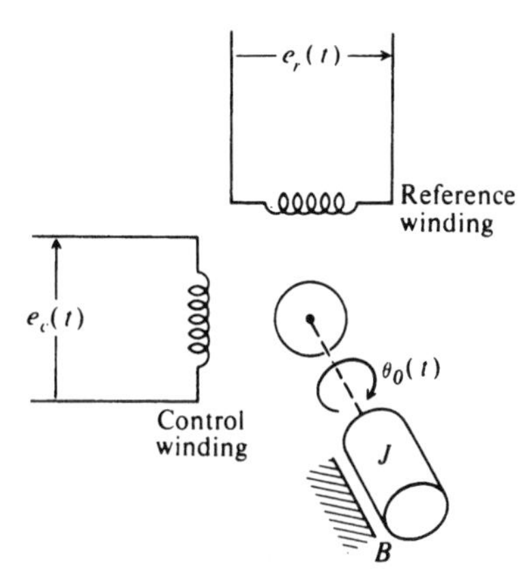

Figure 3.16 Two-phase ac servomotor schematic diagram.

As the control voltage is varied, the torque T_D and speed $n(t)$ vary. A set of torque-speed curves for various values of control voltage are shown in Figure 3.17. Notice that these curves show a very large torque for zero speed which is desirable in developing a very rapid acceleration.

Unfortunately, the torque-speed curves are not straight lines. Therefore, we cannot write a linear differential equation to represent them. However, by linearizing these characteristics, reasonable accuracy can be achieved.* Because the developed torque T_D is a function of the speed $n(t)$ and the control voltage $e_c(t)$, we use a Taylor-series expansion of the nonlinear functin $T_D(n, e_c)$ to get an approximate linear function:

$$\frac{\partial T_D(n, e_c)}{\partial n} n(t) + \frac{\partial T_D(n, e_c)}{\partial e_c} e_c(t) = T_D(n, e_c). \tag{3.97}$$

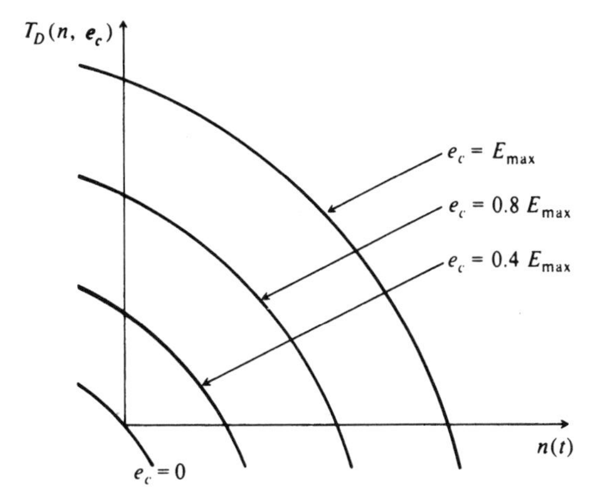

Figure 3.17 Two-phase ac servomotor torque-speed characteristics.

* The subject of linearizing approximations is discussed from a more general viewpoint in Section 10.6.

By defining

$$\frac{\partial T_D(n, e_c)}{\partial e_c} \equiv K_e, \quad \frac{\partial T_D(n, e_c)}{\partial n} \equiv K_n \quad \text{(where } K_n \text{ is a negative number)}$$

and substituting $n(t) = d\theta_0(t)/dt$, we can rewrite Eq. (3.97) as

$$K_n \frac{d\theta_0(t)}{dt} + K_e e_c(t) = T_D(n, e_c). \tag{3.98}$$

The developed torque is used to drive the mechanical load and the torque equation is

$$T_D(n, e_c) = J \frac{d^2\theta_0(t)}{dt^2} + B \frac{d\theta_0(t)}{dt}. \tag{3.99}$$

Substituting Eq. (3.99) into Eq. (3.98) and taking the Laplace transform, we obtain

$$K_n s\theta_0(s) + K_e E_c(s) = Js^2\theta_0(s) + Bs\theta_0(s). \tag{3.100}$$

The transfer function of the device, defined as the output $\theta_0(s)$ divided by the input $E_c(s)$, is given by

$$\frac{\theta_0(s)}{E_c(s)} = \frac{K_m}{s(T_m s + 1)}, \tag{3.101}$$

where

$$K_m = \frac{K_e}{B - K_n} = \text{motor constant,}$$

$$T_m = \frac{J}{B - K_n} = \text{motor time constant,}$$

and its simple block diagram is illustrated in Figure 3.18. Although the torque-speed characteristics are nonlinear, as illustrated in Figure 3.17, the value of K_n is usually obtained graphically by drawing a straight line through the two end points. The accuracy of this approach, compared with drawing a line tangent to the curve at various other points, is analyzed further in Chapter 10, Section 10.6 (see Problem 10.1).

In practice, B is usually very small compared to K_n and can be assumed to be negligible.

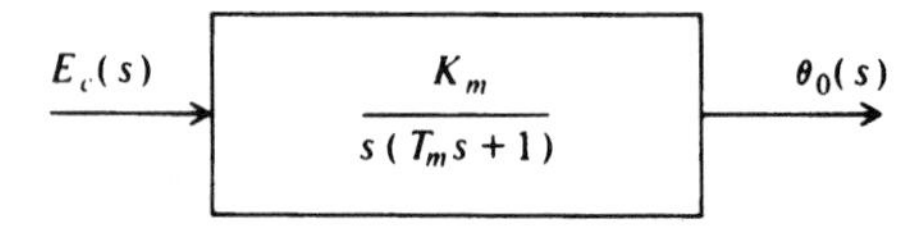

Figure 3.18 Block diagram of a two-phase ac servomotor.

A very important characteristic of the two-phase ac servomotor is the torque-to-inertia ratio which provides a measure of the maximum acceleration that the device can achieve. It is defined as the ratio of the maximum torque at standstill (for the specified control field voltage, $e_c(t)$) to the rotor moment of inertia. The larger the ratio, then the better is the acceleration characteristic of the two-phase ac servomotor.

Equation (3.101) can be written in differential-equation form as follows:

$$T_m \frac{d^2\theta_0(t)}{dt^2} + \frac{d\theta_0(t)}{dt} = K_m e_c(t). \tag{3.102}$$

This second-order differential equation can be written as two first-order differential equations by defining

$$x_1(t) = \theta_0(t), \tag{3.103}$$

$$x_2(t) = \dot{\theta}_0(t). \tag{3.104}$$

The resulting state equations are given by

$$\dot{x}_1(t) = x_2(t), \tag{3.105}$$

$$\dot{x}_2(t) = -\frac{x_2(t)}{T_m} + \frac{K_m}{T_m} e_c(t). \tag{3.106}$$

In phase-variable canonical form, these equations become

$$\dot{\mathbf{x}}(t) = \mathbf{P}\mathbf{x}(t) + \mathbf{B}r(t), \tag{3.107}$$

$$\theta_0(t) = \mathbf{L}\mathbf{x}(t), \tag{3.108}$$

where

$$\mathbf{x}(t) = \begin{bmatrix} x_1(t) \\ x_2(t) \end{bmatrix}, \quad \dot{\mathbf{x}}(t) = \begin{bmatrix} \dot{x}_1(t) \\ \dot{x}_2(t) \end{bmatrix}, \quad \mathbf{P} = \begin{bmatrix} 0 & 1 \\ 0 & -\dfrac{1}{T_m} \end{bmatrix}, \quad \mathbf{B} = \begin{bmatrix} 0 \\ \dfrac{K_m}{T_m} \end{bmatrix},$$

$$\mathbf{L} = [1 \quad 0], \quad r(t) = e_c(t).$$

It is left as an exercise to the reader to prove that the state equations derived from the state-variable diagram agree with Eqs. (3.107) and (3.108) (see Problem 3.30).

F. Servomotors with Gear Reducers

The two-phase ac servomotor is inherently a high-speed, low-torque device. In control-system practice, however, what is usually needed is a low-speed, high-torque device. Therefore, a gear reduction is usually required between the high-speed, low-torque servomotor and the load to obtain speed reduction and torque magnification.

In order to analyze the modifications necessary to the transfer function derived in Eq. (3.101), let us consider the configuration illustrated in Figure 3.19. The sub-

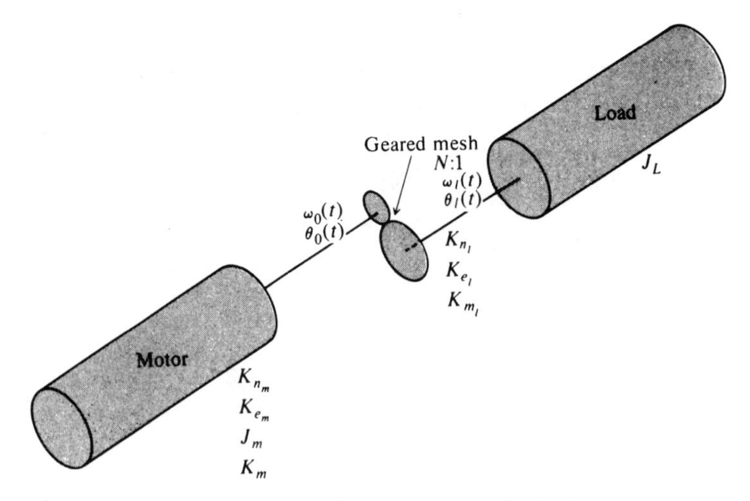

Figure 3.19 Servomotor geared to load.

scripts m and l are used to denote motor and load shafts, respectively. It is assumed initially that the inertias of the gears are negligible, the efficiency of transmission is perfect, and there is only one gear mesh between the load and the motor. It is also assumed that a servomotor and gear train have been selected which will provide the required torque to obtain the desired acceleration, and to drive the load at the maximum required speed. Basically what is desired is a set of relationships to include the effect of the gear reduction and the load inertia.

There are basically two approaches to the problem. In one method, the parameters K_{n_m}, K_{e_m}, and J_m are reflected to the output shaft, θ_l. Another manner of solving the problem is to reflect the load inertia to the motor shaft. Both methods will be illustrated here, and it will be shown that they are equivalent.

Let us first consider the reflection of the paramters K_{n_m}, K_{e_m}, and J_m to the output shaft θ_l. K_{n_m} is the ratio of motor torque to motor speed and the value of K_{n_m} is given by

$$K_{n_m} = \frac{\partial T_D(n, e_c)}{\partial n}. \tag{3.109}$$

Because the driving torque at the load is multiplied by the gear-reduction factor N, and the motor speed is reduced by this factor at the load shaft, the following relationship is obtained:

$$K_{n_l} = \frac{N}{1/N} \frac{\partial T_D(n, e_c)}{\partial n} = N^2 K_{n_m}, \tag{3.110}$$

where K_{n_l} denotes K_n referred to the load shaft.

The parameter K_{e_m} is the ratio of motor torque to control voltage. Because the driving torque of the load is multiplied by the gear reduction factor N, and the control voltage $e_c(t)$ is independent of N,

$$K_{e_l} = NK_{e_m}, \tag{3.111}$$

where K_{e_m} indicates the value of K_e at the motor shaft and K_{e_l} denotes its value referrd to the load shaft. The effect of the motor inertia at the load shaft can be obtained by considering the conservation of energy equation:

$$\tfrac{1}{2} J_m \omega_0^2(t) = \tfrac{1}{2} J_{m_r} \omega_l^2(t), \tag{3.112}$$

where J_{m_r} denotes the value of the motor inertia referred to the output shaft. On substitution, the following is obtained:

$$\tfrac{1}{2} J_m \omega_0^2(t) = \tfrac{1}{2} J_{m_r} \left(\frac{\omega_0(t)}{N} \right)^2. \tag{3.113}$$

Therefore,

$$J_{m_r} = J_m N^2. \tag{3.114}$$

This relationship indicates that the motor inertia referred to the output shaft is amplified by a factor of N^2. This factor must be added to the load inertia J_L in order to obtain the total output inertia:

$$J_{\text{total}_L} = J_m N^2 + J_L.$$

Based on these relationships, the values of K_m and T_m for the overall combination of servomotor, gear train, and load inertia can now be obtained. To do this, it is assumed that the damping factor B in Eq. (3.101) is zero (it is usually negligible compared with the motor damping K_n):

$$K_{m_l} = \frac{N K_{e_m}}{-N^2 K_{n_m}} = \frac{1}{N} K_m, \quad K_{n_m} < 0 \tag{3.115}$$

where K_{m_l} denotes the value of K_m referred to the output shaft. This indicates that the system gain is reduced by a factor of N. Furthermore,

$$T_{m_{\text{total}}} = \frac{J_m N^2 + J_L}{-N^2 K_{n_m}} = T_m - \frac{J_L}{N^2 K_{n_m}}, \quad K_{n_m} < 0. \tag{3.116}$$

This indicates an increase in the servomotor time constant due to the added load inertia. Therefore, the net effect of the gear reduction and the load inertia has been to reduce K_m by a factor of N, and to increase the time constant by the added term $J_L/N^2 K_{n_m}$. Therefore, the transfer function of the two-phase ac servomotor with a gear reduction N and a load inertia J_L is given by

$$\frac{\theta_l(s)}{E_c(s)} = \frac{K_m/N}{s(T_{m_{\text{total}}} s + 1)}. \tag{3.117}$$

Now let us approach the problem by referring all quantities to the motor shaft and check the result. For this case, the only two factors that have to be considered are the effect of the gear reduction N and the load inertia J_L. As we have just observed, the effect of the gear reduction N is basically a loss in system gain N. Therefore, it must be treated as an element having a transfer function $1/N$. What about the load inertia J_L? Let us reconsider Eq. (3.112) and reflect the load inertia to the servomotor:

$$\tfrac{1}{2} J_{L_r}(N\omega_l(t))^2 = \tfrac{1}{2} J_L \omega_l^2(t),$$

where J_{L_r} is the load inertia referred to the motor shaft. Therefore,

$$J_{L_r} = \frac{J_L}{N^2}.$$

This indicates that the load inertia at the motor shaft is reduced by a factor of N^2. We are now ready to determine the values of K_m and T_m of the servomotor and gear-reduction combination. As before, B is assumed to be zero. Therefore,

$$K_{m_l} = \frac{1}{N} K_m, \tag{3.118}$$

and

$$T_{m_{\text{total}}} = \frac{J_m}{-K_{n_m}} + \frac{J_L}{-N^2 K_{n_m}} = T_m - \frac{J_L}{N^2 K_{n_m}}, \quad K_{n_m} < 0. \tag{3.119}$$

Observe the agreement between Eqs. (3.115) and (3.118) and between Eqs. (3.116) and 3.119).

The results derived here can easily be extended to multimesh gear trains as illustrated in Figure 3.20. In this figure, a two-mesh system is illustrated, and the inertia of the respective shafts and gearing combinations are indicated as J_1, J_2, and J_3. It is left as an exercise to the reader to show that the equivalent inertia reflected to the input (motor) shaft is

$$J_{\text{eq}_i} = J_1 + \frac{J_2}{N_1^2} + \frac{J_3}{N_1^2 N_2^2},$$

and the equivalent inertia reflected to the output shaft is given by

$$J_{\text{eq}_0} = J_1 N_1^2 N_2^2 + J_2 N_2^2 + J_3.$$

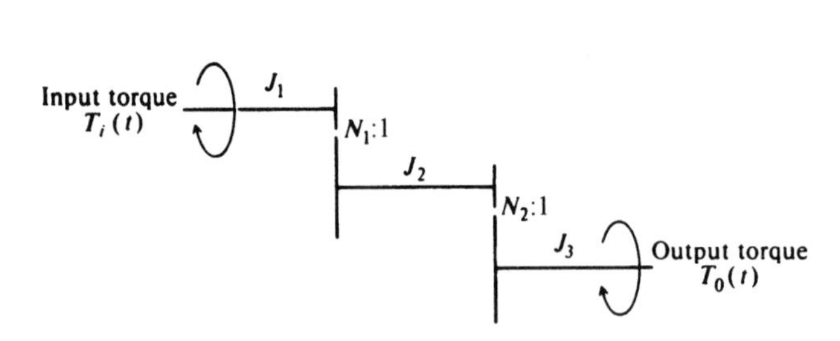

Figure 3.20 A multimesh gear train.

Before leaving the subject of gear ratios, the question is posed as to what is the optimum gear ratio for a given motor and load configuration? This problem can be answered by reconsidering the servomotor, gear train, and load configuration ilustrated in Figure 3.19. The torque developed by the motor (at the motor shaft) in accelerating its own inertia, J_m, and the load inertia, J_L, through the gear ratio N, is given by

$$T_D(n, e_c) = (J_m)\alpha_l(t)N + \left(\frac{J_L}{N^2}\right)\alpha_l(t)N, \tag{3.120}$$

where $\alpha_l(t)$ represents the maximum required acceleration of the load. Equation (3.120) represents the developed torque at the motor shaft with the load inertia component appropriately reflected to it. In order to find the optimum gear ratio N which will minimize the required motor torque and therefore the size of the motor, Eq. (3.120) is differentiated with repect to N and is set equal to zero:

$$\frac{d}{dN}[(J_m)\alpha_l(t)N + (J_L)\alpha_l(t)N^{-1}] = 0$$

$$\alpha_l(t)\left[J_m - \frac{J_L}{N^2}\right] = 0.$$

Therefore, the gear ratio required to minimize the motor torque is given by

$$N = \sqrt{\frac{J_L}{J_m}}, \tag{3.121}$$

where the gear ratio N is the square root of the ratio of the load to motor inertias. Equation (3.121) may also be interpreted from the viewpoint that the value of motor inertia for minimum motor torque required is given by

$$J_m = \frac{J_L}{N^2}. \tag{3.122}$$

Therefore, Eq. (3.122) states that the motor torque is minimized when the motor and load inertias are equal when referred to a common shaft.

Note that this result assumes that the only factor that is important is acceleration of the load. If a maximum load velocity is specified and the motor can only reach a given maximum velocity, the choice of optimum gear ratio is different.

3.5. TRANSFER-FUNCTION AND STATE-VARIABLE REPRESENTATION OF TYPICAL HYDRAULIC DEVICES

Hydraulic components are commonly found in control systems that are either all hydraulic or a combination of electromechanical and hydraulic devices [6, 13]. The procedures used for deriving the transfer function representation of some commonly used hydraulic control system devices from their basic differential equations are illustrated in this section. We specifically consider hydraulic motors, pumps, and valves.

A. Hydraulic Motor and Pump

There is no essential difference between a hydraulic pump and motor, just as there is no essential difference between a dc generator and a dc motor. Basically, the hydraulic device is classified as a motor if the input is hydraulic flow or pressure and the output is mechanical position; or a pump if the input is mechanical torque and the output is hydraulic flow or pressure.

Figure 3.21 illustrates a commonly used hydraulic power transmission system. This device, which is capable of controlling large torques, consists of a variable displacement pump that is driven at a constant speed. A control stroke, which determines the quantity of oil pumped, also controls the direction of fluid flow. The angular velocity of the hydraulic motor is proportional to the volumetric flow and is in the same direction as the oil flow from the pump. A functional diagram of the hydraulic transmission is illustrated in Figure 3.22.

The amount of oil displaced per revolution of the hydraulic pump is a function of the tilt angle $\theta_p(t)$. When $\theta_p(t) = 0°$, there is no flow in the oil lines. As $\theta_p(t)$ is increased in the positive diretion, more oil flows in the lines with the direction shown. When $\theta_p(t)$ is negative, the direction of oil flow reverses.

In order to derive the differential equation relating $\theta_p(t)$ and $\theta_c(t)$, we must define certain hydraulic quantities. The volume of oil flowing from the pump, $Q_p(t)$ is

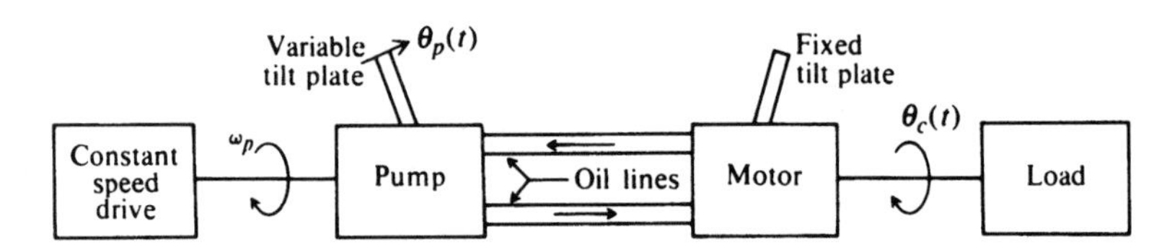

Figure 3.21 Hydraulic power transmission system.

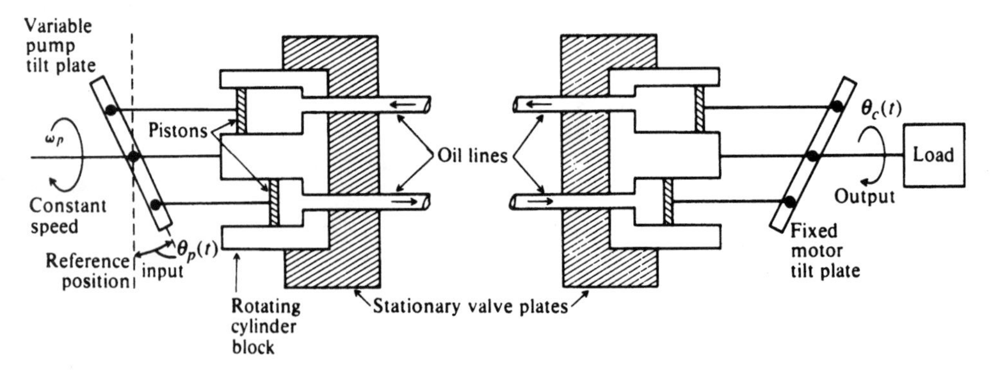

Figure 3.22 Functional block diagram of a hydraulic power transmission system.

composed of flow of oil through the motor, $Q_m(t)$, leakage flow around the motor, $Q_l(t)$, and compressibility flow, $Q_c(t)$ that is,

$$Q_p(t) = Q_m(t) + Q_l(t) + Q_c(t). \tag{3.123}$$

It can be shown that

$$Q_p(t) = K_p\theta_p(t), \tag{3.124}$$

where
$K_p =$ volumetric pump flow per second per angular displacement of $\theta_p(t)$
$\theta_p(t)$ =displacement of the pump stroke;

$$Q_m(t) = V_m\omega_c(t), \tag{3.125}$$

where
$V_m =$ volumetric motor displacement
$\omega_c(t)$ = angular velocity of motor shaft:

$$Q_l(t) = LP_L(t), \tag{3.126}$$

where

$L =$ leakage coefficient of complete system $(\text{ft}^3/\text{sec})/(\text{lb}/\text{ft}^2)$,
$P_L(t) =$ load-induced pressure drop across motor (lb/ft^2); and

$$Q_c(t) = \frac{dV}{dt} = \frac{V}{K_B}\frac{dP_L(t)}{dt}, \tag{3.127}$$

where

$V =$ total volume of liquid under compression (ft^3)
$K_B =$ bulk modulus of oil (lb/ft^2).

Substituting Eqs. (3.124)–(3.127) into Eq. (3.123), we obtain the relationship

$$K_p\theta_p(t) = V_m\omega_c(t) + LP_L(t) + \frac{V}{K_B}\frac{dP_L(t)}{dt}. \tag{3.128}$$

The torque available to drive the motor inertia is

$$\text{torque} = V_m P_L(t) = J\frac{d^2\theta_c(t)}{dt^2}. \tag{3.129}$$

Substituting Eq. (3.129) into (3.128), we obtain

$$K_p\theta_p(t) = V_m\frac{d\theta_c(t)}{dt} + \frac{L}{V_M}J\frac{d^2\theta_c(t)}{dt^2} + \frac{V}{K_B}\frac{J}{V_m}\frac{d^3\theta_c(t)}{dt^3}. \tag{3.130}$$

The Laplace transform of Eq. (3.130) is given by

$$K_p\theta_p(s) = V_m s\theta_c(s) + \frac{LJ}{V_m}s^2\theta_c(s) + \frac{VJ}{K_B V_m}s^3\theta_c(s). \tag{3.131}$$

The transfer function of this device, defined as the ratio of the output $\theta_c(s)$ to the input $\theta_p(s)$, is given by

$$\frac{\theta_c(s)}{\theta_p(s)} = \frac{K_p/V_m}{s[(VJ/K_B V_m^2)s^2 + (LJ/V_m^2)s + 1]}, \tag{3.132}$$

and its simple block diagram is illustrated in Figure 3.23. It is important to emphasize, however, that even a relatively small amount of air in the oil lines would lower K_B. This would cause the resonant frequency of the system to decrease sharply and reduce its capabilities. In addition, a large volume of oil in the lines between the hydraulic pump and motor has a similar effect. Therefore, these lines should be as short and narrow as possible.

Equation (3.132) can be written in differential-equation form as follows:

$$\frac{VJ}{K_B V_m^2}\frac{d^3\theta_c(t)}{dt^3} + \frac{LJ}{V_m^2}\frac{d^2\theta_c(t)}{dt^2} + \frac{d\theta_c(t)}{dt} = \frac{K_p}{V_m}\theta_p(t). \tag{3.133}$$

This third-order differential equation can be written as three first-order differential equations be defining

$$x_1(t) = \theta_c(t), \tag{3.134}$$

$$x_2(t) = \dot{\theta}_c(t), \tag{3.135}$$

$$x_3(t) = \ddot{\theta}_c(t). \tag{3.136}$$

The resulting state equations are given by

$$\dot{x}_1(t) = x_2(t), \tag{3.137}$$

$$\dot{x}_2(t) = x_3(t), \tag{3.138}$$

$$\dot{x}_3(t) = -\frac{K_B V_m^2}{VJ}x_2(t) - \frac{K_B L}{V}x_3(t) + \frac{K_B K_p V_m}{VJ}\theta_p(t). \tag{3.139}$$

In phase-variable canonical form, these equations become

$$\dot{\mathbf{x}}(t) = \mathbf{P}\mathbf{x}(t) + \mathbf{B}r(t), \tag{3.140}$$

$$\theta_c(t) = \mathbf{L}\mathbf{x}(t), \tag{3.141}$$

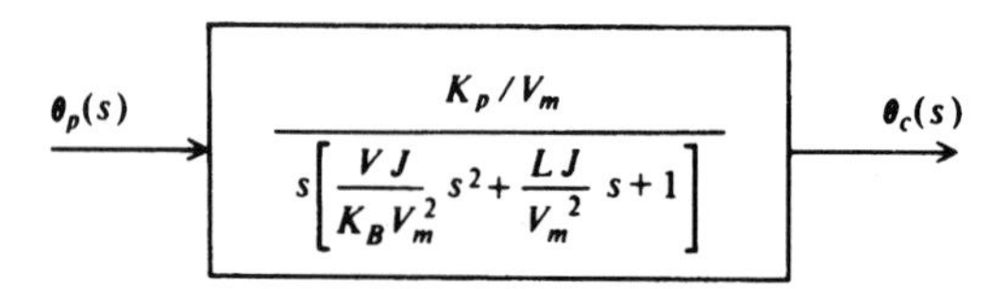

Figure 3.23. Block diagram of a pump-controlled hydraulic transmission system.

where

$$\mathbf{x}(t) = \begin{bmatrix} x_1(t) \\ x_2(t) \\ x_3(t) \end{bmatrix}, \quad \dot{\mathbf{x}}(t) = \begin{bmatrix} \dot{x}_1(t) \\ \dot{x}_2(t) \\ \dot{x}_3(t) \end{bmatrix}, \quad \mathbf{P} = \begin{bmatrix} 0 & 1 & 0 \\ 0 & 0 & 1 \\ 0 & -\dfrac{K_B V_m^2}{VJ} & -\dfrac{K_B L}{V} \end{bmatrix},$$

$$\mathbf{b} = \begin{bmatrix} 0 \\ 0 \\ \dfrac{K_B K_p V_m}{VJ} \end{bmatrix}, \quad \mathbf{L} = [1 \quad 0 \quad 0], \quad r(t) = \theta_p(t).$$

B. Hydraulic Valve-Controlled Motor

Another method of controlling a hydraulic motor is with a constant-pressure source and a valve that controls the flow of oil through it. A valve-controlled hydraulic system is usually smaller than a pump-controlled system and less efficient. Due to the increased losses the time constants are greatly reduced, and the speed of response is greater. Valve-controlled systems also have the disadvantages associated with devices whose characteristics are nonlinear.

Figure 3.24 illustrates a valve-controlled hydraulic system. A fluid source, at constant pressure, is provided at the center of the control valve. Fluid-return lines are located on each side of this pressure source. When the control valve is moved to the right, hydraulic fluid flows through line A into the hydraulic motor. This results in a pressure differential across the piston of the motor which causes it also to move to the right. This action caused fluid to be pushed back into the valve through line B which returns it to the sump through line E. Similar operation occurs when the control valve is moved to the left. Observe that all fluid flows are blocked when the control valve is in the neutral position, as shown in Figure 3.24.

Figure 3.25 represents the characteristics for the valve-controlled hydraulic motor. The pressure between lines going to the motor is denoted by P, the flow

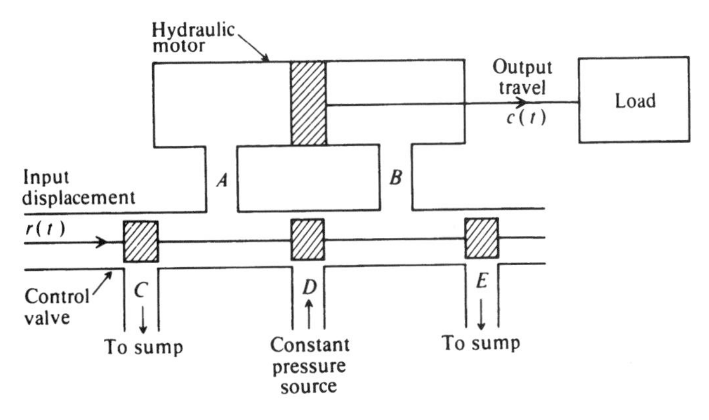

Figure 3.24 A valve-controlled hydraulic system.

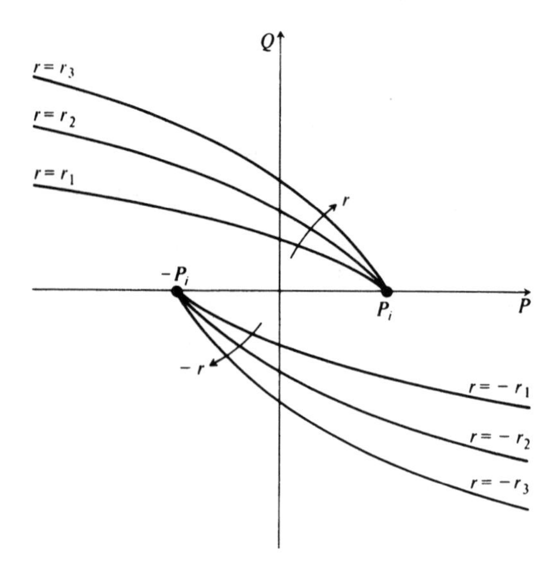

Figure 3.25 Valve characteristics.

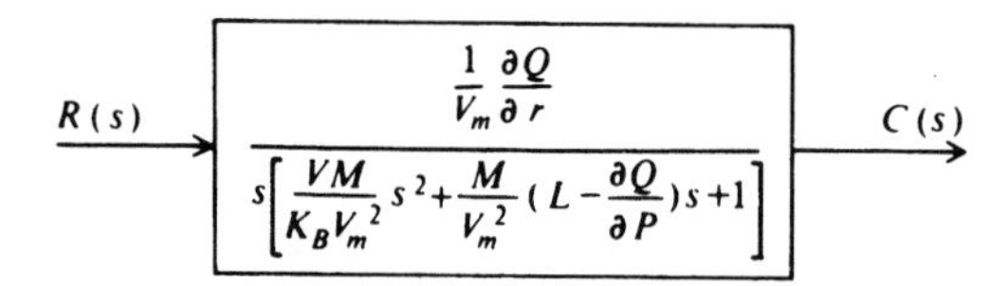

Figure 3.26 Block diagram of a valve-controlled hydraulic transmission system.

through these lines is denoted by Q, and $r(t)$ denotes the displacement of the valve from its neutral position. Although these characteristics are nonlinear, it will be assumed that they are linear for small input displacements. This is basically an application for small-signal theory, used so frequently by circuit designers. For small excursions from a given quiescent operating point,

$$\Delta Q = \frac{\partial Q}{\partial P}\Delta P + \frac{\partial Q}{\partial r}\Delta r. \tag{3.142}$$

At any given quiescent operating point, it will be assumed that $\partial Q/\partial P$ and $\partial Q/\partial r$ are constants.

The transfer function relating the input $r(t)$ and output $c(t)$ can be obtained by comparing the valve-controlled hydraulic system with the pump-controlled hydraulic system. Studying these two systems carefully, it is observed that ΔQ is analogous to Q_p, M (moment of inertia) is analogous to J, ΔP is analogous to P_L, and Δr is analogous to $\theta_p(t)$. Using these analogies, the transfer function can easily be found to be given by

$$\frac{C(s)}{R(s)} = \frac{(1/V_m)(\partial Q/\partial r)}{s[(VM/K_B V_m^2)s^2 + (M/V_m^2)(L - \partial Q/\partial P)s + 1]}. \tag{3.143}$$

The term $L - \partial Q/\partial P$ in the denominator of this equation is always positive, because Figure 3.25 indicates that $\partial Q/\partial P$ is always negative. The simple block diagram for this system is illustrated in Figure 3.26. It is left as an exercise to the reader to determine what the transfer function of the system is if a spring and damper are attached to the control rod (see Problem 3.11), and to determine the state-variable vector form of Eq. (3.143).

3.6. TRANSFER-FUNCTION REPRESENTATION OF THERMAL SYSTEMS

If the assumption is made that the temperature of a body is uniform, then a number of thermal systems can be represented by linear ordinary differential equations [13]. This approximation is reasonably correct for relatively small configurations. This section specifically considers a hot-water heating system as an example of a typical thermal system.

Figure 3.27 illustrates an electric hot-water heating system. The object of this system may be, typically, to supply hot water in a home. Any demand for hot water in the home causes hot water to leave and cold water to enter the tank. In order to reduce heat loss to the surrounding air, the tank is insulated. A thermostatic switch turns an electrical heating element on or off in order to maintain a desired reference temperature.

The law of conservation of energy requires that the heat added to the system equals the heat stored plus the heat lost. This can be expressed by the relationship

$$Q_h(t) = Q_c(t) + Q_o(t) - Q_i(t) + Q_l(t), \tag{3.144}$$

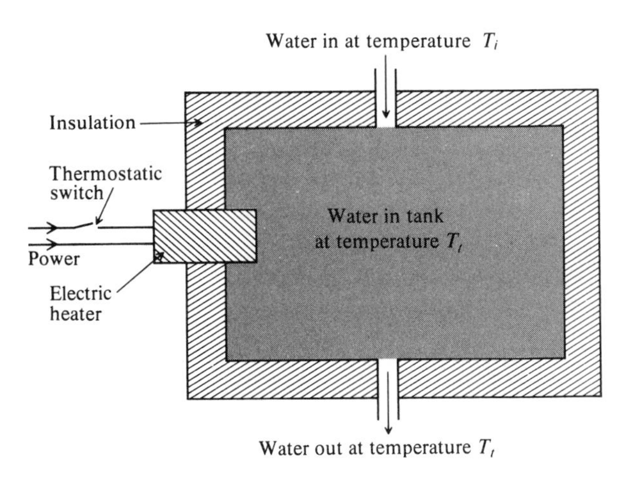

Figure 3.27 Electric hot-water heating system.

where

$Q_h(t) =$ heat flow supplied by heating element,
$Q_c(t) =$ heat flow into storage in the water in the tank,
$Q_o(t) =$ heat flow lost by hot water leaving tank,
$Q_i(t) =$ heat flow carried in by cold water entering tank,
$Q_l(t) =$ heat flow through insulation.

It can be shown that

$$Q_c(t) = C\frac{dT_t(t)}{dt}, \tag{3.145}$$

where

$C =$ thermal capacity of water in tank
$T_t(t) =$ temperature of water in tank;

$$Q_0(t) = V(t)HT_t(t), \tag{3.146}$$

where

$V(t) =$ water flow from tank
$H =$ specific heat of water

$$Q_i(t) = V(t)HT_i(t), \tag{3.147}$$

where

$T_i(t) =$ temperature of water entering tank; and

$$Q_l(t) = \frac{T_t(t) - T_e(t)}{R}, \tag{3.148}$$

where

$T_e(t) =$ temperature of air surrounding tank
$R =$ thermal resistance of insulation, stagnant air film, and stagnant liquid film.

Substitution of Eqs. (3.145)–(3.148) into Eq. (3.144) yields the expression

$$Q_h(t) = C\frac{dT_t(t)}{dt} + V(t)H(T_t(t) - T_i(t)) + \frac{T_t(t) - T_e(t)}{R}. \tag{3.149}$$

The thermal model presented, so far has considered $T_t(t)$, $T_i(t)$, $T_e(t)$, and $V(t)$ as variables. For the specific condition where V is a constant and $T_e(t) = T_i(t)$, Eq. (3.149) reduces to the expression

$$Q_h(t) = C\frac{dT_r(t)}{dt} + \left(VH + \frac{1}{R}\right)T_r(t), \tag{3.150}$$

where $T_r(t)$ = temprature of the water in the tank above the reference $T_e(t)$. The Laplace transform of Eq. (3.150) is given by

$$Q_h(s) = T_r(s)\left(Cs + VH + \frac{1}{R}\right). \tag{3.151}$$

The transfer function of this system, defined as the ratio of the output $T_r(s)$ so the input $Q_h(s)$ is given by

$$\frac{T_r(s)}{Q_h(s)} = \frac{1}{Cs + VH + 1/R} \tag{3.152}$$

and its simple block diagram is illustrated in Figure 3.28.

3.7. A GENERALIZED APPROACH FOR MODELING—THE PRINCIPLES OF CONSERVATION AND ANALOGY

Because of the large variety of control-system components that occur in practice, a generalized approach is useful for obtaining their mathematical model. Therefore, rather than pursue the presentation of further specific control-system components, this section will provide a generalized approach to deriving mathematical models.

There are several general principles that can be useful in serving as guides. The most important are the principle of conservation and the concept of analogous circuits.

A. Principle of Conservation

The principle of conservation is a very important guideline for the derivation of a mathematical model. A statement of this concept is that

$$\text{accumulation} = \text{inflow} - \text{outflow}. \tag{3.153}$$

In terms of rates, the principle of conservation is stated as follows:

$$\text{rate of increase (or storage)} = \text{rate of inflow} - \text{rate of outflow}. \tag{3.154}$$

Exactly what is being conserved depends on the application. However, this principle is usually used to establish a balance or inventory of mass, energy, momentum, or charge.

The principle of conservation has been used several times throughout this chapter. For example, let us reconsider the hydraulic motor and pump power transmission

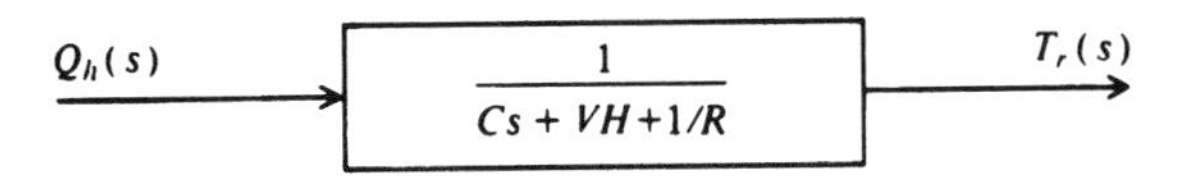

Figure 3.28 Block diagram for the system shown in Figure 3.27.

system illustrated in Figure 3.21 and the electric hot-water heating system of Figure 3.27. In the case of the hydraulic motor and pump power transmission system, Eq. (3.123) was derived. It related the volume of oil flowing from the pump Q_p to its distribution in flow through the motor, $Q_m(t)$, leakage around the motor, Q_l, and compressibility flow (accumulation), $Q_c(t)$. Therefore, this equation was an application of the principle of conservation, where Eq. (3.153) was modified to the form

$$\text{inflow} = \text{outflow} + \text{leakage} + \text{accumulation}. \tag{3.155}$$

Similarly, in the electric hot-water heating system problem, Eq. (3.144) was derived. Let us rearrange it into the following form in order to demonstrate the principle of conservation:

$$Q_h(t) + Q_i(t) = Q_c(t) + Q_o(t) + Q_l(t). \tag{3.156}$$

This equation states that the heat flow supplied by the heating element, $Q_h(t)$, plus the heat flow carried in by cold water entering the tank, $Q_i(t)$, is distributed as heat flow into storage in the water in the tank, $Q_c(t)$, heat flow lost by hot water leaving the tank $Q_o(t)$, and heat flow through the insulation, $Q_l(t)$. Therefore, the principle of conservation was applied, and the form of the principle applied was similar Eq. (3.155).

B. Circuit Concept and Analogy

An alternative viewpoint to the principle of conservation is the concept of an analogous circuit. The basis for applying the principle of analogy is that two different physical systems can be described by the same mathematical model. This permits a generalization of ideas specific to a particular field in order that a broader understanding of a variety of apparently unrelated situations can be achieved.

The analogy concept can best be understood by focusing attention on some of the devices that have been covered in this chapter. For example, let us look at the electrical circuit of the armature-controlled dc servomotor illustrated in Figure 3.11*a* and the hydraulic motor and pump illustrated in Figure 3.21. Equations (3.61) and (3.132) described their respective mathematical models. The direct analogy between Eqs. (3.61) and (3.132) is self-evident. Furthermore, the development of these two models follows a very similar process. It is important to note how their respective relations in both cases enabled one to relate the two coupled sets of physical variables: electrical and mechanical for the armature-controlled dc servomotor, and hydraulic and mechanical for the hydraulic motor and pump transmission system.

This concept of analogy can be extended to relate electrical, mechanical (linear motion), mechanical (rotational), thermal, and hydraulic systems. For purposes of comparison. Table 3.4 illustrates a brief table of analogous quantities in different physical systems. It is important to note that there are two possible analogies between mechanical, and electrical systems. If torque (or force) is chosen to be analogous to current, then the mechanical circuit and electrical circuits look alike with inertia being analogous to capacitance and a spring being analogous to inductance. This is the system illustrated in Table 3.4. Another approach could be to

Table 3.4. Analogous Quantities

Electrical	Mechanical (Linear Motion)	Mechanical (Rotation)	Thermal	Hydraulic
Current	Force	Torque	Heat flow	Flow
Voltage	Linear velocity	Angular velocity	Temperature	Pressure
Inductance	Spring	Spring	—	Inertia
Capacitance	Mass	Inertia	Capacitance	Compression
Resistance	Dashpot	Dashpot	Resistance	Resistance

choose torque (or force) to be analogous to voltage. Then the mechanical and electrical circuits will be analogous, with inertia being analogous to inductance and a spring being analogous to capacitance: it makes no difference which viewpoint is taken.

By using the method of analogs, complex mechanical (or hydraulic, etc.) systems can be drawn as equivalent circuit diagrams, for which Kirchhoff's voltage and current laws can be utilized to obtain the mathematical model of the system. As an example of this approach, let us reconsider the complex mechanical system of Figure 3.5, using Kirchhoff's voltage and current laws. By using the analogs of Table 3.4, the mechanical network of Figure 3.5 is redrawn as an equivalent electrical circuit in Figure 3.29. The node equations for Figure 3.29 are written by inspection as

$$\left(C_2 s + \frac{1}{R_2} + \frac{1}{L_2 s}\right)V_2(s) - \left(\frac{1}{R_2} + \frac{1}{L_2 s}\right)V_1(s) = I(s), \tag{3.157}$$

$$-\left(\frac{1}{R_2} + \frac{1}{L_2 s}\right)V_2(s) + \left(C_1 s + \frac{1}{L_1 s} + \frac{1}{R_1} + \frac{1}{R_2} + \frac{1}{L_2 s}\right)V_1(s) = 0, \tag{3.158}$$

where

$$\begin{aligned} i(t) &= f(t), & L_1 &= 1/K_1, & C_2 &= M_2, \\ R_1 &= 1/B_1, & R_2 &= 1/B_2, & v_1(t) &= \dot{y}_1(t) \\ L_2 &= 1/K_2, & v_2(t) &= \dot{y}_2(t), & C_1 &= M_1. \end{aligned}$$

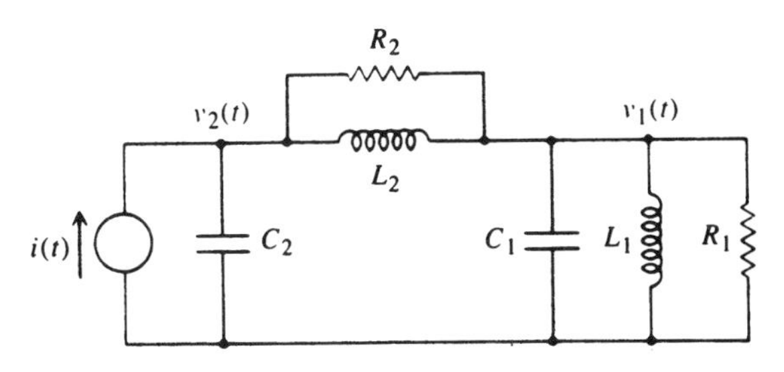

Figure 3.29 Electrical analog of Figure 3.5.

Equations (3.157) and (3.158) are analogous to Eqs. (3.27) and (3.28), respectively. Substituting the analogous quantities into Eqs (3.157) and (3.158), and taking the inverse Laplace transforms we obtain the following set of equations:

$$f(t) = M_2 \frac{d^2 y_2(t)}{dt^2} + B_2 \left[\frac{dy_2(t)}{dt} - \frac{dy_1(t)}{dt} \right] + K_2[y_2(t) - y_1(t)], \tag{3.159}$$

$$0 = B_2 \left[\frac{dy_2(t)}{dt} - \frac{dy_1(t)}{dt} \right] + K_2[y_2(t) - y_1(t)]$$

$$-M_1 \frac{d^2 y_1(t)}{dt^2} - B_1 \frac{dy_1(t)}{dt} - K_1 y_1(t). \tag{3.160}$$

Observe that Eqs. (3.159) and (3.160) are identical to Eqs. (3.27) and (3.28), respectively.

Electrical analogies have the advantage that they can be set up very easily in the laboratory. For example, a change in a particular parameter cna be accomplished very easily in the electric circuit to determine its overall effects and the electric circuit can be appropriately adjusted for the desired response. Afterwards, the parameters in the mechanical (or hydraulic, or thermal, etc.) system can be adjusted by an analogous amount to obtain the same desired response.

3.8. ILLUSTRATIVE PROBLEMS AND SOLUTIONS

This section provides a set of illustrative problems and their solutions to supplement the material in Chapter 3.

I3.1. The torque–speed curve of a two-phase ac servomotor is illustrated where it is assumed that this characteristic can be represented by a straight line:

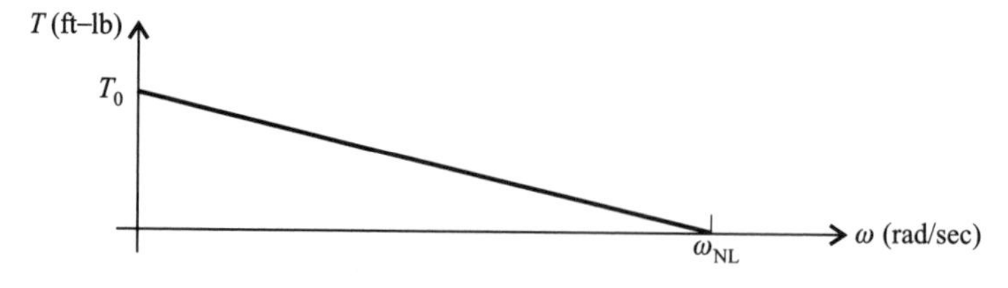

Figure I3.1

(a) Derive an equation for this torque–speed relationship, relating T and ω.

(b) The motor-shaft output power is the product of the motor's derived torque, T, and its speed, ω:

$$P = T\omega.$$

Derive the speed, ω_{max}, where the motor output power is a maximum.

(c) Derive an expression for the corresponding maximum motor-shaft output power, P_{max}.

(d) Determine the maximum motor-shaft output power, P_{max}, if the no-load motor speed, ω_{NL}, is 5000 rev/min and the stall torque, T_0, is 4 in-oz.

SOLUTION: **(a)**

$$T = T_0 - \left(\frac{\omega}{\omega_{NL}}\right) T_o$$

(b) Substituting the equation for T derived in part (a) into the equation for the motor-torque output power

$$P = T\omega$$

we obtain the following:

$$P = \left(T_0 - \left(\frac{\omega}{\omega_{NL}}\right) T_0\right)\omega.$$

To find ω_{max}, we differentiate P with respect to ω as follows:

$$\frac{dP}{d\omega} = T_0 - 2\frac{\omega_{max}}{\omega_{NL}} T_0 = 0.$$

Therefore,

$$\omega_{max} = \frac{1}{2}\omega_{NL}$$

To check that this is a maximum, we take the second derivative of P as follows:

$$\frac{d^2 P}{d\omega^2} = \frac{-2T_0}{\omega_{NL}} < 0.$$

Since the second derivative is negative, ω_{max} is a maximum.

(c) Substituting the expression found for ω_{max} in part (b) into the expression for motor-shaft output power,

$$P = T\omega$$

we obtain the following:

$$P_{max} = T_{max}\left(\frac{\omega_{NL}}{2}\right).$$

Substituting the expression for ω_{max} found in part (b) into the expression for T in part (a), we obtain the following:

$$T_{max} = T_0/2.$$

Substituting this expression for T_{max} into the expression for P_{max}, we obtain the following:

$$P_{max} = \left(\frac{T_0}{2}\right)\left(\frac{\omega_{NL}}{2}\right) = \frac{T_0\omega_{NL}}{4}$$

(d) $$P_{max} = \left(4\ in - oz\frac{1}{12}\frac{ft}{in}\frac{1}{16}\frac{lb}{oz}\right)\left(5000\ \frac{rev}{min}.2\pi\ \frac{rad}{rev}\cdot\frac{1}{60}\ \frac{min}{sec}\right)\frac{1}{4}.$$

Therefore,

$$P_{max} = 2.726 \text{ watts}$$

and P_{max} occurs at a speed of 2500 rev/min.

I3.2. A control-system engineer needs to use an 8 : 1 gear reduction for a two-phase ac servomotor which is to be used for a positioning control system. The 8 : 1 gear reduction is to be made of two gear passes. The stock room has a limited number of gears available which can permit one of the following two configurations:

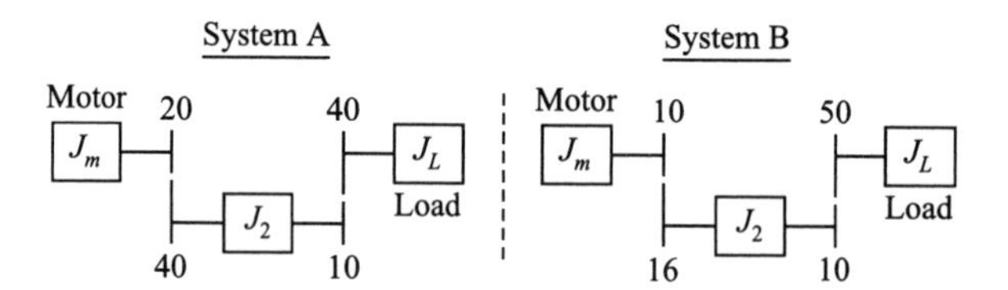

Figure I3.2

Each configuration has the same 8 : 1 gear reduction. The number of teeth on each gear is shown. At the instant of starting, the motor develops a torque T_D. The control-system engineer wants to use the gear configuration which will provide the higher initial load accelertion.

(a) Write one parametric equation expressing the developed motor torque, T_D, in terms of the inertias, the gear reductions, and the load acceleration, a_L, with everything reflected to the motor shaft. (One parametric equation can represent both systems.)

(b) Assuming that $J_L = 16J_2 = 4J_m$, determine which system has the higher initial load acceleration.

SOLUTION: **(a)** Reflecting a_L and all inertias to the motor shaft, we obtain the following:

$$T_D = J_m(N_1N_2a_L) + \frac{J_2}{N_1^2}(N_1N_2a_L) + \frac{J_L}{(N_1N_2)^2}(N_1N_2a_L)$$

where N_1 represents the gear ratio between J_m and J_2, and N_2 represents the gear ratio between J_2 and J_L.

Simplifying this equation, we obtain the following:

$$T_D = a_L\left(J_m N_1 N_2 + \frac{J_2 N_2}{N_1} + \frac{J_L}{N_1 N_2}\right).$$

(b) Setting the initial developed torques, T_D, of system A and system B equal to one another, substituting $J_L = 16J_2 = 4J_m$, and using the gear ratios indicated, we obtain the following:

$$a_{L_A}\left(\frac{8J_L}{4} + \frac{4J_L}{16}\frac{1}{2} + \frac{J_L}{8}\right) = a_{L_B}\left(\frac{8J_L}{4} + \frac{5J_L}{16}\frac{1}{1.6} + \frac{J_L}{8}\right).$$

Factoring out J_L and then canceling it from both sides of this equation, we obtain the following:

$$a_{L_A}(2 + 0.132 + 0.125) = a_{L_B}(2 + 0.1955 + 0.125).$$

Therefore,

$$a_{L_A}(2.257) = (2.3205)a_{L_B}$$

and

$$a_{L_A} = 1.00281a_{L_B}.$$

Therefore, we conclude that System A has the higher initial load acceleration.

PROBLEMS

3.1. **(a)** For the mechanical translational system illustrated in Figure P3.1, write the differential equation relating the position $y(t)$ and the applied force $f(t)$.

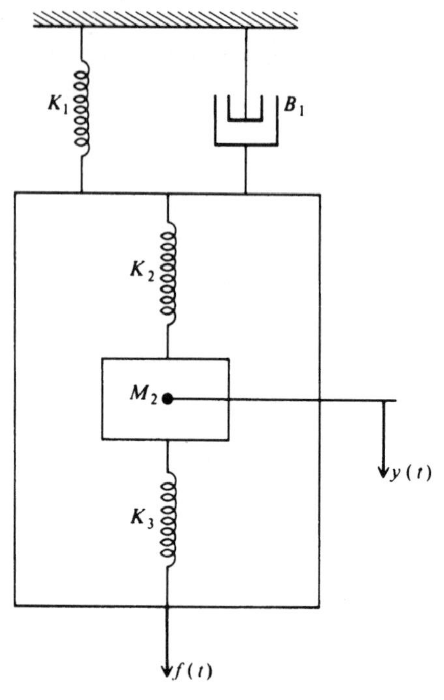

Figure P3.1

(b) Determine the transfer function $Y(s)/F(s)$.

(c) Determine the phase-variable canonical vector form of this system.

3.2. **(a)** For the mechanical translational system illustrated in Figure P3.2, write the differential equation relating the position $y(t)$ and the applied force $f(t)$.

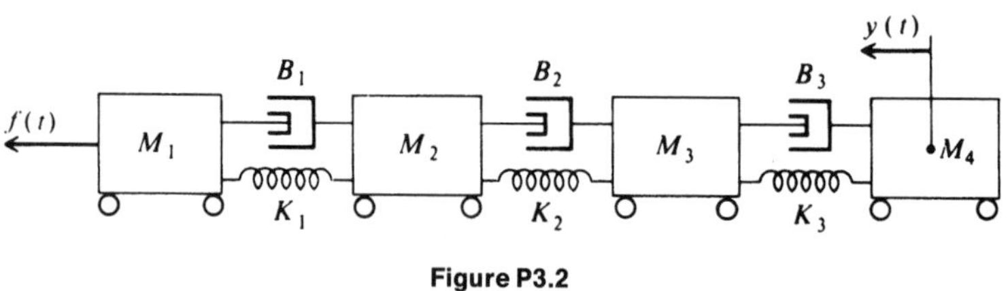

Figure P3.2

(b) Determine the transfer function $Y(s)/F(s)$.

(c) Determine the phase-variable canonical vector form of this system.

3.3. **(a)** For the mechanical rotational system illustrated in Figure P3.3, write the differential equation relating $T(t)$ and $\theta(t)$.

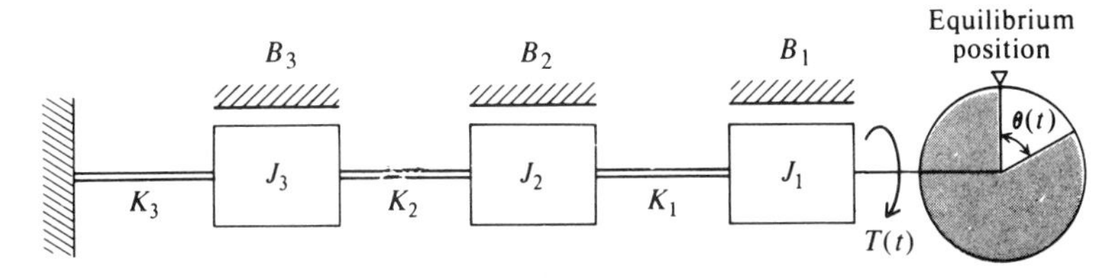

Figure P3.3

(b) Determine the transfer function $\theta(s)/T(s)$.

3.4. Figure P3.4 represents the diagram of a gyroscope which is used quite frequently in autopilots, stabilized fire control systems, and so on. Assume that the rotor speed is constant, that the total developed torque about the output axis is given by

$$T_0 = K' \frac{d\theta_i(t)}{dt},$$

where K' is a constant, and that the inner gimbal's moment of inertia about the output axis is J.

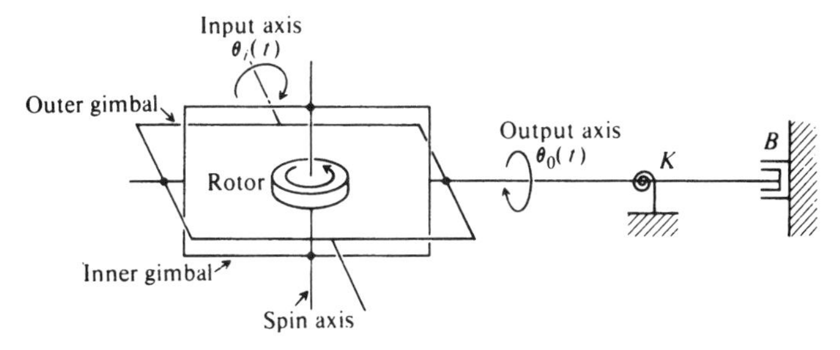

Figure P3.4

(a) Write the differential equation relating $\theta_i(t)$ and $\theta_0(t)$.

(b) Determine the transfer fiunction $\theta_0(s)/\theta_i(s)$.

3.5. The Ward-Leonard system shown in Figure 3.12 has the constants and characteristics shown in Table P3.5. Derive the transfer function for the system $\theta_0(s)/E_f(s)$.

3.6. Repeat Problem 3.5 for the constants and characteristics shown in Table P3.6.

Table P3.5

Generator	Motor
$R_f = 1.1\Omega$	$R_m = 2\Omega$
$L_f = 0.25\text{H}$	$L_m = 0.05\text{ H}$
$R_g = 0.20\Omega$	$K_e = 0.1\text{ V/(rad/sec)}$
$L_g = 0.01\text{H}$	$K_T = 2\text{ ft lb/A}$
$K_g = 1\text{ V/A}$	$J = 0.1\text{ slug ft}^2$
	$B =$ negligible

Table P3.6

Generator	Motor
$R_f = 1\ \Omega$	$R_m = 1\Omega$
$L_f = 0.1\text{ H}$	$L_m = 0.01\text{ H}$
$R_g = 0.1\ \Omega$	$K_e = 10\ V\text{/(rad/sec)}$
$L_g = 0.1\text{ H}$	$K_T = 10\text{ ft lb/A}$
$K_g = 10\text{ V/A}$	$J = 1\text{ slug ft}^2$
	$B =$negligible

3.7. The time constant for highly inductive devices such as the armature circuit of the Ward-Leonard system shown in Figure 3.12 is usually too long for high-performance control systems. A simple technique for decreasing the time constant of such a system uses feedback, which controls the current of the inductive device. Figure P3.7 illustrates how this can be practically accomplished for a Ward-Leonard system by inserting a resistor R in series with the armature circuit. Other notations used in this diagram are the same as those used in Section 3.4. Assume that the armature current is much greater than the field current.

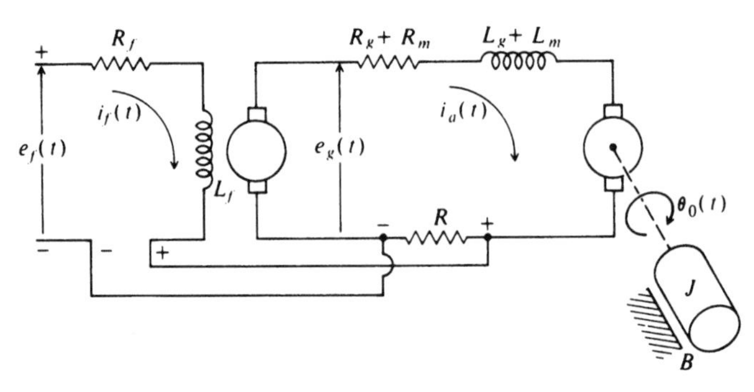

Figure P3.7

(a) Write the differential equation relating $e_f(t)$ and $\theta_0(t)$.

(b) Determine the transfer function $\theta_0(s)/E_f(s)$.

(c) Compare the answer to part (b) with Eq. (3.78) and show that the armature time constant has been reduced by means of feedback.

3.8. Because it is proportional to velocity, the back emf of a motor is sometimes directly used as a stabilizing voltage. Figure P3.8 illustrates a practical bridge-type circuit which can be used for this purpose. Resistors R_1 and R_2, which have very high resistance relative to the armature circuit resistance, are adjustable in order to obtain the desired value of $e_b(t)$. R_{AC} represents the resistance of the armature and commutator voltage drop; R_{CF} and L_{CF} represent the resistance and inductance of the commutating fields, respectively. K_e, K_T, J, and B have the same significance as in the text of this chapter. For the armature-controlled dc servomotor illustrated in Figure P3.8, assume that the field current I_f remains constant.

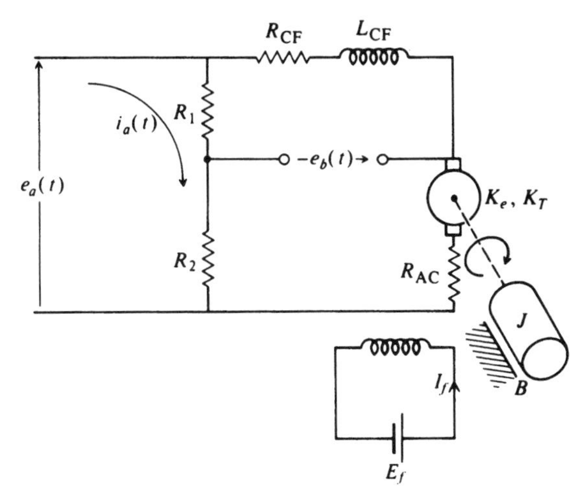

Figure P3.8

(a) Write the differential equation relating $e_a(t)$ and $e_b(t)$.

(b) Determine the transfer function $E_b(s)/E_a(s)$.

3.9. The armature-controlled dc servomotor's transfer function and state equations were developed in Section 3.4. Very often in practice, torque disturbances occur which adversely affect its performance. This may be due to sea waves hitting flaps of a hydrofoil, or wind gusts hitting the reflector of a tracking rada. Figure P3.9 illustrates a modification to the configuration of Figure 3.11c to illustrate a torque disturbance $T_U(s)$ addition. Let us assume for our application that we wish to have the motor velocity $\dot{\theta}(s)$ and position $\theta(s)$ be independent of the torque disturbance $T_U(s)$. This can be accomplished as shown in Figure P3.9 by the addition of a current-feedback loop which is fed back to an added new feedback loop [consisting of an amplifier K_A and $B(S)$] preceding the armature-controlled dc servomotor. Find the required relationship between $A(s)$ and $B(s)$ so that the motor velocity $\dot{\theta}(s)$ and position $\theta(s)$ are independent of the disturbance torque$T_U(s)$. In practice, $K_aB(s)$ is designed to be much greater than 1 over the frequency range of interest.

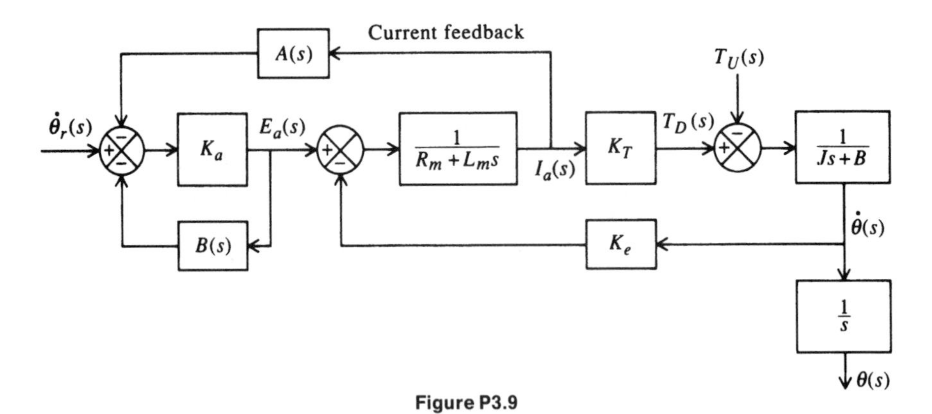

Figure P3.9

3.10. Draw the signal-flow graph for the following devices:

(a) dc generator

(b) the Ward-Leonard system

(c) two-phase ac servomotor.

3.11. Derive the transfer function $C(s)/R(s)$ for the valve-controlled hydraulic transmission system shown in Figure 3.24 if a spring is attached between the right end of the valve control rod and some stationary reference point and viscous damping is added. Assume that the spring stiffness factor is K lb/ft and the viscous damping factor is B lb(ft/sec).

3.12. The control of paper color is a very interesting and important problem in the paper-processing industry. Figure P3.12*a* illustrates a functional diagram of the problem [14]. This color-control method depends on the availability of a precise, reliable, on-line colorimeter. As indicated in the diagram, dyestuff concentrations are added at various stages of the process. In this diagram, the following nomenclature is used:

f = water flow rate, in liters/minute,
α = consistencies (weight of dry fiber/weight of stock),
V_t = header tank volume,
K_v = constant at the dry end,
V_p = pipe volume,
V_s = stirred tank volume,
c = dyestuff concentration, in grams/liter,
v = machine speed, in meters/minute,
m = dyestuff flow rate, in liters/minute.

Figure P3.12*b* illustrates an equivalent block diagram of the color control process [14]. The factor P_r, indicated at several stages of the block diagram, is the retention factor of the dye. It varies with the concentration, with the time allowed for dyeing, and with such factors as pH, alum, resin, temperature,

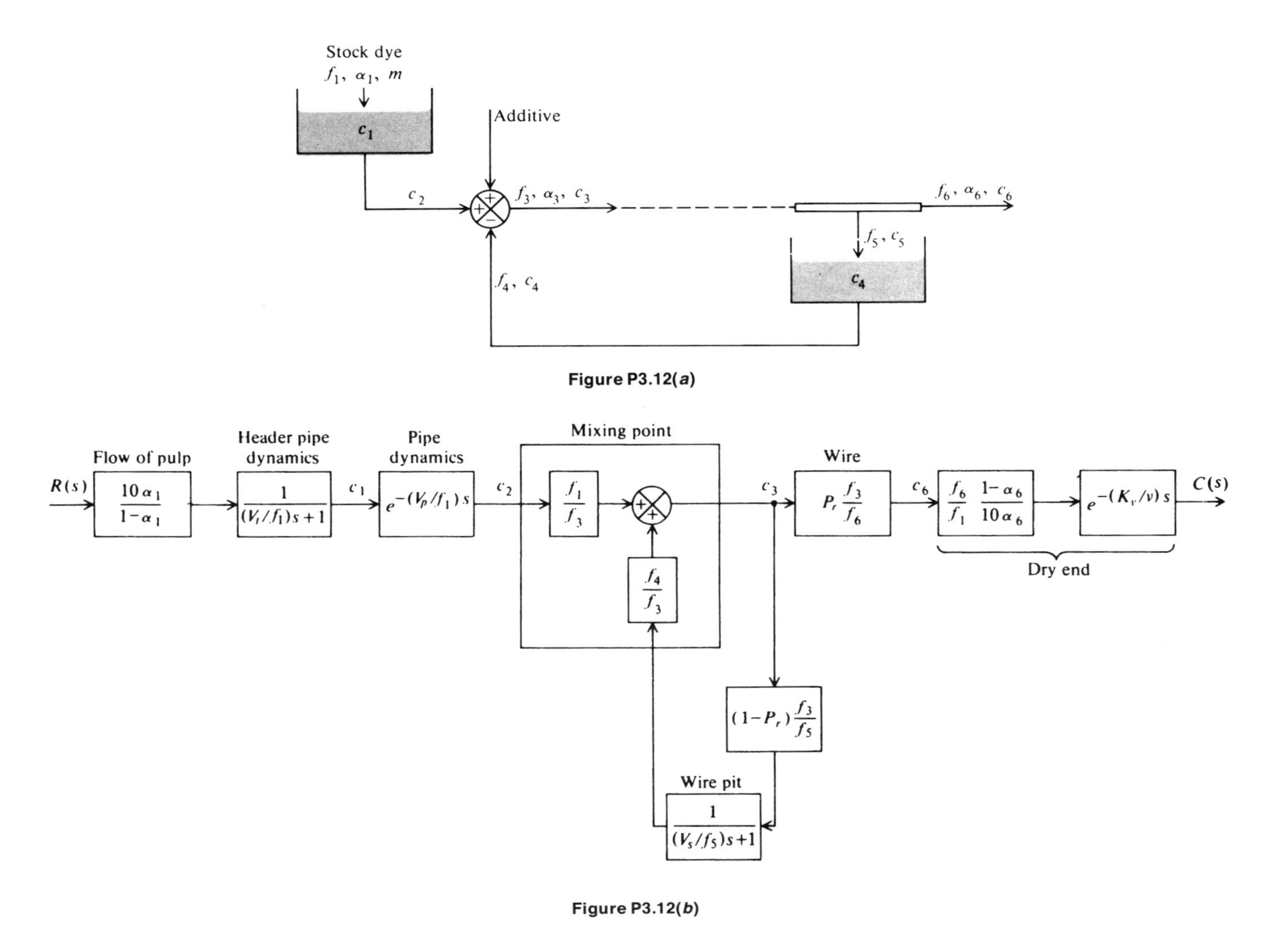

Figure P3.12(a)

Figure P3.12(b)

and dye interaction. Determine the transfer function of this system, $C(s)/R(s)$, where

$$R(s) = \text{dye concentration ratio} = \text{percent of dye/weight of fiber}$$
$$C(s) = \text{weight concentration of dye in the sheet at the output.}$$

3.13. A two-phase ac servomoter has the following specifications: 115/115 V, 60 Hz, 2 phases, 4 poles; rotor moment of inertia: 0.09 oz in^2, rated stalled torque 6 oz in; no-load speed: 2500 rpm; load inertia and coefficient of viscous friction are negligible. Calculate the transfer function for this servomotor, $\theta_0(s)/E_c(s)$.

3.14. It is common practice to place a gear reduction between a servomotor and the load in order to convert the high-speed, low-torque motor characteristics to a low-speed, high-torque device. Assuming a gear reduction of $N:1$ and a load inertia of J_2, derive the transfer function $\theta_0(s)/E_c(s)$ for this system and compare it with Eq. (3.101). Use the same terminology and motor characteristics as in Section 3.4, Part E. Assume B is finite.

3.15. Based on the derivation of Problem 3.14, repeat Problem 3.13 if the load has an inertia of 0.40 oz in^2, and a gear ratio of 36 : 1 is used. Assume $B = 0$.

3.16. Repeat Problem 3.15 with the load inertia doubled. What conclusions can you draw from your results?

3.17. A two-phase ac servomotor has the following specifications: 115/115 V, 400 Hz, 2 phases, 4 poles; rotor moment of inertia: 0.04 oz in^2, rated stall torque: 2 oz in; no-load speed: 5000 rpm; load inertia: 4 oz in^2 geared to the motor through a gear reduction of 9 : 1. Determine the transfer function for the overall servomotor and load. Assume the inertia of the gears is negligible, and the damping factor term B is zero.

3.18. Repeat Problem 3.17 without assuming that the gearing inertia can be neglected. Assume that the gear reduction of 9 : 1 is achieved in one gear pass and each gear has an inertia of 0.01 oz in^2. What conclusions can you reach from your results?

3.19. Repeat Problem 3.17 without assuming that the gearing inertia can be neglected. Assume that the gear reduction of 9 : 1 is achieved in two gear passes of 3 : 1 each, and that each gear has an inertia of 0.01 oz in^2. What conclusions can you draw from your results?

3.20. Derive the transfer function of the armature-controlled dc servomotor from its signal-flow graph.

3.21. The control system which positions the printwheel of a printer connected to a PC is illustrated in Figure P3.21.

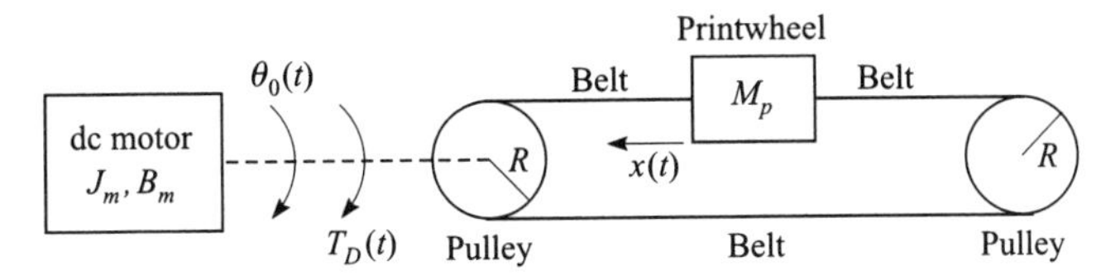

Figure P3.21

The dc motor controls the position of the printwheel and develops a torque, $T_D(t)$. The motor's angular displacement is $\theta_0(t)$, and $x(t)$ represents the linear displacement of the printwheel. Assume that the dc motor inertia is J_m, its motor linear viscous friction is B_m, the radius of the pulley is R, the mass of the printwheel is M_p, and the belts are rigid.

(a) Determine the inertia of the printwheel.

(b) Determine the differential equation relating the developed torque, $T_D(t)$, with the angular motion of the dc motor, $\theta_0(t)$.

(c) Determine the transfer function, $\theta_0(s)/T_D(s)$.

(d) Determine the relation between the linear motion of the printwheel, $x(t)$, and the angular motion of the dc motor, $\theta_0(t)$.

(e) Determine the transfer function, $X(s)/\theta_0(s)$.

(f) Determine the transfer function of this control system, $X(s)/T_D(s)$.

3.22. Repeat Problem I3.2, Part b, for the case where $J_m = J_2 = J_L$. Determine whether system A or system B has the higher initial load acceleration.

3.23. It is desired to determine the transfer function of a two-phase ac servomotor. Its characteristics determined from its specification sheet are as follows: 115/115 V, 400 Hz, 2 phases, 4 poles, rotor moment of inertia: 10 kg m^2, rates stall torque: 1000 N m, no-load speed: 100 rad/sec, load inertia: 1000 kg m^2 geared to the motor through a gear reduction of 10:1. Determine the transfer function for the overall servomotor and load. Assume that the inertia of the gears is negligible, and the damping factor term B is zero.

3.24. A two-phase ac servomotor (Figure P3.24) has the following parameters: 115/115 V, rotor moment of inertia: 0.02 oz in^2, rated stall (developed) torque: 1 oz in, no-load speed: 4000 rpm, load inertia: 2 oz in^2 geared to the motor through a gear reduction of 10:1. Assume the inertia of each gear is 0.01 oz in^2. Determine the transfer function in its simplest form for the overall servomotor and load. Assume that the linear viscous friction term B equals zero.

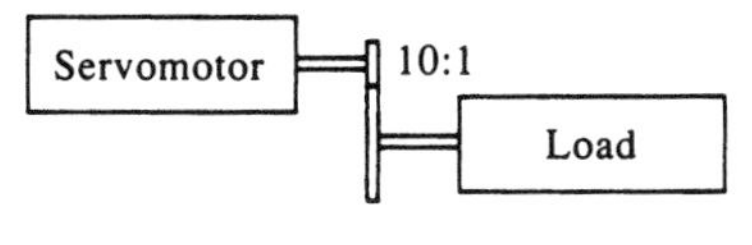

Figure P3.24

3.25. The two-phase ac servomotor shown in Figure P3.24 has the following parameters: 115/115 V, rotor moment of inertia: 0.04 oz in^2, rated stall (developed) torque: 2 oz in, no-load speed: 2000 rpm, load inertia: 4 oz in^2 geared to the motor through a gear reduction of 10 : 1. Assume the inertia of each gear is 0.01 oz in^2. Determine the transfer function in its simplest form for the overall servomotor and load. Assume that the damping term B equals zero.

3.26. A two-phase ac servomotor has the following specifications: 115V/115 V, 400 Hz, 2 phases, 4 poles, rotor moment of inertia : 0.2 oz in^2, rated stall torque : 4 oz in, no-load speed : 2000 rpm, load inertia of 1 oz in^2 geared to the motor through a single stage of gear reduction of 3 : 1. Assume that the inertia of each gear in the gear reduction is 0.1 oz in^2. Determine the transfer function for the overall servomotor and load. Assume that the damping factor term B equals zero.

3.27. A two-phase ac servomotor has the following specifications: 115 V/115 V, 400 Hz, 2 phases, 4 poles, rotor moment of inertia : 0.2 oz in^2, rated stall torque : 8 oz in, no-load speed : 4000 rpm, load inertia of 1 oz in^2 geared to the motor through a single stage of gear reduction of 3 : 1. Assume that the inertia of each gear in the gear reduction is 0.1 oz in^2. Determine the transfer function for the overall servomotor and load. Assume that the damping factor term B equals zero.

3.28. Starting with Eq. (3.62), derive the state equations of the armature-controlled dc servomotor from its state-variable diagram.

3.29. Starting with Eq. (3.78), derive the state equations of the Ward-Leonard system from the state-variable diagram.

3.30. Starting with Eq. (3.101), derive the state equations of the two-phase ac servomotor from the state-variable diagram.

3.31. For the mechanical system illustrated in Figure P3.31, draw an analogous electric circuit and determine the differential equation relating the force $f(t)$ and the position $x(t)$.

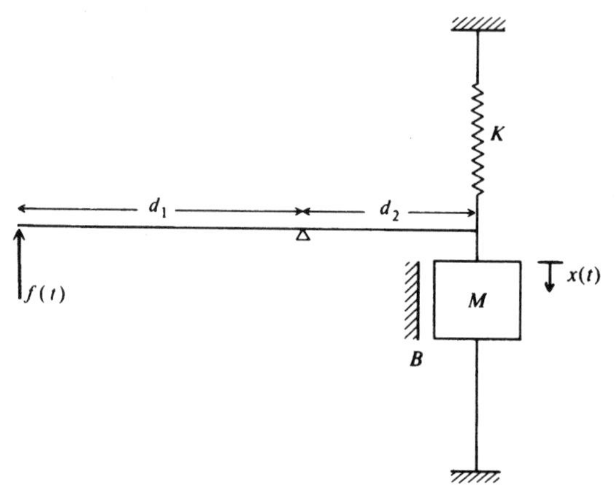

Figure P3.31

3.32. For the mechanical system illustrated in Figure P3.32, draw an analogous electric circuit and determine the differential equation relating $v_1(t)$, $v_2(t)$, and $f(t)$.

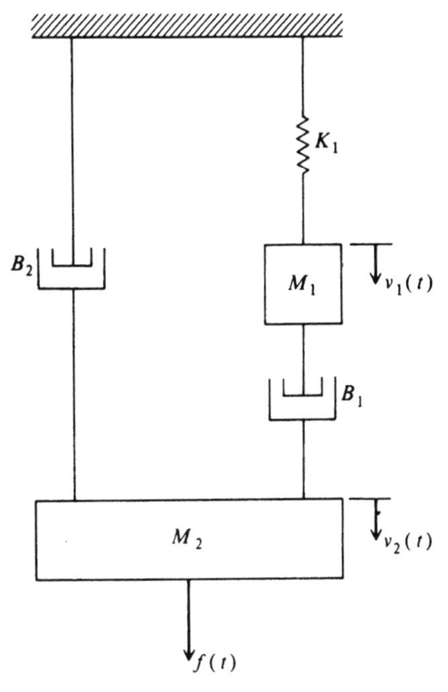

Figure P3.32

3.33. With the advent of modern man-machine control systems utilizing humans to perform various manual tasks, knowledge of the human transfer function is very important in order to enable the prediction of the system's performance. Many researchers have attempted to determine the human transfer function. For manual control at relatively low frequency, the human characteristics indicate a gain K, an anticipation time constant T_A, an operator's error-smoothing lag time constant T_L, an operator's short neuromuscular delay time constant T_N, and an operator's time-delay factor D. Write the form of the human transfer function. What is the meaning of the time-delay factor D [15]?

3.34. The first untethered walk in space was conducted by Astronaut Bruce McCandless II on 7 February 1984. He use a nitrogen-propulsion hand-controlled device called the manned maneuvering unit as shown in the NASA photograph of Figure P3.34*a*. In this photograph, Astronaut McCandless II is a few meters away from the cabin of the Earth-orbiting space shuttle *Challenger*. Figure P3.34*b* is a rare scene of the space shuttle *Challenger* some 50–60 m away from Astronaut McCandless which he took with a fixed camera on his helmet. Observe the robot arm-like remote manipulator system (RMS) that was first introduced in Figure 1.13.

Figure P3.34(*a*) Astronaut Bruce McCandless II, one of two mission specialists participating in a historic extra-vehicular activity, is a few meters away from the cabin of the Earth-orbiting space shuttle *Challenger*. This spacewalk represented the first use of a nitrogen-propelled, hand-controlled device called the manned maneuvering unit.

Figure P3.34(*b*) A fixed camera on Astronaut McCandless's helmet recorded this rare scene of the space shuttle *Challenger* some 50–60 meters away during his extra-vehicular activity. (Photographs courtesy of NASA)

We wish to draw the block-diagram representation of one axis of this manual-control system where the input represents the desired position and the output represents the actual position. Assume that the human model [14] can be represented as defined in Problem 3.33, and assume that the nitrogen-propulsion hand-controlled device converts the astronaut's control action to a force by means of proportional control, where the proportionality constant is K. It will also be assumed that the dynamics of the astronaut in converting the force of the nitrogen-propulsion hand-controlled device to a position in space can be represented as a double integration.

REFERENCES

1. M. E. Van Valkenburg, *Network Analysis*, 3rd ed. Prentice-Hall, Englewood Cliffs, NJ, 1974.
2. K. Furuto, A. Sano, and D. Atherton, *State Variable Methods in Automatic Control*. Wiley, New York, 1988.
3. J. S. Rao and R. V. Dukkipati, *Mechanism and Machine Theory*. Wiley, New York, 1988.
4. R. M. Saunders, "The dynamo electric amplifier—Class A operation," *Trans. Am. Inst. Electr. Eng.* **68**, 1368–1373 (1949).
5. B. Litman, "An analysis of rotating amplifiers," *Trans. Am. Inst. Electr. Eng.* **68**, Pt. 2, 111 (1949).
6. H. Chestnut and R. W. Mayer, *Servomechanisms and Regulating System Design*, 2nd ed., Vol. 1. Wiley, New York, 1959.
7. P. C. Sen, *Principles of Electric Machines and Power Electronics*. Wiley, New York, 1989.
8. G. McPherson and R. D. Laramore, *An Introduction to Electrical Machines and Transformers*, 2nd ed. Wiley, New York, 1989.
9. H. Chestnut and R. W. Mayer, *Servomechanisms and Regulating System Design*, Vol. 2, Chap. 7. Wiley, New York, 1955.
10. R. J. Koopman, "Operating characteristics of two-phase servomotors." *Trans. Am. Inst. Electr. Eng.* **68**, Pt. 1, 319 (1949).
11. A. M. Hopkin, "Transient response of small two-phase servomotors," *Trans. Am. Inst. Electr. Eng.* **70**, Pt. 1, 881 (1951).
12. R. D. Begamudre, *Electro-Mechanical Energy Conversion with Dynamics of Machines*. Wiley, New York, 1988.
13. W. J. Minkowycz, E. M. Sparrow, G. E. Schneider, and R. H. Pletcher, *Handbook of Numerical Heat Transfer*.Wiley, New York, 1988.
14. P. R. Bélanger, "Sensitivity design of a color control system," in *Proceedings of the* 1969 *Joint Automatic Control Conference*, pp. 99–106.
15. S. M. Shinners, "Modelling of human operator performance utilizing time series analysis." *IEEE Trans. Syst. Man Cybernet.*, **SMC-4**(5), 446–458 (1974).

4

SECOND-ORDER SYSTEMS

4.1. INTRODUCTION

From the frequency-domain viewpoint, system order refers to the highest power of s in the denominator of the closed-loop transfer function of a system. In the time domain, system order refers to the highest derivative of the controlled quantity in the equation describing the control system's dynamics. System order is a very signifiicant parameter for characterizing a system.

Second-order systems are very important to the control-system engineer. This type of system characterizes the dynamics of many control-system applications found in the fields of servomechanisms, space-vehicle control, chemical process control, bioengineering, aircraft control systems, ship controls, etc. It is interesting to note that most control-system designs are based on second-order system analysis. Even if the system is of higher order, as it usually is, the system may be approximated by a second-order system in order to obtain a first approximation for preliminary design purposes with reasonable accuracy. A more exact solution can then be obtained in terms of departures from the performance of a second-order system.

Because of the importance of second-order systems, this chapter is devoted to presenting its characteristic response in the time domain and analyzing its state-variable signal-flow graph. In addition, several important control-system definitions are presented. The closed-loop frequency response of second-order systems is presented in Chapter 6, where techniques for obtaining the closed-loop frequency characteristics are derived. A method for modeling the transfer functions of control systems is also presented.

4.2. CHARACTERISTIC RESPONSES OF SECOND-ORDER CONTROL SYSTEMS

The purpose of this section is to describe the transient response of a typical feedback control system. We consider a very common configuratuon in which a two-phase ac servomotor, whose transfer function is given by Eq. (3.101) is enclosed by a simple unity feedback loop. Figure 4.1 illustrates the block diagram of this second-order system. For purposes of simplicity, the gain of the amplifier driving the motor is assumed to be unity.

The closed-loop transfer function of this system is given by

$$\frac{C(s)}{R(s)} = \frac{K_m/T_m}{s^2 + (1/T_m)s + K_m/T_m}. \tag{4.1}$$

By defining the undamped natural frequency ω_n and the dimensionless damping ratio ζ as

$$\omega_n^2 = \frac{K_m}{T_m} \text{ and } \zeta = \frac{1}{2\omega_n T_m}, \tag{4.2}$$

Eq. (4.1) can be rewritten as

$$\frac{C(s)}{R(s)} = \frac{\omega_n^2}{s^2 + 2\zeta\omega_n s + \omega_n^2}. \tag{4.3}$$

The parameters ω_n and ζ are very important for characterizing a system's response. Note from Eq. (4.3) that ω_n turns out to be the radian frequency of oscillation when $\zeta = 0$. As ζ increases from 0, the oscillation decays exponentially and becomes more damped. When $\zeta \geqslant 1$, an oscillation does not occur.

We assume that the initial conditions are zero and the input is a unit step. Therefore, $R(s) = 1/s$, and the Laplace transform of the output can be written as

$$C(s) = \frac{\omega_n^2}{s(s^2 + 2\zeta\omega_n s + \omega_n^2)}. \tag{4.4}$$

Factoring the denominator, we obtain

$$C(s) = \frac{\omega_n^2}{s(s + \zeta\omega_n - \omega_n\sqrt{\zeta^2 - 1})(s + \zeta\omega_n + \omega_n\sqrt{\zeta^2 - 1})}. \tag{4.5}$$

The exact solution for the output in the time domain is dependent on the value of ζ. When $\zeta \geqslant 1$, the second-order system has poles which lie along the negative real axis of the complex plane. When $\zeta < 1$, however, a pair of complex-conjugate poles result. We shall determine the output response to a step input for the three cases: where the damping ratio equals unity, is greater than unity, and is less than unity.

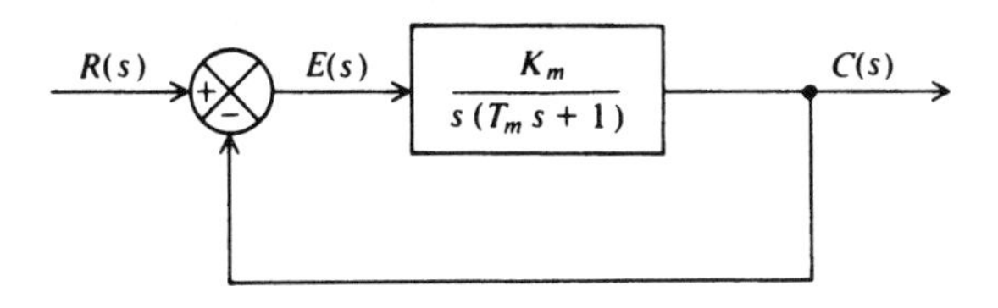

Figure 4.1 Second-order feedback system containing a two-phase ac servomotor.

Case A. Damping Ratio Equals Unity

When $\zeta = 1$, Eq. (4.5) reduces to

$$C(s) = \frac{\omega_n^2}{s(s+\omega_n)^2}. \tag{4.6}$$

The time-domain response can be obtained by utilizing the solution obtained for Eq. (2.82). The partial-fraction expansion of Eq. (4.6) is given by

$$C(s) = \frac{K_1}{(s+\omega_n)^2} + \frac{K_2}{s+\omega_n} + \frac{K_3}{s}. \tag{4.7}$$

Using Eqs. (2.85), (2.87), and (2.88) with $A = D = 0$, $B = \omega_n^2$, $C = \omega_n$, we find that

~~$K_1 = \omega_n$,~~ $K_1 = -\omega_n$, (4.8)

$$K_2 = -1, \tag{4,9}$$

$$K_3 = 1. \tag{4.10}$$

Substituting these constants into Eq. (4.7), we obtain

$$C(s) = \frac{-\omega_n}{(s+\omega_n)^2} - \frac{1}{s+\omega_n} + \frac{1}{s}. \tag{4.11}$$

The time-domain response of the output, $c(t)$, may be obtained by utilizing the table of Laplace transforms given in Appendix A:

$$c(t) = -\omega_n t e^{-\omega_n t} - e^{-\omega_n t} + 1, \quad t \geqslant 0. \tag{4.12}$$

Figure 4.2*a* illustrates the output response together with the unit step input. Notice that the output response exhibits no overshoots when $\zeta = 1$. The response is described as being critically damped.

Case B. Damping Ratio Greater than Unity

When $\zeta > 1$, the time-domain response can be obtained quite simply from its partial fraction expansion. This can be expressed as

$$C(s) = \frac{K_1}{s} + \frac{K_2}{s+\zeta\omega_n - \omega_n\sqrt{\zeta^2-1}} + \frac{K_3}{s+\zeta\omega_n + \omega_n\sqrt{\zeta^2-1}}, \tag{4.13}$$

where $\zeta\omega_n - \omega_n\sqrt{\zeta^2-1}$ and $\zeta\omega_n + \omega_n\sqrt{\zeta^2-1}$ are positive real numbers.

The constants K_1, K_2, and K_3 can be evaluated quite simply by the methods described in Chapter 2. Their values are

$$\begin{aligned} K_1 &= 1, \\ K_2 &= \left[2\left(\zeta^2 - \zeta\sqrt{\zeta^2-1} - 1\right)\right]^{-1}, \\ K_3 &= \left[2\left(\zeta^2 + \zeta\sqrt{\zeta^2-1} - 1\right)\right]^{-1}. \end{aligned} \tag{4.14}$$

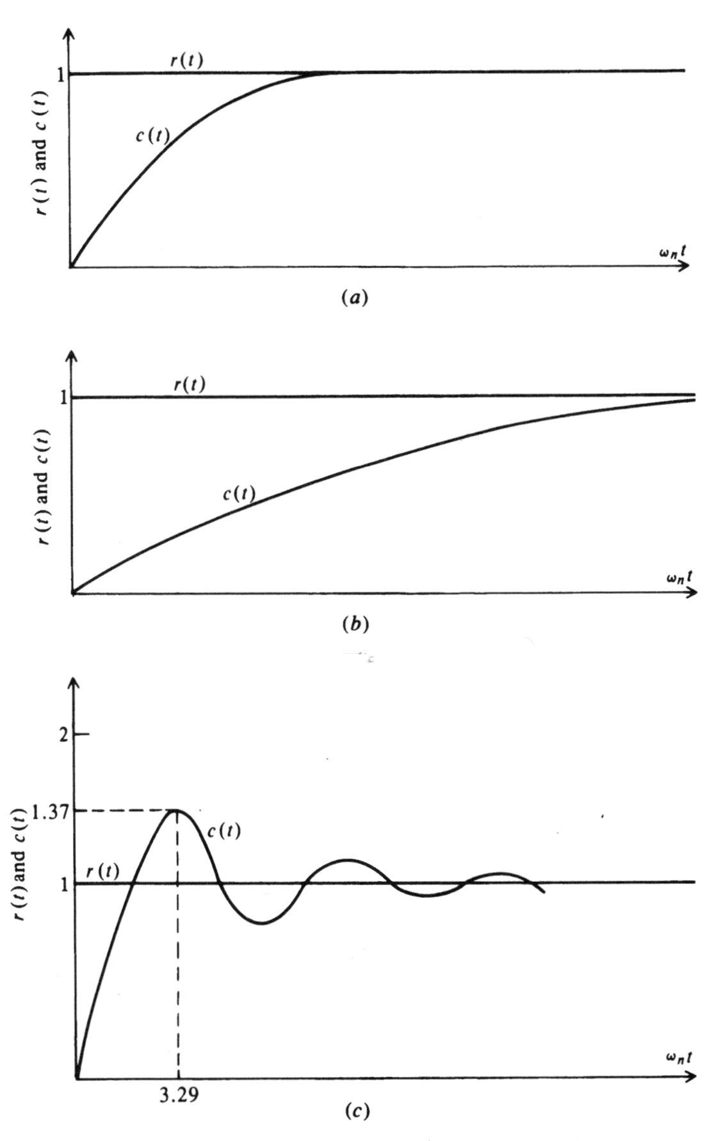

Figure 4.2 (*a*) Input and output response for a critically damped second-order system; (*b*) input and output response for an overdamped second-order system; (*c*) input and output response for an underdamped second-order system ($\zeta = 0.3$).

Therefore Eq. (4.13) can be written as

$$C(s) = s^{-1} + \left[2\left(\zeta^2 - \zeta\sqrt{\zeta^2 - 1} - 1\right)\right]^{-1}\left(s + \zeta\omega_n - \omega_n\sqrt{\zeta^2 - 1}\right)^{-1} + \left[2\left(\zeta^2 + \zeta\sqrt{\zeta^2 - 1} - 1\right)\right]^{-1}\left(s + \zeta\omega_n + \omega_n\sqrt{\zeta^2 - 1}\right)^{-1}. \tag{4.15}$$

The time-domain response of the output, $c(t)$, may be obtained by utilizing the table of Laplace transforms given in Appendix A. It can be expressed as

$$c(t) = 1 + \left[2\left(\zeta^2 - \zeta\sqrt{\zeta^2 - 1} - 1\right)\right]^{-1} e^{-(\zeta - \sqrt{\zeta^2-1})\omega_n t} + \left[2\left(\zeta^2 + \zeta\sqrt{\zeta^2 - 1} - 1\right)\right]^{-1} e^{-(\zeta + \sqrt{\zeta^2-1})\omega_n t}, \quad t \geqslant 0. \tag{4.16}$$

Figure 4.2*b* illustrates the output response together with the unit step input. Notice that when $\zeta > 1$, the output response exhibits no overshoots and takes longer to reach its final value than when $\zeta = 1$. This response is described as bieng overdamped.

Case C. Damping Ratio Less than Unity

When $\zeta < 1$, the time-domain response can be obained in an analogous manner. The solution is slightly more complex, however, because we now have a pair of complex-conjugate poles. The partial fraction expansion of Eq. (4.5) can be written as

$$C(s) = \frac{K_1}{s} + \frac{K_2}{s + \zeta\omega_n - j\omega_n\sqrt{1-\zeta^2}} + \frac{K_3}{s + \zeta\omega_n + j\omega_n\sqrt{1-\zeta^2}}. \tag{4.17}$$

The constants K_1, K_2, and K_3 can be evaluated in an analogous manner by the method described in Chapter 2, but the algebra becomes a little complicated. In order to simplify this situation somewhat, use is made of the relationship between the location of the complex-conjugate poles in the complex plane and the damping ratio ζ. The geometry of the configuration is illustrated in Figure 4.3. Notice that the distance from the origin to either pole equals ω_n. In addition, the angle α has the following trigonometric properties:

$$\cos\alpha = \zeta, \tag{4.18}$$

$$\sin\alpha = \sqrt{1-\zeta^2}. \tag{4.19}$$

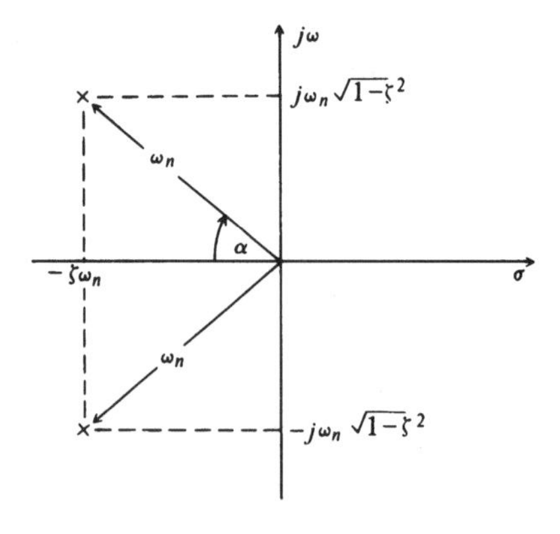

Figure 4.3 Location of the complex-conjugate poles in the complex plane.

Utilizing the relations given by Eqs. (4.18) and (4.19), the constants K_1, K_2, and K_3 can be expressed as

$$\begin{aligned} K_1 &= 1, \\ K_2 &= \frac{e^{-j\alpha}}{2j\sin\alpha}, \\ K_3 &= -\frac{e^{j\alpha}}{2j\sin\alpha}. \end{aligned} \tag{4.20}$$

Therefore Eq. (4.17) can be written as

$$\begin{aligned} C(s) =& \frac{1}{s} + \frac{e^{-j\alpha}}{2j\sin\alpha}\left(s + \zeta\omega_n - j\omega_n\sqrt{1-\zeta^2}\right)^{-1} \\ & - \frac{e^{j\alpha}}{2j\sin\alpha}\left(s + \zeta\omega_n + j\omega_n\sqrt{1-\zeta^2}\right)^{-1}. \end{aligned} \tag{4.21}$$

The time-domain response of the output, $c(t)$, may be obtained by utilizing Appendix A. It can be expressed as

$$\begin{aligned} c(t) =& 1 + \frac{e^{-j\alpha}}{2j\sin\alpha} e^{-(\zeta\omega_n - j\omega_n\sqrt{1-\zeta^2})t} \\ & - \frac{e^{j\alpha}}{2j\sin\alpha} e^{-(\zeta\omega_n + j\omega_n\sqrt{1-\zeta^2})t}. \end{aligned} \tag{4.22}$$

This can be simplified to

$$c(t) = 1 + \frac{e^{-\zeta\omega_n t}}{\sqrt{1-\zeta^2}} \frac{e^{j(\omega_n t\sqrt{1-\zeta^2}-\alpha)} - e^{-j(\omega_n t\sqrt{1-\zeta^2}-\alpha)}}{2j}, \tag{4.23}$$

or

$$c(t) = 1 - \frac{e^{-\zeta\omega_n t}}{\sqrt{1-\zeta^2}} \sin\left(\omega_n\sqrt{1-\zeta^2}t + \alpha\right), \quad t \geqslant 0. \tag{4.24}$$

Figure 4.2*c* illustrates the output response for a value of ζ which has a value of 0.3, together with the unit step input. Notice that the output response exhibits several overshoots and undershoots before finally settling out. This response, which is characteristic of an exponentially damped sinusoid, is described as being underdamped.

Examination of Eq. (4.24) indicates that the term in the exponent, $\zeta\omega$, multiplies t and it controls the exponential decay or rise of the unit step response, $c(t)$. Therefore, $\zeta\omega_n$ determines the "damping" of the system and it is defined as the *damping factor* of the system. Note that the inverse of $\zeta\omega_n$ is proportional to the time constant of the system.

Observe from Eq. (4.12) that when the system is critically damped and $\zeta = 1$, the damping factor equals ω_n. Therefore, we view ζ as the damping ratio which is defined as follows:

$$\zeta = \text{damping ratio} = \frac{\zeta\omega_n}{\omega_n} = \frac{\text{damping factor}}{\text{damping factor at critical damping.}} \tag{4.25}$$

The time to the first overshoot and the value of the first overshoot are two interesting identifying characteristics for this type of response. We shall next derive these values in terms of the undamped natural frequency ω_n and the damping ratio ζ.

Equation (4.24) indicates that the damped natural radian frequency of oscillation of the system, ω_m, is

$$\omega_m = \omega_n\sqrt{1-\zeta^2}. \tag{4.26}$$

The damped natural cyclic frequency of oscillation of the system, f_m, is

$$f_m = \frac{\omega_m}{2\pi} = \frac{\omega_n\sqrt{1-\zeta^2}}{2\pi}. \tag{4.27}$$

The period of oscillation of the underdamped system, t_m, is

$$t_m = \frac{1}{f_m} = \frac{2\pi}{\omega_n\sqrt{1-\zeta^2}}. \tag{4.28}$$

The time at which the peak overshoot occurs, t_p, is found by differentiating $c(t)$, from Eq. (4.24), with respect to time and setting the derivative equal to zero:

$$\begin{aligned}\frac{dc(t)}{dt} &= \zeta\omega_n e^{-\zeta\omega_n t}\left(\sqrt{1-\zeta^2}\right)^{-1}\sin\left(\omega_n\sqrt{1-\zeta^2}t+\alpha\right)\\ &\quad - \omega_n e^{-\zeta\omega_n t}\cos\left(\omega_n\sqrt{1-\zeta^2}t+\alpha = 0\right.\\ \frac{dc(t)}{dt} &= \frac{\omega_n}{\sqrt{1-\zeta^2}}e^{-\zeta\omega_n t}\sin\left(\omega_n\sqrt{1-\zeta^2}t\right) = 0\end{aligned}$$

This derivative is zero when

$$\omega_n\sqrt{1-\zeta^2}t = 0, \pi, 2\pi, \ldots$$

The peak overshoot occurs at the first value after zero, provided there are zero initial conditions. Therefore, the time to the first peak, t_p, and $\omega_n t_p$ are given by

$$t_p = \frac{\pi}{\omega_n\sqrt{1-\zeta^2}}, \tag{4.29}$$

$$\omega_n t_p = \frac{\pi}{\sqrt{1-\zeta^2}}. \tag{4.30}$$

For the case illustrated in Figure 4.2*c*, where $\zeta = 0.3$, the time to the first overshoot is $3.29/\omega_n$.

Substituting Eq. (4.30) into Eq. (4.24) yields the value for the maximum instantaneous value of the output, $c(t)$:

$$c(t) = 1 - \frac{\exp\left(-\zeta\pi/\sqrt{1-\zeta^2}\right)}{\sqrt{1-\zeta^2}} \sin(\pi + \alpha). \tag{4.31}$$

This can be simplified by substituting

$$\sin(\pi + \alpha) = -\sin\alpha \quad \text{and} \quad \sin\alpha = \sqrt{1-\zeta^2}.$$

Therefore,

$$c(t)_{\max} = 1 + \exp\left(-\frac{\zeta\pi}{\sqrt{1-\zeta^2}}\right) \tag{4.32}$$

This is usually expressed as a percentage of the input. Therefore, for a unit step input,

$$\text{Maximum percent overshoot} = \exp\left(-\frac{\zeta\pi}{\sqrt{1-\zeta^2}}\right) \times 100\% \tag{4.33}$$

For the case illustrated in Figure 4.2*c*, where $\zeta = 0.3$, the maximum percent overshoot is 37%.

The second-order system is a very common and popular one. In order for the reader to become more familiar with its typical characteristic responses, Figures 4.4 and 4.5 are shown to illustrate the resulting transient responses and percent maximum overshoots, rexpectively, for several values of damping factor.

It is interesting to compare the sketches of Figure 4.2*a*, *b*, and *c*. The critically damped system appears to be a compromise among the three systems shown. Although it does take a longer time to reach the desired value of unity than the underdamped system, it does not exhibit overshoots. The underdamped system, however, oscillates several times around the desired value before it finally settles to its steady-state value. Depending on the value of the damping ratio, the underdamped system may reach its final value faster than the critically damped system. Overdamped systems are rarely used in practice. As a matter of interest, practical systems are usually designed to be somewhat underdamped. Section 4.5 will discuss the best damping ratio to use for different applications.

4.3. RELATION BETWEEN LOCATION OF ROOTS IN THE s-PLANE AND THE TRANSIENT RESPONSE

We found in the preceding section that complex-conjugate roots located in the left-half of the s-plane, as illustrated in Figure 4.3, resulted in an exponentially damped sinusoidal oscillation as illustrated in Figures 4.2*c* and 4.4. For the case of second-order roots located on the negative real axis, we obtained the transient responses illustrated in Figures 4.2*a* and *b*. It is very interesting to extend our analysis from the preceding section to the location of a second-order control system's roots in other locations of the s-plane.

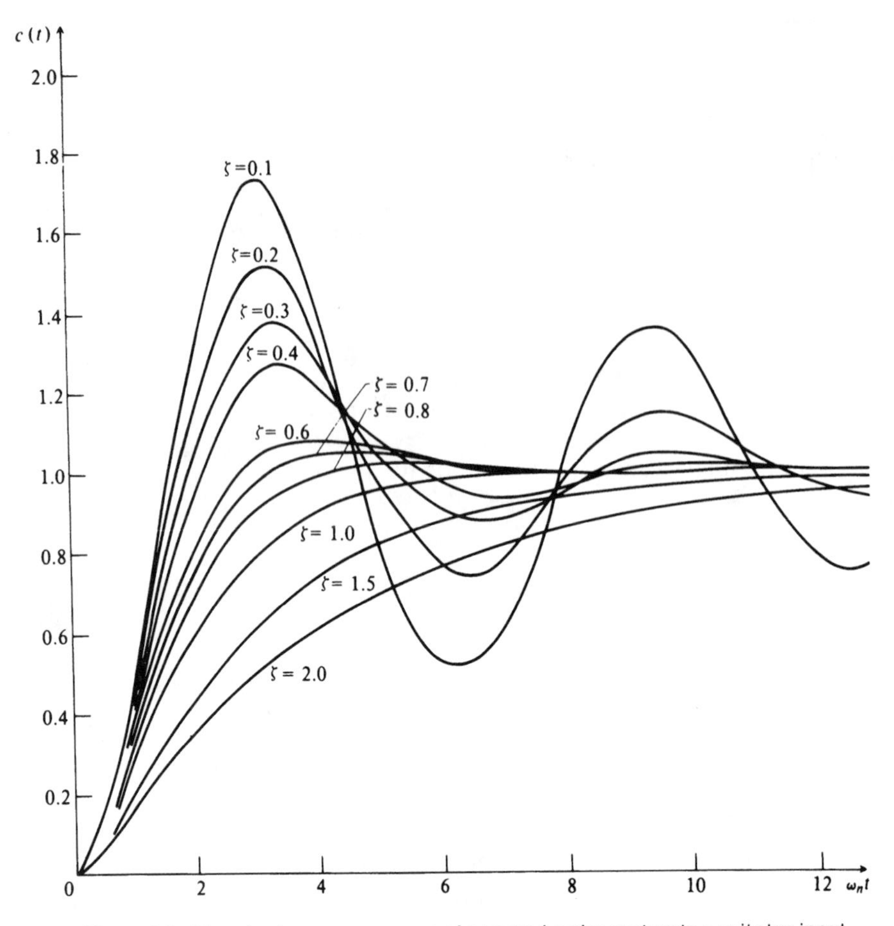

Figure 4.4 Transient response curves of a second-order system to a unit step input.

For the case of a pair of complex-conjugate roots located on the imaginary axis, we obtain an oscillation of fixed amplitude. This is the case of simple harmonic motion. For the case of a pair of complex-conjugate roots located in the right half-plane, we obtain an unbounded exponentially growing oscillation. For the case of real roots located in the right half-plane, we obtain an unbounded growing response. These cases are illustrated in Figure 4.6.

4.4. STATE-VARIABLE SIGNAL-FLOW GRAPH OF A SECOND-ORDER SYSTEM

In the previous section, the derivation of the time response of the second-order system was based on the transfer-function derivation. Therefore, an implicit assumption was made that the initial conditions were zero. This section illustrates how the state-variable signal-flow graph method of the underdamped case yields a more versatile and complete solution in the time domain including the response to the initial condition terms.

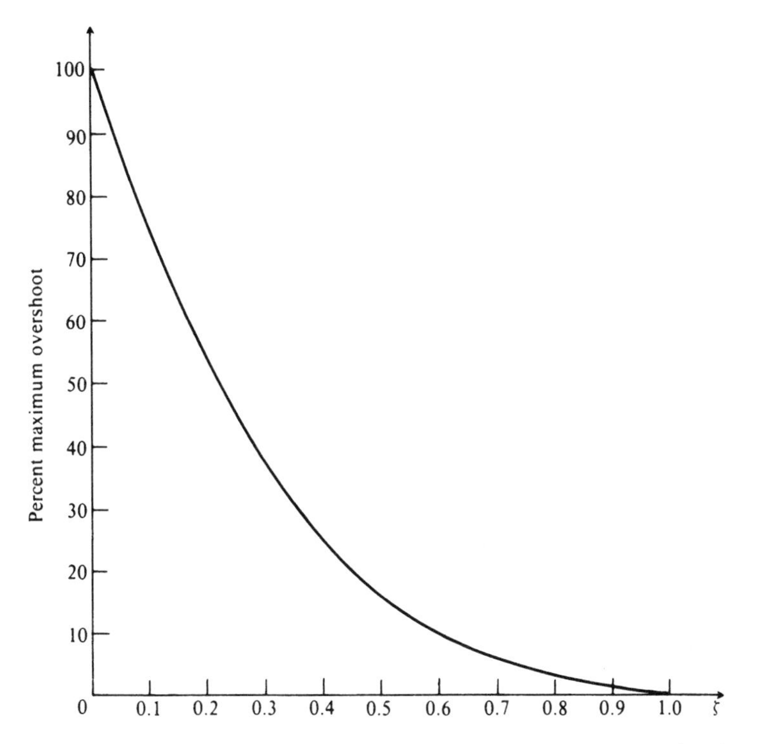

Figure 4.5 Percent maximum overshoot versus damping ratio for a second-order system.

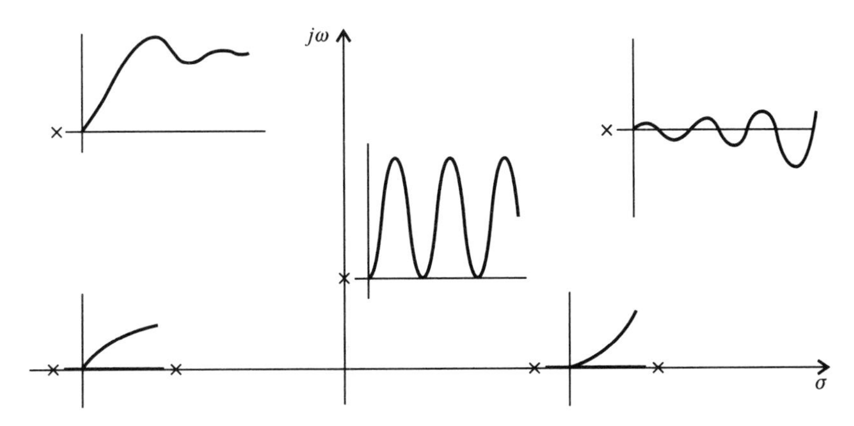

Figure 4.6 Responses of a second-order control system for various root locations in the *s*-plane (conjugate roots are not shown).

The state-variable signal-flow graph for this system can be derived from Eq. (4.3) as follows:

$$\frac{C(s)}{R(s)} = \frac{\omega_n^2}{s^2 + 2\zeta\omega_n s + \omega_n^2}. \tag{4.34}$$

Dividing through by s^2, we obtain

$$\frac{C(s)}{R(s)} = \frac{\omega_n^2 s^{-2}}{1 + 2\zeta\omega_n s^{-1} + \omega_n^2 s^{-2}}.$$

Defining

$$E(s) = \frac{R(s)}{1 + 2\zeta\omega_n s^{-1} + \omega_n^2 s^{-2}}, \tag{4.35}$$

we may rewrite Eq. (4.34) as

$$C(s) = \omega_n^2 s^{-2} E(s). \tag{4.36}$$

From Eq. (4.36) and

$$E(s) = R(s) - 2\zeta\omega_n s^{-1} E(s) - \omega_n^2 s^{-2} E(s), \tag{4.37}$$

the state-variable signal-flow graph for this system can easily be obtained as illustrated in Figure 4.7. The initial conditions are assumed to occur at $t = t_0$. Application of Mason's theorem (Eq. 2.135) to this diagram permits us to write the state equations directly as follows:

$$X_{1s}(s) = \frac{s^{-1}(1 + 2\zeta\omega_n s^{-1})}{\Delta} x_1(t_0) + \frac{s^{-2} x_2(t_0)}{\Delta} + \frac{s^{-2} R(s)}{\Delta}, \tag{4.38}$$

$$X_s(s) = \frac{-\omega_n^2 s^{-2}}{\Delta} x_1(t_0) + \frac{s^{-1} x_2(t_0)}{\Delta} + \frac{s^{-1} R(s)}{\Delta}, \tag{4.39}$$

where

$$\Delta = 1 + 2\zeta\omega_n s^{-1} + \omega_n^2 s^{-2}. \tag{4.40}$$

Simplifying Eqs. (4.38), (4.39), and (4.40), we obtain the following set of equations:

$$\begin{aligned} X_1(s) = {} & \frac{s + 2\zeta\omega_n}{s^2 + 2\zeta\omega_n s + \omega_n^2} x_1(t_0) + \frac{x_2(t_0)}{s^2 + 2\zeta\omega_n s + \omega_n^2} \\ & + \frac{1}{s^2 + 2\zeta\omega_n s + \omega_n^2} R(s) \end{aligned} \tag{4.41}$$

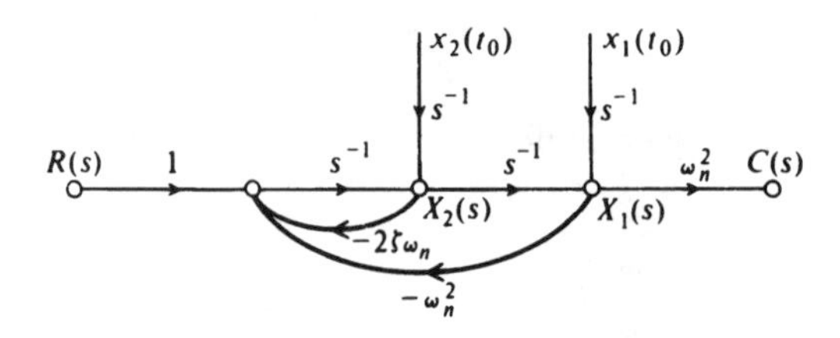

Figure 4.7 State-variable signal-flow graph of a second-order system.

$$X_2(s) = \frac{-\omega_n^2}{s^2 + 2\zeta\omega_n s + \omega_n^2} x_1(t_0) + \frac{s}{s^2 + 2\zeta\omega_n s + \omega_n^2} x_2(t_0) + \frac{s}{s^2 + 2\zeta\omega_n s + \omega_n^2} R(s) \tag{4.42}$$

These two equations can be put into the following vector form:

$$\begin{bmatrix} X_1(s) \\ X_2(s) \end{bmatrix} = \begin{bmatrix} \dfrac{s + 2\zeta\omega_n}{s^2 + 2\zeta\omega_n s + \omega_n^2} & \dfrac{1}{s^2 + 2\zeta\omega_n s + \omega_n^2} \\ \dfrac{-\omega_n^2}{s^2 + 2\zeta\omega_n s + \omega_n^2} & \dfrac{s}{s^2 + 2\zeta\omega_n s + \omega_n^2} \end{bmatrix} \begin{bmatrix} x_1(t_0) \\ x_2(t_0) \end{bmatrix} + \begin{bmatrix} \dfrac{1}{s^2 + 2\zeta\omega_n s + \omega_n^2} \\ \dfrac{s}{s^2 + 2\zeta\omega_n s + \omega_n^2} \end{bmatrix} R(s). \tag{4.43}$$

The inverse Laplace transform of Eq. (4.43) is given by the following expression (it is assumed that $R(s) = 1/s$ and that $\zeta < 1$):

$$\begin{bmatrix} x_1(t) \\ x_2(t) \end{bmatrix} = \underbrace{\begin{bmatrix} \dfrac{1}{\sqrt{1-\zeta^2}} e^{-\zeta\omega_n(t-t_0)} \sin\left[\omega_n\sqrt{1-\zeta^2}(t-t_0) + \phi_1\right] & \dfrac{1}{\omega_n\sqrt{1-\zeta^2}} e^{-\zeta\omega_n(t-t_0)} \sin \omega_n\sqrt{1-\zeta^2}(t-t_0) \\ \dfrac{-\omega_n}{\sqrt{1-\zeta^2}} e^{-\zeta\omega_n(t-t_0)} \sin \omega_n\sqrt{1-\zeta^2}(t-t_0) & \dfrac{1}{\sqrt{1-\zeta^2}} e^{-\zeta\omega_n(t-t_0)} \sin\left[\omega_n\sqrt{1-\zeta^2}(t-t_0) + \phi_2\right] \end{bmatrix}}_{\text{State Transition Matrix}} \begin{bmatrix} x_1(t_0) \\ x_2(t_0) \end{bmatrix}$$

$$+ \begin{bmatrix} \dfrac{1}{\omega_n^2}\left\{1 - \dfrac{1}{\sqrt{1-\zeta^2}} e^{-\zeta\omega_n(t-t_0)} \sin\left[\omega_n\sqrt{1-\zeta^2}(t-t_0) - \phi_2\right]\right\} \\ \dfrac{1}{\omega_n\sqrt{1-\zeta^2}} e^{-\zeta\omega_n(t-t_0)} \sin \omega_n\sqrt{1-\zeta^2}(t-t_0) \end{bmatrix}, t \geq 0 \tag{4.44}$$

where

$$\phi_1 = \tan^{-1}\frac{\sqrt{1-\zeta^2}}{\zeta}, \quad \phi_2 - \tan^{-1}\frac{\sqrt{1-\zeta^2}}{-\zeta}.$$

Notice from this approach that the state transition matrix is readily available.

The corresponding output response of this second-order system having a unit step input is given by the following expression:

$$
\begin{aligned}
c(t) &= \omega_n^2 x_1(t) \\
&= \frac{\omega_n^2}{\sqrt{1-\zeta^2}} e^{-\zeta\omega_n(t-t_0)} \sin\left[\omega_n\sqrt{1-\zeta^2}(t-t_0)+\phi_1\right] x_1(t_0) \\
&+ \frac{\omega_n}{\sqrt{1-\zeta^2}} e^{-\zeta\omega_n(t-t_0)} \sin\left[\omega_n\sqrt{1-\zeta^2}(t-t_0)\right] x_2(t_0) \\
&+ 1 - \frac{1}{\sqrt{1-\zeta^2}} e^{-\zeta\omega_n(t-t_0)} \sin\left[\omega_n\sqrt{1-\zeta^2}(t-t_0)-\phi_2\right], \quad t \geqslant 0. \qquad (4.45)
\end{aligned}
$$

Equation (4.45) is a much more complete solution to the classical second-order system than that previously derived in Eq, (4.24), since the initial conditions are now accounted for. Observe from Eq. (4.45) that the solution, when the initial conditions are zero, reduces to that of Eq. (4.24) with $\alpha = -\phi_2$. The first term in Eq. (4.45) represents the forced response due to the external forcing function, and the last two terms represent the natural or homogeneous solutions due to the initial conditions.

4.5 WHAT IS THE BEST DAMPING RATIO TO USE?

A very important question to answer before leaving this chapter is the best value of damping ζ to choose for a control system. Section 4.2 had indicated that most practical systems are usually designed to be somewhat underdamped. This will be discussed further in Section 5.6 from the viewpoint of various performance criteria. This section will present general guidelines for the selection of damping for three specific cases:

A. Design of a Control System for General Applications

Suppose it is desired to design a linear control system for general applications. It is assumed that the designer does not know the final application of the control system. The selection of the damping ratio for such a general application requires a tradeoff between maximum percent overshoot and the time where the peak overshoot occurs, t_p. A smaller damping ratio decreases t_p (which is desirable), but it increases the maximum percent overshoot (which is undesirable). Final choice of the damping ratio is subjective. It has been my experience that the damping ratio range is usually selected between 0.4 and 0.7 for this general case. It is important to note that Chapter 5, Section 5.8 on "The ITAE Performance Criterion for Optimizing the Transient Response" shows that the damping needed to minimize the integral of time times the absolute value of the error (ITAE) performance criterion is 0.7 for second-order systems.

The selection of the best damping ratio should not be restricted to the limitations imposed by linear system theory. If we analyze Figure 4.4, we recognize that the best choice of the damping ratio is between 0.4 and 0.7 from the viewpoint of linear system theory (tradeoff between maximum percent overshoot and the time at which it peaks). However, it is important to emphasize that a linear system is not the best choice for optimizing transient responses. A nonlinear system can be designed to give a better transient response than a linear system, and an adaptive control system can be designed to give a better transient response than a nonlinear system [1]. Adaptivity refers to a system where the damping factor varies continually with time as a function of the system error, and the damping factor "adapts" to the system error.

Consider the three transient responses illustrated in Figure 4.8 to a unit step input: a linear control system, a nonlinear control system, an adaptive control system. The linear control system's transient response is representative of a second-order system having a damping ratio of 0.4 and an undamped natural frequency of 1 rad/sec.

We can improve this linear control system's transient response with the nonlinear system's response illustrated which can be obtained with the configuration shown in Figure 4.9 where the amplifier gain K is a function of the control-system error [1]. Qualitatively, the amplifier gain K is set at a relatively high value for very large errors

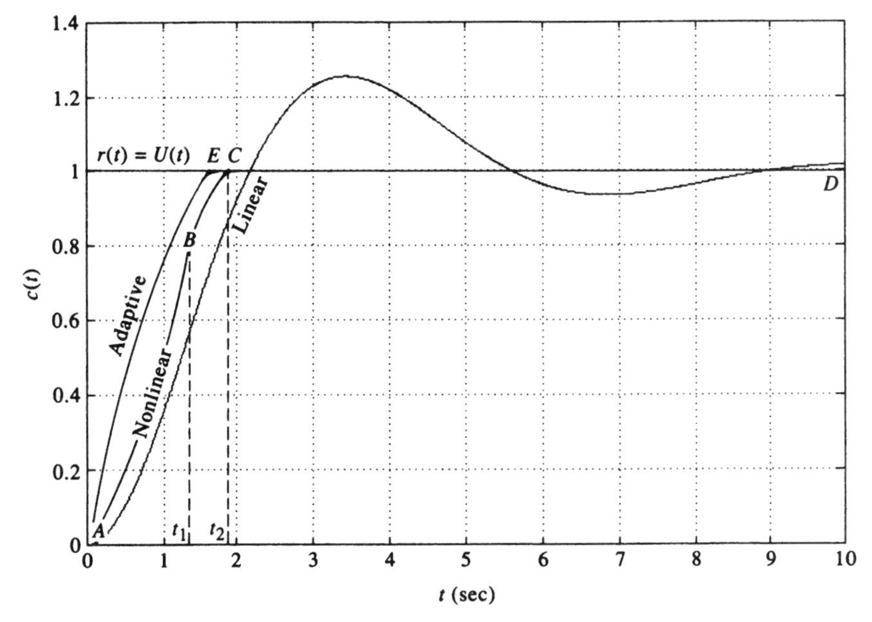

Figure 4.8 Comparison of transient response of a linear control system (where $\zeta = 0.4$ and $\omega_n = 1$), a hypothetical nonlinear control system, and an adaptive control system to a unit step.

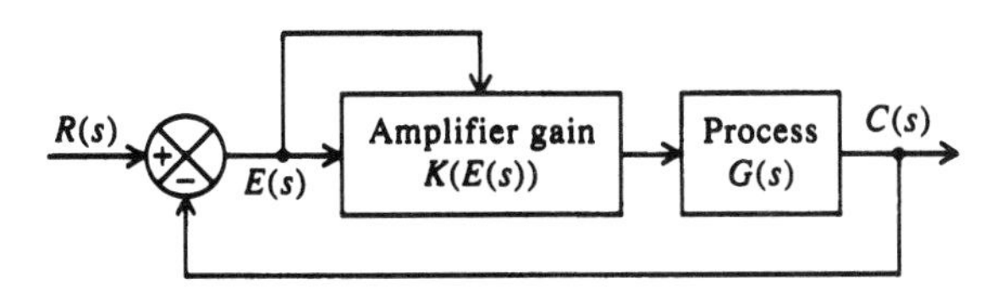

Figure 4.9 A control system where the amplifier gain is a function of the error.

(and the resulting damping factor will be very low), and K is set at a much smaller value for small errors (and the resulting damping factor will be much higher). It is theoretically possible for such a system to have the transient response of A-B-C-D shown in Figure 4.8, where the gain is set at a very high value from 0 to t_1, and it is set at a much lower value from t_1 to t_2. The damping factor might be designed for 0.1 from A to B, and it might be designed to be critically damped from B to C. (The gain would be switched at point B.) Such a system has the advantage that it responds to its steady-state value quicker than the linear control system, and without any overshoots.

We can improve the transient response of the nonlinear control system shown with an adaptive control system where the amplifier gain, in the configuration of Figure 4.9 is continually varied as a function of time (not just two gain settings as in the nonlinear control-system case). Therefore, the damping ratio of this configuration "adapts" to the error present continuously as a function of time. For example, near point A, it is theoretically possible for the damping ratio to be set at 0.01 (or undamped), and then it can be gradually increased to unity just below point E. In practice, the control-system engineer must design this adaptive control system very carefully and take into account the inertias and the dynamics of the system—otherwise, overshoots and/or undershoots can occur which might make this system's transient response worse than the nonlinear and linear control-system configurations. Reference 1 illustrates how to design such an adaptive control system and achieve its full benefits.

B. Design of a Very Accurate Linear Control System

Let us assume that it is necessary to design an extremely accurate linear control system whose steady-state errors to position and velocity inputs are extremely small (e.g., to be used for navigation purposes). For such a system, the transient response is not the primary performance criterion to optimize, but minimum steady-state error is the major objective. In Chapter 5, it is proved that the steady-state error of a second-order control system to a velocity input of unit magnitude is given by $2\zeta/\omega_n$ [reciprocal of Eq. (5.37)]. In such a case, it is very important to design the damping ratio to be as small as possible because the steady-state error is proportional to the damping ratio. Values less than 0.1 are not unreasonable for such applications, and the disadvantage of waiting for the relatively long transient response times to settle out must be tolerated.

C. Design of a Slowly Responding Linear Control System

Some applications require a very slow response of the linear control system. An example of this might be an elevator, or a platform that is part of a simulator used to train sailors in the diving and control functions of a submarine. For these applications, which are usually found where people are located on a platform that the control system is positioning (such as found in trainers) or in simulation of slowly varying processes, the control system's damping factor is selected to try to duplicate the dynamics of the vehicle or system that is being simulated. Damping ratios greater than 1 are usual in these applications, and damping ratios as high as 2 may be reasonable.

From the three control-system cases considered, we have justified damping ratios from 0.01 (for part of the transient response in the case of the adaptive control system) to values up to 2.0. In conclusion, the damping ratio selected is determined by the particular control system's application.

4.6. MODELING THE TRANSFER FUNCTIONS OF CONTROL SYSTEMS [2]

In the previous material presented, we designate transfer functions to various elements of the control system. For example in Figure 2.9, some of the elements shown in this block diagram are designated $G_1(s)$, $G_2(s)$, $G_3(s)$, $G_4(s)$, etc. How do we determine their actual transfer functions? If these elements are simple networks as described in Table 2.4, we can then use their corresponding theoretical transfer functions as shown in Table 2.4. If they are electromechanical motors, we can use the transfer functions for these devices that are derived in Section 3.4. However, in the case of motors and other devices, how do we know whether the actual transfer function agrees with the theoretical transfer function? What if we have an unknown device, or "black box," and we do not know what its transfer function is at all?

To address this problem, we can apply a test signal to the device and measure its output. From knowledge of the input test signal and the measured output, we can then determine the device's actual transfer function. The most commonly used test signals are a unit step and a sinusoidal signal. We use the unit step to obtain a transient response of the device. This approach is presented in this section. The frequency-response approach for testing the frequency reponse of a device to a sinusoidal input signal is presented in Chapter 6.

Suppose we have a linear control system consisting of only one element, $G(s)$, as shown in Figure 2.7 and we do not know its transfer function. To determine its transfer function, we will apply a unit step input and measure the output's transient response. Therefore, $r(t) = U(t)$, and $R(s) = 1/s$. The possible transient responses for the outputs, $c(t)$, are shown in Figure 4.2. How do we model this output transient response into a describable function for $c(t)$?

Having determined the function $c(t)$, the procedure for finding $G(s)$ would proceed as follows. The Laplace transfer function of $c(t)$, $C(s)$, is then determined. Since we know the Laplace transform of the unit step input is $R(s) = 1/s$, we can then determine the transfer function of $G(s)$ as follows:

$$G(s) = \frac{C(s)}{R(S)} = \frac{C(s)}{\frac{1}{s}} = sC(s). \tag{4.46}$$

The resulting transient response may take on the characteristics of any of the results illustrated in Figure 4.2. For example, it may result in a critically damped, overdamped, or underdamped system.

Observe from the transient responses in Figure 4.2 that the various responses are smooth and monotonic. Let us assume that $c(t)$ is given by the sum of exponential terms as follows:

$$c(t) = c_{ss} + K_1 e^{-at} + K_2 e^{-bt} + \dots \tag{4.47}$$

where c_{ss} represents the final-value of $c(t)$ as t approaches infinity. Simplifying Eq. (4.47) and focusing only on "$-a$" as the slowest or smallest root, we obtain the following:

$$c(t) - c_{ss} \approx K_1 e^{-at} + \dots . \tag{4.48}$$

Taking the logarithm to the base 10 of Eq. (4.48), we obtain the following:

$$\log_{10}[c(t) - c_{ss}] \approx \log_{10} K_1 - at \log_{10} e, \tag{4.49}$$

$$\log_{10}[c(t) - c_{ss}] \approx \log_{10} K_1 - 0.4343\, at. \tag{4.50}$$

Equation (4.50) is the equation of a straight line, We can fit a straight line to the plot of $\log_{10}[c(t) - c_{ss}]$, or $\log_{10}[c_{ss} - c(t)]$ if K_1 is negative, and we can estimage K_1 and a. Having estimated K_1 and a, we can then plot $c(t) - c_{ss} - K_1 e^{-at}$ which is determined by subtracting the line $K_1 e^{-at}$ from the original data. At this point, the curve is approximately $K_2 e^{-bt}$ and the plot represents $\log_{10} K_2 - 0.434bt$. The process is repeated until the residual (error) in the modeling process approaches zero.

To illustrate this procedure, let us consider the transient data in Table 4.1 which represent the transient response of a closed-loop, second-order, control system given by the following transfer function

$$\frac{C(s)}{R(s)} = \frac{9}{s^2 + 6s + 9} \tag{4.51}$$

to a unit step input. The transient response is plotted in Figure 4.10 as the solid curve. The control system whose transfer function is given by Eq. (4.51) represents a

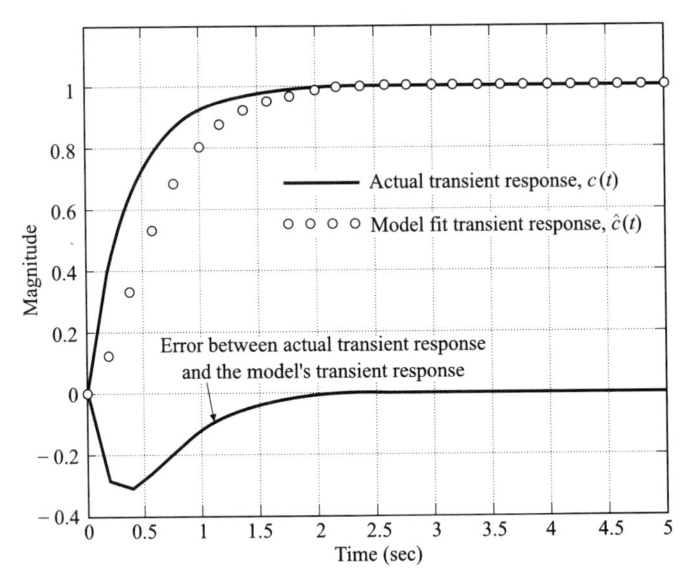

Figure 4.10 Transient response of the control system whose transfer function is given by Eq. (4.51) to a unit step input, and a model fit $\hat{c}(t)$.

Table 4.1. Transient Response of Eq. (4.51) to a Unit Step Input to Obtain the Theoretical Value of $c(t)$, the Model Fit Data $\hat{c}(t)$, and the Error between the Theoretical Value of $c(t)$ and the Model Fit Data

Time (sec)	$c(t)$	$\hat{c}(t)$ (Model fit)	error $= c(t)$–fit of $\hat{c}(t)$
0	0	0	0
0.2000	0.1219	0.3994	−0.2775
0.4000	0.3374	0.6393	−0.3019
0.6000	0.5372	0.7833	−0.2462
0.8000	0.6916	0.8699	−0.1783
1.0000	0.8009	0.9218	−0.1210
1.2000	0.8743	0.9531	−0.0787
1.4000	0.9220	0.9718	−0.0498
1.6000	0.9523	0.9831	−0.0308
1.8000	0.9711	0.9898	−0.0187
2.0000	0.9826	0.9939	−0.0112
2.2000	0.9897	0.9963	−0.0067
2.4000	0.9939	0.9978	−0.0039
2.6000	0.9964	0.9987	−0.0023
2.8000	0.9979	0.9992	−0.0013
3.0000	0.9988	0.9995	−0.0008
3.2000	0.9993	0.9997	−0.0004
3.4000	0.9996	0.9998	−0.0002
3.6000	0.9998	0.9999	−0.0001
3.8000	0.9999	0.9999	−0.0001
4.0000	0.9999	1.0000	0.0000
4.2000	1.0000	1.0000	0.0000
4.4000	1.0000	1.0000	0.0000
4.6000	1.0000	1.0000	0.0000
4.8000	1.0000	1.0000	0.0000
5.0000	1.0000	1.0000	0.0000

critically damped control system whose two roots are equal and lie at −3.0 on the negative real axis. A plot of the absolute value of $\log_{10}[1 - c(t)]$ is illustrated in Figure 4.11. From the straight-line fit obtained using MATLAB, the vertical axis intercept is 0.3436. Therefore,

$$\log_{10} K_1 = 0.3436 \tag{4.52}$$

with the result that $K_1 = -2.2057$, and K_1 is negative because c_{ss} is greater than $c(t)$. The slope of this straight line is −1.104. Therefore,

$$0.4343a = 1.104 \tag{4.53}$$

and $a = 2.5430$.

The next step is to subtract this line from the given data for $c(t)$ and plot the log of this result. This is performed in Figure 4.12. The vertical intercept of this plot is 0.6517. Therefore,

$$\log_{10} K_2 = 0.6517 \tag{4.54}$$

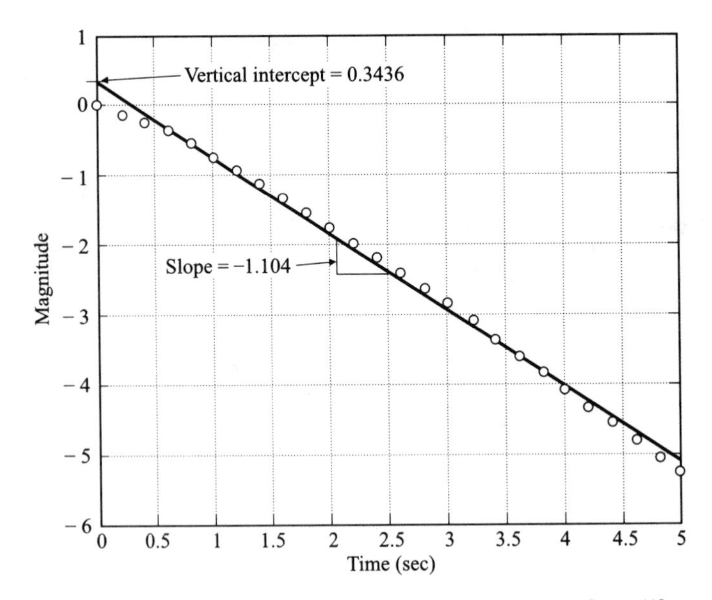

Figure 4.11 Straight-line fit to the absolute value of $\log_{10}[1 - c(t)]$.

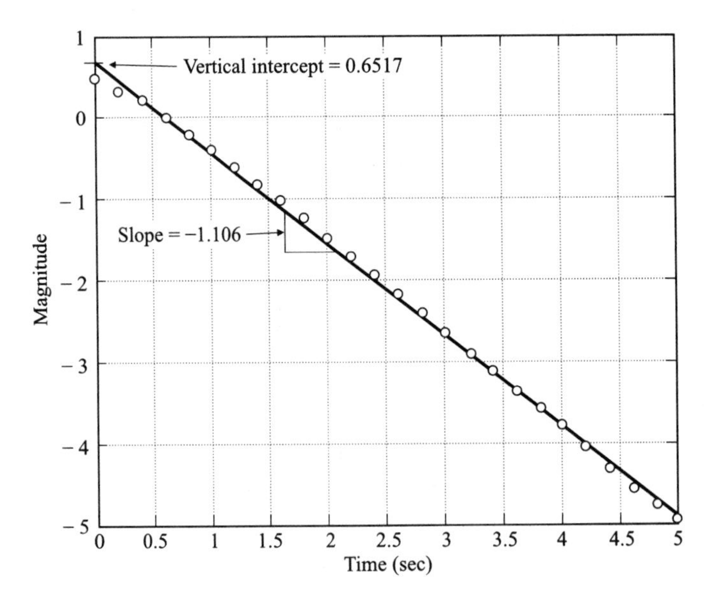

Figure 4.12 Straight-line fit to the subtraction of the straight-line obtained in Figure 4.11 from the given data for $c(t)$.

with the result that $K_2 = -4.4847$, and K_2 is negative because c_{ss} is greater than $c(t)$. The slope of this line is -1.106. Therefore,

$$0.4343b = 1.106 \tag{4.55}$$

and $b = 2.5466$. Combining the results of Eqs. (4.47), (4.52), (4.53), (4.54), and (4.55), we obtain the following estimate for $\hat{c}(t)$:

$$\hat{c}(t) \approx 1 - 2.2057e^{-2.5430t} - 4.4847e^{-2.5466t}. \tag{4.56}$$

Because we want $c(0) = 0$, we must adjust the estimate of Eq. (4.56) by adding the following term which is obtained from the steady-state value and the coefficients of K_1 and K_2:

$$adjustment = -\frac{1 - 2.2057 - 4.4847}{2} = +2.8452. \tag{4.57}$$

Therefore, we add +2.8452 to the coefficients K_1 and K_2 and obtain the following estimate for $c(t)$:

$$\hat{c}(t) \approx 1 + 0.6395e^{-2.5430t} - 1.6395e^{-2.5466t}. \tag{4.58}$$

The result of this estimate from Eq. (4.58) is plotted in Figure 4.10 and superimposed on the result of the actual unit step response of $c(t)$. This figure also illustrates the error between $c(t)$ and the estimate to $\hat{c}(t)$. Table 4.1 also contains the actual data for $c(t)$, the estimated value for $\hat{c}(t)$ and the error between $c(t)$ and the estimated value of $\hat{c}(t)$.

Therefore, the estimated or modeled transfer function can be obtained as illustrated in Eq. (4.46). Taking the Laplace transform of (4.58), we obtain the following:

$$\hat{C}(s) = \frac{1}{s} + \frac{0.6395}{s + 2.543} - \frac{1.6395}{s + 2.5466}. \tag{4.59}$$

Dividing $\hat{C}(s)$ from Eq. (4.59) by $R(s) = 1/s$, we obtain the following:

$$\frac{\hat{C}(s)}{R(s)} = \frac{\frac{1}{s} + \frac{0.6395}{s + 2.5430} - \frac{1.6395}{s + 2.5466}}{\frac{1}{s}}. \tag{4.60}$$

This can be simplified to the following result:

$$\frac{\hat{C}(s)}{R(s)} = \frac{2.5492s + 6.476}{s^2 + 5.089s + 6.475}. \tag{4.61}$$

The resulting modeled value of $\hat{C}(s)/R(s)$ from Eq. (4.61) contains a zero term in the left half-plane, while the known transfer function of $C(s)/R(s)$ in Eq. (4.51) does not contain a zero term.

The resulting estimated value of $c(t)$ is fairly close to the actual value of $c(t)$ as shown in Figure 4.10. This example illustrates that this procedure is very sensitive, and great care must be taken to ensure accuracy. Therefore, MATLAB was used in solving this example. If this procedure were done with hand calculations, a poorer fit would undoubtedly result. Another factor which must be brought to the attention of the reader is that noise can corrupt the results and lead to erroneous models. Therefore, in obtaining the test data, ensure that the data have a very good signal-to-noise ratio. More sophisticated and accurate modeling procedures can be found in Reference 2.

4.7. ILLUSTRATIVE PROBLEMS AND SOLUTIONS

This section provides a set of illustrative problems and their solutions to supplement the material presented in Chapter 4.

I4.1 A control sysem whose two closed-loop poles and one closed-loop zero are located in the s-plane is illustrated:

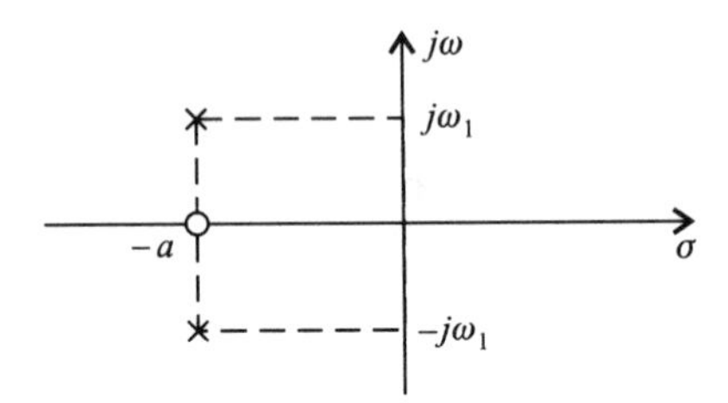

Figure I4.1

(a) Determine the closed-loop transfer function in its simplest form.

(b) Determine the *impulse response* of this system using the table of Laplace transforms in Appendix A.

SOLUTION: (a)

$$\frac{C(s)}{R(s)} = \frac{K(s+a)}{(s+a+j\omega_1)(s+a-j\omega_1)} = \frac{K(s+a)}{(s+a)^2+\omega_1^2}.$$

(b) Since $R(s) = 1$, then

$$C(s) = \frac{K(s+a)}{(s+a)^2+\omega_1^2}.$$

From Appendix A, we obtain the following inverse Laplace transform of this equation:

$$c(t) = (Ke^{-at}\cos(\omega_1 t))U(t).$$

I4.2 A control-system engineer is trying to reduce costs in his development project to produce a positioning system using a two-phase ac servomotor. The engineer finds a discarded two-phase ac servomotor on a shelf in the parts room, and wants to know its characteristics so that he can determine its usefulness for the project. It is decided to observe the response of this motor in a unity-gain feedback loop to a unit step input as illustrated:

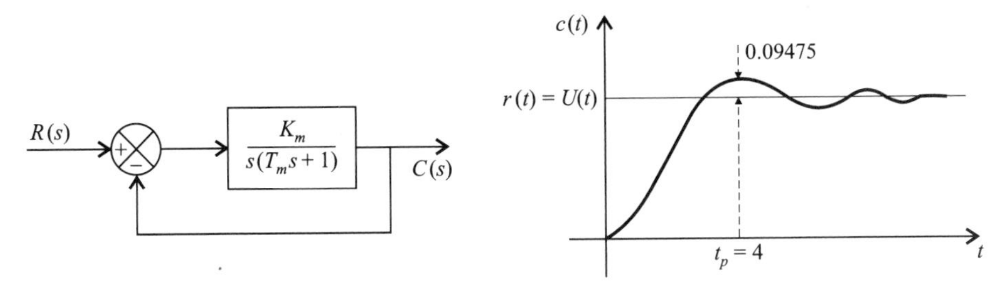

Figure I4.2

From the observed characteristics, determine:

(a) The undamped natural resonant frequency of this system, ω_n.

(b) The motor time constant, T_m.

(c) The motor constant, K_m.

SOLUTION: **(a)** From Eq. (4.32), we determine that a 9.475% overshoot corresponds to a damping factor, ζ, of 0.6. From Eq. (4.29),

$$t_p = \frac{\pi}{\omega_n\sqrt{1-\zeta^2}}.$$

Substituting $t_p = 4.0$ sec and $\zeta = 0.6$ into this equation, we can solve for ω_n:

$$\omega_n = 0.982.$$

(b) From Eq. (4.2),

$$\omega_n^2 = \frac{K_m}{T_m}$$

$$\zeta = \frac{1}{2\omega_n T_m}.$$

We substitute $\zeta = 0.6$ and $\omega_n = 0.982$ into the last equation, and we obtain that

$$T_m = 0.849.$$

(c) Substituting into the following equation from Eq. (4.2):

$$\omega_n^2 = \frac{K_m}{T_m}$$

for $\omega_n = 0.982$ and $T_m = 0.849$, we find that

$$K_m = 0.819.$$

PROBLEMS

4.1. The two-phase ac servomotor of Problem 3.16 is used in a simple positioning system, as shown in Figure P4.1. Assume that a difference amplifier, whose gain is 10, is used as the error detector and also supplies power to the control field.

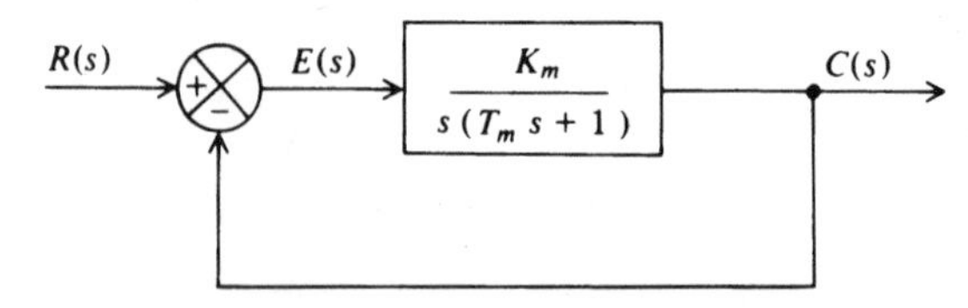

Figure P4.1

(a) What are the undamped natural frequency ω_n and the damping ratio ζ ?

(b) what are the percent overshoot and time to peak resulting from the application of a unit step input?

(c) Plot the error as a function of time on the application of a unit step input.

4.2. Repeat Problem 4.1 with the gain of the difference amplifier increased to 20. What conclusions can you draw from your result?

4.3. The two phase ac servomotor and load, in conjunction with the gear train specified in Problem 3.19, is used in a simple positioning system as shown in Figure P4.1. Assume that a difference amplifier, whose gain is 20, is used as the error detector and also supplies power to the control field.

(a) What are the undamped natural frequency and damping ratio ζ?

(b) What are the percent overshoot and time to peak resulting from the application of a unit step input?

(c) Plot the error as a function of time on the application of a unit step input.

4.4. Repeat Problem 4.3 with the gain of the difference amplifier increased to 40. What conclusions can you draw from your results?

4.5. Repeat Problem 4.3 with the gain of the difference amplifier decreased to 10. What conclusions can you draw from your result?

4.6. A typical aerodynamically controlled missile control system is synthesized by means of the appropriate application of moments to the airframe. These moments are generated by the deflection of control surfaces placed at large distances from the center of gravity. The result is that large moments are created with relatively small surface loads. The design of this type of control system requires a sufficiently high control loop gain in order to minimize the response time to input commands. In addition, it must not be so high as to cause high-frequency instabilities. Figure P4.6 illustrates an acceleration-control steering system of a missile. The command acceleration is compared with the output of an accelerometer to develop the basic error signal which drives the control system. The output of a rate gyro is utilized for damping [3].

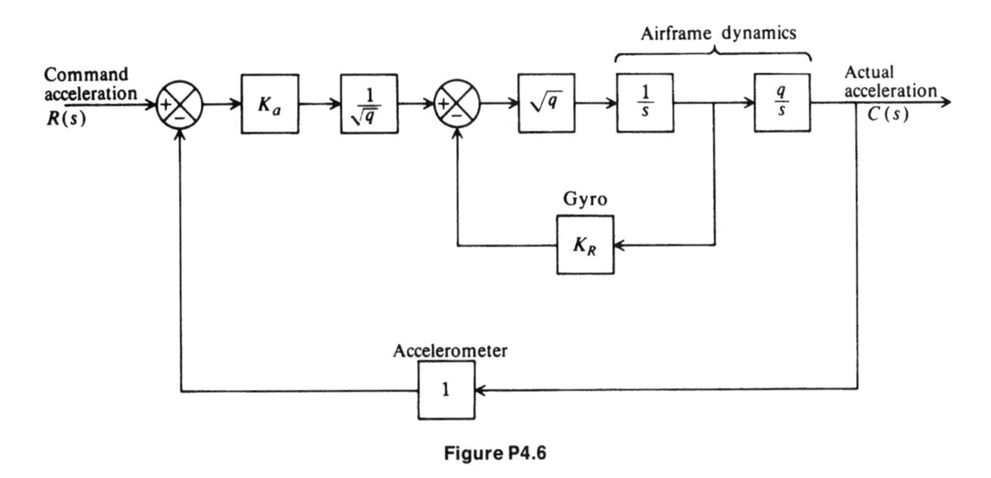

Figure P4.6

(a) Determine the transfer function $C(s)/R(s)$ of this system.

(b) Determine the undamped natural frequency ω_n of this system and the damping ratio ζ for the following set of parameters:

amplifier gain $= K_a = 16$, aircraft gain factor $= q = 4$, $K_R = 4$.

(c) Determine the percent overshoot and time to peak resulting from an input command of a unit step of acceleration.

4.7. Figure P4.7 illustrates the control system of a map drive system used to display a portion of the map of an area to the commander of a task force. A main requirement of this control system is that a constant tension be maintained on the continuous sheet of paper between the wind-up and wind-off rolls in order that the paper does not tear. The system illustrated contains four rollers, and a spring which provices a restoring angular torque of $K_1\theta(t)$. As the radius R of the rollers varies, the tension changes and an adjustment in the wind-up motor speed is required. The map leaving the wind-off roll is assumed to have a velocity $v_1(t)$ and a tension $T_2(t)$; that approaching the wind-up roll is assumed to have a velocity $v_2(t)$ and a tension $T_1(t)$. A synchro-type sensing device, whose transfer function is K_e, is mounted on the pivot point of the tension arm and is used to sense angular deviations from $\theta = 0$. This error signal is amplified by a factor K_a by an amplifier. In addition, assume that the relationship for motor control voltage $E_c(t)$ caused by tension variations $T(t)$ is given by

$$E_c(t) = K_2 T(t).$$

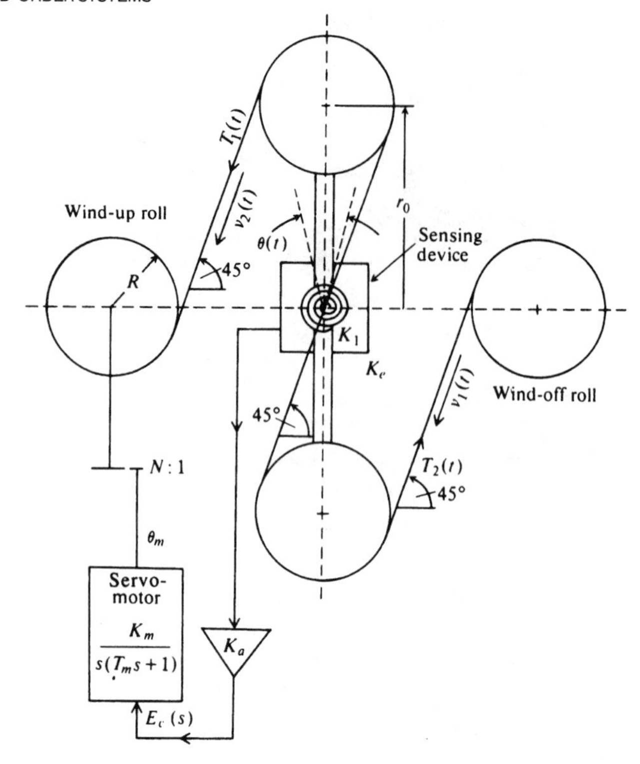

Figure P4.7

The two-phase ac servomotor is assumed to have a transfer function given by Eq. (3.101). It also contains a gear reduction of $N:1$.

(a) Determine the transfer function relating the change in tension $T(s)$ due to an input change velocity $v_1(s)$ of the wind-off roll in terms of the system parameters.

(b) Determine the undamped natural frequency ω_n and the damping ratio ζ in terms of the system parameters.

(c) Calculate the damping ratio ζ, for the case of a fully loaded wind-up roll, with the following parameters:

$$\begin{aligned}
\text{inertia of wind-up roller} &= 0.101 \text{ oz in sec}^2, \\
\text{gear ratio} &= N = 10.4:1 \\
\text{radius of wind-up roller} &= R = 1.5 \text{ in}, \\
\text{motor inertia} &= J_m = 4.43 \times 10^{-4} \text{ oz in sec}^2, \\
\text{motor constant} &= K_m = 0.417 \text{ in oz/(rad/sec)}, \\
\text{synchro sensitivity} &= K_e = 0.4 \text{ V/degree}, \\
\text{amplifier gain} &= K_a = 1000, \\
K_1 &= 10.5 \text{ V/oz/tension}, \\
K_2 &= 0.04 \text{ in oz/degree}.
\end{aligned}$$

(d) How can the damping ratio be increased?

4.8. The Viking Mission conducted by the National Aeronautics and Space Administration included two launches in 1975 of a Viking Spacecraft by a Titan-Centaur launch vehicle consisting of an Orbiter and a Lander. The Orbiter had the capability of orbiting the planet Mars and of separating the Lander capsule, an automated laboratory in the search of signs of life, that entered the Martian atmosphere for soft landing on the surface of Mars. Looking over Mars from orbit, Viking cameras and other instruments aided in the confirmation of a suitable landing site. After this confirmation, the Lander separated from the Orbiter and began its descent to the Martian surface. The descent occurred when Mars was about 225 million miles from Earth and nearly on the other side of the Sun. Therefore, this required a completely automated deorbit and landing operation because two-way communication at that distance was almost 45 minutes. Viking featured a series of scientific experiments in the areas of biology, geology, and meteorology. In order to collect material for these experiments, the Lander had a retractable claw with a 10-ft reach, as shown in Figure P4.8*a*. It was used for

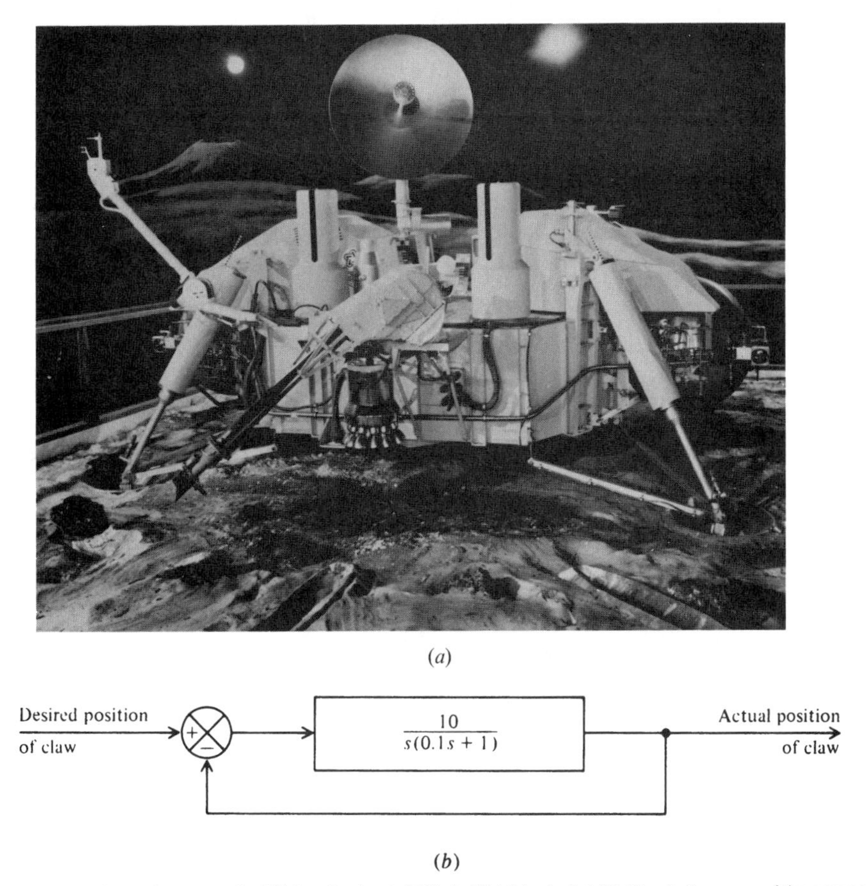

(*a*)

(*b*)

Figure P4.8 (*a*) Lander capsule, Viking Project. (Official NASA photo) (*b*) Block diagram of the second-order control system for positioning the retractable claw.

scooping out soil samples and placing them in its automated chemical laboratory for analysis. For purposes of this example, assume that the control system which positions the retractable claw can be represented by the second-order control system shown in Figure P4.8*b*.

(a) Determine the undamped natural frequency and the damping ratio of this control system.

(b) Determine the maximum percent overshoot and time to peak resulting from the application of a unit command signal.

4.9. A two-phase ac induction motor is used to position a device in a feedback configuration represented by Figure P4.1. The time constant of the motor and load, T_m, is 0.5 sec.

(a) Determine the combined amplifier and motor constant gain K_m which will result in a damping ratio of 0.5.

(b) What is the resulting undamped natural frequency for the value of gain determined in part (a)?

4.10. A simulator used for training the crew of an aircraft is to be designed. The simulator will have a two-degree-of-freedom capability (roll and pitch), which will give the trainees a feel of the dynamics which the aircraft crew will experience during operating conditions. A typical aircraft trainer of this type is usually driven by a digital computer which provides the control inputs/logic of the system. The controls and interior of the simulator will be designed to be a very close replica of the actual aircraft, and give the trainees a real-world feel of the aircraft—both esthetically and dynamically. For purposes of this analysis, assume that the pitch loop can be represented by Figure P4.10.

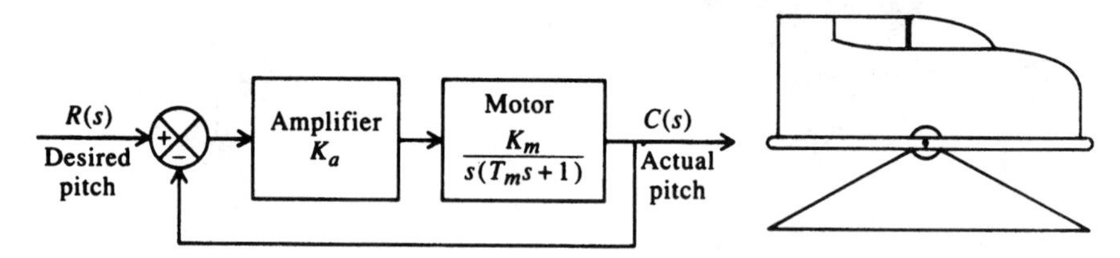

Figure P4.10

Test data of the aircraft dynamics show that pitch can closely be approximated by a critically damped loop. Assuming that $\zeta = 1$, $K_m = 5$ and $T_m = 1$, determine the following:

(a) The resulting undamped natural frequency of the system, ω_n.

(b) Amplifier gain K_a required to achieve critical damping.

4.11. Robotics have revolutionized the manufacturing industry, and this is especially true in the manufacture of automobiles. Figure P4.11*a* shows a photograph of the GMFanuc Robotics Corporation Model P-150 six-axis (a seventh axis is optional) articulated arm, electric-servo-driven robot painting an automobile. All axis motions are controlled simultaneously [4]. All axes are driven by state-of-the-art compact ac servomotors providing fast acceleration and deceleration, precision painting, and no brush maintenance.

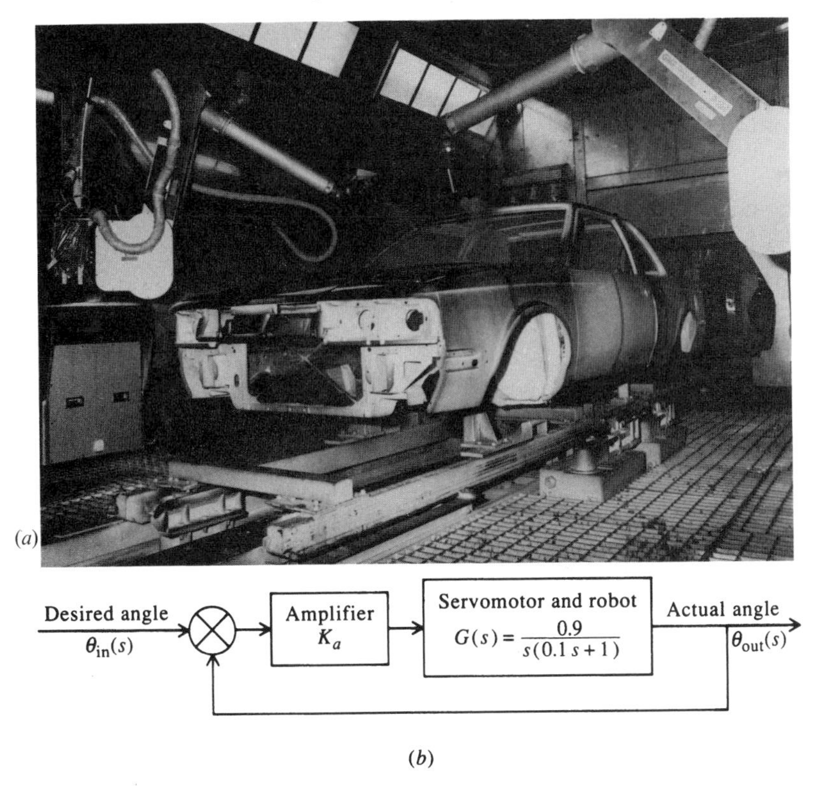

Figure P4.11 (*a*) Photograph of the GMFanuc Robotics Corporation Model P-150 six-axis (a seventh axis is optional) articulated arm, electric-servo-driven robot painting an automobile. (Courtesy of GMFanuc Robotics Corporation) (*b*) Block diagram.

Let us assume that we have to design a robot such as this one, and that one of the axes can be represented by the block diagram shown in Figure P4.11*b*. For the purposes of our application, a critically damped system will be assumed to be desirable. Find the amplifier gain K_a that will achieve a critically damped system.

4.12. It is desired to analyze the response of the feedback configuration shown in Figure P4.12 used to position a device in response to an input. The motor selected for this application has a motor constant $K_m = 0.4$ and a motor time constant $T_m = 0.1$. For an amplifier gain $K_a = 20$, determine the following:

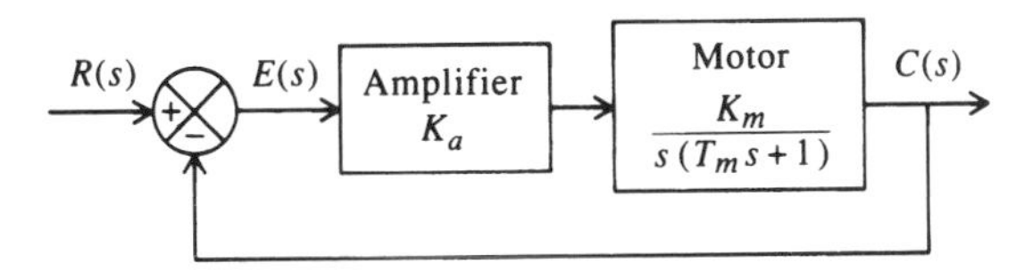

Figure P4.12

(a) What is the undamped natural frequency ω_n?

(b) What is the damping ratio ζ ?

(c) Determine the maximum percent overshoot resulting from the application of a unit step input?

(d) Find the time to peak, t_p, for the unit step input applied in part (c).

(e) Sketch the output $c(t)$ and error $e(t)$ resulting from the application of a unit step input $r(t)$. Indicate the time to peak and the maximum overshoot on your sketch.

4.13. For Figure P4.13 determine the following:

(a) Undamped natural frequency, ω_n.

(b) Damping ratio ζ.

(c) Maximum percent overshoot.

(d) Time to peak, t_p.

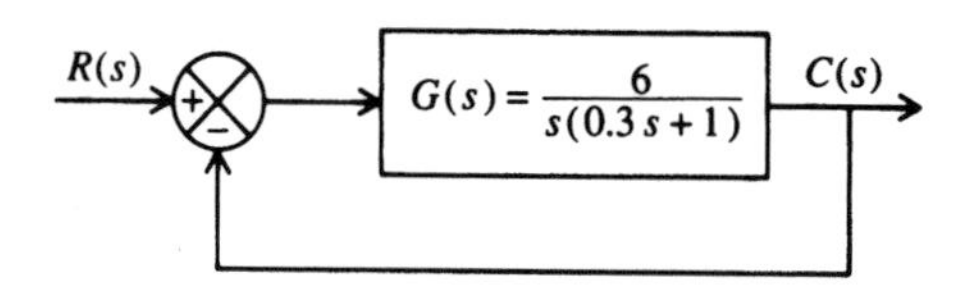

Figure P4.13

4.14. Repeat Problem 4.13 for

$$G(s) = \frac{4}{s(0.2s + 1)}.$$

4.15. A robot is used on an airplane assembly line to drill some holes. The block diagram of the control system used to control the drill in one axis is shown in Figure P4.15.

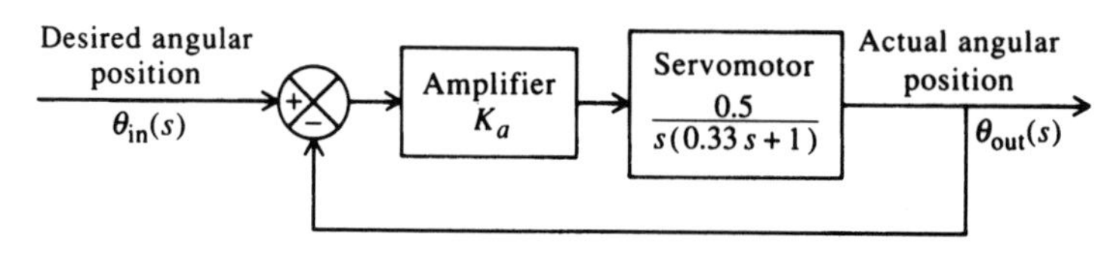

Figure P4.15

(a) Determine the amplifier gain K_a required to achieve a damping ratio in the system of 0.7.

(b) Calculate the corresponding time to peak.

(c) Calculate the corresponding maximum percent overshoot.

4.16. A control-system engineer needs to use a two-phase ac servomotor for an application on a project which is running short of money. The stock room has a spare motor, but the name plate is missing. The control-system engineer needs to determine the motor constant, K_m, and the motor time constant, T_m, before this motor can be used in the application. To accomplish this, the engineer hooks up the simple positioning control system shown in Figure P4.16*a* containing this two-phase ac servomotor and applies a unit step input.

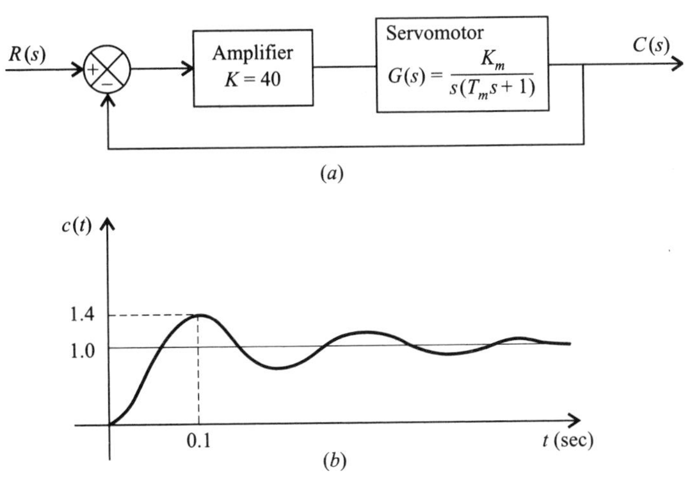

Figure P4.16

The response of this control system to a unit step is illustrated in Figure P4.16*b*.

From this response, determine, K_m and T_m.

4.17. The exact transfer function of a second-order system is unknown. We wish to determine its transfer function by applying a unit step to its input, $r(t)$, and analyzing its output response, $c(t)$, shown in Figure P4.17.

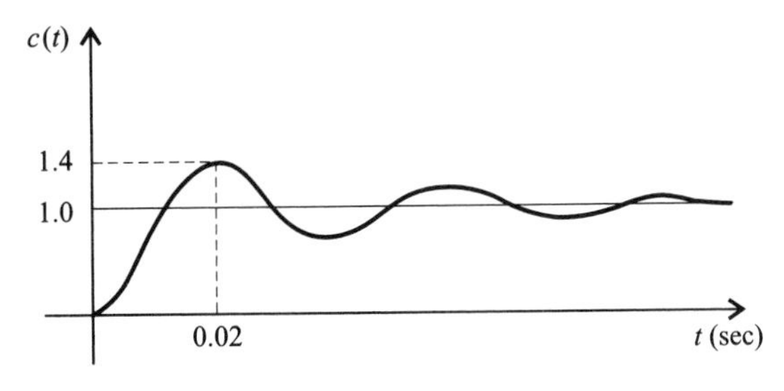

Figure P4.17

Determine it transfer function, $C(s)/R(s)$.

4.18. A robot is used to drill holes in the body of an automobile on an assembly line in response to commands as in Figure P4.18. For this application, a critically damped system is desired.

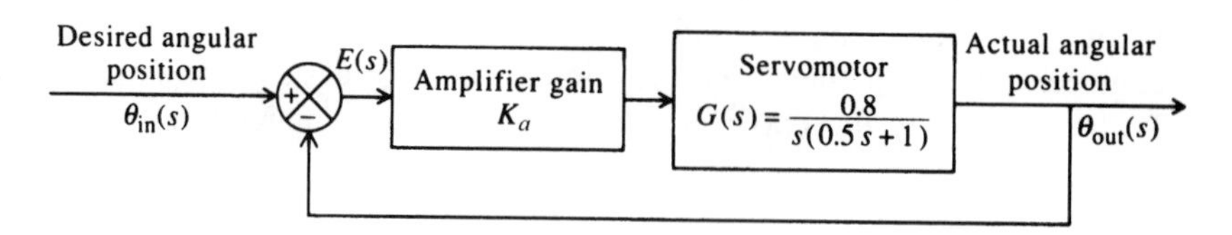

Figure P4.18

(a) Determine the amplifier gain K_a required to achieve a critically damped system.

(b) *Sketch* $r(t)$, $c(t)$, and $e(t)$ on the same set of axes when $r(t)$ is a unit step input.

(c) Determine the time t when the error $e(t)$ and the output $c(t)$ are equal.

4.19 We wish to fit the data contained in Table P4.19 to a model. These data represent the transient response of the following closed-loop transfer function to a unit step input:

$$\frac{C(s)}{R(s)} = \frac{16}{s^2 + 10s + 16}.$$

We wish to fit this data with the following model:

$$c(t) = 1 + K_1 e^{-at} + K_2 e^{-bt}$$

(a) Focus only on the slowest or smallest root, and determine

$$K_1 e^{-at}$$

by subtracting the steady-state value of one from the data, and fitting a straight line to the $\log_{10}$ of the difference. Determine K_1 and a.

(b) Plot the $\log_{10}$ of $c(t) - c_{ss} - K_1 e^{-at}$, and determine K_2 and b from the resulting straight line.

(c) Having determined K_1, K_2, a, and b, determine the estimated value of $\hat{c}(t)$.

(d) Compare the resulting estimated value of $\hat{c}(t)$ with the theoretical value of $c(t)$ provided in Table P4.19.

(e) Plot the theoretical and graphical values of the theoretical value of $c(t)$ with the estimated value of $\hat{c}(t)$ on the same set of axes. In addition, plot the error between the two sets of data on this same set of axes.

(f) How well does your fit of $\hat{c}(t)$ compare with the theoretical value of $c(t)$?

(g) Determine the modeled value of $\hat{C}(s)/R(s)$.

Table P4.19. Set of data representing the theoretical value of $c(t)$.

Time	$c(t)$
0	0
0.2000	0.1735
0.4000	0.4145
0.6000	0.6012
0.8000	0.7314
1.0000	0.8197
1.2000	0.8791
1.4000	0.9189
1.6000	0.9457
1.8000	0.9636
2.0000	0.9756
2.2000	0.9836
2.4000	0.9890
2.6000	0.9926
2.8000	0.9951
3.0000	0.9967
3.2000	0.9978
3.4000	0.9985
3.6000	0.9990
3.8000	0.9993
4.0000	0.9996
4.2000	0.9997
4.4000	0.9998
4.6000	0.9999
4.8000	0.9999
5.0000	0.9999

REFERENCES

1. R. P. Hatcher and S. M. Shinners, "Adaptive control techniques for reducing settling time." In *Proceedings of the 1982 American Control Conference*, Vol. 2, pp. 658–659.
2. N. K. Sinha and B. Kuszta, *Modeling and Identification of Dynamic Systems*. Van Nostrand Reinhold Co., New York, 1983.
3. W. K. Waymeyer and R. W. Sporing, "Closed loop adaption applied to missile control." In *Proceedings of the 1962 Joint Automatic Control Conference*, New York, p. 18–3.
4. GMFanuc *Robotics Corporation Specifications Sheeet on* GMC *Model* P-150 *Robot*. GMFanuc Robotics Corporation, Auburn Hills, MI, 1987.

5

PERFORMANCE CRITERIA

5.1. INTRODUCTION

In the early days of control-system theory, engineers were generally less rigid in defining performance criteria. They were more apt to look on the feedback control system rather qualitatively and center attention primarily on stability and static accuracy. However, modern complex control systems have demanded the development of accurate criteria of performance.

The performance of a feedback control system is generally described in terms of stability, sensitivity, accuracy, transient response, and residual noise jitter. The exact specifications are usually dictated by the required system performance. Certain characteristics are more important in some systems than in others.

The great amount of literature that has appeared on the subject in recent years is evidence of the increasing importance that performance criteria have been given in feedback control-system design. In order to keep pace with the requirements of modern feedback control systems, several new criteria of performance have been developed [1]. It is the purpose of this chapter to review and study several classical performance criteria together with more recent and sophisticated approaches. The control literature abounds with various criteria of performance. Most significant are several performance criteria that have been postulated that are functions of time and error. After examining the literature, we find that the integral of time multiplied by the absolute value of error (ITAE) criterion for optimizing the transient response stands out as being useful; the integral of the square of the error (ISE) criterion is also important. Performance criteria presented in this chapter is extended to Chapters 7 and 8, where pole placement using linear-state-variable feedback and estimation are discussed, and to Chapter 6 of the accompanying volume, where optimal control theory is presented.

5.2. STABILITY

The degree of stability is an extremely important part of describing a control system's performance criteria. It is a very complex subject, and five chapters are devoted to it. In Chapter 6 we address the question of whether a linear system is stable or not. If it is stable, we determine how stable it is. If it is unstable, we determine how unstable it is. No effort is made in Chapter 6 to make an unstable system stable, or improve the stability or a marginally stable system. That is reserved for Chapters 7 and 8, where we design linear control systems and make the systems meet the stability specifications required. The question of stability of nonlinear control systems is addressed in the accompanying volume, where we extend the technique of linear system stability analysis and design to nonlinear systems. The stability of digital control systems is also presented there. The design of control systems to meet required stability specifications is further illustrated in Chapter 7 of the accompanying volume, where case studies of control-system designs are presented.

In this chapter, linear system stability is briefly addressed in this section to focus attention on its importance in the overall concept of control-system performance criteria considerations. A feedback control system must be stable even when the system is subjected to command signals, extraneous inputs anywhere within the loop, power-supply variations, and changes in parameters of the feedback loop.

The qualitative statement that a system is stable is meaningless. The question of how stable the system is must also be determined. In order to answer this question adequately, we must return to Eq. (2.121), which is repeated here:

$$\frac{C(s)}{R(s)} = \frac{G(s)}{1 + G(s)H(s)}. \tag{5.1}$$

As shown in Chapter 6, stability is determined by evaluating the denominator of this equation for $s = j\omega$. If $G(j\omega)H(j\omega) = -1$, the denominator would vanish and the system would oscillate indefinitely or the resulting response would grow with time. The margin by which $G(j\omega)H(j\omega)$ is shy of unity magnitude when its phase is 180° is known as the *gain margin*, and the phase by which it is shy of 180° when its magnitude is unity is known as the *phase margin*. These quantities indicate the degree to which the system is stable. They are used by the control engineer to determine how stable the feedback system is. Useful, qualitative, desirable design values are 30°–60° for the phase margin and 4–12 dB for the gain margin. These numbers indicate that when $G(j\omega)H(j\omega)$ equals unity, its phase is 120°–150°, and when $G(s)H(s)$ has 180° phase shift, its magnitude is 0.25–0.63. In addition, the magnitude of the closed-loop resonant peak M_p is also a measure of the degree of stability. As shown in Chapter 6, a desirable value is between 1.0 and 1.4.

5.3. SENSITIVITY

Sensitivity is a measure of the dependence of a system's characteristics on those of a particular element. The differential sensitivity of a system's closed-loop transfer function $H(s)$ with respect to the characteristics of a given element $K(s)$ is defined as

$$S_K^H(s) = \frac{d\ln H(s)}{d\ln K(s)} = \frac{\%\text{change in } H(s)}{\%\text{change in } K(s)}, \tag{5.2}$$

where

$$H(s) = C(s)/R(s).$$

A more meaningful definition can be obtained by rewriting Eq. (5.2) as

$$S_k^H(s) = \frac{dH(s)/H(s)}{dK(s)/K(s)}. \tag{5.3}$$

Equation (5.2) states that the differential sensitivity of $H(s)$ with respect to $K(s)$ is the percentage change in $H(s)$ divided by that percentage change in $K(s)$ that has caused the change in $H(s)$ to occur. This definition is valid only for small changes. It is important to note that sensitivity is a function of frequency and an ideal system has zero sensitivity with respect to any parameter.

In order to illustrate the concept of sensitivity, consider the general control system shown in Figure 5.1*a*. Here K_1 represents the transfer function of the input transducer, K_2 represents the transfer function of the feedback transducer, and $G(s)$ represents the combined transfer function of an amplifier, stabilizing network, motor, and gear train in the forward part of the feedback loop.

The overall system transfer function $H(s)$ is given by

$$H(s) = \frac{C(s)}{R(s)} = \frac{K_1 G(s)}{1 + K_2 G(s)}. \tag{5.4}$$

Let us now determine the sensitivity of the overall system transfer function with respect to changes in K_1, K_2 and $G(s)$.

A. Sensitivity of *H(s)* with respect to K_1

This is

$$S_{K_1}^H(s) = \frac{dH(s)/H(s)}{dK_1/K_1} = \frac{K_1}{H(s)} \frac{dH(s)}{dK_1},$$

where

$$\frac{dH(s)}{dK_1} = \frac{G(s)}{1 + K_2 G(s)} = \frac{H(s)}{K_1}.$$

Therefore,

$$S_{K_1}^H(s) = \frac{K_1}{H(s)} \frac{H(s)}{K_1} = 1. \tag{5.5}$$

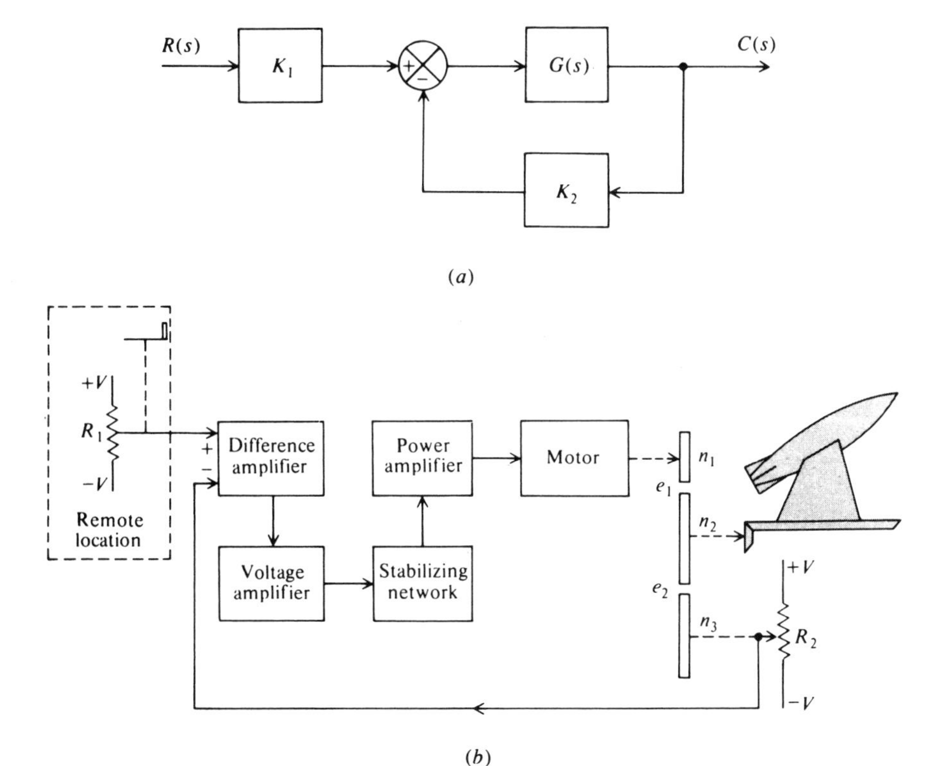

Figure 5.1 (*a*) A representative control system. (*b*) An automatic positioning system for a missile launcher.

B. Sensitivity of *H*(*s*) with respect to K_2

This is

$$S_{K_2}^H(s) = \frac{dH(s)/H(s)}{dK_2/K_2} = \frac{K_2}{H(s)}\frac{dH(s)}{dK_2},$$

where

$$\frac{dH(s)}{dK_2} = \frac{0 - K_1 G^2(s)}{[1 + K_2 G(s)]^2} = \frac{-K_1^2 G^2(s)}{K_1[1 + K_2 G(s)]^2}.$$

Therefore

$$S_{K_2}^H(s) = \frac{K_2}{H(s)}\frac{-K_1^2 G^2(s)}{K_1[1 + K_2 G(s)]^2} = \frac{-K_2}{H(s)}\frac{H^2(s)}{K_1} = \frac{-K_2 G(s)}{1 + K_2 G(s)}.$$

For frequencies where $K_2 G(s) \gg 1$, this reduces to

$$S_{K_2}^H(s) \simeq -1. \tag{5.6}$$

C. Sensitivity of *H*(*s*) with respect to *G*(*s*)

$$S_{G(s)}^{H}(s) = \frac{dH(s)/H(s)}{dG(s)/G(s)} = \frac{G(s)}{H(s)} \frac{dH(s)}{dG(s)},$$

where

$$\frac{dH(s)}{dG(s)} = \frac{(1 + K_2G(s))K_1 - K_1G(s)K_2}{[1 + K_2G(s)]^2} = \frac{K_1}{(1 + K_2G(s))^2}.$$

Therefore,

$$S_{G(s)}^{H}(s) = \frac{G(s)}{H(s)} \frac{K_1}{[1 + K_2G(s)]^2} = \frac{1}{1 + K_2G(s)}. \tag{5.7}$$

The results obtained in Eqs. (5.5), (5.6), and (5.7) are quite interesting. The symbols K_1 and K_2 represent input and feedback transducers, respectively, and Eqs. (5.5) and (5.6) illustrate that they are very critical. Any changes in their characteristics are directly reflected in an overall system transfer-function change. Elements used for K_1 and K_2 must, therefore, possess precise and stable characteristics with temperature and time. Equation (5.7) shows that the sensitivity of the overall system transfer function with respect to $G(s)$ is divided by $1 + K_2G(s)$. From a sensitivity viewpoint, it appears desirable to design $K_2G(s)$ to have as large a value as possible. However, it need not be very precise.

It is also important to recognize that because sensitivity is a function of frequency, we should think of systems as being sensitive or insensitive only over certain frequency bands. In this example, $K_2G(s)$ is a function of frequency and will be large over only a limited range of frequencies. Therefore, $H(s)$ is insensitive to $G(s)$ only over a certain range of frequencies. This point is further illustrated in Problems 5.1 through 5.14.

Let us now try to extend the results derived in this section in order to determine qualitatively the requirements of the various elements shown in the simple missile launcher positioning device of Figure 5.1*b*. In this system R_1, R_2, $\pm V$, and the difference capability of the difference amplifier must all be precise. The gain characteristics of the difference amplifier, voltage, stabilizing network, power amplifier, and motor need not be precise. Any changes in the characteristics of these elements will be divided by $1 + K_2G(s)$. Let us now consider the gear train, which is composed of three gears n_1, n_2, n_3. It is assumed that gears n_2 and n_3 have the same number of teeth and each has 10 times as many teeth as gear n_1. The output is taken off gear n_2. The prime purpose of gear n_3 is to enable the output transducer to be coupled off another shaft. Gear meshes can result in system errors because the tooth space exceeds the thickness of an engaging tooth. This phenomenon is commonly referred to as backlash. It can be measured by holding one gear fast and observing the amount of motion in the other gear. Figure 5.1*b* denotes the backlash between gears n_1 and n_2 as e_1, and that between n_2 and n_3 as e_2. A sensitivity analysis shows that the error produced by backlash e_1 is reduced by $1 + K_2G(s)$, whereas the error produced by backlash e_2 is coupled into the feedback element K_2 and, therefore, its effect is seen through $S_{K_2}^{H}(s)$. Therefore, there is a need for precision gearing for n_2 and n_3, but not for n_1. Chapter 10 illustrates that backlashes at e_1 and

e_2, however, are very important from a stability viewpoint. From a sensitivity viewpoint, however, the backlash at e_1 is not as critical as that at e_2.

5.4. STATIC ACCURACY

Static accuracy ranks as the next most important characteristics for a feedback control system. The designer always strives to design the system to minimize error for a certain anticipated class of inputs. This section considers techniques that are available for determining the system accuracy.

Theoretically, it is desirable for a control system to have the capability of responding to changes in position, velocity, acceleration, and changes in higher-order derivatives with zero error. Such a specification is very impractical and unrealistic. Fortunately, the requirements of practical systems are much less stringent. For example, let us consider the automatic positioning system of the missile launcher illustrated in Figure 5.1*b*. Its functioning is similar to the missile launcher positioning system in Figure 1.9. Realistically, it would be desirable for this system to respond well to inputs of position and velocity, but not necessarily to those of acceleration. In addition, it probably would be desirable for this system to respond with zero error for positional-type inputs. However, a finite tracking error could probably be tolerated for inputs of velocity. In contrast to this system, where the stakes are quite high, let us consider a simpler positioning system, which perhaps is only required to reproduce the angular position of a dial at some remote location. Such a control system would probably be only required to reproduce any positional inputs, but not any higher-order inputs such as those of velocity and acceleration.

A method for determining the steady-state performance of any control system is to apply the final-value theorem of the Laplace transform as given by Eq. (2.74). Let us consider the unity-feedback system shown in Figure 5.2. The relation between the resulting system error, $E(s)$, for a given input $R(s)$ is given by

$$\frac{E(s)}{R(s)} = \frac{1}{1 + G(s)}. \tag{5.8}$$

The steady-state error can be expressed as

$$e_{ss} = \lim_{t \to \infty} e(t) = \lim_{s \to 0} \frac{sR(s)}{1 + G(s)}. \tag{5.9}$$

The control engineer is usually interested in test inputs of position, velocity, and acceleration. A step, ramp, and paraboloid are simple mathematical expressions

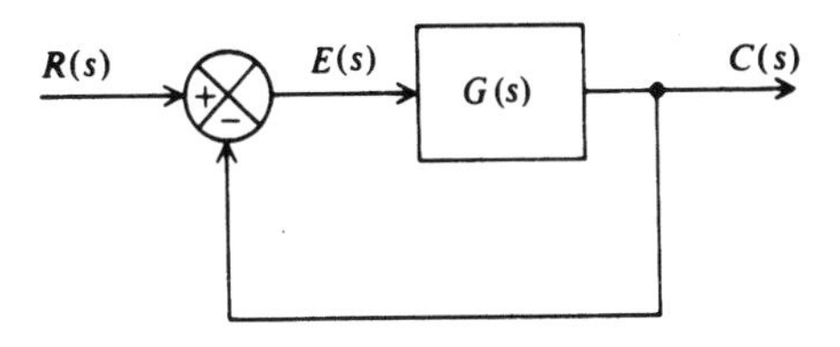

Figure 5.2 A unity-feedback system.

which represent these physical quantities, respectively, and is illustrated in Figure 5.3. They are defined in Eqs. (5.10)–(5.12), where the notation $U(t)$ means a unit step for $t \geqslant 0$:

$$\text{Position input: } r(t) = U(t), \qquad R(s) = 1/s, \tag{5.10}$$

$$\text{Velocity input: } r(t) = tU(t), \qquad R(s) = 1/s^2, \tag{5.11}$$

$$\text{Acceleration input: } r(t) = \frac{1}{2}t^2 U(t), \qquad R(s) = 1/s^3. \tag{5.12}$$

We next determine the steady-state error of several types of systems for each of these three inputs: the unit step, unit ramp, and paraboloid. It is assumed that the loop transfer function $G(s)$ has the general form

$$G(s) = \frac{K(1 + T_1 s)(1 + T_2 s)\cdots(1 + T_m s)}{s^n[(T_o s)^2 + 2\zeta\omega_n s + 1](1 + T_a s)(1 + T_b s)\cdots(1 + T_q)}, \tag{5.13}$$

where

s^n = a multiple pole at the origin of the complex plane,
K = gain factor of the expression.

The term s^n in the denominator is of particular significance. It represents the number of multiple poles in the denominator. Based on these number of pure integrations in the denominator of the open-loop transfer function, we classify the system by system "type". A control system is defined as type 0, type 1, type 2, type 3,..., if $n = 0$, $n = 1$, $n = 2$, $n = 3, \ldots$, respectively. We will show that as the system type is increased, the accuracy is improved. However, the stability problem becomes more difficult as the system type is increased. In practice, we usually never design a control system greater than type 2 because it is very difficult to stabilize a control system containing more than two pure integrations (although it can be done). Therefore, a tradeoff is required between steady-state accuracy and relative stability.

A. Unit Step (Position) Input

The steady-state error can be obtained by substituting $R(s) = 1/s$ into Eq. (5.9):

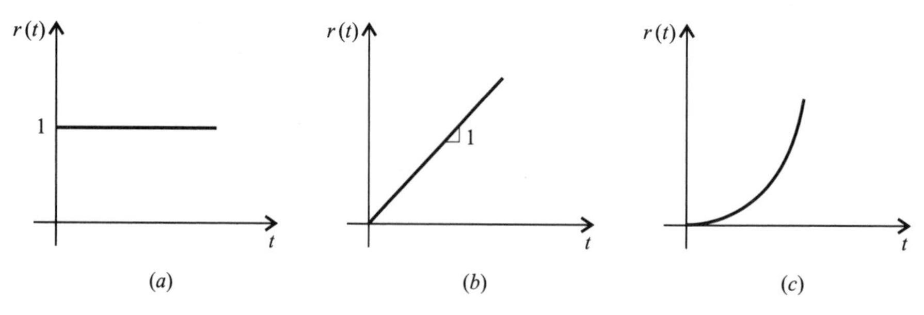

Figure 5.3 Test inputs (*a*) Unit step representing a position input. (*b*) Unit ramp representing a velocity input. (*c*) Parabolic input representing an acceleration input.

$$e_{ss} = \lim_{s\to 0} \frac{s(1/s)}{1+G(s)} = \frac{1}{1+\lim_{s\to 0} G(s)}. \tag{5.14}$$

The quantity $\lim_{s\to 0} G(s)$ is defined as the position constant and is denoted by K_p:

$$K_p = \lim_{s\to 0} G(s). \tag{5.15}$$

Therefore, the steady-state error in terms of the position constant is given by

$$e_{ss} = \frac{1}{1+K_p}. \tag{5.16}$$

Equation (5.16) states that the steady-state tracking error of a feedback control system having a unit step input equals $1/(1+$ the position constant). Table 5.1 summarizes the values of K_p and the resulting steady-state error as a function of the number of pure integrations of the open-loop transfer function $G(s)$.

Table 5.1 indicates that the position constant is infinite for all systems which contian one or more pure integration(s) in the open-loop transfer function $G(s)$. Therefore, Eq. (5.16) implies that all systems containing at least one pure integration result in a theoretical steady-state positional response error of zero. Table 5.1 indicates that the position constant is finite for a system containing no pure integrations and, therefore, the response error for a unit position input is $1/(1+K)$.

B. Unit Ramp (Velocity) Input

The steady-state error can be obtained by substituting $R(s) = 1/s^2$ into Eq. (5.9):

$$e_{ss} = \lim_{s\to 0} \frac{s(1/s)^2}{1+G(s)} = \frac{1}{\lim_{s\to 0} sG(s)}. \tag{5.17}$$

The quantity $\lim_{s\to 0} sG(s)$ is defined as the velocity constant and is denoted by K_v:

$$K_v = \lim_{s\to 0} sG(s). \tag{5.18}$$

Therefore, the steady-state error in terms of the velocity constant is given by

$$e_{ss} = 1/K_v. \tag{5.19}$$

Table 5.1.

Pure integrations of $G(s)$	K_p	e_{ss}
0	K	$1/(1+K)$
1	∞	0
2	∞	0
.	.	.
.	.	.
.	.	.
n, where $n > 0$	∞	0

Equation (5.19) states that the steady-state response error of a feedback control system having a unit ramp equals the reciprocal of the velocity constant. Table 5.2 summarizes the values of K_v and the resulting steady-state error as a function of the number of pure integrations of the open-loop transfer function $G(s)$.

Table 5.2 indicates that the velocity constant is infinite for all systems that contain more than one pure integration in the open-loop transfer function $G(s)$. Therefore, Eq. (5.19) implies that al systems containing at least two pure integrations have a theoretical steady-state velocity response error of zero. Table 5.2 indicates that a system containing no pure integrations cannot follow a velocity input. Table 5.2 also indicates that a system containing one pure integration has a response of 1/K due to a unit ramp input.

C. Parabolic (Acceleration) Input

An expression for the steady-state error due to $r(t) = \frac{1}{2}t^2U(t)$ can be obtained by substituting $R(s) = 1/s^3$ into Eq. (5.9):

$$e_{ss} = \lim_{s\to 0}\frac{s(1/s^3)}{1+G(s)} = \frac{1}{\lim_{s\to 0} s^2G(s)}. \tag{5.20}$$

The quantity $\lim_{s\to 0} s^2G(s)$ is defined as the acceleration constant and is denoted by K_a:

$$K_a = \lim_{s\to 0} s^2G(s). \tag{5.21}$$

Therefore, the steady-state error in terms of the acceleration constant is

$$e_{ss} = 1/K_a. \tag{5.22}$$

Equation (5.22) states that the steady-state response error of a feedback control system having this parabolic input equals the reciprocal of the acceleration constant. Table 5.3 summarizes the values of K_a and the resulting steady-state error as a function of the number of pure integrations of the open-loop transfer function $G(s)$. Table 5.3 inciates that the acceleration constant is infinite for all systems that contain three or more pure integrations in the open-loop transfer function $G(s)$, and the resulting steady-state error is zero. Therefore, Eq. (5.22) implies that all systems containing at last three pure integrations have a theoretical steady-state acceleration response error

Table 5.2

Pure integrations of $G(s)$	K_v	e_{ss}
0	0	∞
1	K	$1/K$
2	∞	0
.	.	.
.	.	.
.	.	.
n, where $n > 1$	∞	0

Table 5.3

Pure integrations of $G(s)$	K_a	e_{ss}
0	0	∞
0	0	∞
2	K	$1/K$
3	∞	0
.	.	.
.	.	.
n, where $n > 2$	∞	0

of zero. Table 5.3 indicates that systems containing less than two pure integrations cannot follow an acceleration input. Table 5.3 also indicates that a system containing two pure integrations has a response error of 1/K for this acceleration input.

A summary of the results derived appears in Table 5.4. It is quite general and enables the reader to compare the capabilities of various types of systems. Notice from this table that the steady-state constants are zero, finite, or infinite. It is important to emphasize at this time that if the inputs are other than unit quantities the steady-state errors are proportionally increased because we are analyzing linear systems. For example, should the input to a system containing one pure integration be a ramp whose value is B position units (ft/sec, rad/sec, etc.), then the steady-state error as given by Eq. (5.19) would be modified to read

$$e_{ss} = B/K_v.$$

It should be noted that the unit of the velocity constant is 1/sec and that of the acceleration constant is $1/\text{sec}^2$. The position constant K_p has no dimensions.

Let us now consider an input composed of position, velocity, and acceleration components which equal $AU(t)$ ft, $BU(t)$ ft/sec, and $C/2U(t)$ ft/sec^2, respectively. The form of the input can be represented as

$$r(t) = AU(t) + BtU(t) + \frac{1}{2}Ct^2U(t).$$

The steady-state response of the system may be obtained by considering each component of the input separately, and then adding the results by means of superposition. The resulting steady-state error is of the following form:

Table 5.4 Summary of Steady-State Constants

Number of Pure Integrations	Constants: Position	Constants: Velocity	Constants: Acceleration
0	K_p	0	0
1	∞	K_v	0
2	∞	∞	K_a
3	∞	∞	∞

$$e_{ss} = \frac{A}{1+K_p} + \frac{B}{K_v} + \frac{C}{K_a}.$$

It is interesting to see how the various types of systems summarized in Table 5.4 would respond to this input.

1. System Containing No Pure Integration. The steady-state error is

$$e_{ss} = \frac{A}{1+K_p} + \infty + \infty.$$

The result indicates that a system containing no pure integration will be able to follow the position input component of $AU(t)$ ft, but not velocity or acceleration inputs of $BU(t)$ ft/sec and $C/2U(t)$ ft/sec^2, respectively. Figure 5.4 illustrates a typical step response for this system (when there are no velocity or acceleration components in the input).

2. System Containing One Pure Integration. The steady-state error is

$$e_{ss} = 0 + B/K_v + \infty.$$

The result indicates that a system containing one pure integration will follow the position input component with zero error and the velocity input component with a finite error of B/K_v. This system will not, however, be able to follow the acceleration input component. The units of B/K_v are feet or radians. This should be interpreted

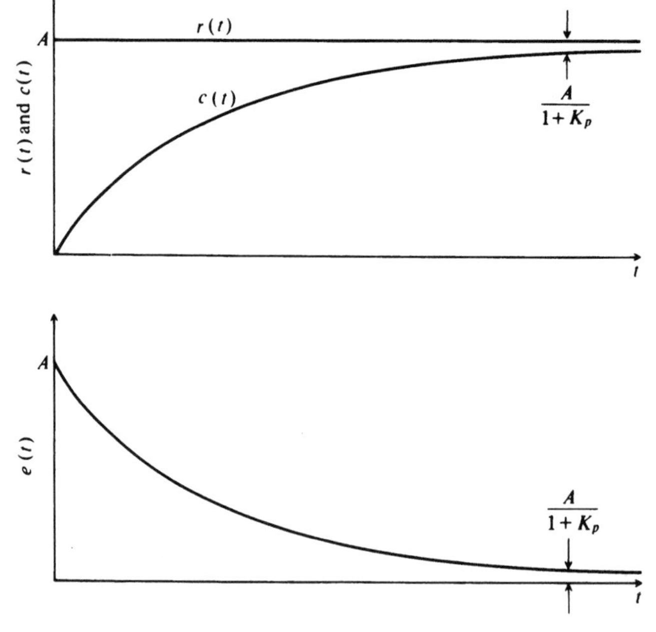

Figure 5.4 Response of a system containing no pure integrations to a step input.

to mean that there is a fixed positional error due to the constant velocity input component. Figure 5.5 illustrates typical step and ramp responses of this system.

3. System Containing Two Pure Integrations. The steady-state error is

$$e_{ss} = 0 + 0 + C/K_a.$$

The result indicates that a system containing two pure integrations will follow the position and velocity input components with zero error and the acceleration input component with a finite error of C/K_a. The units of C/K_a are feet or radians. This should be interpreted to mean that there is a fixed positional error due to the constant acceleration input component. Figure 5.6 illustrates typical step, ramp, and acceleration responses of this system.

D. Relationships of Static Error Constants to Closed-Loop Poles and Zeros

It is often important to relate the position, velocity, and acceleration constants to the *closed-loop* poles and zeros. Let us consider a general single-loop, unity-feedback system having a forward transfer function, $G(s)$, where the closed-loop transfer function is given by

$$\frac{C(s)}{R(s)} = \frac{G(s)}{1 + G(s)}, \tag{5.23}$$

and the relationship between input and error is given by

$$\frac{E(s)}{R(s)} = \frac{1}{1 + G(s)}. \tag{5.24}$$

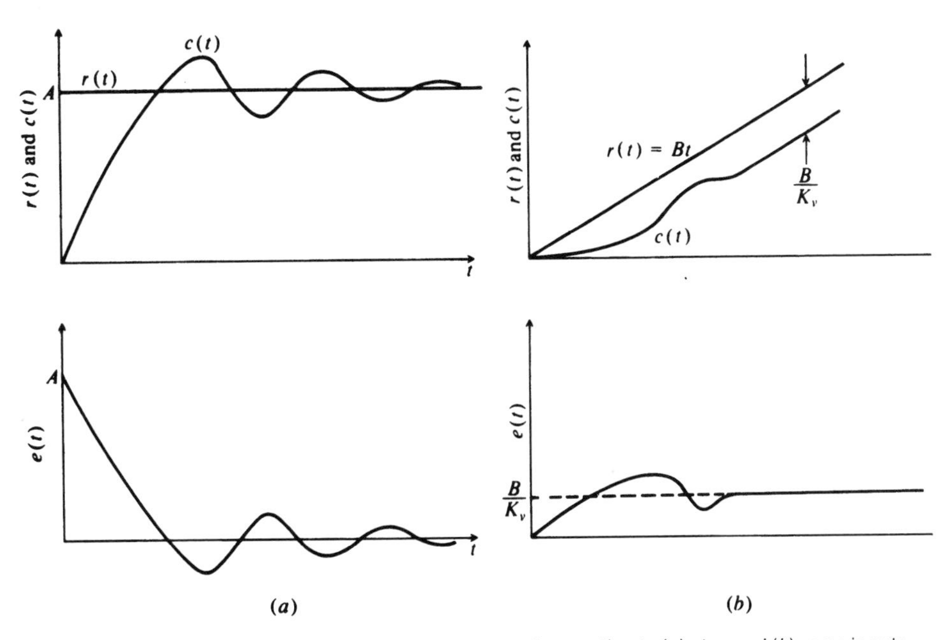

Figure 5.5 Response of a system containing one pure integration to (*a*) step and (*b*) ramp inputs.

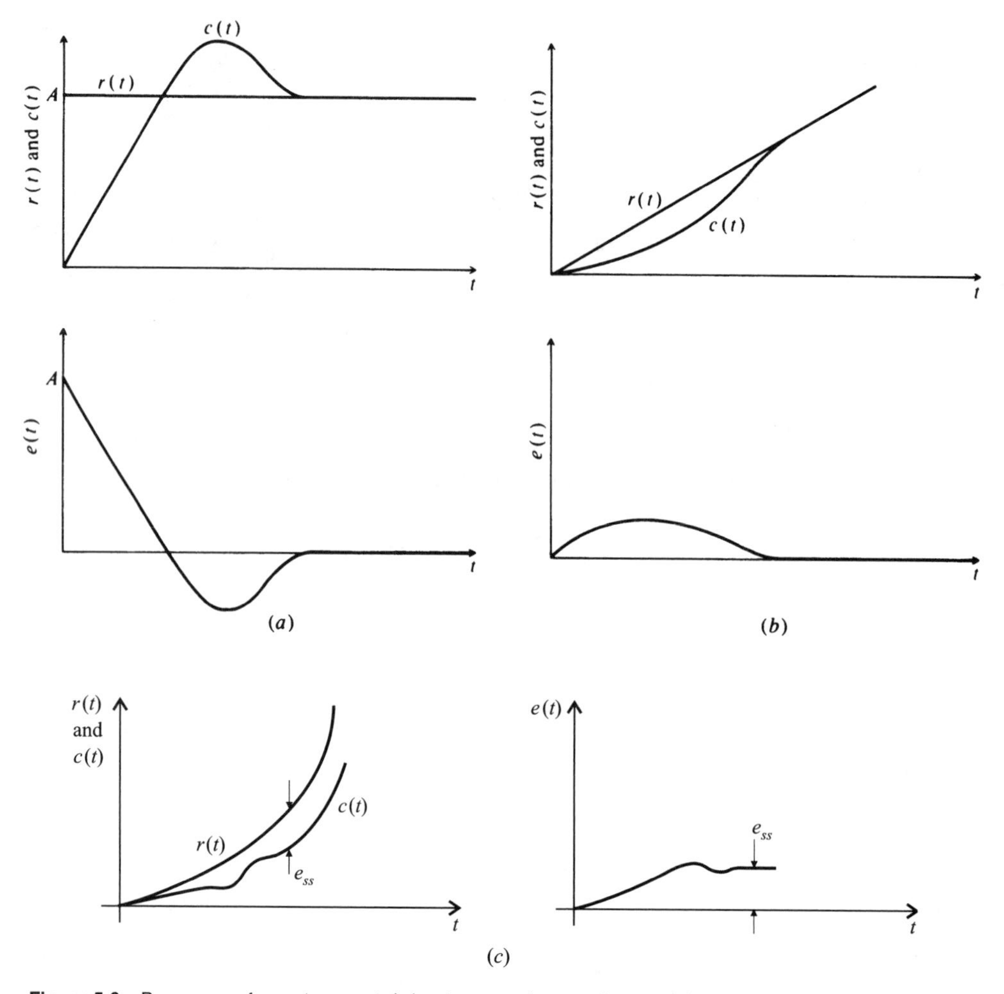

Figure 5.6 Response of a system containing two pure integrations to (*a*) step, (*b*) ramp and (*c*) acceleration inputs.

Because $E(s) = R(s) - C(s)$, it is evident that

$$\frac{C(s)}{R(s)} = 1 - \frac{E(s)}{R(s).} \tag{5.25}$$

It is assumed that $C(s)/R(s)$ can be represented by the rational function

$$\frac{C(s)}{R(s)} = \frac{K(s+z_1)(s+z_2)\cdots(s+z_m)}{(s+p_1)(s+p_2)\cdots(s+p_n)} = \frac{K\Pi_{j=1}^{m}(s+z_j)}{\Pi_{j=1}^{n}(s+p_j)}. \tag{5.26a}$$

In addition, if $1/[1+G(s)]$ in Eq. (5.24) is expanded as a power series in s, the error constants are defined in terms of the successive coefficients:

$$\frac{E(s)}{R(s)} = \frac{1}{1+G(s)} = \frac{1}{1+K_p} + \frac{1}{K_v}s + \frac{1}{K_a}s^2 + \cdots. \tag{5.26b}$$

Now the relationships between K_p, K_v, K_a, and the closed-loop poles and zeros can be determined.

1. Position Constant. Letting s approach zero in Eq. (5.24), we have

$$\frac{E(0)}{R(0)} = \lim_{s\to 0} \frac{1}{1+G(s)}. \tag{5.27}$$

Based on the definition of K_p in Eq. (5.15), we can rewrite Eq. (5.27) as

$$\frac{E(0)}{R(0)} = \frac{1}{1+K_p}. \tag{5.28}$$

Substituting Eq. (5.28) into Eq. (5.25) with s set at zero, we obtain the following:

$$\frac{C(0)}{R(0)} = \frac{K_p}{1+K_p}.$$

Solving for K_p in terms of $C(0)/R(0)$, we have

$$K_p = \frac{C(0)/R(0)}{1 - C(0)/R(0)}. \tag{5.29}$$

Using Eq. (5.26*a*) to represent $C(s)/R(s)$ and letting s approach zero in Eq. (5.26*a*), we have

$$\frac{C(0)}{R(0)} = \frac{K\Pi_{j=1}^{m} z_j}{\Pi_{j=1}^{n} p_j}, \tag{5.30}$$

where

$$\prod_{j=1}^{m} z_j = \text{product of zeros,}$$

$$\prod_{j=1}^{n} p_j = \text{product of poles.}$$

Substituting Eq. (5.30) into (5.29), the following expression for K_p in terms of the closed-loop poles and zeros is obtained:

$$K_p = \frac{K\Pi_{j=1}^{m} z_j}{\Pi_{j=1}^{n} p_j - K\Pi_{j=1}^{m} z_j}. \tag{5.31}$$

2. Velocity Constant. In order to derive the velocity constant in terms of the closed-loop poles and zeros, let us substitute Eq. (5.26*b*) into Eq. (5.25):

$$\frac{C(s)}{R(s)} = 1 - \frac{1}{1+K_p} - \frac{1}{K_v}s - \frac{1}{K_a}s^2 - \cdots.$$

Taking the derivative of this expression with respect to s, and then letting s equal zero, we obtain

$$\left[\frac{d}{ds}\left(\frac{C(s)}{R(s)}\right)\right]_{s=0} = -\frac{1}{K_v}. \tag{5.32}$$

In addition, we make use of the property that

$$\left[\frac{C(s)}{R(s)}\right]_{s=0} = \frac{C(0)}{R(0)} = 1 \tag{5.33}$$

in unity-feedback systems containing one or more pure integrations. Equation (5.33) says that a closed-loop unity-feedback system behaves with an ideal closed-loop transfer function of one at zero frequency. Dividing Eq. (5.32) by (5.33),

$$\frac{1}{K_v} = \frac{-\left[\frac{d}{ds}\left(\frac{C(s)}{R(s)}\right)\right]_{s=0}}{[C(s)/R(s)]_{s=0}} = -\left[\frac{d}{ds}\ln\frac{C(s)}{R(s)}\right]_{s=0}.$$

Substituting Eq. (5.26*a*) into this equation, we have the following:

$$\frac{1}{K_v} = -\left\{\frac{d}{ds}[\ln K + \ln(s+z_1) + \cdots + \ln(s+z_m) - \ln(s+p_1) - \cdots - \ln(s+p_n)]\right\}_{s=0}.$$

This can also be written as

$$\frac{1}{K_v} = -\left(\frac{1}{s+z_1} + \cdots + \frac{1}{s+z_m} - \frac{1}{s+p_1} - \cdots - \frac{1}{s+p_n}\right)_{s=0}, \tag{5.34}$$

or

$$\frac{1}{K_v} = \sum_{j=1}^{n}\frac{1}{p_j} - \sum_{j=1}^{m}\frac{1}{z_j}. \tag{5.35}$$

Therefore, $1/K_v$ equals the sum of the reciprocals of the closed-loop poles minus the sum of the reciprocals of the closed-loop zeros.

3. Acceleration Constant. The acceleration constant in terms of the closed-loop poles and zeros can be derived in a similar manner. We know from Eq. (5.26*b*) that

$$\frac{E(s)}{R(s)} = \frac{1}{1+K_p} + \frac{1}{K_v}s + \frac{1}{K_a}s^2 + \cdots.$$

From Eq. (5.25), this can be rewritten as

$$\frac{C(s)}{R(s)} = 1 - \frac{1}{1+K_p} - \frac{1}{K_v}s - \frac{1}{K_a}s^2 - \cdots.$$

It is obvious from this equation that $-2/K_a$ equals the zero-frequency value of the second derivative of $C(s)/R(s)$. Writing this in terms of the logarithmic derivative:

$$\frac{d^2}{ds^2}\left[\ln\frac{C(s)}{R(s)}\right] = \frac{(C(s)/R(s))''}{C(s)/R(s)} - \left[\frac{(C(s)/R(s))'}{C(s)/R(s)}\right]^2.$$

Setting $C(0)/R(0)$ equal to one, then

$$-\frac{2}{K_a} = \left\{\frac{d^2}{ds^2}\left[\ln\frac{C(s)}{R(s)}\right]\right\}_{s=0} + \frac{1}{K_v^2}.$$

Differentiating the right-hand side of Eq. (5.34) and letting s equal zero yields the following expression:

$$-\frac{2}{K_a} = \frac{1}{K_v^2} + \sum_{j=1}^{n}\frac{1}{p_j^2} - \sum_{j=1}^{m}\frac{1}{z_j^2}, \tag{5.36}$$

where K_v is defined by Eq. (5.35).

4. An Example. As an example, let us consider the second-order system illustrated in Figure 4.1 whose characteristics are defined by Eqs. (4.1), (4.2), and (4.3):

$$\frac{C(s)}{R(s)} = \frac{K_m/T_m}{s^2 + (1/T_m)s + K_m/T_m},$$

or

$$\frac{C(s)}{R(s)} = \frac{\omega_n^2}{s^2 + 2\zeta\omega_n s + \omega_n^2},$$

where

$$\omega_n^2 = K_m/T_m \quad \text{and} \quad \zeta = 1/2\omega_n T_m.$$

This is representative of a wide class of control systems that was thoroughly analyzed in Section 4.2. The problem is to determine K_v in terms of the parameters ζ and ω_n. K_m is the system gain and T_m is the time constant of the open-loop transfer function. The velocity constant of the simple system illustrated in Figure 4.1 is

$$K_v = K_m$$

by inspection. Let us relate this velocity constant to ζ and ω_n from the basic definitions of K_m and T_m in terms of ζ and ω_n:

$$K_v = K_m = \omega_n^2 T_m = \omega_n^2\left(\frac{1}{2\zeta\omega_n}\right) = \frac{\omega_n}{2\zeta}.$$

The fact that

$$K_v = \omega_n/2\zeta \tag{5.37}$$

is very important to remember for all second-order control systems that are characterized by a pair of complex-conjugate poles. Therefore, in order to obtain very accurate responses to velocity inputs, the damping ratio ζ is made very small. An example of this is in inertial navigation systems where accuracy is much more important than transient responses settling quickly.

5.5. TRANSIENT RESPONSE

In addition to stability, sensitivity, and accuracy, we are always concerned with the transient response of a feedback system. Transient response characteristics are defined on the basis of a step input. The response of a second-order system, containing one pure integration, to a unit step is useful for purposes of defining the various transient parameters. Should a problem arise where the system is higher than second order, a reasonably good approximation can be made by assuming that the system is second order if one pair of complex-conjugate roots dominates. This point is amplified during the discussion of the root locus in Chapters 6, 7 and 8. For purposes of illustration, let us consider the second-order system analyzed in Section 4.2. There we had a unity-feedback system whose closed loop transfer function $C(s)/R(s)$ was given by [see Eq. (4.3)]

$$\frac{C(s)}{R(s)} = \frac{\omega_n^2}{s^2 + 2\zeta\omega_n s + \omega_n^2}. \tag{5.38}$$

Its response to a unit step input, for the case of $\zeta < 1$, was given by [see Eq. (4.24)]

$$c(t) = 1 - \frac{e^{-\zeta\omega_n t}}{\sqrt{1-\zeta^2}}\sin\left(\omega_n\sqrt{1-\zeta^2}\,t + \alpha\right) \quad t \geqslant 0, \tag{5.39}$$

where

$$\alpha = \sin^{-1}\sqrt{1-\zeta^2}.$$

The input and output responses are illustrated in Figure 5.7.

The time for the feedback system to reach its first peak of the overshoot is commonly referred to as the time to peak t_p. As was derived in Section 4.2 [see Eq. (4.30)] and is illustrated in Figure 5.7,

$$\omega_n t_p = \frac{\pi}{\sqrt{1-\zeta^2}},$$

or

$$\text{time to peak} = t_p = \frac{\pi}{\omega_n\sqrt{1-\zeta^2}}. \tag{5.40}$$

The time required for the system to damp out all transients is commonly called the settling time t_s. Theoretically, for a second-order system this is infinity. In practice, however, the transient is assumed to be over when the error is reduced below some minimum value. Typically, the minimum level is set at 2% of the final value. The settling time, which is approximately equal to four time constants of the envelope of the damped sinusoidal oscillation, is illustrated in Figure 5.7 and is given by

$$t_s = 4/\zeta\omega_n. \tag{5.41}$$

The rise time t_r is defined as the time required for the response to rise from 10% to 90% of its final value. Figure 5.7 illustrates the rise time for the response shown. Figure 5.8 illustrates the resulting curve obtained relating the variation of $\omega_n t_r$ and the damping factor, ζ. Using MATLAB, a third-order fit was made to the curve and the following expression was obtained:

$$\omega_n t_r = 1.589\zeta^3 - 0.1562\zeta^2 + 0.9247\zeta + 1.0141.$$

Notice that this fit is so close to the actual curve, it is hard to distinguish between these two curves. Solving for t_r, we obtain the following:

$$t_r = (1/\omega_n)(1.589\zeta^3 - 0.1562\zeta^2 + 0.9247\zeta + 1.0141). \tag{5.42}$$

If a simple expression is desired for approximations, we can obtain the values of t_r over the desirable range of damping factors, from 0.4 to 0.7 (as discussed in Section 4.5), and then determine an average value over this range. The value of $\omega_n t_r$ for a damping of 0.4 is 1.4635; the value of $\omega_n t_r$ for a damping of 0.7 is 2.1299. Therefore, an average value of rise time over this damping factor range is given by

$$t_r = 1.80/\omega_n. \tag{5.43}$$

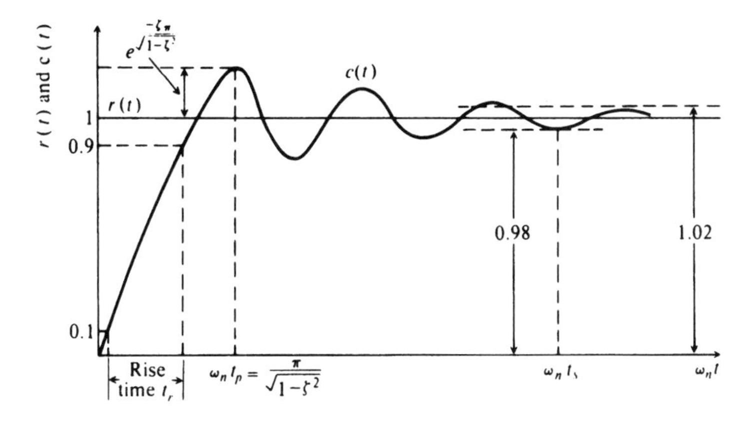

Figure 5.7 Response of a second-order system to a unit step.

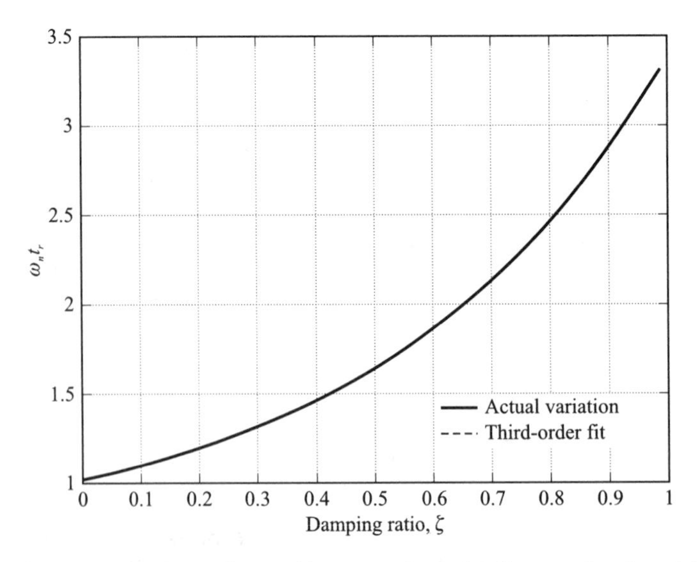

Figure 5.8 Variation of undamped natural frequency times rise time as a function of damping ratio.

This is a reasonable approximation to use for determining the rise time of second-order systems or systems that can be approximated as second-order systems (which is discussed further in Chapter 7).

The rise time is a useful characteristic for underdamped, critically damped, and overdamped systems. The time to peak is useful only for underdamped systems.

As is illustrated in Figure 5.7, a second-order system will have an exponentially decaying sinusoid when the damping factor is between 0 and 1 (this was derived in Section 4.2).

The first overshoot is of particular interest; the ratio of this overshoot's peak value to the steady-state settling value of the system is expressed as a percentage. The amount of overshoot allowable depends entirely on the particular problem. (see Section 4.5.) Notice that the percentage overshoot of the system illustrated in Figure 5.7 is $\exp(-\zeta\pi/\sqrt{1-\zeta^2}) \times 100\%$ [see Eq. (4.33)].

Even though reasonable values for time to peak, rise time, settling time, and the peak overshoot have been chosen, one is not sure whether a good system has been designed. For example, if the rise time is very small, invariably the peak value of the overshoot increases and so does the settling time. On the other hand, if the design is for minimum overshoot, the rise time increases. In order to resolve the conflict that exists between rise time, peak overshoot, and settling time, criteria have been proposed for synthesizing optimum transient performance for linear systems. Most of these criteria consider that the error and/or the time at which the error has occurred during the transient is important [1, 2]. Several of these criteria are presented in the remainder of this chapter.

5.6. PERFORMANCE INDICES

Modern complex control systems usually require more sophisticated performance criteria than those presented so far. As seen in the previous section, error and the time at which it occurs are very important factors that usually must be considered simultaneously. A performance index is a single measure of a system's performance that emphasizes those characteristics of the response that are deemed to be important. The notion of a performance index is very important in estimator design using linear-state-variable feedback, which is presented in Sections 8.1 through 8.6, and in optimal control theory, where the system is designed to optimize this performance index given certain constraints. This subject will be discussed fully in Chapter 11.

Refrences 1 and 2 discuss an entire class of performance indices that are various functions of error and/or time. This section reviews those that are useful in the design of practical linear control systems.

A fairly useful performance index is the *integral of the absolute magnitude of the error* (IAE) criterion, which is

$$S_1 = \int_0^\infty |e(t)| dt. \tag{5.44}$$

By utilizing the magnitude of the error, this integral expression increases for either positive or negative error, and results in a fairly good underdamped system. For a second-order system, S_1 has a minimum for a damping ratio of approximately 0.7.

Another useful performance index is the *integral of the square of the error* (ISE) criterion, which is

$$S_2 = \int_0^\infty e^2(t) dt. \tag{5.45}$$

By focusing on the square of the error function, it penalizes both positive and negative values of the error. For a second-order system, S_2 has a minimum for a damping ratio of approximately 0.5.

A very useful criterion that penalizes long-duration transients is known as the *integral of time multiplied by the absolute value of error* (ITAE). It is given by

$$S_3 = \int_0^\infty t|e(t)| dt. \tag{5.46}$$

This performance index is much more selective than the IAE or the ISE. The minimum value of its integral is much more definable as the system parameters are varied. For a second-order system, S_3 has a minimum for a damping ratio of 0.7.

Other figures of merit which have been proposed are the *integral of time multiplied by the squared error* (ITSE), *the integral of squared time multiplied by the absolute value of error* (ISTAE), and *the integral of squared time multiplied by square error* (ISTSE). These performance indices are

$$\text{ITSE:} \quad S_4 = \int_0^\infty te^2(t)\, dt, \tag{5.47}$$

$$\text{ISTAE:} \quad S_5 = \int_0^\infty t^2|e(t)|\, dt, \tag{5.48}$$

$$\text{ISTSE:} \quad S_6 = \int_0^\infty t^2e^2(t)\, dt. \tag{5.49}$$

The performance indices S_4, S_5, and S_6 have not been applied to any great extent in practice because of the increased difficulty in handling them.

A comparison of this array of performance indices is very interesting. The ISE criterion is not very sensitive to parameter variations, because the minimum is usually broad [3]. In addition, the ISE criterion has the advantage of being easy to deal with mathematically. The IAE criterion gives a slightly better sensitivity than the ISE criterion. The ITAE criterion generally produces smaller overshoots and oscillations than the IAE and ISE criteria. In addition, it is the most sensitive of the three, and sometimes too sensitive—slight parameter variation degrades system performance [1, 2].

In practice, a relatively insensitive criterion may be more useful in those systems where the parameters may not be known very accurately. In addition, even though one tries to optimize a performance criterion, one may also have other performance characteristics in mind. Therefore, it is desirable in some applications to permit moderate deviation from the "optimum" setting of the parameters in order that these other performance characteristics can be achieved without appreciably increasing the performance index. Based on this logic, the ISE criterion may be the most desirable performance index in some practical applications and is considered further in Chapter 6 of the accompanying volume when optimal control theory is discussed. The reader is also referred to Chapter 2 of Reference 3 on the ISE criterion.

A paper by Graham and Lathrop [2] created a great deal of interest in the ITAE criterion. An important aspect of this paper was the detailed discussion and presentation of results of a comparison for various performance criterion with the ITAE. It is more sensitive than the IAE and ISE criteria, and is useful in those practical applications that require a very sensitive criterion. Because the ITAE criterion has practical value and is interesting academically, this performance criterion is studied further. First, however, we are interested in determining the form of the closed-loop system transfer function in order that zero error results for various kinds of inputs. This is necessary in order to determine the relationship of the various coefficients in the numerator and denominator of the closed-loop system transfer function in order to achieve zero error.

5.7. ZERO-ERROR SYSTEMS

The transfer function of the general feedback system, containing a forward transfer function $G(s)$ and a feedback transfer function $H(s)$, is given by the general expression

$$\frac{C(s)}{R(s)} = \frac{G(s)}{1 + G(s)H(s)} = \frac{A_1 s^n + A_2 s^{n-1} + A_3 s^{n-2} + \cdots + A_n s + A_{n+1}}{B_1 s^m + B_2 s^{m-1} + B_3 s^{m-2} + \cdots + B_m s + B_{m+1}}. \tag{5.50}$$

From Eq. (5.26b) of Section 5.4, the error for this system can be expressed as*

$$e_{ss} = \frac{r(t)}{1 + K_p} + \frac{\dot{r}(t)}{K_v} + \frac{\ddot{r}(t)}{K_a} + \cdots. \tag{5.51}$$

It can be shown [2] that $1/(1 + K_p)$ is a function of $B_{m+1} - A_{n+1}$, and it is necessary that $B_{m+1} = A_{n+1}$ for zero steady-state error when the input is a step function. In addition, this implies that the forward transfer function contains at least one pure integrator as discussed previously in Section 5.4. From the general system transfer function of Eq. (5.50), it can be seen that there are many possible forms of $C(s)/R(s)$ that will yield zero steady-state error with a step input. When the numerator consists of the constant B_{m+1}, the system is called a *zero steady-state step error system* [2]:

$$\frac{C(s)}{R(s)} = \frac{B_{m+1}}{B_1 s^m + B_s s^{m-1} + B_3 s^{m-2} + \cdots + B_{m+1}}. \tag{5.52}$$

We can illustrate such a system with an example of a control system we have thoroughly analyzed in Chapter 4, the second-order control system, shown in Figure 4.1 which contains a single pure integrator. We also know from Table 5.1 that this control system has zero steady-state error to a unit step input (since $G(s)$ has one pure integration). Solving for $C(s)/R(s)$ from Figure 4.1, we obtain the following:

$$\frac{C(s)}{R(s)} = \frac{\dfrac{K_m}{s(T_m s + 1}}{1 + \dfrac{K_m}{s(T_m s + 1)}} = \frac{K_m}{T_m s^2 + s + K_m}. \tag{5.53}$$

Equation (5.53) has the same form as Eq. (5.52) and, therefore, the control system shown in Figure 4.1 is a *zero steady-state step error system.*

It can be shown [2] that $1/K_v$ is a function of $B_{m+1} - A_{n+1}$ and $B_m - A_n$. Therefore, for zero steady-state error with a ramp input, $B_m = A_n$ and $B_{m+1} = A_{n+1}$. In addition, this implies that the forward transfer function contains two or more pure integrators, as discussed previously in Section 5.4. A *zero steady-state ramp error system* occurs when the system transfer function is given by

$$\frac{C(s)}{R(s)} = \frac{B_m s + B_{m+1}}{B_1 s^m + B_2 s^{m-1} + B_3 s^{m-2} + \cdots + B_m s + B_{m-1}}. \tag{5.54}$$

To illustrate an example of a zero steady-state ramp error system, let us consider the control system shown in Figure 5.9. It contains two pure integrations and, therefore, $n = 2$ for Table 5.2, where we see that the steady-state error due to a ramp is zero. Let us determine the closed-loop transfer function of the control system shown in Figure 5.9 and verify that it has the same form as Eq. (5.54).

$$\frac{C(s)}{R(s)} = \frac{\dfrac{(1 + T_1 s)}{(1 + T_2 s)} \dfrac{K}{s^2}}{1 + \dfrac{(1 + T_1 s)}{(1 + T_2 s)} \dfrac{K}{s^2}} = \frac{KT_1 s + K}{T_2 s^3 + s^2 + KT_1 s + K}. \tag{5.55}$$

Therefore, the closed-loop transfer function of the control system shown in Figure 5.9 has the same form as the general equation (5.54), and it is, therefore, classified as a zero steady-state ramp error system.

*It is assumed that the input and all of its derivatives do not have any discontinuities in the interval of interest.

5.8. THE ITAE PERFORMANCE CRITERION FOR OPTIMIZING THE TRANSIENT RESPONSE

The ITAE performance index, as defined by Eq. (5.46), is considered further in this section. As discussed in Section 5.6, this performance index does not penalize large initial errors that are unavoidable. However, it does penalize long-duration transients. We now consider the form that the system transfer function should take, for various order systems, in order to achieve zero stedy-state step and ramp error systems and minimize the ITAE.

In the case of the zero steady-state step error system, Eq. (5.52) shows that the form of the system transfer function is given by

$$\frac{C(s)}{R(s)} = \frac{B_{m+1}}{B_1 s^m + B_2 s^{m-1} + B_3 s^{m-2} + \cdots + B_{m+1}}. \tag{5.56}$$

The procedure used to produce a table of standard system transfer functions of the form $C(s)/R(s)$ was to vary each coefficient in Eq. (5.56) separately until the integral of time multiplied by the absolute value of error became a minimum. Then the successive coefficients were varied in sequence to minimize the ITAE value.

If this criterion is applied to the second-order system described by Eq. (5.38), the optimum damping ratio is found to be 0.7. A listing of system transfer functions, $C(s)/R(s)$, has been prepared by Graham and Lathrop [2]. They show the optimum form of the denominator, for systems whose transfer functions are of the form given by Eq (5.56), which will minimize the integral of Eq. (5.46). For example, the optimum form for a second-order system is given by

$$\frac{C(s)}{R(s)} = \frac{\omega_n^2}{s^2 + 1.4\omega_n s + \omega_n^2}, \tag{5.57}$$

where $\zeta = 1.4/2 = 0.7$. The optimum form for a third-order system is given by

$$\frac{C(s)}{R(s)} = \frac{\omega_n^3}{s^3 + 1.75\omega_n s^2 + 2.15\omega_n^2 s + \omega_n^3}. \tag{5.58}$$

Table 5.5, which has been obtained from Reference 2, shows the optimum denominator transfer function for systems through the eighth order which will minimize the integral of Eq. (5.46). These standard forms provide a quick and simple method for synthesizing an optimum dynamic response.

In the case of the zero steady-state ramp error systems, the system transfer function was shown in Eq. (5.54) to be given by

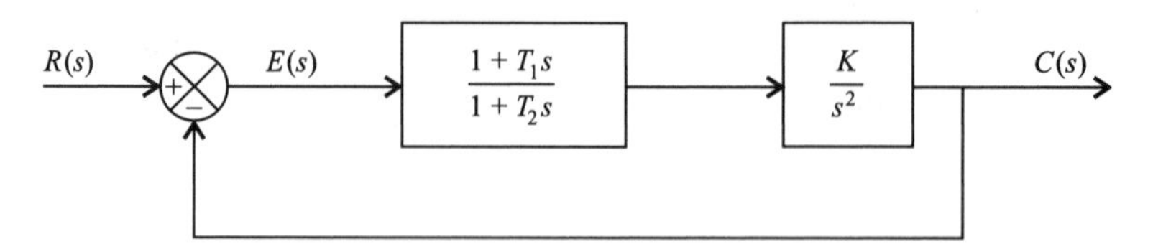

Figure 5.9 Control system containing two pure integrations.

$$\frac{C(s)}{R(s)} = \frac{B_m s + B_{m+1}}{B_1 s^m + B_2 s^{m-1} + B_3 s^{m-2} + \cdots + B_m s + B_{m+1}}. \tag{5.59}$$

The objective here also is to obtain a set of standard forms for the denominators of the system transfer functions given by Eq. (5.59). Table 5.6, obtained from Reference 2 was obtained in a similar way to Table 5.5.

The ITAE criterion is applied to several problems in the problem section of this chapter (see Problems 5.38–5.42, and illustrative problems I5.9 and I5.10.

The ITAE criterion is a straightforward method for optimizing the transient response of a system when the transfer function is known. Generally, it produces smaller overshoots and oscillations than the other criteria presented. It should be emphasized, however, that the ITAE solution is very sensitive and may not be useful for certain systems where most of the system poles and zeros and gains may be specified initially. For the latter case, designers do not have the flexibility for selecting as many of the parameters as they might wish.

5.9. OTHER PRACTICAL CONSIDERATIONS

The control engineer must be concerned with several other practical aspects before becoming able to state intelligently and completely the expected system performance. The concluding section of this chapter qualititatively discusses considerations of feedback system bandwidth, nonlinearities, size, weight, power consumption, and economics. Hopefully, this will aid in giving the reader a complete overall view of the problem.

Table 5.5. The Minimum ITAE Standard Forms for a Zero Steady-State Step Error System [2]

$$s + \omega_n$$
$$s^2 + 1.4\omega_n s + \omega_n^2$$
$$s^3 + 1.75\omega_n s^2 + 2.15\omega_n^2 s + \omega_n^3$$
$$s^4 + 2.1\omega_n s^3 + 3.4\omega_n^2 s^2 + 2.7\omega_n^3 s + \omega_n^4$$
$$s^5 + 2.8\omega_n s^4 + 5.0\omega_n^2 s^3 + 5.5\omega_n^3 s^2 + 3.4\omega_n^4 s + \omega_n^5$$
$$s^6 + 3.25\omega_n s^5 + 6.60\omega_n^2 s^4 + 8.60\omega_n^3 s^3 + 7.45\omega_n^4 s^2 + 3.95\omega_n^5 s + \omega_n^6$$
$$s^7 + 4.475\omega_n s^6 + 10.42\omega_n^2 s^5 + 15.08\omega_n^3 s^4 + 15.54\omega_n^4 s^3 + 10.64\omega_n^5 s^2 + 4.58\omega_n^6 s + \omega_n^7$$
$$s^8 + 5.20\omega_n s^7 + 12.80\omega_n^2 s^6 + 21.60\omega_n^3 s^5 + 25.75\omega_n^4 s^4 + 22.20\omega_n^5 s^3 + 13.30\omega_n^6 s^2 + 5.15\omega_n^7 s + \omega_n^8$$

Table 5.6. The Minimum ITAE Standard Forms for a Zero Steady-State Ramp Error System [2]

$$s^2 + 3.2\omega_n s + \omega_n^2$$
$$s^3 + 1.75\omega_n s^2 + 3.25\omega_n^2 s + \omega_n^3$$
$$s^4 + 2.41\omega_n s^3 + 4.93\omega_n^2 s^2 + 5.14\omega_n^3 s + \omega_n^4$$
$$s^5 + 2.19\omega_n s^4 + 6.50\omega_n^2 s^3 + 6.30\omega_n^3 s^2 + 5.24\omega_n^4 s + \omega_n^5$$
$$s^6 + 6.12\omega_n s^5 + 13.42\omega_n^2 s^4 + 17.16\omega_n^3 s^3 + 14.14\omega_n^4 s^2 + 6.76\omega_n^5 s + \omega_n^6$$

Feedback system bandwidth is usually defined as the frequency at which the open-loop magnitude equals unity. Sometimes it is defined where it is −3 dB. The bandwidth of a system is indicated by its particular application. Usually, the control engineer is interested in designing the system to respond to a certain spectrum of input signal frequencies and to suppress all inputs above a certain frequency. It is important to emphasize that we should not arbitrarily design for a large bandwidth. Although large bandwidths usually result in large error constants, with small resulting system error, they also result in a system that responds to extraneous noise inputs and has considerable jitter due to the noise. The desirable approach is to design the feedback system bandwidth to be just large enough to pass the desired input-signal frequency spectrum and them attenuate all higher-frequency signals [4]. Feedback system bandwidth considerations are discussed in detail in Chapters 6, 7 and 8, where methods for obtaining the frequency response are presented.

Nonlinearities are other factors that affect the performance of a control system. Primary concern is with backlash, stiction (starting friction) and Coulomb friction. Backlash, which was described in Section 5.3, is the amount of free motion of one gear while its mating gear is held fast. Stiction is the frictional force that prevents motion until the driving force exceeds some minimum value. Coulomb friction is a constant frictional drag that opposes motion, but has a magnitude that is independent of velocity. Each of these nonlinearities has an effect on performance. More will be said regarding nonlinearities in Chapter 5 of the accompanying volume.

Other factors of concern are size, weight, power consumption, and economics. The system must conform to certain specifications of size and weight. These are very important factors that usually dictate the design of the system. For example, these specifications may decide the type of power drive to be used. Power is another very important consideration. The system must usually perform within a certain allowable power limitation. Size, weight, and power consumption are usually very critical items for airborne and space applications. Last, but not least, is the question of economics. A basic fact of life is that most engineers work for organizations whose primary purpose is to make profit. Systems must therefore be designed as inexpensively as possible within the framework of good performance. A generally useful rule of thumb is that minimum-bandwidth systems will consume the least power and be the most economical.

5.10. ILLUSTRATIVE PROBLEMS AND SOLUTIONS

This section provides a set of illustrative problems and their solutions to supplement the material presented in Chapter 5.

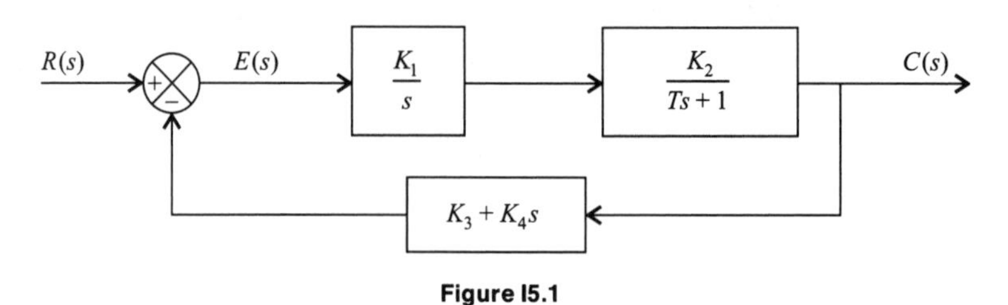

Figure I5.1

I5.1. The control system of Figure I5.1 contains position and rate feedback:

(a) Determine the closed-loop transfer function $C(s)/R(s)$.

(b) Determine the sensitivity of the closed-loop transfer function to K_2 as a function of s.

(c) Determine the sensitivity of the closed-loop transfer function to K_2 at dc.

SOLUTION: **(a)**

$$H(s) = \frac{C(s)}{R(s)} = \frac{\left(\frac{K_1}{s}\right)\left(\frac{K_2}{Ts+1}\right)}{1 + \left(\frac{K_1}{s}\right)\left(\frac{K_2}{Ts+2}\right)(K_3 + K_4 s)} = \frac{K_1 K_2}{Ts^2 + (K_1 K_2 K_4 + 1)s + K_1 K_2 K_3}.$$

(b)

$$S_{K_2}^{H}(s) = \frac{\frac{dH}{H}}{\frac{dK_2}{K_2}} = \left(\frac{K_2}{T}\right)\left(\frac{dH}{dK_2}\right)$$

where

$$\frac{dH}{dK_2} = \frac{(K_1 s)(Ts+1)}{Ts^2 + (K_1 K_2 K_4 + 1)s + K_1 K_2 K_3}.$$

Therefore,

$$S_{K_2}^{H}(s) = \frac{s(Ts+1)}{Ts^2 + (K_1 K_2 K_4 + 1)s + K_1 K_2 K_3}.$$

(c)

$$\left. S_{K_2}^{H}(s) \right]_{s=0} = 0.$$

I5.2. The sensitivity of the following control systems is to be analyzed:

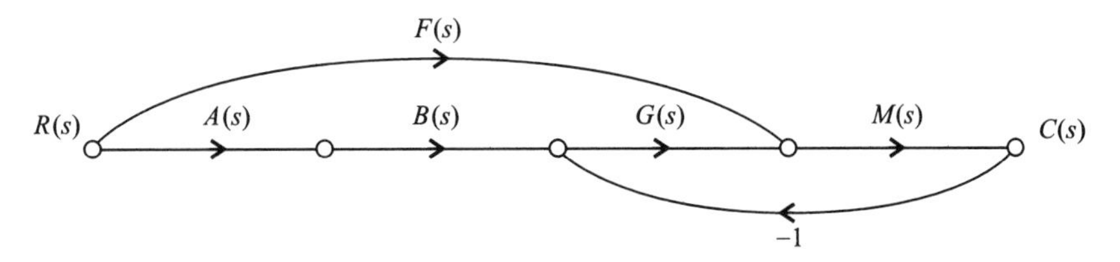

Figure I5.2

(a) Determine the overall transfer function, $H = C(s)/R(s)$.

(b) Determine the sensitivity of the control system's transfer function, $H(s) = C(s)/R(s)$ with respect to the transfer function $M(s)$.

(c) Determine the sensitivity of the control system's transfer function, $H(s) = C(s)/R(s)$ with respect to the transfer function $B(s)$.

(d) What happens to your results in part (c) if the transfer function $A(s)G(s)B(s)$ is much greater than $F(s)$?

SOLUTION: (a)

$$H(s) = \frac{C(s)}{R(s)} = \frac{M(s)[F(s) + A(s)B(s)G(s)]}{1 + G(s)H(s)}$$

(b)

$$H(s) = \frac{\frac{dH}{H}}{\frac{dM}{M}} = \frac{M}{H}\frac{dH}{dM}$$

where

$$\frac{dH}{dM} = \frac{F(s) + A(s)B(S)G(s)}{(1 + G(s)M(s))^2}.$$

Therefore,

$$S_M^H(s) = \frac{1}{1 + G(s)M(s)}.$$

(c)

$$S_B^H(s) = \frac{\frac{dH}{H}}{\frac{dB}{B}} = \frac{B}{H}\frac{dH}{dB}$$

where

$$\frac{dH}{dB} = \frac{A(s)G(s)M(s)}{1 + G(s)M(s)}.$$

Therefore,

$$S_B^H(s) = \frac{A(s)G(s)B(s)}{F(s) + A(s)G(s)B(s)}$$

(d) For $A(s)G(s)B(s)$ much greater than $F(s)$, then

$$S_B^H(s) \approx \frac{A(s)G(s)B(s)}{A(s)G(s)B(s)} = 1$$

15.3. A control-system engineer is designing a feedback amplifier using two stages of gain. He is not certain which of the following two systems will result in a smaller sensitivity to the amplifier gain, K_a:

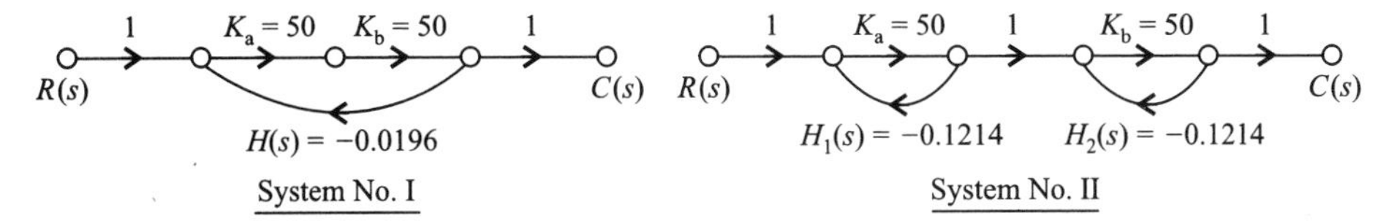

Figure I5.3

(a) Determine the closed-loop transfer functions, $C(s)/R(s)$, for each design, assuming that $K_a = K_b = 50$.

(b) Compare the sensitivities of the two systems with respect to the amplifier gain K_a.

(c) Why do the results come out of the way they do, and which design would you select?

SOLUTION: **(a)** *For System No. I*:

$$H_I(s) = \frac{C(s)}{R(s)} = \frac{K_a K_b}{1 + K_a K_b H(s)} = \frac{(50(50)}{1 + (50)(50(0.196)} = 50.$$

For System No. II:

$$H_{II}(s) = \frac{C(s)}{R(s)} = \frac{K_a}{1 + K_a H_1(s)} \frac{K_b}{1 + K_b H_2(s)} = \frac{50}{1 + 50(0.124)} \frac{50}{1 + 50(0.124)} = 7.07^2.$$

Therefore, $H_{II}(s) = 50$.

(b) *For System No. I*:

$$S_{K_a}^{H_I}(s) = \frac{K_a}{H_1(s)} \frac{dH_{II}(s)}{dK_a} = \frac{K_a}{1 + K_a K_b H(s)} = \frac{1}{1 + (50)(50)(0.0196)} = 0.02.$$

For System No. II:

$$H_{II}(s) = \frac{K_a}{1 + K_a H_1(s)} \text{ (gain of stage 2)} = \frac{K_a}{1 + 0.1214 K_a}(7.07)$$

and

$$S_{K_a}^{H_{II}}(s) = \frac{K_a}{H_{II}(s)} \frac{dH_{II}(s)}{dK_a}.$$

Therefore,

$$S_{K_a}^{H_{II}}(s) = \frac{K_a}{\dfrac{7.072 K_a}{1 + 0.1213 K_a}} \frac{(1 + 0.1214 K_a)(7.07) - 7.07 K_a (0.1214)}{(1 + 0.1214 K_a)^2} = 0.1414.$$

(c)

$$\frac{S_{K_a}^{H_{II}}(s)}{S_{K_a}^{H_I}(s)} = \frac{0.1414}{0.02} = 7.07.$$

The sensitivity of System No. I is less than that of System No. 2 because System No. 1 has feedback around both K_a and K_b, with the result that the gain of System No. 1, $H(s)$, is 50. The gain around K_a in System No. II is only 7.07. Therefore, the sensitivity of Hs) to variations in K_a in System No. I is less than that in System No. II.

I5.4. The accuracy of the following control system is to be analyzed:

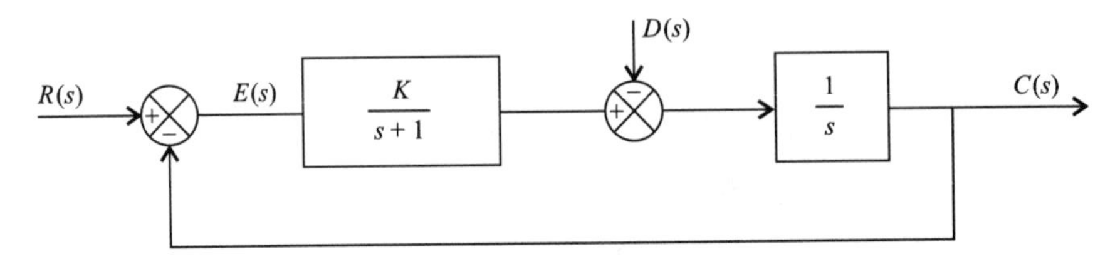

Figure I5.4

(a) Determine the steady-state error, e_{ss}, for a unit step input, $r(t) = U(t)$ with $K = 10$.

(b) Determine the steady-state error, e_{ss}, for a unit disturbance step input, $d(t) = U(t)$ with $K = 10$.

(c) Determine the value of K so that the resulting steady-state error due to a disturbance step input, $d(t) = U(T)$ equals 5% of the disturbance step input value.

SOLUTION: (a) Since the forward transfer function, $C(s)/E(s)$, has one pure integrator in it, the steady-state error, e_{ss}, equals 0.

(b)

$$\frac{E(s)}{D(s)} = \frac{1/s}{1 + \dfrac{10}{s(s+1)}} = \frac{(s+1)}{s^2 + s + 10}.$$

Therefore,

$$e_{ss} = \lim_{s \to 0} sE(s) = \lim_{s \to 0} s \frac{1}{s} \frac{(s+1)}{s^2 + s + 10} = 0.1$$

(c)

$$\frac{E(s)}{D(s)} = \frac{1/s}{1 + \dfrac{K}{s(s+1)}} = \frac{(s+1)}{s^2 + s + K}.$$

Therefore,

$$e_{ss} = \lim_{s\to 0} s\frac{1}{s}\frac{(s+1)}{s^2+s+K} = \frac{1}{K} = 0.05.$$

Therefore, $K = 20$.

I5.5. The control system of a radar-controlled anti-aircraft gun is to be designed. In the illustration shown, the tracking radar measures the distance d to the crossing target aircraft, and the angle of the antenna with respect to the ground, θ_r.

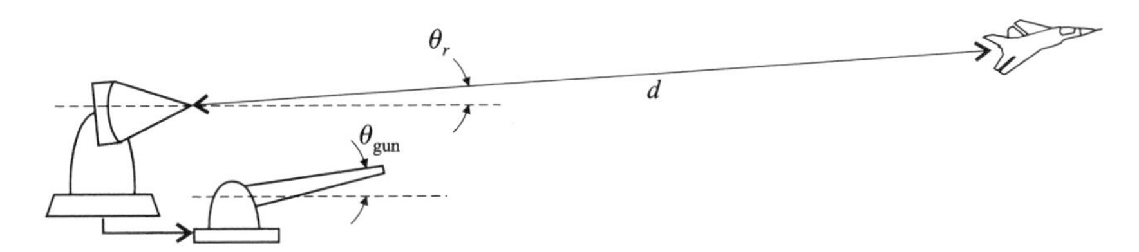

Figure I5.5(i)

For the design of the control system, assume the crossing target aircraft speed is 1500 ft/sec at a range of 25000 ft, and it is assumed to be perpendicular to the line-of-sight from the radar. The following control system for positioning the gun is to be analyzed:

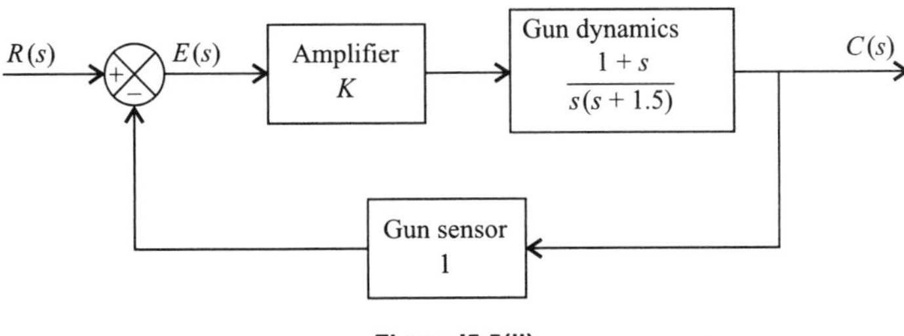

Figure I5.5(ii)

(a) Determine the angular velocity, in rad/sec, the radar sees due to the crossing target.

(b) The maximum desired miss distance of a shell fired from the anti-aircraft gun is 2 ft. How does this translate to an angular steady-state-error, e_{ss}, in radians for the control system?

(c) Determine the velocity constant, K_v, needed from the control sysem to satisfy the answers found in parts (a) and (b).

(d) Determine the value of the amplifier gain. K, in the control system to achieve the velocity constant, K_v, found in part (c).

SOLUTION: **(a)** The problem can be analyzed from the following:

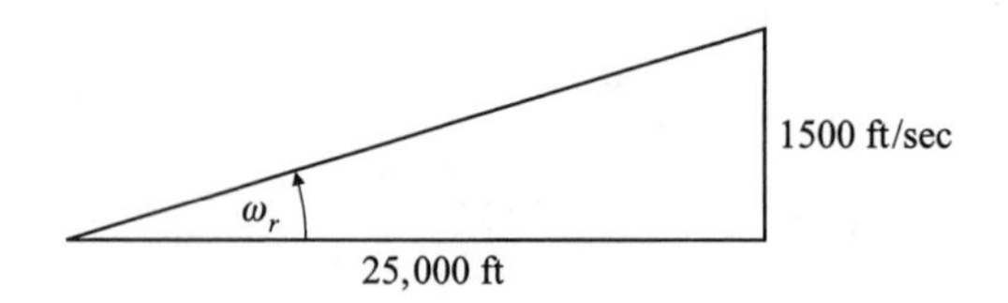

Figure I5.5(iii)

Therefore,

$$\omega_r = \frac{v}{r} = \frac{1500\,\text{ft/sec}}{25,000\,\text{ft}} = 0.06 \text{ rad/sec}.$$

(b) The problem can be analyzed from the following:

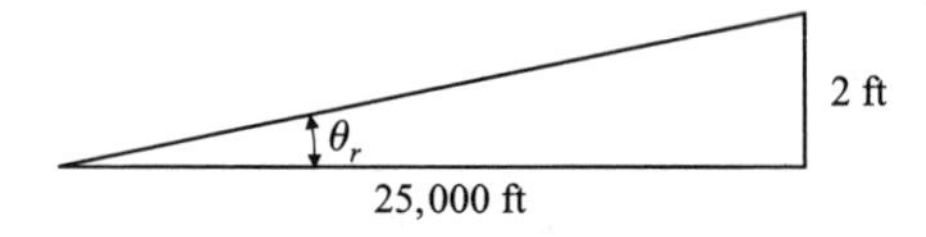

Figure I5.5(iv)

Therefore,

$$e_{ss} = \frac{2\,\text{ft}}{25,000\,\text{ft}} = 0.00008 \text{ rad}.$$

(c)

$$e_{ss} = \frac{\omega_r}{K_v}.$$

Therefore, from the answers in parts (a) and (b), we obtain the following:

$$0.00008 = \frac{0.06}{K_v}.$$

Therefore,

$$K_v = 750.$$

(d) Using the answer from part (c), we obtain the following:

$$K_v = \lim_{s\to 0} sG(s) = \lim_{s\to 0} s\frac{K(1+s)}{s(s+1.5)} = 750.$$

Therefore,

$$\frac{K}{1.5} = 750$$

and

$$K = 1125.$$

I5.6. Determine the values of K and b for the control system shown in order that the damping ratio of the system is 0.5, and the settling time of the unit-step response is 0.2 sec.

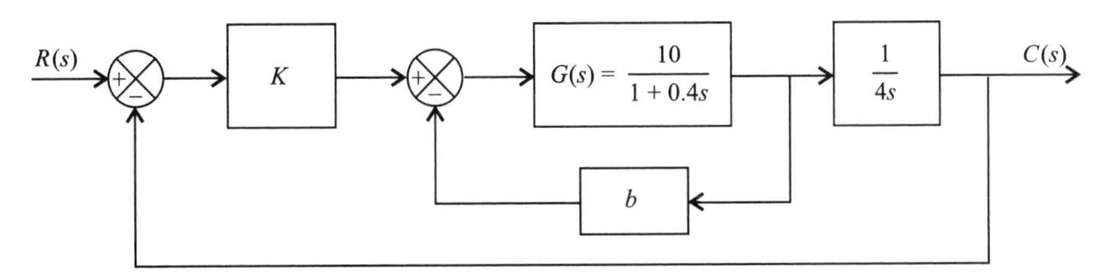

Figure I5.6

SOLUTION: (a) Using Mason's Theorem, we obtain the following:

$$\frac{C(s)}{R(s)} = \frac{\dfrac{K}{4s}\left(\dfrac{10}{1+0.4s}\right)}{1+\dfrac{K}{4s}\left(\dfrac{10}{1+0.4s}\right)+\dfrac{10b}{1+0.4s}}.$$

Simplifying this equation, we obtain the following:

$$\frac{C(s)}{R(s)} = \frac{6.25K}{s^2+(2.5+25b)s+6.25K}.$$

From Eq. (5.41),

$$t_s = \frac{4}{\zeta\omega_n}.$$

Substituting into this equation for t_s,

$$0.2 = \frac{4}{\zeta\omega_n}.$$

Therefore,

$$\zeta\omega_n = 20.$$

We also know from comparing the denominator of Eq. (4.3) and that of $C(s)/R(s)$,

$$2\zeta\omega_n = 2.5 + 25b.$$

Therefore, $b = 1.5$.

Since the problem states that the damping ratio, $\zeta = 0.5$, and since we determined that $\zeta\omega_n = 20$, we can solve for ω_n as follows:

$$0.5\omega_n = 20.$$

Therefore, $\omega_n = 40$. We can now determine K by comparing the equation for $C(s)/R(s)$ and Eq. (4.3):

$$\omega_n^2 = 6.25K.$$

Substituting for $\omega_n = 40$, we solve for K:

$$40^2 = 6.25K.$$

Therefore, $K = 256$.

I5.7. The steering control system of a large ship represents a feedback control system as illustrated in the block diagram shown. The function of the ship's rudder is to compensate for the ship's deviation from its desired heading due to sea waves and wind disturbances (where the ship acts as a sail when wind strikes it).

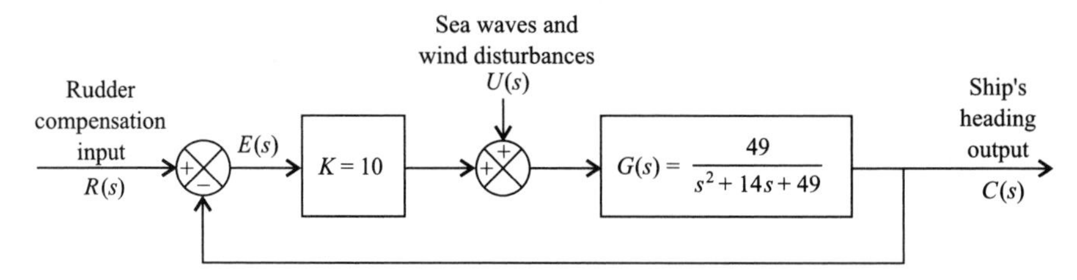

Figure I5.7

(a) Find the steady-state value of the output due to a constant sea wave and wind disturbance which can be approximated as a unit step: $U(s) = 1/s$. Assume that the rudder input is zero: $R(s) = 0$.

(b) Determine the value that the ship's steering officer should set the rudder, $R(s)$, to compensate for the disturbance in part (a).

SOLUTION: (a)

$$\frac{C(s)}{U(s)} = \frac{G(s)}{1 + KG(s)} = \frac{49}{s^2 + 14s + 539},$$

$$C(s) = U(s)\frac{49}{s^2 + 14s + 539} = \left(\frac{1}{s}\right)\left(\frac{49}{s^2 + 14s + 539}\right),$$

$$c(\infty) = \lim_{s\to 0} sC(s) = \lim_{s\to 0} s\left(\frac{1}{s}\right)\left(\frac{49}{s^2 + 14s + 539}\right) = 0.09.$$

(b)

$$C(s) = \frac{G(s)U(s) + KG(s)R(s)}{1 + KG(s)} = \frac{G(s)[U(s) + KR(s)]}{1 + KG(s)}.$$

Therefore, for $C(s) = 0$, it is necessary that we design

$$U(s) + KR(s) = 0$$

Or,

$$R(s) = -\frac{U(s)}{K} = -\frac{1}{10s}.$$

Therefore,

$$r(t) = -0.1U(t).$$

I5.8. A control system containing a reference input $R(s)$ and a disturbance $D(s)$ is represented by the following block diagram:

Assuming that the system is stable, determine the following:

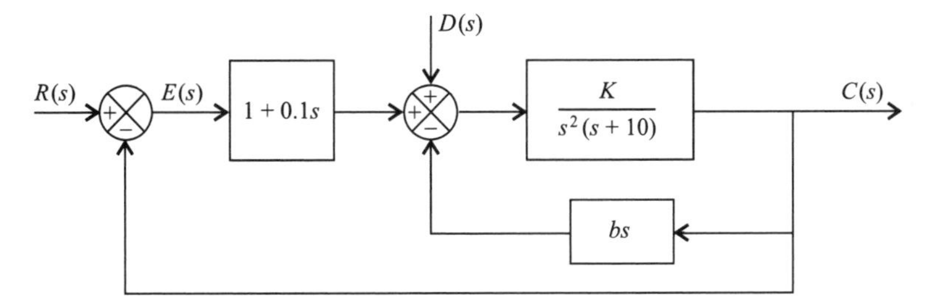

Figure I5.8

(a) The transfer function $C(s)/E(s)$ which represents the transfer function of the control system with the outer loop opened.

(b) The steady-state error of the system in terms of K and b for a unit-ramp input, $r(t) = tU(t)$. Assume that $D(s) = 0$ in this part.

(c) The steady-state value of $c(t)$ when $D(s) = 1/s$, a unit step input. Assume that $R(s) = 0$ in this part.

SOLUTION:

(a)

$$\frac{C(s)}{E(s)} = (1 + 0.1s)\left[\frac{\dfrac{K}{s^2(s+10)}}{1 + \dfrac{(bs)K}{s^2(s+10)}}\right] = \frac{K(1+0.1s)}{s(s^2 + 10s + Kb)}.$$

(b)

$$K_v = \lim_{s\to 0} s\left[\frac{C(s)}{E(s)}\right] = \lim_{s\to 0} s\left[\frac{K(1+0.1s)}{s(s^2+10s+Kb)}\right] = \frac{1}{b}.$$

Therefore,

$$e_{ss} = \frac{1}{K_v} = \frac{1}{1/b} = b$$

(c)
$$\frac{C(s)}{D(s)} = \frac{\dfrac{K}{s^2(s+10)}}{1 + \dfrac{Kbs}{s^2(s+10)} + \dfrac{(1+0.1s)K}{s^2(s+10)}},$$
$$\frac{C(s)}{D(s)} = \frac{K}{s^3 + 10s^2 + K(b+0.1)s + K}.$$

For $D(s) = 1/s$, we obtain the following expression for $C(s)$:

$$C(s) = \left(\frac{1}{s}\right)\left[\frac{K}{s^3 + 10s^2 + K(b+0.1)s + K}\right]$$

$$c_{ss} = \lim_{s\to 0} sC(s) = \lim_{s\to 0} s\left(\frac{1}{s}\right)\left[\frac{K}{s^3 + 10s^2 + K(b+0.1)s + K}\right] = 1$$

I5.9. We wish to analyze the second-order control system illustrated in Figure 4.1, whose closed-loop transfer function was developed in Eq. (5.53). Assuming that $T_m = 1$ sec, what should K_m be designed to be so that the ITAE criterion is achieved for this zero steady-state step error system?

SOLUTION: From Eq. (5.53) with $T_m = 1$, we obtain the following

$$\frac{C(s)}{R(s)} = \frac{K_m}{s^2 + s + K_m}.$$

From Table 5.5, the ITAE form for a second-order control system is given by:

$$\frac{C(s)}{R(s)} = \frac{\omega_n^2}{s^2 + 1.4\omega_n s + \omega_n^2}.$$

Setting like coefficients equal to each other from these two equations, we obtain the following:

$$1.4\omega_n = 1$$

and

$$\omega_n^2 = K_m.$$

Solving these two simultaneous equations, we obtain the following:

$$\omega_n = 0.71$$

and

$$K_m = 0.51.$$

I5.10. We wish to analyze the control system illustrated in Figure 5.9, which contains two pure integrations and whose closed-loop transfer function was given in Eq. (5.55). Assuing that $T_2 = 1$, determine the value sof K and T_1 so that the ITAE criterion is achieved for this zero steady-state ramp error system.

SOLUTION: From Eq. (5.55) with $T_2 = 1$, we obtain the following:

$$\frac{KT_1 s + K}{s^3 + s^2 + KT_1 s + K}.$$

From Table 5.6, the ITAE form for a third-order control system is

$$\frac{\omega_n^3}{s^3 + 1.75\omega_n s^2 + 3.25\omega_n^2 s + \omega_n^3}.$$

Setting like coefficients equal to each other from these two equations, we obtain the following:

$$1.75\omega_n = 1,$$
$$KT_1 = 3.25\omega_n^2,$$
$$K = \omega_n^3.$$

Solving these three equations simultaneously, we obtain:

$$\omega_n = 0.571$$
$$K = 0.186,$$
$$T_1 = 5.697.$$

PROBLEMS

5.1. Assume that the system shown in Figure P5.1 has the following characteristics:

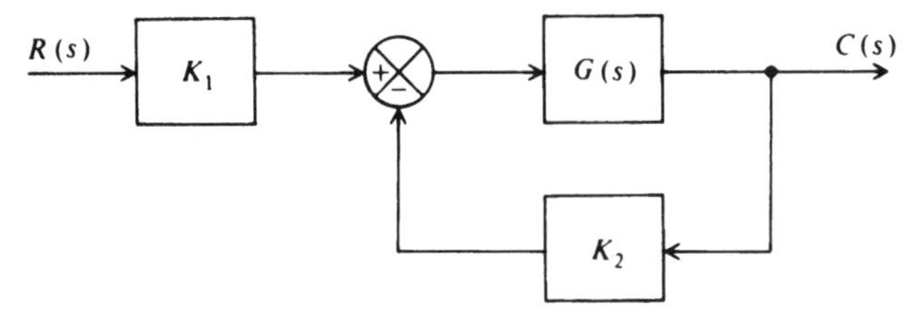

Figure P5.1

$$K_1 = 10 \text{ V/rad},$$
$$K_2 = 10 \text{ V/rad},$$
$$G(s) = \frac{100}{s(s+1)}.$$

(a) Determine the sensitivity of the system's transfer function $H(s)$ with respect to the input transducer, K_1.

(b) Determine the sensitivity of the system's transfer function $H(s)$ with respect to the feedback transducer, K_2.

(c) Determine the sensitivity of the system's transfer function $H(s)$ with respect to $G(s)$

(d) Indicate qualitatively the frequency dependency of $S_G^H(s)$.

5.2. For the system in Figure P5.2, assume the following characteristics:

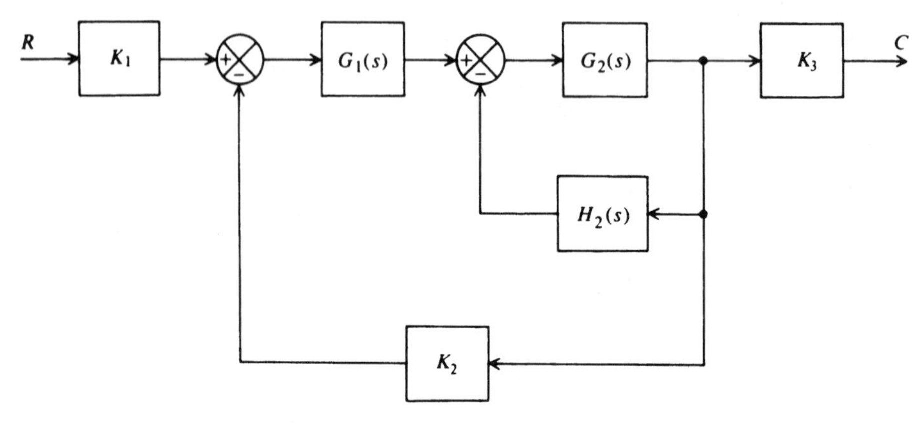

Figure P5.2

$$K_1 = 10 \text{ V/rad}, \qquad G_1(s) = \frac{10}{100 + s}$$

$$K_2 = 10 \text{ V/rad}, \qquad G_2(s) = \frac{20}{s(s+2)},$$

$$K_3 = 2 \text{ V/rad}, \qquad H(s) = 4s.$$

(a) Determine the sensitivity of the system's transfer function $H(s)$ with respect to $G_1(s)$ at $\omega = 1$ rad/sec.

(b) Determine the sensitivity of the system's transfer function $H(s)$ with respect to $G_2(s)$ at $\omega = 1$ rad/sec.

(c) Determine the sensitivity of the system's transfer function $H(s)$ with respect to $H_2(s)$ at $\omega = 1$ rad/sec.

(d) Determine the sensitivity of the system's transfer function $H(s)$ with respect to K_3 at $\omega = 1$ rad/sec.

(e) List the answers of parts (a)–(d) in tabular form with the lowest value first and the largest value last for $\omega = 1$ rad/sec. Normalize your table with repect to the lowest sensitivity found. How could this table change for a different choice of frequency?

5.3. Figure P5.3*a* illustrates the block diagram of a computer-controlled machine tool that utilizes a position loop and a correction loop [5]. A practical problem in machine-tool application is the fact that the desired tool position is ordinarily fed back by the position loop as shown, but this is not indicative of the condition and shape of the finished part. In reality, the finished part is ordinarily removed from the feedback effect due to the tool–work interface. Because these effects are complex and difficult to predict, the process output will not conform precisely to that desired. By means of the correction loop, information is obtained from the process output by means of sensors coupled as closely as possible to the tool–work interface. These sensor signals, which represent variables such as torque, vibration, and temperature, are fed back to improve the performance of the machine tool control system. Figure P5.3*b* illustrates an equivalent block-diagram representation.

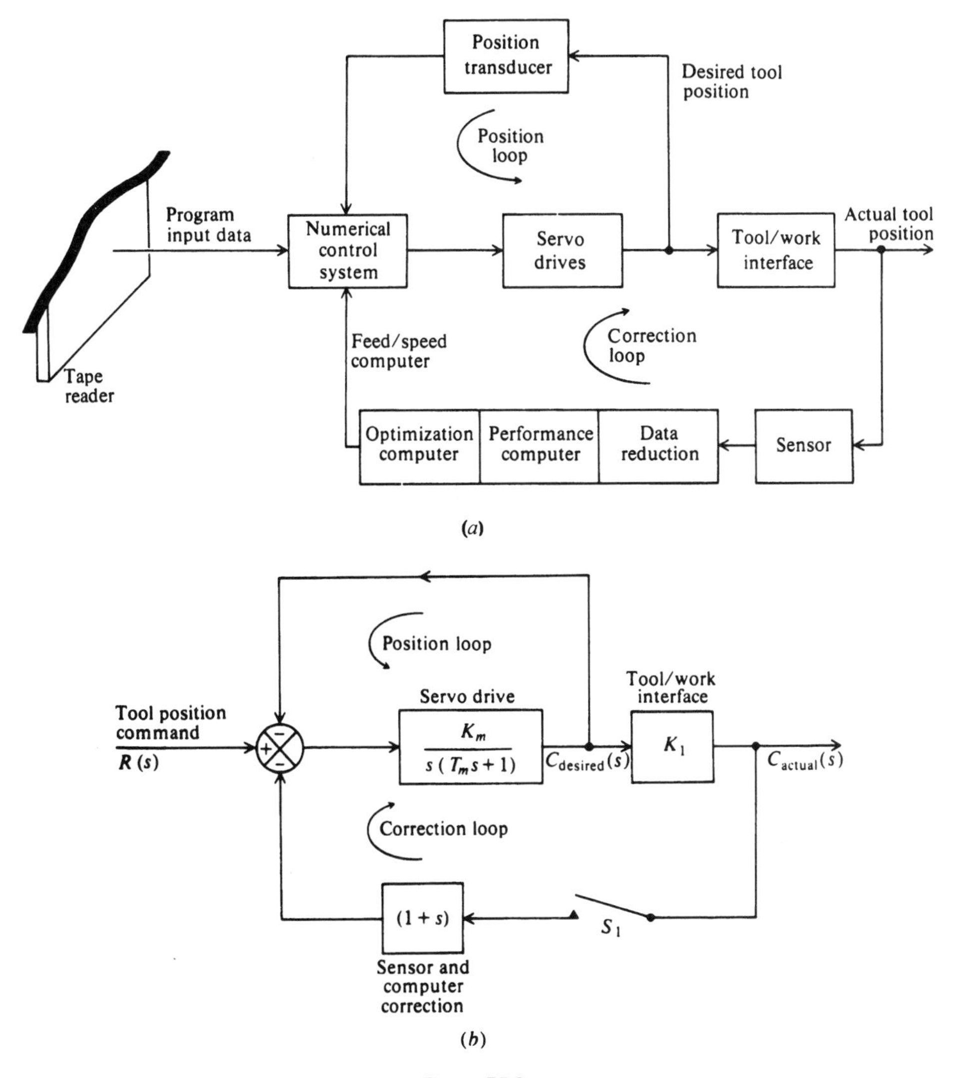

Figure P5.3

(a) With switch S_1 open and the correction loop inoperative, determine the sensitivity of the system's transfer function $T(s)$ to variations of K_m, T_m, and K_1.

(b) Repeat part (a) with switch S_1 closed and the correction loop operative.

(c) What conclusions can you draw from your results?

5.4. For the system illustrated in Figure P5.4 assume the transfer functions indicated.

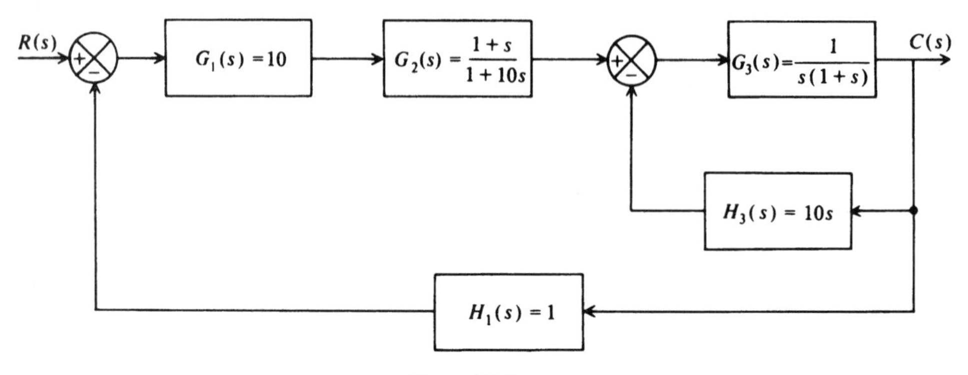

Figure P5.4

(a) Determine the sensitivity of the system's transfer function with respect to $G_1(s)$.

(b) Determine the sensitivity of the system's transfer function with respect to $G_3(s)$.

(c) Determine the sensitivity of the system's transfer function with respect to $H_3(s)$.

(d) Determine the sensitivity of the system's transfer function with respect to $H_1(s)$.

(e) List the answers of parts (a)–(d) in tabular form with the lowest value of sensitivity first and the largest value last at a frequency of 1 rad/sec. Normalize your table with respect to the lowest sensitivity found. How could this table change for a different choice of frequency?

5.5. Simulators for training pilots of commercial and military aircraft are receiving greater emphasis and attention. A simulator used for training the pilot of a new aircraft is to be designed. This simulator will have a three-degree-of-freedom capability (roll, pitch, yaw), which will give the trainees a "feel" of the dynamics that the pilots will experience during flight operation. The controls and interior of the cockpit will be designed to be a very close replica of the actual aircraft, to give the pilot a real-world feel of the aircraft—both esthetically and dynamically. For purposes of this analysis, assume that the pitch loop can be represented by the block diagram in Figure P5.5.

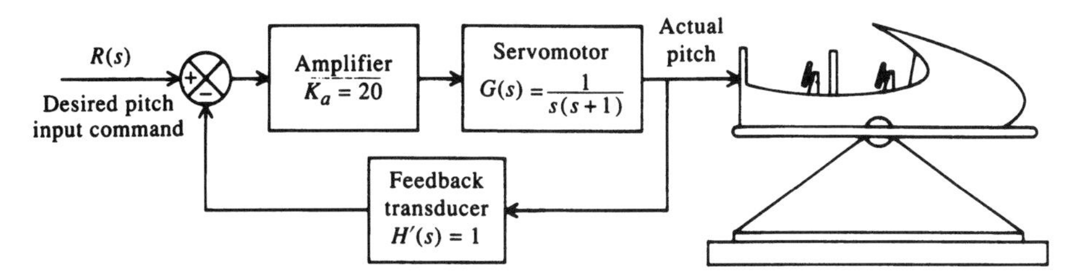

Figure P5.5

(a) Determine the sensitivity of the pitch loop's transfer function, $H(s)$, to the amplifier gain, K_a, in terms of the system parameters.

(b) Determine the sensitivity of the pitch loop's transfer function to the feedback transducer, H_1, in terms of the system parameters.

(c) For the flight environment expected to be encountered, parameter variations at $\omega = 1$ rad/sec are very critical. Determine the absolute values of these two sensitivities at this frequency.

(d) What conclusions can you reach from your results?

5.6. The time constant of a two-phase ac servomotor is not a precise parameter, and is subject to change caused by aging and environmental conditions such as temperature changes. The control system in Figure P5.6 containing a two-phase ac servomotor, is to be analyzed for its sensitivity to the two-phase ac servomotor's time constant, T.

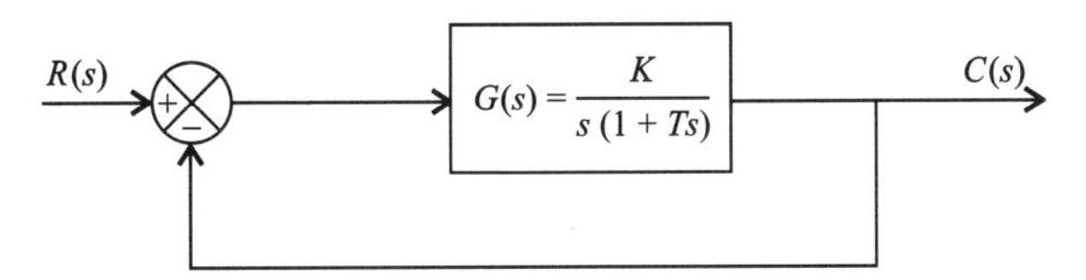

Figure P5.6

Determine the sensitivity of the control system's transfer function, $H(s) = C(s)R(s)$, with respect to the time constant T.

5.7. The signal-flow graph representation of the control system in Figure P5.7 is to be analyzed.

(a) Determine the overall transfer function $C(s)/R(s)$.

(b) The selection of $G_4(s)$ is critical. Determine the sensitivity $S_{G_4(s)}^{H(s)}(s)$.

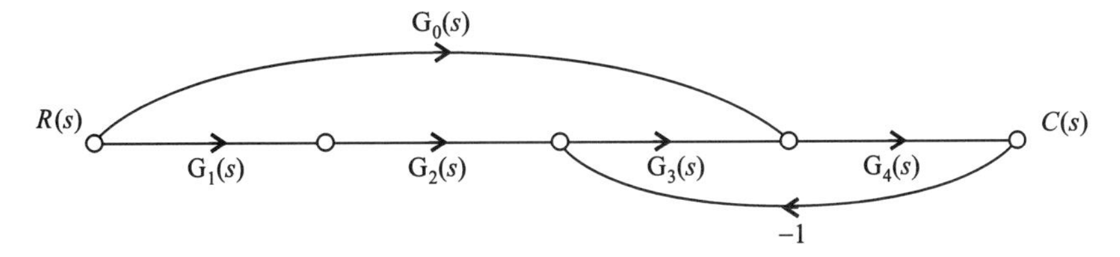

Figure P5.7

5.8. The block diagram for the control system that positions an aircraft's rudder is shown in Figure P5.8.

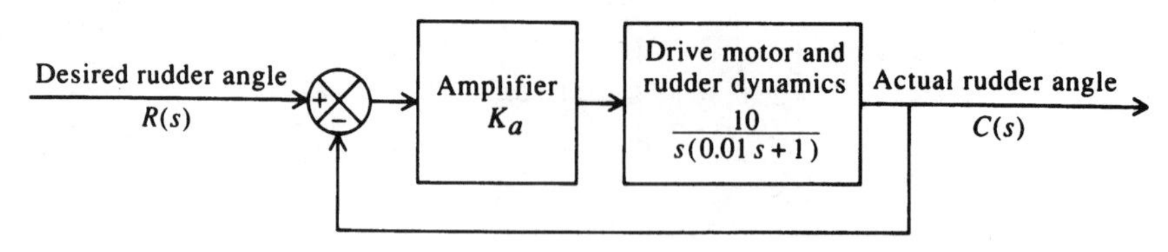

Figure P5.8

(a) What is the sensitivity of the system transfer function, $C(s)/R(s)$, to small changes in K_a? Express your answer for $S_K^H(s)$ as a function of s.

(b) Determine the magnitude of the sensitivity at wind gust disturbances which can be approximated to exist at $\omega = 1$ rad/sec. Assume that $K_a = 10$.

(c) Determine the magnitude of the sensitivity when the aircraft breaks the sound barrier, and the sonic boom causes a vibration frequency disturbance which can be approximated at $\omega = 1000$ rad/sec. Assume that $K_a = 10$.

(d) What conclusions can you reach from your results?

5.9. In the control system illustrated in Figure P5.9, attention is to be focused on the quality of components to be specified for it. Determine the sensitivity of the system's transfer function, $H(s) = C(s)/R(s)$ with respect to the following system parameters:

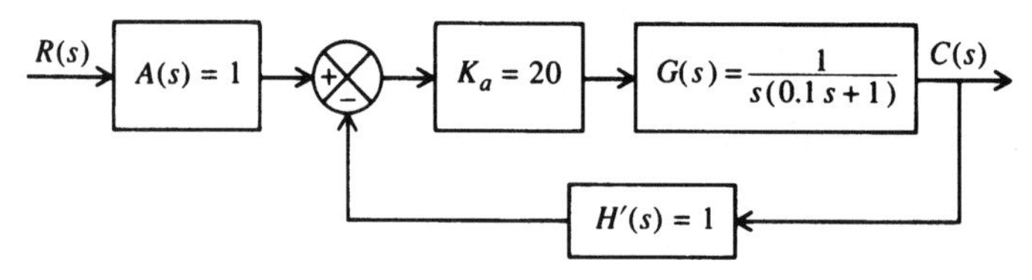

Figure P5.9

(a) $A(s)$

(b) $G(s)$

(c) $H'(s)$

In each case, evaluate the sensitivity function, and then determine its absolute value at $\omega = 1$ rad/sec.

5.10. The sensitivity of the feedback control system illustrated in Figure P5.10 is to be determined.

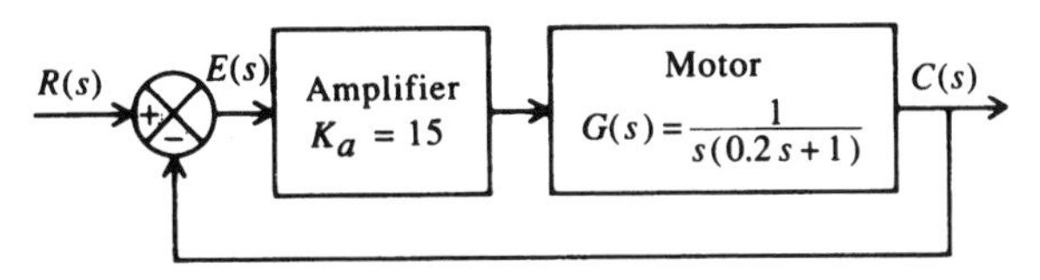

Figure P5.10

(a) Determine the system's transfer function $H(s) = C(s)/R(s)$.

(b) Determine the sensitivity of $H(s)$ with respect to the motor $G(s)$. Show your answer in its simplest possible form as a function of s.

(c) What is the absolute value of the sensitivity of $H(s)$ with respect to $G(s)$ at $\omega = 1$ rad/sec?

(d) If the value of the motor's transfer function $G(s)$ varies by $\pm 25\%$, by how much will the system's transfer function $H(s)$ vary at $\omega = 1$ rad/sec?

5.11. A control system is used to position the rudder of an aircraft. Its block diagram can be adequately represented as shown in Figure P5.11.

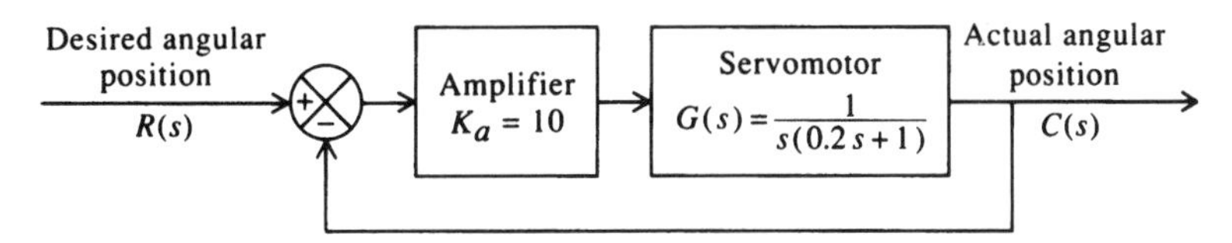

Figure P5.11

(a) Determine the sensitivity of the system's transfer function with respect to $G(s)$.

(b) Determine the sensitivity of the system's transfer function with respect to K_a.

(c) Determine the absolute value of the sensitivities determined in parts (a) and (b) to wind gusts that can be approximated to primarily exist at 1 rad/sec. Which element is more sensitive at this frequency, the amplifier or servo motor?

(d) How much will the system's transfer function vary at 1 rad/sec if K_a changes by 50%?

5.12. A control-system engineer wants to determine the quality of a tachometer needed to be specified in the control system shownin Figure P5.12, which uses both position and rate feedback. In making the decision, it is known that the control system must follow an input at $R(s)$, the dominant frequency component of which is at 0.5 rad/sec.

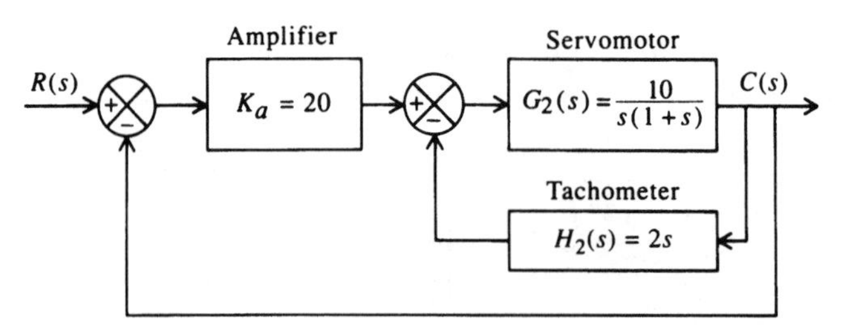

Figure P5.12

(a) Determine the sensitivity of the system's transfer function $H(s)$ with respect to $H_2(s)$ in terms of the Laplace transform s.

(b) Determine the absolute value of this sensitivity at a frequency of 0.5 rad/sec.

(c) What decision regarding the quality of the tachometer should be made?

5.13. Automatic control theory can also be applied to automatic warehousing and inventory control systems [6]. Of particular importance in these systems is the smooth flow of material. Figure P5.13 illustrates a block diagram that is representative of this class of control systems.

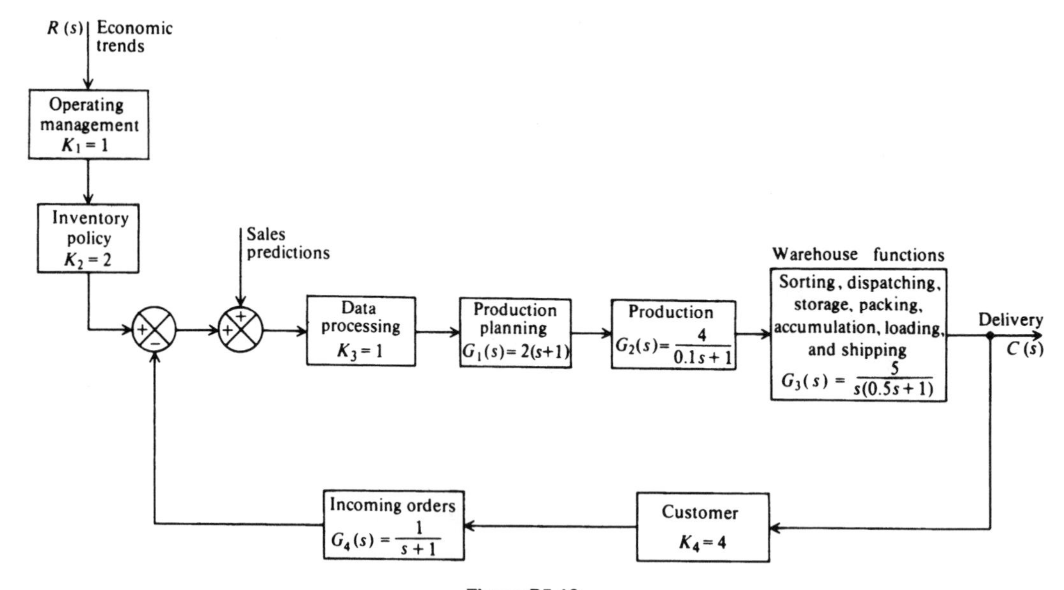

Figure P5.13

(a) Determine the sensitivity of the system's transfer function $C(s)/R(s)$ with respect to production planning. $G_1(s)$.

(b) Determine the sensitivity of the system's transfer function with respect to production, $G_2(s)$.

(c) Determine the sensitivity of the system's transfer function with respect to the various warehouse functions $G_3(s)$.

(d) Determine the sensitivity of the system's transfer function with respect to incoming orders, $G_4(s)$.

(e) Determine the sensitivity of the system's transfer function with respect to the inventory policy, K_2.

(f) For which two parameters is the system most sensitive, and what conclusions can you reach from your results?

(g) Indicate qualitatively the frequency dependence of the sensitivity functions.

5.14. Figure P5.14 illustrates an electronic pacemaker used to regulate the speed of the human heart. Assume that the transfer function of the pacemaker is given by $G_p(s) = K/(0.05s + 1)$ and assume that the heart acts as a pure integrator.

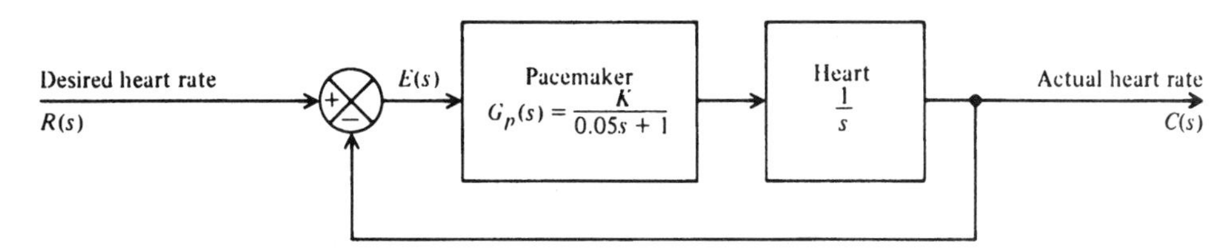

Figure P5.14

(a) For optimum response, a closed-loop damping ratio of 0.5 is desired. Determine the required gain K of the pacemaker in order to achieve this.

(b) What is the sensitivity of the system transfer function. $C(s)/R(s)$, to small changes in K?

(c) Determine this sensitivity at dc.

(d) Find the magnitude of this sensitivity at the normal heart rate of 60 beats/minute.

5.15. A control system used to position a load is shown in Figure P5.15.

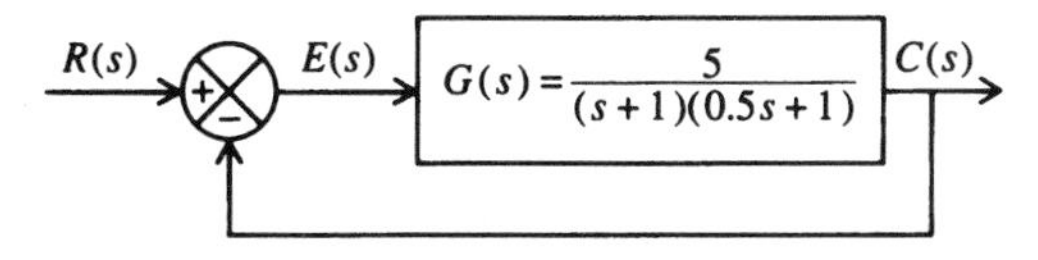

Figure P5.15

(a) Determine the steady-state error for a step input of 10 units.

(b) How should $G(s)$ be modified in order to reduce this steady-state error to zero?

5.16. A unity feedback control system has a forward transfer function given by

$$G(s) = \frac{10(s+1)}{s^2(0.1s+1)(s+5)}.$$

Determine the steady-state error for an input given by

$$r(t) = (4 + 6t + 3t^2)U(t).$$

5.17. The block diagram of a simple instrument servomechanism is shown in Figure P5.17.

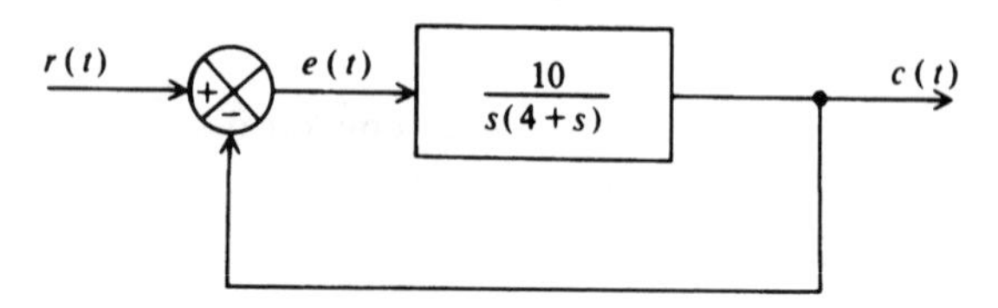

Figure P5.17

(a) Determine the steady-state error resulting from the input of a ramp that may be represented by

$$r(t) = (10t)U(t).$$

(b) Determine the steady-state error resulting from the following input:

$$r(t) = (4 + 6t + 3t^2)U(t).$$

5.18. Repeat Problem 5.17 with the transfer function of the system given by

$$G(s) = \frac{10}{s(1+s)(1+10s)}.$$

5.19. A ground-based tracking radar is used to track aircraft targets very precisely. The block diagram of the azimuth axis of the tracking radar can be adequately represented by the block diagram of Figure P5.19.

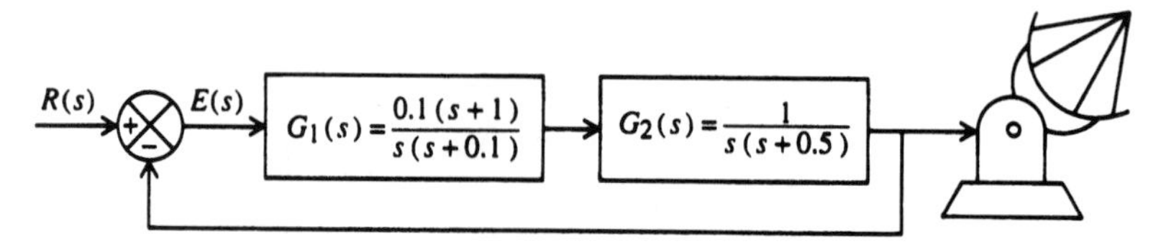

Figure P5.19

Determine the steady-state errors of the tracker for the following types of inputs caused by the aircraft dynamics:

(a) $r(t) = (10t)U(t)$.

(b) $r(t) = (10t + 6t^2)U(t)$.

5.20. A control system containing a reference input, $R(s)$, and a disturbance input, $D(s)$, is illustrated:

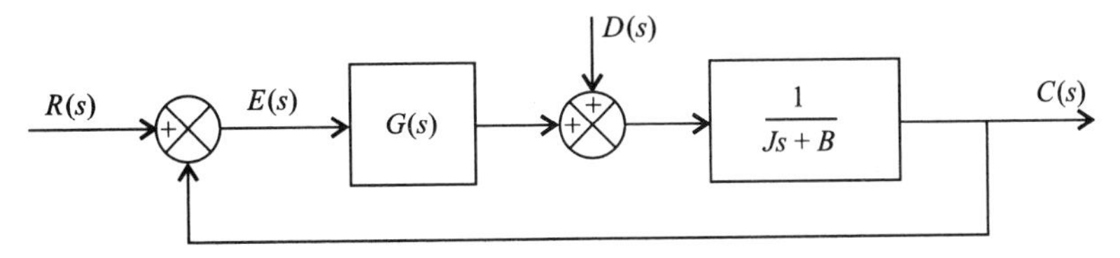

Figure P5.20

(a) Determine the steady-state error, e_{ss}, for a unit-step disturbance at $D(s)$ in terms of the unknown transfer function $G(s)$.

(b) Select the simplest value of $G(s)$ which will result in zero steady-state error for $E(s)$ when $D(s)$ is a unit step input.

5.21. The block diagram of a control system containing a disturbance, $D(s)$, is illustrated in Figure P5.21.

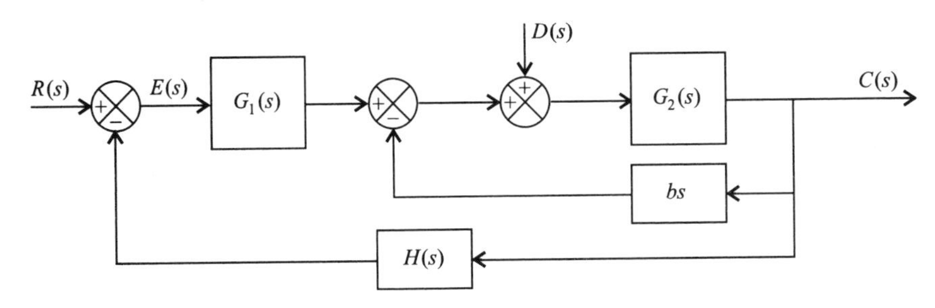

Figure P5.21

(a) Determine the transfer function, $E(s)/D(s)$.

(b) Determine the steady-state error for $E(s)$ when $D(s)$ is a unit step input assuming that $H(s) = b = 1$, and that $G_2(s) = 1/(s+1)$, in terms of the unknown transfer function, $G_1(s)$.

(c) Select the simplest value of $G_1(s)$ which will result in zero steady-state error for $E(s)$ when $D(s)$ is a unit step input.

5.22. We wish to determine the accuracy of the control system illustrated in Figure P5.22.

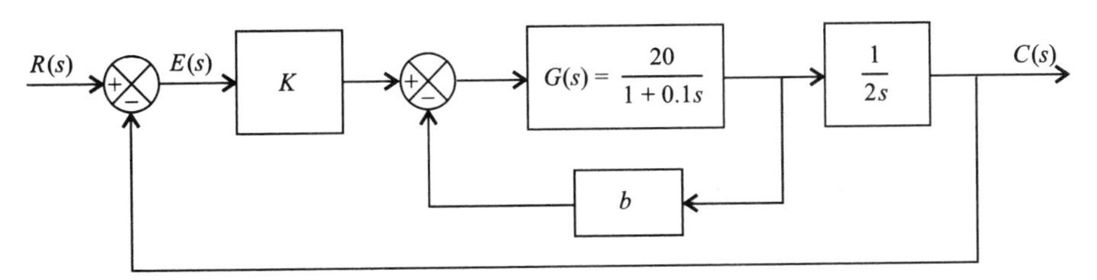

Figure P5.22

Assuming that the system is stable, determine the following:

(a) The transfer function, $C(s)/E(s)$ which represents the open-loop transfer function of the control system with the outer loop opened. Reduce your answer to its simplest form.

(b) The steady-state error, e_{ss}, for $r(t) = U(t)$.

(c) The steady-state error, e_{ss}, for $r(t) = tU(t)$.

(d) The steady-state error, e_{ss}, for $r(t) = (t^2/2)U(t)$.

(e) Assuming that $b = 1$, what must K be designed for in order that the steady-state error, e_{ss}, equals 0.1 when $r(t) = tU(t)$.

5.23. The block diagram of a control system containing a disturbance, $D(s)$, is illustrated in Figure P5.23.

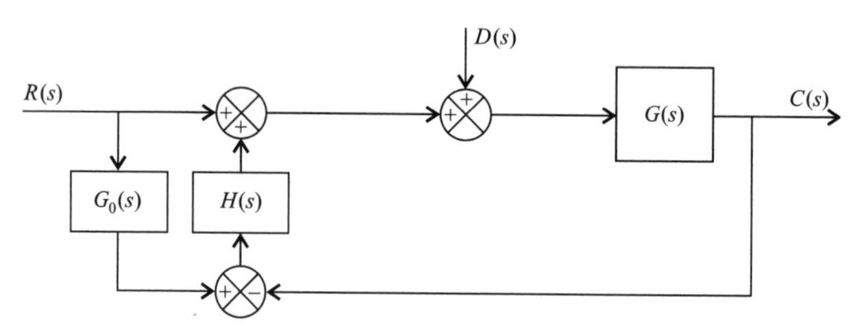

Figure P5.23

The process to be controlled is represented by $G(s)$, and $G_0(s)$ and $H(s)$ represent the controller transfer function.

(a) Determine the transfer function, $C(s)/R(s)$.

(b) Determine the transfer function $C(s)/D(s)$.

(c) Determine $C(s)/R(s)$ when $G_0(s) = G(s)$, assuming $D(s) = 0$.

(d) Assuming that

$$G_0(s) = G(s) = \frac{10}{(s+2)(s+4)}$$

select $H(s)$ from the following choices so that when $D(s) = 1/s$ and $R(s) = 0$, the steady-state value of $c(t)$ is equal to zero:

(i) $H(s) = 1/(s+0.5)$ (iii) $H(s)\ 4/(s+0.5)(s+2)$
(ii) $H(s) = 20(s+0.2)/(s+0.4)$ (iv) $H(s) = 2/s$
(v) $H(s) = 1000,000$ (vi) $H(s) = 1000,000,000$.

5.24. We know from Eq. (5.37) that the steady-state error of a second-order control system to a unit ramp input is given by $2\zeta/\omega_n$ (reciprocal of the velocity constant, K_v). This steady-state error to a unit ramp input can be eliminated if the input, $R(s)$, is introduced into the system through a proportional-plus-derivative filter, as illustrated in Figure P5.24, and the value of A is properly designed.

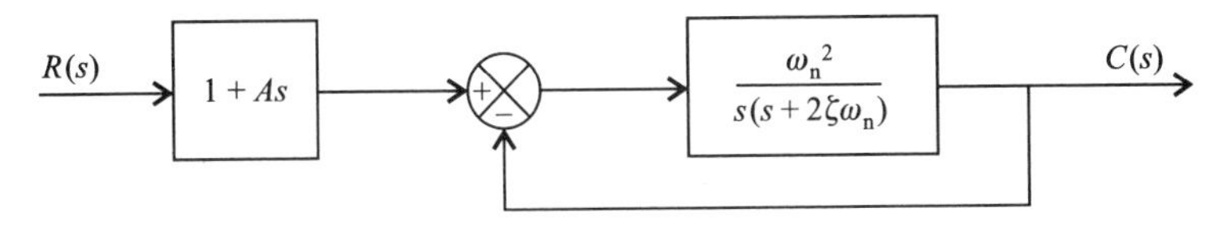

Figure P5.24

Determine the value of A which will result in zero steady-state error to a unit ramp input at the input, $R(s)$, assuming that the error, $e(t)$, is defined as

$$e(t) = r(t) - c(t).$$

5.25. For the system whose block diagram is shown in Figure P5.25, determine the following:

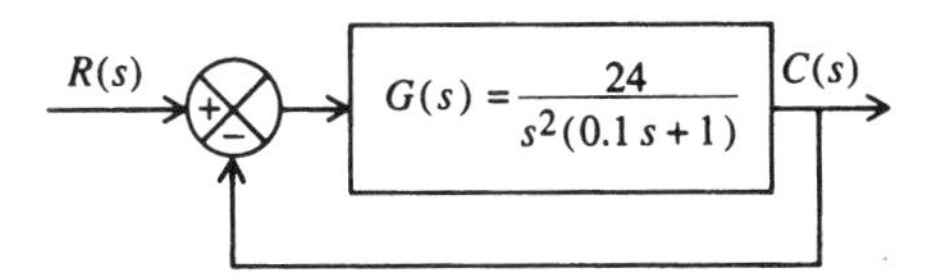

Figure P5.25

(a) Steady-state error resulting from an imput of $r(t) = (100t)U(t)$.

(b) Steady-state error resulting from an input of $r(t) = (12t^2)U(t)$.

(c) Steady-state error resulting from an input of $r(t) = (20+100t + 12t^2)U(t)$.

5.26. For the control system shown in Figure P5.26, determine the steady-state error $e(t)$ resulting from the application of the following inputs:

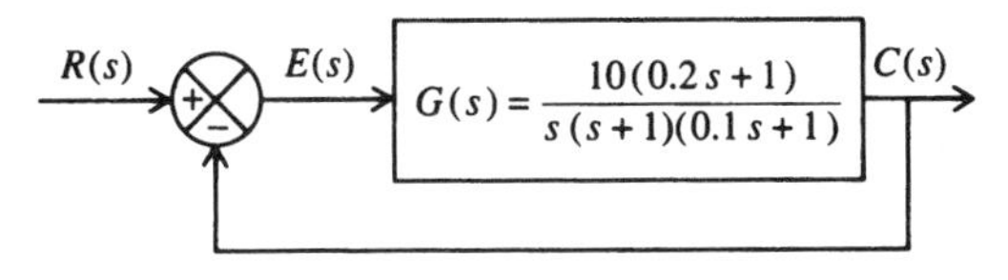

Figure P5.26

(a) $r(t) = (10t)U(t)$.

(b) $r(t) = (2 + 4t + 6t^2)U(t)$.

(c) How would you modify $G(s)$ to reduce the steady-state error in part (b) to zero?

5.27. For the system shown in Figure P5.27, determine the following:

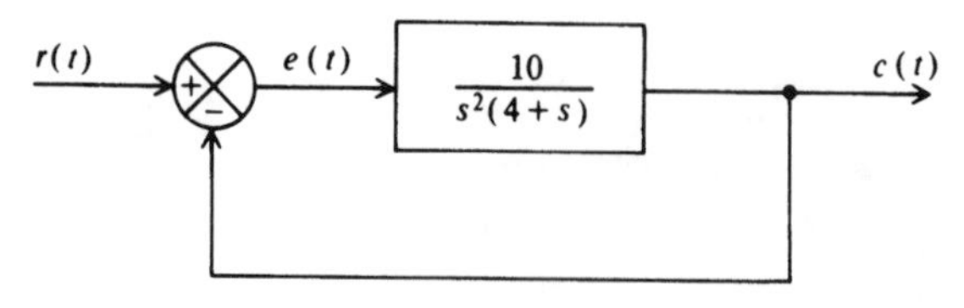

Figure P5.27

(a) Steady-state error resulting from an input

$$r(t) = (10t)U(t).$$

(b) Steady-state error resulting from an input

$$r(t) = (4 + 6t + 3t^2)U(t).$$

(c) Steady-state error resulting from an input

$$r(t) = (4 + 6t + 3t^2 + 1.8t^3)U(t).$$

5.28. Repeat Problem 5.27 with the transfer function of the system given by

$$G(s) = \frac{10}{s^2(1+s)(1+10s)}.$$

5.29 A common problem in the television industry is that of picture wobbling or jumping because of movement of the TV camera while a picture is being taken. The effect of this problem is easily understood by examining Figures P5.29*a* and P5.29*b*. When the camera is at rest (as illustrated in Figure P5.29*a*) a light ray entering the camera lens impinges on point A within the camera. However, if the camera is jolted upward through an angle δ (as in Figure P5.29*b*, the light ray is displaced from its original location at point A to point B. This can be corrected by means of the system illustrated in Figure P5.29*b*. The concept utilizes a device that changes the shape of a fluid lens such that the ray's impinging point does not move [7]. The front transparent plate is rotated in the vertical plane by a torque motor. The rear plate is rotated in the horizontal plane. Two rate gyros are mounted in the camera to detect any disturbances. Their output is fed to a servo amplifier that adjusts the driving current to the torque motors. Tachometers close the rate feedback loops. The equivalent block diagram of one such axis is illustrated in Figure P5.29*c*. Observe that the feedback loop uses speed from the rate gyro as the reference input and speed from the tachometer to indicate speed of the bellows.

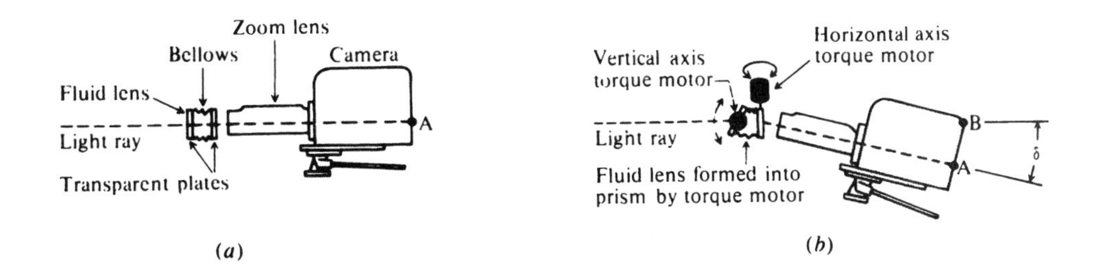

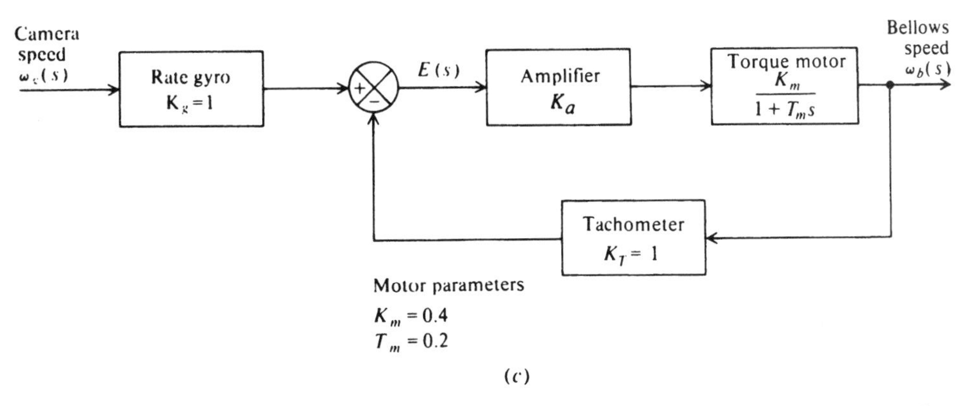

Figure P5.29 (Reprinted from *Control Engineering*, May 1965 © 1965 by Cahners Publishing Company)

(a) Determine the required value of amplifier gain K_a in order that the steady-state error resulting from a camera scanning speed of 50°/sec is only 1°/sec.

(b) What is the steady-state error of this system resulting from camera accelerations of any magnitude?

5.30 A servomechanism, shown in Figure P5.30, is used to drive an inertia load through gearing of negligible inertia.

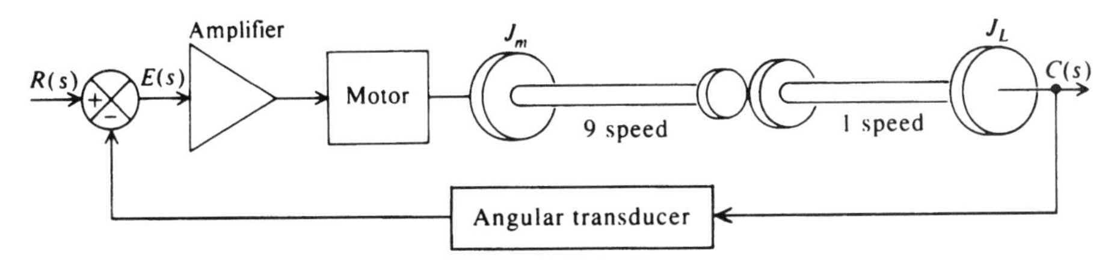

Figure P5.30

1. ac motor characteristics: Torque-speed slope $= 4.5 \times 10^{-6}$ lb ft/(rad/sec). Stall torque constant $= 8.5 \times 10^{-6}$ (lb ft)/V.
2. Load inertia $= J_L = 40 \times 10^{-6}$ lb ft sec^2 (assume $J_m N^2 \ll J_L$).
3. Amplifier gain $= 10$.

4. Gear ratio = 9 : 1 (steps motor speed down).
5. Angular transducer transfer function = 1.

(a) What is the transfer function $G(s)$ relating $C(s)$ and $E(s)$ of the system?

(b) What are the undamped natural frequency ω_n and the damping ratio ζ?

(c) What is the percent overshoot and time to peak resulting from the application of a unit step input?

(d) What is the steady-state error resulting from application of a unit step input?

(e) What is the steady-state error resulting from application of a unit ramp input?

(f) What is the steady-state error resulting from application of a parabolic input?

5.31. Repeat Problem 5.30 with the assumption that the motor inertia J_m is 1.11×10^{-6} lb ft sec^2.

5.32. Repeat Problem 5.30 with the amplifier's gain increased to 100.

5.33. Automatically controlled machine tools form an important aspect of control-system application. The major trend has been towards the use of automatic numerically controlled machine tools using tape inputs [8]. The justification of using tape has been the elimination of costly contour templates and the reduction of the machine set-up procedure required. In addition, it eliminates the tedium of repetitive operations required of human operators, and the possibility of human error. Figure P5.33 illustrates the block diagram of an automatic numerically controlled machine-tool position control system, using a punched tape reader, to supply the reference signal.

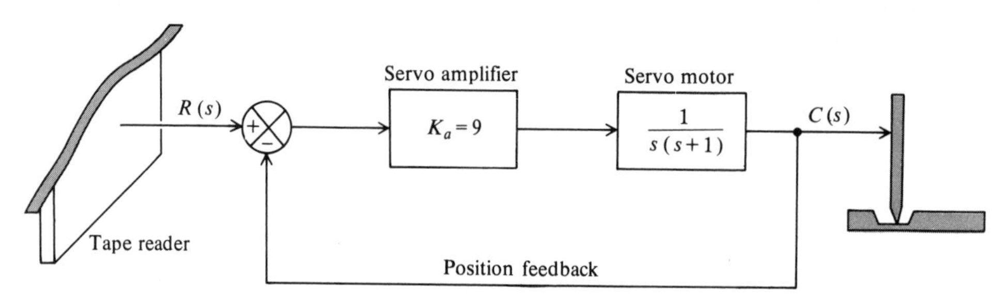

Figure P5.33

(a) What are the undamped natural frequency ω_n and damping ratio ζ?

(b) What are the percent overshoot and time to peak resulting from the application of a unit step input?

(c) What is the steady-state error resulting from the application of a unit step input?

(d) What is the steady-state error resulting from the application of a unit ramp input?

5.34. The transient response of the control system in Figure P5.34 is to be analyzed.

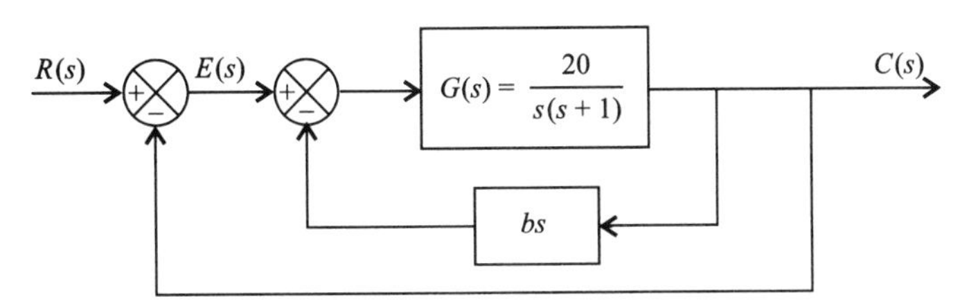

Figure P5.34

(a) Determine $C(s)/R(s)$.

(b) Determine the rate feedback constant b so that this control system has a damping ratio of 0.7.

(c) Determine the rise time, t_r.

(d) Determine the time to peak, t_p.

(e) Determine the settling time, t_s.

5.35. The transfer function of a closed-loop control system is given by

$$\frac{C(s)}{R(s)} = \frac{10(s+1)(s+2)(s+z)}{(s+0.3)(s+4)(s+8)(s+10)}.$$

Based on test data, it has been determined that the system exhibits a position constant K_p of 8. Determine the value of the unknown zero z.

5.36. Modern ocean-going ships utilize stabilization techniques in order to minimize the effects of oscillations due to waves. By utilizing hydrofoils or fins, stabilizing torques can be generated in order to stabilize the ship [9]. Figure P5.36*a* illustrates this concept employing stabilizing fins. An equivalent block diagram of the system is illustrated in Figure P5.36*b*. The reference input signal, θ_{desired}, is normally set equal to zero. A vertical gyro, which senses deviations from $\theta = 0$, feeds back a correction signal that activates the fins in order to drive the error signal to zero. The disturbance signal $U(s)$ represents the disturbance torque due to the waves.

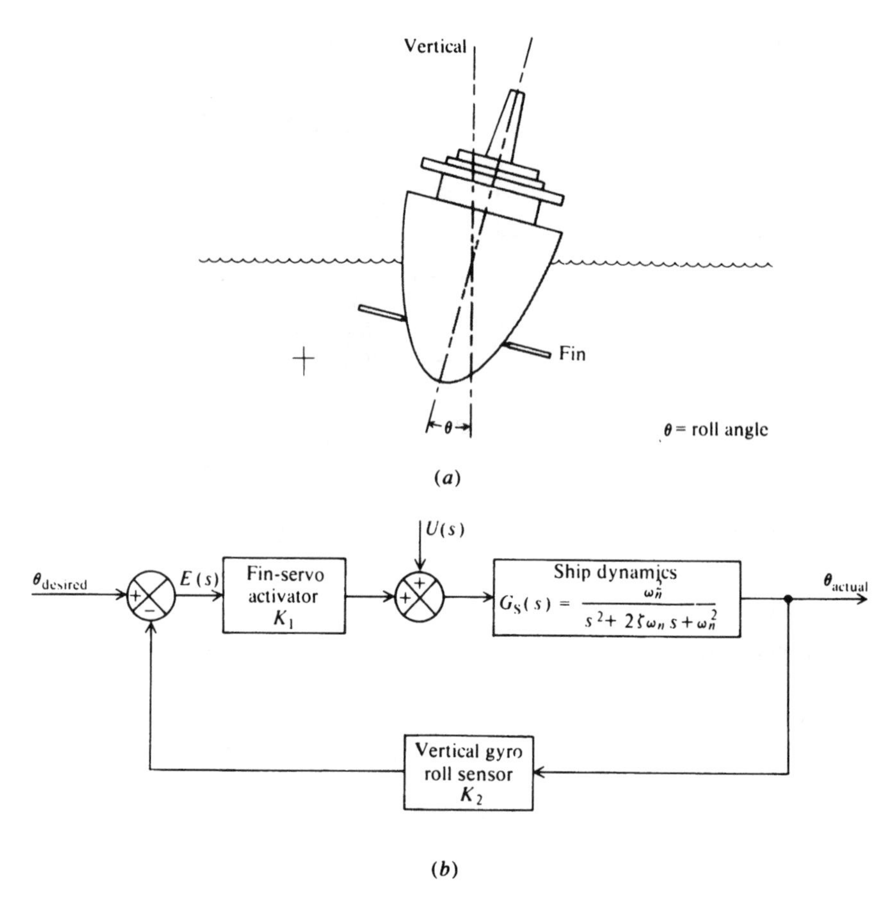

Figure P5.36

(a) Determine the effect of the disturbance torque $U(s)$ on system error $E(s)$ if it is assumed that $K_1 K_2 G_s(s) \gg 1$. What conclusions can you draw from your result?

(b) Based on the approximation of part (a), what should K_1 be set at in order to reduce a disturbance torque input at $U(s)$ of 10° to an equivalent system steady-state error of 0.1°?

5.37. As digital computers become more common in every phase of industry and the scientific communities, there is an ever-increasing demand for the storage and speedy retrieval of data [10]. The resulting control-system requirements are usually quite stringent and usually require new techniques. Figure P5.37*a* illustrates a system used to control the position of the read/write heads in a random-access magnetic-drum memory system [10]. The particular application had 64 data heads mounted in vertical pairs at 2-in intervals along a "headbar." Although the controlling action is obtained by utilizing digital components, an equivalent representation of the control system is given in Figure P5.37*b*.

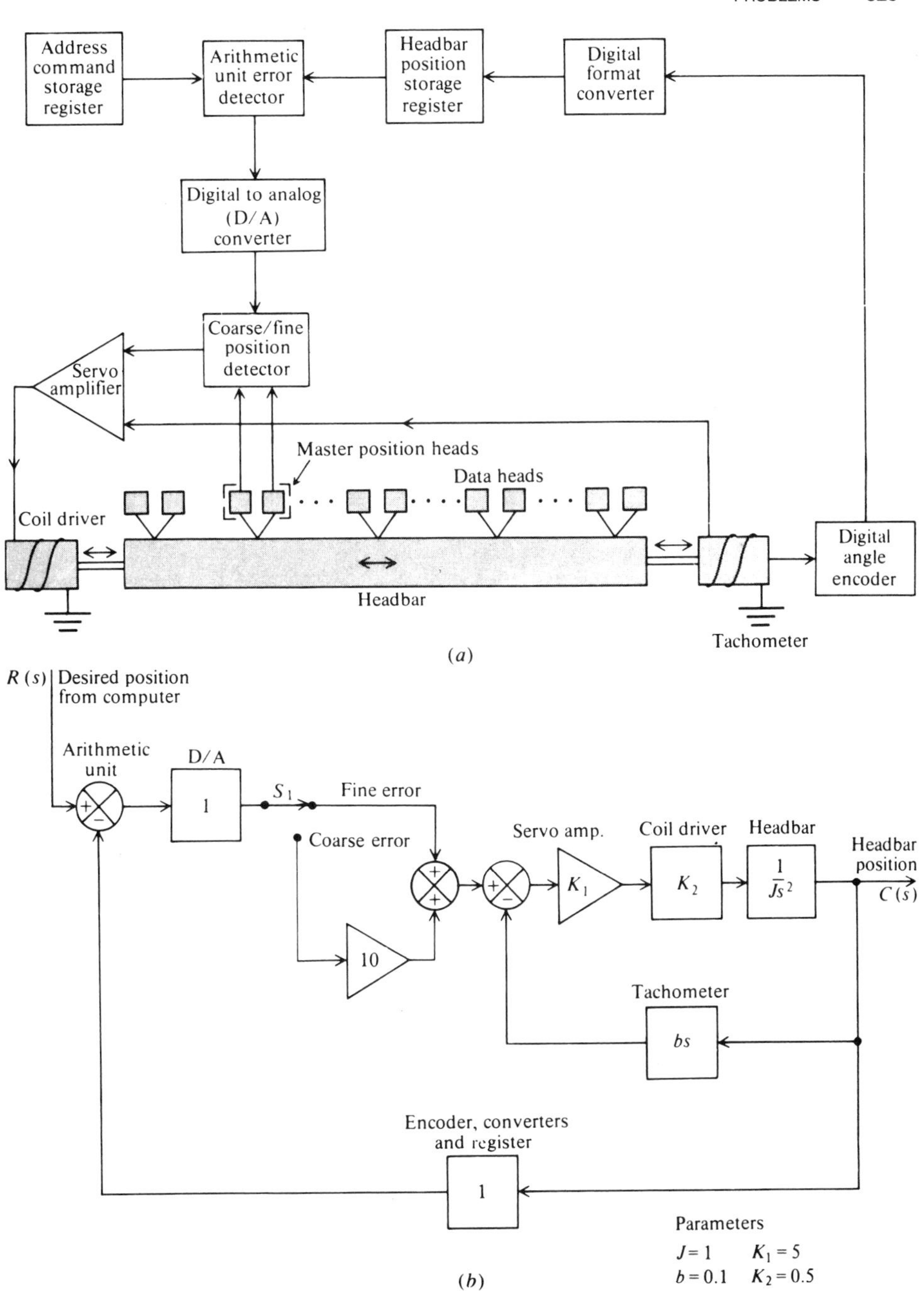

Address command storage register
Arithmetic unit error detector
Headbar position storage register
Digital format converter
Digital to analog (D/A) converter
Coarse/fine position detector
Servo amplifier
Master position heads
Data heads
Coil driver
Headbar
Tachometer
Digital angle encoder
(a)
R (s)
Desired position from computer
Arithmetic unit
D/A
1
S1
Fine error
Coarse error
10
Servo amp.
K1
Coil driver
K2
Headbar
1/Js2
Headbar position
C (s)
Tachometer
bs
Encoder, converters and register
1
Parameters
J = 1 K1 = 5
b = 0.1 K2 = 0.5
(b)

Figure P5.37

The headbar is free to move longitudinally. It is located between two magnetic drums whose axes of rotation are parallel with the longitudinal axis of the headbar. Each of the data heads in this system has access to a total of 200 tracks. The track accessed depends on the placement of the headbar within the limits of its 2-in travel. A fine and a coarse positioning loop are used to obtain the desired accuracy. An electronic switch S_1 switches the system so that it has a large amount of gain for large errors so that the system responds rapidly. For this condition, a large amount of overshoot is tolerated in order to achieve a fast rise time. For small errors, the fine system is activated which has a larger damping ratio, smaller overshoot, and a somewhat longer response time. Assume that the system switches when $e(t) = 0.1$ unit.

(a) Determine the undamped natural frequency, damping ratio, overshoot, and time to peak for the "coarse" loop.

(b) Repeat part (a) for the "fine" loop.

(c) Determine the error of the coarse and fine loops to a unit step input.

(d) Determine the error of the coarse and fine loops to a unit ramp input.

(e) Determine the error of the coarse and fine loops to a unit parabolic input.

5.38. A fourth-order feedback system is illustrated in Figure P5.38. In order to satisfy the ITAE criterion, determine the values of K_1, β_1, and β_2 for a zero steady-state step error system.

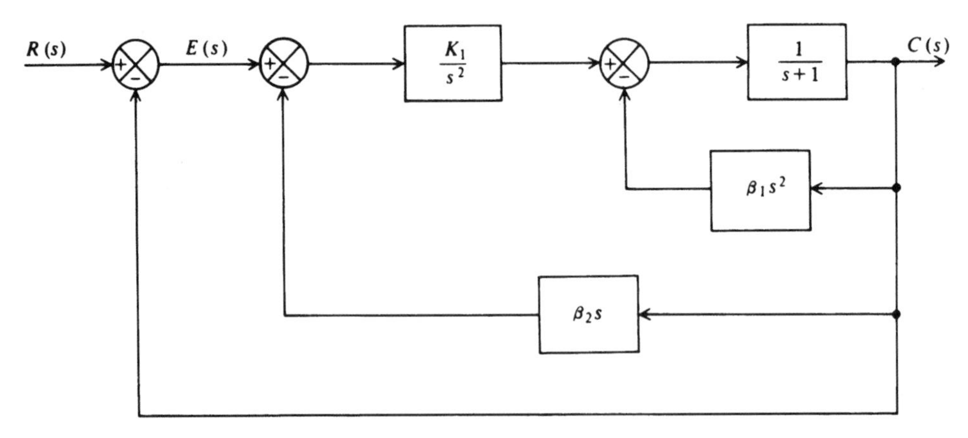

Figure P5.38

5.39. Determine the values of K_1, K_2, ω_n and b, in the feedback system illustrated in Figure P5.39 which will satisfy the ITAE criterion for a zero steady-state step error system.

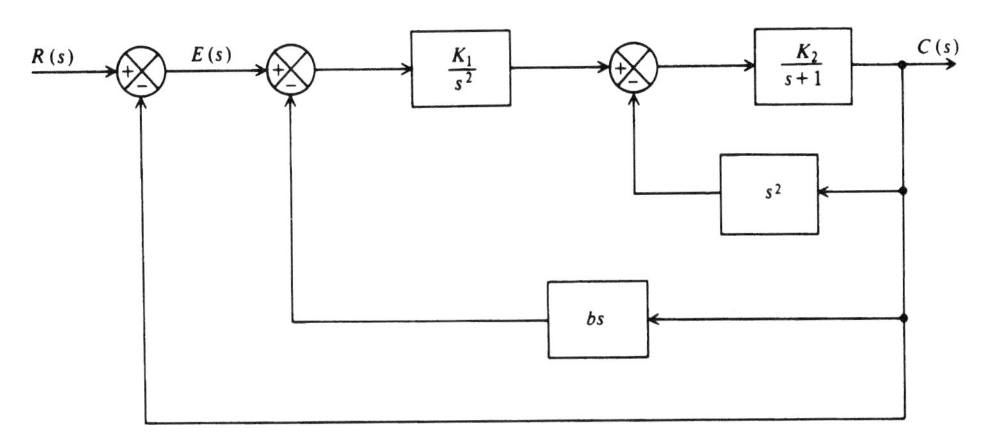

Figure P5.39

5.40. Repeat Problem 5.39 with the acceleration feedback element attentuated by a factor of 0.5. What conclusions can you draw from your result?

5.41. Repeat Problem 5.39 if the gain of the acceleration feedback element is doubled. What conclusions can you draw from your result?

5.42. A position-following system which can respond to position and ramp inputs with zero steady-state errors is shown in Figure P5.42. Determine the values of K_a and T that will satisfy the ITAE criterion for a zero steady-state ramp error system.

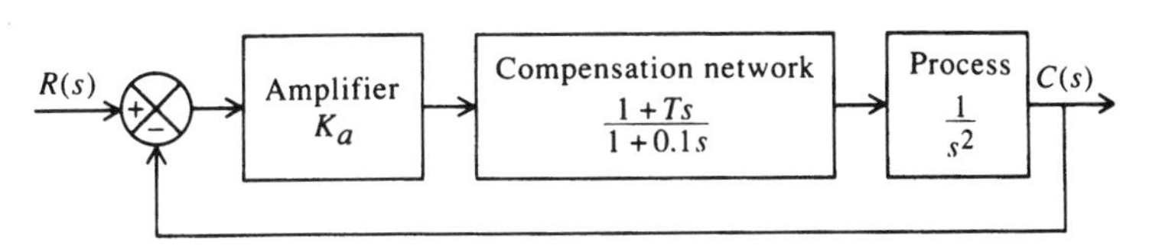

Figure P5.42

5.43. Robots are being used to a great extent in automating arc welding processes in manufacturing. This is especially true in the manufacture of automobiles. Figure P5.43 shows a photograph of the GMFanuc Robotics Corporation GMF ARC Mate Sr™ six-axis robot designed specifically for automating arc welding processes in a cost-effective manner. It uses ac servomotors that are brushless, which eliminates brush-related maintenance. A digital servo system enables high-speed movement, shorter positioning time, and improved path accuracy [11].

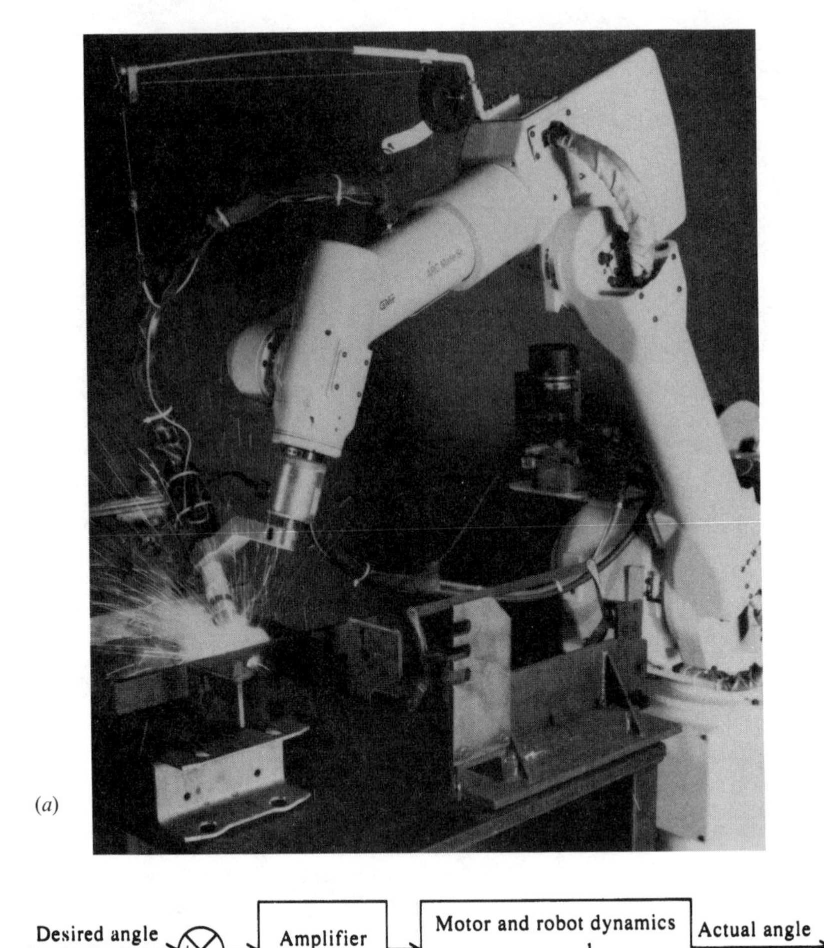
(a)

Figure P5.43 (*a*) Photograph of the GMFanuc Robotics Corporation GMF ARC Mate Sr.^{TM} six-axis robot designed specifically for automating arc welding processes in a cost-effective manner. (Courtesy of GMFanuc Robotics Corporation) (*b*) Block diagram.

Let us assume that we have to design a robot such as this one, but it will be based on a continuous design and not on digital techniques, which are discussed in Chapter 9. It is also assumed that one of the axes can be represented by the block diagram shown in Figure P5.43*b*. For the purposes of our application, a critically damped system will be assumed to be desirable.

(a) Determine the amplifier gain K_a so that the robot is criticaly damped in this axis.

(b) Find the sensitivity function of the robot's transfer function due to amplifier variations, $S_{K_a}^{H(s)}(s)$, where $H(s) = \theta_{out}(s)/\theta_{in}(s)$. Express the sensitivity as a function of s.

(c) Determine the absolute value of the sensitivity at dc.

(d) Determine the steady-state error for an input command signal of

$$r(t) = (4 + 6t)U(t).$$

5.44 The accuracy of the following control system is to be analyzed:

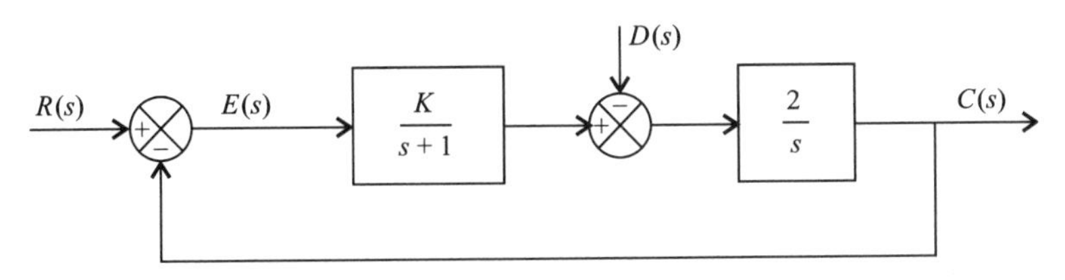

Figure P5.44

(a) Determine the steady-state error, e_{ss}, for a unit step input, $r(t) = U(t)$ with $K = 10$.

(b) Determine the steady-state error, e_{ss}, for a unit disturbance step input, $d(t) = U(t)$ with $K = 10$.

(c) Determine the value of K so that the resulting steady-state error due to a disturbance step input, $d(t) = U(T)$ equals 5% of the disturbance step input value.

5.45 We wish to analyze the second-order control system illustrated in Figure 4.1, whose closed-loop transfer function was developed in Eq. (5.53). Assuming that $T_m = 2$ seconds, what should K_m be designed for so that the ITAE criterion is achieved for this zero steady-state step error system?

REFERENCES

1. W. C. Schultz and C. V. Rideout, "Control system performance measures: Past, present, and future," *IRE Trans. Autom. Control* **AC-6**, 22 (1961).
2. D. Graham and R. C. Lathrop, "The synthesis of optimum transient response: Criteria and standard forms." *Am. Inst. Electr. Eng.*, **72**, 273 (1953).
3. G. C. Newton, Jr., L. A. Gould, and J. F. Kaiser, *Analytical Design of Linear Feedback Control*. Wiley, New York, 1961.
4. S. M. Shinners, "Minimizing servo load resonance error with frequency selective feedback." *Control Eng.* **51**, 51–56 (1962).
5. R. M. Centner and J. M. Idelsohn, "Adaptive controller for a metal cutting process." In *Proceedings of the 1963 Joint Automatic Control Conference*, pp. 262–271.
6. R. Dallimonti, "Developments in automatic warehousing and inventory control." In *Proceedings of the 1965 Joint Automatica Control Conference*, pp. 281–285.
7. J. de la Cierva, "Rate servo keeps TV picture clear." *Control Eng.* **12**, 112 (1965).
8. S. M. Shinners, "Which computer—Analog, digital, or hybrid?" *Mach. Des.* **43**, 104–111 (1971).

9. J. Bell, "Control for ship stabilization." In *Proceedings of the 1st International Federation of Automatic Control Congress 1960*, pp. 208–217.
10. R. Tickell, "A high performance position control." In *Proceedings of the 1966 Joint Automatic Control Conference*, pp. 230–242.
11. GMFanuc *Robotics Corporation Specification Sheet on* ARC Mate Sr.™ *Robot, 1990.* GMFanuc Robotics Corporation, Auburn Hills, MI, 1990.

6

TECHNIQUES FOR DETERMINING CONTROL-SYSTEM STABILITY

6.1. INTRODUCTION

For proper controlling action, a feedback system must be stable. Previous chapters have indicated that feedback systems have the serious disadvantage that they may inadvertently act as oscillators. A feedback control system must maintain stability when the system is subjected to commands at its input, extraneous inputs anywhere within the feedback loop, power-supply variations, and changes in the parameters of the elements comprising the feedback loop.

In the ensuing discussion, if a control system has zero initial conditions, then for every bounded input, the output is bounded and the system is stable. This is popularly referred to by control engineers as bounded input-bounded output (BIBO) stability. In this chapter, analysis is limited to linear time-invariant systems, that is, systems for which the principle of superposition is valid and which may be described by an ordinary linear differential equation with constant coefficients. The analysis of nonlinear systems is presented in Chapter 5, and digital control system stability is presented in Chapter 4 of the accompanying volume.

We showed in Section 4.4 that a control system's total response was the sum of the forced response due to the external forcing function and the natural or homogeneous response due to the initial conditions. The analysis was performed for the second-order control system illustrated in Figure 4.7, and the resulting response to a unit-step input for this control system was given by Eq. (4.45).

Therefore, in general, the total response of a control system, c(t), is composed of the natural or homogeneous response due to the initial conditions, and the forced response due to the external input:

$$c(t)_{\text{total}} = c(t)_{\text{homogeneous}} + c(t)_{\text{forced}} \tag{6.1}$$

A linear, time-invariant control system is stable if the natural or homogeneous response approaches zero, and it is unstable if the natural or homogeneous response grows without bound as time approaches infinity. If the linear, time-invariant system neither grows nor decays, and an oscillation of constant magnitude results, which represents simple harmonic motion, then this is an indication of a marginally stable system. Such a system, however, is considered to be an unstable control system from a practical viewpoint.

Let us consider the stability of a linear, time invariant, control system which only has an input from its initial conditions, and then consider a control system which has a reference input, $r(t)$. For the first case, we will assume that $r(t)$ equals zero, and the control system is an nth-order system. Therefore, the output, $c(t)$, can be expressed as follows:

$$c(t) = \sum_{k=0}^{n-1} g_k(t) c^{(k)}(0) \tag{6.2}$$

where

$$c^{(k)}(0) = \frac{d^k c(t)}{dt^k}$$

and where $g_k(t)$ represents the zero-input response due to $c^{(k)}(0)$. Therefore, the control system is said to be *zero-input stable* if the zero-input response $c(t)$ due to the finite initial conditions, $c^{(k)}(0)$, approaches zero as t approaches infinity. The zero-input stability is also defined to be *asymptotically stable* because it is required that $c(t)$ reaches zero as time approaches infinity. Linear, time-invariant, control systems which are BIBO, zero-input, stable, and asymptotically stable all require that the roots of the characteristic equation be located in the left half of the s-plane. Therefore, if a system is BIBO stable, then it must also be zero-input stable or asymptotically stable. Figure 6.1 illustrates the stable and unstable regions of the s-plane. Roots located on the imaginary axis result in a stable oscillation (e.g., simple harmonic motion), but this is considered to be an unstable s-plane condition to the practicing control-system engineer.

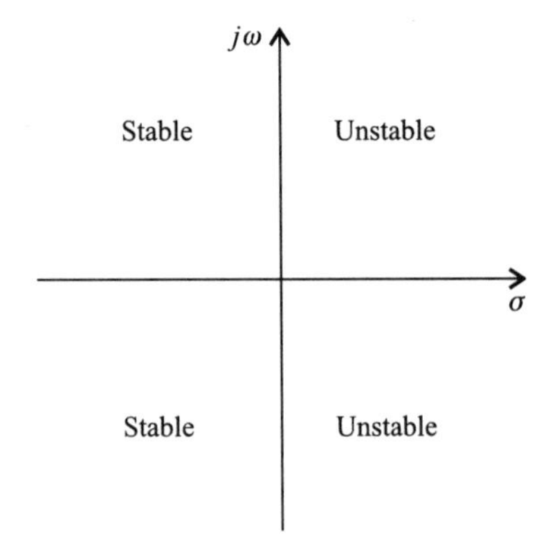

Figure 6.1 Stable and unstable regions of the *s* plane.

To illustrate the second case, where the linear time-invariant control system has a reference input, consider the following closed-loop transfer function of a control system which contains first-order real poles and zeros, and second-order quadratic poles:

$$H(s) = \frac{C(s)}{R(s)} = \frac{K\prod_{j=1}^{v}(s+z_j)}{\prod_{i=1}^{n}(s+p_i)\prod_{m=1}^{q}(s^2+2\zeta_m\omega_m s+\omega_m^2)}. \tag{6.3}$$

Let the reference input be a unit step input. Therefore, $R(s) = 1/s$ in Eq. (6.3), and expanding the resulting equation for $C(s)$ into its partial fractions gives us the following:

$$C(s) = \frac{\alpha}{s} + \sum_{i=1}^{n}\frac{k_i}{s+p_1} + \sum_{m=1}^{q}\frac{a_m(s+\zeta_m\omega_m)+b_m\omega_m\sqrt{1-\zeta_m^2}}{s^2+2\zeta_m\omega_m s+\omega_m^2} \tag{6.4}$$

where k_i is the residue of the pole at $s = -p_i$. Taking the inverse Laplace transform of Eq. (6.4), we obtain the following:

$$c(t) = \alpha + \sum_{i=1}^{n}k_i e^{-p_i t} + \sum_{m=1}^{q}a_m e^{-\zeta\omega_n t}\cos\omega_n\sqrt{1-\zeta_m^2}\,t + \sum_{m=1}^{q}b_m e^{-\zeta\omega_m t}\sin\omega_m\sqrt{1-\zeta_m^2}\,t. \tag{6.5}$$

If all the closed-loop roots lie in the left half of the s-plane, then the exponential terms and the exponentially damped sinusoidal terms in Eq. (6.5) approach zero as time approaches infinity.

Therefore, it is necessary that the poles of the closed-loop transfer function must be in the left-hand part of the s-plane in order to obtain a *bounded response.* We can generalize this result as follows. A necessary and sufficient condition of control-system stability is that all its poles have negative real parts. A control system which has all its roots in the left-half plane is denoted as a stable system. If the control system does not have all the roots in the left-half plane, then the control system is denoted as being unstable. If the control system has roots on the imaginary axis, and all other roots are in the left-half plane, then the output response will have an oscillation of constant amplitude for a bounded input, and is denoted as being marginally stable. This case is considered to be unstable in practical control systems.

This chapter focuses attention on the stability of the general feedback system, which was illustrated in Figure 2.10. The closed-loop transfer function of this system, given by Eq. (2.121), is repeated below:

$$\frac{C(s)}{R(s)} = \frac{G(s)}{1+G(s)H(s)}. \tag{6.6}$$

The characteristic equation for this generalized system can be obtained by setting the denominator of the system transfer function equal to zero:

$$1 + G(s)H(s) = 0. \tag{6.7}$$

In linear systems, stability is independent of the input excitation, and this is the equation that determines system stability. All the methods of stability analysis investigate this equation in some manner. One can show that if the roots of Eq. (6.7) lie in the left half-plane, the system is stable. However, the system is considered unstable if any of the roots of this equation have positive real parts or lie on the imaginary axis.

To illustrate this point, the transient response of the second-order system in Chapter 4 will be analyzed. Reconsider the underdamped case whose complex-conjugate poles lie in the left half of the complex plane, as shown in Figure 4.3. For that case, the transient response to a unit step input was shown to be an exponentially damped sinusoid in Figure 4.4. If the complex-conjugate poles of the second-order system were in the right half-plane, the transient response would be an exponentially increasing sinusoid and the system would be unstable. For the case of complex-conjugate poles that lie on the imaginary axis, the response would be a sinusoidal oscillation of fixed amplitude (simple harmonic motion). From a practical viewpoint, this would be considered to be unstable, also. These three cases are illustrated and compared in Figure 6.2 for three second-order systems labeled S1, S2, and S3 whose transfer functions and roots are given as follows:

(a) S1:
$$G_{S1}(s) = \frac{1}{s^2 + 0.4s + 1}.$$

Complex-conjugate roots are located at $-0.2 \pm 0.9798j$.

(b) S2:
$$G_{S2}(s) = \frac{1}{s^2 - 0.6s + 150}.$$

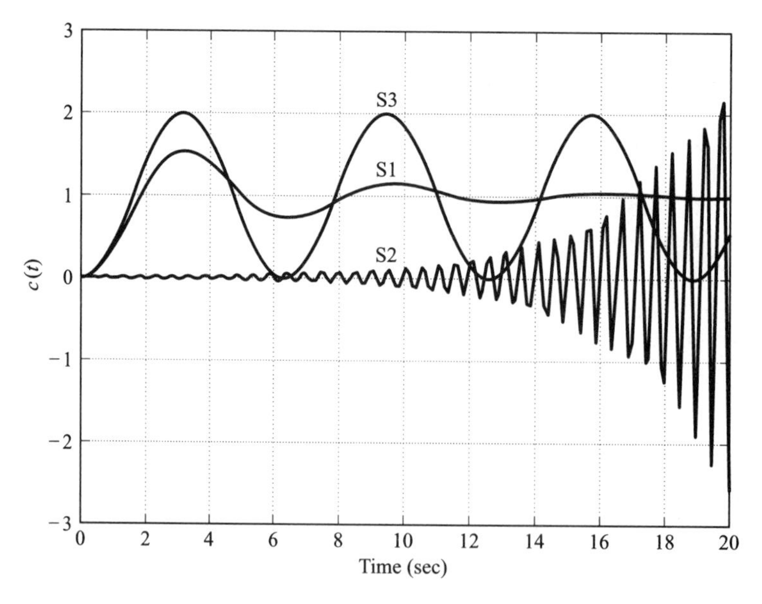

Figure 6.2 Unit step response of three second-order systems (S1, S2, S3)

Complex-conjugate roots are located at $0.3 \pm 12.2438j$.

(c) S3:

$$G_{S3} = \frac{1}{(s^2 + 1)}.$$

Complex-conjugate roots are located on the imaginary axis at $\pm j$.

Therefore, the system S1 has an exponentially decaying sinusoid and is stable; system S2 has an exponentially increasing sinusoid and is unstable; system S3 oscillates at a fixed amplitude and is unstable from a practical viewpoint.

In general, the following three approaches exist for determining stability:

1. Calculating the exact roots of Eq. (6.7)
2. Determination of the region of system parameters which guarantees that the roots of Eq. (6.7) have negative real parts.
3. State-variable method.

Using the first approach, the control engineer has at his or her disposal the following two methods:

(a) Direct-solution utilizing the classical approach
(b) Root-locus method

Using the second approach, the control engineer has at his or her disposal the following criteria:

(a) Routh–Hurwitz criterion
(b) Nyquist criterion.

Clearly, calculation of the exact roots using the classical approach can be extremely tedious. Usually, the designer is interested in the root-locus method and the criteria of the second and third approaches. This chapter represents each of these methods, except the classical approach, together with their relative merits. Additional graphical approaches based on the Nyquist criterion are also discussed. These include the use of the Bode diagram and Nichols chart. The state-variable approach permits the determination of the characteristic equation directly from knowledge of the state equations, and is illustrated in the following section. Application of these methods to actual design problems is deferred to Chapters 7 and 8.

6.2. DETERMINING THE CHARACTERISTIC EQUATION USING CONVENTIONAL AND STATE-VARIABLE METHODS

The characteristic equation can be defined in terms of the state-variable equation of the control system. We have stated in the previous section that stability of linear systems is independent of the input. Therefore, the condition $\mathbf{x}(t) = 0$, where $\mathbf{x}(t)$ is the state vector, can be viewed as the equilibrium state of the system. Let us assume that a linear system is subjected to a disturbance at $t = 0$, resulting in an initial state $\mathbf{x}(0)$ that is finite. If it returns to its equilibrium state as t approaches infinity, the system is considered to be stable. If it does not, in terms of our definition, it is considered to be unstable. These concepts of stability can be generalized. In the state-variable approach [1, 2], a linear system is considered to be stable if, for a finite initial state $\mathbf{x}(0)$, there is a positive number A that depends on $\mathbf{x}(0)$, where

$$\|\mathbf{x}(t)\| < A, \quad \text{for } t \geqslant 0, \tag{6.8}$$

$$\lim_{t \to \infty} \|\mathbf{x}(t)\| = 0. \tag{6.9}$$

The value $\|\mathbf{x}(t)\|$ denotes the norm of the state vector $\mathbf{x}(t)$. It is defined as

$$\|\mathbf{x}(t)\| = \left[\sum_{j=1}^{n} x_j^2(t) \right]^{1/2}. \tag{6.10}$$

Equation (6.8) can be interpreted to mean that the transition of state for positive time, as represented by the norm of the vector $\mathbf{x}(t)$, is bounded. Equation (6.9) can be interpreted to mean that the system must reach its equilibrium point as t approaches infinity. Figure 6.3 shows the state-variable stability criterion for a second-order system having states $x_1(t)$ and $x_2(t)$. Observe from this figure that a cylinder, the radius of which is A, forms the bound for the trajectory as time increases. As time approaches infinity, the linear system reaches the equilibrium point $\mathbf{x}(t) = 0$. Note that, strictly speaking, the above definition corresponds to asymptotic stability.* For simplicity, we will call the system stable if Eqs. (6.8) and (6.9) hold.

Let us next develop analytically, and apply, the state-variable approach for determining the stability of a linear system. It will be shown that this method determines the location of the roots of the characteristic equation, and restricts the roots of the characteristic equation to the left half-plane for stability.

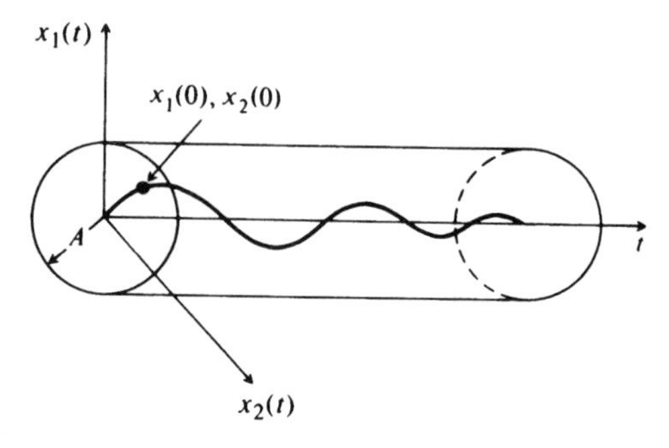

Figure 6.3 State-space stability concept.

*According to Liapunov, this is asymptotic stability—the type of stability preferred by control-system engineers. This is defined and discussed in detail in Section 10.20 on Liapunov's stability criterion.

Consider the general differential equation of a linear system in scalar form:

$$A_n \frac{d^n c(t)}{dt^n} + A_{n-1} \frac{d^{n-1} c(t)}{dt^{n-1}} + \ldots + A_0 c(t) = r(t), \tag{6.11}$$

where all the coefficients are constants, $A_n \neq 0$, $r(t)$ represents the input to the system, and $c(t)$ represents the system output. The Laplace transform of Eq. (6.11) is given by

$$C(s)[A_n s^n + A_{n-1} s^{n-1} + \ldots + A_0] = R(s). \tag{6.12}$$

Stability can be determined from the characteristic equation of this system:

$$A_n s^n + A_{n-1} s^{n-1} + \ldots + A_0 = 0. \tag{6.13}$$

It remains now to determine the roots of this equation.

The nth-order differential equation, given by Eq. (6.11), may also be specified by n first-order differential equations. By defining

$$\begin{aligned} x_1(t) &= c(t), \\ x_2(t) &= \dot{x}_1(t) = \frac{dc(t)}{dt}, \\ &\ldots \\ x_n(t) &= \frac{d^{n-1} c(t)}{dt^{n-1}} \\ \dot{x}_n(t) &= \frac{d^n c(t)}{dt^n}, \end{aligned} \tag{6.14}$$

the linear system can be specified by the following set of first-order differential equations:

$$\begin{aligned} \dot{x}_1(t) &= A_{11} x_1(t) + \ldots + A_{1n} x_n(t) + B_{11} r_1(t), \\ &\ldots \\ \dot{x}_n(t) &= A_{n1} x_1(t) + \ldots + A_{nn} x_n(t) + B_{n1} r_1(t), \end{aligned} \tag{6.15}$$

which is equivalent to Eq. (6.11).

The set of equations (6.15) is solved for the case where the input $r(t)$ equals zero; assume an exponential solution of the form

$$x_i(t) = f_i e^{st}, \tag{6.16}$$

where f_i is an unknown constant. To check the form of the assumed solution, Eq. (6.16) is differentiated and the result is substituted into the set of equations (6.15). Differentiating (6.16), we obtain

$$\dot{x}_i(t) = f_i(s e^{st}). \tag{6.17}$$

Substitution of Eqs. (6.16) and (6.17) into (6.15) results in

$$\begin{aligned} f_1 s e^{st} &= A_{11} f_1 e^{st} + \ldots + A_{1n} f_n e^{st}, \\ &\ldots \\ f_n s e^{st} &= A_{n1} f_1 e^{st} + \ldots + A_{nn} f_n e^{st}. \end{aligned} \tag{6.18}$$

Equating coefficients of e^{st} and rearranging, we obtain the following:

$$\begin{aligned} (A_{11} - s) f_1 + \ldots + A_{1n} f_n = 0, \\ \ldots \\ A_{n1} f_1 + \ldots + (A_{nn} - s) f_n = 0. \end{aligned} \tag{6.19}$$

This set of equations can be rewritten in the following matrix algebra form:

$$[\mathbf{A} - s\mathbf{I}]\mathbf{f} = \mathbf{0}. \tag{6.20}$$

where $\mathbf{f}$ = column matrix, or vector, and $\mathbf{I}$ = identity matrix. Observe that matrix $\mathbf{A}$ used in Eq. (6.20) is analogous to the $\mathbf{P}$ matrix introduced in Chapter 2 when state variables were introduced [see Eq. (2.197]. Equation (6.20) will have a nontrivial solution only if

$$|\mathbf{A} - s\mathbf{I}| = 0. \tag{6.21}$$

By expanding the determinant $|\mathbf{A} - s\mathbf{I}|$ and solving for the roots of the equation $|\mathbf{A} - s\mathbf{I}| = 0$, eigenvalues of matrix $\mathbf{A}$ are obtained. Expanding the determinant

$$|\mathbf{A} - s\mathbf{I}| = 0 \tag{6.22}$$

results in the following expression:

$$s^n + B_1 s^{n-1} + \ldots + B_n = 0. \tag{6.23}$$

Notice that the B_i terms in Eq. (6.23) are equivalent to the A_i/A_n terms in Eq. (6.13). Therefore, the two equations are equivalent. The n values of s that satisfy Eq. (6.22) are called the characteristic values, or eigenvalues, of the matrix.

As an example for comparing the stability determination of a system utilizing conventional Laplace-transform and state-variable techniques, consider the unity-feedback system shown in Figure 6.4. Using the Laplace transform, we could easily obtain the closed-loop transfer function of the system as

$$\frac{C(s)}{R(s)} = \frac{G(s)}{1 + G(s)} = \frac{K}{T_1 T_2 s^3 + (T_1 + T_2)s^2 + s + K}. \tag{6.24}$$

The characteristic equation of this simple linear system is given by

$$T_1 T_2 s^3 + (T_1 + T_2)s^2 + s + K = 0. \tag{6.25}$$

System stability can be determined by locating the roots of this equation if the classical approach is to be used. The same problem will next be analyzed from the state-variable viewpoint.

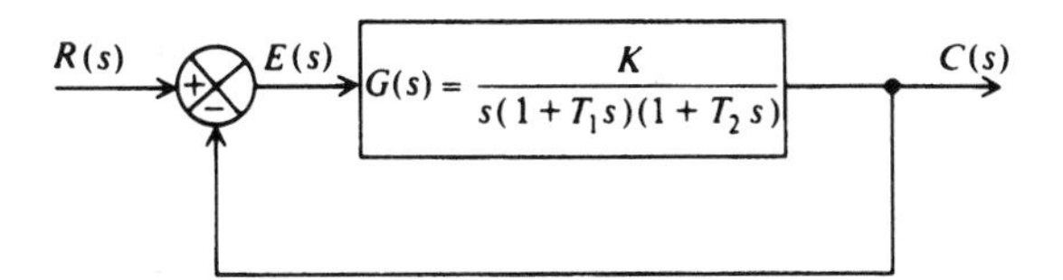

Figure 6.4 Third-order feedback control system.

From Eq. (6.24), we obtain

$$[T_1 T_2 s^3 + (T_1 + T_2)s^2 + s + K]C(s) = KR(s). \tag{6.26}$$

The time-domain expression equivalent to Eq. (6.26) is given by

$$T_1 T_2 \dddot{c}(t) + (T_1 + T_2)\ddot{c}(t) + \dot{c}(t) + Kc(t) = Kr(t). \tag{6.27}$$

As discussed previously, this third-order differential equation may be written as three first-order differential equations as follows:

$$\begin{aligned} c(t) &= x_1(t) \\ \dot{c}(t) &= \dot{x}_1(t) = x_2(t) \\ \ddot{c}(t) &= \dot{x}_2(t) = x_3(t) \\ \dddot{c}(t) &= \dot{x}_3(t) = \frac{1}{T_1 T_2}[-(T_1 + T_2)\ddot{c}(t) - \dot{c}(t) - Kc(t) + Kr(t)]. \end{aligned} \tag{6.28}$$

This can easily be transformed into matrix form:

$$\begin{bmatrix} \dot{x}_1(t) \\ \dot{x}_2(t) \\ \dot{x}_3(t) \end{bmatrix} = \begin{bmatrix} 0 & 1 & 0 \\ 0 & 0 & 1 \\ \dfrac{-K}{T_1 T_2} & \dfrac{-1}{T_1 T_2} & \dfrac{-(T_1 + T_2)}{T_1 T_2} \end{bmatrix} \begin{bmatrix} x_1(t) \\ x_2(t) \\ x_3(t) \end{bmatrix} + r(t) \begin{bmatrix} 0 \\ 0 \\ \dfrac{K}{T_1 T_2} \end{bmatrix}. \tag{6.29}$$

Using vector and matrix notation, the above equation becomes

$$\dot{\mathbf{x}}(t) = \mathbf{A}\mathbf{x}(t) + \mathbf{B}r(t),$$

where

$$\mathbf{x}(t) = \begin{bmatrix} x_1(t) \\ x_2(t) \\ x_3(t) \end{bmatrix}, \quad \dot{\mathbf{x}}(t) = \begin{bmatrix} \dot{x}_1(t) \\ \dot{x}_2(t) \\ \dot{x}_3(t) \end{bmatrix},$$
$$\mathbf{A} = \begin{bmatrix} 0 & 1 & 0 \\ 0 & 0 & 1 \\ \dfrac{-K}{T_1 T_2} & \dfrac{-1}{T_1 T_2} & \dfrac{-(T_1 + T_2)}{T_1 T_2} \end{bmatrix}, \quad \mathbf{B} = \begin{bmatrix} 0 \\ 0 \\ \dfrac{K}{T_1 T_2} \end{bmatrix}. \tag{6.30}$$

In order to obtain the characteristic equation, the input will be set equal to zero and the solution x_i will be assumed equal to $f_i e^{st}$. As outlined previously [see Eq. (6.22)], this procedure results in

$$|\mathbf{A} - s\mathbf{I}| = 0. \tag{6.31}$$

The resulting determinant is given by

$$\begin{vmatrix} -s & 1 & 0 \\ 0 & -s & 1 \\ \dfrac{-K}{T_1T_2} & \dfrac{-1}{T_1T_2} & \dfrac{-(T_1+T_2)}{T_1T_2} & -s \end{vmatrix} = 0. \tag{6.32}$$

Expansion of the determinant by means of minors along the first row gives

$$-s\begin{vmatrix} -s & 1 \\ \dfrac{-1}{T_1T_2} & \dfrac{-(T_1+T_2)}{T_1T_2} & -s \end{vmatrix} - 1\begin{vmatrix} 0 & 1 \\ \dfrac{-K}{T_1T_2} & \dfrac{-(T_1+T_2)}{T_1T_2} & -s \end{vmatrix} = 0, \tag{6.33}$$

which reduces to

$$T_1T_2s^3 + (T_1+T_2)s^2 + s + K = 0. \tag{6.34}$$

Equation (6.34), obtained utilizing the state-variable approach, is the same characteristic equation as (6.25), which was obtained by using conventional Laplace-transform techniques. Although the mathematics involved in obtaining Eq. (6.34) was more laborious than for Eq. (6.25) for this simple problem, many important features and characteristics of the state-variable approach have been presented and applied. It is important to emphasize that for more complex problems involving multiple inputs and outputs, the state-variable approach greatly simplifies the solution. In addition, it greatly facilitates computation utilizing digital computers.

6.3. ROUTH–HURWITZ STABILITY CRITERION

The Routh–Hurwitz stability criterion is an algebraic procedure for determining whether a polynomial has any zeros in the right half-plane. It involves examining the signs and magnitudes of the coefficients of the characteristic equation without actually having to determine its roots. This method does not indicate the relative degree of stability or instability.

Routh [3] and Hurwitz [4] independently determined the necessary and sufficient conditions for stability from the signs and magnitudes of the coefficients of the characteristic equation. A useful form of their approach is described below.

Let us represent the general form of the characteristic equation by

$$B_1s^m + B_2s^{m-1} + B_3s^{m-2} + \ldots + B_{m+1} = 0, \quad B_{m+1} \neq 0. \tag{6.35}$$

If a coefficient is negative or zero when at least one of the other coefficients is positive, then a root exists which is in the right half-plane or is imaginary. For this case, the system is unstable and one can stop here. If all the coefficients are present, real, and positive, then the coefficients are arranged in two rows:

$$\begin{array}{c|ccccc} s^m & B_1 & B_3 & B_5 & B_7 & \cdots \\ s^{m-1} & B_2 & B_4 & B_6 & B_8 & \cdots \end{array}. \tag{6.36}$$

We obtain additional rows of coefficients from these two rows as follows:

$$\begin{array}{c|ccccc} s^m & B_1 & B_3 & B_5 & B_7 & \cdots \\ s^{m-1} & B_2 & B_4 & B_6 & B_8 & \cdots \\ s^{m-2} & U_1 & U_3 & U_5 & U_7 & \cdots \\ s^{m-3} & U_2 & U_4 & U_6 & U_8 & \cdots \\ s^{m-4} & V_1 & V_3 & V_5 & V_7 & \cdots \\ s^{m-5} & V_2 & V_4 & V_6 & V_8 & \cdots \\ \vdots & \vdots & \vdots & \vdots & \vdots & \\ s^0 & Z_1 & & & & \end{array} \tag{6.37}$$

where

$$\begin{aligned} U_1 &= \frac{B_2B_3 - B_1B_4}{B_2}, & U_3 &= \frac{B_2B_5 - B_1B_6}{B_2}, & U_5 &= \frac{B_2B_7 - B_1B_8}{B_2}, \\ U_2 &= \frac{U_1B_4 - B_2U_3}{U_1}, & U_4 &= \frac{U_1B_6 - B_2U_5}{U_1}, & U_6 &= \frac{U_1B_8 - B_2U_7}{U_1}, \\ V_1 &= \frac{U_2U_3 - U_1U_4}{U_2}, & V_3 &= \frac{U_2U_5 - U_1U_6}{U_2}, & V_5 &= \frac{U_2U_7 - U_1U_8}{U_2}. \end{aligned} \tag{6.38}$$

This pattern will continue until all the terms in a row are zero. The rows are indexed downwards, the first row being numbered m, the degree of the original polynomial; the last row being numbered 0. The number of rows obtained will be $m + 1$, where m is the order of the characteristic equation. Note that there is one exceptional case where this will not be so; this is discussed later on. The criterion of stability is to check that all the terms in the left-hand column $(B_1, B_2, U_1, U_2, V_1, V_2, \ldots)$ have the same sign. If so, there are no roots in the right half-plane. If there are X changes of sign, then X roots exist in the right half-plane.

Let us illustrate the approach with a simple example. Consider the characteristic equation

$$1 + G(s)H(s) = s^3 + 4s^2 + 100s + 500 = 0. \tag{6.39}$$

Using the procedure described, the resulting array is

$$\begin{array}{c|cc} s^3 & 1 & 100 \\ s^2 & 4 & 500 \\ s & -25 & 0 \\ s^0 & 500 & 0 \end{array} \tag{6.40}$$

There are two changes of sign in the first column: 4 to −25 and −25 to 500; therefore, there are two roots in the right half-plane, and the system is unstable.

If the first term in any row is zero, and the other terms of the row are not zero, the array of Eq. (6.37) may be continued by replacing the first column zero by an arbitrary small positive constant ϵ. The process is then continued in the usual manner. Let us illustrate the procedure for this particular case with a simple example. Consider the following characteristic equation:

$$1 + G(s)H(s) = s^5 + s^4 + 4s^3 + 4s^2 + 2s + 1 = 0. \tag{6.41}$$

Using the procedure described, the resulting array is

$$\begin{array}{l|lccc}
s^5 & & 1 & 4 & 2 \\
s^4 & & 1 & 4 & 1 \\
s^3 & \text{Replace this row} \rightarrow & (0 & 1 & 0) \\
 & \text{with} \rightarrow & \epsilon & 1 & 0 \\
s^2 & & \frac{4\epsilon-1}{\epsilon} & 1 & 0 \\
s & & \frac{-\epsilon^2+4\epsilon-1}{4\epsilon-1} & 0 & 0 \\
s^0 & & 1 & 0 & 0
\end{array} \tag{6.42}$$

As ϵ approaches zero, the limiting value of the term in the left-hand column, fourth row, is negative. The limiting value of the term in the left-hand column, fifth row, is positive. Therefore there are two changes of sign, two roots must lie in the right half-plane, and the system is unstable.

The exceptional case referred to above occurs when all the terms in a row are zero before the $(m+1)$th row is reached. This means that there are pairs of real roots existing that are negatives of each other located on the real axis, pairs of conjugate roots on the imaginary axis, or quadruples of roots symmetrically located with respect to the origin. For this special case, the array of Eq. (6.37) can be completed by obtaining a subsidiary polynomial from the preceding row. The subsidiary polynomial equation is formed by constructing a polynomial whose coefficients are the coefficients of the last nonzero row. To determine the degree of the subsidiary polynomial, the rows are indexed downwards, the first row being numbered m, the degree of the original polynomial. Then the index of the last nonzero row is the degree of the subsidiary polynomial. This polyomial, which is always even, is then differentiated and the resulting coefficients are used to complete the array. The zeros of the subsidiary polynomial are actual roots of the characteristic equation. This procedure is illustrated with a simple example. Consider the following characteristic equation:

$$1 + G(s)H(s) = s^3 + 10s^2 + 16s + 160 = 0. \tag{6.43}$$

Using the procedure described, the resulting array is

$$\begin{array}{l|cc}
s^3 & 1 & 16 \\
s^2 & 10 & 160 \\
s & 0 & 0
\end{array} \tag{6.44}$$

The presence of zeros in the third row indicates the exceptional case. Using the coefficients of the second row for the subsidiary equation, we obtain

$$F(s) = 10s^2 + 160 = 0. \tag{6.45}$$

In order to complete the array, Eq. (6.45) is differentiated and the resulting coefficients are then inserted into the array as follows:

$$\begin{array}{c|cc} s^3 & 1 & 16 \\ s^2 & 10 & 160 \\ s & 20 & 0 \\ s^0 & 160 & 0 \end{array} \tag{6.46}$$

No roots lie in the right half-plane because there are no changes of sign in the left-hand column. The roots that are negatives of each other can be obtained from Eq. (6.45) as $\pm j4$, indicating a pair of imaginary roots. Although the Routh–Hurwitz criterion does not consider such a system to be unstable (because its roots do not lie in the right half plane), it is unstable from practical control considerations.

As a concluding example of the application of the Routh–Hurwitz criterion, let us determine the maximum value of gain K that a feedback system could have before the system becomes unstable. Another way of looking at this is to find K where the roots exist on the imaginary axis. Consider a system whose open-loop transfer function is given by

$$G(s)H(s) = \frac{K}{s(s+1)^2}. \tag{6.47}$$

Its characteristic equation is given by

$$1 + G(s)H(s) = s^3 + 2s^2 + s + K = 0. \tag{6.48}$$

The resulting Routh–Hurwitz criterion is given by

$$\begin{array}{c|cc} s^3 & 1 & 1 \\ s^2 & 2 & K \\ s^1 & U_1 = \frac{2-K}{2} & \\ s^0 & K & \end{array}$$

By setting the term U_1 equal to zero, we find that the maximum value of K before the system becomes unstable is given by

$$\frac{2-K}{2} = 0$$

or

$$K_{\max} = 2. \tag{6.49}$$

If $K > 2$, then there are two sign changes in the first column, indicating two roots in the right half-plane, and the system is unstable.

Although the Routh–Hurwitz criterion gives a relatively quick determination of absolute stability, it does not show how to improve the design. In addition, it does not give an indication of relative system performance. Its main attribute is to serve as a check of other design criteria. The greatest difficulty with using the Routh–Hurwitz criterion, however, is that it assumes that the characteristic polynomial is known.

6.4. MAPPING CONTOURS FROM THE *s*-PLANE TO THE *F*(*s*)-PLANE

Prior to introducing the Nyquist stability criterion which is very useful for determining the degree of stability or instability of feedback control systems, it is important to understand how to map contours from the s-plane to the $F(s)$-plane where $F(s)$ is some rational function of s. A *contour map* is defined as a contour in the s-plane which is then mapped or transferred into the $F(s)$-plane by the definition of the rational function $F(s)$. Because s is the complex variable $s = \sigma + j\omega$ and $F(s)$ is a complex function, then we can represent $F(s)$ as $F(s) = u + jv$. Therfore, the $F(s)$-plane is also a complex plane having coordinates of u and v.

To illustrate the mapping process from the s-plane to the $F(s)$-plane, consider the unit-square contour in the s-plane illustrated in Figure 6.5a. We wish to map this unit-square contour from the s-plane to the $F(s)$-plane where $F(s)$ is a rational function of s and is defined by the following expression:

$$F(s) = \frac{s}{s+3}. \tag{6.50}$$

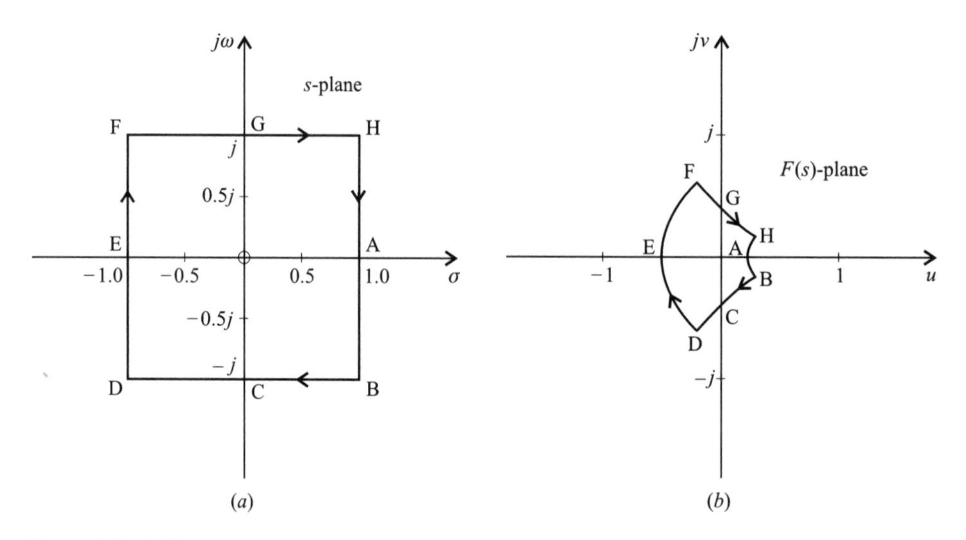

Figure 6.5 (a) Contour in the s-plane to be mapped for $F(s) = s/(s+3)$. (b) Resulting mapping in the $F(s)$-plane for $F(s) = s/(s+3)$ for the contour shown in the s-plane in Figure 6.5(a).

Since we are defining $F(s) = u + jv$, then

$$F(s) = u + jv = \frac{\sigma + j\omega}{\sigma + j\omega + 3} \tag{6.51}$$

Table 6.1 provides several values of $F(s)$ as s traverses the unit-square contour of the s-plane illustrated in Figure 6.5*a*. Observe that for each point in the s-plane, there is only one corresponding point in the $F(s)$-plane. Therefore, the mapping from the s-plane into the $F(s)$-plane is one-to-one. The resulting contour in the $F(s)$-plane is illustrated in Figure 6.5*b*. The points A, B, C, D, E, F, G, and H illustrated in the contour map in the s-plane map into the corresponding points A, B, C, D, E, F, G, and H in the $F(s)$-plane. We conclude that the unit-square contour of the s-plane maps into a different-shaped cotour in the $F(s)$-plane as illustrated in Figure 6.5*b*. We also conclude that a closed contour in the s-plane results in a closed contour in the $F(s)$-plane.

Another important point to recognize is the direction of the contour. Arrows are shown in Figure 6.5*a* of the traversal of the unit-square in the s-plane in going from A to B to C to D to E to F to G to H and back to A. A corresponding traversal occurs in the $F(s)$-plane contour as illustrated in Figure 6.5*b*. By convention in control systems, the area within a contour to the right of the traversal (as viewed by an observer standing on the contour and facing in the direction of the traversal) is defined as the area enclosed by the contour. Another convention in control systems is to consider a clockwise traversal of the contour to be positive. The direction and number of encirclements of the origin in the $F(s)$-plane by the closed contour is very significant in the Nyquist diagram which is discussed in the following section where we will correlate the number and direction of encirclements of the origin of the $F(s)$-plane with the stability of the control system.

It is significant that the path we chose for s in Figure 6.5*a* did not go through any singular points such as the pole of $F(s)$ at -3 but it did contain the zero of $F(s)$ at the origin. The result was a closed contour in the $F(s)$-plane which enclosed the origin of the $F(s)$-plane in the clockwise direction. The number of encirclements of the origin of the $F(s)$-plane depends on the closed contour in the s-plane. For example, if we have a contour that encloses one pole in the s-plane and the contour encircles the origin in the s-plane in the clockwise direction, then the contour in the $F(s)$-plane

Table 6.1. Corresponding Values in the s-Plane and $F(s)$-Plane for Figures 6.5*a* and *b*, respectively

Point	$s = \sigma + j\omega$	$F(s) = u + jv$
A	1	0.25
B	$1 - j$	$0.29 - 0.18j$
C	$-j$	$0.1 - 0.3j$
D	$-1 - j$	$-0.2 - 0.6j$
E	-1	-0.5
F	$-1 + j$	$-0.2 + 0.6j$
G	j	$0.1 + 0.3j$
H	$1 + j$	$0.29 + 0.18$

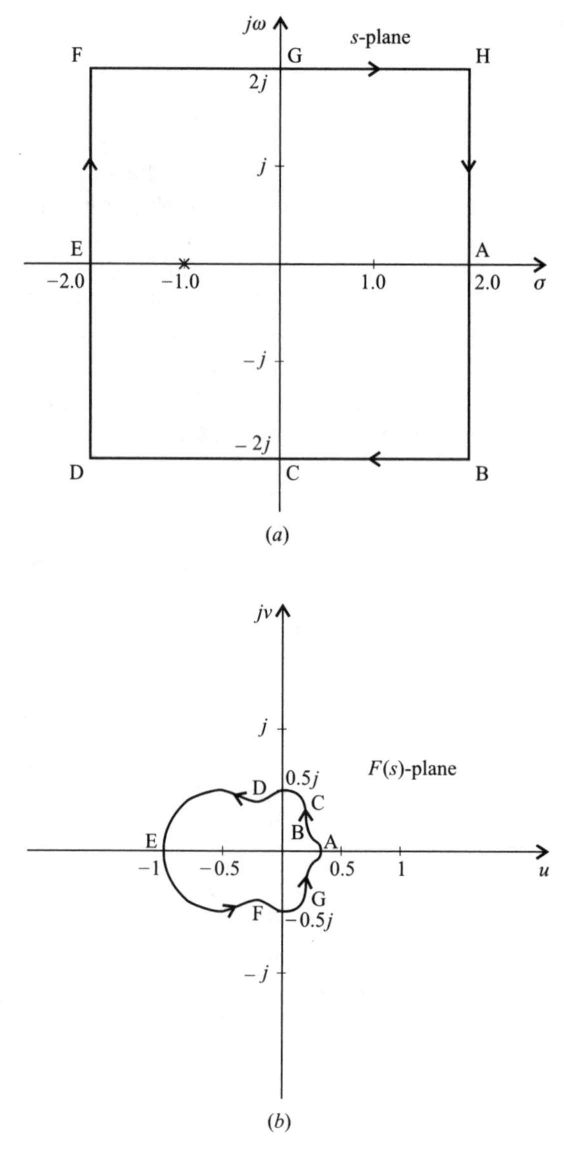

Figure 6.6 (*a*) Contour in the *s*-plane to be mapped for $F(s) = 1/(s+1)$. (*b*) Resulting mapping in the $F(s)$-plane for $F(s) = 1/(s+1)$ for the contour shown in the *s*-plane in Figure 6.6 (*a*).

encloses the origin in the $F(s)$-plane in the counterclockwise direction. This is illustrated in Figure 6.6*a* and *b* for the following function:

$$F(s) = \frac{1}{s+1}. \tag{6.52}$$

The clockwise contour illustrated follows the path ABCDEFGHA and encloses the pole at $s = -1$. The corresponding contour in the $F(s)$-plane, illustrated in Figure

6.6b, shows that the closed contour ABCDEFGHA encircles the origin of the $F(s)$-plane in the counter-clockwise direction. Table 6.2 tabulates several values of $F(s)$ as s traverses the contour of the s-plane illustrated in Figure 6.6a.

As a third illustration on mapping contours from the s-plane to the $F(s)$-plane, let us consider the contour ABCDEFGHA traversed in Figure 6.7a which encloses the pole and zero terms for the following function:

$$F(s) = \frac{s}{s+1}. \tag{6.53}$$

The corresponding contour in the $F(s)$-plane is illustrated in Figure 6.7b. Table 6.3 provides several values of $F(s)$ as s traverses the contour illustrated in Figure 6.7a. Figure 6.7b shows that the resulting contour does not enclose the origin in the $F(s)$-plane. This clearly illustrates that the encirclement of a pole and a zero in the s-plane does not result in an encirclement of the origin of the $F(s)$-plane because the zero cancels the effect of the pole. (The clockwise encirclement of a zero alone in the s-plane would result in a clockwise encirclement of the origin in the $F(s)$-plane, while a clockwise encirclement of a pole alone in the s-plane would result in a counterclockwise encirclement of the origin in the $F(s)$-plane.)

This analysis illustrates that if there is an encirclement of the origin of the $F(s)$-plane, its direction depends on whether the contour in the s-plane encloses a pole(s) and/or zero(s). The results derived are now summarized:

Table 6.2. Corresponding Values in the s-Plane and $F(s)$-Plane for Figures 6.6a and b, respectively

Point	$s = \sigma + j\omega$	$F(s) = u + jv$
A	2	0.33
B	$2 - 2j$	$0.23 + 0.15j$
C	$-2j$	$0.2 + 0.4j$
D	$-2 - 2j$	$-0.2 + 0.4j$
E	-2	-1
F	$-2 + 2j$	$-0.2 - 0.4j$
G	$2j$	$0.2 - 0.4j$
H	$2 + 2j$	$0.23 - 0.15j$

Table 6.3. Corresponding Values in the s-Plane and $F(s)$-Plane for Figures 6.7a and b, respectively

Point	$s = \sigma + j\omega$	$F(s) = u + jv$
A	2	0.67
B	$2 - 2j$	$0.77 - 0.15j$
C	$-2j$	$0.8 - 0.4j$
D	$-2 - 2j$	$1.2 - 0.4j$
E	-2	2
F	$-2 + 2j$	$1.2 + 0.4j$
G	$2j$	$0.8 + 0.4j$
H	$2 + 2j$	$0.77 + 0.15j$

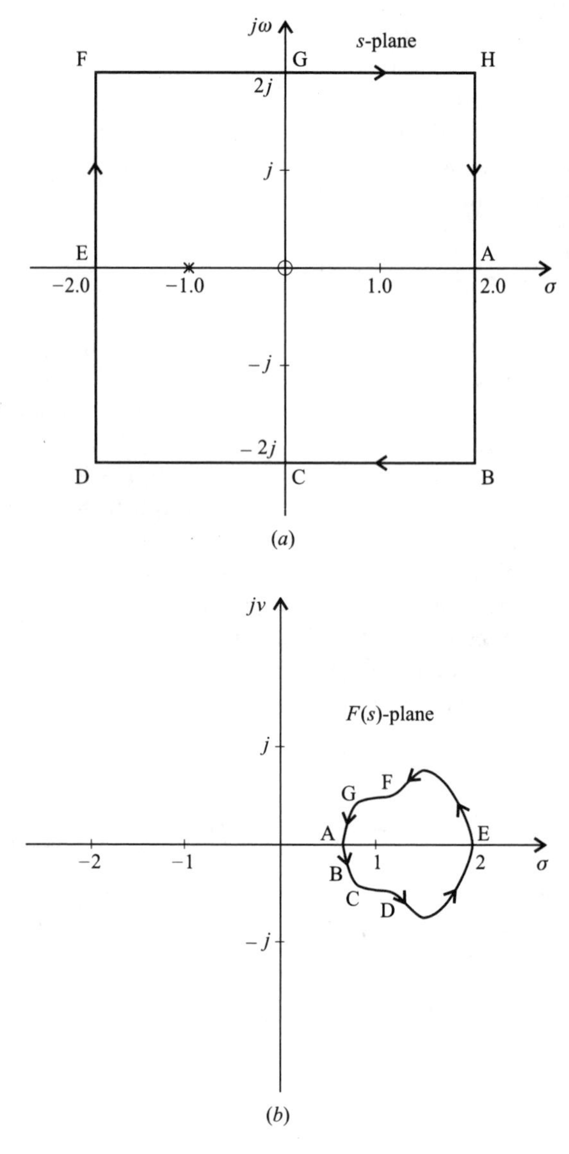

Figure 6.7 (*a*) Contour in the *s*-plane to be mapped for $F(s) = s(s + 1)$. (*b*) Resulting mapping in the $F(s)$-plane for $F(s) = s/(s + 1)$ for the contour shown in the *s*-plane in Figure 6.7(*a*).

1. If a clockwise contour in the *s*-plane encloses one zero as illustrated in Figure 6.5*a*, then the origin of the $F(s)$-plane will have a contour encircling it once in the clockwise direction as illustrated in Figure 6.5*b*. If there are two (or more) zeros contained within the contour in the *s*-plane, then the contour in the $F(s)$-plane will encircle in the clockwise direction the origin two (or more) times.

2. If a clockwise contour in the *s*-plane encloses one pole as illustrated in Figure 6.6*a*, then the origin of the $F(s)$-plane will have a contour encircling it in the counter clockwise direction as illustrated in Figure 6.6*b*. If there are two (or more) poles

contained within the contour in the s-plane, then the contour in the $F(s)$-plane will encircle in the counterclockwise direction the origin two (or more) times.

3. If a clockwise contour in the s-plane encloses both poles and zeros, then the origin of the $F(s)$-plane will have a contour encircling it whose net counterclockwise (or clockwise) encirclements will equal the difference of the poles and zeros (zeros and poles) enclosed by the contour in the s-plane. If the number of poles and zeros contained within the contour of the s-plane are equal, as illustrated in Figure 6.7*a*, then the contour in the $F(s)$-plane will not encircle the origin as illustrated in Figure 6.7*b*.

This graphical analysis of mapping from the s-plane to the $F(s)$-plane is the basis of the Nyquist stability criterion discussed in Section 6.5.

6.5. NYQUIST STABILITY CRITERION

The Nyquist stability criterion [5] is a very valuable tool that determines the degree of stability, or instability, of a feedback control system. In addition, it is the basis for other methods that are used to improve both the steady-state and the transient response of a feedback control system. Application of the Nyquist stability criterion requires a polar plot of the open-loop transfer function, $G(j\omega)H(j\omega)$, which is usually referred to as the Nyquist diagram.

The Nyquist criterion determines the number of roots of the characteristic equation that have positive real parts from a polar plot of the open-loop transfer function, $G(j\omega)H(j\omega)$, in the complex plane. Let us consider the characteristic equation

$$F(s) = 1 + G(s)H(s) = 0. \tag{6.54}$$

System stability can be determined from Eq. (6.54) by identifying the location of its roots in the complex plane. Assuming that $G(s)$ and $H(s)$, in their general form, are functions of s which are given by

$$G(s) = \frac{N_A(s)}{D_A(s)}. \tag{6.55}$$

and

$$H(s) = \frac{N_B(s)}{D_B(s)}, \tag{6.56}$$

then we can say that

$$G(s)H(s) = \frac{N_A(s)N_B(s)}{D_A(s)D_B(s)}. \tag{6.57}$$

Substituting Eq. (6.57) into Eq. (6.54), we obtain the following equivalent expression for $F(s)$:

$$F(s) = 1 + G(s)H(s) = 1 + \frac{N_A(s)N_B(s)}{D_A(s)D_B(s)} = \frac{D_A(s)D_B(s) + N_A(s)N_B(s)}{D_A(s)D_B(s)} = 0. \tag{6.58}$$

In terms of factors, we may rewrite Eq. (6.58) as

$$F(s) = \frac{(s+z_1)(s+z_2)(s+z_3)\dots(s+z_j)\dots}{(s+p_1)(s+p_2)(s+p_3)\dots(s+p_i)\dots}. \tag{6.59}$$

The factors $s + z_j$ are called the zero factors of $F(s)$. This terminology is due to the fact that $F(s)$ vanishes when $s = -z_1, -z_2$, and so on. The factors $s + p_i$ are called the pole factors of $F(s)$. This terminology is due to the fact that $F(s)$ is infinite when $s = -p_1 - p_2$, and so on.

Because $F(s)$ is the denominator of the closed-loop system transfer function given by Eq. (6.6), we see that the zeros of $F(s)$ are the poles of Eq. (6.6). Therefore, for a stable system, it is necessary the the zeros $z_1, z_2, z_3, \dots$ have negative real parts. The roots p_1, p_2, p_3, have no real restrictions on them. As we shall shortly see, however, if we try to determine stability based on $G(s)H(s)$ above, then a knowledge of the roots $p_1, p_2, p_3, \dots$ is also required. The only limitations of the Nyquist criterion are that the system be describable by a linear differential equation having constant coefficients.

The Nyquist diagram is a polar plot in the complex plane of $G(s)H(s)$ as s follows the contour shown in Figure 6.8. Notice that the locus of s avoids poles of $G(s)H(s)$ that lie anywhere on the imaginary axis by small semicircular paths passing to the right. These semicircular paths are assumed to have radii of infinitesimal magnitude. Any roots of the characteristic equation having positive real parts will lie within the contour shown in Figure 6.8.

Use is made of Cauchy's [6] *principle of the argument*, which states that if function $F(s)$ is analytic on and within a closed contour, except for a finite number of poles and zeros within the contour, then the number of times the origin of $F(s)$ in the F plane is encircled as s traverses the closed contour once is equal to the number of zeros minus the number of poles of $F(s)$, the poles and zeros being counted according to their multiplicity. In our particular case, the function $F(s)$ equals $1 + G(s)H(s)$. Therefore, if $1 + G(s)H(s)$ is sketched for the contour defined by Figure 6.8, the number of times that $1 + G(s)H(s)$ encircles the origin equals the number of zeros minus the number of poles of $1 + G(s)H(s)$ for s in the right half-plane.

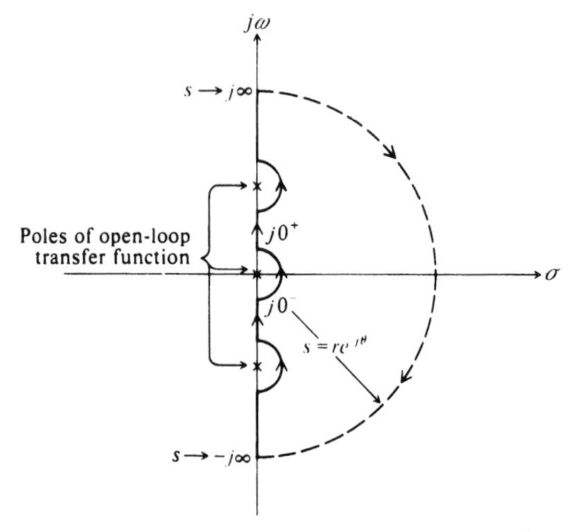

Figure 6.8 Locus of s in the complex plane for determining the Nyquist diagram.

Before we proceed to draw the Nyquist diagram, let us consider the polar plot of a sinusoidal transfer function $G(j\omega)$. It is a plot of the vector $G(j\omega)$, comprising the magnitude and phase angle of $G(j\omega)$, on polar coordinates as ω is varied from zero to infinity. Figure 6.9 illustrates the plot of

$$G(j\omega) = \frac{K}{j\omega(j\omega + 1)(j\omega + 2)} \tag{6.60}$$

as ω varies from zero to infinity. Observe from Figure 6.9 that each point on the polar plot of $G(j\omega)$ represents the end point of a vector for a specific ω.

To illustrate the procedure using the Nyquist diagram, consider the system whose characteristic equation is given by

$$1 + G(s)H(s) = 1 + \frac{K}{(1 + s)^3} = 0. \tag{6.61}$$

Its Nyquist diagram is shown in Figure 6.10. In part (a) of this figure, the Nyquist path for this system is illustrated. Note that its direction is clockwise, and it is defined by three sections: A, B, and C. Section A is defined from $0 \leqslant \omega \leqslant \infty$; section B is defined from $-\infty \leqslant \omega \leqslant 0$; section C is defined from $-\infty \leqslant \omega \leqslant \infty$. For part ($b$) of this figure, the Nyquist plot in the $F(s) = 1 + G(s)H(s)$ plane corresponding to the Nyquist path defined in part (a) of the figure is illustrated. To obtain the direct polar plot of this function, we substitute $s = j\omega$ into Eq. (6.61) as follows:

$$1 + G(j\omega)H(j\omega) = 1 + \frac{K}{(1 + j\omega)^3}. \tag{6.62}$$

Figure 6.10b represents the polar plot of $1 + G(s)H(s)$ for values of s along the imaginary axis. Sections A and B represent the polar plot of $1 + G(s)H(s)$ for values

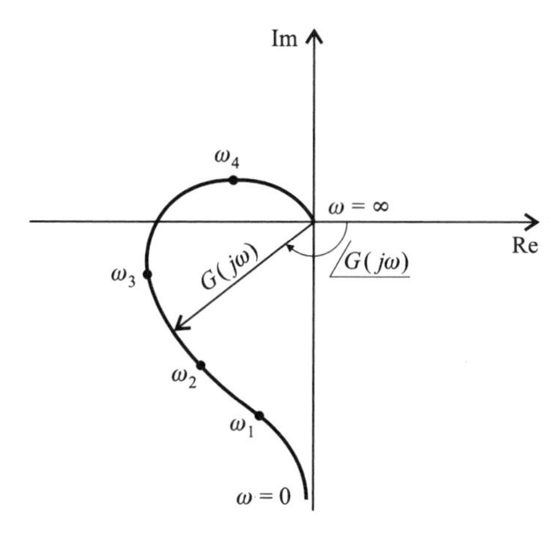

Figure 6.9 Polar plot of

$$G(j\omega) = \frac{K}{j\omega(j\omega + 1)(j\omega + 2)}$$

as ω varies from zero to infinity.

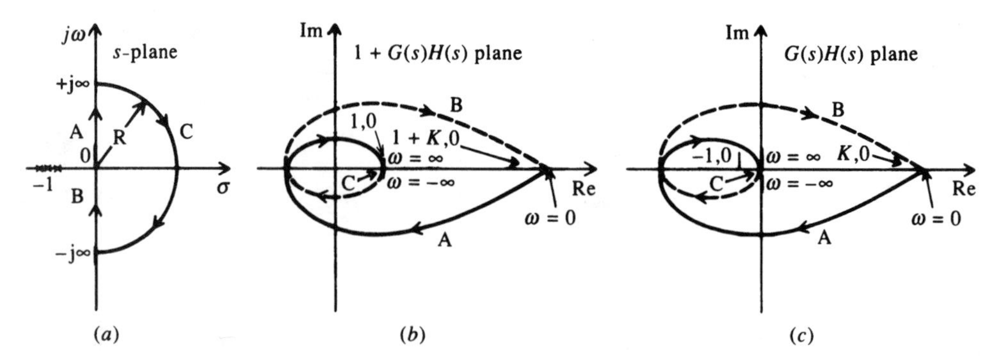

Figure 6.10 The Nyquist path (*a*) in the *s*-plane, and the Nyquist plots in the (*b*) $1 + G(s)H(s)$ and (*c*) $G(s)H(s)$ planes for a system whose open-loop transfer function is given by

$$G(s)H(s) = \frac{K}{(1+s)^3}.$$

of s along the imaginary axis when $-j\infty \leqslant j\omega \leqslant j\infty$. The dashed portion of the curve denotes the negative-frequency portion. The three sections corresponding to A, B, and C of both the Nyquist path (Figure 6.10*a*) and the Nyquist plot of Figure 6.10*b* are indicated. Notice that the semicircular section C of Figure 6.10*a* is only a point in Figure 6.10*b*. Observe that the $F(s)$ locus encircles the origin twice. The $G(s)H(s)$ locus is drawn in Figure 6.10*c*. Notice that the encirclement of origin by the plot of $1 + G(j\omega)H(j\omega)$ is equivalent to the encirclement of the $-1, 0$ point of the $G(j\omega)H(j\omega)$ locus alone. Because it is easier to just draw the locus of $G(j\omega)H(j\omega)$ rather than $1 + G(j\omega)H(j\omega)$, in practice we draw only the $G(j\omega)H(j\omega)$ locus as in Figure 6.10*c* and determine the encirclement of the $-1, 0$ point. We start from $\omega = -\infty$, go through $\omega = 0$, and end at $\omega = \infty$, and we count the number of clockwise rotations of the vector in the $G(j\omega)H(j\omega)$ plane. Observe that the plot of $G(j\omega)H(j\omega)$ and the plot of $G(-j\omega)H(-j\omega)$ are always symmetrical with each other about the real axis in the $G(j\omega)H(j\omega)$ plane.

It is interesting to examine the $1 + G(s)H(s)$ and $G(s)H(s)$ planes further as illustrated in Figure 6.11. Notice that the $1 + G(j\omega)H(j\omega)$ locus in the $1 + G(s)H(s)$ plane is measured from the origin to the locus. In the $G(s)H(s)$ plane, the $1 + G(j\omega)H(j\omega)$ locus is the vector sum of the unit vector and the vector $G(j\omega)H(j\omega)$. Therefore, observe that the $1 + G(j\omega)H(j\omega)$ vector drawn from the origin of the $1 + G(s)H(s)$ plane, and the $1 + G(j\omega)H(j\omega)$ vector drawn from the $-1, 0$ point in the $G(s)H(s)$ plane are identical. The encirclement of the origin in the $1 + G(s)H(s)$ plane is, therefore, equivalent to encirclement of the $-1 + j0$ point in the $G(s)H(s)$ plane.

With the Nyquist diagram sketched on a set of axes where the origin is shifted to the $-1 + j0$ point, the Nquist stability criterion can be stated algebraically as*

$$N = Z - P, \tag{6.63}$$

*The proof of the Nyquist stability criterion is contained in Appendix B.

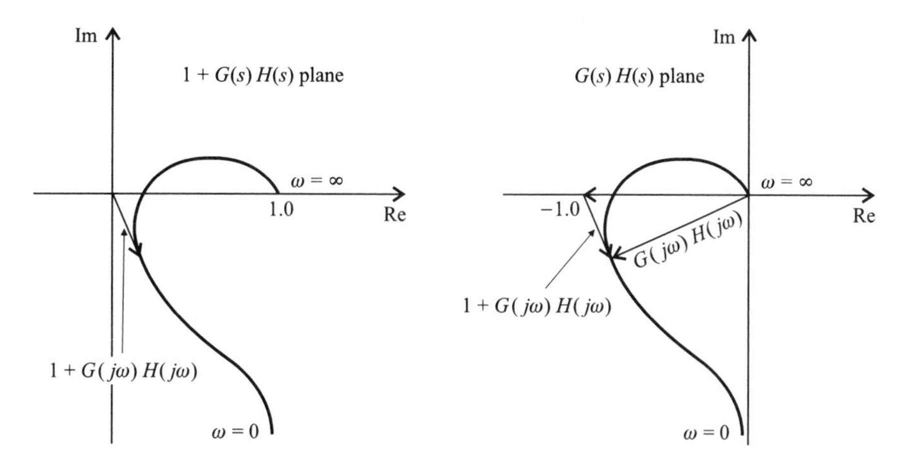

Figure 6.11 Polar plots of $1 + G(j\omega)H(j\omega)$ in the $1 + G(s)H(s)$ plane and $G(j\omega)H(j\omega)$ in the $G(s)H(s)$ plane for

$$G(j\omega)H(j\omega) = \frac{K}{j\omega(j\omega + 1)(j\omega + 2)}.$$

where N is the number of clockwise encirclements of the $-1 + j0$ point by the Nyquist locus, P is the number of poles of the open-loop transfer function $G(s)H(s)$ having positive real parts, and Z is the number of roots of the characteristic equation having positive real parts (Z must equal zero for stability). In most practical cases, the open-loop transfer function is in itself stable and P would equal zero. Because Z must equal zero for stability, N must equal zero for these practical systems.

For the example of Eq. (6.61) analyzed in Figure 6.10, $G(s)H(s)$ does not have any poles in the right half-plane. Therefore, $P = 0$, and $Z = N$. Because the plot of Figure 6.10*c* shows two clockwise encirclements of the $-1, 0$ point, $N = 2$, and Z also equals 2, which means that there are two roots of the characteristic equation in the right half-plane. Therefore, the system is unstable. To stabilize this system, the gain K would have to be reduced, or a frequency-sensitive network would be added, until there were no encirclements of the $-1, 0$ point.

For cases where there exist one or more poles at the origin of the s-plane, the mapping from the s-plane to the $G(s)H(s)$ plane is a little more difficult. As indicated in Figure 6.12*a*, all poles on the imaginary axis must be bypassed with small semicircular paths. To illustrate the technique for this case, an open-loop transfer function containing a pure integration will be analyzed.

Figure 6.12*a* illustrates a polar plot, corresponding to the contour defined by Figure 6.8, for a system whose open-loop transfer function is given by

$$G(s)H(s) = \frac{K}{s(1 + T_1 s)(1 + T_2 s)}. \tag{6.64}$$

To obtain the direct polar plot of this function, we substitute $s = j\omega$ into Eq. (6.64) as follows:

$$G(j\omega)H(j\omega) = \frac{K}{j\omega(1+T_1 j\omega)(1+T_2 j\omega)}. \tag{6.65}$$

Figure 6.12*b* represents the polar plot of $G(s)H(s)$ for the path shown in Figure 6.12*a*. The dashed portion of the curve denotes the negative-frequency portion. Notice that the polar plot for negative frequencies is the conjugate of the positive portion.

In part (a) of Figure 6.12, the Nyquist path for this system is illustrated. It is defined by four sections. Section AB is defined from $0^+ \leqslant \omega \leqslant \infty$; section EFGHA is defined from $0^- \leqslant \omega \leqslant 0^+$; section DE is defined from $-\infty \leqslant \omega \leqslant 0^-$; section DCB is defined from $-\infty \leqslant \omega \leqslant \infty$. The plots of sections AB, DE, and DCB are similar in Figure 6.12*a* to that in Figure 6.10*a*. Notice that semicircular section DCB in Figure 6.12*a* is only a point in Figure 6.12*b*. However, the detour of the origin by path EFGHA in Figure 6.12*a* is new, and the procedure for obtaining it is as follows. The path along the imaginary axis in the vicinity of the origin is modified to be a semicircle of infinitesimal radius δ in the positive half-plane in order to avoid passing through this pole on the imaginary axis at the origin (see Figure 6.8). For this semicircular portion of the path, from $j0^-$ to $j0^+$, $s = \delta e^{j\theta}$, where $\delta \to 0$ and $-\pi/2 \leqslant \theta \leqslant \pi/2$. Therefore, for $s \to 0$, Eq. (6.64) becomes

$$G(s)H(s)\Big]_{s\to 0} = K/s = K/\delta e^{j\theta} = (K/\delta)e^{-j\theta} = (K/\delta)e^{j\alpha}. \tag{6.66}$$

Observe that the magnitude of $K/\delta \to \infty$ as $\delta \to 0$, and $\alpha = -\theta$ goes from $\pi/2$ to $-\pi/2$ as the directed segment s goes from $\delta\underline{/-\pi/2}$ to $\delta\underline{/\pi/2}$. This implies that the

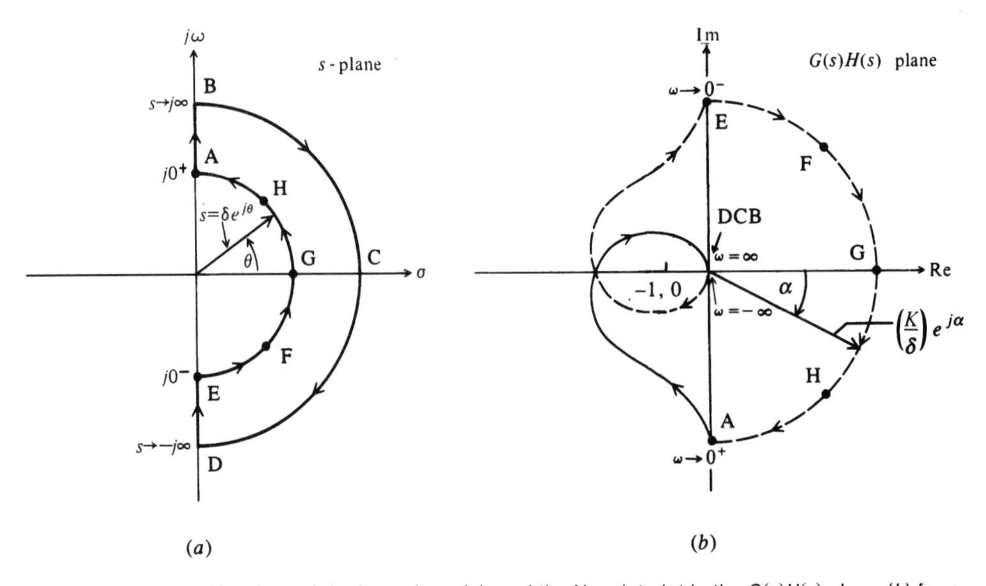

Figure 6.12 The Nyquist path in the *s*-plane (*a*), and the Nyquist plot in the $G(s)H(s)$ plane (*b*) for a system whose open-loop transfer function is given by

$$G(s)H(s) = \frac{K}{s(1+T_1 s)(1+T_2 s)}.$$

end points from $\omega \to 0^-$ and $\omega \to 0^+$ in Figure 6.12*b* are joined by a semicircle of infinite radius in the first and fourth quadrants because as ω goes from 0^- to 0^+ in the *s*-plane (θ goes 180° counterclockwise), in the $G(s)H(s)$ plane the locus has to go 180° clockwise in going from $\omega = 0^-$ to 0^+ because $\theta = -\alpha$ in Eq. (6.66).

Analysis of the Nyquist plot in Figure 6.12*b* indicates that there are two clockwise encirclements of the $-1, 0$ point. Therefore, $N = 2$ in Eq. (6.23), where

$$N = Z - P.$$

Because none of the poles of the open-loop transfer function are in the right half of the *s*-plane ($P = 0$), then

$$2 = Z - 0, \quad Z = 2,$$

and there are two roots of the characteristic equation in the right half-plane. Therefore, this system is unstable for the value of K chosen. We could stabilize this system by reducing its gain K or adding a frequency-sensitive network until there were no encirclements of the $-1, 0$ point.

Figure 6.13 illustrates several examples of application of the Nyquist stability criterion using the relationship given by Eq. (6.63). In all cases, the values of the open-loop transfer function $G(s)H(s)$ are shown together with the values of N, P, and Z. Notice that those systems whose Nyquist diagrams are illustrated in parts (*a*)–(*d*) are open-loop stable, whereas those in parts (*e*) and (*f*) are open-loop unstable. The systems illustrated in parts (*a*), (*d*), and (*e*) are closed-loop stable, whereas the systems illustrated in parts (*b*), (*c*) and (*f*) are closed-loop unstable.

It is quite clear that a good picture of a system's margin of stability can be obtained from the Nyquist diagram. The proximity of the $G(s)H(s)$ locus to the $-1 + j0$ point is an indication of relative stability. The farther away the locus is from this point, the greater the margin of stability. One conventional measure of the relative degree of stability is the distance two points on the $G(s)H(s)$ locus are from the point $-1 + j0$. This is illustrated in Figure 6.14 by the points A and B.

Point A is defined by the intersection of the $G(s)H(s)$ locus and the unit circle. Obviously, the magnitude of $G(s)H(s)$ at point A is unity and is denoted by $[G(j\omega)H(j\omega)]_1$ in Figure 6.14. The angle of $[G(j\omega)H(j\omega)]_1$ with respect to the positive real axis is designated as θ. The phase margin γ is defined as the angle which $[G(j\omega)H)(j\omega)]_1$ makes with respect to the negative real axis. It is related to θ (which is a negative angle) by the equation

$$\gamma = 180^\circ + \theta. \tag{6.67}$$

A positive value of phase margin indicates stability, and a negative value of phase margin indicates instability. A zero value of phase margin indicates that the $G(s)H(s)$ locus passes through the $-1 + j0$ point. The phase margin tells us how shy $G(s)H(s)$ is of -180° when its magnitude is 1. (Remember, from Eq. (6.7), that we do not want $G(s)H(s) = -1$.) The magnitude of $+\gamma$ indicates the relative degree of stability. Usually, a desirable value of γ is between 30° and 60°. It has been shown that the phase margin is approximately related to the damping ratio in second-order systems by the following approximation [6]

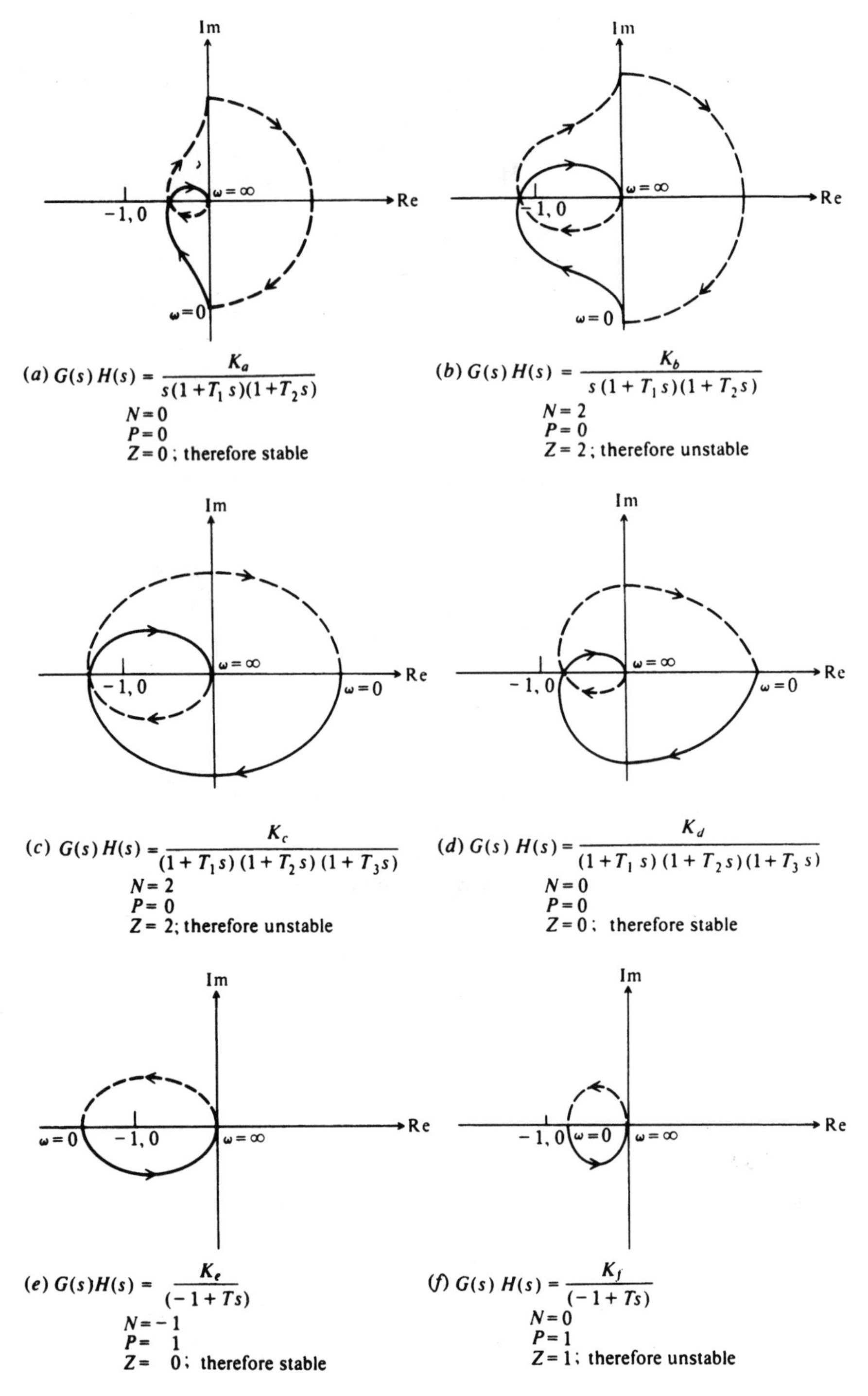

Figure 6.13 Examples of typical Nyquist diagrams.

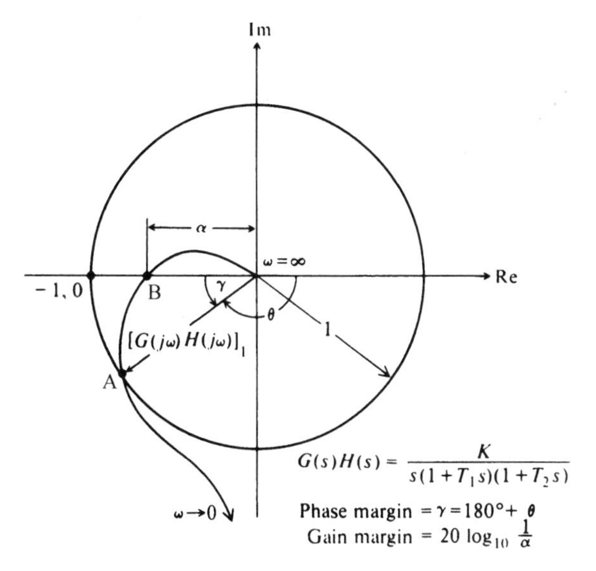

Figure 6.14 Definition of phase and gain margins.

$$\text{phase margin} \approx 100\zeta. \tag{6.68}$$

Therefore, a damping factor of 0.5 requires a phase margin of approximately 50°.

The point B is defined by the intersection of the $G(s)H(s)$ locus and the negative real axis. The gain margin is defined as the reciprocal of the magnitude of the $G(s)H(s)$ locus at point B. For the configuration illustrated in Figure 6.14, the gain margin equals $1/\alpha$. The significance of the gain margin is that the system gain could be increased by a factor of $1/\alpha$ before the $G(s)H(s)$ locus would intersect the $-1 + j0$ point. A value of gain margin greater than unity indicates stability; a gain margin smaller than unity indicates instability. The magnitude of the gain margin indicates the relative degree of stability. Gain margin is conventionally expressed in decibels as follows:

$$\text{gain margin in dB} = 20 \log_{10}(1/\alpha). \tag{6.69}$$

For example, if $\alpha = 0.5$, the gain margin is 6 dB. Usually, a desirable value of gain margin is between 4 and 12 dB. It is important to recognize that if the $G(j\omega)H(j\omega)$ locus intersects the $-1, 0$ point, then the roots of the characteristic equation are located on the imaginary axis, and the phase and gain margins are zero. This should be avoided in practical control systems. The gain margin tells us how shy $G(s)H(s)$ is of 1 when its phase is $-180°$. (Remember, from Eq. (6.7), that we do not want $G(s)H(s) = -1$.)

One important issue to be considered is that of conditionally stable systems. If Figure 6.14 is examined, the erroneous impression can be obtained that decrease in gain will always cause a system to become stable. This is not always true, as illustrated in Figure 6.15. Here, if the gain is decreased, point B will move to enclose the $-1 + j0$ point and the system will become unstable.

On the other hand, if the gain is increased, then point A will move to enclose the $-1 + j0$ point and the system will become unstable. Observe from the Nyquist diagram illustrated in Figure 6.15 that this control system is stable even though it appears that the $-1, 0$ point is enclosed. This is because the inner encirclement of the $-1, 0$ point, BCAD, is counterclockwise while the outer encirclements of the $-1, 0$ point, EHGF, is clockwise. Therefore, the net encirclements of the $-1, 0$ point is zero.

Some control systems have two or more phase crossover frequencies as illustrated in the Nyquist diagram of Figure 6.16. The phase margin for such a control system is measured at the highest gain crossover frequency.

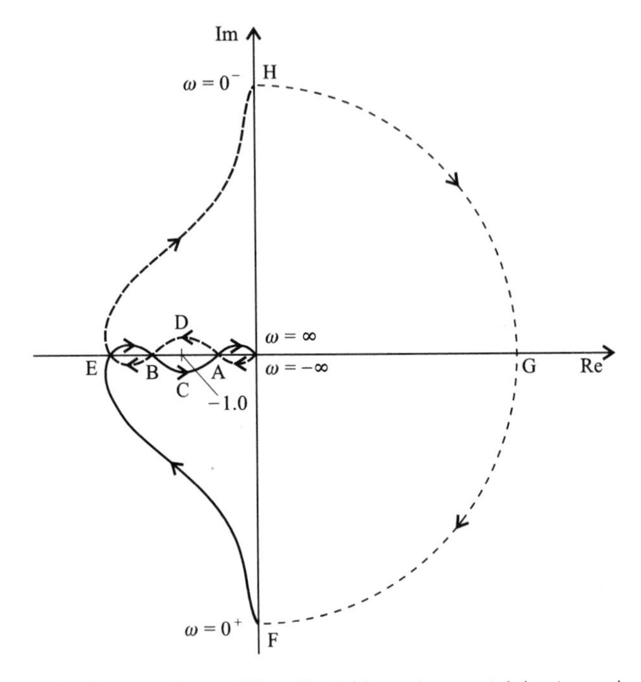

Figure 6.15 Nyquist diagram of a conditionally stable system containing two gain margins.

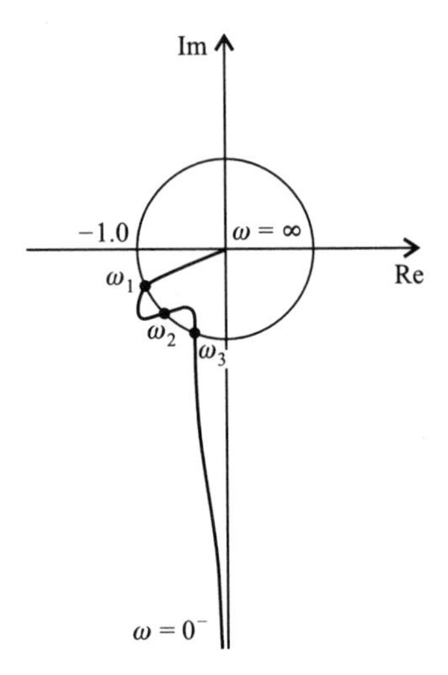

Figure 6.16 The Nyquist diagram of a control system having three gain crossover frequencies.

Conditionally stable control systems containing multiple gain margins are discussed further in Section 6.14 from the viewpoints of the Bode diagram, and the root locus in Section 6.19.

Section 6.6 describes the approach for obtaining the Nyquist diagram using MATLAB. These Nyquist diagrams are contained in the M-files (that are part of my *Modern Control System Theory and Design Toolbox*) that can be retrieved free from The MathWorks, Inc. anonymous FTP server at ftp://ftp.mathworks.com/pub/books/shinners.

6.6. NYQUIST DIAGRAMS USING MATLAB [7]

In the previous section, several Nyquist diagrams were illustrated which were hand drawn. They were based on the transformation of the contour drawn in the s-plane to the $G(s)H(s)$ plane. In this section, the MATLAB utility used for obtaining the Nyquist diagram is identified, and it is applied.

A. Drawing Nyquist Diagrams of Systems Defined by Transfer Functions

If the system is defined by a transfer function, the Control System Toolbox command which determines the Nyquist plot is:

nyquist.

The Nyquist diagram can be used with polynomial transfer functions, defined as $G(s) = \text{num}(s)/\text{den}(s)$ where "num" and "den" contain the polynomial coefficients, if invoked with three right-hand arguments as follows

[re,im] = nyquist(num,den,w)

where

re = real part on the Nyquist diagram
im = imaginary part on the Nyquist Diagram
num = row matrix format representing the numerator of the polynomial
den = row matrix format representing the denominator of the polynomial
w = frequency in rad/sec.

As shown in the previous section on the Nyquist diagram, it will occasionally go to infinity. Without special precautions in the MATLAB program, an erroneous Nyquist plot may occur. We can avoid this with MATLAB by specifying the finite area we want plotted. This will be illustrated next.

Let us consider the Nyquist diagram of a unity feedback control system whose forward transfer function is given by

$$G(s) = \frac{1}{s(1+s)^2}. \tag{6.70}$$

Since the Nyquist diagram goes to infinity, we can avoid this with MATLAB by using a manually determined range. For example, we can limit it from −1.8 to 0 on the real axis, and from −2.5 to 2.5 on the imaginary axis by entering the commands

```
v = [-1.8   0   -2.5   2.5]
axis(v)
```

Alternatively, we can use the single command

```
axis([-1.8   0   -2.5   2.5])
```

Another important point to emphasize in this problem is that the numerator is clearly defined as a constant. However, the denominator is given as the product of terms, with the result that we must multiply these terms to get a polynomial in s. We can use the *convolution* command in MATLAB to perform this multiplication as follows. Defining

$$a = s(s+1) = s^2 + s \quad : \quad a = [1 \quad 1 \quad 0]$$
$$b = s + 1 \quad : \quad b = [0 \quad 1 \quad 1]$$

and using the MATLAB command

$$c = \text{conv}(a,b)$$

will result in the product of the terms in the denominator. The resulting MATLAB program for multiplying the factors in the denominator is illustrated in Table 6.4.

We are now ready to develop the MATLAB program for plotting this Nyquist diagram which is shown in Table 6.5.

Table 6.4. MATLAB Program for Multiplying Factors

```
a = [1   1   0]
b = [0   1   1]
c = conv(a,b)
c =
    1  2  1  0
```

Table 6.5. MATLAB Program for Obtaining the Nyquist Diagram

```
num = [0   0   0   1];
den = [1   2   1   0];
nyquist(num,den)
Warning: Divide by zero
axis([-1.8   0   -2.5   2.5])
grid
title('Nyquist Plot of G(s) * H(s) = 1/[(s * (s + 1)^2]')
```

The resultant Nyquist plot obtained is illustrated in Figure 6.17. Observe that the Nyquist diagram illustrated in Figure 6.17 is only a partial Nyquist diagram because we limited the real and imaginary axes range with the MATLAB command "axis." If we could let the real and imaginary axes approach infinity with MATLAB, this Nyquist diagram would be similar to the Nyquist diagram shown in Figure 6.13*a*. However, enlarging the real and imaginary axes with MATLAB results in losing observability of the portion of the Nyquist diagram closest to the origin which intersects the real axis at -0.5. This is a very important value to know, because it means that the gain could be doubled from 1 to 2 before the Nyquist diagram intersects the $-1, 0$ point.

B. Drawing Nyquist Diagrams of Systems Defined in State Space

If the control system is defined in state-space form, the *Control System Toolbox* command

nyquist(A,B,C,D)

will produce the Nyquist diagram, where the matrices **A** (same as **P**), **B**, **C** (same as **L**), and **D** were previously defined and discussed in Section 2.21. We will show how to obtain the Nyquist diagram for the same control system illustrated earlier in this section, but we will now obtain it from its state-space form.

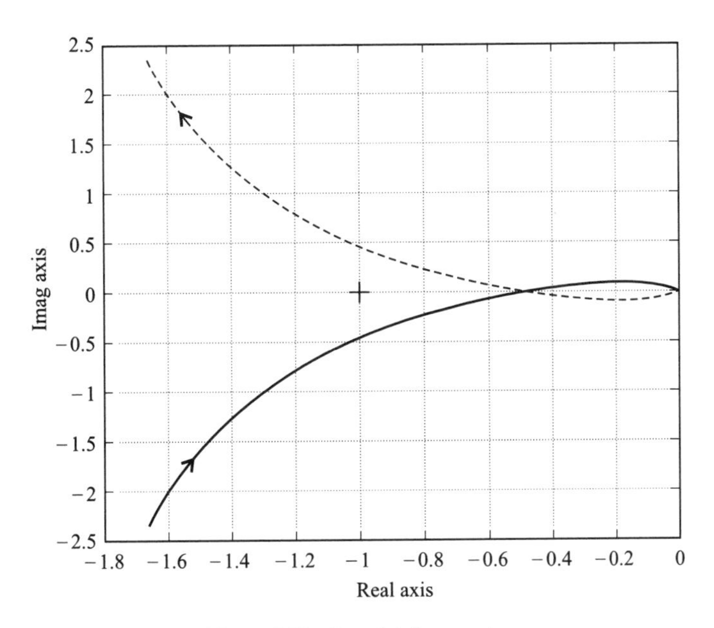

Figure 6.17 Nyquist diagram for

$$G(s)H(s) = \frac{1}{s(s+1)^2}.$$

Let us use the approach we used in Section 2.21 to convert from the transfer function to the state-space form. The resulting MATLAB program for accomplishing this is shown in the MATLAB Program in Table 6.6.

Therefore, the MATLAB program for determining the Nyquist diagram from the state-space form is given by the MATLAB program in Table 6.7.

The result is the same Nyquist diagram shown in Figure 6.17.

6.7. BODE-DIAGRAM APPROACH

The Bode-diagram approach [8] is one of the most commonly used methods for the analysis and synthesis of linear feedback control systems. This method, which is basically an extension of the Nyquist stability criterion, has the same limitations and uses as the Nyquist diagram. The presentation of information in the Bode-diagram approach, however, is modified to permit relatively quick determinations of the effects of changes in system parameters without the laborious calculations associated with the Nyquist diagram.

Table 6.6. MATLAB Program for Converting from the Transfer Function to the State-Space Form

```
num = [0  0  0  1];
den = [1  2  1  0];
[A, B, C, D] = tf2ss(num,den)
A =
   -2  -1  0
    1   0  0
    0   1  0
B =
    1
    0
    0
C =
    0  0  1
D =
    0
```

Table 6.7. MATLAB Program for Determining the Nyquist Diagram from the State-Space Form

```
A = [-2  -1  0; 1  0  0; 0  1  0];
B = [1; 0; 0]
C = [0  0  1];
D = [0];
nyquist(A,B,C,D)
axis([-1.8  0  -2.5  2.5])
grid
title('Nyquist plot of G(s) * H(s) = 1/[s * (s + 1)^2]')
```

The Nyquist diagram gives the amplitude and phase of the open-loop transfer function $G(s)H(s)$ as s traverses a contour that encloses the right half-plane. As s traverses the positive imaginary axis, it has the value of real frequency ω, and the plot corresponds to $G(j\omega)H(j\omega)$. We can illustrate the same amount of information by means of two diagrams which have ω as a common axis. These two diagrams, illustrated in Figure 6.18, are usually referred to as the Bode diagram. The magnitude of $G(j\omega)H(j\omega)$ is $20\log_{10} G(j\omega)H(j\omega)$; the unit used to represent this is decibels (dB). For this logarithmic representation, the amplitude and phase curves are drawn on semilog paper with the log scale used for frequency. The linear scales are used for magnitude in decibels, and phase degrees. The number of cycles needed for the frequency range is based on the frequency range of the problem being analyzed. It is important to emphasize that the Bode diagram provides information only of s corresponding to the positive imaginary axis, and therefore represent the frequency response.

Let us consider the system illustrated in Figure 6.19. It is assumed that the input $r(t)$ is a sinusoidal waveform ω and unity amplitude given by

$$r(t) = \sin \omega t. \tag{6.71}$$

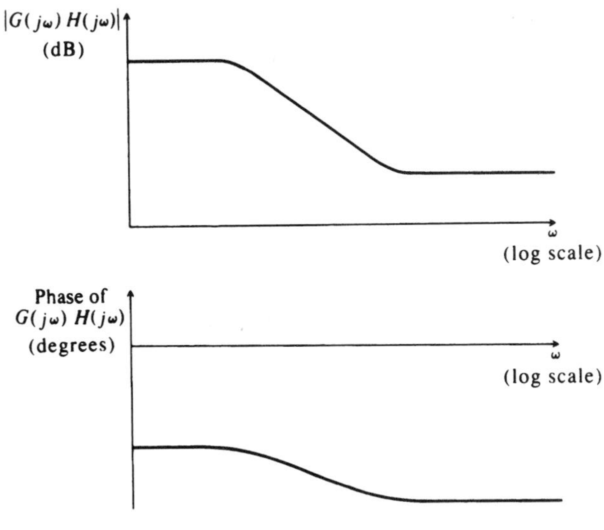

Figure 6.18 Typical Bode diagram.

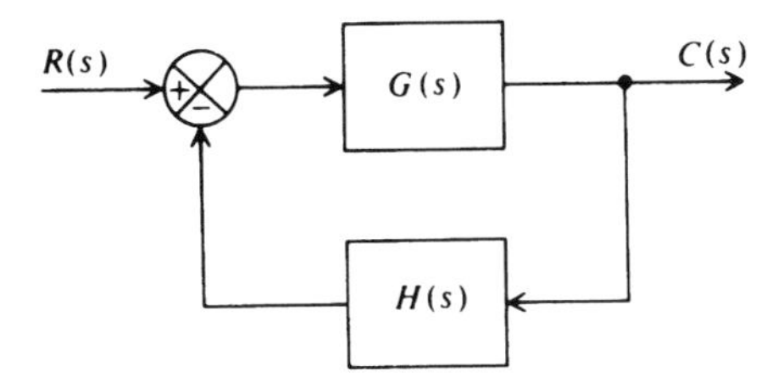

Figure 6.19 Representative feedback control system.

Assuming a linear system, the output would have the general form

$$c(t) = a \sin(\omega t - \phi), \tag{6.72}$$

where $a =$ gain of system and $\phi =$ phase shift of the system. Let us represent the system transfer function as

$$H(j\omega) = \frac{C(j\omega)}{R(j\omega)} = a(j\omega)e^{-j\phi(j\omega)}. \tag{6.73}$$

Here $H(j\omega)$ is a complex function which can be represented by an amplitude $a(j\omega)$ and a phase shift of $\phi(j\omega)$. In the Bode-diagram approach, the amplitude is expressed in decibels and is plotted versus frequency on semilogarithmic graph paper. The amplitude, in decibels, is given by

$$\text{amplitude in decibels} = 20 \log_{10} |a(j\omega)|. \tag{6.74}$$

The phase shift $\phi(j\omega)$ is conventionally expressed in degrees.

By introducing the logarithm concept, the tedious process of multiplying two complex numbers is simplified to one of addition. For example, let us consider two complex numbers: $G_a(j\omega)$ and $G_b(j\omega)$. The logarithm of the product of two complex quantities,

$$G_a(j\omega) = a_a(j\omega)e^{-j\phi_a(j\omega)} \tag{6.75}$$

and

$$G_b(j\omega) = a_b(j\omega)e^{-j\phi_b(j\omega)}, \tag{6.76}$$

is given by

$$\begin{aligned} \log_{10} |G_a(j\omega)G_b(j\omega)| &= \log_{10} |a_a(j\omega)| + \log_{10} |a_b(j\omega)|, \\ \phi(\omega) &= -[\phi_a(j\omega) + \phi_b(j\omega)]. \end{aligned} \tag{6.77}$$

Equation (6.77) illustrates that the logarithm of the product of two complex numbers is the sum of the logarithms of the magnitude components. The total phase equals the sum of the individual phase-angle components. A minus appears before the phase contribution term in Eq. (6.77) because the individual phase components were assumed to be negative in Eqs. (6.75) and (6.76). In addition, Eq. (6.77), shows that the logarithm of the magnitude and phase components, respectively, are separate functions of the common parameter ω and can be sketched separately, as was shown in Figure 6.18.

Bode's theorems are presented in Chapter 7 when we discuss the Bode-diagram approach for the design of feedback control systems in order to meet certain specifications. In this chapter, however, it is important to emphasize the fact that the Bode-diagram approach applies primarily only to *minimum-phase systems*. The basic definition of a minimum-phase system [9] is a system whose phase shift is the minimum possible for the number of energy storage elements in the network. A system is denoted as being minimum phase if all its poles and zeros lie in the left half of the

s-plane. If at least one of its poles or zeros lies in the right half-plane, then the system is denoted as being a nonminimum-phase system. This definition restricts the poles and zeros of minimum-phase systems to the left half of the complex-plane. A little thought indicates that when we specify either the amplitude or phase of a minimum-phase system, we have also automatically specified the other. This concept is the basis of one of Bode's theorems. However, the Bode diagram can be applied to nonminimum-phase systems if we know where the nonminimum-phase characteristics come from.

Understanding the basic concepts of the Bode diagram, let us now gain the facility for constructing them. The laborious procedure of plotting the amplitude and phase of $G(j\omega)H(j\omega)$ by means of substituting several values of $j\omega$ is not necessary when drawing the Bode diagram, because we can use several shortcuts. These shortcuts are based on simplifying the approximations that allow us to represent the exact, smooth plots with straight-line asymptotes. The difference between actual amplitude characteristics and the asymptotic approximations is only a few decibels. Now we shall demontrate the application of this approximating technique to seven common, representative transfer functions: a constant, a pure integration, a pure differentiation, a simple phase-lag network, a simple phase-lead network, a quadratic phase-lage network and a time-delay factor. The basic concepts illustrated are then used to draw the Bode diagram of the transfer function for representative systems. We will then show how the Bode diagram can be obtained using MATLAB in Section 6.8, and other commercial programs presented in Section 6.21.

A. Bode Diagram of a Constant

Using the definition of Eq. (6.74), the logarithm of a constant K, or $1/K$, where $K > 1$, is given by

$$(K)_{\mathrm{dB}} = 20\log_{10} K, \tag{6.78}$$

$$\left(\frac{1}{K}\right)_{\mathrm{dB}} = -20\log_{10} K. \tag{6.79}$$

The corresponding phase angle of a constant is 0°. Figure 6.20 illustrates the Bode diagram of a constant.

Figure 6.21 provides an easy way to obtain a decibel value for different numbers. Observe from this conversion line that ratios of two (an octave) corresponds to 6 dB; ratios of ten (decade) correspond to 20 dB. In addition, observe that the reciprocal of a number differs form its value only by the sign:

$$20\log K = -20\log\frac{1}{K} \tag{6.80}$$

B. Bode Diagram of a Pure Integration

The Bode diagram of a pure integration

$$G(j\omega) = 1/j\omega \tag{6.81}$$

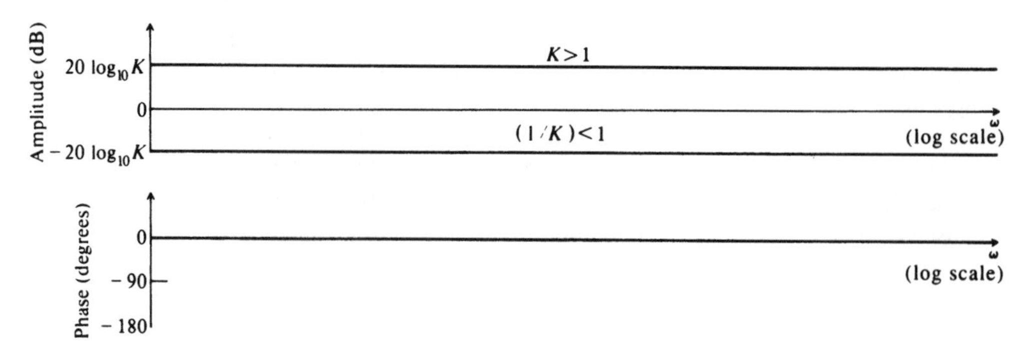

Figure 6.20 Bode diagram of a constant.

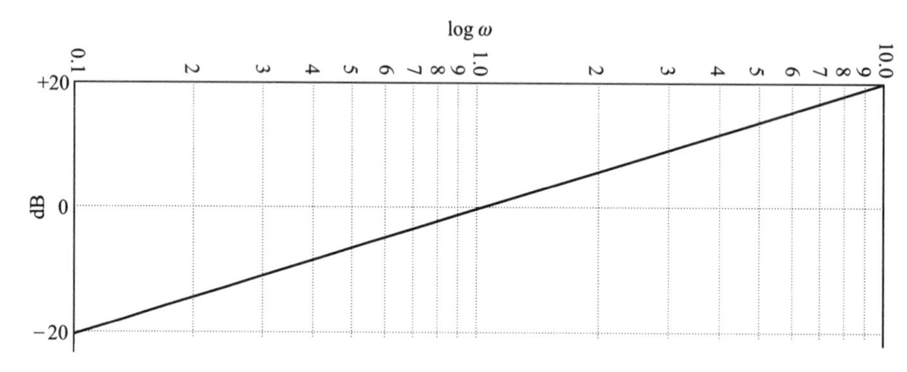

Figure 6.21 Conversion curve between numbers and decibels.

can be obtained by taking the logarithm, yielding

$$20 \log_{10} \left| \frac{1}{j\omega} \right| = -20 \log_{10} \omega \text{ in dB.} \tag{6.82}$$

Figure 6.22 illustrates the Bode diagram of a pure integration. Notice that the resulting amplitude curve is linear when the amplitude is plotted on a linear scale and the frequency is plotted on a logarithmic scale. The slope of the amplitude characteristic is a constant and equals −20 dB/decade. A slope of −20 dB/decade also corresponds to a slope of −6 dB/octave. The phase characteristic for the pure integration is constant and equals −90°.

C. Bode Diagram of a Pure Differentiation

The Bode diagram of a pure differentiation

$$G(j\omega) = j\omega \tag{6.83}$$

can be obtained in a manner similar to that used for the pure integration. The major differences are that the amplitude characteristics now have a positive slope and the phase characteristic is positive. Figure 6.23 illustrates the Bode diagram of a pure differentiation.

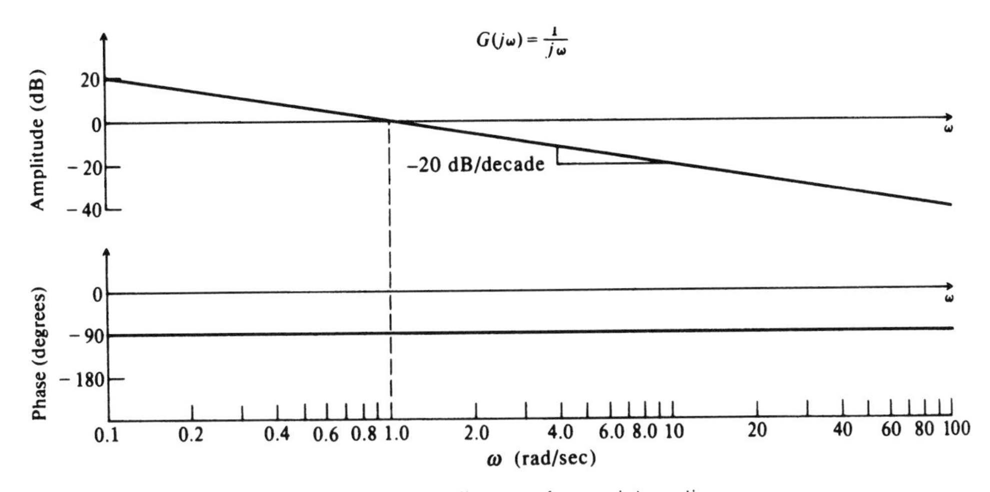

Figure 6.22 Bode diagram of a pure integration.

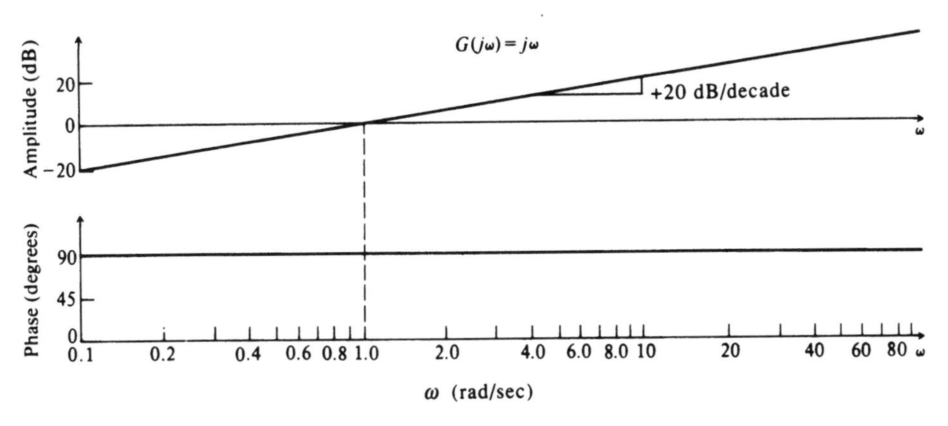

Figure 6.23 Bode diagram of a pure differentiation.

D. Bode Diagram of a Simple Phase-Lag Network

A phase-lag network produces a phase lag that is a function of frequency. The transfer function of such a network is given by

$$G(j\omega) = \frac{a}{j\omega + a} = \left(j\frac{\omega}{a} + 1\right)^{-1}. \tag{6.84}$$

The Bode diagram for this transfer function can be obtained as follows:

$$20\log_{10}\left(j\frac{\omega}{a} + 1\right)^{-1} = -20\log_{10}\left(\frac{\omega^2}{a^2} + 1\right)^{1/2} \angle -\tan^{-1}\frac{\omega}{a}. \tag{6.85}$$

Figure 6.24 illustrates the Bode diagram of a simple phase-lag network. The dashed portion of the amplitude characteristic represents the exact plot, and the heavy-line segments represent the straight line asymptotic approximation. They differ by a maximum of 3 dB at $\omega/a = 1$ and by 1 dB an octave away.

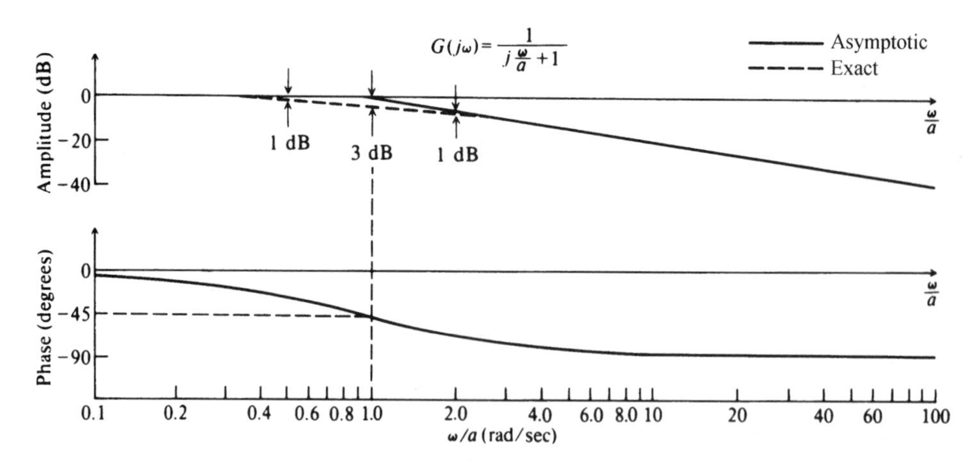

Figure 6.24 Bode diagram of a simple phase-lag network.

The asymptotic approximation can be drawn quite easily. For example, when ω/a is much less than unity, the imaginary component is very much smaller than the real component (unity), and the imaginary component can be neglected. Therefore,

$$20\log_{10} 1 = 0\,\text{dB} \quad \text{for } \frac{\omega}{\text{a}} \ll 1. \tag{6.86}$$

When ω/a is much greater than unity, the imaginary component is much greater than the real component (unity) and the real component may be neglected. Therefore

$$20\log_{10}\left|\left(j\frac{\omega}{a}\right)^{-1}\right| = -20\log_{10}\frac{\omega}{a}\ \text{dB for } \frac{\omega}{a} \gg 1. \tag{6.87}$$

Equation (6.87) is very similar to the amplitude characteristic of a pure integrator, given by Eq. (6.82), and has a slope of −20 dB/decade. At $\omega/a = 1$, the two asymptotes join each other. This frequency ($\omega = a$) is referred to as the break or corner frequency. Ordinarily, the straight-line asymptotic approximation is accurate for most applications. If further correction is needed, the exact curve can be obtained from the approximate curve by using the following corrections: −1 dB at $\omega/a = 0.5$ and 2; −3 dB at $\omega/a = 1$. Additional corrections can be obtained from Figure 6.25 which illustrates the corrections in decibels as a function of a.

The phase shift produced by the simple phase-lag network can be obtained from the expression

$$\text{phase lag} = -\tan^{-1}\frac{\omega}{a}. \tag{6.88}$$

The phase shift at the break frequency is −45°; at $\omega = 0$ it is 0°; at $\omega = \infty$ it is −90°. Notice that this phase-lag network is a minimum-phase network, because the pole is in the left half-plane, and it has the minimum phase shift possible for the number of energy storage elements in the network (one).

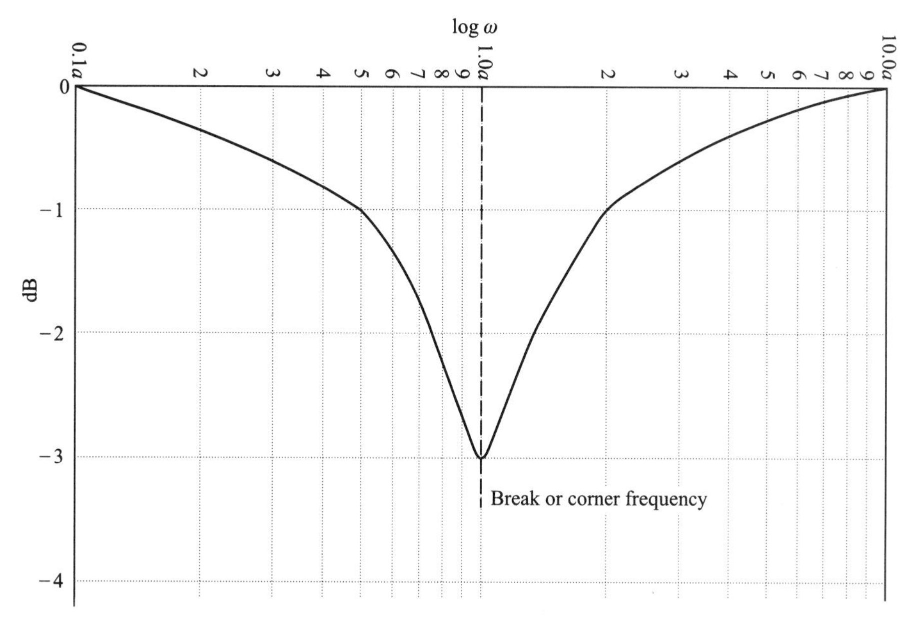

Figure 6.25 Errors in the asymptotic plot for the simple phase-lag network $a/(j\omega + a)$.

E. Bode Diagram of a Simple Phase-Lead Network

The Bode diagram of a simple phase-lead network

$$G(j\omega) = j\frac{\omega}{a} + 1 \tag{6.89}$$

can be obtained in a manner similar to that used for the simple phase-lag network. The major differences are that the amplitude characteristic has a positive slope, and the phase characteristic has a positive (phase-lead) value. Figure 6.26 illustrates the Bode diagram of a simple phase-lead network.

F. Bode Diagram of a Quadratic Phase-Lag Network

Let us consider the second-order quadratic, phase-lag transfer function given by

$$G(j\omega) = \frac{\omega_n^2}{(j\omega)^2 + 2\zeta\omega_n(j\omega) + \omega_n^2}, \tag{6.90}$$

or

$$G(j\omega) = \left[\left(j\frac{\omega}{\omega_n}\right)^2 + 2\zeta\left(j\frac{\omega}{\omega_n}\right) + 1\right]^{-1}. \tag{6.91}$$

The Bode diagram for this transfer function can be obtained as follows:

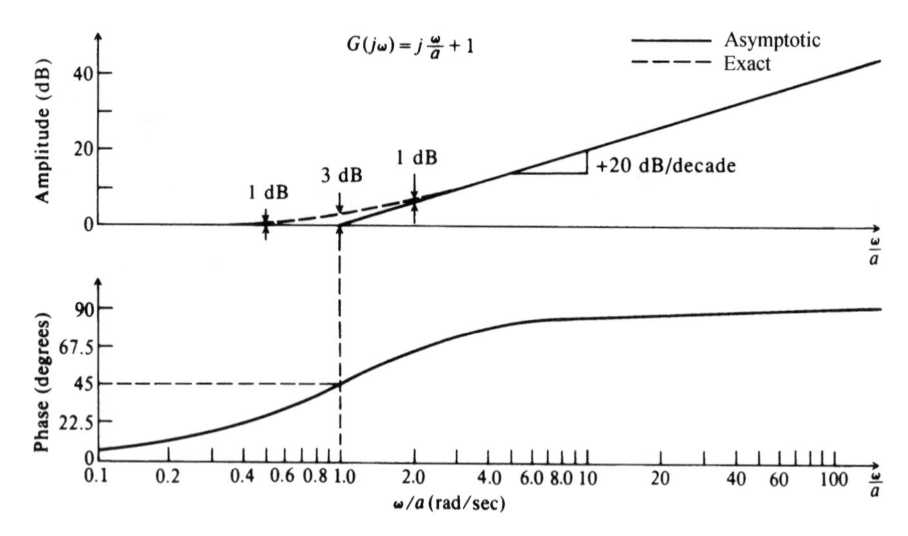

Figure 6.26 Bode diagram of a simple phase-lead network.

$$20\log_{10}\left[\left(j\frac{\omega}{\omega_n}\right)^2 + 2\zeta\left(j\frac{\omega}{\omega_n}\right) + 1\right]^{-1}$$

$$= -20\log_{10}\left[\left(\frac{2\zeta\omega}{\omega_n}\right)^2 + \left(1 - \frac{\omega^2}{\omega_n^2}\right)^2\right]^{1/2} \angle - \tan^{-1}\frac{2\zeta\omega_n\omega}{\omega_n^2 - \omega^2}. \qquad (6.92)$$

Figures 6.27 and 6.28 illustrate the Bode diagram of the quadratic phase-lag network for various values of damping ratio, ζ. In Figure 6.27, the dashed portion of the

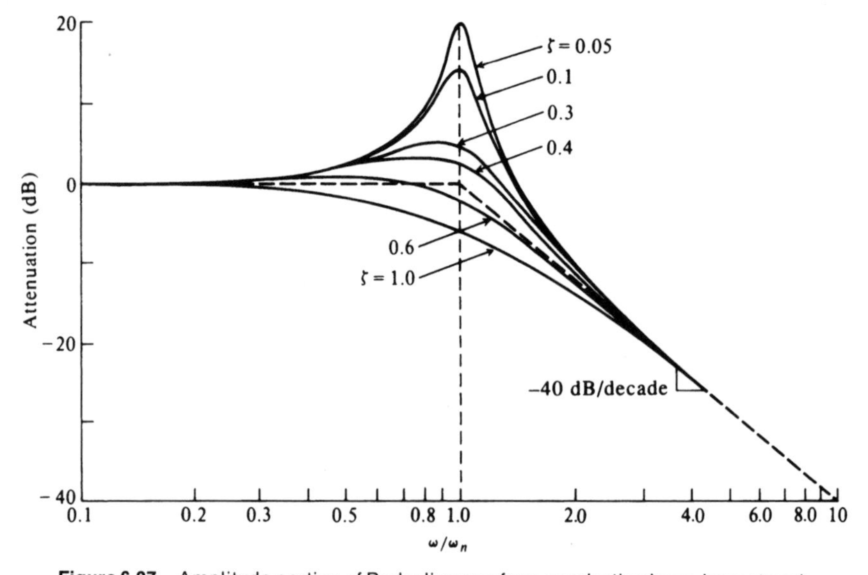

Figure 6.27 Amplitude portion of Bode diagram for a quadratic phase-lag network.

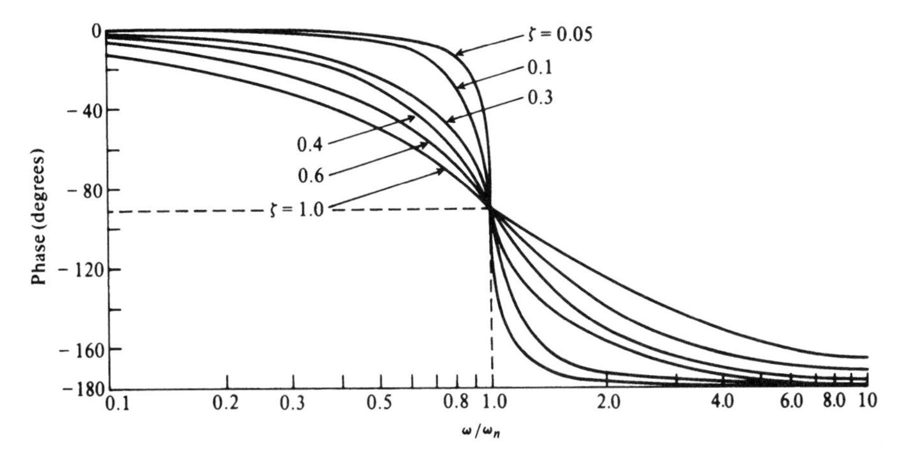

Figure 6.28 Phase portion of Bode diagram for a quadratic phase-lag network.

amplitude characteristic represents the straight-line asymptotic approximation and the heavy-line portions represent the exact plot. For $\zeta = 1$, the two curves differ by a maximum of 6.02 dB at ω/ω_n equal to unity.

The asymptotic approximations can be drawn quite simply. For example, when ω/ω_n is much less than unity, the imaginary component is very much smaller than the real component, and the imaginary component can be neglected. Therefore,

$$20 \log_{10} \sqrt{1} = 0 \text{ dB/decade} \quad \text{for } \frac{\omega}{\omega_n} \ll 1. \tag{6.93}$$

When ω/ω_n is much greater than unity, the dominant term is

$$\begin{aligned} 20 \log_{10} \left| \left[\left(j \frac{\omega}{\omega_n} \right)^2 \right]^{-1} \right| &= -20 \log_{10} \left(\frac{\omega^2}{\omega_n^2} \right) \\ &= -40 \log_{10} \frac{\omega}{\omega_n} \text{ dB} \quad \text{for } \frac{\omega}{\omega_n} \gg 1. \end{aligned} \tag{6.94}$$

This indicates that the slope of the straight-line asymptotic approximation for $\omega/\omega_n \gg 1$ is twice that of a phase-lag network. The difference between the approximate curve and the exact curve depends on the damping factor ζ. Similar analysis indicates the phase shift possible is twice that of the simple phase-lag network (e.g., $-180°$). The phase shift characteristics are shown in Figure 6.28.

G. Bode Diagram of a Time-Delay or Transportation Lag Factor

The time-delay factor [10] occurs in systems which are characterized by the movement of mass that requires a finite time to pass from one point to another. The transfer function of a pure time-delay or transportation lag factor, without attenuation, is given by

$$G(j\omega) = e^{-j\omega T}. \tag{6.95}$$

The delay factor $e^{-j\omega T}$ results in a phase shift

$$\phi(\omega) = -\omega T \text{ rad}, \tag{6.96}$$

which has to be added to the phase shift resulting from the rest of the system. Observe that the magnitude of the time-delay or transportation lag factor is always unity and, therefore, does not affect the magnitude characteristics. Figure 6.29 illustrates the Bode diagram of the pure time-delay factor on a logarithmic scale.

Time-delay or transportation lag factors are very common in hydraulic, pneumatic, thermal, chemical, and manual (human in the loop) control systems. They have nonminimum-phase characteristics with very large phase lags and no attenuation at high frequencies.

H. Bode Diagram of a Composite Transfer Function—Example 1

Let us next apply the basic concepts developed in this section for the transfer function of a simple system. Consider a unity-feedback system whose open-loop transfer function is given by

$$G(s) = \frac{10}{s(1 + 0.1s)}. \tag{6.97}$$

This system contains a gain of 10, a pure integration, and a pole term whose break frequency is at 10 rad/sec. The step-by-step procedure for drawing this Bode diagram is as follows. We will describe graphical techniques for drawing the Bode diagram in this section. Working computer programs are provided in Section 6.9. Commercially available computer programs for obtaining the Bode diagram are presented in Section 6.8. The MATLAB program, which is described in Section 6.8, was used for drawing the actual Bode diagrams for Examples 1 (Figure 6.30), 2 (Figure 6.31),

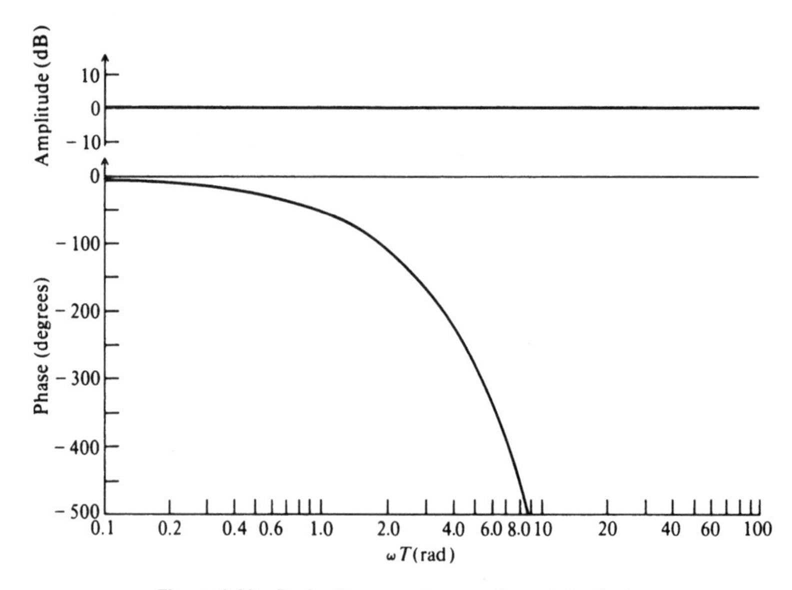

Figure 6.29 Bode diagram of a pure time-delay factor.

3 (Figure 6.33), and 4 (Figure 6.34) of this section. These Bode diagrams are contained in the M-files (that are part of my *Modern Control System Theory and Design Toolbox*) that can be retrieved free from The MathWorks, Inc. anonymous FTP server at ftp://ftp.mathworks.com/pub/books/shinners.

Step 1. We will plot the amplitude characteristics first. A frequency is picked at least 1 decade below the lowest break frequency and the gain and slope are fixed at that frequency. In this example, that frequency is 1 rad/sec, where the pole term is approximated as one [see Eq. (6.86)]. Therefore, the composite gain is 10 (or 20 dB) and the slope is −20 dB/decade due to the pure integration.

Step 2. The slope of −20 dB/decade is continued until the first break at 10 rad/sec occurs, where we add another −20 dB/decade and make the slope −40 dB/

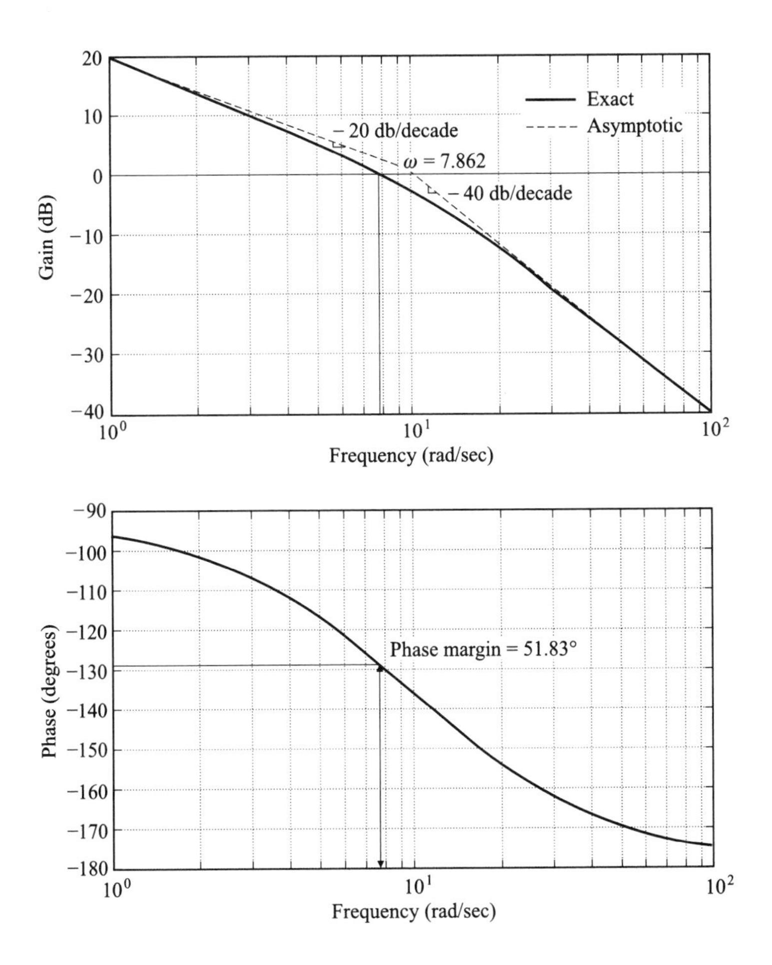

Figure 6.30 Bode diagram for system whose open-loop transfer function is

$$G(s)H(s) = \frac{10}{s(0.1s + 1)}.$$

decade, which continues to infinity. The result is shown in Figure 6.30. Notice that if we use the straight-line asymptotic approximation (dashed line in Figure 6.30), a gain crossover frequency (defined as the frequency where the $G(s)H(s)$ curve crosses the 0-dB line) of 10 rad/sec occurs where a break frequency also occurs. Therefore, we must correct the straight-line asymptotic approximation by the 3-dB error at the break frequency as shown in Figure 6.30 (solid line). Therefore, the actual gain crossover frequency is at 7.862 rad/sec. Whenever a break frequency occurs at the gain crossover frequency, or close to it, we must use the corrected curve to determine the actual gain crossover frequency. Otherwise, large errors will occur in determining the phase margin from the Bode diagram (see Step 4).

Step 3. The phase characteristic is obtained by adding the phase of each component of the composite transfer function. The gain factor contributes zero degrees of phase shift (see Figure 6.20). The pure integration provides a constant phase shift of -90° at all frequencies (see Figure 6.22). The pure integration provides a constant phase shift of -90° at all frequencies (see Figure 6.22). The pole term provides a phase contribution of $-\tan^{-1} 0.1\omega$ (see Eq. (6.88). Therefore, in this example, the composite phase is computed from

$$\phi(\omega) = -90^\circ - \tan^{-1} 0.1\omega \tag{6.98}$$

Step 4. The phase and gain margins, defined in Eqs. (6.67) and (6.69) for the Nyquist diagram, can also be determined from the Bode diagram in this step. Let us determine next how this can be accomplished.

I. Relationship Between Bode and Nyquist Diagrams

At this point in the development of the Bode diagram, it is appropriate to relate the Nyquist stability criterion to the Bode amplitude and phase diagrams. When we discussed the Nyquist diagram, the degree of stability was defined in terms of the phase and gain margins present (see Figure 6.14). This can easily be related to the Bode diagrams by making use of the following two facts.

1. The unit circle of the Nyquist diagram transforms into the unity or 0-dB line of the amplitude plot for all frequencies.
2. The negative real axis of the Nyquist diagram transforms into a negative 180° phase line for all frequencies.

Therefore, the phase margin can be determined on the Bode diagram by determining the phase shift present when $G(j\omega)H(j\omega)$ crosses the 0-dB line. The gain margin can be obtained on the Bode diagram by determining the gain present when $G(j\omega)H(j\omega)$ crosses the -180° line.

Returning to Step 4 of Example 1, the Bode diagram which is illustrated in Figure 6.30, the gain crossover frequency is at 7.862 rad/sec where the phase shift is 128.17° (this corresponds to the θ of Figure 6.14), and the resulting phase margin is 51.83° (this corresponds to the γ of Figure 6.14). Because this example never reaches -180° (except at $\omega = \infty$), its gain margin is infinity. Therefore, this system is stable.

The values of gain crossover frequency, phase, and gain margins determined in this section and the rest of the book where the Bode diagram is applied, use the function "margins," which is part of my *Modern Control System Theory and Design Toolbox*. It calculates their values based on their definitions, analytically and very accurately, rather than using graphical techniques. The disk that contains my toolbox can be retrieved free fromThe MathWorks, Inc. anonymous FJP server at ftp://ftp.mathworks.com/pub/books/shinners.

J. Bode Diagram of a Composite Transfer Function—Example 2

For the second example, let us consider a unity-feedback control system whose forward transfer function is given by

$$G(s) = \frac{10}{s(1+0.1s)^2}. \tag{6.99}$$

This system contains a gain, a pure integration, and a double pole term whose break frequency is at 10. The Bode diagram of this system, shown in Figure 6.31, was also obtained using MATLAB and is contained in the M-file that is part of my *Modern Control System Theory and Design Toolbox*.

The gain and slope at $\omega = 1$ rad/sec (two decades below the lowest and only break frequency) is fixed, where the gain is 40 dB and the slope is -20dB/decade due to the pure integration. This slope is continued using the straight-line approximation (dashed line in Figure 6.31) to the double break at 10 rad/sec. There, we add -40 dB/decade to the slope due to the double pole, and the straight-line approximation continues at a slope of -60 dB/decade thereafter. Because there is a break at the gain crossover frequency (a double one in this example), we must again correct for this to find the actual gain crossover frequency. The solid-line curve in the amplitude plot of Figure 6.31 indicates that the actual crossover frequency is at 6.825 rad/sec (not 10 rad/sec). The composite phase is computed from

$$\phi(\omega) = -90^\circ - 2\tan^{-1} 0.1\omega. \tag{6.100}$$

The phase shift at 6.825 rad/sec is 158.61°. Therefore, the phase margin is 21.39°. At a frequency of 10.02 rad/sec, the phase shift is -180°. This frequency is defined as the phase-crossover frequency. Therefore, the gain margin for this system is 6.039 dB as indicated in Figure 6.31. Therefore this system is stable.

K. The Phase-Shift Scale

For, quick approximate phase-shift calculations, the phase characteristics can easily be determined using superposition of the individual characteristics of each component using the phase-shift scale shown in Figure 6.32. This scale, which can easily be derived from the tangent relationship, illustrates the effect on the resultant phase shift occurring at frequency ω for a first-order factor, due to a break frequency occurring at ω_0. For example, at $\omega/\omega_0 = 5$, the phase shift contributed by the break occurring at $\omega_0 = 1$ is 78.7°, whereas at $\omega/\omega_0 = 0.2$ the phase shift contributed by this break is only 11.3°.

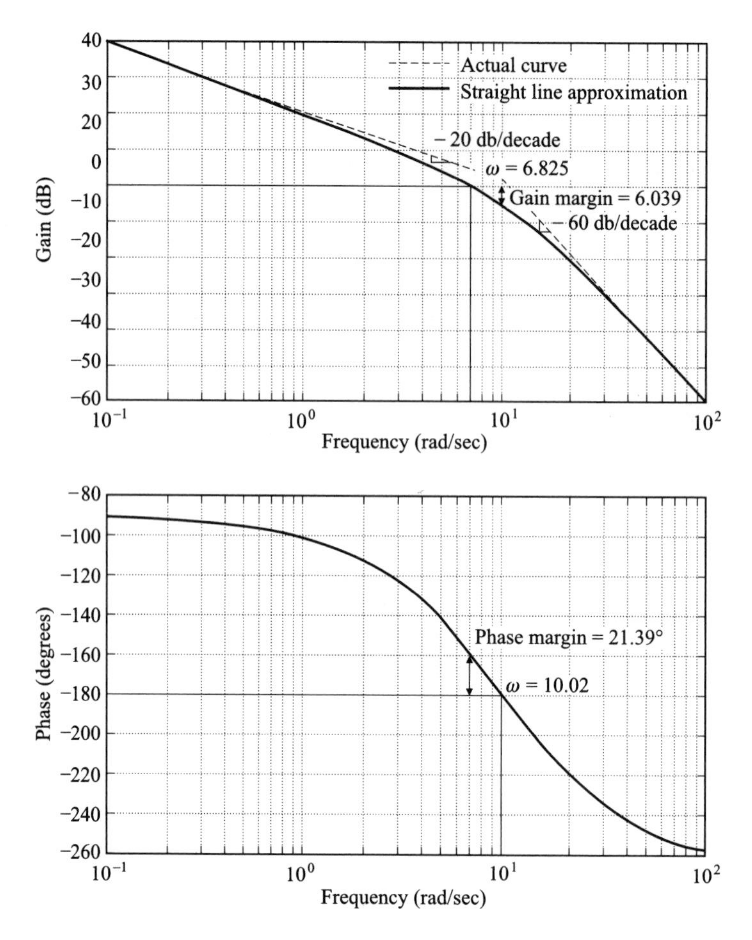

Figure 6.31 Bode diagram for system whose open-loop transfer function is

$$G(s)H(s) = \frac{10}{s(1+0.1s)^2}.$$

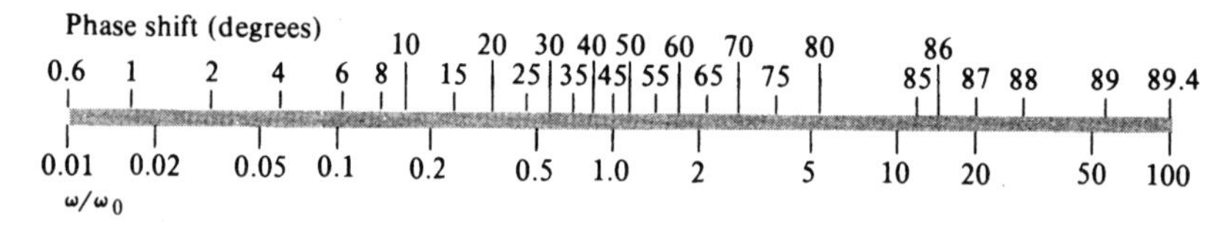

Figure 6.32 Phase-shift scale illustrating the effect on the resultant phase shift occurring at ω due to a break at ω_0.

H. Bode Diagram of a Composite Transfer Function—Example 3

Let us next apply the basic notions developed in this section for the transfer function of a unity-feedback system whose forward transfer function is given by

$$G(j\omega) = \frac{775}{j\omega}\frac{0.1j\omega + 1}{0.5j\omega + 1}\frac{0.2j\omega + 1}{j\omega + 1} \times [0.655 \times 10^{-4}(j\omega)^2 + 6.55 \times 10^{-3}j\omega + 1]^{-1}. \tag{6.101}$$

This transfer function contains a pure integration, two simple phase lags, two simple phase leads, and a quadratic phase-lag network. Each of these characteristics has previously been considered individually. The simplest procedure for plotting the amplitude characteristic for the entire transfer function is to start by locating one point, $\omega = 0.1$, which is one decade below the lowest break ($\omega = 1$). At the same time, the slope at $\omega = 0.1$ is determined. For the problem at hand, at $\omega = 0.1$,

$$[G(j\omega)]_{\omega=0.1} = \left(\frac{775}{j0.1}\right)\left(\frac{1}{1}\right)\left(\frac{1}{1}\right)\left(\frac{1}{1}\right). \tag{6.102}$$

Equation (6.102) indicates that the gain of the straight-line asymptotic curve is 7750 (77.8 dB) and the slope is -20 dB/decade at $\omega = 0.1$.

The next step is to locate all the break frequencies (frequencies at which the slopes change because a real or imaginary component starts or stops dominating). The break frequencies for this transfer function are at $\omega = 1$ (-20 dB/decade to -40 dB/decade); $\omega = 2$ (-40 dB/decade to -60 dB/decade); $\omega = 5$ (-60 dB/decade to -40 dB/decade); $\omega = 10$ (-40 dB/decade to -20 dB/decade); $\omega = 123.8$ (-20 dB/decade to -60 dB/decade). The corresponding phase characteristics can most easily be determined using superposition of the individual phase characteristics of each component as was illustrated in the previous two examples.

Figure 6.33 illustrates the composite Bode diagram for the transfer function given by Eq. (6.101). Notice that the peaking due to the quadratic phase-lag component occurs at $\omega_n = 123.8$, and corresponds to $\zeta = 0.405$. The Bode diagram was drawn using MATLAB (see Section 6.8) and is contained in the M-File that is part of my *Modern Control System Theory and Design Toolbox* and can be retrieved free from The MathWorks, Inc. anonymous FJP server at ftp://ftp.mathworks.com/pub/books/shinners. The straight-line asymptotic approximations (dashed line) were added for illustrative purposes. The results of this Bode diagram show a gain crossover frequency of 34.33 rad/sec, a phase margin of 56.74°, and a gain margin of 9.622 dB (the phase crossover frequency where the phase is $-180°$, occurs at 118.7 rad/sec).

M. Bode Diagram of a Composite Transfer Equation Containing a Time-Delay Factor—Example 4

As a fourth example of a Bode diagram, let us consider a unity-feedback system whose forward transfer function is given by

$$G(j\omega) = \frac{20(1 + 0.2j\omega)}{j\omega(1 + 0.5j\omega)}e^{-0.1j\omega}. \tag{6.103}$$

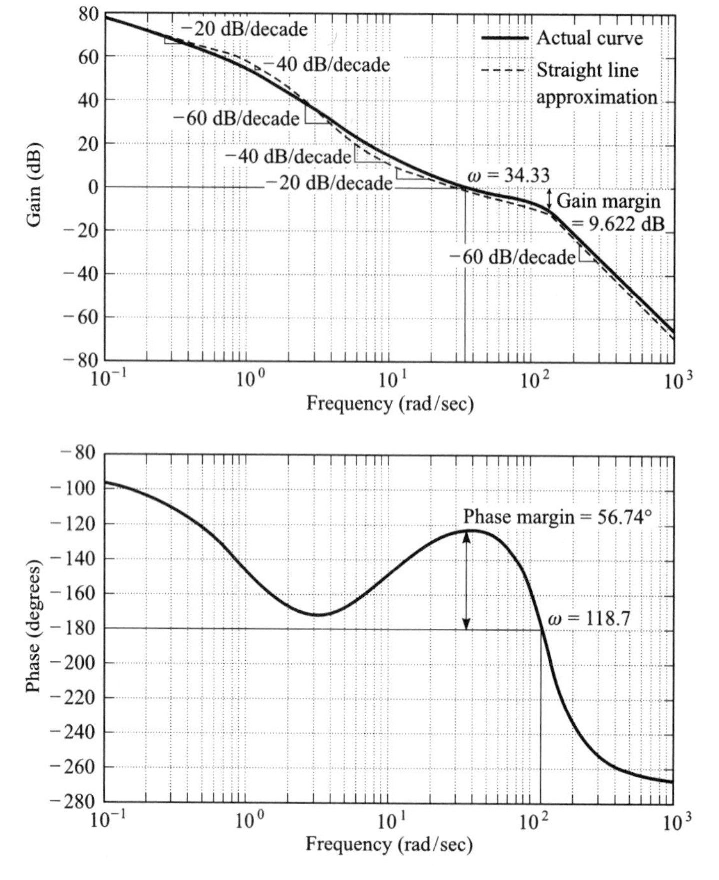

Figure 6.33 Composite Bode diagram for a unity-feedback system whose $G(j\omega)$ is given by Eq. (6.101).

This transfer function contains a pure integration, one pole, one zero, and a time delay factor. This represents the transfer function in a steel mill, and the time-delay factor is caused by the finite time it takes the steel to move from one point to another. In this system, a motor adjusts the separation l of two rolls so that the thickness error of the steel is minimized. If the steel is traveling at a velocity v of 10 ft/sec, and the nominal separation d is 1 ft, then the time delay between the roll thickness measurement and thickness adjustment is given by

$$T = \frac{d}{v} = \frac{1 \text{ ft}}{10 \text{ ft/sec}} = 0.1 \text{ sec.}$$

The overall transfer function $G(j\omega)$, given by Eq. (6.103), relates the transfer function of the system error, defined as system error = desired thickness − actual thickness, and the output which represents the actual thickness.

Each of the other characteristics in Eq. (6.103) has been previously considered separately. Again, our procedure for plotting the amplitude characteristics is to start by finding the gain and slope at a frequency below the lowest break ($\omega = 2$).

Therefore, we will perform the calculation at $\omega = 0.1$. For this problem, the amplitude of $G(j\omega)$ at $\omega = 0.1$ is

$$[G(j\omega)]_{\omega = 0.1} = \left(\frac{20}{j0.1}\right)\frac{(1)}{(1)}1. \tag{6.104}$$

Equation (6.104) indicates that the gain is 200 (46 dB) and the slope is −20 dB/decade at $\omega = 0.1$. The break frequencies occur at $\omega = 2$, where the slope changes from −20 dB/decade to −40 dB/decade, and $\omega = 5$ when the slope changes from −40 dB/decade to −20 dB/decade. The corresponding phase characteristic can be readily determined using superposition of the individual phase characteristics of each component. Figure 6.34 illustrates the composite Bode diagram for the transfer function given by Eq. (6.103). For interest, the phase characteristic is illustrated with and without the time-delay factor. Observe from this curve that the gain crossover frequency is 8.967 rad/sec and the phase margin of the system without the time-delay factor is 73.42°. For this case, the gain margin is infinity because the phase never reaches −180°. However, with the time-delay factor in the system, the phase

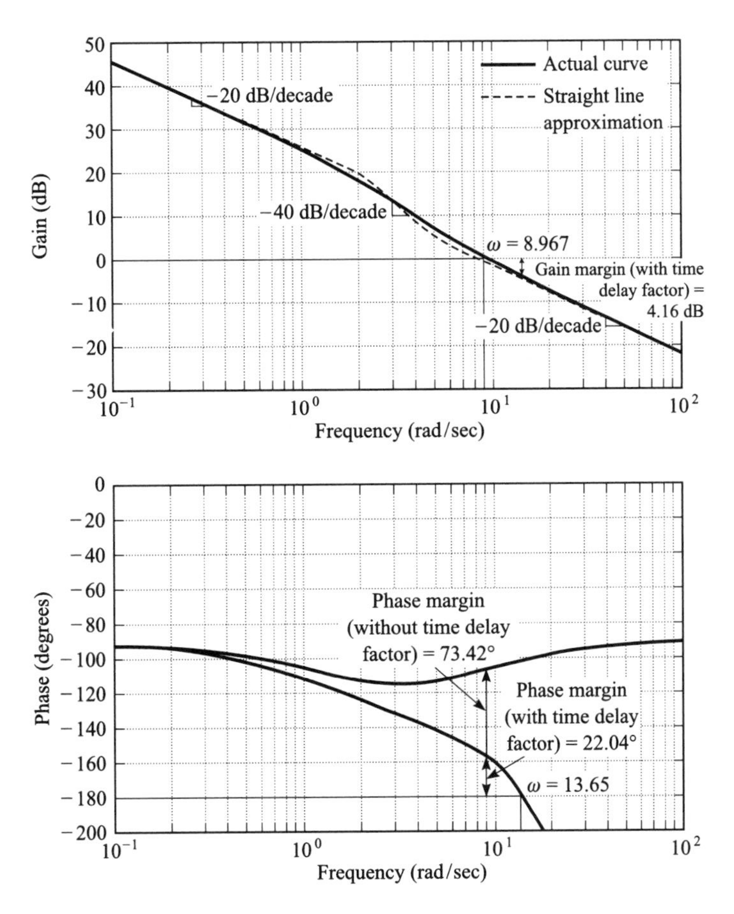

Figure 6.34 Composite Bode diagram for a unity-feedback system where $G(j\omega)$ is given by Eq. (6.103).

margin drops to 22.04°, the gain margin becomes 4.16 dB (the phase crossover frequency equals 13.65 rad/sec), and the system has a smaller margin of stability than before. The control-system engineer should always give particular attention to time-delay factors and avoid them if possible, because they produce a phase lag and decrease the phase margin. The Bode diagram was drawn using MATLAB (see Section 6.8). The straight-line asymptotic approximation (dashed line) was added for illustrative purposes.

The Bode-diagram method has the very practical virtue that the amplitude and phase frequency response can easily be measured in the laboratory. The relative ease of synthesizing feedback control systems with the Bode-diagram approach is further demonstrated in Chapters 7, 8, 9, and 12.

6.8. BODE DIAGRAMS USING MATLAB [7]

In the previous section, several Bode diagrams were illustrated whose amplitude plots were obtained from hand-drawn straight-line asymptotic slopes and whose phase characteristics were calculated from the appropriate trigonometric fucntions. Additionally, several Bode diagrams were also illustrated which were obtained using MATLAB. In this section, the reader will be shown how to obtain the Bode diagrams very easily and accurately using MATLAB.

Let us practice creating Bode diagrams using MATLAB. There are many examples to practice with for creating Bode diagrams on the *Modern Control System Theory and Design* (MCSTD) Toolbox. Two functions exist that assist in Bode diagrams:

1. "bode" returns/plots the Bode response of a system.
2. "margins" is described in the MCSTD toolbox. This, in my opinion, has an advantage over the professional version of the Control System toolbox's "margin" routine. Margins analytically calculates, with analytic precision, the gain and phase margins and their associated frequencies (versus interpolating a single value from a plot).

When trying to find the proper syntax to call the "bode" utility, either use the help feature or look in the reference manual. I personally prefer the help feature, unless I need an example.

Valid syntax for the "bode" utility, for transfer functions, is:

1. [mag,phase,w] = bode(num,den)
2. [mag,phase,w] = bode(num,den,w)
3. [mag,phase] = bode(num,den,w)
4. bode(num,den,w)
5. bode(num,den)

The left-hand arguments (mag, phase and w) are optional for this function, as described on the second page of the Bode function in *The Student Edition* book. The result, which can be manually typed, is listed below so that you may try the Bode example in the book:

$$a = [0, 1; -1, -0.4];$$

$$b = [0; 1];$$

$$c = [1, 0];$$

$$d = 0;$$

A major short-coming of the Bode diagram is that the margins (gain and phase) are not put onto the plot when it generates the plots. The effect of this is compounded when you want to put them onto the plot, and you discover that you can only modify the phase plot with reasonable ease. Adding to the magnitude plot is almost impossible (prior to MATLAB version 4.0). This is why they have the left-hand arguments, so that you can generate the Bode plots yourself with whatever customization on the plot that you desire. This is what is accomplished in the MCSTD Toolbox when Bode diagrams are obtained in the DEMO, figures, or problems directory.

A. Drawing the Bode Diagram if the System is Defined by a Transfer Function

To illustrate the use of MATLAB for obtaining the Bode diagram, let us consider the following example. The open-loop transfer function of a control system is given by the following:

$$G(s)H(s) = \frac{(s+4)(s+40)}{s^3(s+200)(s+900)}. \tag{6.105}$$

In order to ease its transformation to MATLAB notation, we multiply all terms in the numerator and denominator as follows:

$$G(s)H(s) = \frac{s^2 + 44s + 160}{s^5 + 1100s^4 + 180,000s^3} \tag{6.106}$$

Therefore, the row matrices for the numerator and denominator are as follows:

$$\text{num} = [0 \quad 0 \quad 0 \quad 1 \quad 44 \quad 160]$$

$$\text{den} = [1 \quad 1100 \quad 180,000 \quad 0 \quad 0 \quad 0].$$

The resulting MATLAB program will first be provided by the listing in Table 6.8, and new commands will then be explained. The resulting Bode diagram is shown in Figure 6.35.

MATLAB and the MCSTD Toolbox automatically select the frequency range used in Figure 6.35 when using the MATLAB program in Table 6.8. If the control-system engineer wants to select a different frequency range, such as from 0.01 to 10,000 rad/sec instead of from 0.1 to 10,000 rad/sec, then we have to use the "log-space" command which is defined as follows:

Table 6.8. MATLAB Program for Obtaining Bode Diagram of System Defined in Eq. (6.106)

```
num = [0  0  0  1  44  160];
den = [1  1100  180000  0  0  0];
w = logspace(−2, 4);
[mag,ph] = bode(num,den);
grid
title('Bode Diagram of
G(s)H(s) = (s + 4)(s + 40)/s^3(s + 200)(s + 900)')
```

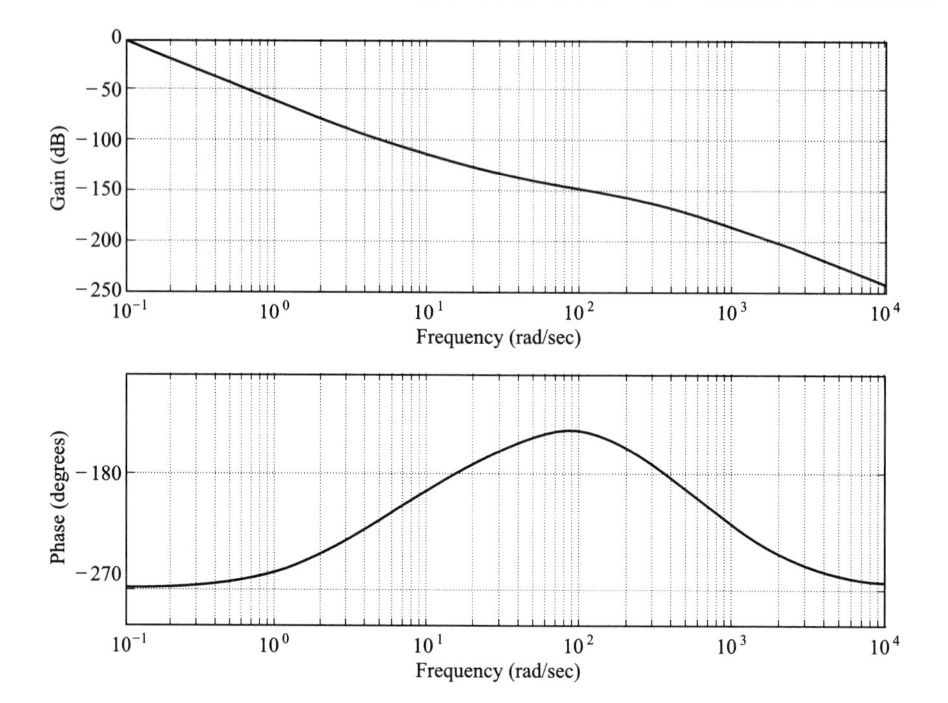

Figure 6.35 Bode diagram for control system whose transfer function is defined by Eq. (6.105).

w = logspace(−2, 4): Generates 50 points equally spaced between $\omega = 10^{-2}$ and 10^4 rad/sec

In addition, we now have to use the MATLAB command "bode(num,den,w)" which is defined as follows:

bode(num,den,w): Generates the Bode diagram from the user-supplied num, den, and the frequency vector **w** which specifies the frequencies at which the Bode diagram will be calculated.

Using the MATLAB commands "logspace" and "bode(num,den,w)", in the MATLAB Program in Table 6.8, is modified as shown in the MATLAB Program in Table 6.9.

Table 6.9. MATLAB Program for Obtaining Bode Diagram of System Defined in Eq. (6.106) with the Logspace Command

```
num = [0  0  0  1  44  160];
den = [1  1100  180000  0  0  0];
w = logspace(-2,4);
[mag,ph] = bode(num,den,w);
grid
title ('Bode Diagram of
G(s)H(s) = (s + 4)(s + 40)/s^3(s + 200)(s + 900)')
```

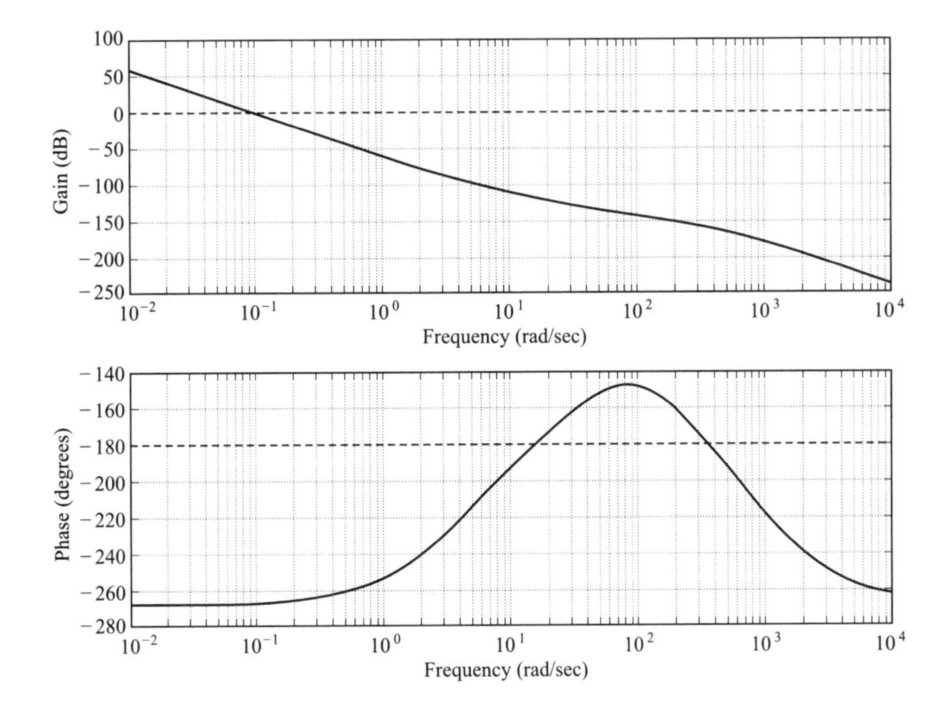

Figure 6.36 Bode diagram of $G(s)H(s) = \dfrac{(s+4)(s+40)}{s^3(s+200)(s+900)}$ with the logspace command added.

The resulting Bode diagram is shown in Figure 6.36. Observe that the frequency range of this Bode diagram is 0.01 to 10,000 rad/sec, compared to 0.1 to 10,000 rad/sec in Figure 6.35.

The MCSTD Toolbox command *margins* provide the following:

‡ gain margin (gm)

‡ phase margin (pm)

‡ frequency (rad/sec) where the phase equals -180° (wcg: ω_{cg}, gain crossover frequency)

‡ frequency (rad/sec) where the gain equals zero dB (wcp: ω_{cp}, phase crossover frequency).

Therefore, adding the following MCSTD Toolbox command to the MATLAB Program in Table 6.9,

$$[\text{gm, pm, wcg, wcp}] = \text{margins(num,den)}$$

will result in the Bode diagram of Figure 6.37 which contains everything previously shown on Figure 6.36 plus the phase margin, the gain margin, the frequency where the gain equals zero dB, and the frequencies where the phase is $-180°$. Observe from Figure 6.37 that this control system has two gain margins and one phase margin.

Therefore, the MCSTD Toolbox enhances MATLAB. The MCSTD Toolbox DEMO M-file elaborates further on the creation of the Bode diagram with several on-screen examples.

B. Drawing the Bode Diagram if the System is Defined in State-Space Form

The MATLAB command for this case is given by

$$\text{bode(A,B,C,D)}.$$

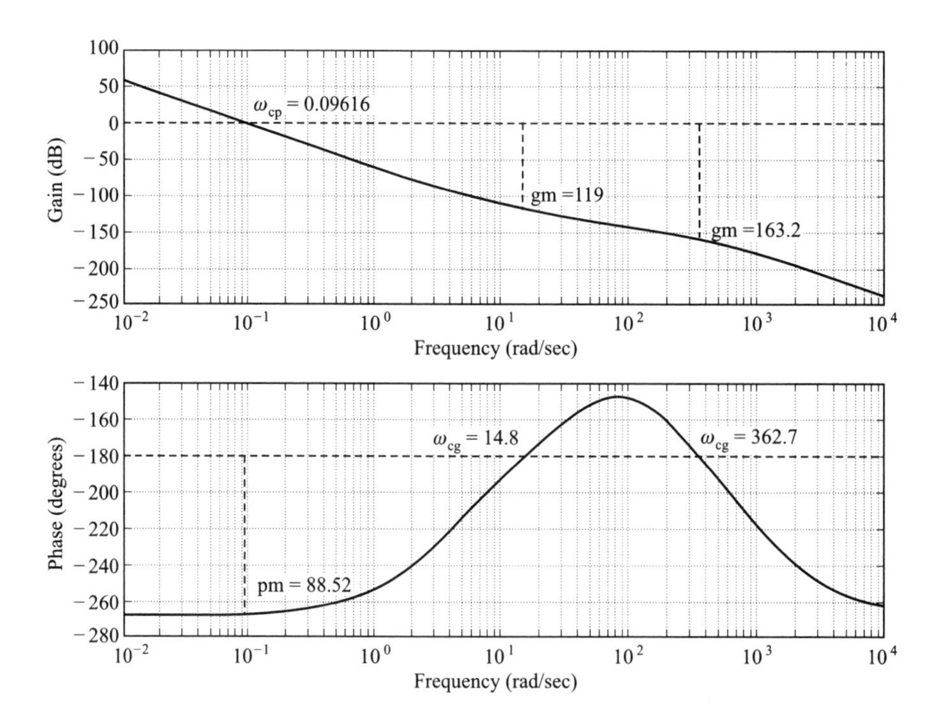

Figure 6.37 Bode diagram of $G(s)H(s) = \dfrac{(s+4)(s+40)}{s^3(s+200)(s+900)}$ with the logspace command added.

As was demonstrated previously in this chapter (e.g., Section 6.6 on the Nyquist Diagram), we must first obtain the state-space form and then use this new command.

To demonstrate this procedure, let us obtain the state-space form from the transfer function given by Eq. (6.106) using the MATLAB command

$$[A,B,C,D] = \text{tf2ss}(\text{num},\text{den})$$

whose application has been demonstrated earlier in the section on the Nyquist diagram (Section 6.6). The resulting MATLAB program for accomplishing this is shown in the MATLAB program in Table 6.10.

Therefore, the resulting MATLAB program for obtaining the Bode diagram of this problem from the state-space formulation is given by the MATLAB program in Table 6.11. The resulting Bode diagram is identical to that shown in Figure 6.36.

Table 6.10. MATLAB Program for Converting from the Transfer Function to the State-Space Form

```
num = [0  0  1  44  160]
den = [1  1100  180000  0  0]
[A, B, C, D] = tf2ss(num,den)
A =
   -1100   - 180000  0  0  0
       1          0  0  0  0
       0          1  0  0  0
       0          0  1  0  0
       0          0  0  1  0
B =
       1
       0
       0
       0

C =
       0  0  1  44  160
D =
       0
```

Table 6.11. MATLAB Program for Determining the Bode Diagram from the State-Space Form.

```
A = [-1100   - 180000  0  0  0; 1  0  0  0  0; 0  1  0  0  0;
    0  0  1  0  0; 0  0  0  1  0]
B = [1; 0; 0; 0; 0]
C = [0  0  1  44  160];
D = [0];
w = logspace(-2,4);
[mag, ph] = bode(A,B,C,D,w)
grid
title('Bode Diagram of
G(s)H(s) = (s + 4)(s + 40)/s^3(s + 200)(s + 900)')
```

6.9. DIGITAL COMPUTER PROGRAMS FOR OBTAINING THE OPEN-LOOP AND CLOSED-LOOP FREQUENCY RESPONSES AND THE TIME-DOMAIN RESPONSE

For those who do not have access to MATLAB, which has now become an industry standard (more or less), this book also provides other digital computer programs for obtaining the Bode diagram. This approach for providing both MATLAB and other digital computer programs which are not dependent on COTS (commercial-off-the-shelf software) is used throughout this book.

The digital computer is a very valuable tool for computing the gain and phase characteristics as a function of frequency [11, 12]. Several program languages can be used to perform this computation. Perhaps the simplest languages are Basic (Beginner's All-purpose Symbolic Instruction Code) [13, 14] and Fortran (FORmula TRANslator) [13,14]. Several problems are solved in this chapter and in the rest of this book utilizing these languages with working programs provided. Commercially available computer programs for solving control-system problems are presented in Section 6.21. One of these, MATLAB, has been used for the four illustrative problems shown in Section 6.7, and in Section 6.8 which is dedicated to MATLAB.

Consider a unity-feedback control system where

$$G(s) = \frac{99(1 + 0.1s)}{(1 + 0.01s)^2(1 + 0.2s)(1 + 1.5s)}. \tag{6.107}$$

It is desired to determine the phase and gain margins of this linear control system, and a Basic program will be utilized in this problem [13, 14]. The coding symbols used are as follows:

$$\begin{aligned} W &= \omega, \\ \mathrm{G2} &= |G(s)|^2, \\ \mathrm{P} &= \text{phase of } G(s), \\ \mathrm{PM} &= \text{phase margin}, \\ \mathrm{GOSUB} &\rightarrow \text{Compute G2, P}. \end{aligned}$$

Figure 6.38 illustrates the logic flow diagram for developing the program that determines the phase and gain margins. Table 6.12 illustrates the actual program. Figure 6.38 and Table 6.12 should be compared in order to obtain a thorough understanding of the method. Table 6.13 illustrates the computer's output for phase and gain margin. It indicates that unity gain occurs at $\omega = 31.2$ rad/sec where the phase shift is 132.1°. Therefore, the phase margin is $180^\circ - 132.1^\circ = 47.9^\circ$. It also indicates that at $\omega = 95.7$ rad/sec, the phase shift is 180°, and the gain margin is 149 dB. To utilize the conventional straight-line asymptotic Bode-diagram technique as a check, Figure 6.39 is drawn. It indicates a phase margin of 48° at a crossover of 31 rad/sec, which is in good agreement. The frequency where 180° phase shift occurs is at 96 rad/sec (compared with 95.7 in the computer run). The gain margin using the straight-line asymptotic curve is only 9 dB and is quite far from the computer run (14.9 dB)

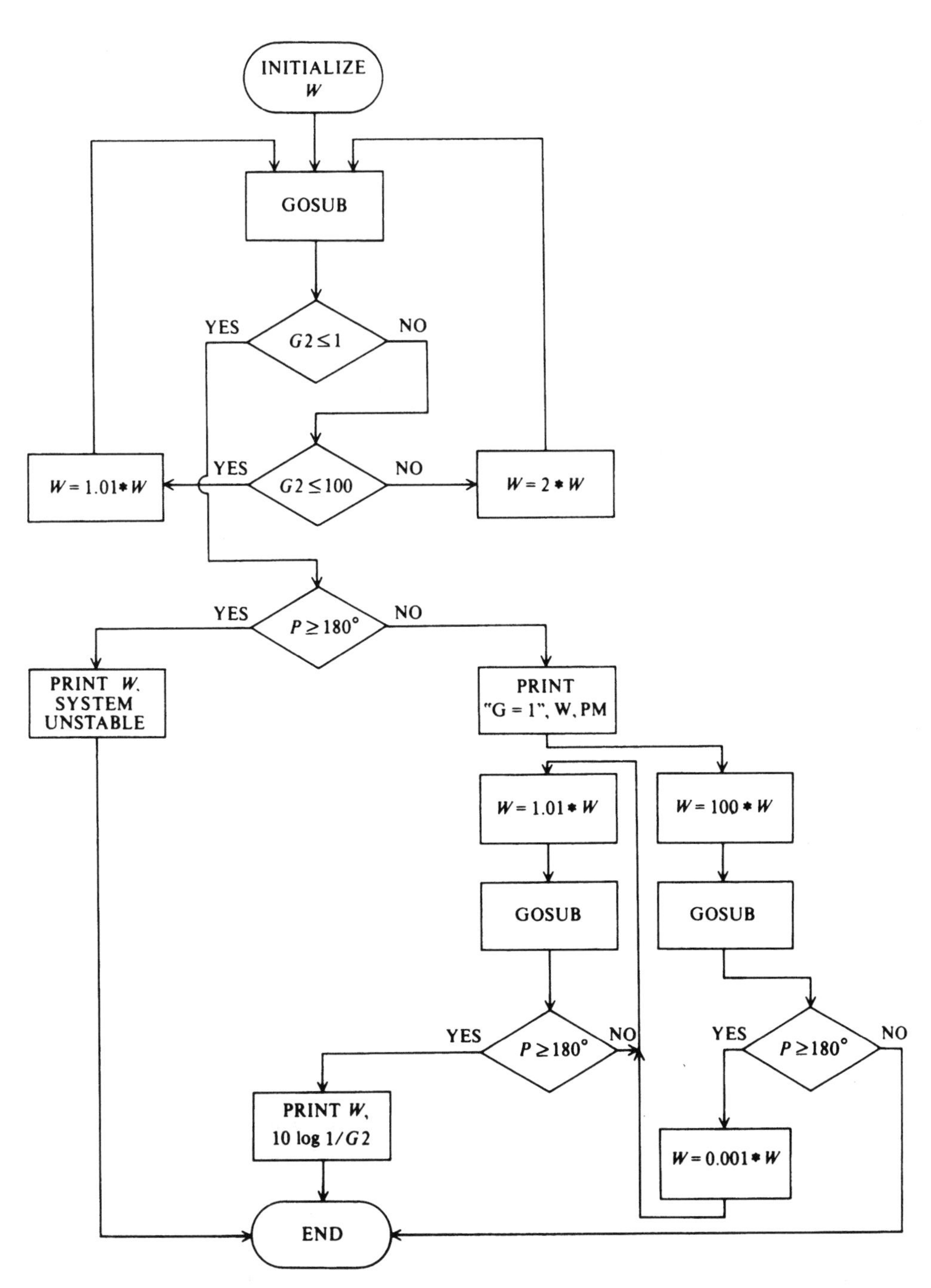

Figure 6.38 Logic flow diagram for Bode-diagram analysis.

Table 6.12. Computer Program for Bode-Diagram Analysis (Basic program) to Determine Phase and Gain Margins

```
LIST

1 REM GEN PHASE AND GAIN MARGIN COMPUTATION FOR GENERAL
RESPONSE
10 LET W=.01
20 GOSUB 210
30 IF G2<=1 THEN 90
40 IF G2<=100 THEN 70
50 LET W=2*W
60 GOTO 20
70 LET W=1.01*W
80 GOTO 20
90 IF P>=180 THEN 250
100 PRINT "UNITY GAIN," "W=" W, "P=" P
110 LET W=100*W
120 GOSUB 210
130 IF P<180 THEN 260
140 LET W=.001*W
150 LET W=1.01*W
160 GOSUB 210
170 IF P>=180 THEN 190
180 GOTO 150
190 PRINT "W="W,"GAIN MARGIN=" 4.3429448*LOG(1/G2)
200 GOTO 260
210 LET P=57.29578*(2*ATN(.01*W)+ATN(.2*W)+ATN(1.5*W)-ATN(.1*W))
220 LET X=W*W
230 LET G2=99*99*(1+.01*X)/((1+.0001*X)↑2*(1+.04*X)*(1+2.25*X))
240 RETURN
250 PRINT "W="W, "SYSTEM UNSTABLE"
260 END
OK
```

Table 6.13. Phase and Gain Margin Results of Computer Analysis for $G(s)H(s) = 99(1+0.1s)/[(1+0.01s)^2(1+0.2s)(1+1.5s)]$

```
RUN
UNITY GAIN,W=31.20995   P=132.1068
W=95.6849   GAIN MARGIN=14.8573
OK
```

because a double break occurs at 100 rad/sec. However, if we correct each of these breaks by the 3-dB error factor, we obtain a gain margin of approximately 15 dB, which is fairly close to the computer solution of 14.7 dB.

Notice the precision, speed, and simplicity of the digital computer's solution. Its usefulness as an aid to the control-system engineer is quite evident, and it will be used extensively in this book with either programs presented in this book or MATLAB.

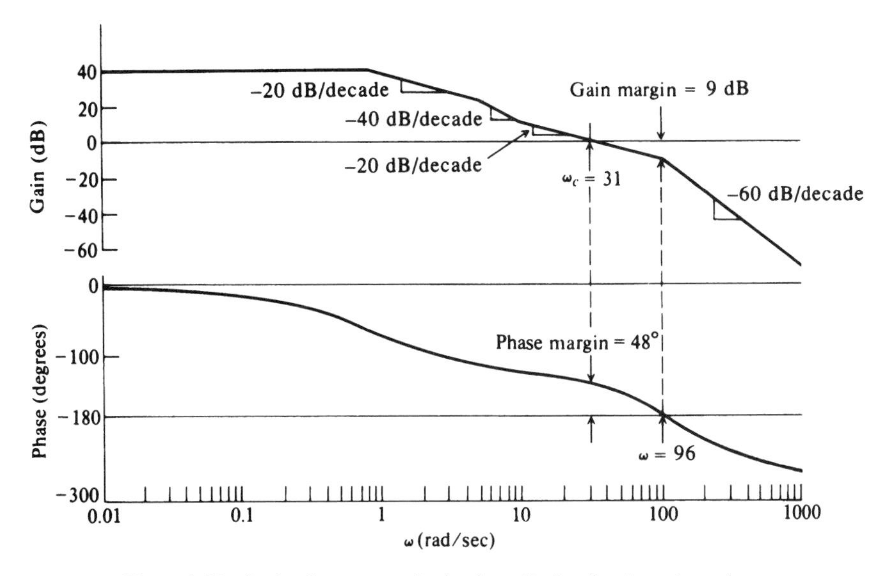

Figure 6.39 Bode-diagram analysis of a unity-feedback system where

$$G(s) = \frac{99(1 + 0.1s)}{(1 + 0.01s)^2(1 + 0.2s)(1 + 1.5s)}.$$

We next extend the digital computer as a tool for also determining the open- and closed-loop frequency response, and the time-domain response to a number of standard input signals. Consider a unity-feedback control system where

$$G(s) = \frac{60(1 + 0.5s)}{s(1 + 5s)}. \tag{6.108}$$

A Fortran [14–17] program will be utilized in this example in order to determine the open- and closed-loop frequency response, and the Basic [13, 14] program will be utilized to determine the time-domain response.

Figure 6.40 shows the logic flow chart of a Fortran program, called FRECOM, which computes the following relations for various frequencies:

(a) $20 \log |G(s)H(s)|$
(b) Phase of $G(s)H(s)$
} for open-loop Bode diagram

(c)
$$20 \log \left|\frac{C(s)}{R(s)}\right| = \left|\frac{G(s)}{1 + G(s)H(s)}\right|$$

(d) Phase of
$$\frac{C(s)}{R(s)}$$
} for closed-loop frequency response

The Fortran program for computing these quantities is shown in Table 6.14 and should be compared with the flow chart of Figure 6.40 to obtain a thorough under-

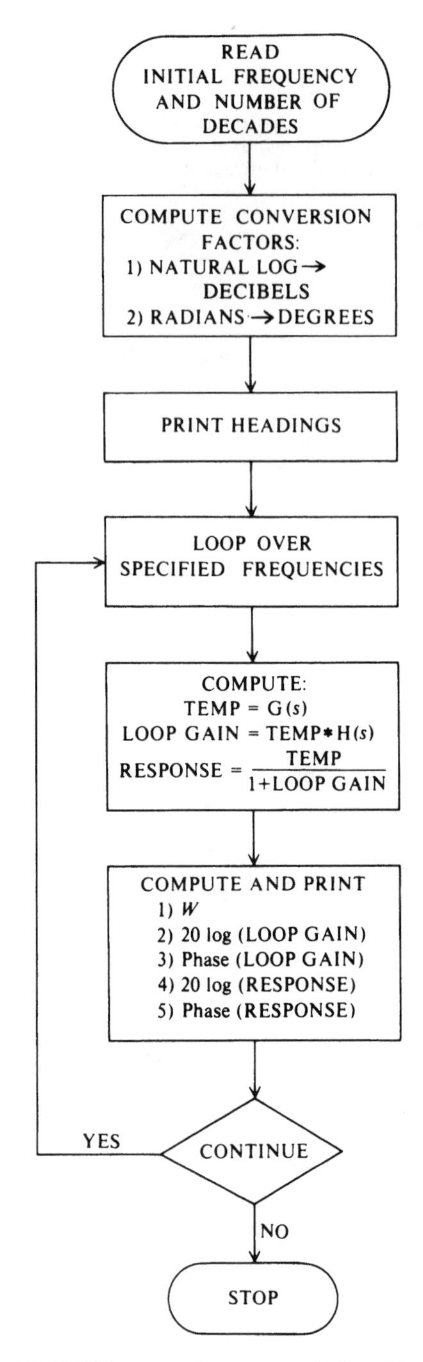

Figure 6.40 Flow diagram FRECOM for computing open- and closed-loop frequency responses.

standing of the technique. The column labeled "loop again" in Table 6.14 provides the open-loop gain, and the column labeled "response" provides the closed-loop frequency response. Note how simple the complex specification statements become in this language. One should also observe the use of complex function subprograms

Table 6.14. Computer Program FRECOM for Determining the Bode Diagram, Open-Loop Frequency Response and Closed-Loop Frequency Response

```
      COMPLEX S,TFRFCT,LOOPG
      COMPLEX G,H
      EXTERNAL G,H
      COMPLEX TEMP
      PRINT *,'INPUT INITIAL FREQUENCY AND # OF DECADES,'
      READ *,WMIN,JM
      PRINT *,WMIN,JM
      PRINT *,'FREQUENCY LOOP GAIN            RESPONSE'
      PRINT *,'RAD/SEC     DB      DEGREES  DB  DEGREES'
      PRINT*,'-------   ---------------   --------------'
      FAC1=20./ALOG(10.)
      FAC2=180./3.141592654
      WINC=WMIN
      DO 1 J-1,JM
      IF(J.GT.1)WINC=WMIN*(10**(J-1))
      DO 1 K=1,9
      W=K*WINC
      S=CMPLX(0.,W)
      TEMP=G(S)
      LOOPG=TEMP*H(S)
      TFRFCT=TEMP/(1+LOOPG)
      LOOPG=CLOG(LOOPG)
      TFRFCT=CLOG(TFRFCT)
      V1=REAL(LOOPG)*FAC1
      V2=AIMAG(LOOPG)*FAC2
      V3=REAL(TFRFCT)*FAC1
      V4=AIMAG(TFRFCT)*FAC2
      PRINT 900,W,V1,V2,V3,V4
900   FORMAT (F7.2,2X,2F10.2,1X,2F10.2)
1     CONTINUE
      STOP
      END
      COMPLEX FUNCTION G(S)
      COMPLEX S
      G=60.*(1.+0.5*S)/(S*(1+5.*S))
      RETURN
      END
      COMPLEX FUNCTION H(S)
      COMPLEX S
      H=1.
      RETURN
      END
```

$G(s)$ and $H(s)$. Once written, the program can now be revised for different systems merely by replacing the two lines containing the arithmetic statements for G and H.

The actual use of the program is extremely simple. During program execution, one enters the starting frequency value and the number of decades of frequency range required. Table 6.15 shows the computer run for the open-(labeled "loop gain") and

Table 6.15. Open and Closed-Loop Frequency Response of $G(s)H(s) = (60(1 + 0.5s))/(s(1 + 5s))$

INPUT INITIAL FREQUENCY AND # OF DECADES
.10000000,4

FREQUENCY	LOOP GAIN		RESPONSE	
RAD/SEC	DB	DEGREES	DB	DEGREES
0.10	54.60	−113.70	0.01	−0.10
0.20	46.58	−129.29	0.03	−0.21
0.30	41.00	−137.78	0.06	−0.35
0.40	36.70	−142.13	0.10	−0.52
0.50	33.24	−144.16	0.15	−0.74
0.60	30.37	−144.87	0.22	−1.02
0.70	27.94	−144.76	0.29	−1.37
0.80	25.84	−144.16	0.36	−1.79
0.90	24.01	−143.24	0.44	−2.28
1.00	22.38	−142.13	0.53	−2.84
2.00	12.51	−129.29	1.21	−12.17
3.00	7.60	−119.88	1.20	−24.53
4.00	4.48	−113.70	0.57	−35.73
5.00	2.22	−109.51	−0.34	−44.55
6.00	0.45	−106.53	−1.34	−51.26
7.00	−1.00	−104.31	−2.32	−56.39
8.00	−2.24	−102.60	−3.24	−60.39
9.00	−3.31	−101.26	−4.11	−63.57
10.00	−4.27	−100.16	−4.91	−66.14
20.00	−10.41	−95.14	−10.57	−77.99
30.00	−13.96	−93.43	−14.03	−81.98
40.00	−16.47	−92.58	−16.51	−83.99
50.00	−18.41	−92.06	−18.43	−85.19
60.00	−20.00	−91.72	−20.01	−85.99
70.00	−21.34	−91.47	−21.35	−86.56
80.00	−22.50	−91.29	−22.51	−86.99
90.00	−23.52	−91.15	−23.53	−87.33
100.00	−24.44	−91.03	−24.44	−87.59
200.00	−30.46	−90.52	−30.46	−88.80
300.00	−33.98	−90.34	−33.98	−89.20
400.00	−36.48	−90.26	−36.48	−89.40
500.00	−38.42	−90.21	−38.42	−89.52
600.00	−40.00	−90.17	−40.00	−89.60
700.00	−41.34	−90.15	−41.34	−89.66
800.00	−42.50	−90.13	−42.50	−89.70
900.00	−43.52	−90.11	−43.52	−89.73

closed-loop frequency response (labeled "response") of this system. Figure 6.41 is a plot of these results. It indicates a phase margin of 74.5° at a gain crossover frequency of 6.4 rad/sec and a closed-loop peaking of 1.8 dB at 2.5 rad/sec. It is immediately evident from examination of the open-loop frequency characteristics that the system is always stable because the phase shift never exceeds −144.87°. Note that the gain margin is infinity because the phase never reaches −180° (except at $\omega = \infty$).

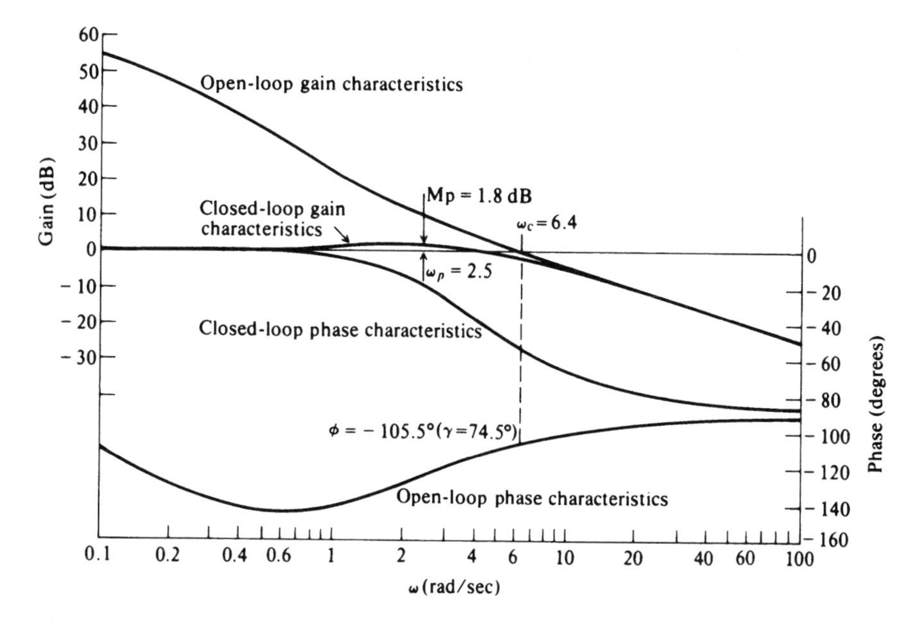

Figure 6.41 Open- and closed-loop frequency response of $G(s)H(s) = \dfrac{60(1+0.5s)}{s(1+5s)}$.

In order to obtain the time-domain response, the differential equations that describe the system must be derived from the given frequency-domain description. Because, in the running example,

$$G(s) = \frac{60(1 + 0.5s)}{s(1 + 5s)}$$

and

$$H(s) = 1,$$

then

$$\frac{C(s)}{R(s)} = \frac{G(s)}{1 + G(s)}.$$

Upon substitution of $G(s)$ into the expression for $C(s)/R(s)$, the following is obtained:

$$\frac{C(s)}{R(s)} = \frac{30s + 60}{5s^2 + 31s + 60}$$

or

$$C(s)[5s^2 + 31s + 60] = R(s)[30s + 60].$$

This is equivalent to the following differential equation:

$$5\frac{d^2c(\tau)}{d\tau^2} + 31\frac{dc(\tau)}{d\tau} + 60c(\tau) = 60r(\tau) + 30\frac{dr(\tau)}{d\tau}. \tag{6.109}$$

In order to obtain the state equations, let

$$x_1(\tau) = c(\tau),$$

$$x_2(\tau) = \frac{dc(\tau)}{d\tau},$$

$$y(\tau) = 12r(\tau) + 6\frac{dr(\tau)}{d\tau}.$$

Therefore, Eq. (6.109) is equivalent to the following two first-order differential equations:

$$\frac{dx_1(\tau)}{d\tau} = x_2(\tau), \tag{6.110}$$

$$\frac{dx_2(\tau)}{d\tau} = -12x_1(\tau) - \frac{31}{5}x_2(\tau) + y(\tau). \tag{6.111}$$

Table 6.16 illustrates the coding of a Basic program called INTER 2, which applies a second-order Runge-Kutta numerical integration method to the solution of two simultaneous differential equations [16, 18, 20]:

Table 6.16. Coding of a Basic Program for Solving Two Simultaneous Differential Equations

STEP 1	Read in $x_1(\tau_0)$, $x_2(\tau_0)$, τ_0, $\Delta\tau$, τ_f Set $n = 0$, $N_{max} = \frac{(\tau_f - \tau_0)}{\Delta\tau}$
STEP 2	Compute $f = f(x_1(\tau_n), x_2(\tau_n), \tau_n)$ $g = g(x_1(\tau_n), x_2(\tau_n), \tau_n)$
STEP 3	Compute $x_1' = x_1(\tau_n) + f * \Delta\tau$ $x_2' = x_2(\tau_n) + g * \Delta\tau$ $\tau_{n+1} = \tau_n + \Delta\tau$
STEP 4	Compute $f' = f(x_1', x_2', \tau')$ $g' = g(x_1', x_2', \tau')$
STEP 5	Compute $x(\tau_{n+1}) = (x_1' + x_1)/2 + \Delta\tau * f'/2$ $y(\tau_{n+1}) = (x_2' + x_2)/2 + \Delta\tau * g'/2$
STEP 6	If the final time has not yet been reached, set $n = n + 1$ and go back to Step 2. Otherwise stop.

$$\dot{x}_1 = f(x_1, x_2, \tau)$$
$$\dot{x}_2 = g(x_1, x_2, \tau).$$

Table 6.16 shows how the values of $x_1(\tau)$ and $x_2(\tau)$ are obtained for $\tau = \tau_0 + n\Delta\tau$, for $n = 1, 2, \ldots$. A flow chart of the program is shown in Figure 6.42, and the Basic language program is shown in Table 6.17.

In order to use this program, one must define the specific functions $f(x_1, x_2, \tau)$, $g(x_1, x_2, \tau)$, and $y(\tau)$ on the specified lines. One must then enter the initial values of

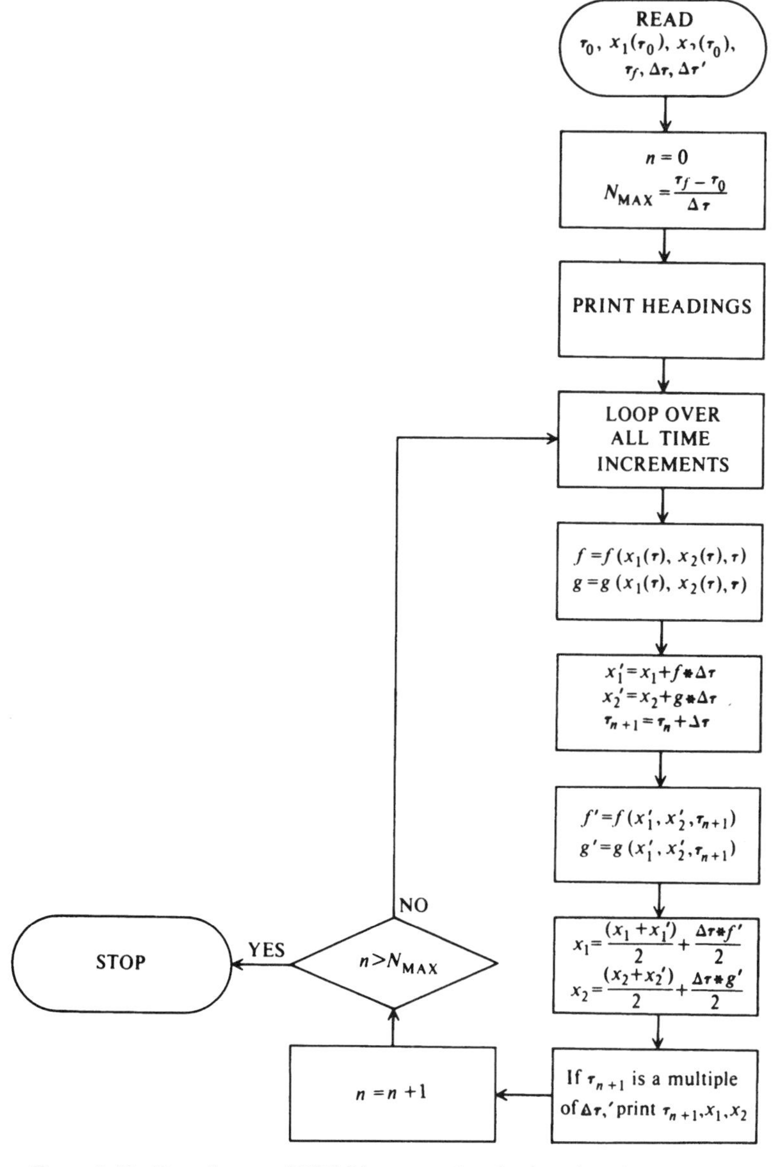

Figure 6.42 Flow diagram INTER 2 for computing the time-domain response.

Table 6.17. Computer Program INTER 2 for Obtaining the Time-Domain Response

```
READY
LIST
  1     REM  THIS ROUTINE APPLIES THE RUNGE-KUTTA METHOD WITH
  2     REM  SECOND-ORDER ACCURACY TO THE SOLUTION OF THE
  3     REM  SYSTEM OF DIFFERENTIAL EQUATIONS X' = F(X, Y, T), Y' =
  4     REM  G(X,Y,T) WITH THE INITIAL CONDITIONS X(TO)=X0, Y(TO)=Y0.
  5     REM  THE INTEGRATION STEP IS H AND THE SOLUTIONS X(T) AND
  6     REM  Y(T) ARE TABULATED ON THE INTERVAL TO < =T< =B IN
  7     REM  STEPS OF SIZE L.
  8     REM    THE FUNCTIONS F(X,Y,T), G(X,Y,T) ARE ENTERED IN LINES
  9     REM  509 AND 510 RESPECTIVELY AS FOLLOWS:
 10     REM                        509 LET F=F(X,Y,T)
 11     REM                        510 LET G=G(X,Y,T)
 12     REM    THE NUMBERS T0, X0, Y0, B, L, AND H ARE ENTERED AS
 13     REM  DATA IN LINE 900.
 14     REM
 15     REM
100     READ T,X,Y,B,L,H
110     LET M-INT(L/H)
120     LET N=INT(B−T)/L)
130     PRINT "VALUE OF T," "VALUE OF X," "VALUE OF Y"
140     PRINT
150     PRINT
160     PRINT T,X,Y
170     FOR J=1 TO N
180     FOR I=1 TO M
190     LET X1=X
200     LET Y1=Y
210     GOSUB 500
220     LET X=X+H*F
230     LET Y=Y+H*G
240     LET T=T+H
250     GOSUB 500
260     LET X=(X1+X)/2+0.5*H*F
270     LET Y=(Y1+Y)/2+0.5*H*G
280     NEXT I
290     PRINT, T,X,Y
300     NEXT J
310     STOP
500     LET Y9=12
509     LET F=Y
510     LET G=−(31/5)*Y-12*X+Y9
520     RETURN
900     DATA 0,0,1,10,1,0.01
999     END
```

τ_0, $x_1(\tau_0)$, $x_2(\tau_0)$, and the time increment $\Delta\tau$. The tabulation step size is a multiple of $\Delta\tau$ for which a printout of results is required (this cuts down on a voluminous output).

The solution for a unit step input is shown in Table 6.18. Similarly, the solution for a unit ramp input is shown in Table 6.19. For both cases, the initial input parameters were:

$\tau_0 = 0$	(initial time)	$\tau_f = 10$	(final time)
$x_1(\tau_0) = 0$	[initial value of $x_1(\tau)$]	$\Delta\tau = 0.01$	(time increment)
$x_2(\tau_0) = 1$	[initial value of $x_2(\tau)$]	$\Delta\tau' = 0.5$	(tabulation step size)

Figures 6.43 and 6.44 plot the results of the tabulation in Tables 6.18 and 6.19.

Table 6.18. Unit Step Response of Unity-Feedback System where $G(s) = 60(1 + 0.5s)/[s(1 + 5s)]$

```
READY
510 LET G=-(31/5)*Y-12*X+Y9
RUN
```

VALUE OF T	VALUE OF X	VALUE OF Y
0	0	1
1.	0.937669	0.260442
2.	1.00189	−1.43078E − 3
3.	1.00013	−5.31968E − 4
4.	0.999996	1.68118E − 6
5.	1.	1.08411E − 6
6.	1.	−4.45157E − 10
7.	1.	−1.49581E − 9
8.	1.	3.54907E − 10
9.	1.	3.56945E − 10
10.	1.	3.56945E − 10

```
OK
```

Table 6.19. Unit Ramp Response of Unity-Feedback System where $G(s) = 60(1 + 0.5s)/[s(1 + 5s)]$

```
READY
500 LET Y9=12*T+6
RUN
```

VALUE OF T	VALUE OF X	VALUE OF Y
0	0	1
1.	0.984858	0.994173
2.	1.9833	0.999987
3.	2.98333	1.00001
4.	3.98333	1.
5.	4.98333	1.
6.	5.98333	1.
7.	6.98333	1.
8.	7.98333	1.
9.	8.98333	1.
10.	9.98333	1.

```
OK
```

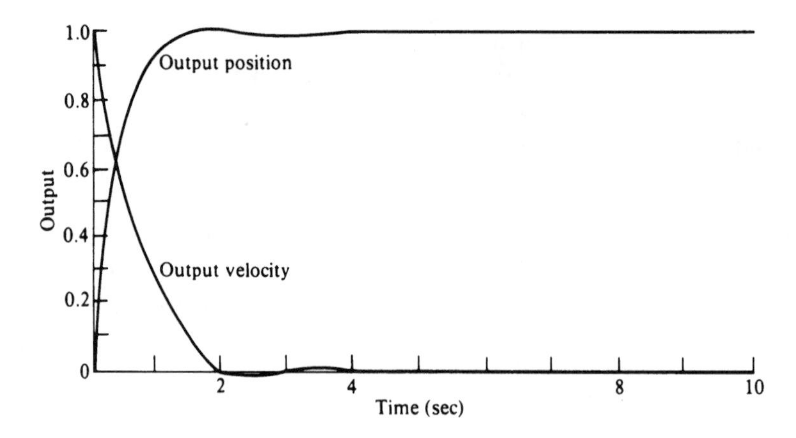

Figure 6.43 Unit step response of unity-feedback system where $G(s)H(s) = \dfrac{60(1 + 0.5s)}{s(1 + 5s)}$.

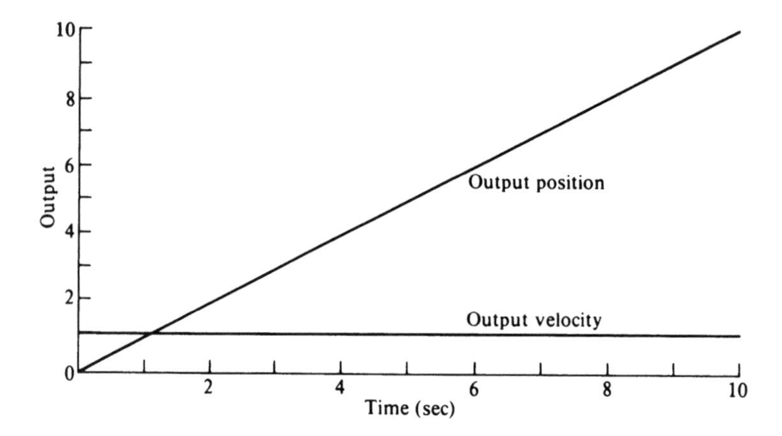

Figure 6.44 Unit ramp response of unity-feedback system where $G(s)H(s) = \dfrac{60(1 + 0.5s)}{s(1 + 5s)}$.

6.10. NICHOLS CHART

The Nichols chart [21] is a very useful technique for determining stability and the closed-loop frequency response of a feedback system. Stability is determined from a plot of the open-loop gain versus phase characteristics. At the same time, the closed-loop frequency response of the system is determined by utilizing contours of constant closed-loop amplitude and phase shift which are overlaid on the gain-phase plot.

In order to derive the basic Nichols chart relationships, let us consider the unity-feedback system illustrated in Figure 6.45. The closed-loop transfer function is given by

$$\frac{C(j\omega)}{R(j\omega)} = \frac{G(j\omega)}{1 + G(j\omega)}, \tag{6.112}$$

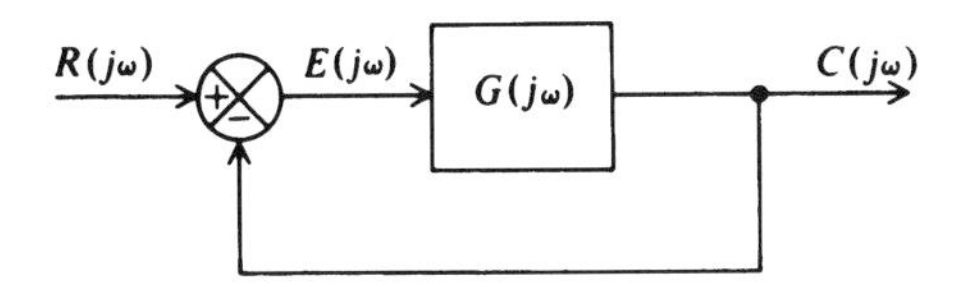

Figure 6.45 Block diagram of a simple feedback control system.

or

$$\frac{C(j\omega)}{R(j\omega)} = M(\omega)e^{j\alpha(\omega)}, \tag{6.113}$$

where $M(\omega)$ represents the amplitude component of the transfer function and $\alpha(\omega)$ the phase component of the transfer function. The radian frequency at which the maximum value of $C(j\omega)/R(j\omega)$ occurs is called the resonant frequency of the system, ω_p, and the maximum value of $C(j\omega)/R(j\omega)$ is denoted by M_p. For the system illustrated in Figure 6.45, we would expect a typical closed-loop frequency response to have the general form shown in Figure 6.46.

From Section 4.2 we know that a small margin of stability would mean a relatively small value for ζ and a relatively large value for M_p. This can be further clarified if we examine the Nyquist diagrams illustrated in Figure 6.47. Notice that $M(\omega)$ is much less than unity for the case of a system having a large degree of stability, as shown in part (*a*) of the figure. In contrast, $M(\omega)$ is much greater than unity for the case of a system having a very small degree of stability, as shown in part (*b*) of the figure. We next develop the Nichols-chart contours in the complex plane in order to be able to determine M_p and ω_p quantitatively for a feedback control system.

Reconsidering the basic relationships of the unity-feedback control system illustrated in Figure 6.45, let us represent the complex vector $G(j\omega)$ by an amplitude and phase as follows:

$$G(j\omega) = |G(j\omega)|e^{j\theta}. \tag{6.114}$$

Substituting Eq. (6.114) into the system transfer function, as given by Eq. (6.112), we obtain

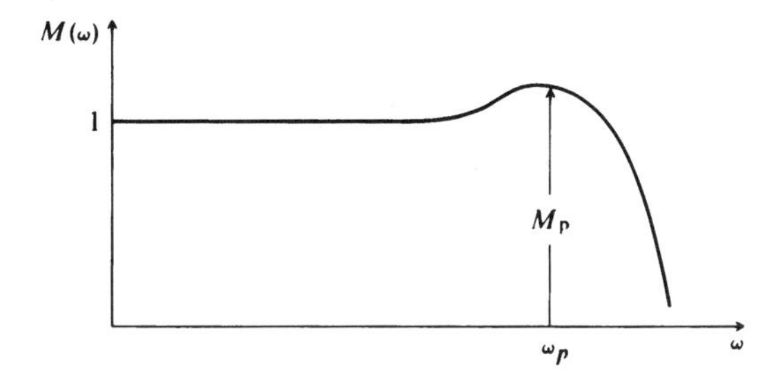

Figure 6.46 A typical closed-loop frequency-response curve for the system shown in Figure 6.45.

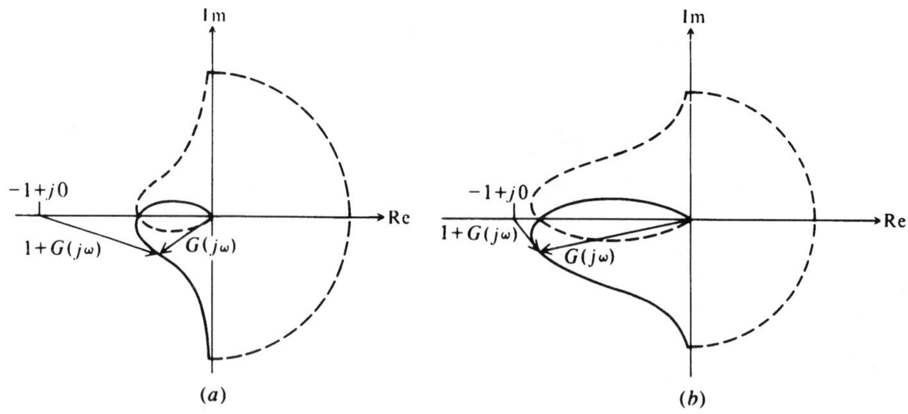

Figure 6.47 Qualitative stability comparison of two systems illustrating corresponding values of $M(\omega)$. (*a*) Stable system with large degree of stability:

$$\left|\frac{C(j\omega)}{R(j\omega)}\right| = \left|\frac{G(j\omega)}{1+G(j\omega)}\right| = M(\omega) \ll 1.$$

(*b*) Stable system with very small degree of stability:

$$\left|\frac{C(j\omega)}{R(j\omega)}\right| = \left|\frac{G(j\omega)}{1+G(j\omega)}\right| = M(\omega) \gg 1.$$

$$\begin{aligned}\frac{C(j\omega)}{R(j\omega)} &= \frac{G(j\omega)}{1+G(j\omega)} \\ &= \frac{|G(j\omega)|e^{j\theta}}{1+|G(j\omega)|e^{j\theta}}.\end{aligned} \tag{6.115}$$

Dividing through by $|G(j\omega)|e^{j\theta}$, we obtain

$$\frac{C(j\omega)}{R(j\omega)} = \left[\frac{e^{-j\theta}}{|G(j\omega)|} + 1\right]^{-1} \tag{6.116}$$

Using the trigonometric relationship for the exponential term, we obtain the expression

$$\frac{C(j\omega)}{R(j\omega)} = \left[\frac{\cos\theta}{|G(j\omega)|} - \frac{j\sin\theta}{|G(j\omega)|} + 1\right]^{-1} \tag{6.117}$$

The expression of Eq. (6.117) has a magnitude of $M(\omega)$ and a phase angle $\alpha(\omega)$ which are given by

$$M(\omega) = \left\{\left[1 + \frac{1}{|G(j\omega)|^2} + \frac{2\cos\theta}{|G(j\omega)|}\right]^{1/2}\right\}^{-1}, \tag{6.118}$$

$$\alpha(\omega) = -\tan^{-1}\frac{\sin\theta}{\cos\theta + |G(j\omega)|}. \tag{6.119}$$

A detailed plot of G in decibels versus θ in degrees, with M and α as parameters, known as the Nichols chart, is illustrated in Figure 6.48. Notice the symmetry of these curves about the $-180°$ and 0 dB point.

The stability criterion for the Nichols chart is quite simple, because the $-1+j0$ point of the Nyquist diagram corresponds to the 0-dB, $-180°$ point of Figure 6.48.

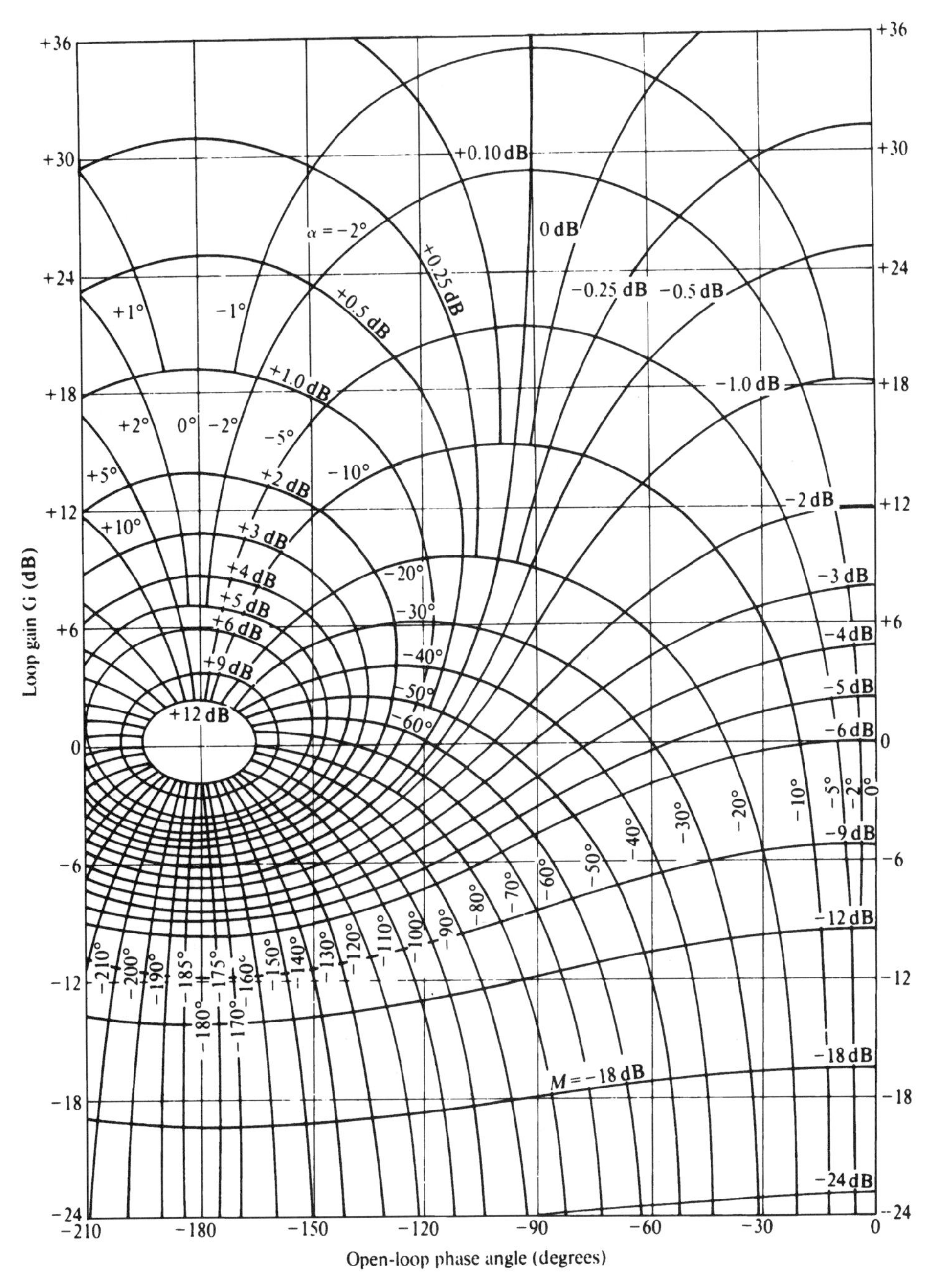

Figure 6.48 Nichols chart.

Therefore, for minimum-phase systems, the phase margin and gain margin can be determined.

The Nichols chart can be used for the analysis and/or synthesis of a feedback control system. For analysis, we can use the Nichols chart to obtain the closed-loop frequency response from the Bode diagram and determine the maximum value of peaking M_p and the frequency at which it occurs, ω_p. From a synthesis viewpoint, we can use the Nichols chart to meet certain requirements as to the values of M_p and ω_p. We demonstrate the use of this method as a total for analysis in this section. Its value as a tool for synthesis is illustrated in Section 7.8 of Chapter 7.

Let us determine the closed-loop frequency response for the third-order feed-back control system which is illustrated in Figure 6.49. Specifically, we are now interested in obtaining the values of M_p and ω_p using the Nichols chart. They can be obtained by first drawing the Bode diagram as shown in Figure 6.50. Then, for each value of ω, the magnitude and phase of $G(j\omega)$ are then plotted onto a Nichols chart as shown in Figure 6.51*a*. The Bode diagram and Nichols chart were obtained using MATLAB, and are contained in the M-file that is part of my MCSTD Toolbox and can be retrieved free from The MathWorks, Inc. anonymous FTP server at ftp://ftp.mathworks.com/pub/books/shinners. It indicates a phase margin of 27.7° and a gain margin at 8.175 dB which agrees with the values obtained from the Bode diagram of Figure 6.50. The intersections of $G(j\omega)$ with $M(\omega)$ on the Nichols chart give the closed-loop response from which M_p and ω_p can be obtained. The resulting response, shown in Figure 6.51*b*, indicates a value of $M_p = 6.758$ dB (2.2) and $\omega_p = 8.989$ rad/sec. This too was obtained using MATLAB. It is important to note that the use of MATLAB and the MCSTD Toolbox permits one to obtain the Nichols chart **directly** without having to first obtain the Bode diagram.

In practice, a value of $M_p = 2.2$ is a little too high. Usually, M_p is chosen somewhere between 1.0 and 1.4 (see Section 7.8). The technique of compensation of this system utilizing the Nichols chart is illustrated in Section 7.8 of Chapter 7. Let us here, however, indicate how the constant M loci on the Nichols chart may be used to limit M_p to some maximum value, say 2.2 dB (1.3). Because the interior of the $M = 2.2$ dB locus consists entirely of constant-magnitude loci which represent M greater than 2.2 dB, we must confine the Nichols locus to the exterior of the $M = 2.2$ dB locus. The system illustrated in Figure 6.49 is compensated, to meet certain maximum requirements of M_p, in Section 7.8 of Chapter 7 utilizing the Nichols chart.

The values of ω_p and M_p determined in this section and the rest of the book where the Nichols chart is applied, uses the function "wpmp" which is part of my MCSTD Toolbox. It calculates ω_p and M_p analytically and very accurately rather than using graphical techniques. My toolbox is available to the reader free, and can

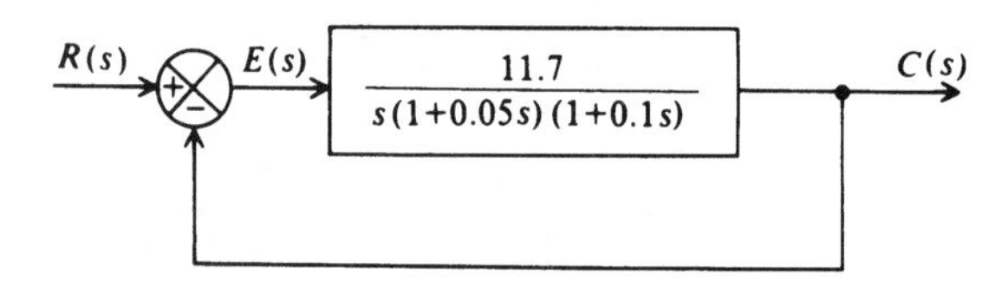

Figure 6.49 A third-order feedback control system.

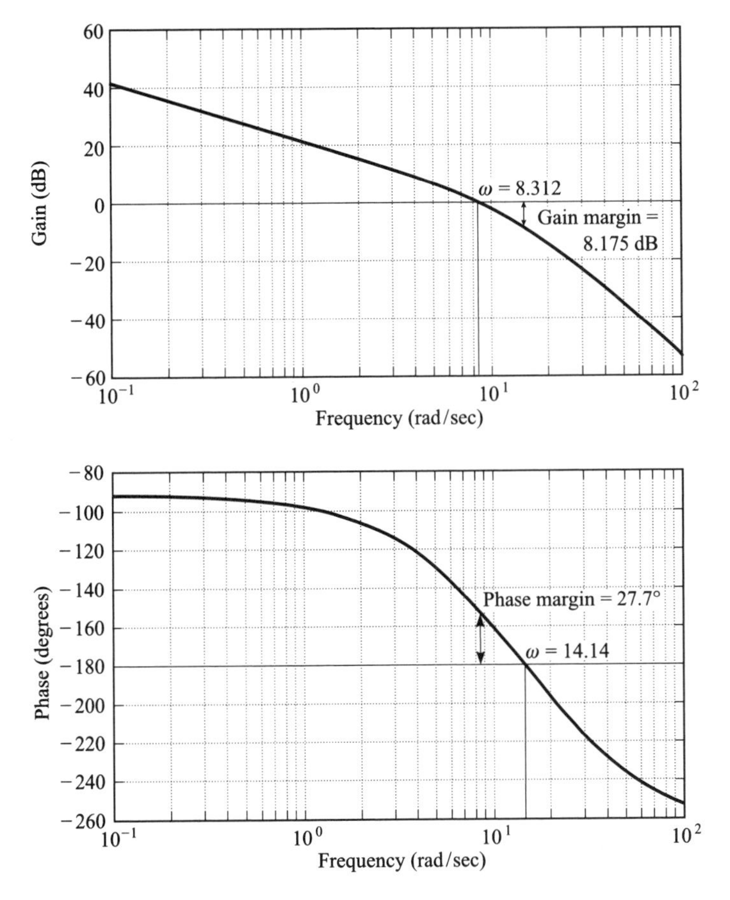

Figure 6.50 Bode diagram for the system shown in Figure 6.49, where

$$G(s)H(s) = \frac{11.7}{s(1+0.05s)(1+0.1s)}.$$

be retrieved from The Mathworks, Inc. anonymous FTP server at ftp://ftp.mathworks.com/pub/books/shinners.

In the following section, 6.11, the procedure for obtaining the Nichols chart using MATLAB is provided. This is of great assistance to the control-system engineer in obtaining very powerful results with a few MATLAB statements.

6.11. NICHOLS CHART USING MATLAB [7]

In the previous section, the Nichols chart was illustrated. We superimposed a hand-drawn plot of the open-loop transfer function $G(s)H(s)$ on a pre-drawn and supplied Nichols chart. In this section, the MATLAB utility used for obtaining the Nichols chart is identified, and the reader will be shown how to obtain the Nichols chart very easily and accurately using MATLAB. The control-system engineer does not have to

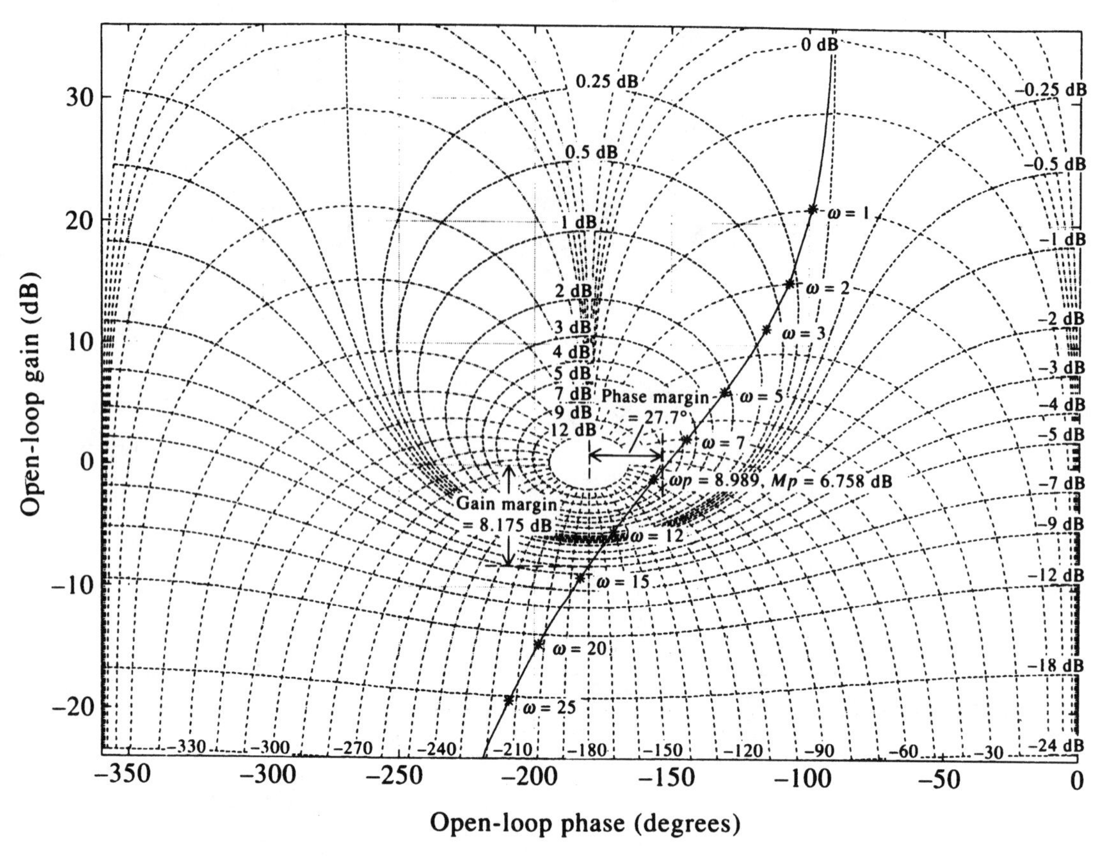

Figure 6.51(*a*) Nichols chart for the system shown in Figure 6.49.

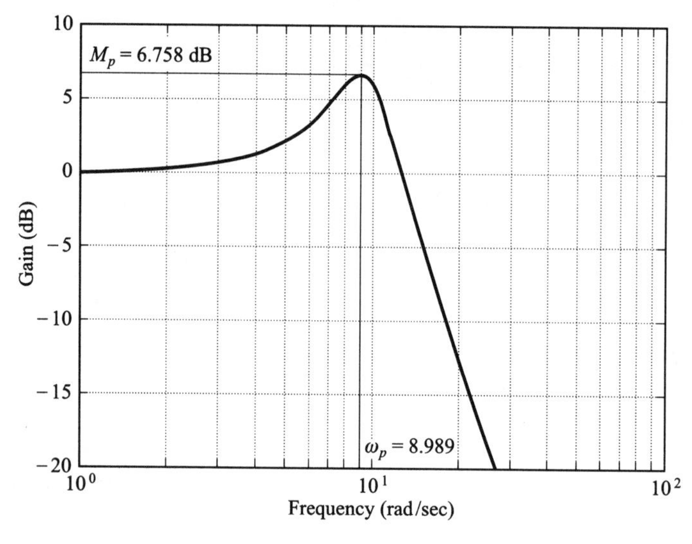

Figure 6.51(*b*) Closed-loop frequency response of the system shown in Figure 6.49 from the Nichols chart of Figure 6.51*a*.

go and find a pre-drawn Nichols chart when using MATLAB. The Nichols chart and the superimposed plot of $G(s)H(s)$ are both created and drawn at the same time.

The MATLAB Control System Toolbox uses the command *nichols* to create the Nichols frequency response plot, and it uses the command *ngrid* to generate the grid lines for a Nichols chart. The MCSTD Toolbox simplifies the process with the comman *nichgrid*.

The definition of the nichgrid command is as follows:

$$\text{function}[x,y] = \text{nichgrid}(g,a,b,n)$$

where

g = grid values for axis (in degrees and dB)

a = angles (in degrees, $-180^\circ < a < 0$)

b = gains (in dB)

n = interpolation value (optional, default = none)

x = phase values

y = magnitude (in dB).

The MCSTD Toolbox [38] discusses the command nichgrid in great detail.

Let us consider drawing the Nichols chart for a unity-feedback control system whose open-loop transfer function is given by

$$G(s)H(s) = \frac{11.7(1 + 0.58s)}{s(1 + 0.05s)(1 + 0.1s)(1 + 1.74s)}. \tag{6.120}$$

For purposes of MATLAB formulation, we will put this transfer function in the following format:

$$G(s)H(s) = \frac{780s + 1344.8273}{s^4 + 30.5747s^3 + 217.2414s^2 + 114.9425s} \tag{6.121}$$

The resulting MATLAB program is shown in Table 6.20 which develops the Nichols chart and also superimposes this transfer function. Comments made previously in Section 6.8 on the use of MATLAB for the Bode diagram shown in the MATLAB Program Table 6.8 apply here too. This illustration is contained in DEMO 3 in the MCSTD Toolbox.

The resulting Nichols chart obtained from this program is plotted in Figure 6.52.

6.12. RELATIONSHIP BETWEEN CLOSED-LOOP FREQUENCY RESPONSE AND THE TIME-DOMAIN RESPONSE

Section 6.10 has illustrated how the closed-loop frequency-domain response may be obtained from the open-loop transfer function. The next logical question to ask is

Table 6.20. MATLAB Program to Obtain the Nichols Chart for the System whose Open-Loop Transfer Function is shown in Eq. (6.121)

```
num = [780  1344.8273]
den = [1  30.5747  217.2414  114.9425  0]
a = -[.25:.25:1  2  5  10:10:170  179.99];
b = [-24  - 18  - 12  - 9  - 7  - 5: - 1  - .5:.25:.5  1:5  7  9  12];
[x,y] = nichgrid([-360  0  - 24  36],a,b,3);
[mag,ph] = bode(num,den,logspace(-1,2));
plot(ph,20 * log10(mag));
W = [.5  1  2  5  7  9  12];
[mag, ph] = bode(num,den,w);
plot(ph,20 * log10(mag),'* g')
title('Nichols Frequency Response Plot')
```

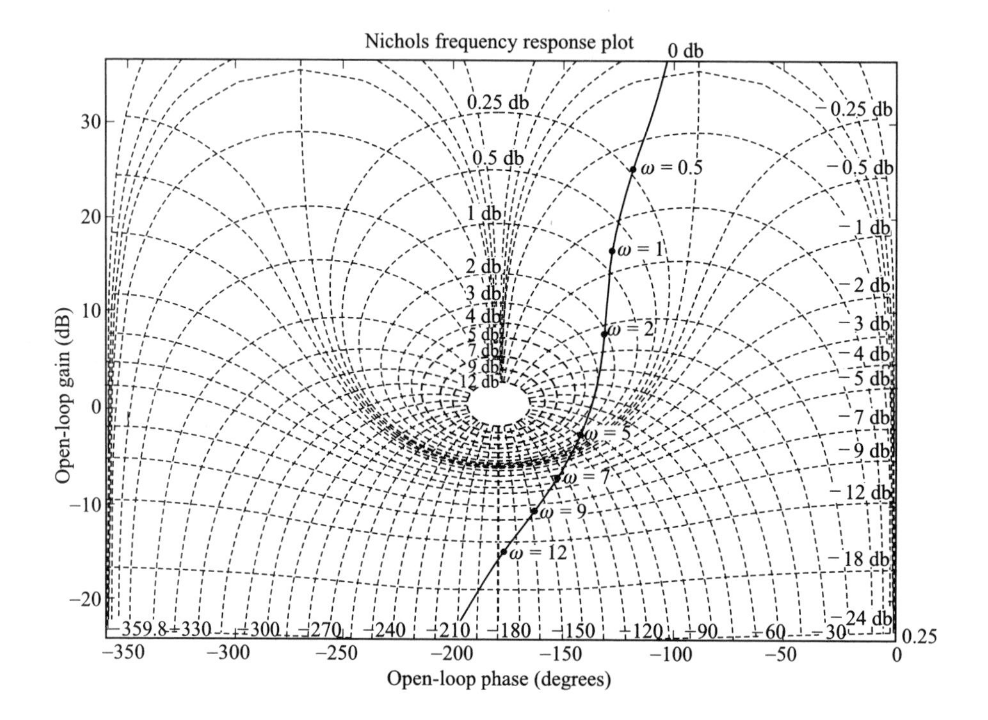

Figure 6.52 Nichols chart with $G(s)H(s) = \dfrac{11.7(1+0.58s)}{s(1+0.05s)(1+0.1s)(1+1.74s)}$ superimposed.

how to determine the relationship between M_p and the peak overshoot one obtains in the time domain. In Chapter 4, we defined the time at which the peak overshoot occurs as t_p in terms of ζ and ω_n [see Eq. (4.29)]. For example, does an M_p of 1.3 mean a 30% transient overshoot in the time domain?

This problem has been analyzed for the general, unity-feedback system of Figure 6.45 [22]. After determining the Fourier transform of the input, $R(\omega)$, and that of the output, $C(\omega)$, the value of $C(\omega)$ was then related to the transient response in the time domain. The resulting approximate general relationship between M_p and the peak overshoot of the transient in the time domain has been shown [22] to be given by

$$c(t)_{\max} \leqslant 1.18 M_p. \tag{6.122}$$

Therefore, the overshoot in the time domain is, in general, related to M_p by some factor equal to or less than 18%. This approximation is extremely important, because it permits the control-system engineer to correlate peak overshoots in the frequency and time domains.

For second-order systems, exact relationships can be obtained for M_p and ω_p in terms of the damping factor ζ. In order to derive this expression, let us reconsider the closed-loop transfer function of a second-order system in the following form:

$$\frac{C(j\omega)}{R(j\omega)} = \frac{1}{(1 - \omega^2/\omega_n^2) + j2\zeta(\omega)/\omega_n} = M(\omega)e^{j\alpha(\omega)}. \tag{6.123}$$

Therefore, the magnitude of the closed-loop response is given by

$$M(\omega) = \frac{1}{\left[(1 - \omega^2/\omega_n^2)^2 + 4\zeta^2(\omega^2/\omega_n^2)\right]^{1/2}}. \tag{6.124}$$

In order to find the maximum value M_p of M, and the frequency at which it occurs, ω_p, Eq. (6.124) is differentiated with respect to frequency and set equal to zero. The frequency ω_p at which M_p occurs is found to be given by

$$\omega_p = \omega_n\sqrt{1 - 2\zeta^2}. \tag{6.125}$$

If this value of ω_p is substituted into Eq. (6.124), the value of M_p is found to be given by

$$M_p = \frac{1}{2\zeta\sqrt{1 - \zeta^2}}. \tag{6.126}$$

A plot of M_p versus ζ is illustrated in Figure 6.53. Observe from this figure that M_p increases very rapidly for $\zeta < 0.4$. The resulting transient oscillatory response is excessively large in this region, and is undesirable from practical considerations. Therefore, systems having a $\zeta < 0.4$ are not normally desired. In electro-mechaical systems, values of M_p in the range of

$$1 < M_p < 1.4$$

are usually specified, which requires $\zeta > 0.4$ if the system is a pure second-order system. In industrial process control, values of M_p up to 2.0 are often used.

Let us carry the analysis of the second-order system one step further by correlating its frequency- and time-domain response. In Section 4.2, we found that the peak value of $c(t)$ for the step response of a second-order system is given by [see Eq. (4.32)]

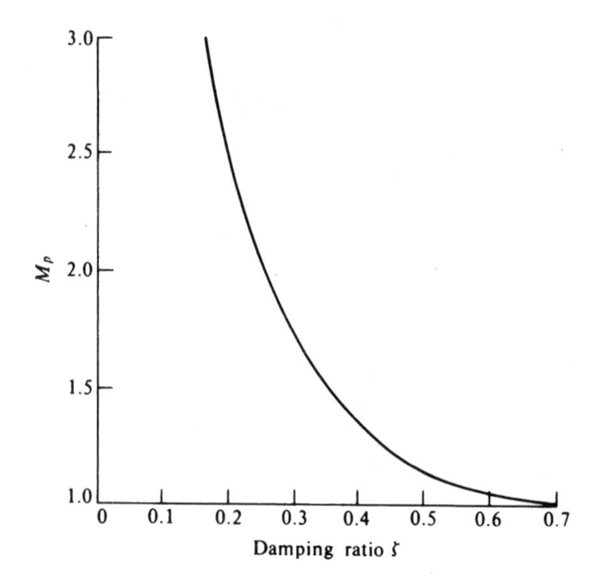

Figure 6.53 M_p versus the damping ratio.

$$c(t)_{\max} = 1 + e^{-\zeta\pi/\sqrt{1-\zeta^2}} \tag{6.127}$$

To check the correspondence between M_p [from Eq. (6.126)] and $c(t)_{\max}$ [from Eq. (6.127)], consider the case where $\zeta = 0.4$. Substituting into Eqs. (6.126) and (6.127),

$$M_p = 1.364, \tag{6.128}$$

$$c(t)_{\max} = 1.254, \tag{6.129}$$

and they are within 18% of each other as stated in Eq. (6.122). Actually, the difference is only 8.76%. As the damping factor ζ increases, the correspondence gets better. For example, if $\zeta = 0.6$, then $M_p = 1.04$ and $c(t)_{\max} = 1.09$; a difference of only 4.8%. In general, when $\zeta > 0.4$, there is a close correspondence between M_p and $c(t)_{\max}$. For values of $\zeta < 0.4$, the correspondence between M_p and $c(t)_{\max}$ is only qualitative. For example, for the second-order system and at $\zeta = 0.1$, we find that $M_p = 5.03$ and $c(t)_{\max} = 1.73$. However, because practical second-order systems usually do not use a $\zeta < 0.4$, the 18% relationship given in Eq. (6.122) is quite accurate for systems of interest.

6.13. CLOSED-LOOP FREQUENCY BANDWIDTH AND CUTOFF FREQUENCY

Let us consider the closed-loop frequency response shown in Figure 6.46 and 6.51*b* so that we can define the closed-loop frequency bandwidth and cutoff frequency. We define the bandwidth of a closed-loop control system in a manner similar to other electronic equipment such as amplifiers. The bandwidth of a closed-loop control system is defined as the frequency range where the magnitude of the closed loop

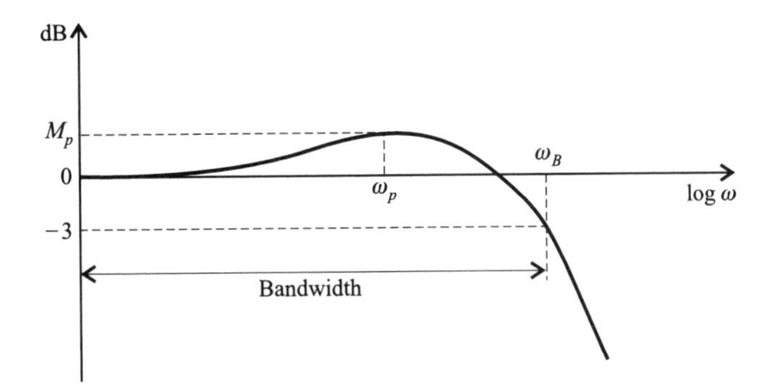

Figure 6.54 Closed-loop frequency response of a control system indicating the bandwidth and the cutoff frequency, ω_B.

gain does not drop below −3 dB as shown in Figure 6.54. Therefore, the bandwidth of the control system, ω_B, is defined to be that frequency range in which the magnitude of the closed-loop frequency response is greater than −3 dB. The frequency ω_B is defined as the cutoff frequency. At frequencies greater than ω_B, the closed-loop frequency response is attenuated by more than −3 dB.

The question arises as to the desired bandwidth and cutoff frequency of a control system. We would like to have a very large bandwidth so that we can reproduce the input signal (external forcing function) very accurately. Unfortunately, very large bandwidths pass high-frequency noise which can be a detriment to the control-system performance. Therefore, there are conflicting requirements on the bandwidth and cutoff frequency, and a tradeoff is necessary in order to achieve a good design.

6.14. ROOT-LOCUS METHOD FOR NEGATIVE-FEEDBACK SYSTEMS

The root-locus method is a technique for determining the roots of the closed-loop characteristic equation of a system as a function of the static gain. This method is based on the relationship that exists between the roots of the closed-loop transfer function and the poles and zeros of the open-loop transfer function. The root-locus method, which was conceived by Evans [23, 25], has several distinct advantages.

Knowledge of the location of the closed-loop roots permits the very accurate determination of a control system's relative stability and transient performance. Alternatively, approximate solutions may be obtained, with a considerable reduction of labor, if very accurate solutons are not required. This and the following section present the graphical method of constructing the root locus and of interpreting the results for negative- and postive-feedback systems, respectively. Working digital computer programs for obtaining the root locus are presented in Section 6.17 (using MATLAB) and 6.18. The technique for synthesizing a system utilizing the root-locus method is discussed in Section 7.9. The method is a useful one and should be part of the designer's bag of tricks.

Let us consider the general feedback control system illustrated in Figure 6.55. In order to find the poles of the closed-loop transfer function, we require that

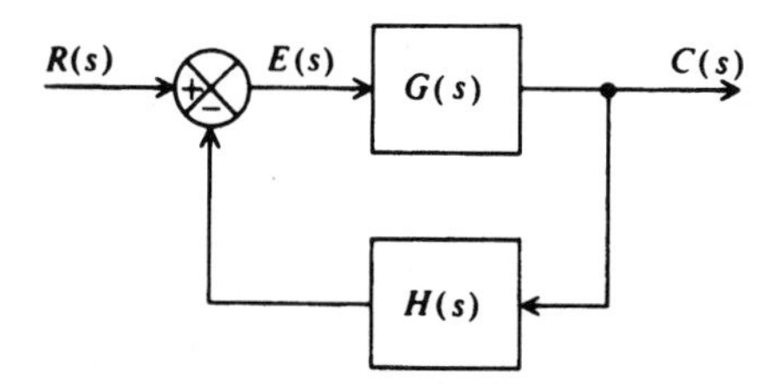

Figure 6.55 A nonunity-feedback control system.

$$1 + G(s)H(s) = 0 \tag{6.130}$$

$$G(s)H(s) = -1 = 1 \angle(2n+1)\pi \tag{6.131}$$

where $n = 0, \pm 1, \pm 2, \ldots$ Equation (6.131) specifies two conditions that must be satisfied for the existence of a closed-loop pole.

1. The angle of $G(s)H(s)$ must be an odd multiple of π:

$$\text{angle of } G(s)H(s) = (2n+1)\pi, \tag{6.132}$$

   where $n = 0, \pm 1, \pm 2, \ldots$
2. The magnitude of $G(s)H(s)$ must be unity:

$$|G(s)H(s)| = 1. \tag{6.133}$$

The construction of the root locus for a particular system can start by locating the open-loop poles and zeros in the complex plane. Other points on the locus can be obtained by choosing various test points and determining whether they satisfy Eq. (6.132). The angle of $G(s)H(s)$ can be easily determined at any test point in the complex plane by measuring the angles contributed to it by the various poles and zeros. For example, consider a feedback control system where

$$G(s)H(s) = \frac{K(s + s_A)(s + s_C)}{s(s + s_B)(s + s_D)}. \tag{6.134}$$

At some exploratory point s_E, $G(s)H(s)$ has the value

$$G(s_E)H(s_E) = \frac{K(s_E + s_A)(s_E + s_C)}{s_E(s_E + s_B)(s_E + s_D)}. \tag{6.135}$$

Graphically, Eq. (6.135) can be represented by Figure 6.56 where the vectors

$$\boldsymbol{\alpha} = s_E + s_A,$$

$$\boldsymbol{\beta} = s_E + s_B,$$

$$\gamma = s_E + s_C,$$

$$\boldsymbol{\delta} = s_E + s_D,$$

$$\epsilon = s_E.$$

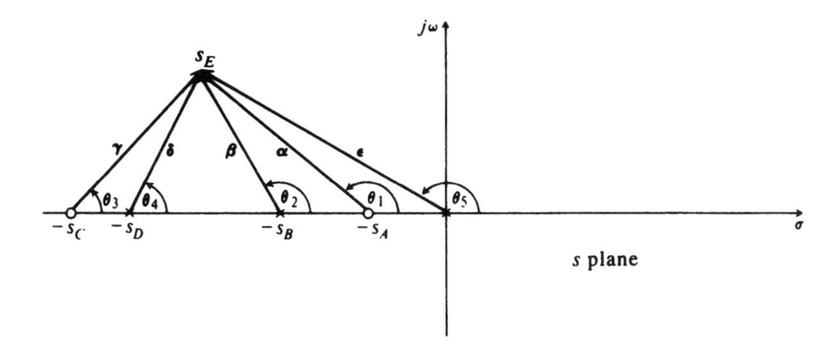

Figure 6.56 Vector representation of $G(s_E)H(s_E) = \dfrac{K(s_E + s_A)(s_E + s_C)}{s_E(s_E + s_B)(s_E + s_D)}$ at the exploratory point s_E.

The angle of $G(s_E)H(s_E)$ is the sum of the angles $\theta_1, \theta_2, \theta_3, \theta_4$, and θ_5, determined by the vectors α, β, γ, δ, and ϵ, respectively:

$$\begin{aligned}\text{Angle of } G(s_E)H(s_E) &= \sum \text{angles of vectors } \boldsymbol{\alpha}, \boldsymbol{\beta}, \boldsymbol{\gamma}, \boldsymbol{\delta}, \text{ and } \boldsymbol{\varepsilon} \text{ to point } s_E \\ &= \theta_1 - \theta_2 + \theta_3 - \theta_4 - \theta_5. \end{aligned} \tag{6.136}$$

If this angle equals $(2n + 1)\pi$ where $n = 0, \pm 1, \ldots,$ then s_E lies on the root locus. If it does not, the point s_E does not lie on the locus and a new point must be tried. When a point is found which does satisfy Eq. (6.132), the vector magnitudes are determined and are substituted into Eq. (6.133) in order to find the value of the gain constant K at the exploratory point s_E:

$$|G(s_E)H(s_E)| = \frac{K|(s_E + s_A)||(s_E + s_c)|}{|s_E||(s_E + s_B)||(s_E + s_D)|} = 1 \tag{6.137}$$

Fortunately, the actual construction of a root locus does not entail an infinite search through the complex plane. Because the zeros of the characteristic equation are continuous functions of the coefficients, the root locus is a continuous curve. Therefore, the root locus must have certain general patterns that are governed by the location and number of open-loop zeros and poles. Once these governing rules are established, the drawing of a root locus is not a tedious and lengthy trial-and-error procedure. We next present 12 basic rules that aid in determining the approximate location of the root locus.

Rule 1. The number of branches of the locus equals the order of the characteristic equation. This is because there are as many roots (and branches) of the root locus as the order of the characteristic equation. Each segment, or branch, of the root locus describes the motion of a particular pole of the closed-loop system as the gain is varied.

Rule 2. The open-loop poles define the start of the root locus ($K = 0$), and the open-loop zeros define the termination of the root locus ($K = \infty$). This can easily be shown by considering Eq. (6.137). At open-loop zeros, K must approach infinity because there is a zero in the numerator due to either s_A or s_C and the expression equals 1. However, K must approach zero when open-loop poles occur because there is a zero in the denominator due to the pole at the origin, s_B or s_D and the expression equals 1. When the order of the denominator of $G(s)H(s)$ is greater than the numerator, the root locus ends at infinity, whereas if the order of the numerator is greater than the denominator, the root locus starts at infinity.

This can be illustrated by considering the following characteristic equation of the control system:

$$1 + \frac{K(s+z_1)(s+z_2)\ldots(s+z_m)}{(s+p_1)(s+p_2)\ldots(s+p_n)} = 0. \tag{6.138}$$

Rearranging Eq. (6.138), the following is obtained:

$$\frac{(s+z_1)(s+z_2)\ldots(s+z_m)}{(s+p_1)(s+p_2)\ldots(s+p_n)} = -\frac{1}{K}. \tag{6.139}$$

To demonstrate that the points on the root locus which correspond to $K = 0$ are open-loop poles, let us let K approach zero in Eq. (6.139) as follows:

$$\lim_{K \to 0} \left| \frac{(s+z_1)(s+z_2)\ldots(s+z_m)}{(s+p_1)(s+p_2)\ldots(s+p_n)} \right| = \lim_{K \to 0} \frac{1}{K} = \infty. \tag{6.140}$$

Therefore, as K approaches zero, a pole must exist (s must approach one of the open-loop poles). Therefore, each branch of the root locus must start at a pole of the open-loop transfer function, $G(s)H(s)$. Now let us consider what happens as K is increased to infinity. For this case, each branch of the root locus approaches either a zero of the open-loop transfer function, $G(s)H(s)$, or infinity in the complex plane. To demonstrate that the points on the root locus which correspond to K equals to infinity are open-loop zeros, let K approach infinity in Eq. (6.139) as follows:

$$\lim_{K \to \infty} \left| \frac{(s+z_1)(s+z_2)\ldots(s+z_m)}{(s+p_1)(s+p_2)\ldots(s+p_n)} \right| = \lim_{K \to \infty} \frac{1}{K} = 0. \tag{6.141}$$

Therefore, as K approaches infinity, a zero must exist (s must approach one of the open-loop zeros). Note that there are the same number of poles and zeros in a root locus, if we include the zeros at infinity in the count. This is at it should be since every branch of the root locus starts at a pole ($K = 0$) and ends (terminates) at a zero ($K = \infty$).

Rule 3. Complex portions of the root locus always occur as complex-conjugate pairs if the characteristic equation is a rational function of s having real coefficients. Therefore, the root locus for this condition is symmetrical with respect to the real axis.

Rule 4. Sections of the real axis are part of the root locus if the number of poles and zeros to the right of an exploratory point along the real axis is odd. This is easily demonstrated from Eq. (6.132). Because the angular contribution along the real axis due to complex-conjugate poles cancels, the total angle of $G(s)H(s)$ is due only to the contributions of the real poles and zeros. Therefore, at any exploratory point along the real axis, the angular contribution due to a pole or zero to its right is 180°, whereas that due to a pole or zero to its left is zero.

Rule 5. Angles of asymptotes to the root locus α_m are given by [25]

$$\alpha_m = \pm \frac{(2m+1)\pi}{p-z}, \tag{6.142}$$

where

$$p = \text{number of finite open-loop poles,}$$
$$z = \text{number of finite open-loop zeros,}$$
$$m = 0, 1, 2, \ldots, \quad \text{up to } m = p - z \text{ (exclusively).}$$

Although m can have an infinite number of values as m is increased, the angle α_m repeats itself, and the number of distinct asymptotes is $p - z$.

To determine the asymptotes of the root locus, consider a test point s_E which is located very far from the origin. For this condition, the angle of each complex quantity can be considered to be the same. Open-loop poles then cancel the effect of open-loop zeros. Therefore, the root loci for very large values of s are asymptotic to straight lines. Because the total angular component must add up to $\pm 180°$ or some odd multiple, Eq. (6.142) follows. Figure 6.57 illustrates the asymptotes of a third-order system where

$$G(s)H(s) = \frac{K}{s(s+4)(s+5)}. \tag{6.143}$$

The asymptotic angles for the root locus illustrated in Figure 6.57 are

$$\alpha_0 = \pm\pi/3 = \pm 60°$$
$$\alpha_1 = \pm 3\pi/3 = \pm 180°.$$

Rule 6. The intersection of the asymptotes and the real axis occurs along the real axis at s_r, where [25]

$$s_r = \frac{\sum_{\text{poles}} - \sum_{\text{zeros}}}{\text{number of finite poles} - \text{number of finite zeros}}. \tag{6.144}$$

The value of s_r is the centroid of the open-loop pole and zero configuration.

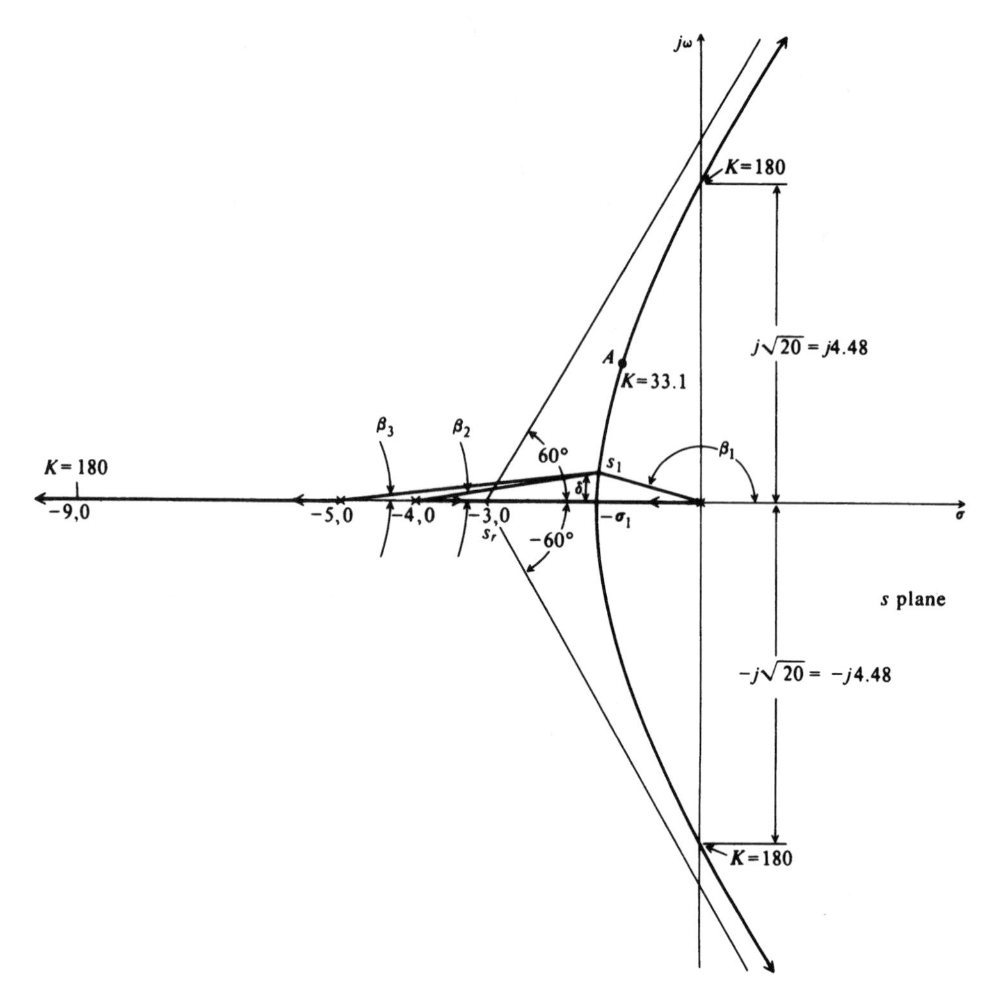

Figure 6.57 Root locus of a system where $G(s)H(s) = \dfrac{K}{s(s+4)(s+5)}$.

To determine the intersection of the asymptotes and the real axis, let us expand the numerator and denominator of the following open-loop transfer function:

$$\frac{K(z+z_1)(s+z_2)\ldots(s+z_m)}{(s+p_1)(s+p_2)\ldots(s+p_n)} = \frac{K[s^m+(z_1+z_2+z_3+\ldots+z_m)s^{m-1}+\ldots+z_1z_2\ldots z_m]}{s^n+(p_1+p_2+p_3+\ldots+p_n)s^{n-1}+\ldots+p_1p_2\ldots p_n}. \tag{6.145}$$

We are interested in a test point which is located very far from the origin of the s-plane. For this case, Eq. (6.145) can be rewritten as follows:

$$G(s)H(s) = \frac{K}{s^{n-m} + [(p_1 + p_2 + p_3 + \ldots + p_n) - (z_1 + z_2 + z_3 + \ldots + z_m)]\, s^{n-m-1} + \ldots}. \tag{6.146}$$

Substituting Eq. (6.146) into the characteristic equation

$$G(s)H(s) = -1 \tag{6.147}$$

we obtain the following result:

$$s^{n-m} + [(p_1 + p_2 + p_3 + p_n) - (z_1 + z_2 + z_3 + \ldots + z_m)]\, s^{n-m-1} + \ldots = -K. \tag{6.148}$$

For very large values of s, Eq. (6.144) can be approximated as follows:

$$\left[s + \frac{(p_1 + p_2 + p_3 + \ldots + p_n) - (z_1 + z_2 + z_3 + \ldots + z_m)}{n - m}\right] \ldots = 0 \tag{6.149}$$

Let us denote the intersection of the asymptotes and the root locus by s_r. Therefore,

$$s_r = \frac{(p_1 + p_2 + p_3 + \ldots + p_n) - (z_1 + z_2 + z_3 + \ldots z_m)}{n - m}. \tag{6.150}$$

Note that s_r is always a real number because complex zeros and poles always occur in complex-conjugate pairs.

The intersection of the asymptotes for the root locus illustrated in Figure 6.57 is given by

$$s_r = \frac{(-4 - 5) - 0}{3 - 0} = -3.$$

Rule 7. Consider the simple case where the locus has a branch on the real axis between two poles as shown in Figure 6.58*a*. A point must exist where the two branches *break away* from the real axis and enter the complex region of the s-plane in order to approach zeros which are finite or are located at infinity. Because K has a value of zero at the two poles and increases in value as the locus moves along the real axis away from the poles, the K's for the two branches simultaneously reach a maximum value at the breakaway point. A plot of $|K|$ versus σ is shown in Figure 6.58*b*. For the case where the locus has branches on the real axis between the two zeros as illustrated in Figure 6.58*c*, branches come from poles in the complex region and *break in* onto the real axis. The variation in value of $|K|$ along the real axis locus between these two zeros is shown in Figure 6.58*d*. The two poles that enter the real axis and then move to zeros on the real axis will enter simultaneously with a value of K, which is a minimum because the gain is increasing continuously as the loci approach the zeros. Thus the breakaway and

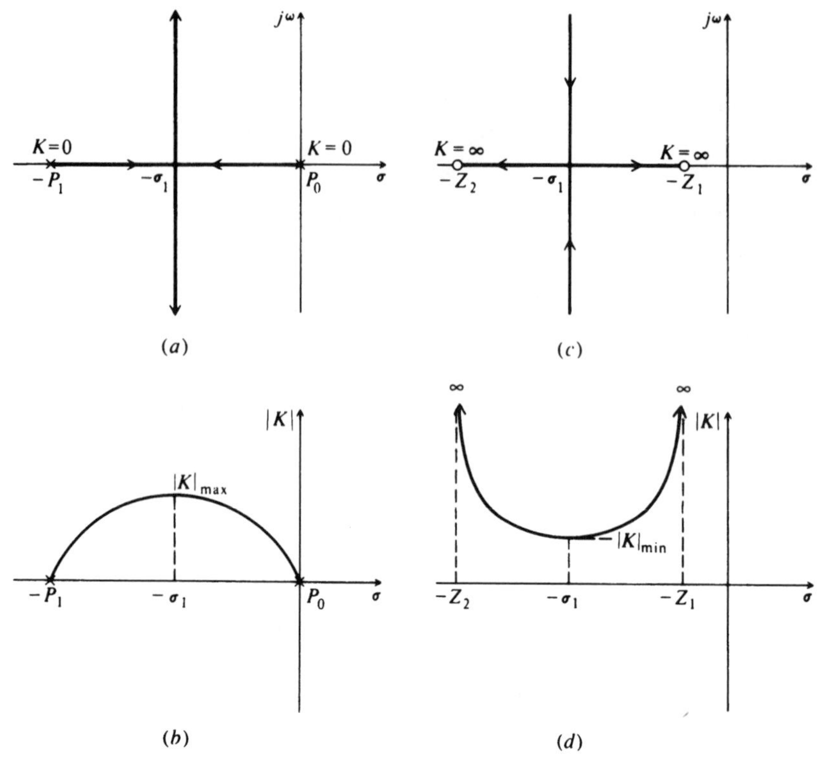

Figure 6.58 Loci for two consecutive poles and zeros on the real axis and the corresponding relation of $|K|$ versus s.

break-in-points can be evaluated from the magnitude condition for Re $s = \sigma$ and solving for $K(\sigma)$. This can be accomplished graphically or by solving

$$\frac{dK(\sigma)}{d\sigma)} = 0 \tag{6.151}$$

to find all the maxima and minima of $K(\sigma)$ and their locations. The most straightforward method for isolating the factor K is to rearrange the characteristic equation. An example will illustrate the procedure.

For the system analyzed in Figure 6.57

$$G(s)H(s) = \frac{K}{s(s+4)(s+5)}.$$

The characteristic equation of this system is given by

$$1 + G(s)H(s) = 1 + \frac{K}{s(s+4)(s+5)} = 0.$$

Alternatively, this can be written as

$$K = -s(s+4)(s+5).$$

For the case of Re $s = \sigma$, we obtain

$$K(\sigma) = -\sigma(\sigma + 4)(\sigma + 5).$$

Multiplying the factors together, we have

$$K(\sigma) = -\sigma^3 - 9\sigma^2 - 20\sigma.$$

Taking the derivative of this function and setting it equal to zero, we can determine the breakaway point:

$$\frac{dK}{d\sigma} = -3\sigma^2 - 18\sigma - 20 = 0.$$

The roots are

$$\sigma_1 = -1.47,$$

$$\sigma_2 = -4.53.$$

The breakway point of σ_1 is indicated in Figure 6.57; the value of σ_2 is not a possible solution in this negative-feedback example because the root locus does not exist on the real axis at this point. It is interesting to note that the point -4.53 is a breakaway point for this system when there is positive feedback present (see Section 6.16).

Points of breakaway of the root locus from the real axis can also be obtained by considering the transition from the real axis to a point s_1, which is a small distance δ off the axis. The basis of this method is that the transition from the real axis to s_1 must result in a zero net change of the angle of $G(s)H(s)$. This is illustrated for the root locus considered in Figure 6.57. For this example,

$$-(\beta_1 + \beta_2 + \beta_3) = (2n + 1)\pi, \quad n = 0, \pm 1, + \ldots \tag{6.152}$$

The very small angles we are considering are equal to their tangents, in radians, as follows:

$$-\left(\pi - \frac{\delta}{\sigma_1}\right) - \frac{\delta}{4 - \sigma_1} - \frac{\delta}{5 - \sigma_1} = -\pi.$$

This can be rewritten as

$$-\frac{\delta}{\sigma_1} + \frac{\delta}{4 - \sigma_1} + \frac{\delta}{5 - \sigma_1} = 0.$$

Canceling δ, and simplifying, results in the equation

$$-\frac{1}{\sigma_1} + \frac{1}{4 - \sigma_1} + \frac{1}{5 - \sigma_1} = 0. \tag{6.153}$$

Solution of Eq. (6.153) yields the absolute values of σ_1 equal to -1.47 and -4.53. As indicated before, the value of -4.53 is impossible for negative feedback.

Both techniques presented for determining the breakaway and break-in points are utilized in this book. They are referred to as the *maximization* (or *minimization*) of $K(\sigma)$ and the *transition* methods.

Rule 8. The intersection of the root locus and the imaginary axis can be determined by applying the Routh–Hurwitz stability criterion to the characteristic equation. The characteristic equation for the system illustrated in Figure 6.57 is given by

$$s^3 + 9s^2 + 20s + K = 0. \tag{6.154}$$

The resulting Routh–Hurwitz array is given by

$$\begin{array}{c|cc} s^3 & 1 & 20 \\ s^2 & 9 & K \\ s & \frac{180-K}{9} & \\ s^0 & K & \end{array}$$

For this simple array, a zero in the third row indicates a pair of complex-conjugate poles crossing the imaginary axis. The corresponding maximum value of gain and the value of s at which this occurs can be obtained as follows: For the third row to equal zero,

$$\frac{180 - K}{9} = 0$$

or

$$K_{\text{max}} = 180. \tag{6.155}$$

Therefore, this system is stable for all gains up to a value of 180. The corresponding value of s occurring at the crossing of the imaginary axis can be obtained from the characteristic equation, Eq. (6.154), with $K = 180$ and $s = j\omega$. We can either set the real part of Eq. (6.154) equal to zero, or we can set the imaginary part of Eq. (6.153) equal to zero. Either approach will give us the correct value for the crossing of the imaginary axis. Using the real parts, we find that

$$-9\omega^2 + 180 = 0,$$

or

$$s = \pm j\sqrt{20} = \pm j4.48. \tag{6.156}$$

These values are illustrated in Figure 6.57.

Rule 9. The angles made by the root locus leaving (or entering) a complex pole can be evaluated by applying the principle of Eq. (6.132). This is illustrated by considering the system shown in Figure 6.59. As indicated in Figure 6.59, Rule 5 shows that the asymptotes are at $\pm 60°$, and Rule 6 shows that the intersection of

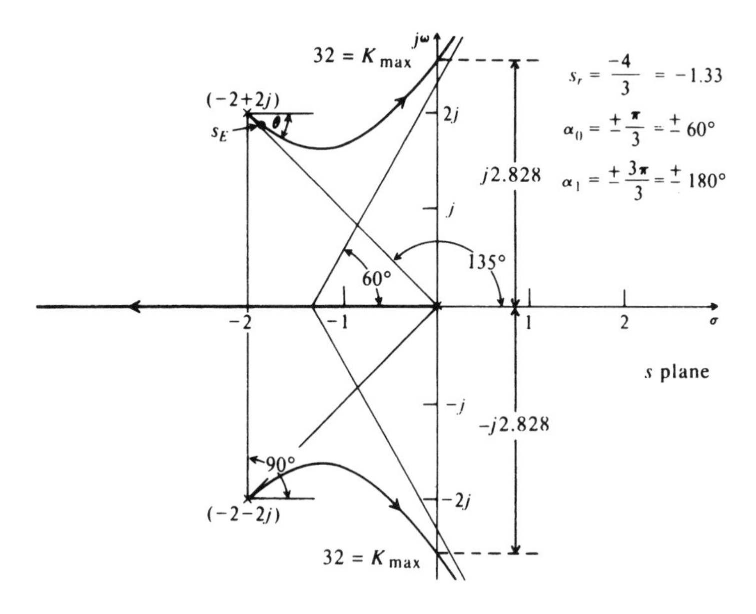

Figure 6.59 Root locus of system where $G(s)H(s) = \dfrac{K}{s(s^2+4s+8)}$.

the asymptotes and the real axis is at -1.33. We can find K_{max} and the crossing of the imaginary axis from Rule 8 as follows. The characteristic equation for the system shown in Figure 6.59 is given by

$$s^3 + 4s^2 + 8s + K = 0. \tag{6.157}$$

Therefore, its Routh–Hurwitz array is given by

$$\begin{array}{c|cc} s^3 & 1 & 8 \\ s^2 & 4 & K \\ s^1 & \frac{32-K}{4} = U_1 & \\ s^0 & K & \end{array}$$

Setting $U_1 = 0$, we find that $K_{max} = 32$. The crossing of the imaginary axis can be found by setting the real or imaginary terms of Eq. (6.157) equal to zero with $K = 32$. Using the real terms,

$$-4\omega^2 + 32 = 0. \tag{6.158}$$

Therefore, we find that the root locus crosses the imaginary axis at $s = \pm j2.828$ as shown in Figure 6.59.

Let us calculate the angle that the root locus makes with the complex pole located at $-2 + 2j$. An exploratory point s_E will be assumed slightly displaced from this pole.

The angle made by the root locus leaving the pole at $-2+2j$ to the point s_E is assumed to be $-\theta$, as illustrated in Figure 6.59. This angle is defined as the angle of departure. The angles contributed to the point s_E, due to various open-loop poles of the system, must conform with Eq. (6.132) and are given by (see Figure 6.59)

$$-135^\circ - 90^\circ + \theta = -180^\circ,$$
$$\theta = 45^\circ.$$

Therefore, the angle contributed by the branch of the root locus leaving the pole at $-2+2j$ must be sufficient to satisfy the basic relationshp given by Eq. (6.132). The negative sign appears before aech of the angles in this angular equation, because they are in the denominator of the expression given by Eq. (6.132). If zeros were present, however, they would contribute positive angles, because they are in the numerator of the expression given by Eq. (6.132). If we had assumed θ to be in the wrong direction, then our answer for θ would be negative.

Rule 10. In order to derive a useful relation between the poles of the open-loop transfer function and the roots of the characteristic equation, consider the following form of the open-loop transfer function for the system illustrated in Figure 6.55, where z_i represents the open-loop zeros and p_a represents the open-loop poles excluding those at the origin:

$$G(s)H(s) = \frac{K\prod_{i=1}^{x}(s-z_i)}{s^n \prod_{a=1}^{y}(s-p_a)}. \tag{6.159}$$

In practical physical systems

$$n + y > x,$$

and the denominator of $C(s)/R(s)$ is of the form

$$1 + G(s)H(s) = \frac{\prod_{j=1}^{m}(s-r_j)}{s^n \prod_{a=1}^{y}(s-p_a)}, \tag{6.160}$$

where

$$m = n + y$$
$$r_j = \text{roots described by the root locus.}$$

Substituting Eq. (6.159) into Eq. (6.160) and equating the expressions on each side of the resulting equation, gives the following:

$$s^n \prod_{a=1}^{y}(s-p_a) + K\prod_{i=1}^{x}(s-z_i) = \prod_{j=1}^{m}(s-r_j).$$

Expanding the product terms of this equation yields

$$\left(s^m - \sum_{a=1}^{y} p_a s^{m-1} + \ldots\right) + K\left(s^x - \sum_{i=1}^{x} z_i s^{x-1} + \ldots\right) = s^m - \sum_{j=1}^{m} r_j s^{m-1} + \ldots \tag{6.161}$$

For those open-loop transfer functions where the denominator of $G(s)H(s)$ is at least of degree 2 higher than that of the numerator (which is often the case in practice)

$$x \leqslant m - 2,$$

and the following is obtained by equating the coefficients of s^{m-1} in Eq. (6.161):

$$\sum_{a=1}^{y} p_a = \sum_{j=1}^{m} r_j.$$

By defining p_j to represent all of the open-loop poles, incuding those at the origin, this equation can be written as

$$\sum_{j=1}^{m} p_j = \sum_{j=1}^{m} r_j. \tag{6.162}$$

Equation (6.162), known as Grant's rule, indicates the sum of the system roots is a constant as the gain is varied from zero to infinity. Therefore, the sum of the system roots is conserved and is independent of gain. This rule, sometimes also referred to as the conservation of the sum of the roots, aids in the drawing the root locus, because it implies that as certain loci turn to the right, others must turn to the left in order that the sum of the closed-loop poles may be constant. In addition, this rule, as described by Eq. (6.162), aids in determining the gain along the root locus. For interest, we can determine the location of the third root of the system illustrated in Figure 6.57 when the root locus crosses the imaginary axis as follows:

$$\sum(-4 - 5) = \sum(+j4.48 - j4.48 + r).$$

Therefore, at $r = -9$, the gain is also 180.

Rule 11. The gain along the root locus can be determined in a number of ways. One of the two fundamental rules of the root locus, Eq. (6.133), can be used to determine this as indicated previously in Figure 6.56 and Eq. (6.137). Basically, for any point along the root locus, the control-system engineer can substitute the distance of the various poles and zeros to the point into Eq (6.133) and solve for K. As an example, let us reconsider the root locus illustrated in Figure 6.57. We have already determined that the gain when the root locus intersects the imaginary axis is 180 (using rule 8). Now, let us determine the value of gain at point A. To do this, we need to solve the following equation:

$$\left|\frac{K}{s_A(s_A + 4)(s_A + 5)}\right| = 1. \tag{6.163}$$

Measuring the distances $|s_A|$, $|s_A + 4|$, and $|s_A + 5|$ to be 2.2, 3.5, and 4.3, respectively, and substituting these values into Eq. (6.163), we obtain

$$\frac{K}{(2.2)(3.5)(4.3)} = 1$$

$$K = 33.1. \tag{6.164}$$

Gains along the rest of the root locus can be similarly determined.

Rule 12. The root locus never crosses itself. The accuracy of the root locus constructed can be greatly improved by use of the digital computer to implement the rules of construction. In Section 6.17, the application of MATLAB to obtaining the root locus is demonstrated. In Section 6.18, a working program for constructing the root locus is presented for those readers who do not have MATLAB available. Other commercially available software packages for constructing the root locus are presented in Section 6.18. Many of the root-locus plots presented hereafter in this book are constructed using either this digital computer program, or the MATLAB program.

This section will conclude with illustrative examples of the root-locus procedure. The techniques of stabilizing systems utilizing the root-locus method are illustrated also in Chapters 7, 8, 9 and 12.

Example 1

As the first example, consider a unity negative-feedback system whose open-loop transfer function is given by

$$G(s) = \frac{K}{(s+1)(s-1)(s+4)^2}. \tag{6.165}$$

This system has four poles: three on the negative real axis and one on the positive real axis. In addition, it has four zeros at infinity. The root locus of this system, illustrated in Figure 6.60, can be drawn on the basis of the 12 rules, as follows.

Rule 1. There are four separate loci, as the characteristic equation, $1 + G(s)H(s)$, is a fourth-order equation.

Rule 2. The root locus starts ($K = 0$) from the poles located at 1, -1, and a double pole at -4. All loci terminate ($K = \infty$) at zeros, which are located at infinity for this problem.

Rule 3. Complex portions of the root locus occur in complex-conjugate pairs, and the root locus is symmetrical with respect to the real axis.

Rule 4. The portions of the real axis between -1 and 1, and the point -4 are part of the root locus.

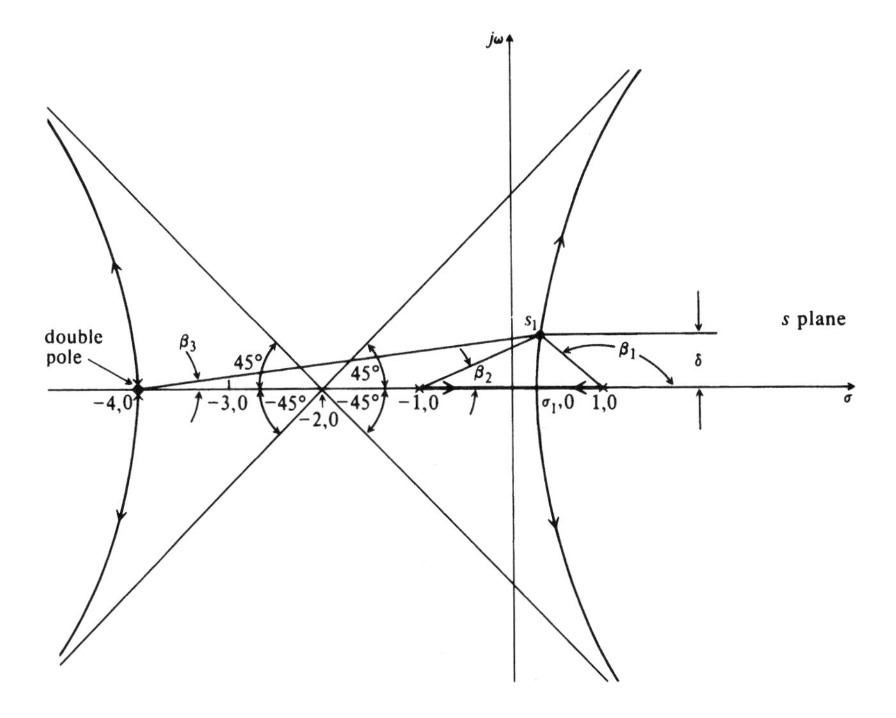

Figure 6.60 Root locus of system where $G(s)H(s) = \dfrac{K}{(s+1)(s-1)(s+4)^2}$.

Rule 5. The four loci approach infinity as K becomes large at angles given by

$$\alpha_0 = \pm\pi/4 = \pm 45^\circ$$

and

$$\alpha_1 = \pm 3\pi/4 = \pm 135^\circ.$$

Rule 6. The intersections of the asymptotic lines and real axis occur at

$$s_r = \frac{-8-0}{4-0} = -2.$$

Rule 7. The point of breakaway from the real axis is determined using the two techniques presented: maximization of $K(\sigma)$ and the transition methods.

Maximization of $K(\sigma)$ Method. From the equation

$$1 + G(s)H(s) = 1 + \frac{K}{(s+1)(s-1)(s+4)^2} = 0,$$

we obtain

$$K = -(s+1)(s-1)(s+4)^2,$$

$$K(\sigma) = -(\sigma^2-1)(\sigma+4)^2,$$

$$\frac{dK(\sigma)}{d\sigma} = -2(\sigma+4)(2\sigma^2+4\sigma-1) = 0.$$

The roots are

$$\sigma_1 = 0.22, \quad \sigma_2 = -2.22, \quad \sigma_3 = -4.$$

Of these, σ_1 represents the breakaway point from the positive real axis, and σ_3 represents the breakaway from the double pole at $-4, 0$ for this negative feedback system. The value of σ_2 represents the breakaway point for positive feedback (see Problem 6.48).

Transition Method. The point of a breakaway from the real axis occurring between -1 and 1 will be assumed to lie along the positive real axis at σ_1. The angles contributed from the various poles to a point s_1 that lies a small distance δ off the positive real axis are

$$-(\beta_1+\beta_2+2\beta_3) = (2n+1)\pi$$

$$-\left(\pi - \frac{\delta}{1-\sigma_1}\right) - \frac{\delta}{1+\sigma_1} - \frac{2\delta}{4+\sigma_1} = -\pi,$$

or

$$\frac{-\delta}{1-\sigma_1} + \frac{\delta}{1+\sigma_1} + \frac{2\delta}{4+\sigma_1} = 0.$$

Solving, we obtain roots at 0.22, -2.22 and -4. The interpretation of these roots is the same as with the first method.

Rule 8. This particular root locus intersects the imaginary axis only at the origin.

Rule 9. This rule does not apply to this problem.

Rule 10. This rule shows that as certain of the loci turn to the right, others turn to the left to ensure that the sum of the roots is a constant.

Rule 11. This rule does not apply to this problem, because the problem does not require the calculation of gains along the locus.

Rule 12. The root locus does not cross itself.

It is interesting to observe from the root locus illustrated in Figure 6.60 that the system is always unstable because at least one root of the characteristic equation always lies in the right half-plane. We illustrate in Section 7.9 the stabilization of this system by means of a lead network, and by using a minor loop containing rate feedback.

Example 2

As a second example, consider a unity negative-feedback system where

$$G(s) = \frac{k(1 + 0.1s)}{s(1 + s)(1 + 0.25s)^2} = \frac{1.6k(s + 10)}{s(s + 1)(s + 4)^2}. \tag{6.166}$$

It is important to recognize in Eq. (6.166) that the gain in this equation is $1.6k$. By defining $1.6k = K$, we obtain the following equation:

$$G(s) = \frac{K(s + 10)}{s(s + 1)(s + 4)^2}. \tag{6.167}$$

The gain defined on the root locus is K, and not k. Therefore, to find the system gain from the root locus, it is necessary to divide the gain determined from the root locus plot, K, by 1.6.

This system has four poles (two being a double pole) and one zero, all on the negative real axis. In addition, it has three zeros at infinity. The root locus of this system, illustrated in Figure 6.61a can be drawn on the basis of the 12 rules presented as follows:

Rule 1. Three are four separate loci, as the characterisic equation, $1 + G(s)H(s)$, is a fourth-order equation.

Rule 2. The root locus starts ($K = 0$) from the poles located at zero, -1, and a double pole located at -4. One pole terminates ($K = \infty$) at the zero located at -10, and three loci terminate at zeros, which are located at infinity.

Rule 3. Complex portions of the root locus occur as complex-conjugate pairs, and the root locus is symmetrical with respect to the real axis.

Rule 4. The portions of the real axis between the origin and -1, the double poles at -4, and between -10 and $-\infty$ are part of the root locus.

Rule 5. The loci approach infinity as K becomes large at angles given by

$$\alpha_0 = \pm\frac{\pi}{4 - 1} = \pm 60^\circ$$

and

$$\alpha_1 = \pm\frac{3\pi}{4 - 1} = \pm 180^\circ.$$

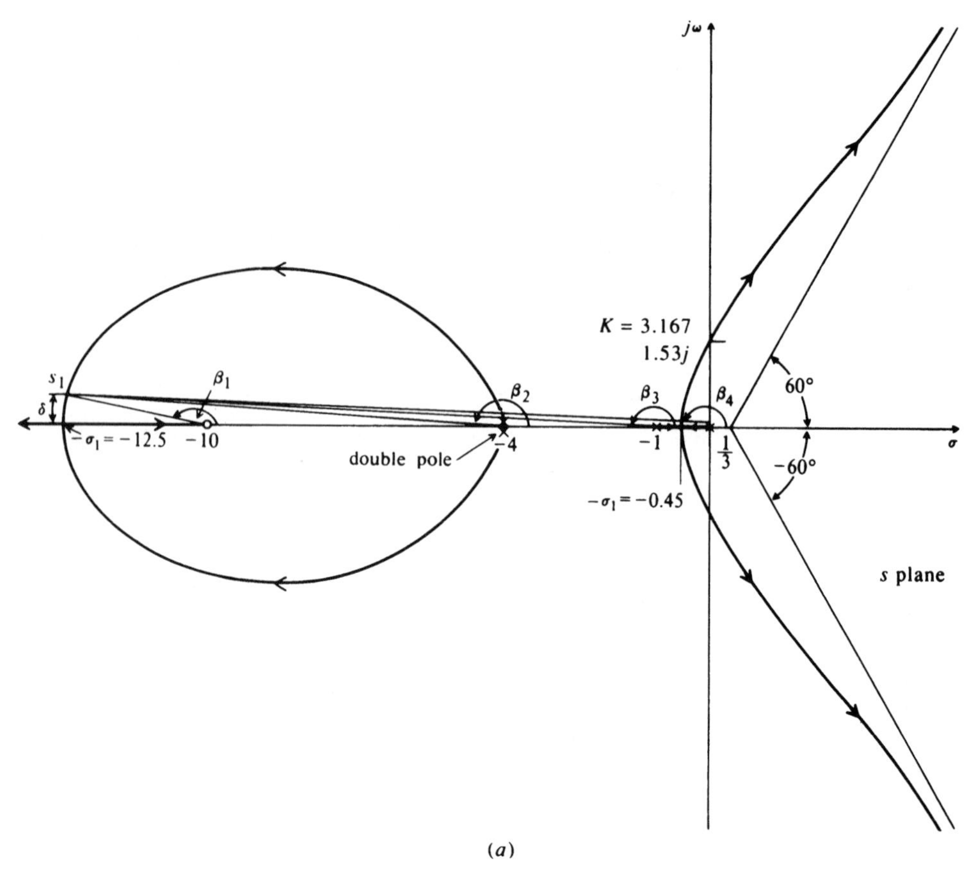

Figure 6.61(a) Root locus obtained using 12 rules of construction of negative-feedback system where

$$G(s)H(s) = \frac{1.6K(+10)}{s(s+1)(s+4)^2}.$$

Rule 6. The intersection of the asymptotic lines and the real axis occur at

$$s_r = \frac{-9 - (-10)}{4 - 1} = 0.33.$$

Rule 7. The point of breakaway from the real axis is determined using the maximization of $K(\sigma)$ and the transition methods.

Maximization of $K(\sigma)$ Method. From the relation

$$1 + G(s)H(s) = 1 + \frac{1.6K(s+10)}{s(s+1)(s+4)^2} = 0,$$

we have

$$K = -\frac{s(s+1)(s+4)^2}{1.6(s+10)},$$

$$K(\sigma) = -\frac{\sigma(\sigma+1)(\sigma+4)^2}{1.6(\sigma+10)}.$$

Taking the derivative of $K(\sigma)$ with respect to σ and solving, we obtain roots at -0.45, -2.25, -4, and -12.5. The root at -2.25 is impossible for the negative-feedback case, because the root locus does not lie here. In Section 6.16 we shall find that -2.25 is the breakaway point for a positive-feedback system. Breakaways occur at -0.45 and -4; a break in occurs at -12.5.

Transition Method. The point of breakaway from the real axis occurring between the origin and -1, and the break occuring between -10 and $-\infty$, are evaluated by summing the angles contributed from the various poles and zero to a point s_1, located a small distance δ off the negative real axis. In addition, there must be a breakaway point at the double pole at -4, because sections of the negative real axis on both sides of this double pole are not part of the root locus. The point chosen, as illustrated in Figure 6.61a, is to the left of $-10, 0$. Actually, it does not matter if the point is chosen on this segment of the root locus or between the $-1, 0$ point and the origin. The resulting equation, in either case, will result in the other points of breakaway:

$$[\beta_1 - 2\beta_2 - \beta_3 - \beta_4] = (2n+1)\pi,$$

$$\left(\pi - \frac{\delta}{\sigma_1 - 10}\right) - 2\left(\pi - \frac{\delta}{\sigma_1 - 4}\right) - \left(\pi - \frac{\delta}{\sigma_1 - 1}\right) - \left(\pi - \frac{\delta}{\sigma_1}\right) = -3\pi.$$

Solving, we obtain values of σ_1 at -0.45, -2.25, -4, and -12.5. the interpretation of these roots is the same as with the first method.

Rule 8. The intersection of the root locus and the imaginary axis can be determined by applying the Routh–Hurwitz stability criterion to the characteristic equation:

$$s^4 + 9s^3 + 24s^2 + (16 + 1.6K)s + 16K = 0.$$

Unfortunately, the Routh–Hurwitz method for a fourth-order system can given two possible answers. For example, in this problem, possible values of K where the root locus intersects the imaginary axis are 125 (from the second row) and 3.167 (from the third row). An alternative technique, which overcomes this problem, is to first solve for the frequencies where the locus intersects the imaginary axis, and then obtain the maximum values of gain from this expression. Letting $s = j\omega$, the characteristic equation becomes

$$\omega^4 - j9\omega^3 - 24\omega^2 + j\omega(16 + 1.6K) + 16K = 0. \tag{6.168}$$

The frequencies where the locus crosses the imaginary axis can be calculated from the real or the imaginary part of Eq. (6.168). Using the real part,

$$\omega^4 - 24\omega^2 + 16K = 0 \tag{6.169}$$

and using the imaginary part,

$$\omega = \sqrt{\frac{16 + 1.6K}{9}}. \tag{6.170}$$

Solving Eqs. (6.169) and (6.170) simultaneously, we obtain

$$K_{max} = 3.167$$

or

$$k_{max} = 1.9794.$$

and the system is stable for $0 < K < 3.167$. Substituting $K_{max} = 3.167$ into Eq. (6.170), we obtain

$$\omega = \sqrt{\frac{16 + 1.6(3.167)}{9}} = 1.53 \text{ rad/sec}$$

as the frequency of crossover of the imaginary axis.

Rule 9. This rule does not apply to this problem.

Rule 10. This rule shows that as certain of the loci turn to the right, others turn to the left to ensure that the sum of the roots is a constant.

Rule 11. This rule does not apply to this problem, because the problem does not require the calculation of gains along the locus.

Rule 12. The root locus does not cross itself.

The root locus of this system obtained using the 12 rules of construction is illustrated in Figure 6.61*a*. The root locus of this system obtained using the program MATLAB (see Section 6.17) is shown in Figure 6.61*b*, and is contained in the M-file that is part of my MCSTD Toolbox and it can be retrieved free from The MathWorks, Inc., anonymous FTP server at ftp://ftp.mathworks.com/pub/books/shinners.

Pertinent values of the root locus determined in this section and the rest of the book where the root locus is applied (for both positive and negative feedback) use the following special functions, which are part of my MCSTD Toolbox and are discussed in greater detail in Section 6.17: rlaxis; rlpoba; rootmag; rootangl. They determine the following:

1. rlaxis determines portions of the real axis that are part of the root locus.

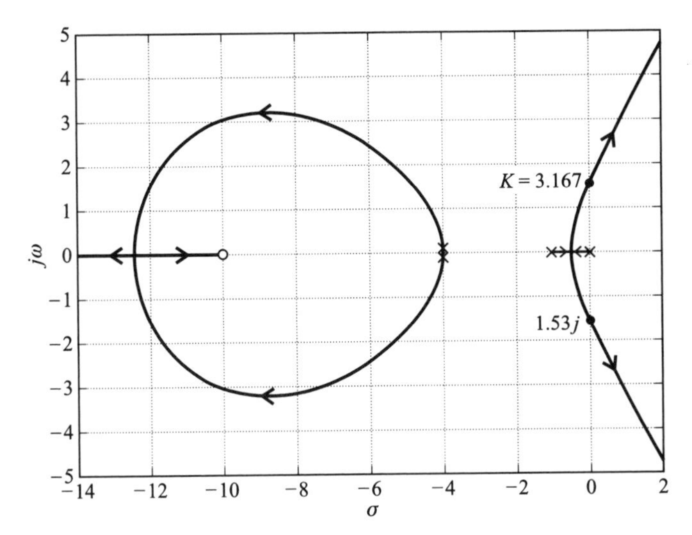

Figure 6.61(b) Root locus obtained using MATLAB for same system illustrated in figure (a).

2. rlpoba determines the points of breakaway and break-in, and the values of gain at these points.
3. rootmag determines gain and values of the roots at a particular distance from the origin. (This is very valuable to us in Chapter 9 where the root locus is extended to discrete systems and we wish to find the gain and root values when the root locus crosses a unit circle that is the stability boundary for discrete systems.)
4. rootangl determines the gain and values of the roots at a particular angle from the origin. For example, to find K_{max} on the root locus and the value of s when it crosses the imaginary axis, the angle would be made equal to 90°. Another example is to determine the value of gain resulting for a particular value of ζ through the application of $\alpha = \cos^{-1}(\zeta)$.

These calculations are based on the root locus definitions, and are analytical and very accurate rather than using graphical techniques. My MCSTD Toolbox can be retrieved free from The MathWorks, Inc. anonymous FTP server at ftp://ftp.mathworks.com/pub/books/shinners.

Example 3

As a concluding third example for obtaining the root locus of a negative-feedback system, let us consider a unity-feedback system wehre

$$G(s) = \frac{K(s+4)}{(s+0.5)^2(s+2)}. \tag{6.171}$$

This system has three poles (two being a double pole) and one zero. The root locus of this system, illustrated in Figure 6.62, can be drawn on the basis of the rules of construction presented as follows:

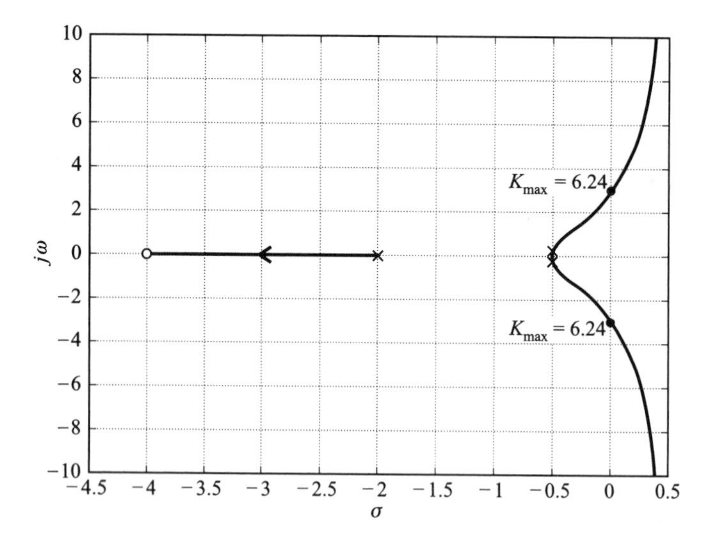

Figure 6.62 Root locus of feedback system where $G(s)H(s) = \dfrac{K(s+4)}{(s+0.5)^2(s+2)}$.

Rule 1. There are three separate loci, as the characteristic equation is third order.

Rule 2. The root locus starts ($K = 0$) from the double poles located at −0.5, and at −2. One pole terminates ($K = \infty$) at the zero located at −4, and two loci terminate at zeros, which are located at infinity.

Rule 3. Complex portions of the root locus occur as complex-conjugate pairs, and the root locus is symmetrical with respect to the real axis.

Rule 4. The portions of the real axis between −2 and −4, and the point where the double poles exist at −0.5 are part of the root locus.

Rule 5. The loci approach infinity as K becomes large at angles given by

$$\alpha_0 = \pm\frac{\pi}{3-1} = \pm 90^\circ$$

Rule 6. The intersection of the asymptotes and the real axis occur at

$$s_r = \frac{(-0.5 - 0.5 - 2) - (-4)}{3-1} = 0.5.$$

Rule 7. The point of breakaway is obviously at the double pole at −0.5. It is not necessary to apply this rule to find that fact.

Rule 8. The value of K_{max} and the intersection of the root locus and the imaginary axis can be found using the Routh–Hurwitz method as follows. The characteristic equation of this system is given by

$$s^3 + 3s^2 + (2.25 + K)s + (4K + 0.5) = 0;$$

the resulting Routh–Hurwitz array is given by

$$\begin{array}{l|cc} s^3 & 1 & 2.25 + K \\ s^2 & 3 & 4K + 0.5 \\ s^1 & U_1 = \dfrac{3(2.25 + K) - (4K + 0.5)}{3} & \\ s^0 & 4K + 0.5 & \end{array}$$

Setting the term in the third row $U_1 = 0$, we find that

$$K_{max} = 6.24.$$

Therefore, this system is stable for gains of

$$0 < K < 6.24.$$

To find the intersection of the root locus and the imaginary axis, we substitute $s = j\omega$ and $K = 6.24$ into the characteristic equation. By setting the real or imaginary parts of the resulting equation equal to zero, we find that the intersection occurs at $\pm j2.92$.

Rules 9, 10, 11, and 12 are not needed for this problem. Observe that the root locus plot of Figure 6.62 was obtained using the commercially available software package known as MATLAB (see Section 6.17), and is contained in the M-file that is part of my MCSTD Toolbox and can be retrieved free from The MathWorks, Inc. anonymous FTP server at ftp://ftp.mathworks.com/pub/books/shinners.

6.15. ROOT LOCUS OF TIME-DELAY FACTORS [26–28]

We previously considered time-delay factors in Section 6.7 when we discussed the Bode-diagram approach. The transfer function of a time-delay factor was defined in Eq. (6.95), and an example was solved using the Bode-diagram approach for a unity feedback control system whose forward transfer function was defined in Eq. (6.103). How would we draw the root locus of a control system containing a time-delay factor?

To answer this question, let us consider a unity-feedback control system whose forward transfer function, which contains a time-delay factor, is given by the following

$$G(s) = \frac{Ke^{-Ts}}{s + 2} \tag{6.172}$$

and is illustrated in Figure 6.63.

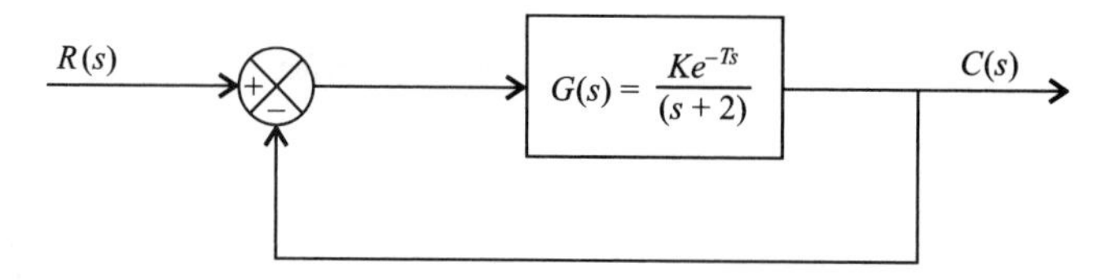

Figure 6.63 Control system containing a time-delay factor.

The characteristic equation of the control system shown in Figure 6.63 is given by

$$1+\frac{Ke^{-Ts}}{s+2}=0 \tag{6.173}$$

or

$$\frac{Ke^{-Ts}}{s+2}=-1. \tag{6.174}$$

Therefore, the angular condition is given by

$$\angle\frac{Ke^{-Ts}}{s+2}=\angle e^{-Ts}-\tan^{-1}\frac{\omega}{2}=(2n+1)\pi, \quad n=0,\pm1,\pm2,\ldots. \tag{6.175}$$

The angular condition of e^{-Ts} is given by

$$\angle e^{-Ts}=\angle e^{-T(\sigma+j\omega)}=\angle e^{-T\sigma}+\angle e^{j\omega T}. \tag{6.176}$$

Since the term $e^{-T\sigma}$ is real, its angular value is zero. Therefore,

$$\angle e^{-Ts}=\angle e^{-j\omega T}=-\omega T \text{ rad}=-\frac{180}{2\pi}\omega T \text{ degrees}=-57.3\omega T \text{ degrees} \tag{6.177}$$

as shown in Eq. (6.96). Therefore, the angular condition of Eq. (6.175) is given by

$$-57.3\omega T-\tan^{-1}\frac{\omega}{2}=(2n+1)\pi, \quad n=0,1,\pm2,\ldots. \tag{6.178}$$

The value of T in this example is assumed to equal one second and, therefore, the angle of e^{-Ts} is a function of ω only:

$$-57.3\omega-\tan^{-1}\frac{\omega}{2}=(2n+1)\pi, \quad n=0,\pm1,\pm2,\ldots. \tag{6.179}$$

For the case of $\omega=0$, the angular contribution of -57.3ω is zero. Therefore, the real axis from -2 to inus infinity forms a branch of the root locus. We can prove that $(Ke^{-Ts})/(s+2)$ approaches minus infinity as s approaches minus infinity:

$$\lim_{s\to-\infty}\frac{Ke^{-Ts}}{s+2}=\left.\frac{\frac{d}{ds}(Ke^{-Ts})}{\frac{d}{ds}(s+2)}\right]_{s\to-\infty}=-KTe^{-Ts}\Big]_{s\to-\infty}=-\infty \tag{6.180}$$

Therefore, minus infinity represents a pole with the time-delay factor.

The root locus for this control system is illustrated in Figure 6.64.

Let us consider the angular condition of Eq. (6.175). It has an infinite number of values depending on n ($n = 0, \pm 1, \pm 2, \ldots$). Therefore, there are an infinite number of root loci. To illustrate this, suppose $n = 1$. Therefore, the angular condition from Eq. (6.179) is:

$$-57.3\omega - \tan^{-1}\frac{\omega}{2} = +540^\circ. \tag{6.181}$$

The root locus for $n = 1$ is the same as for $n = 0$, except that it is shifted along the imaginary axis by 2π. The imaginary axis crossings of the asymptotes are given by

$$\omega = \frac{(1 + 2n)\pi}{T} \tag{6.182}$$

where $T = 1$ in this example and where $n = 0, \pm 1, \pm 2, \ldots$. Figure 6.65 illustrates the root locus for this system for $n = 0$ and $n = \pm 1$. Observe that although there are an

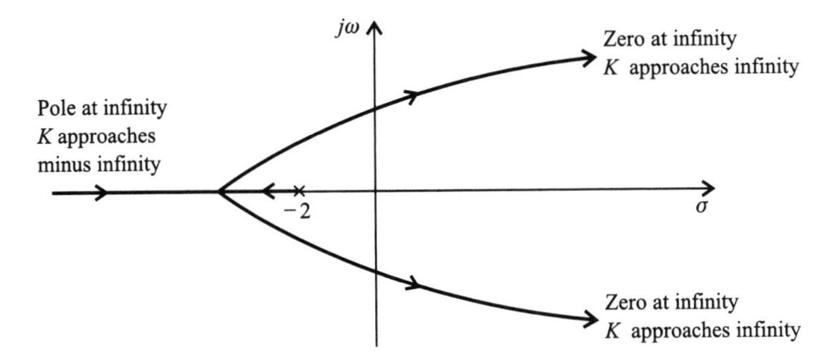

Figure 6.64 Root locus for system in Figure 6.63 for $n = 0$.

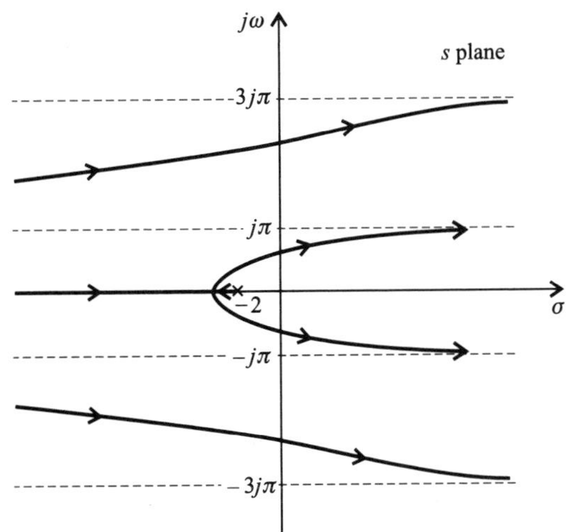

Figure 6.65 Root locus of system of Figure 6.63 for $n = 0$ and $n = \pm 1$.

infinite number of root locus branches, the only branch of interest lies between $-j\pi$ and $j\pi$. The other branches corresponding to $n = \pm 1, \pm 2, \ldots$ are not as important and can be neglected.

6.16. ROOT-LOCUS METHOD FOR POSITIVE-FEEDBACK SYSTEMS

The rules presented in the previous section for constructing the root locus were directed toward negative-feedback systems. For positive-feedback systems, several of these rules must be modified. The purpose of this section is to indicate the changes to the rules and apply them to the second problem considered in Section 6.14 for positive, instead of negative, feedback.

For positive feedback, Eq. (6.130) becomes

$$1 - G(s)H(s) = 0 \tag{6.183}$$

or

$$G(s)H(s) = 1 = 1 \underline{/2n\pi} \tag{6.184}$$

where $n = 0, \pm 1, \pm 2, \pm 3, \ldots$. Equation (6.184) specifies two conditions that must be satisfied for the existence of a closed-loop pole in positive-feedback systems.

1. The angle of $G(s)H(s)$ must be an even multiple of π:

$$\text{angle of } G(s)H(s) = 2n\pi, \tag{6.185}$$

   where $n = 0, \pm 1, \pm 2, \pm 3, \ldots$.
2. The magnitude of $G(s)H(s)$ must be unity:

$$|G(s)H(s)| = 1. \tag{6.186}$$

Based on Eqs. (6.185) and (6.186), it is necessary to modify Rules 4, 5, 7, and 9 given in Section 6.14 for construction of the root locus with negative feedback as follows:

Rule 4. This rule is modified for positive feedback so that sections of the real axis are part of the root locus if the number of poles and zeros to the right of an exploratory point along the real axis is even.

Rule 5. For positive feedback, the numerator in Eq. (6.142) is changed to $2m\pi$.

Rules 7 and 9. The transition method of Rule 7 and the angle-of-departure method of Rule 9 are modified in both cases so that the sum of the angles in the calculations is $2n\pi$ instead of $(2n + 1)\pi$ where $n = 0, \pm 1, \pm 2, \ldots$. The procedure of Rule 7 using the maximization of $K(\sigma)$ method is exactly the same for negative- and positive-feedback systems.

These changes will now be considered in view of the second example presented in Section 6.14 with positive instead of negative feedback. The open-loop transfer function of the unity-feedback linear system with positive feedback is given by [see Eq. (6.166)]

$$G(s) = \frac{1.6k(s+10)}{s(s+1)(s+4)^2}. \tag{6.187}$$

Let us construct the root locus for this system by reconsidering the changes required to Rules 4, 5, 7, and 9.

Rule 4. The portions of the real axis between $+\infty$ and the origin, and between -1 and -10, are part of the root locus as indicated in Figure 6.66.

Rule 5. The loci approach infinity as K becomes large at angles given by

$$\alpha_0 = 0$$

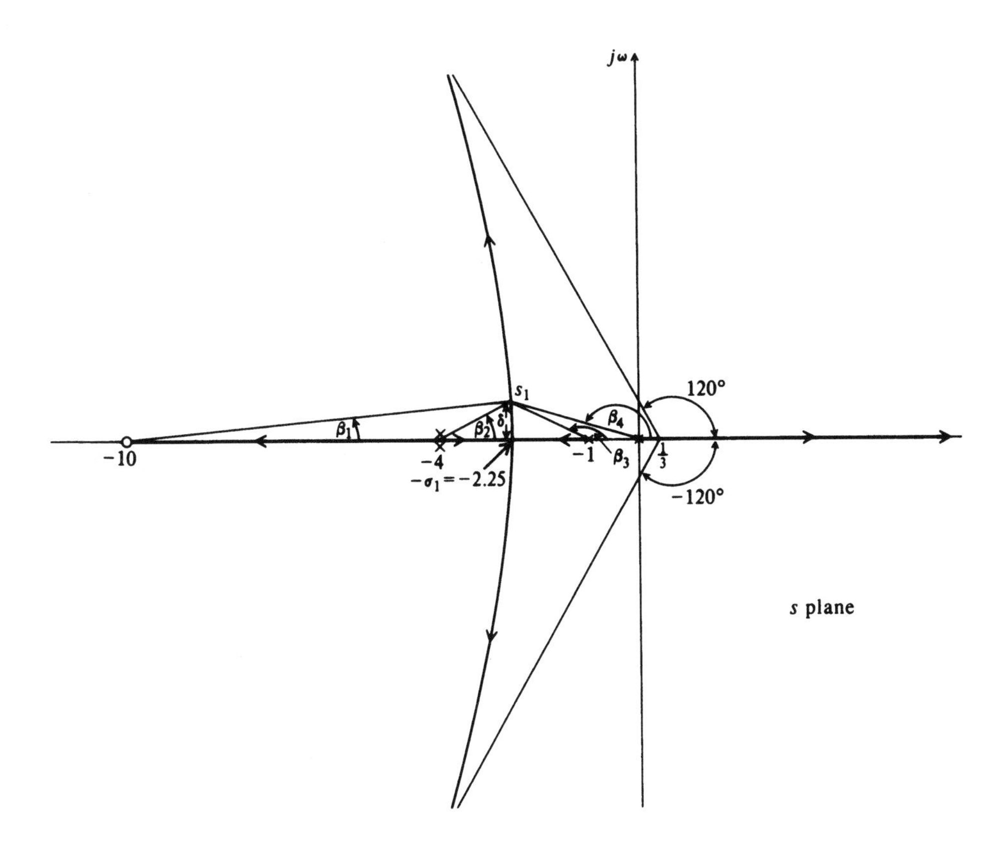

Figure 6.66 Root locus of a positive-feedback system where $G(s)H(s) = \dfrac{1.6k(s+10)}{s(s+1)(s+4)^2}$.

and

$$\alpha_1 = \pm \frac{2\pi}{4-1} = \pm 120^\circ.$$

Rule 7. The point of breakaway from the real axis is determined using the maximization of $K(\sigma)$ and the transition from the real-axis to the complex-plane methods.

Maximization of $K(\sigma)$ Method. The procedure and resulting equations are exactly the same as for the negative-feedback case presented in Section 6.14. The only difference is in the interpretation of the resulting roots. For the positive-feedback case only the root at −2.25 is possible; the roots at −0.45, −4 and −12.5 are impossible.

Transition Method. The point of breakaway from the real axis can be computed from the following equation:

$$[\beta_1 - 2\beta_2 - \beta_3 - \beta_4] = 2n\pi, \quad n = 0, \pm 1, + \ldots,$$

$$\left[\frac{\delta}{10-\sigma_1} - \frac{2\delta}{4-\sigma_1} - \left(\pi - \frac{\delta}{\sigma_1 - 1}\right) - \left(\pi - \frac{\delta}{\sigma_1}\right)\right] = -2\pi.$$

This equation is similar to the one obtained for the negative-feedback-system case. Solving, we obtain the values of σ_1 as before: −0.45, −2.25, −4, and −12.5. In this case, however, only the root at −2.25 is possible. The other solutions at −0.45, −4, and −12.5 are impossible.

Rule 9. This rule does not apply to this problem.

The complete root locus for this system is illustrated in Figure 6.66. It is interesting to observe from this figure that the system is unstable for all values of gain, as compared to the negative-feedback case (see Figure 6.61), where we found that the system was stable for $0 < K < 3.167$.

A useful interpretation is possible if we combine the analysis of Figure 6.61 and 6.66 and study the behavior of the feedback system where

$$G(s)H(s) = \frac{1.6k(s+10)}{s(s+1)(s+4)^2}$$

for

$$-\infty \leqslant k \leqslant \infty.$$

Figure 6.67 illustrates this combination of negative- $(0 \leqslant K \leqslant \infty)$ and positive-feedback $(-\infty \leqslant K \leqslant 0)$ behavior. Note that the root locus is a continuous curve as K varies from minus infinity to plus infinity.

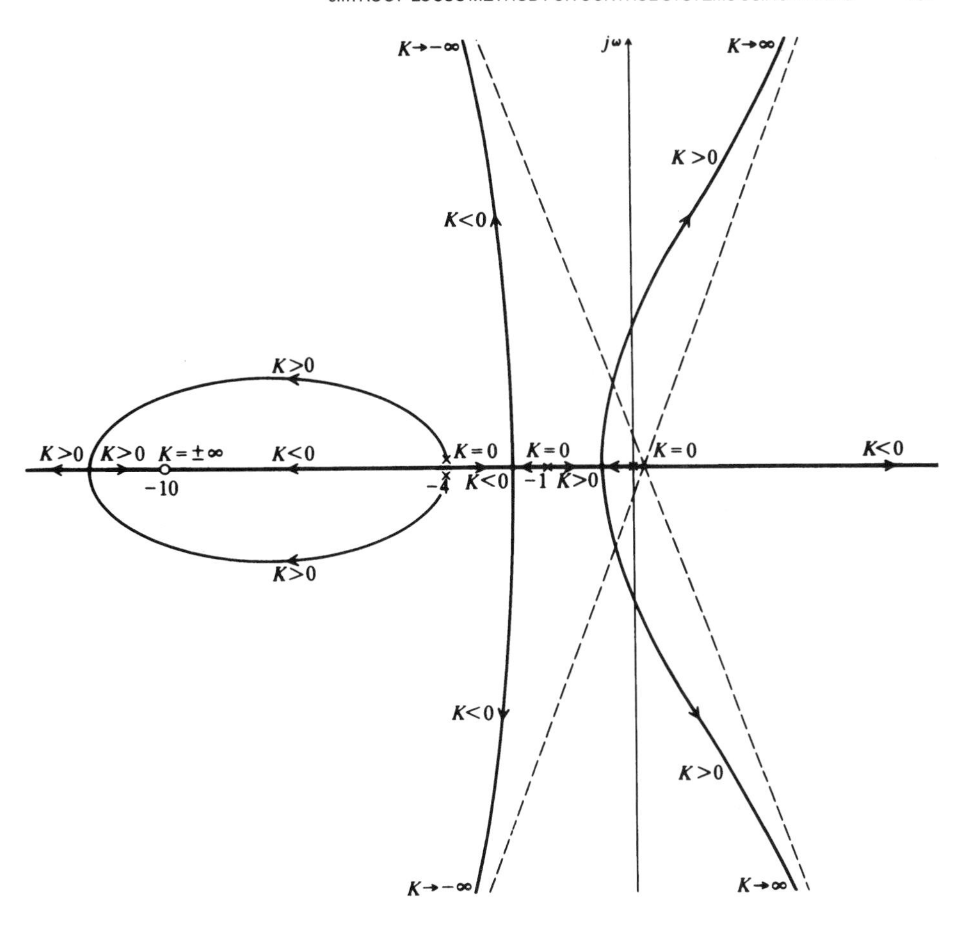

Figure 6.67 Root locus of a feedback system where $G(s)H(s) = \dfrac{1.6k(s+10)}{s(s+1)(s+4)^2}$, for $-\infty \leqslant k \leqslant \infty$.

6.17. ROOT-LOCUS METHOD FOR CONTROL SYSTEMS USING MATLAB [7]

In Sections 6.14–6.16, several root-locus plots were generated using the 12 rules of construction and were hand drawn. Additionally, several root-locus plots were also illustrated which were obtained using MATLAB. In this section, the reader will be shown how to obtain the root-locus diagrams very easily and accurately using MATLAB.

A. Drawing the Root Locus if the System is Defined by a Transfer Function

The MATLAB command used for plotting the root locus is

```
rlocus(num,den)
```

The *Modern Control System Theory and Design* (MCSTD) Toolbox enhances the professional toolbox by not only plotting the root locus with "rlocus" but also calculating almost any point of interest on the root-locus plot directly (the professional version has only "rootfind," a graphical method, to locate these points). The most commonly required value, K_{max}, can now be obtained directly using the MCSTD Toolbox.

What is left now is to understand how to create root-locus diagrams, and to practice creating them. There are many examples of creating root-locus diagrams on the MCSTD diskette (try some). Several functions exist that assist with root-locus diagrams, of which I have listed ones that I believe are important:

1. "rlocus" is described in Reference 7. This function returns/plots the root-locus response of the system. A single example is given in the text manual.
2. "rlaxis" is described in the MCSTD Toolbox. This function returns a set of data representing the portion of the root locus that is on the real axis.
3. "rlpoba" is described in the MCSTD Toolbox. This function returns the value(s) (with their associated gains) of the breakaway/break-in points of the root locus.
4. "rootmag" is described in the MCSTD Toolbox. This function returns the value(s) (with their associated gains) of any point on the root locus that corresponds to the specified magnitude. In the discrete root locus, a magnitude of 1 corresponds to K_{max}.
5. "rootangl" is described in the MCSTD Toolbox. This function returns the value(s) (with their associated gains) of any point on the root locus that corresponds to the specified angle. In the continuous root locus, an angle of ± 90 corresponds to K_{max}. Also, desired gains that accomplish specified damping factors (corresponding to particular angles), can easily be picked off (usually presented in compensation and design using the root locus).

These MCSTD features/utilities alone make the MCSTD Toolbox well worth any design engineer's review. Most of the critical values that normally have to be picked off the plot can be calculated directly using the MCSTD Toolbox. The K_{max} value, indicating the maximum gain before instability occurs, is calculated. Breakaway and break-in points of the root locus are also calculated in the MCSTD Toolbox, instead of by doing them by hand.

When trying to find the proper syntax to call the "rlocus" utility, either use the help feature or look in the reference manual. Valid syntax for the "rlocus" utility is described in Reference 7. Like the Bode utility, the left-hand arguments (r and k) are optional for this function.

Other functions dealing with the root locus are also available only in the MCSTD Toolbox. The MCSTD demo M-file does elaborate further on the creation of the root locus and the important associated points of interest, with on-screen examples. The significance of these utilities will be understood as you progress to the appropriate topics (e.g., discrete time system analysis).

To illustrate the use of MATLAB for obtaining the root locus, let us reconsider the root locus plotted in Figure 6.61*b* for the negative-feedback control system defined by Eq. (6.166) which is repeated here:

$$G(s) = \frac{1.6k(s+10)}{s(s+1)(s+4)^2}. \tag{6.188}$$

In order to ease its transformation to MATLAB notation, we multiply all terms in the numerator and denominator (or use the MATLAB "conv" command):

$$G(s) = \frac{k(1.6s+16)}{s^4+9s^3+24s^2+16s}. \tag{6.189}$$

Therefore, the row matrices for the numerator and denominator are as follows:

$$\text{num} = [0 \quad 0 \quad 0 \quad 1.6 \quad 16]$$

$$\text{den} = [1 \quad 9 \quad 24 \quad 16 \quad 0].$$

Because the root locus goes to infinity, it can lead to difficulties when using MATLAB. Therefore, we have to limit the area of the root locus drawn in the same manner we did for the Nyquist diagram which was discussed in Section 6.6. Therefore, we also add the command "v = [−14 4 − 5 5]; axis (v)." The resulting MATLAB program to draw the root locus is given in Table 6.21.
The resulting root locus diagram obtained using MATLAB is illustrated in Figure 6.68 and Fig. 6.61*b*.

Table 6.21. MATLAB Program for Obtaining the Root Locus for the System Defined by Eq. (6.189)

```
num = [0  0  0  1.6  16]
den = [1  9  24  16  0]
rlocus(num,den)
v = [-14  4  -5  5]; axis(v)
title('Root Locus of Unity Feedback System where G(s) is given
by Eq. (6.189)')
```

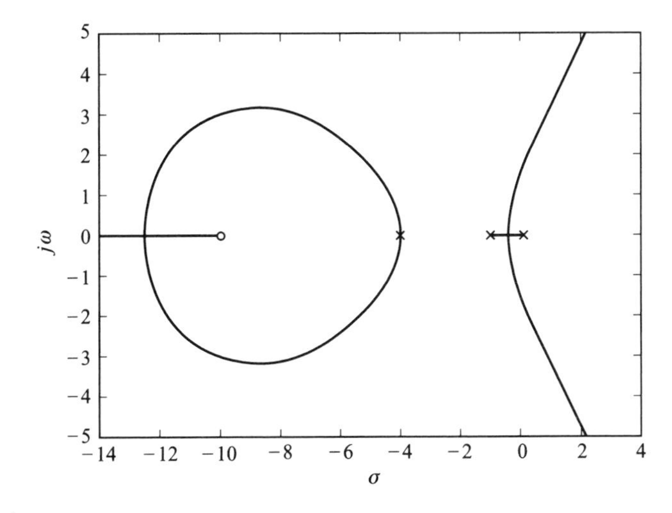

Figure 6.68 Root locus of unity-feedback system where $G(s)$ is given by Eq. (6.189).

The MCSTD Toolbox contains many additional commands which can greatly aid the control-system engineer in finding additional important information on the root locus. These are illustrated as follows:

(a) *Determining portions of the real axis which are part of the root locus*: Use the command *rlaxis(num,den)*. The program output for this problem is

$$\begin{array}{cc} -\infty & -10.0000 \\ -4.0000 & -4.0000 \\ -1.0000 & 0 \end{array}$$

(b) *Determining the point of breakaway and or break-in of the root locus, and the value of the gain K at these points*: Use the command *rlpoba(num,den)*. The program output for this problem is

```
kbreak =
        2.5964e + 003
              0
        2.0417e – 001
sbreak =
        – 12.4812
         – 4.0000
         – 0.4435
```

(c) *Determining the gain at the crossing of the imaginary axis, and the imaginary axis value*: Use the command *rootangl(num,den,angle)*, and set the angle portion equal to 90 as follows:

```
rootangl(num,den,90)
```

The program output for this problem is

```
kimag =
        3.1692
simag =
        0.0000 + 1.5301i
```

The MATLAB Program of Table 6.21 is modified to add and display the additional commands of rlaxis, rlpoba, and rootangl in the MATLAB Program in Table 6.22.

The resulting root locus containing the original root-locus plot shown in Figure 6.68 and the additional information obtained from the three commands rlaxis, rlpoba, and rootangl is shown in Figure 6.69. This problem is also shown in

Table 6.22. MATLAB Program of Table 6.21 with Additional Command of rlaxis, rlpoba, and rootangl

```
num = [0  0  0  1.6  16]
den = [1  9  24  16  0]
rlocus(num,den)
v = [-14  4  - 5  5]; axis(v)
realaxis = rlaxis(num, den)
[kbreak,sbreak] = rlpoba(num,den)
[kimag,simag] = rootangl(num,den,90)
disp('realaxis = rlaxis(num,den);' )
disp('[kbreak,sbreak] = rlpoba(num,den);' )
disp('[kimag,simsag] = rootangl(num,den,90;' )
title('Root locus of Figure 6.68 with Additional Points of
Interest Labeled')
```

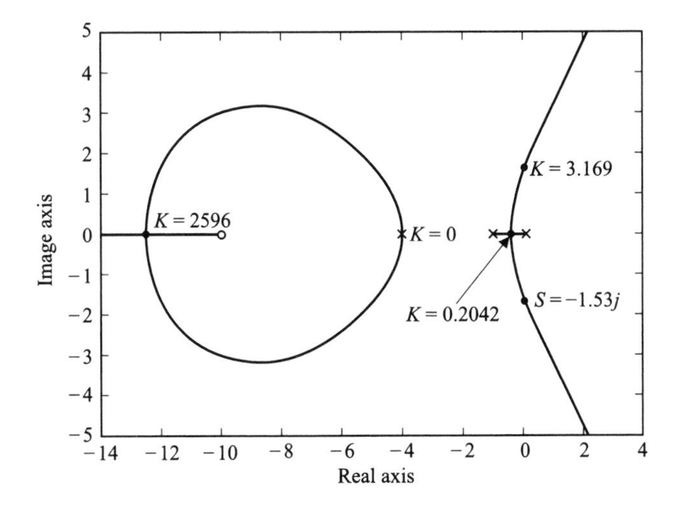

Figure 6.69 Root locus of Figure 6.68 with additional points of interest labeled.

DEMO 4 of the MCSTD Toolbox where additional features for drawing the root locus from commands found in the MCSTD Toolbox are illustrated.

B. Drawing the Root Locus if the System is Defined in State-Space Form

The MATLAB command for this case is given by

rlocus(A,B,C,D).

As was demonstrated previously in this chapter (e.g., Section 6.6 on the Nyquist Diagram, and Section 6.8 on the Bode diagram), we must first obtain the state-space form and then use this new command.

To demonstrate this procedure, let us obtain the state-space form from the transfer function given by Eq. (6.189) using the MATLAB command

$$[A,B,C,D] = \text{tf2ss(num,den)}$$

whose application has been demonstrated before with the sections on the Nyquist diagram (Section 6.6) and the Bode diagram (Section 6.8). The resulting MATLAB program for accomplishing this is shown in the MATLAB program in Table 6.23.

Therefore, the resulting MATLAB program for obtaining the root locus of this problem from the state-space formulation is given by the MATLAB program in Table 6.24. The resulting root locus diagram is identical to that shown in Figure 6.68.

Table 6.23. MATLAB Program for Transforming the System of Eq. 6.189 from Transfer Function to State-Space Form

```
num[0  0  0  1.6  16]
den = [1  9  24  16  0]
[A,B,C,D] = tf2ss(num,den)
A =
      -9   - 24   - 16  0
       1     0      0  0
       0     1      0  0
       0     0      1  0
B =
       1
       0
       0
       0
C =
       0    0    1.6000    16.0000
D =
       0
```

Table 6.24. MATLAB Program for Obtaining the Root Locus for the System Defined by Eq. 6.169 from its State-Space Form determined in Table 6.23

```
A = [-9  - 24  - 16  0; 1  0  0  0; 0  1  0  0; 0  0  1  0];
B = [1; 0; 0; 0];
C = [0  0  1.6  16];
D = [0];
rlocus(A,B,C,D)
v = [-14  4  - 5  5]; axis(v)
title('Root Locus of Unity Feedback System where G(s) is given
by Eq. (6.188)'
```

6.18. DIGITAL COMPUTER PROGRAM FOR OBTAINING THE ROOT LOCUS

For those who do not have access to MATLAB, which has now become an industry standard (more or less), this book also provides another digital computer program for obtaining the root locus. This approach for providing both MATLAB and other digital computer programs which are not dependent on COTS (commercial-off-the-shelf software) is used throughout this book.

The root locus can be plotted automatically using a variety of methods, including utilizing a digital computer [29–31]. This section presents a working digital computer program for obtaining the root locus. In our discussions, we will only consider the case of negative feedback. The program can be modified with a few simple changes for positive feedback.

The digital computer is a very versatile and flexible tool that is easily adaptable for automatically determining the roots in the complex plane. The method preseted, based on the material of References [29–31], starts at the poles of the control system and searches for the overall root locus in a segmented manner.

Reference [29] discusses this conceptual algorithm for obtaining the root locus using a digital computer. It presents the logic that can be used to code a digital computer in order to determine the locus of points which satisfy Eqs. (6.132), (6.133), (6.151). The program presented in this section is written in Fortran IV.

The computer logic flow diagrams of the technique are illustrated in Figures 6.70*a–d* and *f*. The basis of the conceptual algorithm is a convergent trial-and-error procedure for progressing from a known point on the root locus to a succeeding point. Initial points for the algorithm can be determined by realizing that the locus begins at the poles where the loop gain equals zero. For poles located on the real axis, the locus will remian on it until some breakaway point is reached. For these loci, it will be necessary to determine the breakaway points and begin a search in the complex plane from these points. Terminating points of the root locus are zeros that are located at some finite value or at infinity. Therefore, it is necessary to test and deterine if a locus point is near a zero or if it appears to be heading away from the origin of the complex plane.

The procedure determines the root locus in segments that are ultimately joined together. The process is composed of the following six phases:

(a) Input and initialization—Figure 6.70*a*.
(b) Determine locus segments on real axis—Figure 6.70*a*.
(c) Determine breakaway and break-in points of root locus from real axis in positive (upper) half of complex plane—Figure 6.70*b*.
(d) Search for root loci which begin at poles/zeros in positive (upper) half of complex plane—Figure 6.70*c*.
(e) Search for root loci which begin at poles/zeros on real axis and subsequently enter the positive (upper) half of complex plane—Figure 6.70*f*.
(f) Reflect the root-locus segments of the positive (upper) half-plane into the negative (lower) half-plane—Figure 6.70*f*.

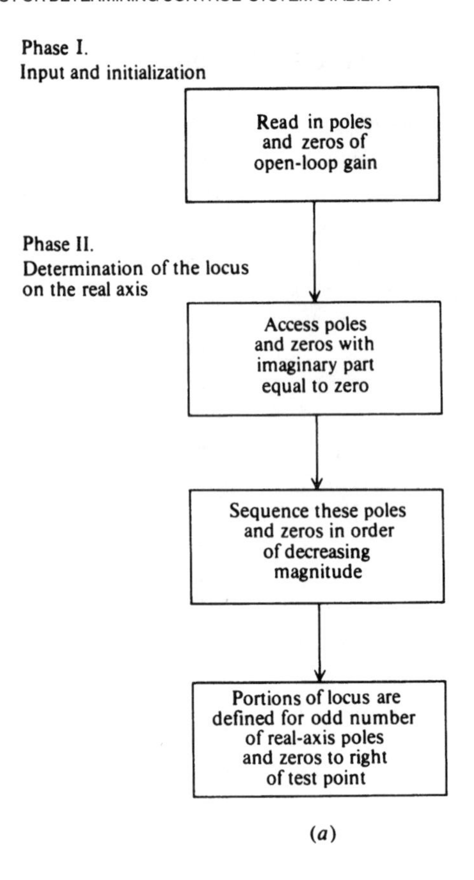

Figure 6.70(*a*) Digital computer logic flow chart, phases I and II.

Figure 6.70*d* illustrates a computer subroutine associated with Figure 6.70*c* and denoted as "next point." Let us next consider some of the details associated with each of these computer steps.

The *input and initialization phase*, illustrated in Figure 6.70*a*, is concerned with reading in poles and zeros of the open-loop gain.

The second phase of the procedure, denoted as the *determination of the locus on the real axis* in Figure 6.70*a*, consists of three sequences. Poles and zeros with imaginary parts equal to zero are determined, and are then sequenced in order of decreasing magnitude. Portions of the locus are then defined for an odd number of real-axis poles and zeros, which exist to the right of a test point.

The third phase, illustrated in Figure 6.70*b* is concerned with the *determination of breakaway and break-in points of the root locus from the real axis in the positive (upper) half of the complex plane*. The procedure involves determining the location of the maxima and minima of the function $K(\sigma)$ as discussed in Rule 7 for construction of the root locus.

Search for root loci which begin/end at poles/zeros in the positive (upper) half of the complex plane, the fourth phase, is illustrated in Figure 6.70*c*. Successive points of

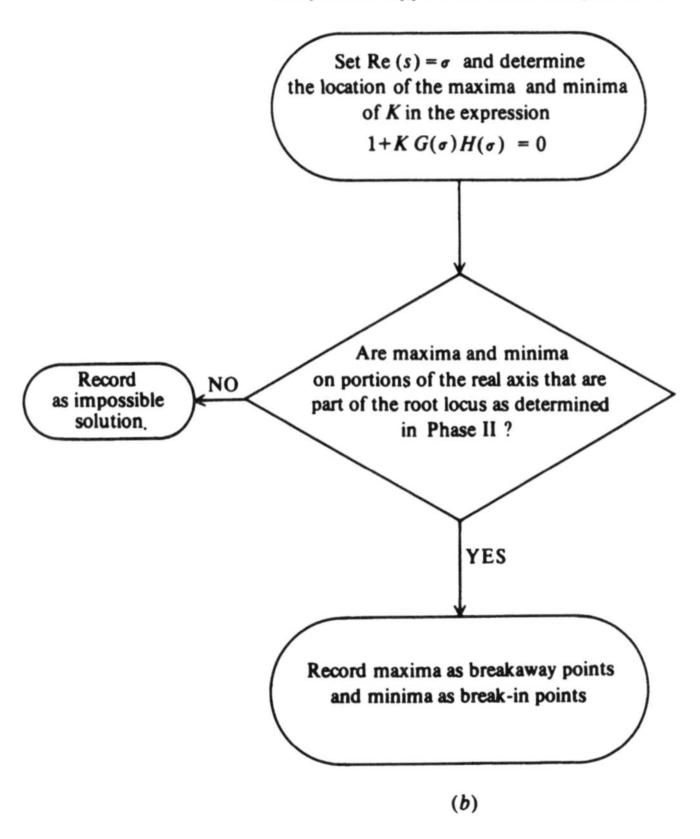

Figure 6.70(*b*) Digital computer logic flow chart, phase III.

the root locus are determined and the search for additional points are terminated if the following conditions occur:

(a) A zero/pole is recognized as the termination of the current locus.
(b) The termination of the current locus occurs at infinity/zero.
(c) The current locus intercepts the real axis.

If the last condition occurs, then it is necessary to determine the succeeding behavior of the root locus. In general, the following three situations can occur:

(d) The locus continues onto the real axis and terminates at a zero/pole on the real axis that is located at a finite value or at infinity.
(e) the locus continues on the real axis until a breakaway point occurs and it then reenters the full complex plane.
(f) The locus immediately reenters the imaginary portion of the complex plane.

Phase VI
Search for root loci which begin/end at poles/zeros on the positive (upper) half of the complex plane.

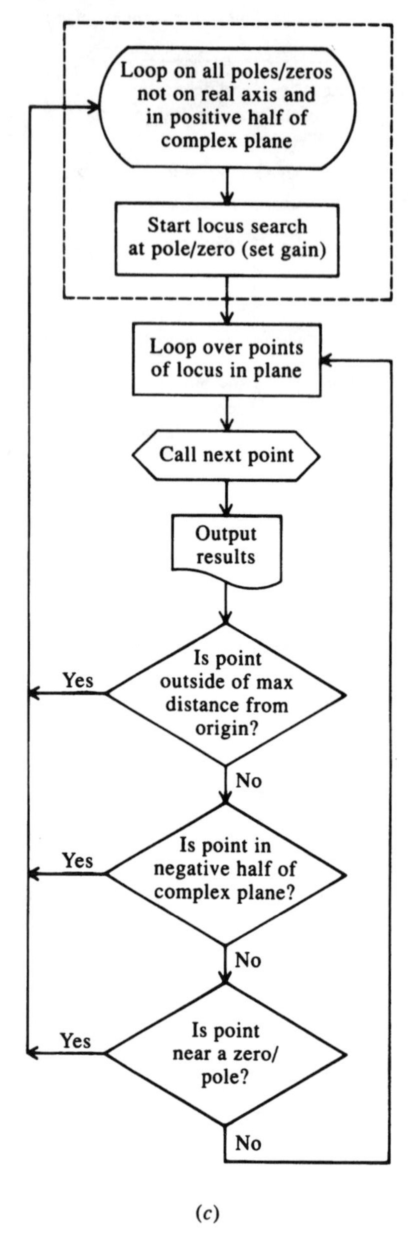

Figure 6.70(*c*) Digital computer logic flow chart, phase IV.

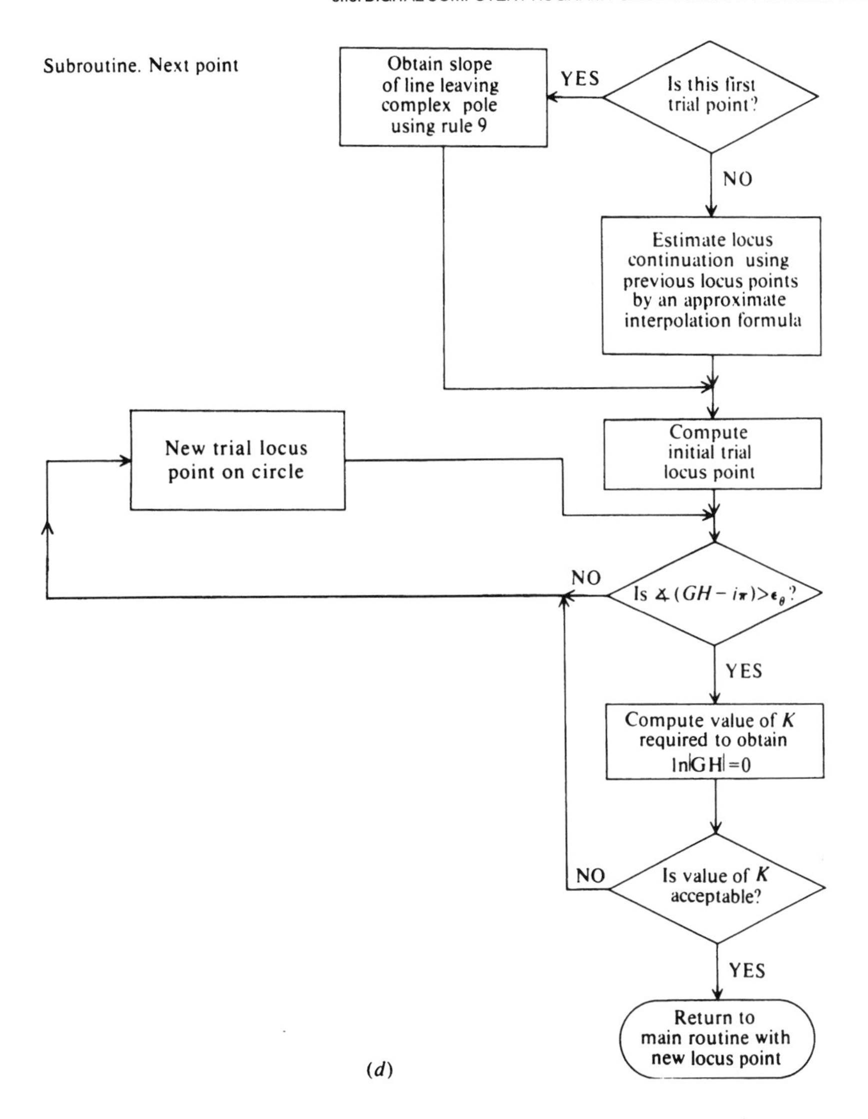

Figure 6.70(*d*) Digital computer logic flow chart, subroutine.

As each new locus point is obtained, it is examined to determine whether conditions (a), (b), or (c) occur.

Each real-axis intercept should be tested to determine if it lies between the bounds of real-axis root-locus segments determined in the second phase. If it does not, computer error should be noted, and the computer scan should be redone with a higher-accuracy set. Similarly, the real-axis intercepts should be compared with the breakaway points determined in the third phase. If the intercept is not within some small error, then an error condition should be recognized. If more than one break-in point exists at the breakaway point, tests should be made to deterimne which of them has the same slope as the current root locus. If no correspondence is obtained, then an error condition should be recognized.

The root-locus continuation is examined next. A test is made to determine if the root locus can be continued at a nearby zero/pole or a zero/pole located at infinity. If it can, the full locus from start to completion is recorded and displayed. If the root locus reenters the imaginary portion of the complex plane, it will occur at the nearest unassigned breakaway point. The repeated use of "call next point" can then be applied.

The "next point" flow diagram is illustrated in Figure 6.70*d*. The real and imaginary parts of the point in the complex plane are treated in a real two-dimesional space. An estimate is made of the continuation of the root locus. The initial trial locus point is obtained by computing the intersection of the estimated root-locus continuation with a circle having a small radius r_Δ, as illustrated in Figure 6.70*e*. There will be two intersecting points and the computer logic must be able to differentiate between them. The net angle contribution for all poles and zeros is computed at the trial locus point. The difference $\Delta\theta$ is computed and compared with the maximum permissible angular deviation of GH from $i\pi$, ϵ_θ. If $\Delta\theta \leqslant \epsilon_\theta$, then the angle criterion is satisfied and the value of K required to obtain $|GH| = 1$ is determined. However, if $\Delta\theta > \epsilon_\theta$, a new trial locus point must be found. The procedure used is to search for an acceptable solution that falls on the previously defined circle illustrated in Figure 6.70*e*. In the figure, a test point δ_θ is chosen. The difference of the angular contribution of the poles and zeros from $i\pi$, evaluated at the new test point, is denoted as $\Delta'\theta$ in Figure 6.70*e*. The change in the angular error is used to determine a new trial locus point on the circle. For computer simplicity, a simple scan of the unit circle is used. After the location of the point and the value of gain are determined, the computer returns to the main routine.

The fifth phase, illustrated in Figure 6.70*f*, is to *search for root loci that begin/end at poles/zeros on the real axis and subsequently enter the positive (upper) half of the complex plane*. Only those as-yet-unassigned breakaway (break-in) points in the same segments as the poles (zeros) which lead into the right half-plane and which are also unassigned, are examined. If an assignment can be made, the continuation of

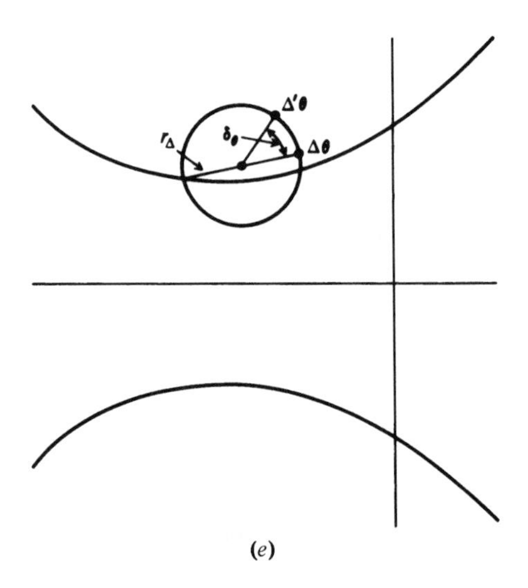

Figure 6.70(*e*) Estimated root locus point.

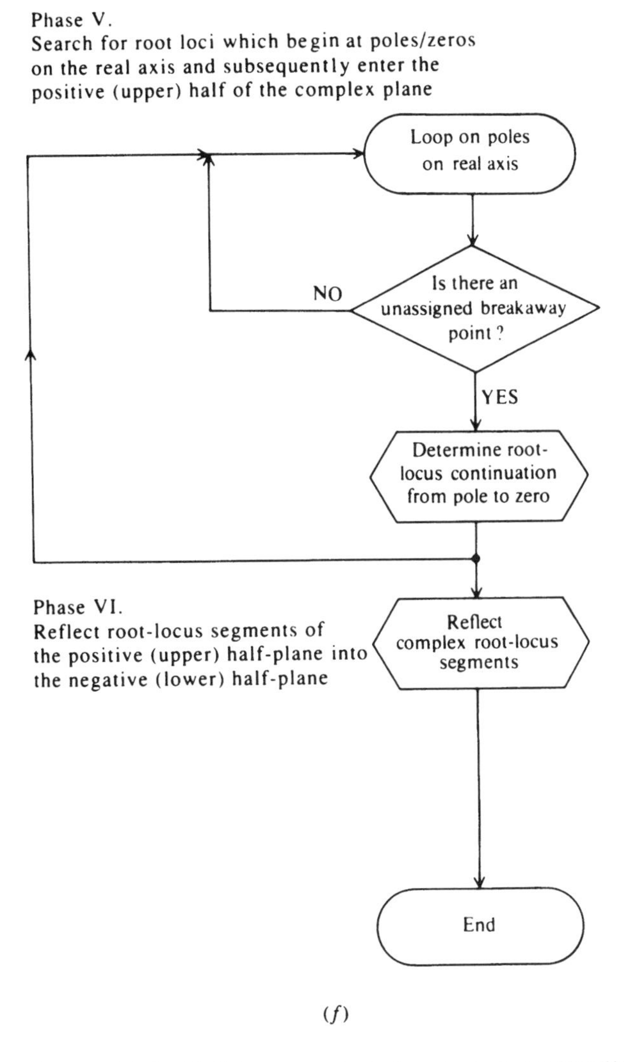
Phase V.
Search for root loci which begin at poles/zeros on the real axis and subsequently enter the positive (upper) half of the complex plane
Loop on poles on real axis
NO
Is there an unassigned breakaway point ?
YES
Determine root-locus continuation from pole to zero
Phase VI.
Reflect root-locus segments of the positive (upper) half-plane into the negative (lower) half-plane
Reflect complex root-locus segments
End
(f)

Figure 6.70(*f*) Digital computer logic flow chart, phases V and VI [29].

the current root locus is determined. The coding in the box labeled "determine root-locus continuation from pole to zero" is exactly the same as that used in the fourth phase (see Figure 6.70*c*) that exists outside the dotted lines.

The sixth phase, illustrated in Figure 6.70*f* is concerned with *reflecting the root-locus segments of the positive (upper) half-plane into the negative (lower) half-plane.* The fourth and fifth phases of the computer routine are sufficient to have determined all of the root-locus segments in the positive half-plane. Similar segments are also part of the root locus in the negative half-plane.

The computer program written in Fortran for computing the root locus is shown in Table 6.25 and should be compared with the flow charts of Figure 6.70*a*–*f* to obtain a thorough understanding of the technique. This program was applied to the root-locus problem analyzed in Figure 6.57 for a system whose transfer function was given by Eq. (6.143). Table 6.26 shows the computer run for the root locus of this

Table 6.25. FORTRAN Program for Generating the Root Locus Diagram

```
 1  C *** THIS PROGRAM LISTS THE ROOT LOCUS OF AN EQUATION
 2  C ***
 3  C *** MAXZP IS THE MAXIMUM # OF POLES OR ZEROS ALLOWED IN
 4         THIS PROGRAM PARAMETER (MAXZP=5)
 5  C *** BRKACC IS THE ACCURACY OF THE BREAK POINTS OF THE REAL-
         AXIS
 6  C *** DAXIS IS THE MOVEMENT DISTANCE OF THE PLOT POINTS
 7  C *** AXSMX IS THE MAXIMUM DISTANCE, FROM THE ORIGIN, TO PLOT
 8  C *** MAXTRY IS THE # OF PARTS THE UNIT CIRCLE IS BROKEN INTO
 9  C ***                FOR FINDING THE NEXT POINT ON THE LOCUS
10         PARAMETER    (BRKACC=0.01,    DAXIS=0.2,    AXSMX=6.0,
           MAXTRY=360)
11  C ***
12         PARAMETER (MAXA=2*MAXZP)
13         COMPLEX GAINK,ZP(0:1,MAXZP),FREQ,G
14         DIMENSION NZP(0:1),AXIS(MAXA),IZP(MAXA)
15         CHARACTER*4 NAME(0:1)/'ZERO','POLE'/
16  C ************** PHASE 1
17  C *** INITIALIZE THE INPUT, & LIST IT OUT
18         DO 100 I=0,1
19           READ(5,*)NZP(I),(ZP(I,J),J=1,NZP(I))
20           WRITE(6,*)' THERE ARE INPUT',NZP(I),' ',NAME(I),'(S) AT : '
21           DO 100 J=1,NZP(I)
22  100      WRITE(6*)ZP(I,J)
23  C *********** PHASE 2
24  C *** FIND THE REAL AXIS ZEROS & POLES
25         NAXIS=0
26         DO 200 I=0.1
27           DO 200 J=1,NZP(I)
28             IF(IMAG(ZP(I,J)).NE.0)GO TO 210
29             NAXIS=1+NAXIS
30             AXIS(NAXIS)=REAL(ZP(I,J))
31             IZP(NAXIS)=I
32  210      CONTINUE
33  200    CONTINUE
34  C *** SORT THE LIST OF ZEROS AND POLES ON THE AXIS (LIST IN PHASE
         3)
35         DO 220 I=1,NAXIS-1
36           DO 220 J=I+1,NAXIS
37             IF(AXIS(I).GE.AXIS(J))GO TO 220   @ NUMBER IS IN ORDER
38             TEMP=AXIS(I)
39             AXIS(I)=AXIS(J)
40             AXIS(J)=TEMP
41             ITEMP=IZP(I)
42             IZP(I)=IZP(J)
43             IZP(J)=ITEMP
44  220    CONTINUE
45  C ************* PHASE 3
46  C *** FIND THE BREAKS ON THE REAL AXIS
```

Table 6.25. ***(Continued)***

```
47          WRITE(6,*)
48          WRITE(6,*)' THE ROOT LOCUS SEGMENTS ON THE REAL AXIS ARE:'
49          DO 300 I=1,NAXIS-1,2
50            WRITE(6,*)
51            WRITE(6,*)NAME(IZP(I))&' AT',AXIS(I),', '&
52           +                    NAME(IZP(I+1)&' AT',AXIS(I+1)
53            IF(IZP(I)+IZP(I+1).EQ.1)GO TO 300 @ NO BREAKS, POLE TO ZERO
54            FREQ=CMPLX(AXIS(I))
55            IF(AXIS(I).EQ.AXIS(I+1))GO TO 390  @ TWO POINTS ARE THE
              SAME
56            OLDK=(1-IZP(I))*1.OE+30
57  310       ESTK=ABS(GAINK(FREQ-BRKACC,NZP,ZP))
58            IF((IZP(I).EQ.0).AND.(ESTK.GE.OLDK))GO TO 390
59            IF((IZP(I).EQ.1).AND.(ESTK.LE.OLDK))GO TO 390
60            FREQ=FREQ-BRKACC
61            OLDK=ESTK
62            IF(SNGL(FREQ-BRKACC).GT.AXIS(I+1))GO TO 310
63            STOP 'ERROR IN DETERMINING BREAK LCATION'
64  390       WRITE(6,*)1 WITH A BREAK AT ',SNGL(FREQ),' CONTINUING AS :'
65  C ************* PHASE 5
66  C *** CONTINUE THE ROOT LOCUS FROM THE BREAK, TO WHERE-EVER
67            WRITE(6,'(/42H     REAL(s)     IMAG(s)     ABS(K))')
68            ANGLE=90.0*(3.1416/180)                @ BREAK ANGLE
69            G=GAINK(FREQ+BRKACC/4,NZP,ZP)    @ GAIN NEAR BREAK
70  510       CALL NEXT(FREQ,G,ANGLE,IZP(I),NZP,ZP)
71  C************* PHASE 6
72  C *** PRINT OUT THE POSITIVE COMPLEX PLANE PORTION
73  C *** REFLECTION ONTO THE NEGATIVE COMPLEX PLANE IS ASSUMED
74            WRITE(6,*)FREQ,ABS(G)
75            IF(ABS(FREQ).GE.AXSMX) GO TO 300
76            IF(AIMAG(FREQ).LE.0) GO TO 300
77            J=IZP(I)
78            DO 320 K=1,NZP(1-J)
79  320         IF(ABS(FREQ-ZP(1-J,K)).LE.DAXIS/2)GO TO 510
80            GO TO 510
81  300       CONTINUE
82          WRITE(6,*)
83          IF(MOD(NAXIS,2).EQ.1)
84         +  WRITE(6,*)NAME(IZP(NAXIS))&' AT',AXIS(NAXIS) 'TO -INFINITY'
85  C ************* PHASE 4
86  C *** DO ALL NON-REAL AXIS POLES/ZEROS IN THE POSITIVE COMPLEX
          PLANE
87  C *** REFLECTION ONTO THE NEGATIVE COPMPLEX PLANE IS
          ASSUMED
88          DO 400 I=0,1
89            DO 410 J=1,NZP(I)
90              FREQ=ZP(I,J)
91              IF(IMAG(FREQ).LE.0) GO TO 410
92              WRITE(6,*)' FOUND POSITIVE COMPLEX ',NAME(I),' AT :',ZP(I,J)
```

Table 6.25. (*Continued*)

```
 93              G=(1-I)*1.03+30
 94              ANGLE=0.0
 95              WRITE(6,'(/42H   REAL(S)   IMAG(S)   ABS(K))')
 96   420        CALL NEXT(FREQ,G,ANGLE,I,NZP,ZP)
 97              WRITE(6,*)FREQ,ABS(G)
 98              IF(ABS(FREQ).GE.AXSMX)GO TO 410
 99              IF(AIMAG(FREQ).GT.0)GO TO 410
100              DO 430 K=1,NZP(1-I)
101   430          IF(ABS(FREQ-ZP(1-I,K)).LE.DAXIS/2)GO TO 410
102              GO TO 420
103   410      CONTINUE
104   400    CONTINUE
105          STOP
106   C ********************************************
107   C ********** FUNCTION TO CALCULATE THE REQUIRED GAIN (K)
         @ THIS FREQ.
108          COMPLEX FUNCTION GAINK(FREQ,NZP,ZP)
109          COMPLEX FREQ,ZP(0:1,MAXZP)
110          DIMENSION NZP(0:1)
111          GAINK=(-1,0)
112          DO 110 IP=1,NZP(1)
113   110    GAINK=GAINK*(FREQ-ZP(1,IP))
114          DO 120 IZ=1,NZP(0)
115   120    GAINK=GAINK/(FREQ-ZP(0,IZ))
116          RETURN
117   C ************ SUBROUTINE TO FIND THE NEXT POINT ON THE
LOCUS
118          SUBROUTINE NEXT(FREQ,G,ANGLE,IZP,NZP,ZP)
119          COMPLEX GAINK,FREQ,G,ZP(0:1,MAXZP),GG,DELTA
120          DIMENSION NZP(0:1)
121          NTRY=-1
122          DA=2*3.14159/MAXTRY
123          ANGLE=ANGLE-DA
124   100    NTRY=NTRY+1
125          IF(NTRY.GE.MAXTRY) STOP 'NEXT POINT PROBLEM, INCREASE
             MAXTRY!'
126          ANGLE=ANGLE+DA
127          DELTA=DAXIS*CMPLX(COS(ANGLE),SIN(ANGLE))
128          GG=GAINK(FREQ+DELTA,NZP,ZP)
129          GANGLE=ATAN2(AIMAG(GG),SNGL(GG))*180/3.14159
130          IF(ABS(GANGLE).GT.0.5)GO TO 100
131          IF((IZP.EQ.1).AND.(ABS(G).GT.ABS(GG)))GO TO 100
132          IF((IZP.EQ.0).AND.(ABS(G).LT.ABS(GG)))GO TO 100
133          FREQ=FREQ+DELTA
134          G=GG
135          RETURN
136          END
```

Table 6.26. Computer Run for Generating the Root-Locus Plot for the System Shown in Figure 6.57

```
    THERE ARE INPUT        0 ZERO(S) AT:
    THERE ARE INPUT        3 POLE(S) AT:
( .00000000   , .00000000 )
(-4.0000000   , .00000000)
(-5.0000000   , .00000000)

    THE ROOT LOCUS SEGMENTS ON THE REAL AXIS ARE:

POLE AT .00000000     POLE AT -4.0000000
    WITH A BREAK AT -1.4699994     CONTINUING AS:
```

REAL(S)		IMAG(S)		ABS(K)
(−1.4700001	,	.20000000	)	13.312023
(−1.4700008	,	.40000000	)	13.862932
(−1.4148716	,	.59225181	)	14.781479
(−1.3904957	,	.79076080	)	16.116799
(−1.3661199	,	.98926979	)	17.872451
(−1.3143477	,	1.1824527	)	20.081075
(−1.2659548	,	1.3765097	)	22.781750
(−1.2175620	,	1.5705667	)	26.004418
(−1.691691	,	1.7646237	)	29.782210
(−1.0910084	,	1.9487185	)	34.155326
(−1.0258799	,	2.1378171	)	39.192993
(−.96075132	,	2.3269157	)	44.919437
(−.89562277	,	2.5160142	)	51.375926
(−.80792807	,	2.6957631	)	58.602841
(−.72976083	,	2.8798551	)	66.677787
(−.65159360	,	3.0639471	)	75.627802
(−.57342636	,	3.2480392	)	85.497834
(−.49525912	,	3.4321312	)	96.333202
(−.41709188	,	3.6162233	)	108.17956
(−.33892464	,	3.8003153	)	121.08286
(−.22997188	,	3.9680332	)	134.98466
(−.14227079	,	4.1477789	)	150.16485
(−.54569695 − 001	,	4.3275245	)	166.55115
(.33131399 − 001	,	4.5072702	)	184.19051
(.12083249	,	4.6870159	)	203.13002
(.20853359	,	4.8667616	)	223.41687
(.29623468	,	5.0465072	)	245.09832
(.38393577	,	5.2262529	)	268.22173
(.47163687	,	5.4059986	)	292.83450
(.55933796	,	5.5857443	)	318.98413
(.64703905	,	5.7654899	)	346.71817
(.73474014	,	5.9452356	)	376.08419
(.86063277	,	6.1006417	)	406.54803

POLE AT −5.0000000 TO -INFINITY

system. The results of the computer run are plotted in Figure 6.71. Compare the agreement between the root locus of Figure 6.57 obtained using the construction rules, and the computer-generated root-locus plot of Figure 6.71. It is important to understand the rules of construction and work several examples out using them. After having mastered the technique, it is recommended that this program, or one of the many commercially available programs for plotting the root locus or MATLAB, which was presented in Section 6.17, be used to plot the root locus. Using the computer programs will save you time and give you a root-locus plot of greater accuracy.

6.19. CONTROL SYSTEMS CONTAINING MULTIPLE GAIN MARGINS

It was shown in Figure 6.16 during the discussion of the Nyquist diagram that some control systems can have more than one phase margin. Figure 6.15 illustrated the Nyquist diagram of a control system which contained two gain margins. This section will address how multiple gain margins appear on both the Bode diagram and the root locus for the same control system.

Let us consider a unity-feedback control system whose forward transfer function, $G(s)$, is given by the following:

$$G(s) = \frac{K(1+s)(1+0.1s)}{s^3(1+0.01s)(1+0.002s)}. \tag{6.190}$$

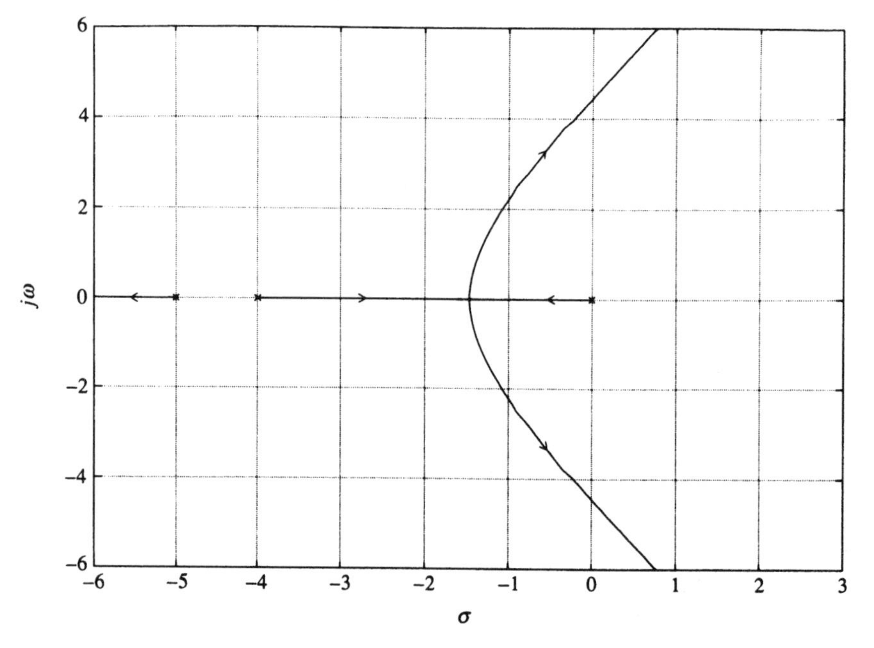

Figure 6.71 Computer-program-generated root locus plot for system defined by Eq. (6.143), and whose root locus was previously obtained using the rules of construction and shown in Figure 6.57.

The Bode diagram for this control system is plotted in Figure 6.72 with K set equal to one. It illustrates that this control system has a gain crossover frequency of 1.0 rad/sec, and a phase margin of -40°. There are two phase crossover frequencies, one at 3.4 rad/sec with a gain margin of 22 dB and one at 210.0 rad/sec with a gain margin of 64 dB. Since the phase margin is negative, this control system is unstable.

Let us now view this control system on the root locus illustrated in Figure 6.73, and we let K vary from zero to infinity. We see that the system is stable between the gains of 22 dB (12.59) and 64 dB (1590). This is a very interesting result as we now go back to the Bode diagram in Figure 6.72 and try to correlate this result. If we increase the gain by 22 dB, or from 1 to 12.59, then the new gain crossover frequency will be at 3.4 rad/sec and its phase margin will be zero. Increasing the gain more than 22 dB or more than 12.59, results in a positive phase margin and the control system is stable. If the gain is increased by 64 dB, or to 1590, then the phase margin will revert back to zero degrees phase margin. Increasing the gain to more than 1590 will cause negative phase margins, and the control system will be unstable.

Therefore, both the Bode diagram and the root locus agree that this control system is unstable for gains less than 12.59 (or 22 dB), it is stable for gains between 12.59 and 1590 (or between 22 dB and 64 dB), and it is unstable for gains greater than 1590 (or 64 dB). The analysis in this section using both the Bode diagram and the root locus for the same control system is used to introduce the next section where the Nyquist diagram, the Bode diagram, the Nichols chart, and the root locus are all used to analyze 12 commonly used transfer functions. The synergism among these methods complements each other method's results and message.

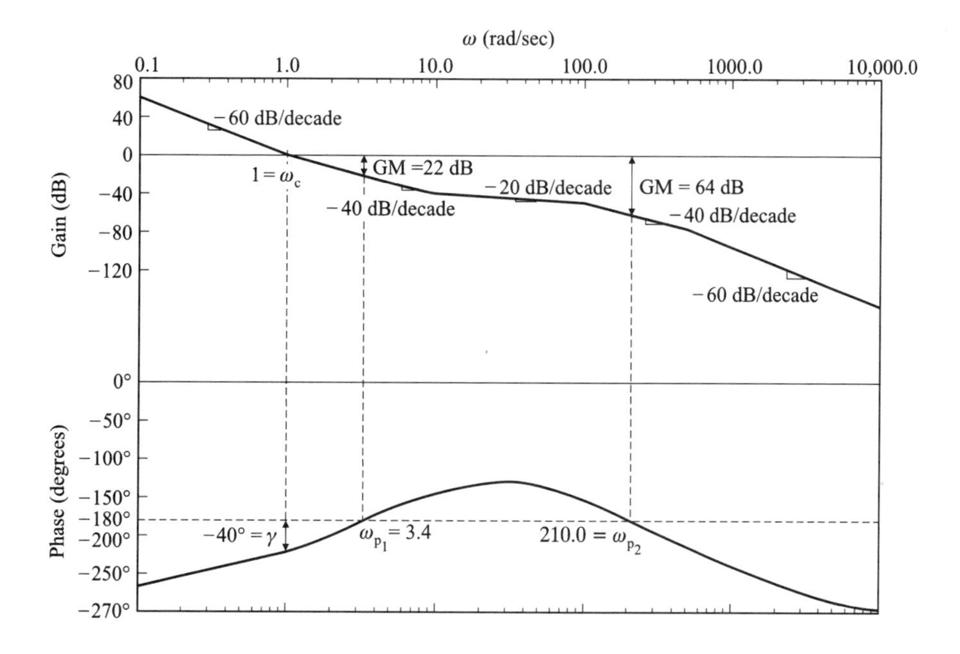

Figure 6.72 Bode diagram for a unity-feedback control system whose forward transfer function is given by Eq. (6.190).

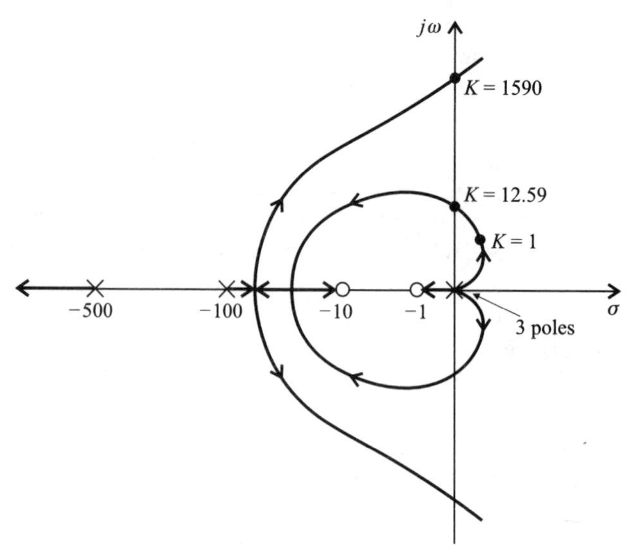

Figure 6.73 Root-locus diagram for a unity-feedback control system whose forward transfer function is given by Eq. (6.190).

6.20. COMPARISON OF THE NYQUIST DIAGRAM, BODE DIAGRAM, NICHOLS CHART, AND ROOT LOCUS FOR 12 COMMONLY USED TRANSFER FUNCTIONS

As shown in this chapter, the control-system engineer can analyze linear control systems for stability by applying the following four graphical methods: the Nyquist diagram, Bode diagram, Nichols chart, and the root-locus method. How do these methods complement each other? This section helps to answer this question by comparing the methods for 12 commonly used transfer functions.

For a perspective on the various methods and how each method complements the others, Table 6.27 [32] compares the information provided by the Nyquist diagram, Bode diagram, Nichols chart, and the root-locus method for 12 commonly used transfer functions. This table defines time constants, and phsae and gain margins where applicable. For simplicity, the M and α contours that represent the magnitude and phase-angle contours have been omitted from the Nichols chart.

A comparison of these 12 transfer functions shows that these four graphical methods provide complementary information regarding control-system stability. It is important to emphasize that the stability results obtained by the four graphical methods should be checked by the Routh–Hurwitz method's mathematical test.

6.21. COMMERCIALLY AVAILABLE SOFTWARE PACKAGES FOR COMPUTER-AIDED CONTROL-SYSTEM DESIGN

We have shown in this chapter how to determine the results for obtaining stability analysis analytically and by means of a digital computer. The two approaches shown for obtaining digital computer results were the use of MATLAB (which has now become, more or less, the industry standard) or a dedicated programs (written in

Table 6.27. A comparison of the Nyquist Diagram, Bode Diagram, Nichols Chart, and the Root Locus Method for 12 Common Transfer Functions [28]

Case No.	Open Loop Transfer Function G(S)H(S) in a Negative Feedback System	Nyquist Diagram γ = Phase Margin	Bode Diagram	Nichols Chart	Root Locus
1. A system containing a pure double integration is a borderline case of stability, and is considered unstable from a practical viewpoint. It has a phase margin of zero degrees and its roots lie on the imaginary axis. This system requires compensation in the form of a phase-lead network, or rate feedback.	$\frac{K}{s^2}$	Im ω→0⁻ ω = -∞ Re ω→0⁺ -1,0 ω = ∞	db -40 db/dec log ω φ log ω -90° -180° Phase margin = 0	db ω→0 Phase margin = 0° φ -180° -90° ω→-∞	jω σ
2. This system reflects compensation of the previous system with rate feedback. This system is always stable, as indicated by the positive phase margin, and roots which always lie in the left-half plane. In addition, it has an infinite gain margin.	$\frac{K(T_s + 1)}{s^2}$	Im ω = 0⁻ ω = -∞ Re ω = 0⁺ ω = ∞ γ -1,0	db -40 db/dec log ω -20 db/dec φ log ω -90° -180° Phase margin	db ω→0 -90° -180° φ Phase margin ω→∞	jω σ
3. This represents the type of transfer function commonly found in instrument servomechanisms. The second-order system is always stable (from a linear viewpoint), as indicated by the positive phase margin and roots which always lie in the left-half plane. Its gain margin is infinite. Stability can be increased by adding a cascade network (phase-lead, phase lag, or phase-lag-lead) or rate feedback.	$\frac{K}{S(T_1 S + 1)}$	ω→0⁻ Im -1,0 γ ω = ∞ Re ω→0⁺	db -20 db/dec log ω -40 db/dec φ log ω -90° -180° Phase margin	db ω→0 Phase margin φ -180° -90° 0° ω→∞	jω σ $-\frac{1}{T}$
4. Modification of the preceding transfer function by adding rate feedback. The phase margin is greater than in the preceding case, indicating improved stability characteristics and greater damping. The gain margin remains infinite.	$\frac{K(T_1 S + 1)}{S(T_2 S + 1)}$, $T_1 > T_2$	ω→0⁻ Im -1,0 ω = -∞ ω = ∞ γ Re ω→0⁺	db -20 db/dec -40 db/dec log ω $\frac{1}{T_1}$ $\frac{1}{T_2}$ -20 db/dec φ log ω -90° -180° Phase margin	db ω→0 φ -180° -90° 0° Phase margin ω→∞	jω σ $-\frac{1}{T_2}$ $-\frac{1}{T_1}$

Table 6.27. A comparison of the Nyquist Diagram, Bode Diagram, Nichols Chart, and the Root Locus Method for 12 Common Transfer Functions [28] (*Continued*)

Case No.	Open Loop Transfer Function G(S)H(S) in a Negative Feedback System	Nyquist Diagram γ = Phase Margin	Bode Diagram	Nichols Chart	Root Locus
5. This system is always stable, as indicated by the positive phase margin and roots which always lie in the left-half plane. In addition, it has an infinite gain margin.	$\frac{K}{(T_1S+1)(T_2S+1)}$, $T_1>T_2$	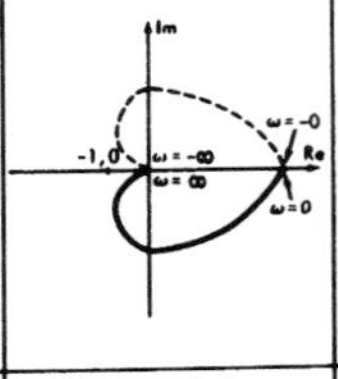 Im; -1,0; ω = -∞; ω = ∞; ω = -0; ω = 0; Re	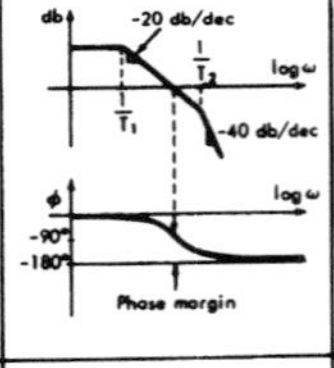 db; -20 db/dec; 1/T₁; 1/T₂; log ω; -40 db/dec; φ; -90°; -180°; log ω; Phase margin	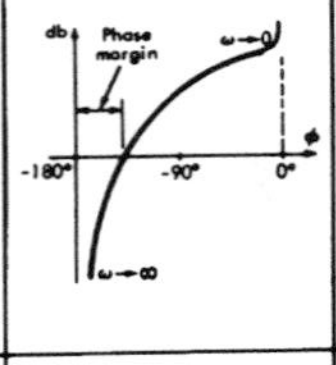 db; Phase margin; ω→0; -180°; -90°; 0°; φ; ω→∞	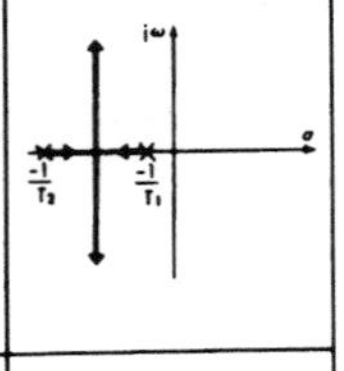 jω; -1/T₂; -1/T₁; σ
6. A system contain three pure integrations is always unstable. As indicated, it has a negative phase margin of minus ninety degrees, and two roots always lie in the right-half plane (for K > 0).	$\frac{K}{S^3}$	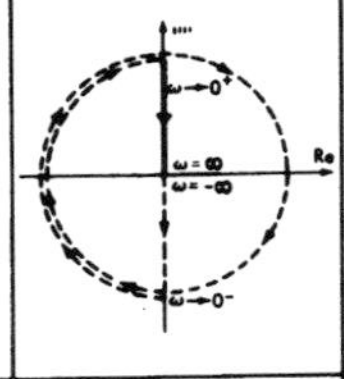 ω→0⁺; ω = ∞; ω = -∞; ω→0⁻; Re	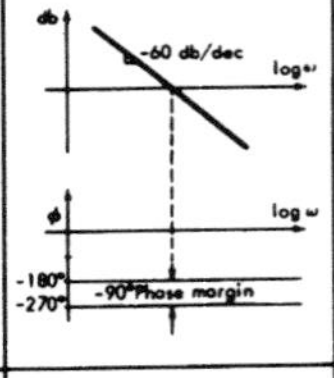 db; -60 db/dec; log ω; φ; log ω; -180°; -270°; -90° Phase margin	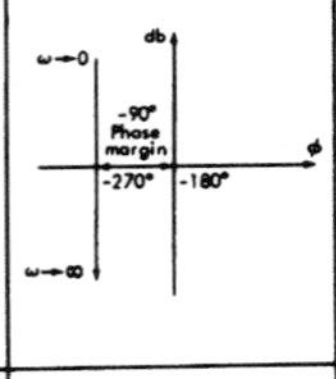 ω→0; db; -90° Phase margin; -270°; -180°; φ; ω→∞	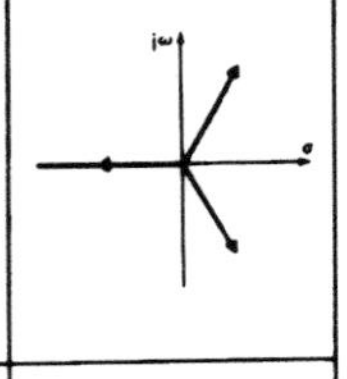 jω; σ
7. This type of transfer function is found in field-controlled servomotors, and in power servomechanisms. Although the system is stable for the gains shown, it would become unstable at higher values of gain. Stability can be improved by the addition of a cascade network (phase lead, phase lag, or phase lag-lead) or rate feedback.	$\frac{K}{S(T_1S+1)(T_2S+1)}$, $T_1>T_2$	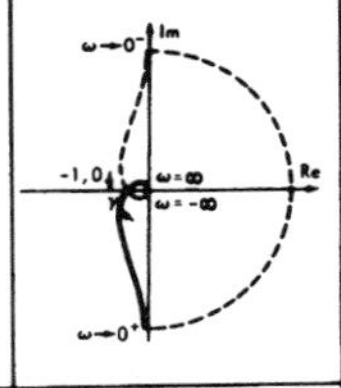 ω→0⁻; Im; -1,0; ω = ∞; ω = -∞; Re; ω→0⁺	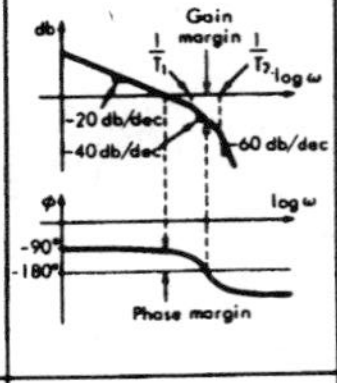 db; Gain margin; 1/T₁; 1/T₂; log ω; -20 db/dec; -40 db/dec; -60 db/dec; φ; log ω; -90°; -180°; Phase margin	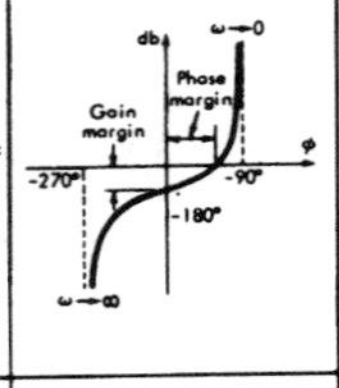 db; ω→0; Phase margin; Gain margin; -270°; -90°; -180°; φ; ω→∞	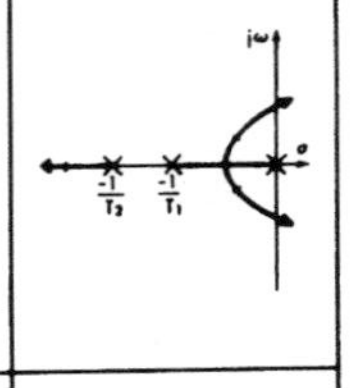 jω; -1/T₂; -1/T₁; σ
8. Modification of the preceding transfer function, with higher gain for greater accuracy and the addition of rate feedback, results in this transfer function. As shown, the system is stable, as indicated by the phase margin and roots which always lie in the left-half plane.	$\frac{K(T_1S+1)}{S(T_2S+1)(T_3S+1)}$, $T_2>T_1>T_3$	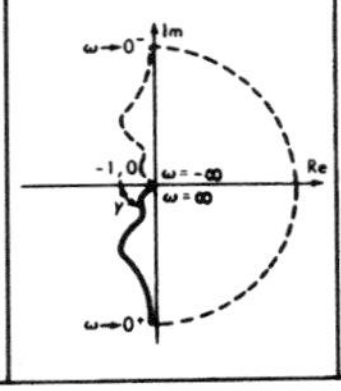 ω→0⁻; Im; -1,0; ω = -∞; ω = ∞; γ; Re; ω→0⁺	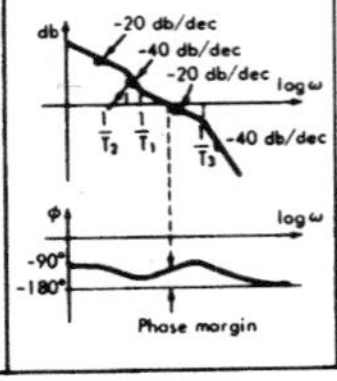 db; -20 db/dec; -40 db/dec; -20 db/dec; log ω; 1/T₂; 1/T₁; 1/T₃; -40 db/dec; φ; log ω; -90°; -180°; Phase margin	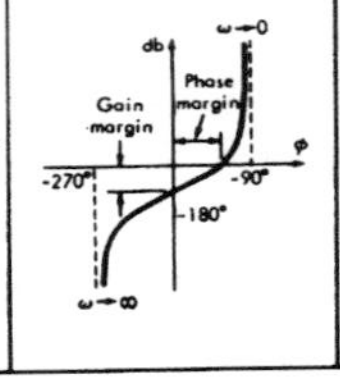 db; ω→0; Phase margin; Gain margin; -270°; -90°; -180°; φ; ω→∞	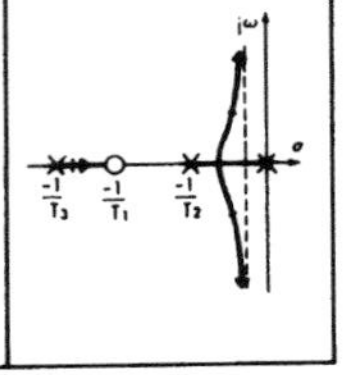 jω; -1/T₃; -1/T₁; -1/T₂; σ

Case No.	Open Loop Transfer Function G(S)H(S) in a Negative Feedback System	Nyquist Diagram γ = Phase Margin	Bode Diagram	Nichols Chart	Root Locus
5. This system is always stable, as indicated by the positive phase margin and roots which always lie in the left-half plane. In addition, it has an infinite gain margin.	$\frac{K}{(T_1S+1)(T_2S+1)}$, $T_1 > T_2$	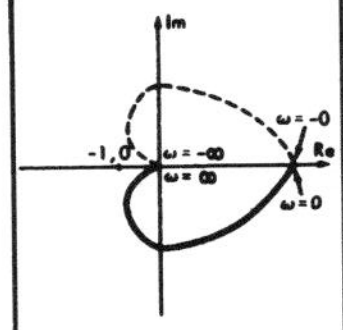 Im -1, 0 ω = -∞ ω = ∞ ω = -0 ω = 0 Re	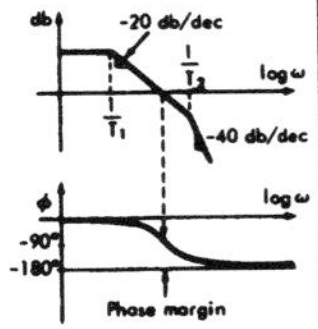 db -20 db/dec $\frac{1}{T_1}$ $\frac{1}{T_2}$ log ω -40 db/dec φ -90° -180° Phase margin	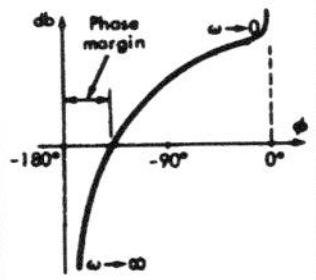 db Phase margin ω → 0 -180° -90° 0° φ ω → ∞	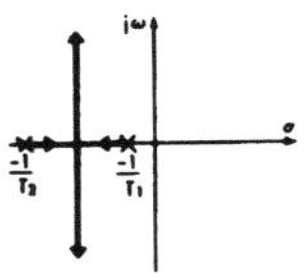 jω σ $\frac{-1}{T_2}$ $\frac{-1}{T_1}$
6. A system contain three pure integrations is always unstable. As indicated, it has a negative phase margin of minus ninety degrees, and two roots always lie in the right-half plane (for K > 0).	$\frac{K}{S^3}$	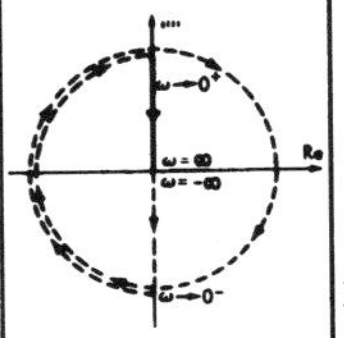 ω → 0⁺ ω = ∞ ω = -∞ ω → 0⁻ Re	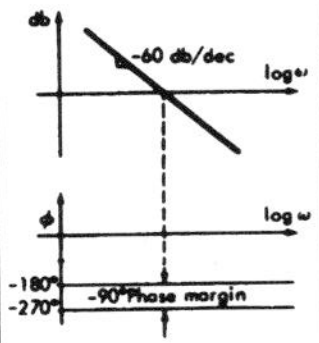 db -60 db/dec log ω φ log ω -180° -270° -90° Phase margin	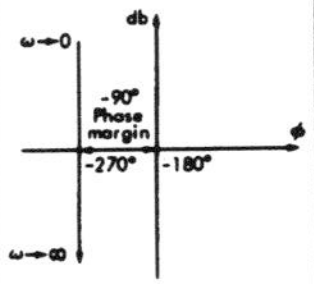 ω → 0 db -90° Phase margin -270° -180° φ ω → ∞	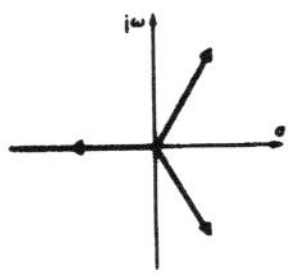 jω σ
7. This type of transfer function is found in field-controlled servomotors, and in power servomechanisms. Although the system is stable for the gains shown, it would become unstable at higher values of gain. Stability can be improved by the addition of a cascade network (phase lead, phase lag, or phase lag-lead) or rate feedback.	$\frac{K}{S(T_1S+1)(T_2S+1)}$, $T_1 > T_2$	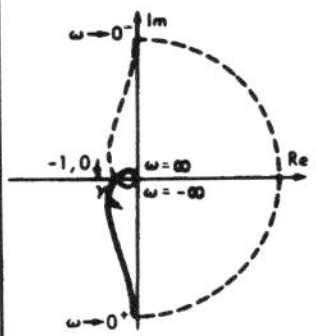 ω → 0⁻ Im -1, 0 ω = ∞ ω = -∞ Re ω → 0⁺	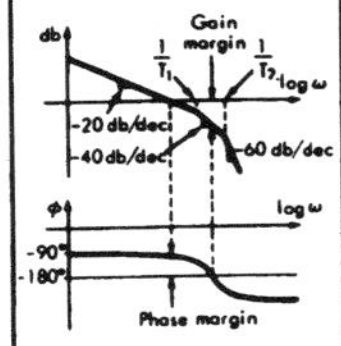 db Gain margin $\frac{1}{T_1}$ $\frac{1}{T_2}$ log ω -20 db/dec -40 db/dec -60 db/dec φ log ω -90° -180° Phase margin	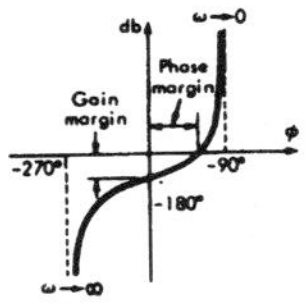 db ω → 0 Phase margin Gain margin -270° -90° -180° φ ω → ∞	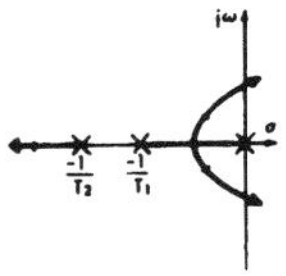 jω σ $\frac{-1}{T_2}$ $\frac{-1}{T_1}$
8. Modification of the preceding transfer function, with higher gain for greater accuracy and the addition of rate feedback, results in this transfer function. As shown, the system is stable, as indicated by the phase margin and roots which always lie in the left-half plane.	$\frac{K(T_1S+1)}{S(T_2S+1)(T_3S+1)}$, $T_2 > T_1 > T_3$	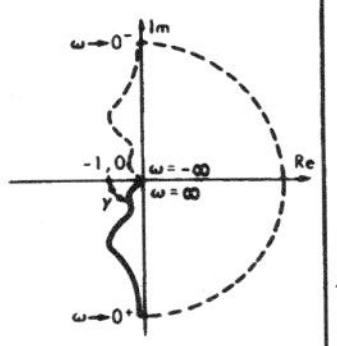 ω → 0⁻ Im -1, 0 ω = -∞ ω = ∞ γ Re ω → 0⁺	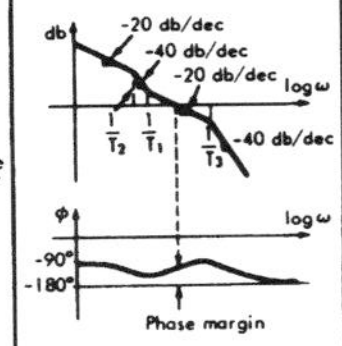 db -20 db/dec -40 db/dec -20 db/dec log ω $\frac{1}{T_2}$ $\frac{1}{T_1}$ $\frac{1}{T_3}$ -40 db/dec φ log ω -90° -180° Phase margin	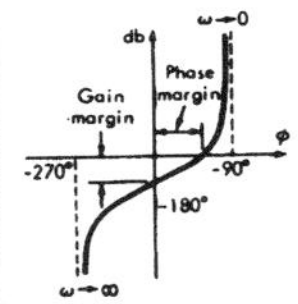 db ω → 0 Phase margin Gain margin -270° -90° -180° φ ω → ∞	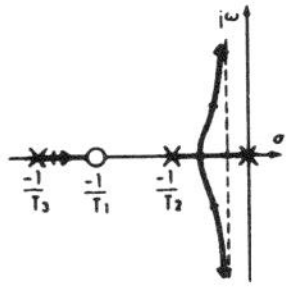 jω σ $\frac{-1}{T_3}$ $\frac{-1}{T_1}$ $\frac{-1}{T_2}$

Fortran or Basic) provided in this book for those readers who do not have access to MATLAB. For example, the following sections were dedicated to providing MATLAB programs:

- Section 6.6—Nyquist diagram
- Section 6.8—Bode diagram
- Section 6.11—Nichols chart
- Section 6.17—root-locus diagram.

The Modern Control System Theory and Design Toolbox, available to the readers of this book through The MathWorks anonymous FTP server at ftp://ftp.mathworks.com/pub/books/shinners, also contains the MATLAB m-files to most of the figures and problems in this book, a demonstration m-file, tutorial file, synopsis file, and on-line help. In addition, the following sections contained dedicated programs written in Fortran and/or Basic:

- Section 6.9—Bode diagram; closed-loop frequency response; time-domain response
- Section 6.18—root-locus diagram.

The computer programs shown in Section 6.9 and 6.18 of this chapter for linear system analysis were generated with programs written in Basic and Fortran. They included the following (tables containing the associated program listings are shown in parentheses):

- Generation of phase and gain margins (Table 6.12)
- Generation of Bode diagrams (open-loop frequency response) and closed-loop frequency response (Table 6.14)
- Time-domain response (Table 6.17)
- Generation of root-locus diagrams (Table 6.25)

For each of these programs, the accompanying logic flow diagram is shown, together with application problems containing the program output and graphical results. As a preview to nonlinear system analysis in Chapter 5 of the accompanying volume, digital computer programs for determining the describing function analysis are presented in Sections 5.11 (using MATLAB) and 5.12 (using a dedicated program), and for the phase plane in Section 5.19 (using a dedicated program). The reader can use these tested programs for application to typical control problems that the student and practicing engineer may encounter.

For those who prefer to use commercially available programs, this section discusses some of them, (inculding MATLAB) and their capabilities. Several software "packages" now exist for control-system engineers that integrate basic algorithms with data management, an interactive user interface, and graphical outputs [33–37]. A good summary of the various packages available is found in the "Extended List of Control Software" (ELCS), which is a collection of one-page summaries of subroutine libraries and software packages [34]. Another good summary on the availability of computer-aided control-system design packages is found in Reference [36]. We will focus attention in this section on the MATLAB with their Control System Toolbox, Ctrl-C®, CODAS-II, Program CC, and ACET™.

*MATLAB and the accompanying Control System Toolbox (The MathWorks, Inc., 24 Prime Park Way, Natick, MA 01760)**. MATLAB and the accompanying Control System Tookbox are very useful for the analysis and design of control systems [7,38]. (The name MATLAB stands for "matrix laboratory.") It has the capability to do all of the analyses presented in this book from classical transfer-function techniques to the modern state-variable methods. MATLAB integrates numerical analysis, matrix computation, signal processing, and graphics. PC-MATLAB runs on IBM and other MS-DOS compatible personal computers. PRO-MATLAB runs on large computers such as Sun Workstations and VAX computers, and the Macintosh MATLAB runs on the Macintosh.

The basic data element in MATLAB is a matrix that does not require dimensioning. This allows the solution to many numerical problems in a much shorter amount of time than it would take if a program were written in Basic or Fortran. MATLAB is a completely integrated system that includes graphics, programmable macros, IEEE arithmetic, and many analytical commands. The Control System Toolbox uses the MATLAB matrix functions to provide specialized functions for control-system analysis, design, and modeling techniques.

MATLAB permits extension of control-system analyses to signal processing, system identification, robust control, and simulation using the following toolboxes.

1. Signal Processing Toolbox for digital filtering and spectral analyses
2. System Identification Toolbox for parametric modeling, and model-based and adaptive signal processing
3. Robust-Control Toolbox for loop transfer recovery and optimal control synthesis
4. SIMULAB™ for describing, analyzing, and simulating dynamic systems.

In addition to these toolboxes, users can write their own functions/toolboxes to suit their needs, and freely exchange them on a MathWorks maintained Bulletin Board (MUG). For information on this software and other noncontrol-based toolboxes, contact The MathWorks, Inc.

Besides being used extensively in this book, the application of MATLAB to control-system problems is also found in References [39–44].

Ctrl-C® (Systems Control Technology, Inc., 2300 Geng Road, P.O. Box 10180, Palo Alto, CA 94303-0888). Ctrl-C® is also very useful for the analysis and design of control systems [45]. (The name Ctrl-C is pronounced "control-see'.) It can perform the control analyses presented in this book from classical transfer-function techniques to the modern state-variable methods. Ctrl-C® provides integrated

*The majority of the solutions in this book have been obtained using MATLAB. The corresponding M-files that are part of my Modern Control System Theory and Design Toolbox can be retrieved free from The MathWorks, Inc. anonymous FTP server at ftp://ftp.mathworks.com/pub/books/shinners.

software that can perform control-system design and analysis, digital signal processing, matrix analysis, and engineering graphics. Systems may be described in modern state-variable or classical transfer-function format, and in continuous- or discrete-time forms.

Ctrl-C® can be viewed as an interactive computer language that provides an operating environment for the analysis and design of multivariable systems. It has syntax commands that form a complete, very high-level programming language that permits direct interfaces to users' preexisting Fortran and C programs, and to two commercially available software packages, ACSL and NASATRAN®. The ACSL Interface can be used to access ACSL for nonlinear simulation and for finding linearized, small-perturbation system models. NASTRAN® can be used for finite-element modeling of continuous systems. Other specialized functions are provided for multivariable control systems, incuding dynamic system simulation, computation of multivariable zeros, decomposition of systems into observable and unobservable parts (observability is discussed in Chapter 8), and the analysis of optimal control systems (optimal control is discussed in Chapter 11).

Application of Ctrl-C® to a control-system problem is found in Reference [39].

CODAS-II (Dynamical Systems, Inc., P.O. Box 35241, Tucson, AZ 85740). CODAS-II is also very useful for the analysis and design of control systems on the PC [46]. It too has the capability to perform all of the analyses presented in this chapter. CODAS for the PC evolved from a version for the British ACORN computer, which incorporates the enhanced memory and graphics capabilities of PCs. Transfer functions can be entered in pole-zero form or as polynomials or a mixture of both. The system contains the controller (or compensator), the plant, an overall system gain and an optional transportation delay factor. It has the capability of analyzing discrete-time/sampled-data systems and nonlinear systems. It also has the capability of switching the nonlinearity in or out in order to determine its effect on the control system's performance. CODAS-II can also analyze nonlinear systems using the describing function in the frequency domain. (The describing function is discussed in Chapter 10.)

For the analysis of process control systems on the PC, Dynamical Systems has available PCS. It has many useful features found in practical process control systems, and it can be applied for analyzing controller operation, tuning, and plant behavior. PCS also permits the simulation of direct digital control, which allows for determining the effects of varying the sampling time on control system performance.

Program CC (Systems Technology, Inc., 13766 S. Hawthorne Blvd., Hawthorne, CA 90250-7083). Program CC is also very useful for the analysis and design of control systems [47]. It can be used for the analysis and design of classical state-variable, sampled-data, and optimal control systems. In addition, it can be used for the analysis of stochastic systems (which are beyond the scope of this book). However, it does not have the capability for simulation of nonlinear systems. It can be used on IBM-PCs and compatible personal computers (with a coprocessor recommended).

Program CC is written entirely in Microsoft Basic, compiled with the Microsoft QB compiler, and it executes on DOS operating systems [47]. It uses double-precision variables for all arithmetic, except data files and graphics. Program CC is actually a collection of many programs that are tied together at run time using the chaining operation.

All of the system elements, time histograms, and frequency data are stored on disk as ASCII files. Program CC has the nice capability of being augmented by the user, who can write new algorithms in the form of Basic programs, and then tie them in with Program CC at run time. The Program CC file structure is open and, therefore, available for communication with other prorams. Data bridges exist for related software products including PC-MATLAB (discussed previously in this section) for additional matrix algorithms, particularly those involving looping and branching.

ACET™ (Advanced Control Engineering Techniques), (Information & Control Systems, 28 Research Drive, Hampton, Virginia 23666). ACET™ is also useful for the analysis and design of control systems [48]. It permits the analysis of control systems using such tools as Bode, Nyquist, and root-locus plots for both continuous and sampled-data systems. ACET™ analyzes the frequency domain characteristics of single-input/single-output systems as well as multi-input/multi-output systems. For the case of multiple-input/multiple-output systems, ACET™ generates one Nyquist or Bode plot for each loop closing one loop at a time to determine the phase and gain margins.

This software package contains modules on control design, model building, system analysis, filter design, simulation matrix operations and plotting. ACET™ can be used to perform analysis, simulate a control system, and design a control law. It has no limits on the order of the transfer functions or the number of system blocks except for hardware memory limitations, and it automatically transforms the block diagram entered into state-space form. ACET™ contains a menu-driven Graphical User Interface, and it can be used on IBM compatible MS-DOS personal computers with math coprocessor.

6.22. WHAT IS THE "BEST" STABILITY ANALYSIS TECHNIQUE? GUIDELINES FOR USING THE ANALYSIS TECHNIQUES PRESENTED

We have presented the following methods in this chapter for analyzing linear control systems: Routh–Hurwitz method, Nyquist diagram, Bode diagram, Nichols chart, and the root-locus method. Which of these stability analyses methods presented is the "best" to use? This is a logical question to ask at this point in the presentation of this chapter. For a particular application, one of these methods may be better than the others. However, in general, each method presented in this chapter complements the other. In general, there is no "best" method. The reader should regard all of the methods presented in this chapter as control-system analysis tools that can be used. The practicing control-system engineer should not restrict himself or herself to one method.

Guidelines are presented here for selecting the best method for particular cases [32]. For systems of first or second order, all of the methods presented reveal about the same amount of information, and they are of similar complexity. For systems from about the third to seventh order, there are two primary approaches, and these two approaches complement each other. One approach uses the Bode-diagram method in conjunction with the Nichols chart. The Routh–Hurwitz method can be used as a quick check. The other method uses the root locus. Using the first approach, the control-system engineer can determine relative stability from the phase and gain margins of the Bode diagram. The closed-loop frequency response can then be found from the Nichols chart, and bounds on the time-domain response can be determined as was illustrated in Section 6.12. The root-locus method is the basis of the second approach. Using this method, the control-system engineer can determine relative stability, and the transient response can be found from the location of the closed-loop roots of the characteristic equation—in particular, the location of the dominant pair of complex-conjugate roots. In conclusion, the information provided by these two basic approaches complement each other.

For systems greater than the seventh order, I have found that the root-locus method is more complex than the Bode-diagram–Nichols-chart approach. This guideline, however, is general, and as in all general statements is not true all of the time. For example, we would not use the Bode-diagram–Nichols-chart approach if the open-loop control system did not satisfy the minimum-phase requirement regardless of the system order.

I want to make one additional important point before leaving this discussion. The Bode-diagram approach has the great advantage that it is a fequency-response approach that can be easily measured in the laboratory using readily available instruments. The procedure involves opening the feed-back loop and applying a sinusoidal signal to the input of the open-loop system, and measuring the magnitude of the input and output quantities and the phase relationship between them. (Be careful of avoiding saturation when testing the open-loop control system.) From this procedure, the phase and gain margins can be easily measured. Therefore, the control-system engineer can check his or her design directly using experimental techniques.

On the other hand, an experimental direct determination of the closed-loop roots to check the root-locus method is not easy. We can only test the root locus indirectly in the laboratory based on system performance measures such as the control system's transient response, and then infer the location of the dominant pair of complex-conjugate roots. Therefore, many practicing control-system engineers prefer the Bode-diagram–Nichols-chart approach over the root-locus method in cases where either approach is applicable. However, in the final analysis, all methods of stability analysis should be considered before selecting the technique or techniques to be used in the final design.

6.23. ILLUSTRATIVE PROBLEMS AND SOLUTIONS

This section provides a set of illustrative problems and their solutions to supplement the material presented in Chapter 6.

I6.1.

A control system can be represented by the following companion matrix:

$$\mathbf{A} = \begin{bmatrix} 0 & 1 & 0 \\ 0 & 0 & 1 \\ 0 & -2 & -2 \end{bmatrix}.$$

(a) Determine the characteristic equation.

(b) Determine the location of the route of the characteristic equation.

(c) Is the control system operating in a stable region?

SOLUTION: **(a)**

$$[\mathbf{A} - s\mathbf{I} = \begin{bmatrix} 0 & 1 & 0 \\ 0 & 0 & 1 \\ 0 & -2 & -2 \end{bmatrix} - \begin{bmatrix} s & 0 & 0 \\ 0 & s & 0 \\ 0 & 0 & s \end{bmatrix} = \begin{bmatrix} -s & 1 & 0 \\ 0 & -s & 1 \\ 0 & -2 & -2-s \end{bmatrix}.$$

Using minors along the first column, we obtain the following:

$$-s(-1)^2 \begin{vmatrix} -s & 1 \\ -2 & -2-s \end{vmatrix} = -s[(2s + s^2) + 2] = -s^3 - 2s^2 - 2s = 0.$$

Therefore, the characteristic equation is given by the following:

$$s(s^2 + 2s + 2) = 0$$

(b) Since

$$s(s^2 + 2s + 2) = s(s + 1 + j)(s + 1 - j) = 0$$

the roots of the characteristic equation are given by the following:

$$s_1 = 0,$$

$$s_2 = -1 - j,$$

$$s_3 - 1 + j.$$

(c) Because the complex-conjugate roots have negative real parts, and the third root is located at the origin, none of the roots are located in the right half-plane and the system is operating in a stable region.

I6.2. A second-order control system can be represented by the following companion matrix:

$$\mathbf{A} = \begin{bmatrix} 0 & -2 \\ 1 & -1 \end{bmatrix}.$$

(a) Determine the characteristic equation of this control system.

(b) Determine the location of the roots of the characteristic equation.

(c) Is this control system operating in a stable region?

SOLUTION: **(a)** From Eq. (6.21):

$$|\mathbf{P} - s\mathbf{I}| = 0$$

Therefore,

$$\begin{vmatrix} 0 & -2 \\ 1 & -1 \end{vmatrix} - \begin{vmatrix} s & 0 \\ 0 & s \end{vmatrix} = 0$$

or,

$$\begin{vmatrix} -s & -2 \\ 1 & -1-s \end{vmatrix} = 0.$$

Therefore, the characteristic equation is given by:

$$s^2 + s + 2 = 0.$$

(b)

$$\begin{aligned} s^2 + s + 2 &= (s + 0.5 + 1.325j)(s + 0.5 - 1.325j), \\ s_1 &= -0.5 - 1.325j, \\ s_2 &= -0.5 + 1.325j. \end{aligned}$$

(c) Since the complex-conjugate roots are located in the left half-plane, the system is stable.

I6.3. A control system containing a tachometer, which provides rate feedback, is illustrated. Determine the range of the tachometer constant, b. in order that the system is always stable.

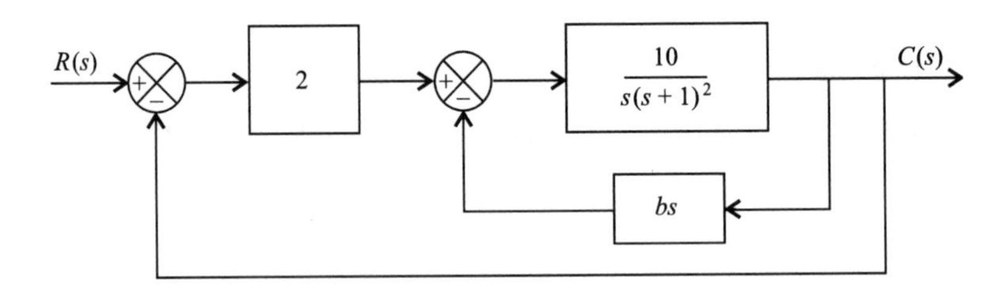

Figure I6.3

SOLUTION: Using Mason's Theorem, we obtain:

$$\frac{C(s)}{R(s)} = \frac{\dfrac{2(10)}{s(s+1)^2}}{1 + \dfrac{2(10)}{s(s+1)^2} + \dfrac{10bs}{s(s+1)^2}},$$

$$\frac{C(s)}{R(s)} = \frac{20}{s^3 + 2s^2 + (1 + 10b)s + 20},$$

$$\begin{array}{c|cc} s^3 & 1 & 1+10b \\ s^2 & 2 & 20 \\ s^1 & U_1 = \dfrac{2(1+10b)-20}{2} & \\ s^0 & U_3 = 20 & \end{array}$$

Setting $U_1 = 0$, we obtain:

$$\frac{2(1+10b)-20}{2} = 0.$$

Therefore, $b = 0.9$, and the tachometer constant, b, must be

$$b > 0.9$$

for the system to be stable.

I6.4 The forward-loop transfer function of a unity-feedback control system is given by the following:

$$G(s) = \frac{K(s+1)}{s(s+4)(1+Ts)}.$$

Determine the values of K and T for stability of this control system.

SOLUTION:

$$1 + G(s)H(s) = \frac{K(s+1)}{s(s+4)(1+Ts)} = 0$$

The characteristic equation is given by

$$Ts^3 + (4T+1)s^2 + (4+K)s + K = 0.$$

The Routh–Hurwitz array is given by:

$$\begin{array}{c|cc} s^3 & T & 4+K \\ s^2 & 4T+1 & K \\ s^1 & U_1 = \dfrac{(4T+1)(4+K)-TK}{(4T+1)} & \\ s^0 & U_3 = K & \end{array}$$

Analysis of this Routh–Hurwitz array, shows that the system will be stable if K and T are both greater than zero.

I6.5. The forward-loop transfer function of a unity-feedback control system is given by the following:

$$G(s) = \frac{2}{(s+1)^4}.$$

(a) Sketch the Nyquist diagram, and determine whether the control system is stable.

(b) Determine the intersection of the $G(j\omega)H(j\omega)$ and the negative real axis of the $G(j\omega)H(j\omega)$ plane *analytically*, and find the values of the *frequency*, ω, and *location* of the intersection on the negative real axis.

(c) How much can the system gain, 2, be increased before the roots of the characteristic equation cross the $j\omega$ axis and go into the right-half plane?

SOLUTION: (a)

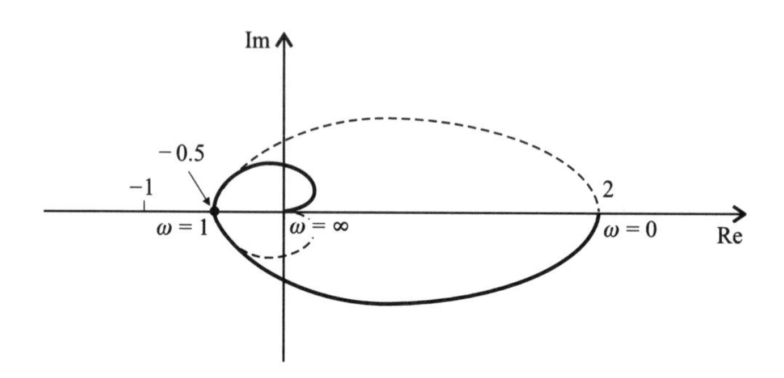

Figure I6.5 Solution.

(b)

$$G(j\omega) = \frac{2}{(j\omega + 1)^4} = \frac{2}{\omega^4 - 4\omega^3 j - 6\omega^2 + 4\omega j + 1}.$$

Separating the real and imaginary terms in the denominator, we obtain the following:

$$G(j\omega) = \frac{2}{(\omega^4 - 6\omega^2 + 1) + (4\omega - 4\omega^3)j}.$$

Therefore,

$$G(j\omega) = \frac{2[(\omega^4 - 6\omega^2 + 1) - (4\omega - 4\omega^3)j]}{(\omega^4 - 6\omega^2 + 1)^2 + (4\omega - 4\omega^3)^2}.$$

Setting the imaginary part of the numerator equal to zero, we obtain:

$$-2(4\omega - 4\omega^3) - 0.$$

Therefore,

$$\omega = 0, \quad 1 \quad -1$$

and we are interested in the condition of $\omega = 1$. The case of $\omega = 0$ refers to the intersection of the Nyquist diagram with the positive real axis.

Substituting $\omega = 1$ into $G(j\omega)$, we obtain that the intersection of the Nyquist diagram and the real axis occurs at -0.5.

(c) Since the intersection of the Nyquist diagram and the negative real axis occurs at -0.5, we can increase the gain, 2, to 4 at which point the Nyquist diagram crosses the negative real axis at -1 and the roots of the characteristic equation cross the $j\omega$ axis.

I6.6. The open-loop transfer function of a feedback control system is given by the following:

$$G(s)H(s) = \frac{20}{s(0.2s+1)(0.4s+1)}.$$

(a) Sketch the Nyquist diagram, and determine whether the control system is stable.

(b) Determine the intersection of the $G(j\omega)H(j\omega)$ and the negative real axis of the $G(j\omega)H(j\omega)$ plane *analytically*, and find the values of the *frequency*, ω, and *location* of the intersection on the negative real axis.

(c) What is the maximum value of gain for this control system before the roots of the characteristic equation cross the $j\omega$ axis and go into the right half-plane?

SOLUTION: (a)

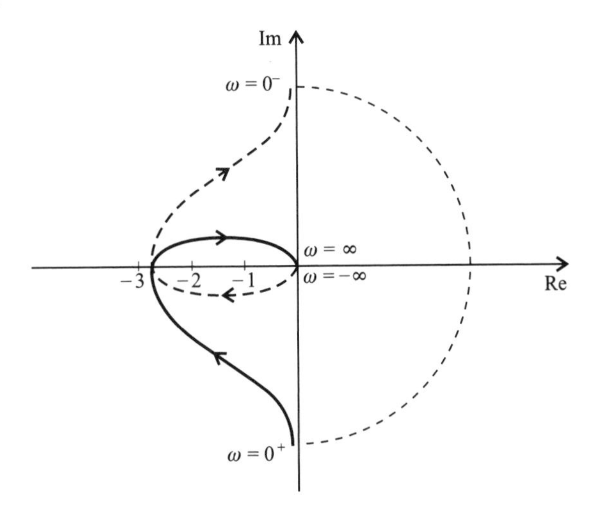

Figure I6.6 Solution.

$$N = Z - P$$

Since $P = 0$, and $N = +2$, $Z = 2$, there are two roots in the right half-plane and the system is unstable.

(b)

$$G(j\omega)H(j\omega) = \frac{20}{(j\omega)(0.2j\omega+1)(0.4j\omega+1)} = \frac{20}{-0.6\omega^2 + (\omega - 0.08\omega^3)j}.$$

Therefore,

$$G(j\omega)H(j\omega) = \frac{20[0.6\omega^2 + (\omega - 0.08\omega^3)j]}{(0.6\omega^2)^2 - (\omega - 0.08\omega^3)^2}.$$

Setting the imaginary part of the numerator equal to zero, we obtain:

$$20j(\omega - 0.08\omega^3) = 0$$

$$\omega^2 = 12.5.$$

Therefore,

$$\omega = 3.536 \text{ rad/sec and } -3.536 \text{ rad/sec}.$$

So, $G(j\omega)H(j\omega)$ intersects the negative real axis at $\omega = 3.536$ rad/sec. Substituting this value of ω into the equation for $G(j\omega)H(j\omega)$, we find that the intersection of $G(j\omega)H(j\omega)$ with the negative real axis occurs at:

$$G(j3.536)H(j3.536) = -2.6667.$$

(c) Therefore, the value of gain for the $G(j\omega)H(j\omega)$ plane to cross the negative real axis at -1 is given by

$$\frac{20}{2.6667} = 7.5.$$

Thus, the maximum value of gain for this control system before the roots of the characteristic equation cross the $j\omega$ axis and go into the right half-plane is 7.5.

16.7. An engineer is called in to consult on a control system in a piece of equipment in the field. No one can find the design report or test results from the original design of the control system. The engineer, therefore, decides to take a frequency response of the control system by opening the outer feedback loop. The resulting asymptotic gain-frequency characteristic is obtained:

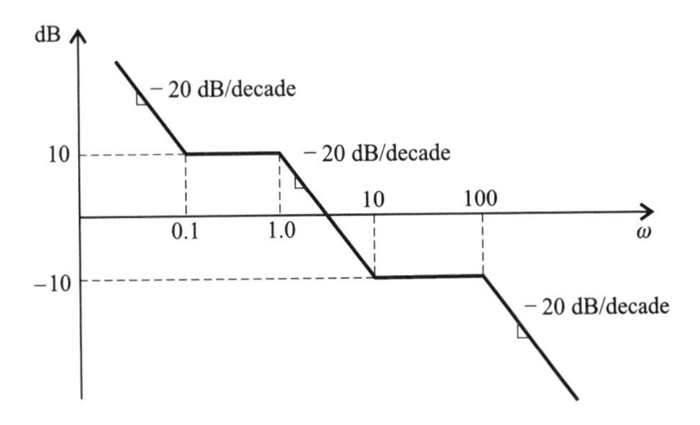

Figure I6.7 Asymptotic gain frequency characteristic solution

Assuming that the system has a minimum phase transfer function, determine the transfer function, $G(s)H(s)$.

SOLUTION: From knowledge of the break points, we know that the transfer function is given by the following transfer function where K is the gain to be determined:

$$G(s)H(s) = \frac{K\left(1+\frac{s}{0.1}\right)\left(1+\frac{s}{10}\right)}{s(1+s)\left(1+\frac{s}{100}\right)}.$$

The value of K can be obtained from knowledge that at $\omega = 0.1$ rad/sec, the gain is 10 dB. Therefore,

$$20\log_{10}\left|\frac{K}{0.1j}\right| = 10.$$

Therefore, $K = 0.316$ and the complete transfer function is given by:

$$G(s)H(s) = \frac{0.316(1+10s)(1+0.1s)}{s(1+s)(1+0.01s)}.$$

I6.8. (a) Determine the root locus for a unity-feedback control system whose forward transfer function is given by

$$G(s) = \frac{K(s+2)}{s^2(s+18)}.$$

(b) Determine the location of the roots when all three roots are all real and equal.

(c) Find the gain when all the roots are real and equal.

SOLUTION: **(a)** See root locus plot (Figure I6.8) using MATLAB.

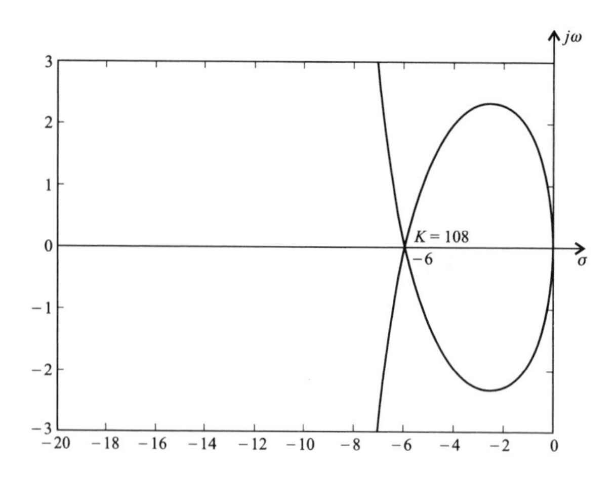

Figure I6.8 Root-locus plot solution.

The angles of the intersection of the asymptotes and the real axis are given by:

$$\alpha_0 = +\frac{\pi}{3-1} = \pm 90^\circ.$$

The intersection of the asymptotes and the real axis is given by:

$$s_r = \frac{-18-(-2)}{3-1} = -8.$$

(b) We can find the location of the roots when they are all equal from Rule 7 as follows:

$$1 + \frac{K(s+2)}{s^2(s+18)} = 0.$$

Solving for K, we obtain the following:

$$K = -\frac{s^2(s+18)}{(s+2)}.$$

Limiting s to the real axis, we let $s = \sigma$. Therefore,

$$K(\sigma) = -\frac{\sigma^3 + 18\sigma^2}{(\sigma+2)}. \qquad (I6.8\text{-}1)$$

Taking the derivative of $K(\sigma)$ with respect to σ, we obtain:

$$\frac{dK(\sigma)}{d(\sigma)} = \frac{(\sigma+2)(3\sigma^2+36\sigma) - (\sigma^3+18\sigma^2)(1)}{(\sigma+2)^2} = 0.$$

Simplifying this equation, we obtain the following:

$$2\sigma(\sigma^2 + 12\sigma + 36) = 0.$$

Solving, we find that $\sigma = 0$ and -6. Therefore, at $\sigma = -6$ all three roots intersect and are real. We can find the value of gain when all the roots are equal at $\sigma = -6$ by substituting $\sigma = -6$ into Eq. (I6.8-1). Therefore, we find that $K = 108$.

I6.9.

(a) Draw the root locus of a negative feedback, unity-feedback control system whose forward-path transfer function is given by the following:

$$G(s) = \frac{K}{(s+2)^4}.$$

(b) At what value of K does the system become unstable, and where does the root locus intersect the $j\omega$ axis when this occurs?

SOLUTION: (a)

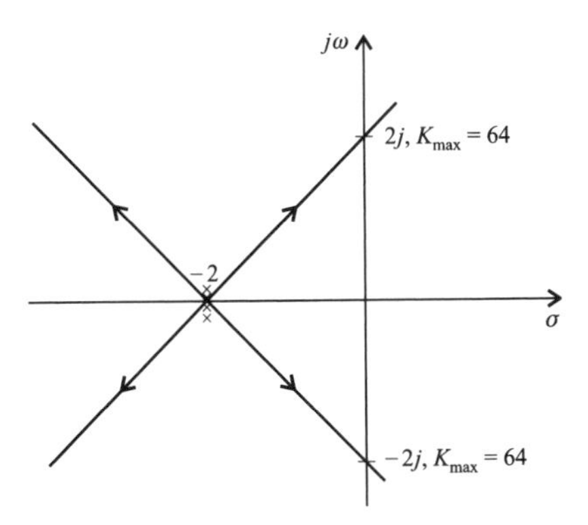

Figure I6.9 Root-locus plot solution

Rule 5: Asymptotes are given by

$$\alpha_0 = \pm\frac{\pi}{4} = \pm 45^\circ$$

$$\alpha_1 = \pm\frac{3\pi}{4} = \pm 135^\circ.$$

Rule 6: Intersection of the asymptotes and the real axis is given by:

$$s_r = \frac{(-2-2-2-2)}{4} = -2$$

(b) *Rule 8*: The intersection of the root locus and the imaginary axis and K_{max} are found from this rule. The characteristic equation is given by:

$$1 + \frac{K}{(s+2)^4} = 0$$

which simplifies to the following:

$$s^4 + 8s^3 + 24s^2 + 32s + 16 + K = 0.$$

The intersection of the root locus and the imaginary axis can be determined by substituting $s = j\omega$ into the characteristic equation:

$$(j\omega)^4 + 8(j\omega)^3 + 24(j\omega)^2 + 32j\omega + 16 + K = 0.$$

Setting the imaginary terms equal to zero, we obtain:

$$-8(j\omega)^3 + 32j\omega = 0$$

$$j\omega[-8\omega^2 + 32] = 0.$$

Therefore, $\omega = \pm 2$, and the root locus crosses the imaginary axis at $\pm 2j$.

The value of K_{max} can be determined from the following Routh–Hurwitz array:

$$\begin{array}{c|ccc} s^4 & 1 & 24 & 16+K \\ s^3 & 8 & 32 & \\ s^2 & 20 & 16+K & \\ s^1 & \frac{512-8K}{20} & & \\ s^0 & 16+K & & \end{array}$$

Setting the term in the fourth row equal to zero, we obtain

$$\frac{512 - 8K_{max}}{20} = 0.$$

Therefore, $K_{max} = 64$.

16.10. Consider a negative-feedback control system whose open-loop transfer function is given by:

$$G(s)H(s) = \frac{Ks(s+6)}{(s+2+j)(s+2-j)}.$$

(a) Sketch the root locus.

(b) Determine when the roots are both equal.

(c) Determine the gain when these two roots are both equal.

(d) Determine the settling time when the roots are equal.

SOLUTION: **(a)**

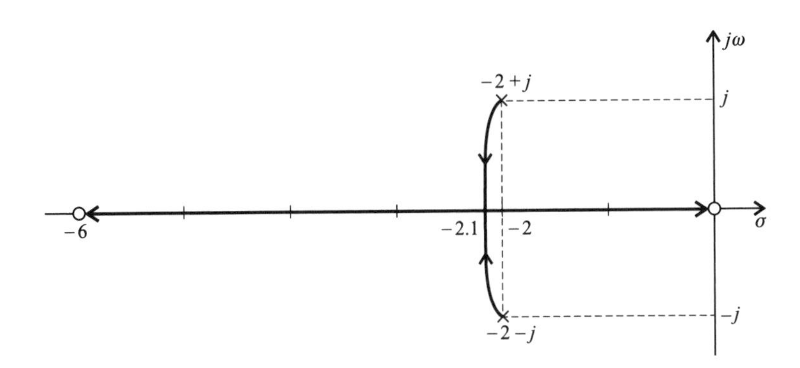

Figure I6.10 Root-locus plot solution.

(b) *Rule 7*: To find the point of breakaway, we start with the characteristic equation:

$$1 + \frac{Ks(s+6)}{(s+2+j)(s+2-j)} = 1 + \frac{Ks^2 + 6Ks}{s^2 + 4s + 5}.$$

We restrict s to real values, let $s = \sigma$, and we solve for K:

$$K = -\frac{\sigma^2 + 4\sigma + 5}{\sigma^2 + 6\sigma}.$$

Taking the derivative of $K(\sigma)$ with respect to σ, we get the following:

$$\frac{dK(\sigma)}{d\sigma} = -\frac{(\sigma^2 + 6\sigma)(2\sigma + 4) - (\sigma^2 + 4\sigma + 5)(2\sigma + 6)}{(\sigma^2 + 6\sigma)^2} = 0.$$

Simplifying, we obtain the following:

$$\frac{dK(\sigma)}{d\sigma} = -\frac{2\sigma^2 - 10\sigma - 30}{(\sigma^2 + 6\sigma)^2} = 0.$$

Therefore:

$$\sigma^2 - 5\sigma - 15 = 0.$$

Therefore, the two resulting roots of this equation are:

$$\sigma_1 = -2.1,$$
$$\sigma_2 = 9.1.$$

Only σ_1 is a valid solution for negative feedback. Therefore, the roots are both equal at $\sigma_1 = -2.1$.

(c) The gain when the two roots are equal can be determined from Rule 11:

$$\left|\frac{K(s)(s+6)}{(s^2+4s+5)}\right|_{s=-2.1} = 1$$

Therefore, $K = 0.134$.

(d) We can determine the settling time from Eq. (5.41):

$$t_s = \frac{4}{\zeta\omega_n}.$$

In this equation, the damping ratio ζ equals one because the two roots are on the real axis, and ω_n equals the intersection value on the real axis. Therefore,

$$t_s = \frac{4}{(1)(2.1)} = 1.9.$$

Therefore, the settling time is 1.9 sec when the roots are equal.

PROBLEMS

6.1. The stability of the feedback control system of Figure P6.1 is to be determined.

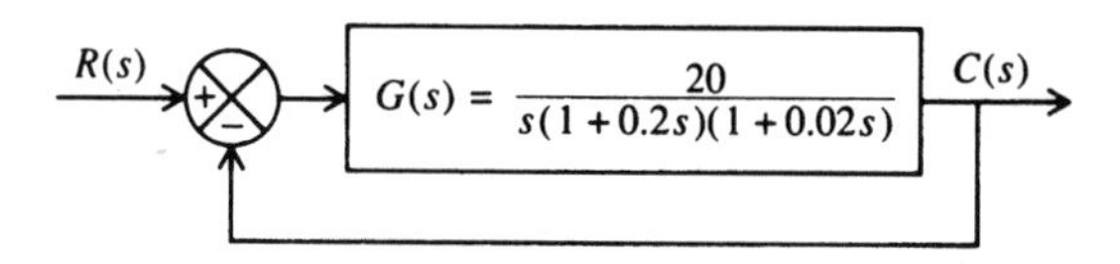

Figure P6.1

(a) Determine the system's **P** matrix from its state equations.

(b) Find the system's characteristic equation from knowledge of the **P** matrix.

(c) Using the Routh–Hurwitz criterion, determine whether this feedback control system is stable.

6.2. Stability of the control system of Figure P6.2 is to be determined.

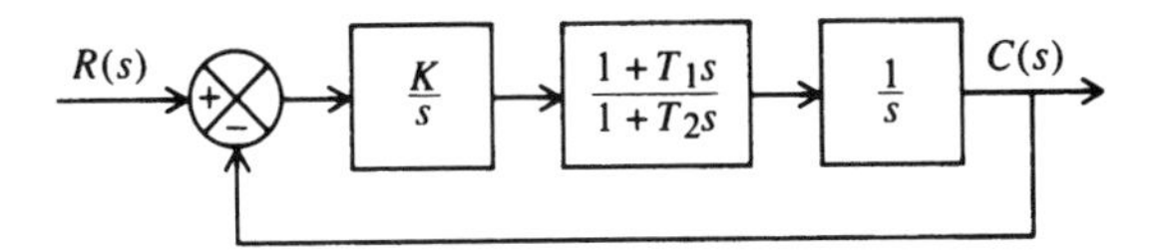

Figure P6.2

(a) Determine the system's **P** matrix from its state equations.

(b) Determine the characteristic equation of this system from knowledge of the **P** matrix.

(c) Utilizing the Routh–Hurwitz criterion, determine the necessary relationship between T_1 and T_2 for this system to be stable.

6.3. A feedback control system can be represented by a state vector differential equation where

$$\mathbf{A} = \begin{bmatrix} 0 & 2 & 0 \\ 0 & 0 & 1 \\ -2 & -K & -1 \end{bmatrix}.$$

(a) Determine the characteristic equation of this control system.

(b) Using the Routh–Hurwitz criterion, determine the range of K where the system is stable.

6.4. Consider the control system of a tracking radar system which operates in two coordinate axes. Its signal-flow graph, which is illustrated in Figure P6.4 indicates that there is electrical coupling between the control systems for the two axes:

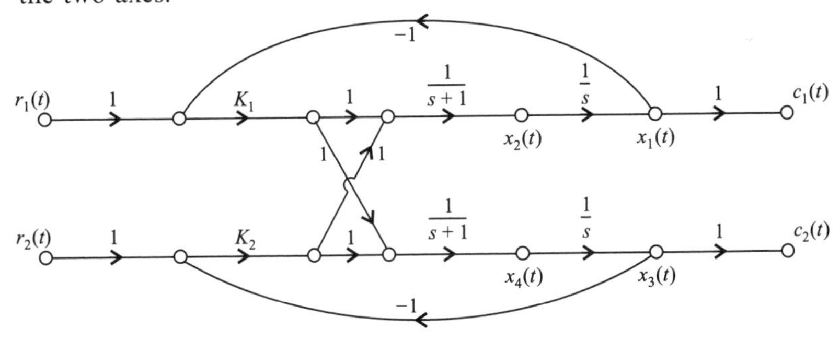

Figure P6.4

(a) Determine the phase-variable canonical equations for this control system.

(b) Determine the characteristic equation of this system.

(c) Is the system stable?

6.5. Phase-locked loops maintain zero difference in phase between an input carrier signal and a voltage-controlled oscillator. They are used in various applications including television, space telemetry, and missile tracking systems. The block diagram in Figure P6.5 is a linearized approximation of a specific phase-locked loop used for a space telemetry application:

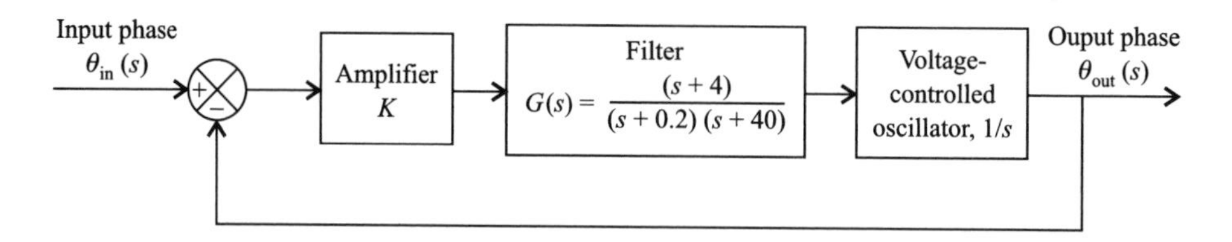

Figure P6.5

(a) Determine the maximum value of K so that this phase-locked loop is stable.

(b) The control system engineer designing this phase-locked loop also has a system requirement in this application that the steady-state error of the system shall be one degree for a ramp signal change of 10 rad/sec. Determine the value of K to meet this specification. Is this value of K within the allowable range of gain found in part (a) for which the system is stable?

6.6. A control system used to position a load is illustrated in Figure P6.6.

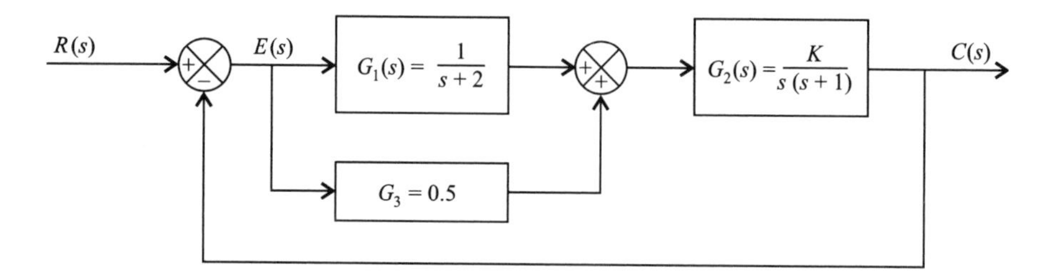

Figure P6.6

(a) Determine the characteristic equation of this control system.

(b) Determine the maximum value of K which can be used before the system becomes unstable using the Routh–Hurwitz criterion.

6.7. The field of fluidics is relatively new [49]. The use of fluids as a power source is not new, but the discovery that this energy can be manipulated and utilized in much the same way as electricity, and without the need for moving parts, has sparked the technology of fluidics. The term *fluidics* refers to the field of technology that is concerned with the use of either liquid or gaseous fluids in motion to perform functions such as amplification, sensing, switching, computation, and control. These systems have the advantages of high reliability, operation under extreme environmental conditions, resistance to radiation, and low cost. Fluidic control systems are especially applicable to fluid-flow systems, such as those using a turbine. Figure P6.7 illustrates the equivalent block diagram of a fluidic speed-control system. Actual speed, derived from a tuning-fork device, is compared with the desired speed. Any difference is amplified by a fluidic amplifier, which then activates a valve used to control the turbine's speed.

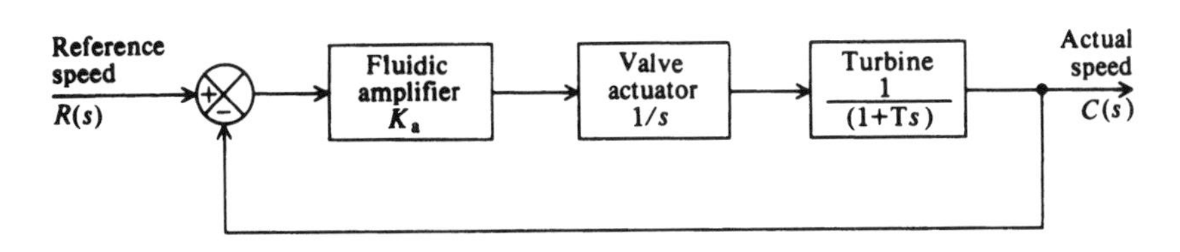

Figure P6.7

(a) Determine the state-variable signal-flow diagram and the phase variable canonical equations of this system when $K_a = 10$ and $T = 100$.

(b) Determine the characteristic equation of this system from the relation $|\mathbf{P} - s\mathbf{I}| = 0$.

(c) Utilizing the Routh–Hurwitz criterion, determine whether the system is stable.

6.8. Using the Routh–Hurwitz stability criterion, determine if the feedback control system shown in Figure P.68 is stable for the following transfer functions:

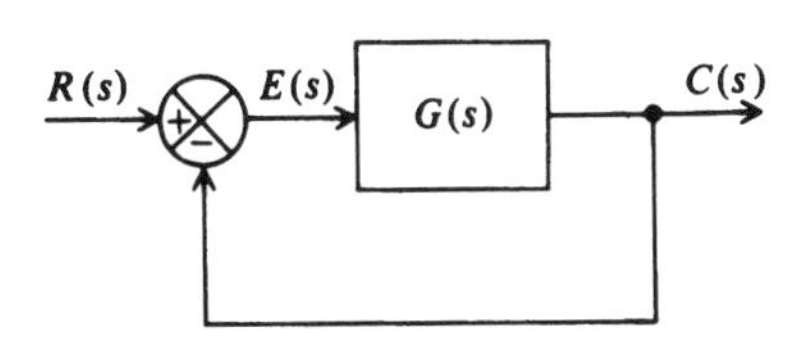

Figure P6.8

(a) $$G(s) = \frac{100}{s(s^2 + 8s + 24)},$$

(b) $$G(s) = \frac{3s + 1}{s^2(300s^2 + 600s + 50)},$$

(c) $$G(s) = \frac{24}{s(s + 2)(s + 4)},$$

(d) $$G(s) = \frac{0.2(s + 2)}{s(s + 0.5)(s + 0.8)(s + 3)}.$$

6.9. Figure P6.9*a* illustrates the submarine depth-control problem. The object is to adjust the actual depth C to equal a desired depth R. The control system depends on a pressure transducer, which is used to measure the actual depth. Any difference between the actual and desired depths is amplified and is used to drive the stern plane actuator through an appropriate angle θ until the actual depth equals the desired depth. An equivalent block diagram of such a system is illustrated in Figure P6.9*b*. Utilizing the Routh–Hurwitz criterion, determine whether this system is stable for the parameters indicated.

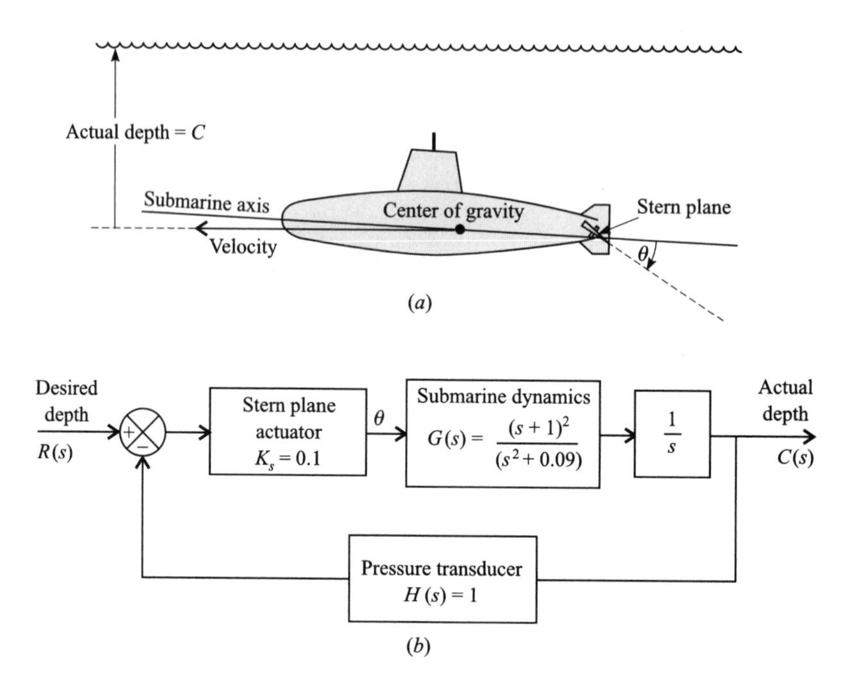

Figure P6.9 (*a*) Submarine depth-control problem. (*b*) Block diagram of system.

6.10. Using the Routh–Hurwitz stability criterion, determine whether the feedback control system illustrated in Figure P6.10 is stable.

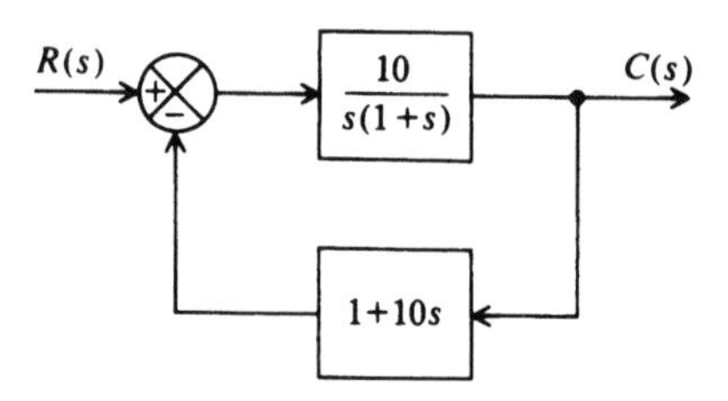

Figure P6.10 Feedback control system.

6.11. Automatic control systems are being used to an ever-increasing degree in the railroad transportation field [50]. A very widely acclaimed high-speed rail transportation system is in operation in Japan [51]. Figure P6.11*a* is a photograph of the Shinkansen Tokyo-to-Osaka railroad, which is commonly referred to as the Tokaido line. Figure P6.11*b* illustrates an equivalent block diagram for the automatic braking system used to regulate this class of high-speed train.

Figure P6.11 (*a*) Shrinkansen Tokaido line. [Courtesy of Japan National Tourist Organization (JNTO)]

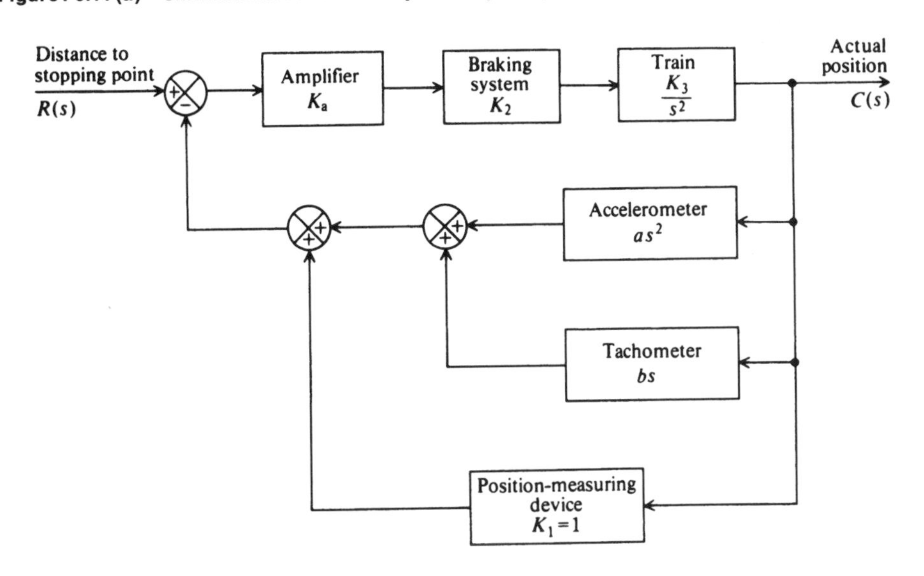

Figure P6.11 (*b*) Block diagram: automatic braking system.

(a) Determine the signal-flow graph and the characteristic equation of this system.

(b) Using the Routh–Hurwitz criterion, determine the allowable values of amplifier gain K_a for system stability. Assume the following parameters:

$$K_1 = 1, \quad K_2 = 1000, \quad K_3 = 0.001,$$

$$a = 0.1, \quad b = 0.1.$$

6.12. Using the Routh–Hurwitz stability criterion, determine whether the feedback control system illustrated in Figure P6.12 is stable.

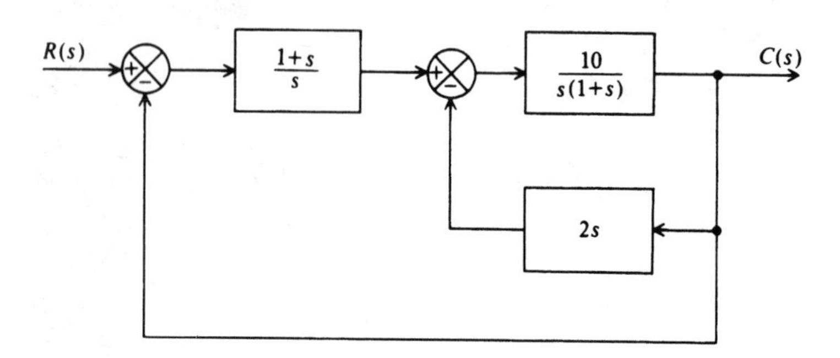

Figure P6.12

6.13. The tool of mathematical modeling can be used to study economic problems of a much larger scope than that found in a single business organization. The economics concerned with national income, government policy on spending, private business investment, business production, taxes, and consumer spending may be represented [49] by the block diagram of Figure P6.13. Although business production lags available funds according to a pure time delay, the representation of $G_s(s)$ by the lag factor $1/(1 + Ts)$, is adequate. Assuming that $E(s)$ and $U(s)$ are related by

$$U(s) = -(A + Bs)E)s),$$

and government policy is represented by

$$G_1(s) = C + Ds,$$

determine the requirements on C and D in terms of A and B for system stability.

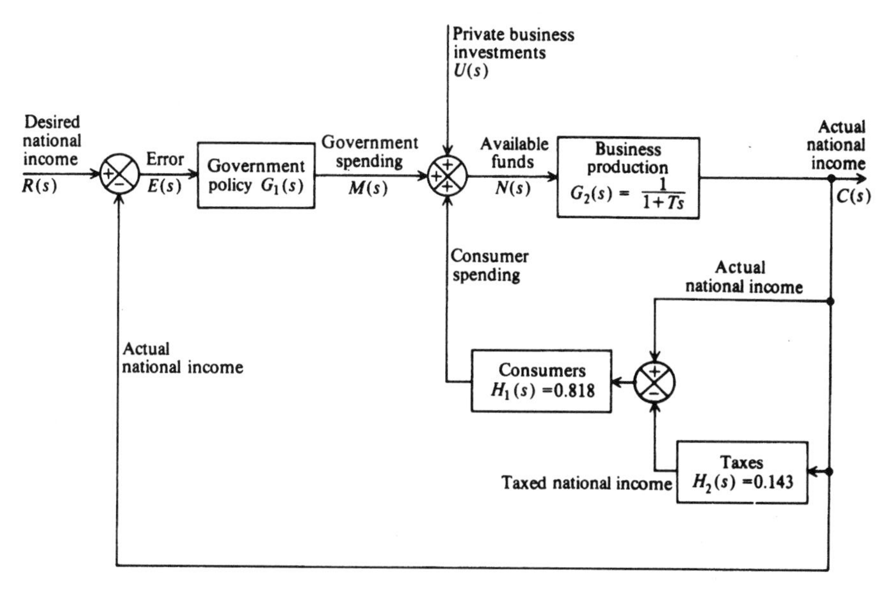

Figure P6.13

6.14. By means of Nyquist-diagram plots, determine whether feedback systems represented by the following values of $G(j\omega)H(j\omega)$ are stable.

(a)
$$G(j\omega)H(j\omega) = \frac{10}{(1+j\omega)(1+2j\omega)(1+3j\omega)},$$

(b)
$$G(j\omega)H(j\omega) = \frac{10}{j\omega(1+j\omega)(1+10j\omega)},$$

(c)
$$G(j\omega)H(j\omega) = \frac{10}{(j\omega)^2(1+0.1j\omega)(1+0.2j\omega)},$$

(d)
$$G(j\omega)H(j\omega) = \frac{2}{(j\omega)^2(1+0.1j\omega)(1+10j\omega)}.$$

Do not attempt to plot the exact values of $G(j\omega)H(j\omega)$ for all values of frequency. It should only be necessary to determine a few values of frequency exactly.

6.15. By means of a Nyquist-diagram plot, determine whether the system illustrated in Figure P6.15 is stable.

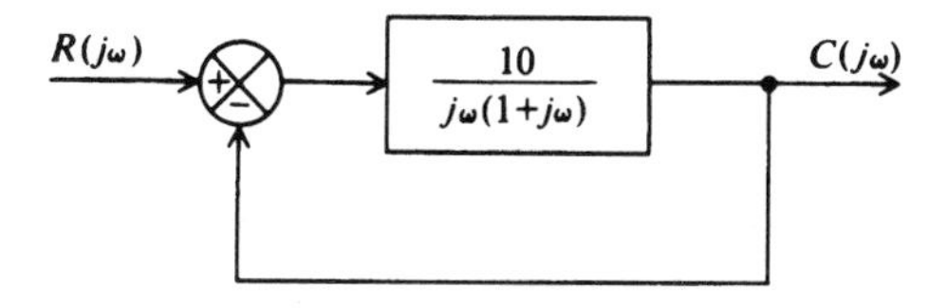

Figure P6.15

6.16. Determine whether the system illustrated in Figure P6.16 is stable by means of a Nyquist-diagram plot.

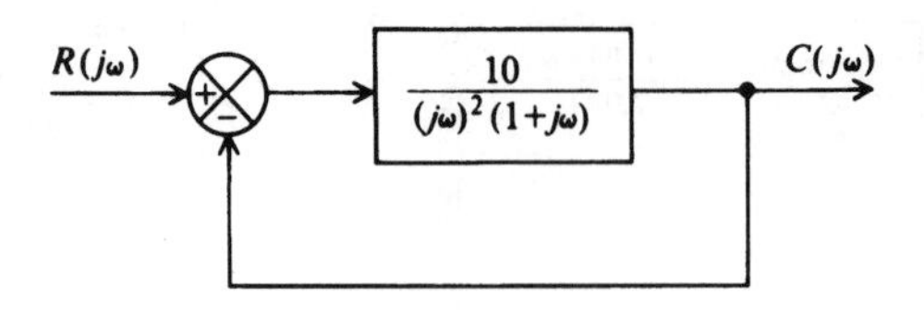

Figure P6.16

6.17. The Nyquist diagram for a control system is as shown in Figure P6.17.

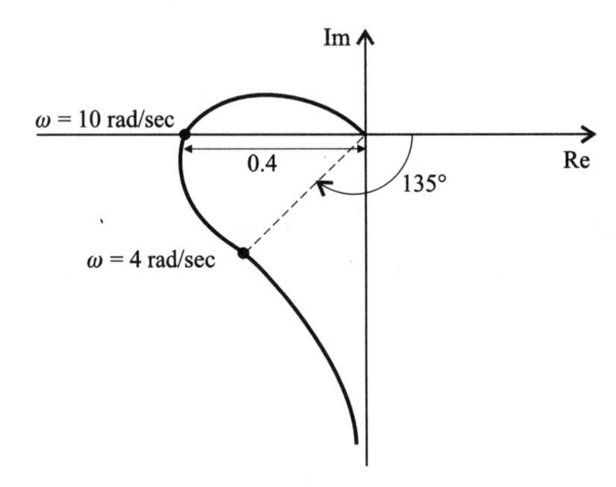

Figure P6.17

The control system is known to have an open-loop transfer function whose form is given by the following:

$$G(s)H(s) = \frac{K_1}{s^3 + K_2 s^2 + K_3 s}.$$

Determine the values of K_1, K_2, and K_3.

6.18. The Nyquist diagram of a control system is illustrated in Figure P6.18. This control systems's open-loop transfer function, $G(s)H(s)$, does not have any poles in the right half-plane. Determine whether the system is stable or not, and determine the number of roots (if any) which are located in the right half-plane.

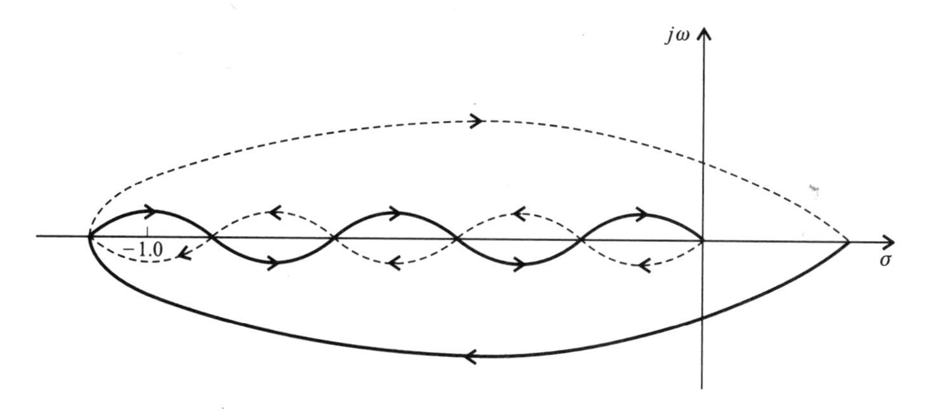

Figure P6.18

6.19. The positive-frequency portions of the Nyquist diagram for several transfer functions are shown in Figure P6.19. Complete the Nyquist diagram and determine stability, assuming that $G(j\omega)H(j\omega)$ has no poles in the right half-plane.

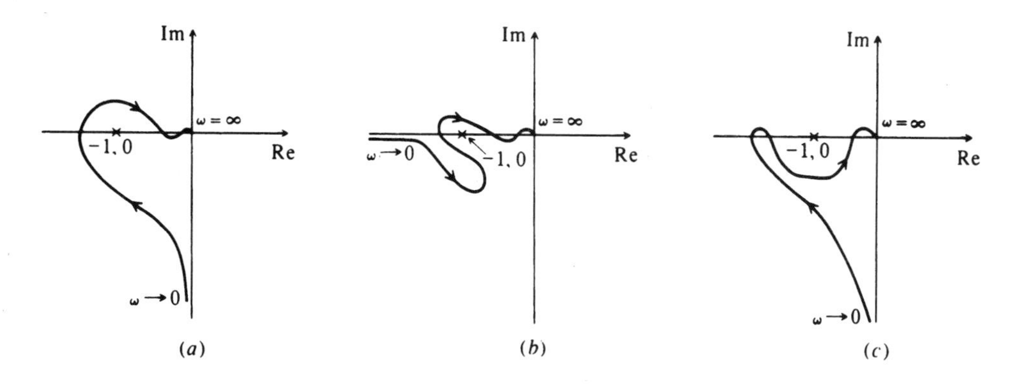

Figure P6.19

6.20. The design of automatic control systems for electronic activation of human limb movements is an interesting example of the application of control theory [52]. By means of electrical pulses, a paralyzed limb can be made to contract, with the result that functional movements of the extremity are performed. Figure P6.20a illustrates the concept implemented using conventional control-system techniques. The electronic controller $G(s)$ feeds electrical signals to the contracting muscles (agonists), which stretch the opposing muscles (antagonists). Figure P6.20b illustrates the equivalent block diagram where the following transfer functions can be assumed:

$$G(s) = A,$$

$$H(s) = \frac{\phi(s)}{I(s)} = \frac{K'e^{-\tau_0 s}}{s(Js + C)}.$$

Typical values of the parameters are as follows:

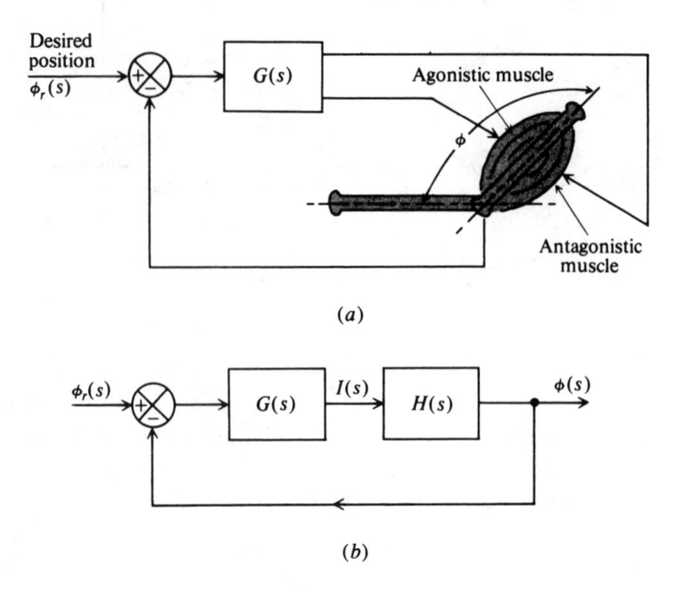

Figure P6.20

$$J = 1, \quad C = 20, \quad \tau_0 = 0.1, \quad K' = 1.$$

(a) Utilizing the Nyquist diagram, determine the phase margin when the electronic gain A is set at 2.

(b) Repeat part (a) when the electronic gain is set at 100.

6.21. A unity negative-feedback control system's forward transfer function is given by the following:

$$G(s) = \frac{10}{s^2(1+s)^3}.$$

(a) Sketch the Nyquist diagram for this control system.

(b) Is the resulting system stable or unstable? If it is unstable, determine the number of roots in the right half-plane.

6.22. A feedback control system has the configuration shown in Figure P6.22.

(a) Draw the Nyquist diagram of the loop gain function.

(b) Draw the Bode diagram showing the magnitude, in decibels, and phase angle, in degrees, as a function of frequency.

(c) Find the phase margin and gain margin of the system. Illustrate these points on the graphs for parts (a) and (b).

(d) Find the values of K_p, K_v, and K_a.

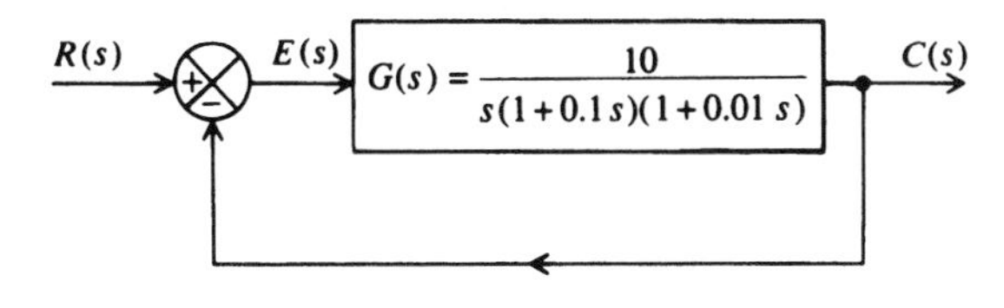

Figure P6.22

(e) What is the steady-state velocity-lag error for a velocity input of 5 rad/sec?

6.23. A unity negative-feedback control system's forward transfer function is given by the following:

$$G(s) = \frac{10}{s^2(1+s)^2}.$$

(a) Sketch the Nyquist diagram for this control system.

(b) Is the resulting system stable or unstable? If it is unstable, determine the number of roots in the right half-plane.

6.24. A feedback control system has the configuration shown in Figure P6.24.

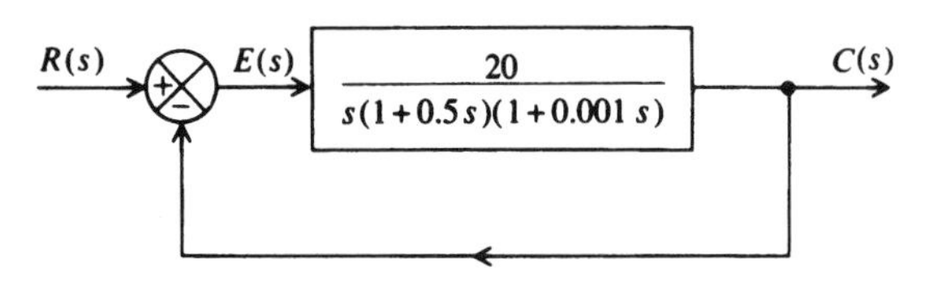

Figure P6.24

(a) Plot the Bode diagram for this system.

(b) What is the gain crossover frequency? What is the phase margin at gain crossover? What is the gain margin? Is the system stable?

(c) Find the values of K_p, K_v, K_a.

(d) What is the steady-state velocity-lag error for a velocity input of 40 rad/sec?

6.25. Draw the straight-line attenuation diagrams, showing the magnitude in decibels, and the phase characteristics, showing the phase angle in degrees, as a function of frequency, for the following transfer functions:

(a)
$$G_A(s) = \frac{20}{s(1+0.5s)(1+0.1s)},$$

(b)
$$G_B(s) = \frac{2s^2}{(1+0.4s)(1+0.04s)},$$

(c)
$$G_C(s) = \frac{50(0.6s + 1)}{s^2(4s + 1)},$$

(d)
$$G_D(s) = \frac{7.5(0.2s + 1)(s + 1)}{s^2(s^2 + 16s + 100)}.$$

Assuming that $G_A(s)$ and $G_C(s)$ represent open-loop transfer functions of unity-gain feedback systems, determine the phase and gain margins in parts (a) and (c).

6.26. We wish to design the control system shown in Figure P6.26 so that the phase margin is 70°.

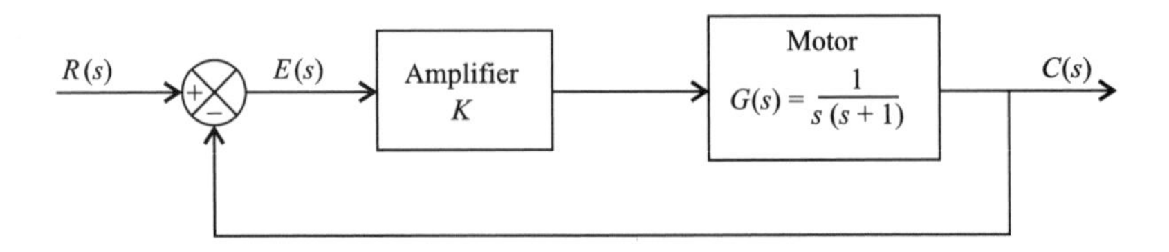

Figure P6.26

Without using semi-log graph paper, determine analytically the value of the amplifier gain K so that the phase margin is 70°.

6.27. An engineer is called in to consult on a control system in a piece of equipment in the field. No one can find the design report or test results from the original design of the control system. The engineer, therefore, decides to take a frequency response of the control system by opening up the outer feedback loop. The resulting asymptotic gain–frequency characteristic is obtained as shown in Figure P6.27.

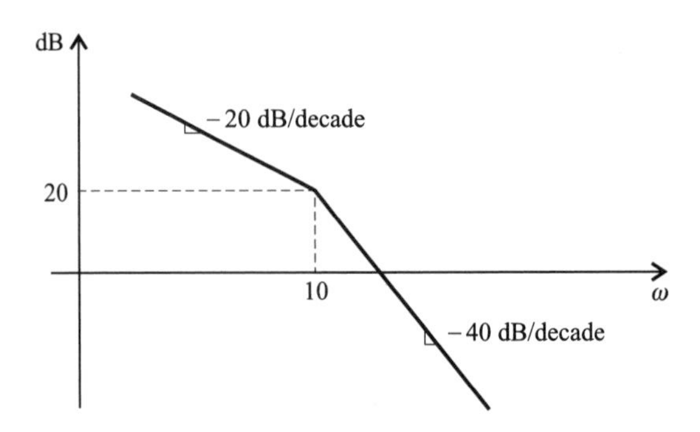

Figure P6.27

Assuming that the system is a minimum-phase transfer function, determine the transfer function, $G(s)H(s)$.

6.28. A tank-level control is shown in Figure P6.28. It is desired to hold the tank level C within limits even though the outlet flow rate V_1 is varied. If the level is not correct, an error voltage E_n is developed which is amplified and applied to a servomotor. This in turn adjusts a valve through appropriate gearing N and thereby restores balance by adjusting the inlet rate, M_2.

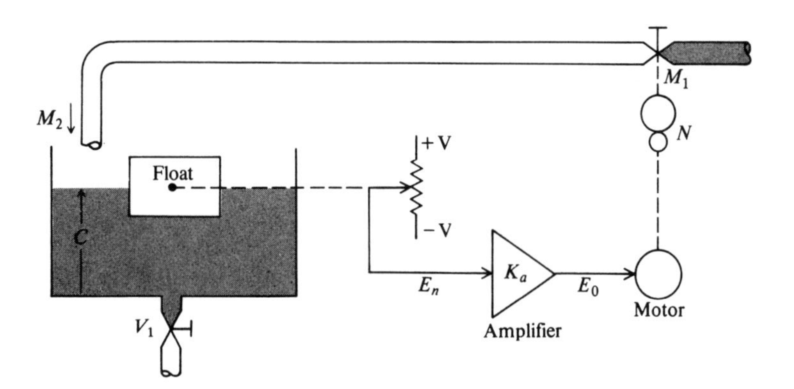

Figure P6.28

The following relations are valid from this figure:

$$
\begin{aligned}
E_n &= R - C = \text{error in level (in)},\\
E_0 &= K_a E_n = \text{voltage applied to motor (volts)},\\
M_1 &= G_1(s)E_0 = \text{valve position (rad)},\\
M_2 &= K_v M_1 = \text{tank feed flow (ft}^3/\text{sec)},\\
C &= G_2(s)M_2 = \text{tank level (in)}.
\end{aligned}
$$

The pertinent constants and transfer functions are as follows:

$$
\begin{aligned}
G_1(s) &= \frac{1}{s(0.1s+1)} \text{ rad of valve motion per volt},\\
G_2(s) &= 0.5/s \text{ in of level per ft}^3/\text{sec},\\
K_v &= 0.1 \text{ ft}^3\text{sec per rad of valve motion},\\
K_a &= \text{amplifier gain to be set},
\end{aligned}
$$

$$\text{Error detector sensitivity} = 10 \text{ V per in of error.}$$

(a) Draw the system block diagram showing all transfer functions.

(b) Using a Bode diagram for the solution, determine the amplifier gain K_a required to meet a required gain crossover frequency of 1.5 rad/sec.

(c) With the gain set at the value determined in part (b), what are the resulting phase and gain margins?

6.29. By utilizing automatic ship steering systems, a ship can maintain a desired heading much more accurately than if it depended on a helmsman correcting the heading at infrequent intervals. Accurate ship heading is a particular crucial problem for minesweepers who must sweep desired, prescribed, straight paths with very little allowable deviations. Figure P6.29 illustrates the overall control problem where δ represents the deviation of the mine sweeper from the desired heading and θ represents the angle of deflection of the steering rudder. The transfer function relating $\theta(s)$ and $\delta(s)$ for a 180-ft minesweeper moving at 13 ft/sec is given by [53]

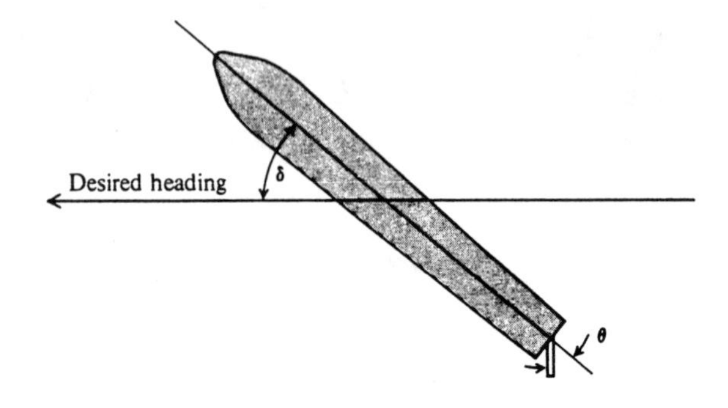

Figure P6.29

$$\frac{\delta(s)}{\theta(s)} = \frac{0.164(s+0.2)(s-32)}{s^2(s+0.25)(s-0.009)}.$$

Determine the Bode diagram for this transfer function.

6.30. Repeat Problem 6.22 for the transfer function

$$G(s) = \frac{2}{s(1+0.1s)(1+0.5s)}.$$

6.31. A feedback control system has the configuration shown in Figure P6.31, where $U(s)$ represents an extraneous signal appearing at the input to the plant.

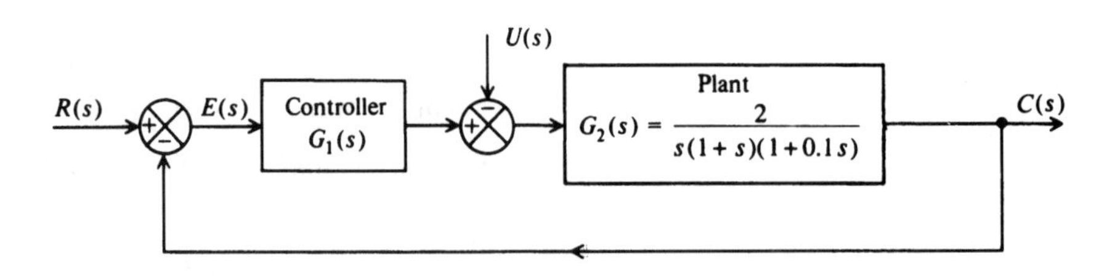

Figure P6.31

(a) Assuming that $G_1(s) = 1$, plot the decibel-log frequency diagram and the phase diagram for this system.

(b) It is desired that the steady-state error resulting from an extraneous unit step input signal at $U(s)$ shall be 0.1 unit. Assuming $G_1(s) = K$, determine K to meet this specification.

(c) Determine the gain crossover frequency, phase margin, and gain margin resulting from part (b).

6.32. A feedback control system has the configuration shown in Figure P6.32.

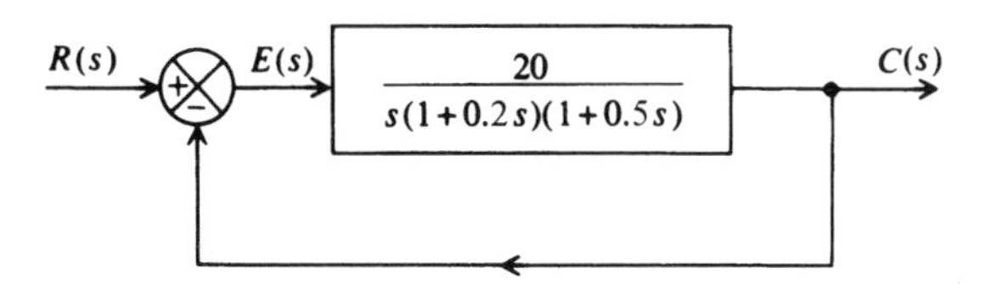

Figure P6.32

(a) Plot the Bode diagram for this system.

(b) What is the gain crossover frequency? What is the phase margin at gain crossover? What is the gain margin? Is the system stable?

(c) Find the values of K_p, K_v, K_a.

(d) What is the steady-state velocity-lag error for a velocity input of 40 rad/sec?

6.33. Repeat problem 6.28 with an error-detector sensitivity of 1 V per inch of error.

6.34. Repeat Problem 6.31 with the transfer function of $G_2(s)$ given by

$$G_2(s) = \frac{10}{(1 + 0.1s)(1 + 0.2s)(1 + 1.5s)}.$$

6.35. A second-order servomechanism has a forward transfer function given by

$$G(s) = \frac{16}{s(2 + s)}.$$

The feedback transfer function is unity.

(a) Draw the Bode diagram showing the magnitude and the phase characteristics as a function of frequency.

(b) Using the Nichols chart, plot a curve of frequency response of the closed-loop system.

(c) What are ω_p and M_p?

(d) Can this system ever be unstable no matter how large the forward gain is made? Explain.

6.36. Repeat problem 6.35 for

$$G(s) = \frac{60(1+0.5s)}{s(1+5s)}.$$

6.37. Repeat problem 6.35 for

$$G(s) = \frac{60(1+s)}{s^2(1+0.1s)}.$$

6.38. Driving an automobile is a feedback process, as indicated in the block diagram [54] of Figure P6.38.

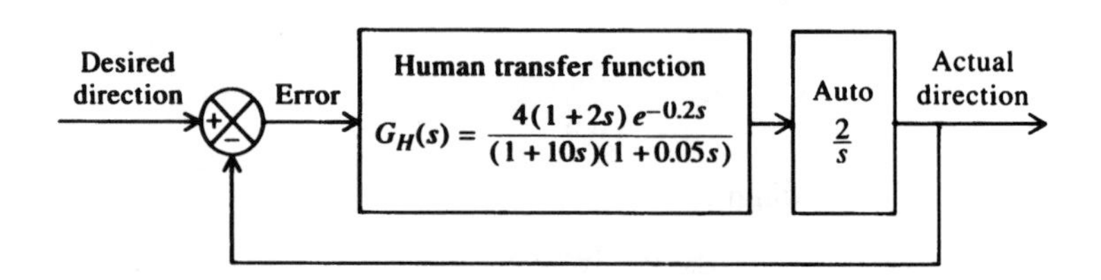

Figure P6.38

The objective of the system shown is for the driver to maintain a desired direction along the road. Any error is sensed by the driver's eyes and is transmitted to the brain, which then activates his muscles appropriately. With his muscles, the driver controls the steering wheel. The human transfer function shown in the block diagram is based on test data for a particular driver, and is of the same form discussed in Problem 6.64.

Using the Bode-diagram method, determine the phase margin and gain margin that this system has.

6.39. Consider the same feedback control system analyzed in Problem 6.1. Its stability is to be analyzed in this problem using the Bode diagram.

(a) Draw the Bode diagram for this system, both the amplitude and phase diagrams.

(b) Determine the phase and gain margins from your Bode diagram. Is this system stable? (Note: Your answer as to whether the system is stable in this problem should agree with your answer regarding stability in Problem 6.1.)

(c) Repeat parts (a) and (b) if the gain is increased to 200. (Hint: Modify your Bode diagram of part (a) to do this part. Drawing a new Bode diagram is not necessary.)

6.40. The stability of a submarine depth-control system illustrated in Figure P6.40*a* is to be analyzed using the Bode diagram. In the block diagram of the system illustrated in Figure P6.40*b*, a pressure transducer measures the actual depth, which is then compared to the desired depth. Any difference between the actual and desired depths is amplified and is used to drive the stern plane actuator through an appropriate angle θ until the actual depth equals the desired depth.

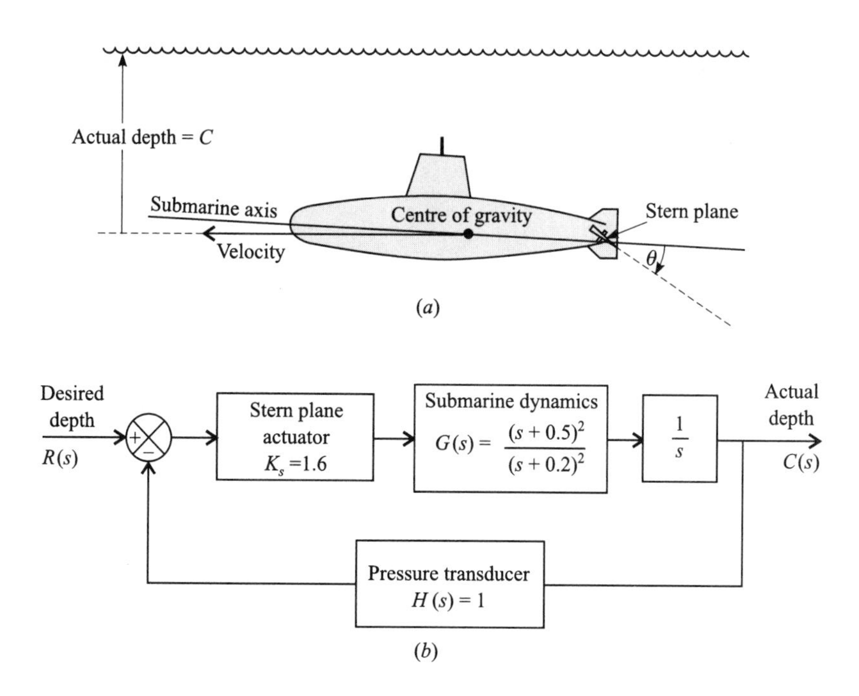

Figure P6.40 (a) Submarine depth-control system (b) Block diagram of the system

(a) Draw the Bode diagram, and find the gain crossover frequency.

(b) Find the phase margin and gain margin of the depth-control system.

6.41. In Figure P6.41*a*, Astronauts George D. Nelson (right) and James D. van Hoften (left) use the mobile foot restraint and the remote manipulator system (a robot arm-like device) as a "cherry picker" for moving about on the Challenger space shuttle mission, Flight 41-C, on April 11, 1984. They shared a repair task on the "captured" Solar Maximum Mission Satellite in the aft end of the Challenger's cargo bay. Later, the remote manipulator system lifted the Solar Maximum Mission Satellite into space once more. An approximate equivalent block diagram of this operation, which is illustrated in Figure P6.41*b*, contains an approximation for modeling the astronaut [54].

(a) Draw the Bode diagram and determine the gain crossover frequency, phase, and gain margins for $K = 10$.

(b) Repeat part (a) if K is reduced in half to 5.

(*a*) (Official NASA photograph)

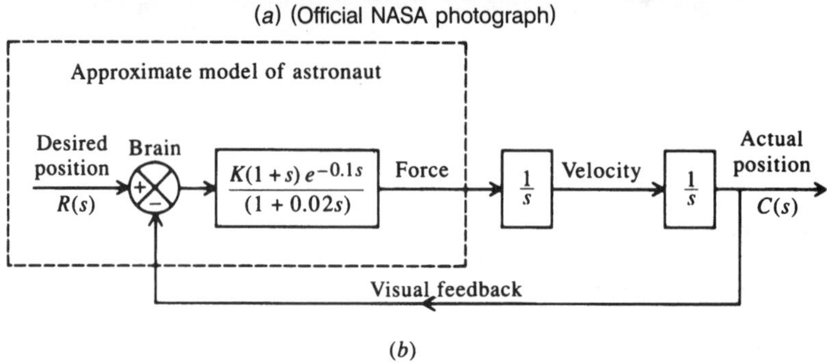

(*b*)

Figure P6.41 (*a*) Mobile foot restraint and remote manipulator: Challenger space-shuttle Flight 41-c, April 1984 (Official NASA photograph). (*b*) Block diagram of operation

(c) Repeat part (a) if K is doubled to 20.

(d) What conclusions can you reach from your results?

6.42. A feedback control system has a block diagram as shown in Figure P6.42.

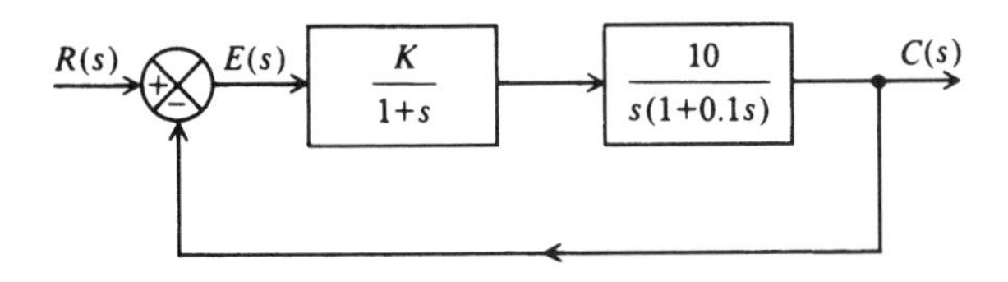

Figure P6.42

(a) Determine the required gain K for a steady-state velocity-lag error of 30° with an input velocity of 10 rad/sec.

(b) What are the values of M_p and ω_p?

6.43. A unity-feedback control system has a forward transfer function given by

$$G(s) = \frac{10}{s(1 + 0.1s)(1 + 0.05s)}.$$

(a) Plot $G(j\omega)H(j\omega)$ on the Nichols chart.

(b) What are the values of M_p and ω_p?

6.44. Repeat problem 6.43 for

$$G(s) = \frac{2}{s(1 + s)(1 + 10s)}.$$

6.45. A unity negative-feedback control system has a forward transfer function given by

$$G(s) = \frac{K}{s(1 + 0.1s)(1 + s)}.$$

(a) Sketch the root locus giving all pertinent characteristics of the locus.

(b) At what value of gain does the system become unstable?

6.46. Repeat Problem 6.45 for the following transfer functions:

(a) $$G(s) = \frac{K}{s^2},$$

(b) $$G(s) = \frac{K(1 + s)}{s^2(1 + 0.1s)},$$

(c) $$G(s) = \frac{K(s + 1)}{s(s^2 + 8s + 16)},$$

(d) $$G(s) = \frac{K(s + 0.1)^2}{s^2(s^2 + 9s + 20)}.$$

6.47. Sketch the root locus for a negative-feedback control system having the following forward and feedback transfer functions:

$$G(s) = \frac{K(s+0.1)}{s^2(s+0.01)}, \quad H(s) = 1 + 0.6s.$$

6.48. Sketch the root locus of a unity positive-feedback system whose transfer function is given by

$$G(s) = \frac{K}{(s+1)(s-1)(s+4)^2}.$$

6.49.

(a) Draw the root-locus diagram of a negative-feedback control system whose feedback-path transfer function is unity and whose forward-path transfer function is given by the following:

$$G(s) = \frac{K}{(s+2)^5}.$$

(b) At what value of K does the system become unstable, and where does the root locus intersect the $j\omega$ axis when this occurs?

6.50. Sketch the root locus for a negative-feedback control system having the following forward and feedback transfer functions:

$$G(s) = \frac{K(s+1)}{s^2(s+2)(s+4)}, \quad H(s) = 1.$$

6.51. Repeat problem 6.50 for positive feedback. What conclusions can you draw from your result?

6.52. Draw the root locus of a positive feedback system where

$$G(s) = \frac{K}{(s+1)^2(s+4)^2}, \quad H(s) = 1.$$

6.53. Determine the root locus for a feedback system whose open-loop transfer function is given by

$$G(s)H(s) = \frac{K(s+2)}{s(s+4)(s+8)(s^2+2s+5)}$$

For $-\infty \leqslant K \leqslant \infty$. Indicate all pertinent values on the root locus.

6.54. A negative-feedback system has an open-loop transfer function given by

$$G(s)H(s) = \frac{K(s+4)(s+40)}{s^3(s+200)(s+900)}.$$

(a) Draw the root locus and label all pertinent values on the root locus.

(b) For what range of values of gain is the system stable?

(c) Draw the Bode diagram of the system and correlate the regions of stability and instability with the root-locus results. Assume that $K = 1$.

6.55. Repeat Problem 6.52 for the case of negative feedback.

6.56. Reconsider the ecological model of rabbits and rabbit-eating foxes presented in Problem 2.72. The state equations of the process were given by

$$\dot{x}_1(t) = Ax_1(t) - Bx_2(t),$$
$$\dot{x}_2(t) = -Cx_2(t) + Dx_1(t),$$

where x_1 represented the number of rabbits and x_2 represented the fox population. Determine the requirements of A, B, C, and D for a stable ecological system. What occurs if A is greater than C?

6.57. Sketch the root locus for the following negative-feedback control system, and determine the range of K for which the system is stable:

$$G(s) = \frac{K(s+6)}{s(s+1)(s+4)}, \quad H(s) = 1.$$

6.58. Reconsider the model depicting the development of unindustrialized nations discussed previously in Problem 2.69. For the coefficients listed in part (a) of Problem 2.69, determine whether the system is stable. Stability in this problem means that the states $x_1(t)$ and $x_2(t)$ decrease to zero as time increases.

6.59. The stability of the feedback control system shown in Figure P6.59 is to be determined.

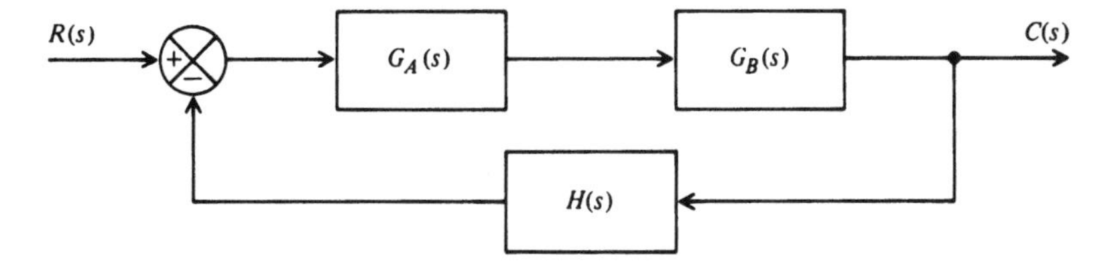

Figure P6.59

(a) Using the Routh–Hurwitz stability criterion, determine whether the system is stable for the following transfer functions:

$$G_A(s) = \frac{45}{s+2},$$

$$G_B(s) = \frac{2}{(s+3)(s+5)},$$

$$H(s) = 1.$$

(b) Repeat part (a) for the following transfer functions:

$$G_A(s) = \frac{45}{s+2},$$

$$G_B(s) = \frac{2}{s+3},$$

$$H(s) = \frac{1}{s+5}.$$

(c) What conclusions concerning stability can you draw from your results in parts (a) and (b)?

(d) Will the outputs, $c(t)$, in parts (a) and (b), differ in response to the same input? Discuss your results.

6.60. Determine the range of positive real gain K that will result in a stable system for the system shown in Figure P6.60 for the following conditions:

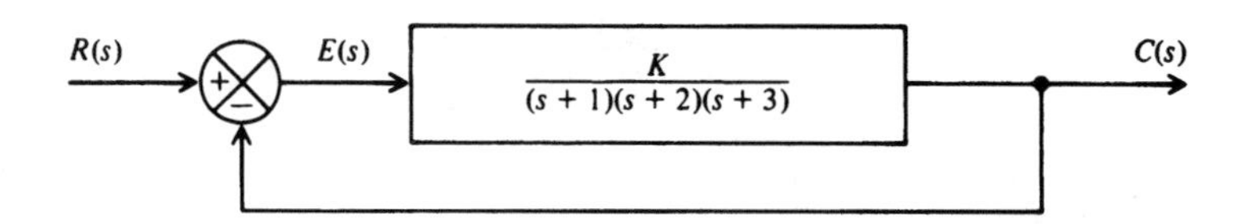

Figure P6.60

(a) The system has negative feedback.

(b) The system has positive feedback.

6.61. The transfer function of an unknown element in a control system can be determined by measuring its frequency response using a sinusoidal input. For a particular element, the frequency and amplitude data shown in Table P6.61 were obtained.

The form of the transfer function is given by

$$G(s) = \frac{K[1 + (s/\omega_a)]}{s[1 + (s/\omega_b)][1 + (s/\omega_c)]}.$$

Determine the values of K, ω_a, ω_b, and ω_c.

Table P6.61. Experimental Data

ω	$G(j\omega)$ in dB
0.1	66
0.2	60
0.4	54
0.7	51
1.0	49
2.0	47
4.0	46
7.0	45
10.0	43
20.0	39
40.0	34
70.0	28
100.0	23
200.0	13
400.0	2
700.0	−8
1000.0	−14

6.62. The transfer function of an element used in a feedback system is not known explicitly, but its frequency response has been measured experimentally and is given in Figure P6.62. Based on your knowledge of Bode-diagram characteristics, determine the transfer function of this element.

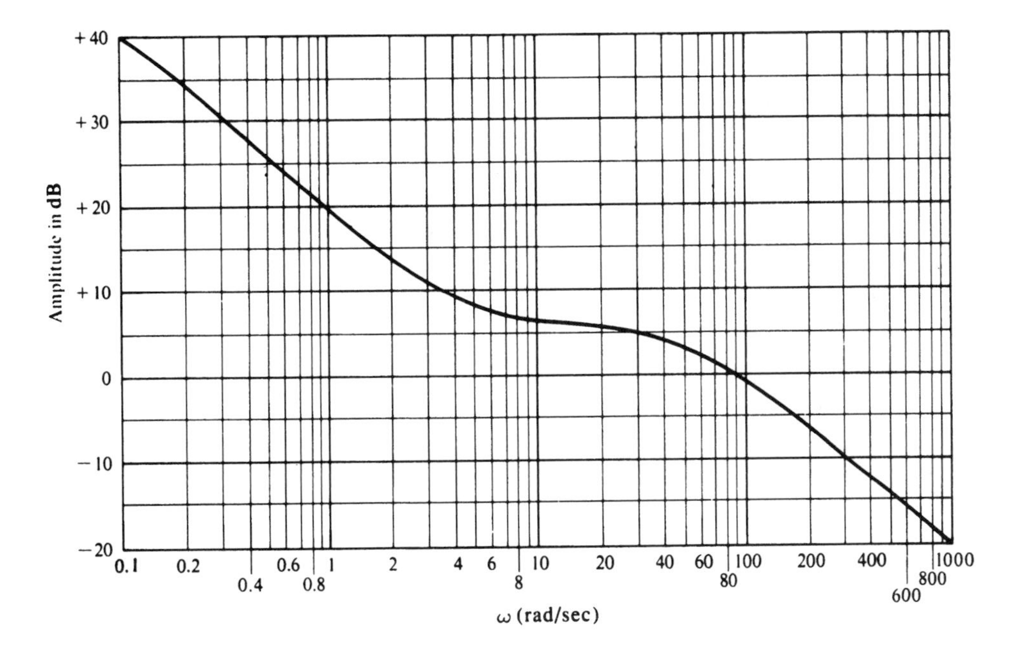

Figure P6.62

6.63. Electromechanical nose-wheel steering systems have been developed that can supply general aviation short-haul aircraft with needed maneuverability and durability [55]. These intermediate-sized aircraft must be able to utilize small airfields, many of which do not have elaborate service and maintenance facilities. During takeoff from such landing strips, the task of keeping the aircraft on the proper heading is achieved by utilizing nose-wheel power steering provided by this device. The block diagram of a nose-wheel steering system is illustrated in Figure P6.63. Pilot command signals are compared with a feedback signal representing the nose wheel's heading. The resulting error signal is amplified and applied to a magnetic particle clutch, which activates the rotation of the wheel heading. Assuming that $L/R = 0.1$, $K_c = 1$, $J = 1$, $C_v = 9$, and $K_e = 9$, utilize the Bode diagram to determine the gain required to achieve a phase margin of 40°.

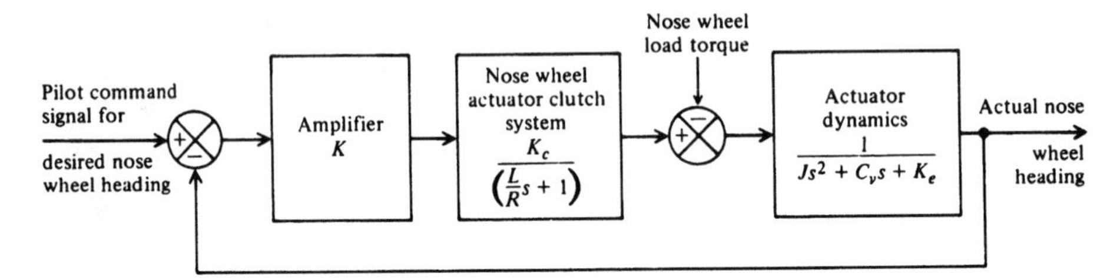

Figure P6.63

6.64. The proper design of man-machine control systems requires as much knowledge of the human element as that of the machine. Therefore, the determination of the human transfer function is very essential in order that the performance of man-machine systems can be evaluated. This problem can be better understood by referring to Figure P6.64, which depicts the classical compensatory manual tracking problem. In this configuration, an operator attempts to maintain a moveable follower coincident with a stationary reference point that represents the target. A very general and useful form of the human transfer function, $G_H(s)$, which can be applied to manual tracking problems, was provided in Problem 3.33. It is given by

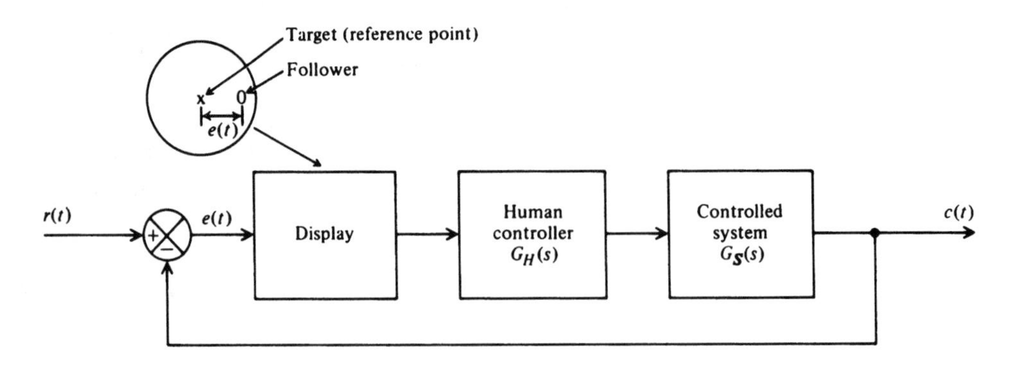

Figure P6.64

$$G_H(s) = \frac{K(1 + T_A s)e^{-Ds}}{(1 + T_L s)(1 + T_N s)},$$

where D represents the operator's transportation lag, T_A represents the operator's anticipation time constant, T_L represents the operator's error-smoothing lag-time constant, and T_N represents the operator's short neuro-muscular delay. Representative values for the elements are as follows [54]:

$$D = 0.2 \text{ sec} \pm 20\%.$$

$$T_A = 0\text{-}2.5 \text{ sec (variable)},$$

$$T_L = 0\text{-}20 \text{ sec (variable)},$$

$$T_N = 0.1 \text{ sec} \pm 20\%,$$

$$K = 1\text{-}100 \text{ (variable)}.$$

The gain K and the time constants T_A and T_L are considered to be variable according to the control task being performed. This transfer function has met with reasonable success for predicting manual tracking control system response where the bandwidth is relatively low. For this problem, assume that the human transfer function is given by the following expression:

$$G_H(s) = \frac{10(1 + s)e^{-0.2s}}{(1 + 10s)(1 + 0.1s)}.$$

Assume that the controlled system has a transfer function given by

$$G_s(s) = \frac{2}{s}.$$

Draw the Bode diagram and determine the phase and gain margin of the resulting system.

6.65. The pitch attitude control system for a booster rocket containing attitude and rate gyros is shown in Figure P6.65. Sketch the root locus and determine the maximum value of K that would permit stable operation.

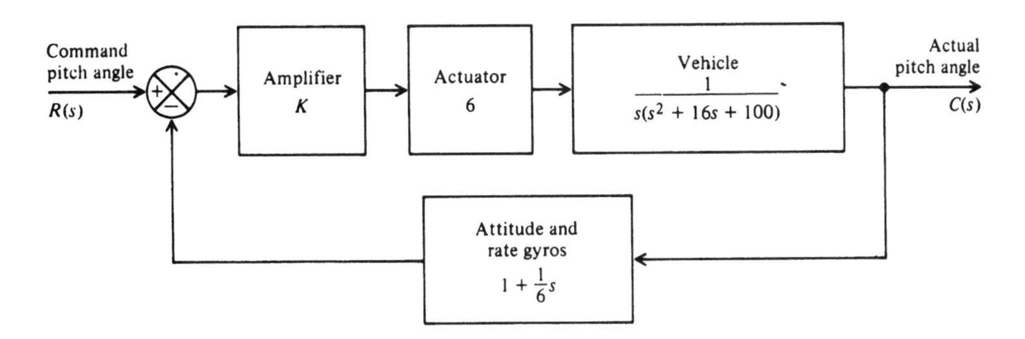

Figure P6.65

6.66. A unity feedback control system has the following forward transfer function:

$$G(s) = \frac{K}{s(s+1)(s+5)}.$$

(a) If the gain is set at 1.1 and one of the closed-loop poles is known to be located at $s_1 = -0.5$, is the system stable?

(b) If the gain is increased to 30 and one of the closed-loop poles is known to be located at $s_1 = -6$, is the system stable?

6.67. The root locus method can be used to study the stability of a motorcycle and rider combination. Analysis of a motorcycle/rider system must include a model of the rider in addition to the handling characteristics of the motorcycle. For this analysis, the approximate human model to be used for the system shown in Figure P6.67 is [54]:

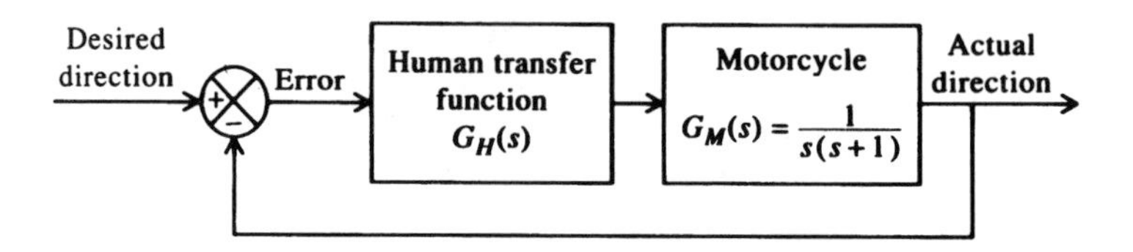

Figure P6.67

$$G_H(s) = \frac{K(s+A)}{(s+0.1)(s+10)}.$$

(a) Draw the root locus of this motorcycle/rider system assuming that the rider is new, his anticipation is very small, and therefore the "zero" term $(s+A)$ can be neglected in the numerator of $G_H(s)$. Show all pertinent values on the root locus including the allowable range of K for stability.

(b) Draw the root locus of this motorcycle/rider system assuming that the rider has gained experience, and the anticipation term is finite. Assume that $A = 1$. Show all pertinent values on the root locus including the allowable range of K for stability.

(c) What conclusions can you reach by comparing the allowable range of gain for stability in parts (a) and (b)?

6.68. The roll attitude-control system for a missile containing attitude and rate gyros is shown in Figure P6.68.

(a) Sketch the root locus and indicate all pertinent characteristics.

(b) Find the maximum value of K that would permit stable operation.

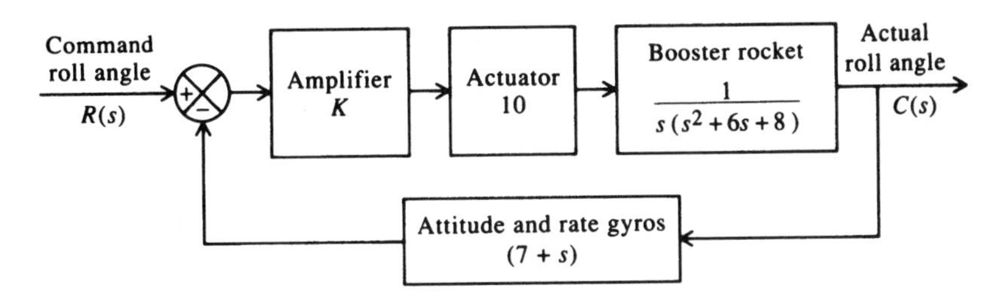

Figure P6.68

6.69. A block diagram of a unity negative-feedback positioning system used in a robot has a forward transfer function given by

$$G(s) = \frac{K(s+4)}{(s+0.5)^2(s+2)}.$$

(a) Sketch the root locus giving all pertinent characteristics of the locus. Identify all asymptotes, points of breakaway from the real axis, and the intersection of the root locus and the imaginary axis.

(b) At what value of gain does the system become unstable?

(c) When the system becomes unstable at the gain found in part (b), determine the location of all three roots of the system.

6.70. Consider the control system in Figure P6.70.

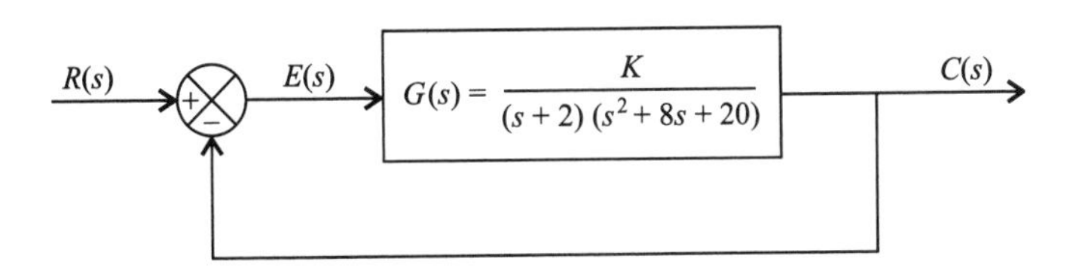

Figure P6.70

(a) Sketch the root locus of this control system. Determine all pertinent values on the root locus including the asymptotes, angles of departure, $K_{\max}$, and the crossing of the imaginary axis.

(b) We wish to operate this system at a gain which will result in the real part of all the closed-loop poles being equal. Determine the value of the gain and the real part of these closed-loop poles for this condition.

6.71. Determine the root locus for a negative-feedback system whose open-loop transfer function is given by

$$G(s)H(s) = \frac{K(s+8)}{s(s+3.125)(s+100)}$$

for $0 \leqslant K \leqslant \infty$. (Hint: One of the points of breakaways from the real axis exists at -1.74). Indicate all pertinent points on the root locus, and determine the range of K for which the system is stable.

6.72. Repeat Problem 6.26 if we wish to design the control system for a phase margin of 45°.

6.73. Repeat Problem I6.9 for the following transfer function:

$$G(s) = \frac{K}{(s+2)^3}$$

REFERENCES

1. R. C. Rosenberg and D. C. Karnopp, *Introduction to Physical System Design.* McGraw-Hill, New York, 1987.
2. M. F. O'Flynn and G. Moriarity, *Linear Systems.* Harper & Row, New York, 1987.
3. E. Routh, *Advanced Dynamics of a System of Rigid Bodies*, Macmillan, London, 1905.
4. A. Hurwitz, "Uber die Bedingungen, unter welchen eine Gleichung nur Wurzeln mit negativen realen Theilen besitzt." *Math. Ann.* **46**, 273 (1895).
5. H. Nyquist, "Regeneration theory." *Bell Syst. Tech. J.* **11**, 126 (1932).
6. R. Sarcedo and E. E. Shiring, *Introduction to Continuous and Digital Control Systems.* Macmillan, New York, 1968.
7. *MATLAB™ for MS-DOS Personal Computers User's Guide,* Control System Toolbox. MathWorks, Inc., Natick, MA, 1997.
8. H. Chestnut and R. W. Mayer, *Servomechanisms and Regulating System Design*, 2nd ed., Vol. 1. Wiley, New York, 1959.
9. S. M. Shinners, "Minimizing servo load resonance error with frequency selective feedback." *Control Eng.* **51**, 51–56 (1962).
10. R. C. Dorf, *Modern Control Systems*, 5th ed. Addison-Wesley, Reading, MA, 1989.
11. C. H. C. Leung, *Quantitative Analysis of Computer Systems.* Wiley, New York, 1988.
12. S. M. Shinners, "Which computer—Analog, digital or hybrid?" *Mach. Des.* **43**, 104–111 (1971).
13. N. Stern and R. A. Stern, *Introducing QuickBASIC 4.0 and 4.5: A Structured Approach.* Wiley, New York, 1989.
14. S. A. Hovanessian and L. A. Pipes. *Digital Computer Methods in Engineering.* McGraw-Hill, New York, 1969.
15. T. Ward and E. Bromhead, *FORTRAN and the Art of PC Programming.* Wiley, New York, 1989.

16. D. D. McCracken and W. J. Dorn, *Numerical Methods and FORTRAN Programming: With Applications in Engineering and Science*. Wiley, New York, 1964.
17. E. I. Organick, *A FORTRAN IV Primer*. Addison-Wesley, Reading, MA, 1966.
18. S. Gill, "A process for the step-by-step integration of differential equations in an automatic digital computing machine" *Proc. R. Soc. London, Ser. A* **193**, 407–433 (1948).
19. H. H. Rosenbrick and C. Storey, *Computational Techniques for Chemical Engineers*, Pergamon, Oxford, 1966.
20. R. W. Hamming, *Numerical Methods for Scientists and Engineers*. McGraw-Hill, New York, 1962.
21. H. M. James, N. B. Nichols, and R. S. Phillips, *Theory of Servomechanisms*. McGraw-Hill, New York, 1947.
22. A Papoulis, *The Fourier Integral and Its Application*. McGraw-Hill, New York, 1962.
23. W. R. Evans, "Graphical analysis of control systems." *Trans. Am. Inst. Elect. Eng.* **67**, 547 (1948).
24. W. R. Evans, "Control system synthesis by root locus method." *Trans. Am. Inst. Electr. Eng.* **69**, 66 (1950).
25. W. R. Evans, *Control System Dynamics*, McGraw-Hill, New York, 1954.
26. C. S. Chang, "Analytical Method for Obtaining the Root Locus with Positive and Negative Gain." *IEEE Trans. Automatic Control*, **AC-10**, 92–94 (1965).
27. G. F. Franklin, J. D. Powell, and A. Emami-Naeini, *Feedback Control of Dynamic Systems,* 3rd ed., Addison-Wesley, Reading, MA, 1994.
28. K. Ogata, Modern Control Engineering, 3rd ed., Prentice Hall, Englewood Cliffs, NJ, 1997.
29. J. Lipow, *A Computer Algorithm for Obtaining the Root Locus*. National Biscuit Co., New York, 1962.
30. M. J. Remec, "Saddle-points of a complete root locus and an algorithm for their easy location on the complex frequency plane." *Proc. Natl. Electron. Conf.* **21**, 605–608 (1965).
31. H. M. Paskin, "Automatic computation of root loci using a digital computer." Unpublished M.S. thesis, Air Force Institute of Technology, Dayton, OH, March 1962.
32. S. M. Shinners, "How to approach the stability analysis and compensation of control systems." *Control Eng.* **25**(5), 62–67 (1978).
33. J. M. Maciejowski, *Multivariable Feedback Design*. Addison-Wesley, Reading, MA, 1989.
34 D. K. Frederick, C. J. Herget, R. Kool, and M. Rimvall, *ELCS: The Extended List of Control Software*, February, No. 3, 1987.
35. *Proceedings of the Third IEEE Symposium on Computer-Aided Control Systems Design*, Arlington, VA, September 1986, IEEE, New York, 1986.
36. M. Jamshidi and C. J. Herget, eds. *Advances in Computer-Aided Control Systems Engineering*, North-Holland, Amsterdam, 1985.
37. C. J. Herget and A. L. Laub, eds. Special issue on computer-Aided control system design. *IEEE Control Systems Magazine*, **2**(4) (Dec.), 2–37 (1982).
38. A. Grace, A. J. Laub, J. N. Little, and C. Thompson, *Control System Toolbox for Use with MATLAB™ User's Guide*. MathWorks, Inc., Natick, MA, 1990
39. R. E. Klein, "Using bicycles to teach system dynamics." *IEEE Control Syst. Mag.* **9**(3), 4–9 (1989).
40. G. F. Franklin and J. D. Powell, "Digital control laboratory courses." *IEEE Control Syst. Mag.* **9**(3), 10–13 (1989).

41. M. Mansour and W. Schaufelberger, "Software and laboratory experiments using computers in control education." *IEEE Control Syst. Mag.* **9**(3), 19–24 (1989).
42. J. M. Boyle, M. P. Ford, and J. M. Maciejowski "Multivariable Toolbox for use with MATLAB." *IEEE Control Syst. Mag.* **9**(1), 59–65 (1989).
43. K. Gustafson, M. Lundh, and M. Lilja, "A set of MATLAB routines for control system analysis and design." In *Proceedings of the 1991 Advances in Control Education Conference*, Boston, MA. IEEE, New York, 1991.
44. D. Atherton and T. C. Yang, "Programs for teaching nonlinear control in MATLAB." In *Proceedings of the 1991 Advances in Control Education Conference*, Boston, MA. IEEE, New York, 1991.
45. *Ctrl-C® User's Guide*. Systems Control Technology, Palo Alto, CA, 1990.
46. *CODAS-II*. Dynamical Systems, Inc., Tucson, AZ.
47. *Programm CC, Version 4*. Systems Technology, Inc., Hawthorne, CA.
48. *ACET™ (Advanced Control Engineering Techniques)*. Information & Control Systems, Hampton, VA.
49. S. M. Shinners, *Techniques of System Engineering*. McGraw-Hill, New York, 1967.
50. B. Blake, "Four views on train control." *Control Eng.* **11**, 62–68 (1964).
51. I. Nakamura and S. Yamazaki, "On the centralized system for train operation and traffic control—Including signaling and routing information." *Railw. Tech. Res. Inst.* **5**, 9–11 (1964).
52. W. Crochetiere, L. Vovovnik, and J. B. Resnick, "The design of control systems for electronic activation of human limb movement." In *Proceedings of the 1967 Joint Automatic Control Conference*, pp. 51–57.
53. J. Goclowski and A. Gelb, "Dynamics of an automatic ship steering system." In *Proceedings of the 1966 Joint Automatic Control Conference*, pp. 294–304.
54. S. M. Shinners, "Modeling of human operator performance utilizing time series analysis." *IEEE Trans. Syst., Man Cybernet.* **SMC-4**(5), 446–458 (1974).
55. J. Camp and M. J. Campbell, "Aircraft power steering." *Sperry Rand Eng. Rev.* **24**, 37–40 (1971).

7

LINEAR CONTROL-SYSTEM COMPENSATION AND DESIGN

7.1. INTRODUCTION

After the stability of a feedback control system has been analyzed, by using any of the tools presented in Chapter 6, it will often be found that system performances is not satisfactory and needs to be modified. It is necessary to ensure that the open-loop gain is adequate for accuracy, and that the transient response is desirable for the particular application. In order for the system to meet the requirements of stability, accuracy, and transient response, certain types of equipment must be added to the basic feedback control system. We use the term *design* to encompass the entire process of basic system modification in order to meet the specifications of stability, accuracy, and transient response. The term *stabilization* is usually used to indicate the process of achieving the requirements of stability alone; the term *compensation* is usually used to indicate the process of increasing accuracy and speeding up the response.

There are several commonly used configurations for compensating (or stabilizing) a control system. The compensating (or stabilizing) device may be inserted into the system either in cascade with the forward portion of the loop (cascade compensation) as shown in Figure 7.1, or as part of a minor feedback loop (feedback compensation) as shown in Figure 7.2 [1, 2]. The cascade-compensation technique is usually concerned with the addition of phase-lag, phase-lead, and phase-lag–lead passive networks. The feedback-compensation technique is primarily concerned with the addition of rate or acceleration feedback. The type of compensation chosen usually depends on the nonlinearities and the location of the noises in the loop, and economic considerations.

Linear-state-variable-feedback compensation, illustrated in Figure 7.3, consists of feeding back the state variables through constant real gains.

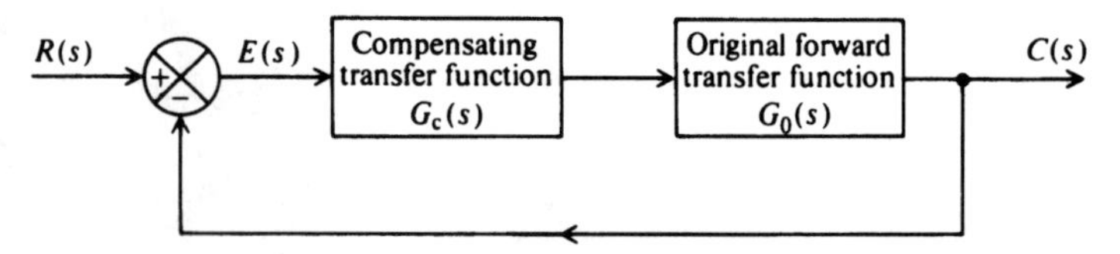

Figure 7.1 Illustration of series or cascade compensation.

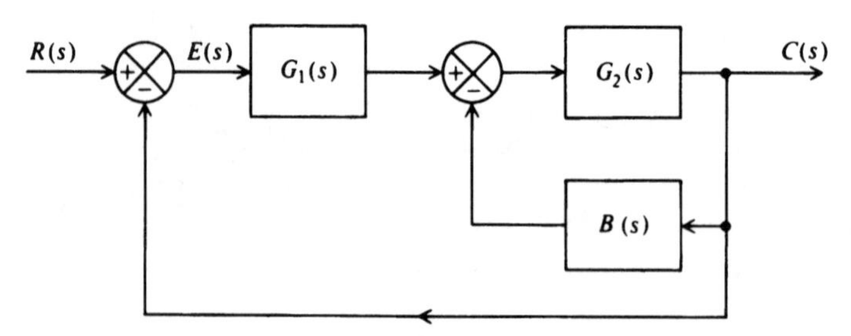

Figure 7.2 Illustration of minor-loop feedback compensation.

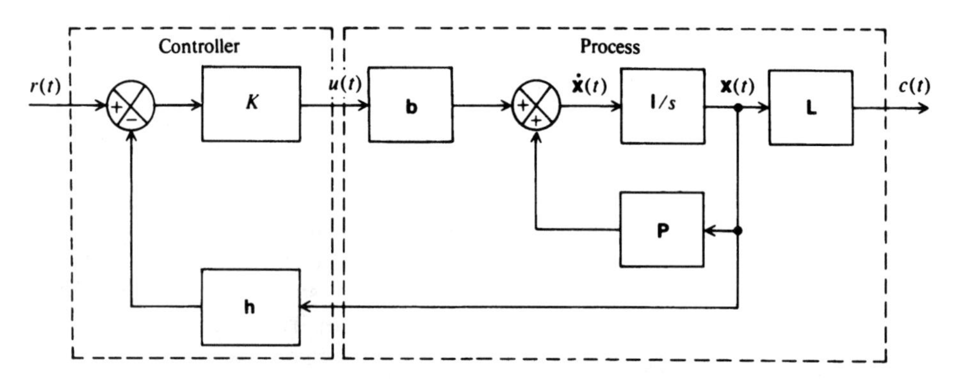

Figure 7.3 Illustration of linear-state-variable-feedback compensation.

Linear-state-variable feedback requires that all state variables be fed back to the controller. For very high-order control systems, it requires a larger number of transducers to sense the state variables for feedback. Therefore, practical applications of this technique can be very costly or impractical. In addition, some state variables may not be directly accessible (e.g., nuclear power plant processes), and this can limit this technique's application even in low-order control systems. For cases where some state variables may not be directly accessible, we use an estimator to estimate the state variables from measurements made of the accessible variables. Figure 7.4 illustrates an estimator system for a regulator system where the reference input $r(t)$, equals zero.

The compensation techniques illustrated in Figures 7.1 through 7.4 all have only one degree of freedom because there is only one controller in each system, even though the controller can have more than one parameter which can be varied. These one-degree-of-freedom controllers have the major disadvantage that the performance criteria which can be achieved is limited. For example, if a control system

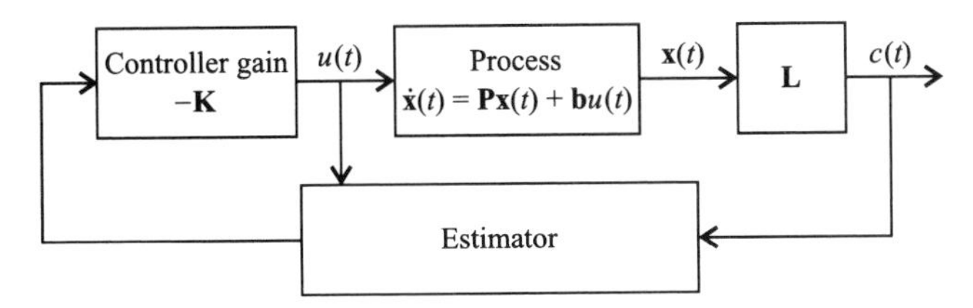

Figure 7.4 Illustration of an estimator system for a regulator system (where the reference input $r(t) = 0$).

is designed to have a desired stability, then it may have poor sensitivity to parameter variations. Let us, therefore, next consider compensation techniques which have two degrees of freedom.

Figure 7.5 illustrates the series-feedback compensation technique in which a series controller and a feedback controller are used. This technique is denoted as having two degrees of freedom compensation because it has two controllers, the series controller $G_c(s)$ and the feedback controller $B(s)$.

Figure 7.6 illustrates forward compensation in which the feedforward controller $G_{c1}(s)$ is in series with the closed-loop control system which also has a controller $G_{c2}(s)$ in the forward part of the control system. Because the controller $G_{c1}(s)$ is not within the feedback loop of the control system, it does not affect the characteristic equation roots of the original control system. Therefore, the poles and zeros of $G_{c1}(s)$ can be designed to cancel (or add) to the poles and zeros of the closed-loop transfer function. This technique is also referred to as having two degrees of freedom compensation because it has two controllers, the series controller $G_{c1}(s)$ and the forward-loop controller $G_{c2}(s)$.

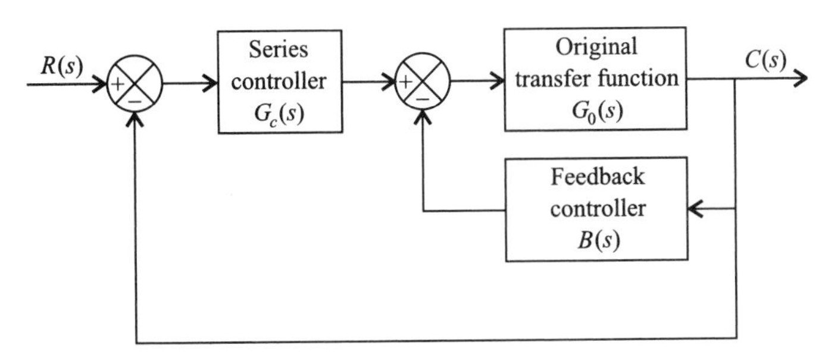

Figure 7.5 Series-feedback compensation technique illustrating two degrees of freedom.

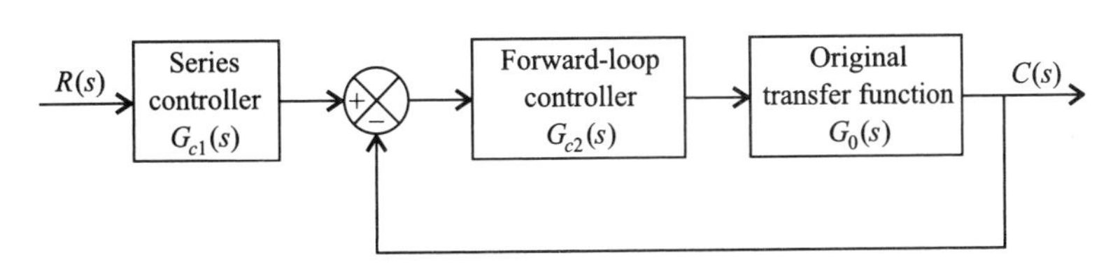

Figure 7.6 Forward compensation with series compensation illustrating two degrees of freedom.

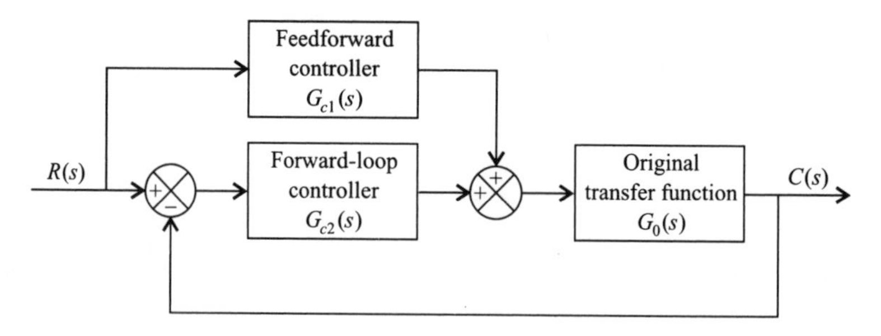

Figure 7.7 Feedforward compensation illustrating two degrees of freedom.

Figure 7.7 illustrates another form of two degrees of freedom compensation system which uses a feedforward controller, $G_{c1}(s)$, which is placed in parallel with the forward path which contains the controller $G_{c2}(s)$.

This chapter focuses attention on the tools presented in Chapter 6 which are of practical and useful interest to the control engineer. We will focus our attention on the application of the techniques to those particular design problems they are most suited to solve. Chapter 8 presents modern control-system design techniques using state-space techniques, Ackermann's formula for pole placement, estimation, robust control, and the H^∞ method for sensitivity reduction. Chapter 6 of the accompanying volume discusses the design of linear feedback control systems from the point of view of modern optimal control theory. Additional linear design problems are presented there in Chapter 7 where actual case studies are presented.

7.2. CASCADE-COMPENSATION TECHNIQUES

Let us consider the system of Figure 7.1 as our basic starting point in order to analyze the effect of cascade compensation. The compensating transfer function $G_c(s)$ is designed in order to provide additional phase lag, phase lead, or a combination of both at certain frequencies, in order to achieve certain specifications regarding stability and accuracy. We will illustrate and derive the transfer functions for representative compensating passive networks [1–6].

A *phase-lag network* is a device that shifts the phase of the control signal in order that the phase of the output lags the phase of the input over a certain range of frequencies. An electrical network performing this function was illustrated in Table 2.4 as item 4. Its transfer function was

$$\frac{E_{\text{out}}(s)}{E_{\text{in}}(s)} = \frac{1 + R_2 C_2 s}{1 + (R_1 + R_2) C_2 s}. \tag{7.1}$$

This can be written in the following more useful form:

$$\frac{E_{\text{out}}(s)}{E_{\text{in}}(s)} = \frac{1 + Ts}{1 + \alpha Ts}, \tag{7.2}$$

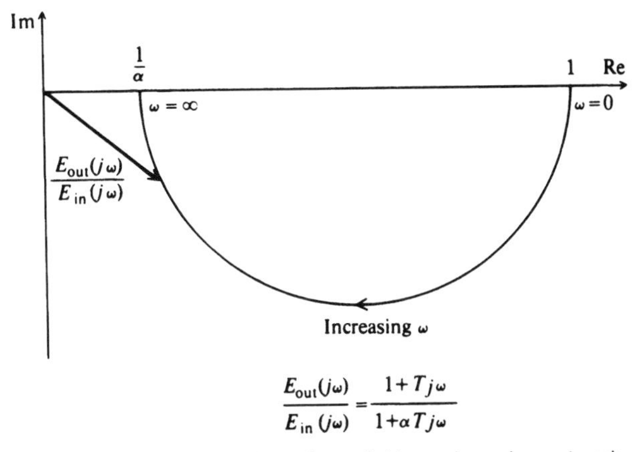

Figure 7.8 A complex-plane plot for a phase-lag network.

where

$$T = R_2 C_2, \quad \alpha = 1 + R_1/R_2.$$

Observe that $\alpha T > T$. A complex-plane plot of this network as a function of frequency is shown in Figure 7.8. Notice that the output voltage lags the input in phase angle for all positive frequencies. In addition, observe that the magnitude of $E_{out}(j\omega)/E_{in}(j\omega)$ decreases from unity at $\omega = 0$ to $1/\alpha$ at $\omega = \infty$. The Bode diagram for the phase-lag network is illustrated in Figure 7.9. The frequency at which the maximum phase lag occurs, ω_{max}, and the maximum phase lag, ϕ_{max}, can be easily derived. The results are

$$\omega_{max} = 1/(T\sqrt{\alpha}) \tag{7.3}$$

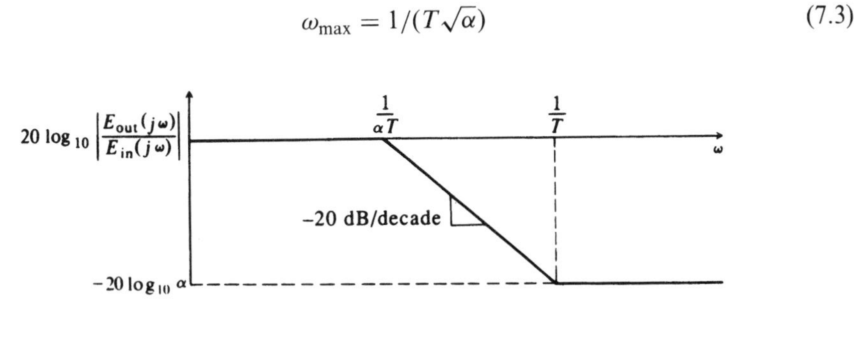

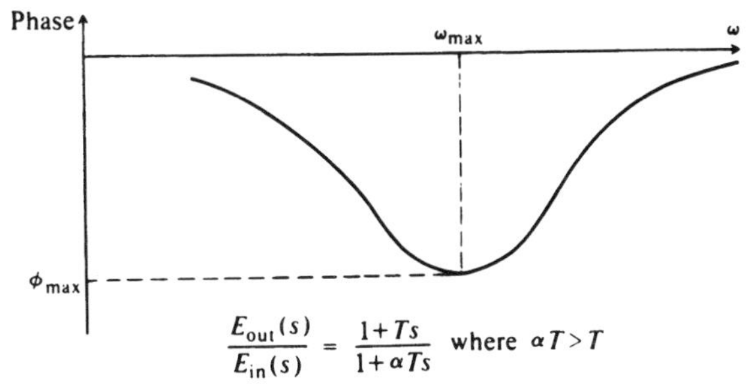

Figure 7.9 Bode diagram of a phase-lag network.

and

$$\phi_{\max} = \sin^{-1}\frac{1-\alpha}{1+\alpha}. \tag{7.4}$$

Values of $\phi_{\max}$ for certain values of α, which are useful for design purposes, are listed in Table 7.1.

A *phase-lead network* is a network which shifts the phase of the control signal in order that the phase of the output leads the phase of the input at certain frequencies. An electrical network performing this function was illustrated in Table 2.4 as item 3. Its transfer function was as follows:

$$\frac{E_{\text{out}}(s)}{E_{\text{in}}(s)} = \frac{R_2}{R_1+R_2}\,\frac{1+R_1C_1s}{1+[R_2/(R_1+R_2)]R_1C_1s}. \tag{7.5}$$

This can be written in the following more useful form:

$$\frac{E_{\text{out}}(s)}{E_{\text{in}}(s)} = \frac{1}{\alpha}\left(\frac{1+\alpha Ts}{1+Ts}\right) \tag{7.6}$$

where

$$T = \frac{R_1R_2}{R_1+R_2}C_1, \qquad \alpha = 1+\frac{R_1}{R_2}.$$

Observe that $\alpha T > T$. A complex-plane plot of this network as a function of frequency is shown in Figure 7.10. Notice that the output voltage leads the input in phase angle for all positive frequencies. In addition, notice that the magnitude of $E_{\text{out}}(j\omega)/E_{\text{in}}(j\omega)$ increases from $1/\alpha$ at $\omega = 0$ to unity at $\omega = \infty$. The Bode diagram for the phase-lead network is illustrated in Figure 7.11. The corresponding values of $\omega_{\max}$ and $\phi_{\max}$ for the phase-lead network are

$$\omega_{\max} = 1/(T\sqrt{\alpha}) \tag{7.7}$$

and

$$\phi_{\max} = \sin^{-1}\frac{\alpha-1}{\alpha+1}. \tag{7.8}$$

Table 7.1. $\phi_{\max}$ **as a Function of** α

α	$\phi_{\max}$ (degrees)
1	0
2	−19.4
4	−36.9
8	−51.0
10	−55.0
20	−64.8

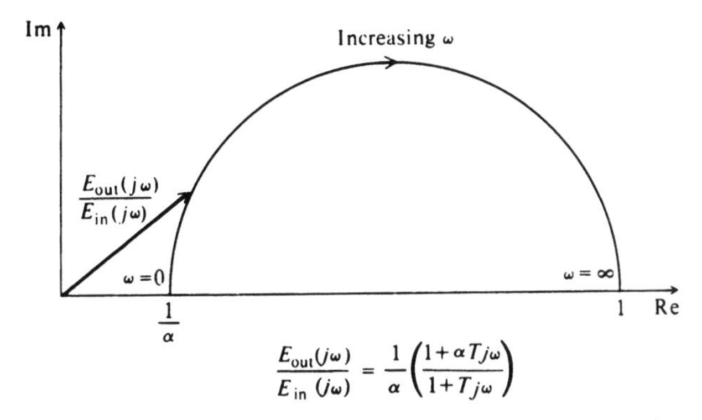

Figure 7.10 A complex-plane plot for a phase-lead network.

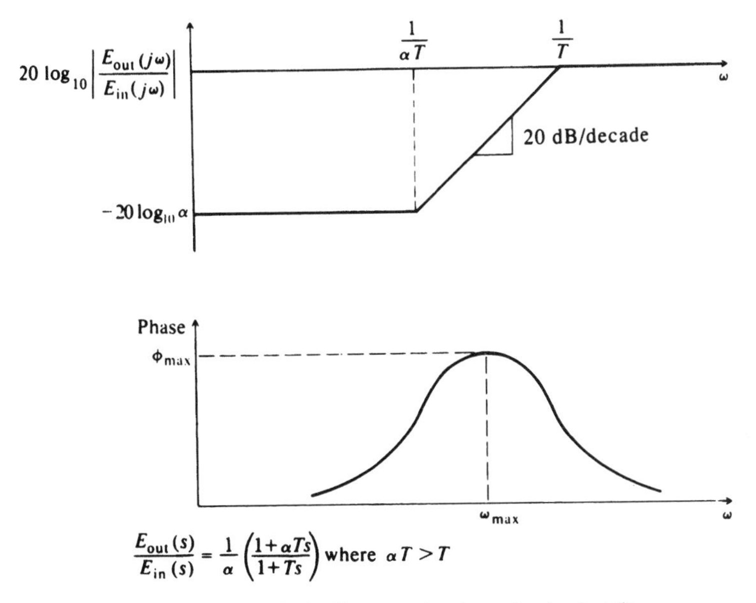

Figure 7.11 Bode diagram of a phase-lead network.

The values shown in Table 7.1 are also true for the phase-lead case except for the sign. An important practical point to emphasize is that the control engineer would not in practice use any ratio of $\alpha > 10$ because the lead network acts as an attenuation, which must be made up for somewhere in the feedback control system, with an amplification whose ratio is α.

A *phase-lag–lead network* is a network that shifts the phase of a control signal in order that the phase of the output lags at low frequencies and leads at high frequencies relative to the input. An electrical network performing this function was illustrated in Table 2.4 as item 5. Its transfer function was as follows:

$$\frac{E_{out}(s)}{E_{in}(s)} = \frac{(1 + R_1C_1s)(1 + R_2C_2s)}{R_1R_2C_1C_2s^2 + (R_1C_1 + R_2C_2 + R_1C_2)s + 1}. \tag{7.9}$$

Defining

$$T_1 = R_1C_1. \quad T_2 = R_2C_2, \quad T_{21} = R_1C_2$$

we can rewrite Eq. (7.9) as

$$\frac{E_{out}(s)}{E_{in}(s)} = \frac{T_1T_2s^2 + (T_1 + T_2)s + 1}{T_1T_2s^2 + (T_1 + T_2 + T_{21})s + 1}.$$

A complex-plane plot of this network as a function of frequency is shown in Figure 7.12. Notice that the output voltage lags the input in phase angle for low frequencies and leads in phase angle for high frequencies. In addition, notice that the magnitude of $E_{out)}(j\omega)/E_{in}(j\omega)$ decreases for intermediate frequencies and increases to unity as ω approaches 0 and ∞. A corresponding Bode diagram for the phase-lag-lead network is illustrated in Figure 7.13.

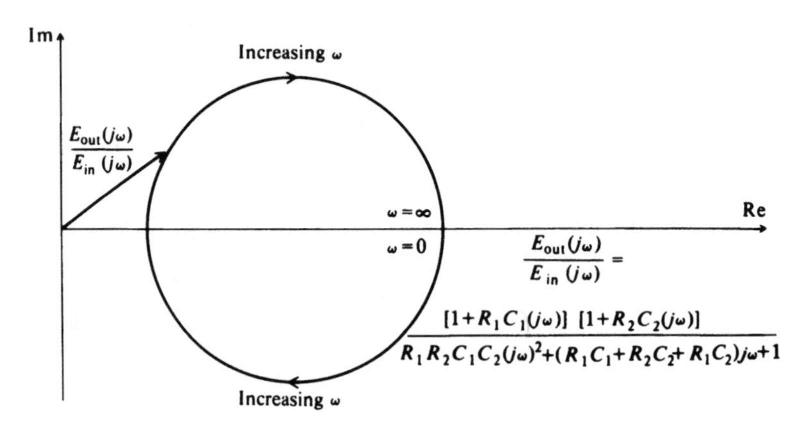

Figure 7.12 A complex-plane plot for a phase-lag-lead network.

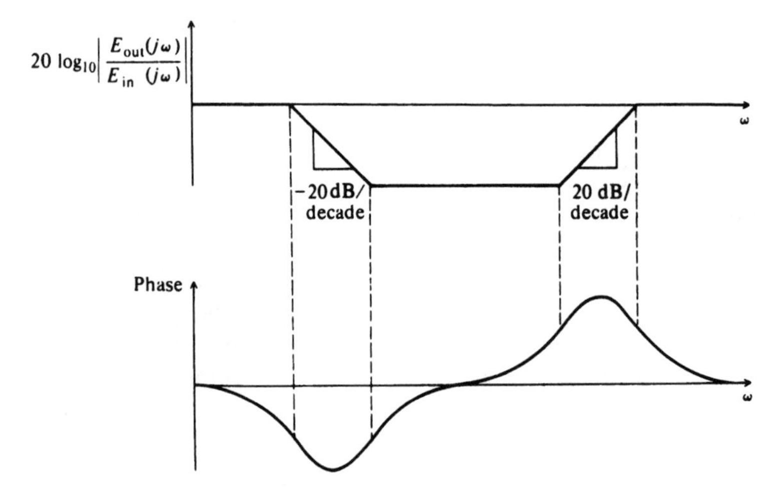

Figure 7.13 Bode diagram of a phase-lag-lead network.

The stabilizing effect of cascaded, phase-shifting networks can easily be demonstrated for a simple second-order system. For example, let us consider the configuration illustrated in Figure 7.1, where the original forward transfer function $G_0(s)$, is given by

$$G_0(s) = \frac{\omega_n^2}{s(s + 2\zeta\omega_n)}. \tag{7.10}$$

If this system were uncompensated $[G_c(s) = 1]$, then the transfer function $G_0(s)$ would result in the familiar second-order system response which was discussed at great length in Section 4.2. The resulting damping ratio of the system would be given by ζ and its undamped natural frequency by ω_n. Let us now assume that we add a lead network to this system whose transfer function is given by

$$G_c(s) = \frac{1 + \alpha Ts}{1 + Ts}. \tag{7.11}$$

It is assumed that the attenuation factor of $1/\alpha$ is compensated for with an amplification increase of α, and the system will maintain the same static error. Let us consider the case where $T \ll \alpha T$ and, therefore, $G_c(s)$ can be approximated by a zero factor (proportional plus derivative control):

$$G_c(s) \approx 1 + \alpha Ts, T \ll \alpha T. \tag{7.12}$$

Another way of looking at this approximation is the viewpoint of adding a rate feedback loop. In the following section, it is shown in Figure 7.14*a* that the addition of rate feedback bs in parallel with unity position feedback is equivalent to adding the pure zero factor term $(1 + bs)$ to the open-loop transfer function, $G(s)H(s)$. Therefore, the approximation in Eq. (7.12) can be viewed as an approximate representation of a phase-lead network, or the exact representation of the addition of rate feedback in parallel with position feedback to the system. In the following analysis, we will use the terminology of Eq. (7.12), which is based on an approximation to the phase-lead network, although it could just as easily represent exactly the addition of rate feedback in parallel with position feedback.

The form of Eq. (7.12) suggests that this lead network (or the addition of rate feedback in parallel with unity position feedback) is equivalent to a proportional plus derivative controller. The resulting system transfer function with the lead network is given by

$$\frac{C(s)}{R(s)} \approx \frac{\omega_n^2(1 + \alpha Ts)}{s^2 + (2\zeta\omega_n + \alpha T\omega_n^2)s + \omega_n^2}. \tag{7.13}$$

Comparing the denominators of Eqs. (7.13) and (4.3), we observe that it is still of second order and ω_n remains the same, but ζ is greater due to the increase in the coefficient of s in the denominator. The equivalent damping ratio with $G_c(s)$ can be obtained as follows:

$$2\zeta\omega_n + \alpha T\omega_n^2 = 2\zeta_{\text{eq}}\omega_n \tag{7.14}$$

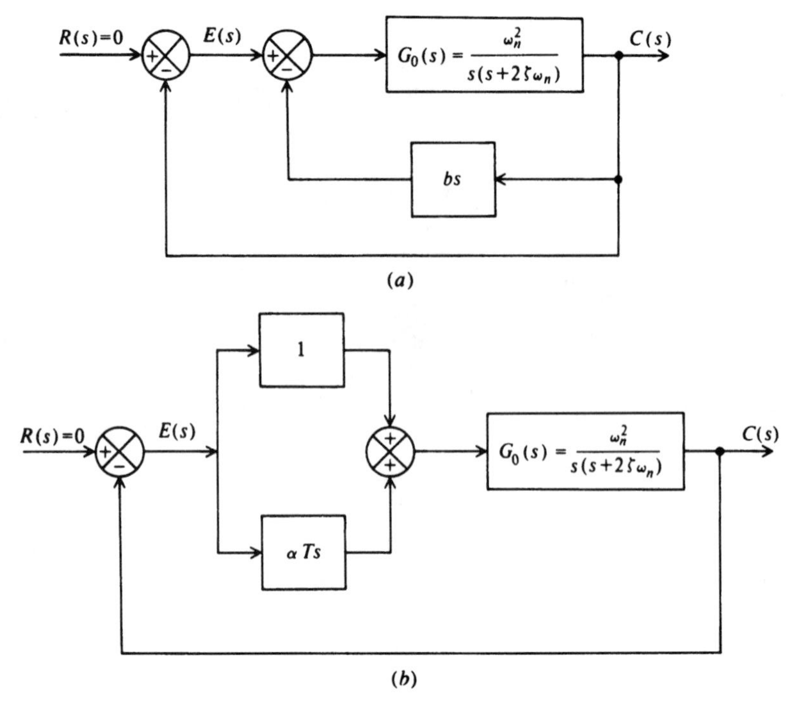

Figure 7.14 The stabilizing effects of the systems illustrated are equivalent. (*a*) Feedback compensation. (*b*) Cascade compensation—proportional plus derivative controller

where ζ_{eq} is an equivalent damping ratio with the addition of a zero factor for compensation. Solving for ζ_{eq}, we obtain

$$\zeta_{\text{eq}} = \zeta + \alpha T \omega_n / 2. \tag{7.15}$$

Therefore, we can conclude that the addition of a zero factor in $G_c(s)$ has increased the damping ratio from ζ to ζ_{eq} by an amount equal to $\alpha T \omega_n / 2$. This assumes that T is positive, or the zero of the factor $(1 + \alpha Ts)$ is in the left half of the s-plane.

The next question is what is the steady-state error resulting from cascade compensation? To answer this, we must find the steady-state errors resulting from the application of a unit ramp input for the cases of no compensation and compare them with those resulting from cascade compensation. We choose a unit ramp as our input because it is the only input which results in a finite response error for a system with a pole at the origin. The transfer function relating error to input for the system shown in Figure 7.1 is given by

$$\frac{E(s)}{R(s)} = \frac{1}{1 + G_c(s) G_0(s)}. \tag{7.16}$$

Assuming that

$$G_c(s) = 1 \quad \text{(no cascade compensation)},$$

$$G_0(s) = \frac{\omega_n^2}{s(s + 2\zeta\omega_n)},$$

and

$$R(s) = 1/s^2 \quad \text{(a unit ramp input)},$$

we find that

$$E(s) = \frac{s + 2\zeta\omega_n}{s(s^2 + 2\zeta\omega_n s + \omega_n^2)}. \tag{7.17}$$

Applying the final-value theorem to Eq. (7.17), we find the steady-state error to be

$$e_{ss(\text{ramp input})} = \lim_{s \to 0} sE(s) = \frac{2\zeta\omega_n}{\omega_n^2} = \frac{2\zeta}{\omega_n}. \tag{7.18}$$

For the case with cascade compensation, a similar analysis yields the following result:

$$G_c(s) = \frac{1 + \alpha Ts}{1 + Ts}, \quad G_0(s) = \frac{\omega_n^2}{s(s + 2\zeta\omega_n)}, \quad R(s) = \frac{1}{s^2}.$$

Therefore,

$$E(s) = \frac{1}{s} \frac{(s + 2\zeta\omega_n)(1 + Ts)}{\left[s(s + 2\zeta\omega_n)(1 + Ts) + \omega_n^2(1 + \alpha Ts)\right]}. \tag{7.19}$$

Applying the final-value theorem to Eq. (7.19), the steady-state error is found to be

$$e_{ss(\text{ramp input})} = \lim_{s \to 0} sE(s) = \frac{2\zeta\omega_n}{\omega_n^2} = \frac{2\zeta}{\omega_n}. \tag{7.20}$$

Comparing the results of Eqs. (7.18) and (7.20), we conclude that the addition of the cascade lead network as given by Eq. (7.11) does not increase or decrease the steady-state response error of the system.

It is important to emphasize that the relationships derived in this analysis apply only to the simple system considered. For example, if a zero factor were contained in the numerator of Eq. (7.10), then these relationships are modified (see Problems 7.5 and 7.6).

If we attempt to extend this analysis of a second-order system to the case of phase-lag compensation, the characteristic equation becomes third order and difficult to factor. For example, if $G_c(s)$ were only to represent the pole factor of the phase-lag network, then

$$G_c(s) = \frac{1}{1 + \alpha Ts}. \tag{7.21}$$

In a similar manner, the closed-loop system transfer function can be found to be given by

$$\frac{C(s)}{R(s)} = \frac{\omega_n^2}{\alpha Ts^3 + (2\zeta\omega_n\alpha T + 1)s^2 + 2\zeta\omega_n s + \omega_n^2}. \tag{7.22}$$

The factorization of this characteristic equation is not trivial and a similar analysis to that performed for the phase lead-network case is more complex. The root-locus method is an excellent tool which can be used for factorization, and the analysis of this problem for third- and higher-order systems is presented in Sections 6.14 and 7.9.

7.3. MINOR-LOOP FEEDBACK-COMPENSATION TECHNIQUES

Let us consider the general system illustrated in Figure 7.2. The compensating element in this case is the transfer function $B(s)$. In order to have a basis of comparison, we will follow an analysis for minor-loop feedback compensation similar to that performed for the case of phase lead-network cascade compensation.

The minor-loop feedback element $B(s)$ usually represents rate feedback or acceleration feedback. In general, phase-lag, -lead, and/or lag–lead networks may also be cascaded with $B(s)$.

The stabilizing effect of minor-loop feedback compensation can easily be demonstrated for a simple second-order system. We assume that the system illustrated in Figure 7.2 contains simple rate feedback. The specific transfer functions for the system are

$$G_1(s) = 1 \tag{7.23}$$

$$G_2(s) = \frac{\omega_n^2}{s(s + 2\zeta\omega_n)}, \tag{7.24}$$

$$B(s) = bs. \tag{7.25}$$

The system is redrawn with these transfer functions and shown in Figure 7.14*a*.

Without any rate feedback, the configuration represents a simple second-order system whose damping ratio is ζ and undamped natural frequency is ω_n. The resulting system transfer function with rate-feedback compensation is given by

$$\frac{C(s)}{R(s)} = \frac{\omega_n^2}{s^2 + (2\zeta\omega_n + \omega_n^2 b)s + \omega_n^2}. \tag{7.26}$$

Comparing the denominator of Eq. (7.26) with that of Eq. (4.3), we observe that it is still of second order and ω_n remains the same, but ζ is greater due to the increase in the coefficient of s in the denominator. The equivalent damping ratio with rate feedback added can be obtained by setting the coefficients of the s terms equal to each other, as follows:

$$2\zeta\omega_n + \omega_n^2 b = 2\zeta_{eq}\omega_n. \tag{7.27}$$

where ζ_{eq} = an equivalent damping ratio with rate feedback added. Solving for ζ_{eq} we obtain

$$\zeta_{eq} = \zeta + \frac{\omega_n b}{2}. \tag{7.28}$$

Therefore, we can conclude that the addition of the minor loop using rate feedback has increased the damping ratio from ζ to ζ_{eq} by an amount equal to $\omega_n b/2$. This assumes that b is positive (negative feedback).

It is important at this point to compare Eqs. (7.15) and (7.28). Note that they are very similar, and they imply that

$$\alpha T = b \tag{7.29}$$

The fact that rate feedback behaves as the approximated phase lead network, as defined by Eq. (7.12) (porportional plus derivative controller), can be easily demonstrated from Figure 7.14a and b. Let us assume that there is zero input to both systems, because we are concerned only with the system poles. Clearly, in both cases, there are two negative-feedback paths in parallel around $G_0(s)$. In the cascade-compensation case, the total feedback around $G_0(s)$ is $1 + \alpha Ts$; in the rate-feedback-compensation case, the total feedback around $G_0(s)$ is $1 + bs$. Therefore, the stabilizing effects of αT and b are equivalent. We discuss proportional plus derivative (PD) controllers fully in the next section, 7.4, on proportional-plus-integral-plus-derivative (PID) compensators.

Let us next determine the steady-state error resulting from the use of minor-loop rate-feedback compensation. We assume that the input to this system is a unit ramp in order to have a finite steady-state response error and a basis for comparison. From our discussion of cascade compensation in Section 7.2 we know from Eq. (7.18) that the resulting steady-state error of this system without any compensation ($b = 0$) is $2\zeta/\omega_n$. For the case of minor-loop rate-feedback compensation, the resulting expression for $E(s)$ is given by

$$E(s) = \frac{1}{s}\left[\frac{s + 2\zeta\omega_n + b\omega_n^2}{s\left(s + 2\zeta\omega_n + b\omega_n^2\right) + \omega_n^2}\right]. \tag{7.30}$$

Applying the final-value theorem, the steady-state error is found to be

$$e_{ss(\text{ramp input})} = \lim_{s\to 0} sE(s) = \frac{2\zeta\omega_n + b\omega_n^2}{\omega_n^2} = \frac{2\zeta}{\omega_n} + b. \tag{7.31}$$

Therefore, the steady-state response error of the system with minor-loop rate-feedback compensation has increased by a factor of b. This unfavorable result can easily be remedied by placing a high-pass filter in cascade with the rate device. Such a filter would block the steady-state value of the rate output. This technique is illustrated in Figure 7.15.

As in the preceding section, it is important to emphasize that the relationships derived apply only to the simple system considered. Problems 7.5 and 7.6 illustrate how these relationships change if a zero factor is added to the basic system transfer function considered.

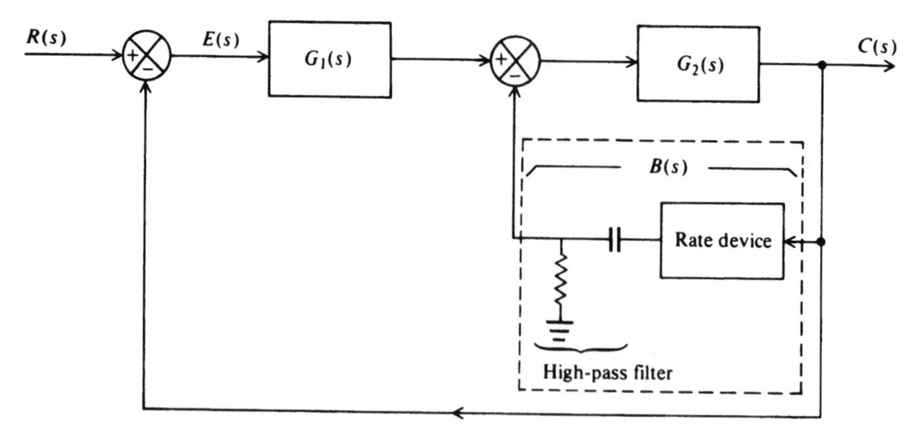

Figure 7.15 Illustration of minor-loop feedback compensation using a rate device in cascade with a high-pass filter

7.4. PROPORTIONAL-PLUS-INTEGRAL-PLUS DERIVATIVE (PID) COMPENSATORS

PID compensators are another form of compensation frequently used in control systems, especially in the industrial process control field. They are very popular in the industrial process control field due to their robust (insensitive) performance over a wide range of operating conditions including plant uncertainty, parameter variation, and external disturbances. Robust control is presented in Section 8.10 of Chapter 8.

Assuming that the input to the PID compensator is $e(t)$ and its output is $u(t)$, the equation defining the operation of the PID compensator is given by:

$$u(t) = K_p e(t) + K_I \int_0^t e(\tau)\, d\tau + K_D \frac{de(t)}{dt}. \tag{7.32}$$

Figure 7.16 illustrates a block diagram representation of Eq. (7.32). The transfer function of the PID compensator is obtained as follows:

$$U(s) = K_P E(s) + \frac{K_I}{s} E(s) + K_D s E(s) \tag{7.33}$$

$$U(s) = \left(K_P + \frac{K_I}{s} + K_D s\right) E(s). \tag{7.34}$$

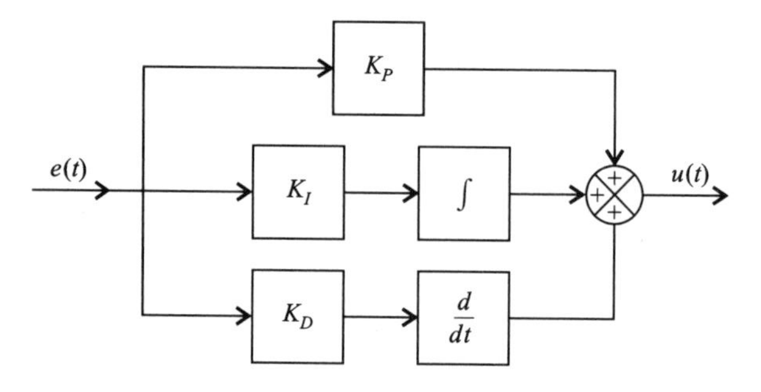

Figure 7.16 PID compensator block-diagram representation of Eq. (7.32).

The resulting transfer function of the PID compensator, $G_c(s)$, is given by

$$G_c(s) = \frac{U(s)}{E(s)} = K_P + \frac{K_I}{s} + K_D s. \tag{7.35}$$

Figure 7.17 illustrates a block diagram representation of Eq. (7.35). Observe from Eqs. (7.32) and (7.35) that the PID compensator provides a proportional term, an integration term, a derivative term, as its name implies.

Very often, only a portion of the very general PID compensator is used. For example, if $K_D = 0$, then we have

$$G_c(s) = K_P + \frac{K_I}{s} \tag{7.36}$$

which is denoted as the proportional plus integral, or PI, compensator. If $K_I = 0$, then we have

$$G_c(s) = K_P + K_D s \tag{7.37}$$

which is denoted as the proportional plus derivative, or PD, compensator. This was illustrated previously in Figure 7.14.

To consider the design of a PID compensator, let us reconsider the transfer function of $G_c(s)$ given by Eq. (7.35):

$$G_c(s) = K_p + \frac{K_I}{s} + K_D s. \tag{7.38}$$

This can be rewritten as

$$G_c(s) = \frac{K_D s^2 + K_P s + K_I}{s} \tag{7.39}$$

or,

$$G_c(s) = \frac{K_D(s^2 + K_1 s + K_2)}{s} \tag{7.40}$$

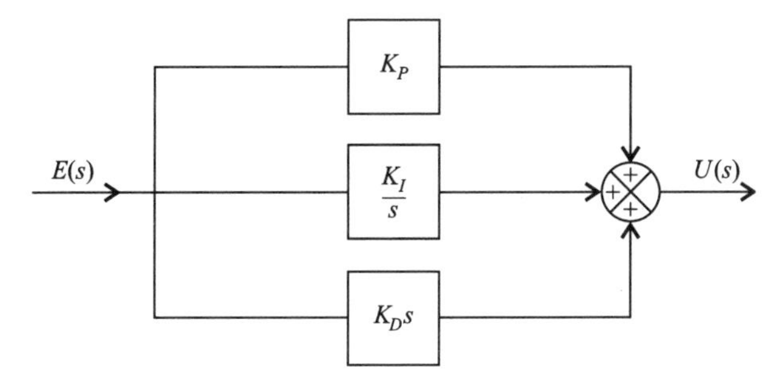

Figure 7.17 PID compensator block diagram representation of Eq. (7.35).

where

$$K_1 = \frac{K_P}{K_D}, \quad K_2 = \frac{K_I}{K_D}.$$

Factoring the numerator of Eq. (7.40), and assuming the quadratic factors as two real zeros, we obtain the following:

$$G_c(s) = \frac{K_D(s + z_a)(s + z_b)}{s}. \tag{7.41}$$

The very pleasing result of Eq. (7.41) is that the PID compensator results in two zero factors which can be located anywhere in the left-hand of the s-plane, in addition to the pole at the origin. We know from the analysis in Sections 7.2 and 7.3 that a zero factor provides a phase lead and aids in the compensation of a control system. In particular, as illustrated in Figure 7.14*b*, a PD is equivalent to one zero factor (which the rate feedback compensator also provides and is illustrated in Figure 7.14*a*). With the complete PID compensator, two zero factors are provided for compensation, in addition to the pole at the origin.

Applications of PI, PD, and PDI compensators are illustrated throughout this book for compensating control systems. These techniques are very useful for compensation of control systems in addition to the phase-lag, phase-lead, and phase-lag–lead networks illustrated in Section 7.2, and minor-loop rate feedback illustrated in Section 7.3.

7.5. EXAMPLE FOR THE DESIGN OF A SECOND-ORDER CONTROL SYSTEM

In this section we consider the design and resulting performance of the second-order system by means of cascade and minor-loop rate-feedback techniques. This problem is useful in unifying concepts which were introduced in Chapters 4, 5, and 6, together with the design techniques illustrated in this chapter.

Let us consider the second-order system illustrated in Figure 7.18*a*. We will assume that the original forward-loop transfer function $G_0(s)$ is given by

$$G_0(s) = \frac{14.4}{s(0.1s + 1)}. \tag{7.42}$$

The closed-loop transfer function, $C(s)/R(s)$, is given by

$$\frac{C(s)}{R(s)} = \frac{G_0(s)}{1 + G_0(s)} = \frac{14.4/[s(0.1s + 1)]}{1 + 14.4/[s(0.1s + 1)]} \tag{7.43}$$

or

$$\frac{C(s)}{R(s)} = \frac{144}{s^2 + 10s + 144}. \tag{7.44}$$

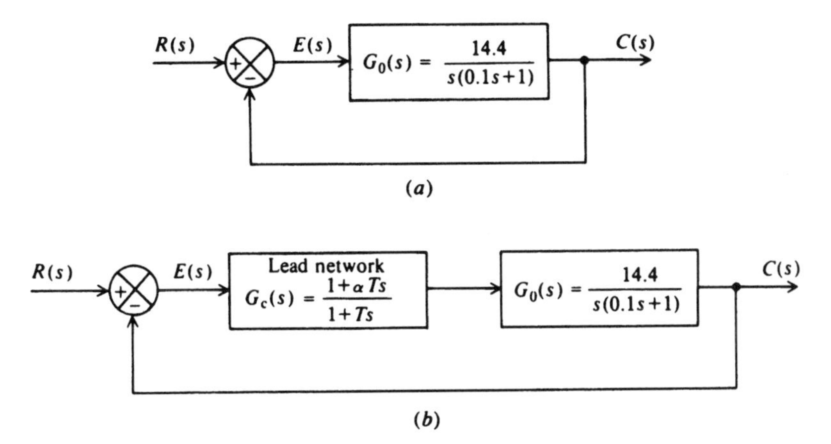

Figure 7.18 Design of a second-order system.

Comparing Eqs. (4.3) and (7.44), we observe that the undamped natural frequency ω_n and damping ratio ζ of the system are given by

$$\omega_n = 12 \text{ rad/sec} \tag{7.45}$$

and

$$\zeta = 0.417.$$

If the system is subjected to a unit step input, the transient response will have the form shown in Figure 4.4 (interpolate between $\zeta = 0.4$ and $\zeta = 0.6$). The maximum percent overshoot can be obtained from Eq. (4.33) and is found to be 23.5%.

Let us assume, for this application, that it is desired to have a critically damped system ($\zeta = 1$). We will demonstrate how this can be achieved using a cascaded network and minor-loop rate feedback.

Figure 7.18*b* illustrates the form that the system illustrated in Figure 7.18*a* would have if a cascaded network were used. We attempt to achieve a damping ratio equal to 1 using a phase-lead network, where

$$G_c(s) = \frac{1 + \alpha Ts}{1 + Ts} \tag{7.47}$$

As was assumed previously, in Section 7.2, we assume that $T \ll \alpha T$, and therefore $G_c(s)$ can be approximated by (proportional plus derivative control)

$$G_c(s) \approx 1 + \alpha Ts. \tag{7.47}$$

The closed-loop system transfer function for this case was derived in Section 7.2 [see Eq. (7.13)] as

$$\frac{C(s)}{R(s)} \approx \frac{\omega_n^2 + \alpha T\omega_n^2 s}{s^2 + (2\zeta\omega_n + \alpha T\omega_n^2)s + \omega_n^2}. \tag{7.49}$$

The equivalent damping ratio for this situation was derived in Section 7.2 [see Eq. (7.15)] as

$$\zeta_{\mathrm{eq}} = \zeta + \frac{\alpha T \omega_n}{2}. \tag{7.50}$$

The object in this problem is to design $\zeta_{\mathrm{eq}} = 1$ for the case where

$$\zeta = 0.417 \quad \text{and} \quad \omega_n = 12. \tag{7.50}$$

Substituting the values into Eq. (7.50), we find that

$$\alpha T = 0.0972.$$

We know that the resulting system will be stable and critically damped, and will have a steady-state response error for a unit step input of zero. The steady-state response error of the system to a unit ramp input was derived [see Eqs. (7.18) and (7.20)] as

$$e_{ss(\text{ramp input})} = \frac{2\zeta}{\omega_n} = \frac{2(0.417)}{12} = 0.0695 \text{ units}. \tag{7.51}$$

Let us next attempt to achieve the same type of performance using minor-loop rate feedback. Figure 7.19 illustrates the form that the system illustrated in Figure 7.18*a* would have if minor-loop feedback were used. Our goal is to achieve a damping ratio of $\zeta = 1$. The closed-loop system transfer function for this case was derived previously, in Section 7.2 [see Eq. (7.26)] as

$$\frac{C(s)}{R(s)} = \frac{\omega_n^2}{s^2 + (2\zeta\omega_n + \omega_n^2 b)s + \omega_n^2}. \tag{7.52}$$

The equivalent damping ratio for this configuration was derived in Section 7.2 [see Eq. (7.28)] as

$$\zeta_{\mathrm{eq}} = \zeta + \omega_n b/2. \tag{7.53}$$

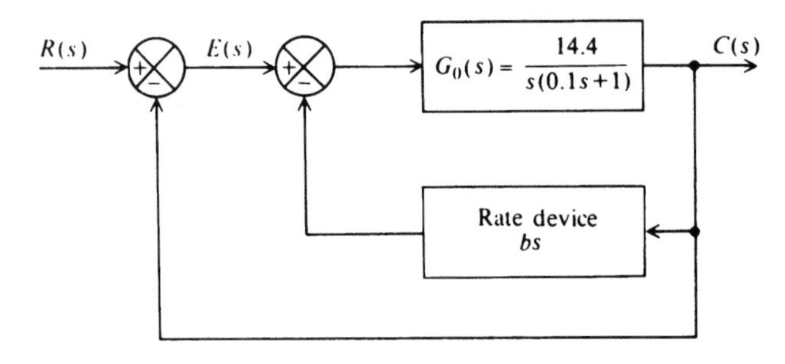

Figure 7.19 Minor-loop feedback added to the system shown in Figure 7.18*a*.

The object in this problem is to design $\zeta_{eq} = 1$ for the case where

$$\zeta = 0.417 \tag{7.54}$$

and

$$\omega_n = 12. \tag{7.55}$$

Substituting these values into Eq. (7.53), we find that

$$b = 0.0972. \tag{75.6}$$

Notice that b and αT are identical.

We know that the resulting system will be stable and critically damped, and will have a steady-state response error of zero for a unit step input. The steady-state response error of the system to a unit ramp input was derived [see Eq. (7.31)] as

$$e_{ss(\text{ramp input})} = \frac{2\zeta}{\omega_n} + b = 0.0695 + 0.0972 = 0.1667 \text{ units.} \tag{7.57}$$

The increase has been accounted for in Section 7.3; by using a high-pass filter in the feedback path to block a steady-state output from the rate feedback, the steady-state error can be reduced to 0.0695.

7.6. COMPENSATION AND DESIGN USING THE BODE-DIAGRAM METHOD

The techniques necessary to construct and analyze the open-loop frequency response of a feedback control system utilizing the Bode-diagram approach were presented in Section 6.7. This section illustrates how the Bode diagram can be used for designing a feedback-control system in order to meet certain specifications regarding relative stability, transient response, and accuracy. It is important to emphasize that the Bode-diagram approach is used very frequently by the practicing control engineer. Its use is due to the fact that the anticipated theoretical results may be relatively simply checked with actual performance in the laboratory just by opening the feedback loop and obtaining an open-loop frequency response of the system.

Bode's primary contribution to the control art is summarized in two theorems [7]. We introduce the concepts embodied in these theorems first in a qualitative manner, and then the mathematical statements are given.

A. Bode's Theorems

Bode's first theorem essentially state that the slopes of the asymptotic amplitude–log-frequency curve implies a certain corresponding phase shift. For example, in Section 6.7 it was shown that a slope of $20n$ dB/decade (or $6n$ dB/octave) corresponded to a phase shift of $90n^\circ$ for $n = 0, \pm 1, \pm 2, \ldots$. Furthermore, this theorem states that the slope at crossover (where the attenuation–log-frequency curve crosses the 0-dB line) is weighted more heavily toward determining system stability than a slope further removed from this frequency. This results in a rather complex weighting factor which is a measure of relative importance toward determining system stability.

From what has been presented so far, this theorem is intuitively seen to be valid. Gain crossover frequency is one of the two points that is checked to determine the degree of stability when using the Bode diagram. Specifically, the phase shift is measured at this particular frequency in order to determine the phase margin. A feedback system whose slope at gain crossover is −20 dB/ decade, and whose other slope sections are relatively far away from crossover in accordance with the relative weighting function, implies a phase shift of approximately −90° in the vicinity of crossover and a corresponding phase margin of about 90°. This value of phase margin certainly implies a stable system. A system, however, whose slope at crossover is −40 dB/decade, and whose other slope sections are relatively far away from crossover in accordance with the relative weighting function, implies a phase shift of approximately −180° and a corresponding phase margin of about 0°. This value of phase margin implies a system which is on the verge of being unstable and would probably be so when actually tested. Steeper slopes would indicate negative phase margins and definitely unstable systems. Therefore, one strives to maintain the slope of the amplitude–log-frequency curve in the area of gain crossover at a slope of −20 dB/decade. Notice that the system, illustrated in Figure 6.33, has slopes of −60 dB/ decade that are relatively far from the gain crossover frequency. Therefore, this system has a fairly respectable phase margin of 56.74° by maintaining the 20 dB/ decade slope for about an octave below, and about 2 octaves above crossover.

Bode's second theorem essentially states that the amplitude and phase characteristics of linear, minimum-phase-shift systems are uniquely related. When we specify the slope of the amplitude–log-frequency curve over a certain frequency interval, we have also specified the corresponding phase-shift characteristics over that frequency interval. Conversely, if we specify the phase shift over a certain frequency interval, we have also specified the corresponding amplitude–log-frequency characteristic over that frequency interval. The theorem emphasizes the fact that we can specify the amplitude–log-frequency characteristic over a certain interval of frequencies together with the phase-shift–log-frequency characteristics over the remaining frequencies. It should be emphasized that these conclusions apply only if the transfer function is minimum phase.

The second theorem may appear quite trivial at first glance. Its implications however, are quite important. We will make further use of this this theorem when designing feedback control systems using the Bode-diagram approach.

The formal *mathematical statement of Bode's first theorem* is given by the following expression:

$$\phi(\omega_d) = \frac{\pi}{2}\left|\frac{dG}{dn}\right|_d + \frac{1}{\pi}\int_{-\infty}^{\infty}\left[\left|\frac{dG}{dn}\right| - \left|\frac{dG}{dn}\right|_d\right] \ln\, \coth\left|\frac{n}{2}\right| dn, \tag{7.58}$$

where $\phi(\omega_d)$ is the phase shift of the system in radians at the desired (e.g., crossover) frequency ω_d, G represents the gain in nepers (1 neper $=$ $\ln|e|$), $n = \ln(\omega/\omega_d)$, $|dG/dn|$ represents the slope of the amplitude–log-frequency curve in nepers per unit change of n (1 neper/unit change of n is equivalent to 20 dB/decade), $|dG/dn|_d$ is the slope of the amplitude–log-frequency curve at the desired (crossover) frequency ω_d, and $\ln\, \coth|n/2|$ is the weighting function which is plotted in Figure

7.20. The first term of Eq. (7.58) represents the phase shift contributed by the slope of the amplitude–log-frequency curve at the reference frequency ω_d. For example, it yields a phase shift of 90° for every neper per unit of n (20 dB/decade). The second term of Eq. (7.58) is proportional to the integral of the product of the weighting function and the difference in slope of the amplitude–log-frequency curve at a frequency ω as compared to its value at the reference frequency, ω_d. Attention is drawn to the fact that it is the weighting function that determines the phase-shift contribution at ω_d due to the amplitude–log-frequency curve which exists at some frequency ω. Because the second term of Eq. (7.58) is zero for large values of n and where $n = 0$, the value of the integral will be relatively small compared with the first term if the slope of dG/dn is constant over a relatively wide range of frequencies above ω_d. Therefore, under these conditions, the phase shift would be determined primarily by the first term of Eq. (7.58). Following this line of reasoning, the slope of the amplitude–log-frequency curve should be less than −2 nepers per unit of n (−40 dB/decade) over a relatively wide range of frequencies at crossover in order to ensure stability.

The formal *mathematical statement of Bode's second theorem* is given by the following expression:

$$\int_0^{\omega_s} \frac{G\,d\omega}{\sqrt{\omega_s^2 - \omega^2}\left(\omega^2 - \omega_d^2\right)} + \int_{\omega_s}^{\infty} \frac{\phi\,d\omega}{\sqrt{\omega^2 - \omega_s^2}(\omega^2 - \omega_d^2)} = \begin{cases} \dfrac{\pi}{2}\dfrac{\phi(\omega_d)}{\omega_d\sqrt{\omega_s^2 - \omega_d^2}} & (\text{for } \omega_d < \omega_s) \\[2ex] -\dfrac{\pi}{2}\dfrac{G(\omega_d)}{\omega_d\sqrt{\omega_d^2 - \omega_s^2}} & (\text{for } \omega_d > \omega_s) \end{cases}, \tag{7.59}$$

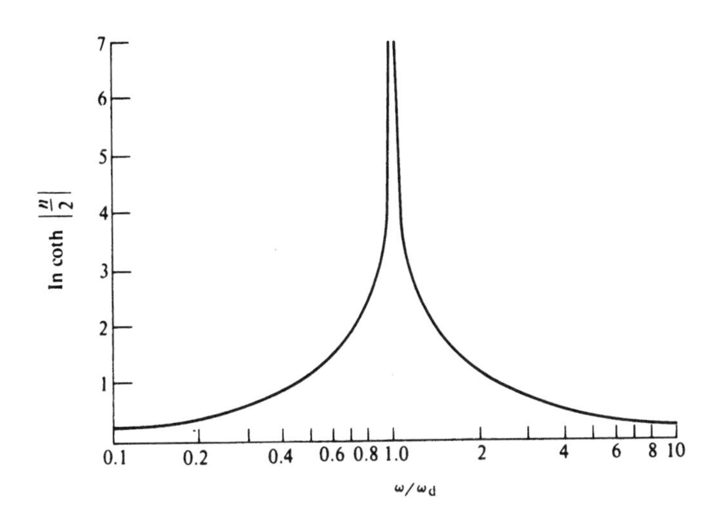

Figure 7.20 Plot of the weighting function used in Bode's first theorem.

where ω_s represents the frequency in radians per second below which the amplitude–log-frequency characteristics is specified and above which the phase characteristic is specified. This theorem emphasizes the interdependence of amplitude and phase shift over the entire range of positive frequencies. In addition, notice that although it is possible to specify amplitude or phase in one range of frequencies, and the other quantity in the remaining frequencies, these quantitites reflect their presence back into the other range of frequencies. Therefore the integration with respect to frequency is performed over the entire range of positive frequencies.

The design of several systems using the Bode-diagram approach is considered next. We shall illustrate a method that determines steady-state accuracy from the Bode diagram as well as meeting relative stability requirements.

B. Example of Phase-Lead and Phase-Lag Network Compensation, and Rate-Feedback Compensation

Let us first consider the third-order system illustrated in Figure 7.21*a*. Its open-loop transfer function $G_0(s)$ is given by

$$G_0(s) = \frac{80}{s(1+0.02s)(1+0.05s)}. \tag{7.60}$$

The Bode diagram for the uncompensated system $[G_c(s) = 1]$ is illustrated in Figure 7.22. It shows a gain crossover frequency of 33.78 rad/sec, a phase margin of $-3.402°$, and a gain margin of -1.1422 dB. The frequency where the phase shift equals $-180°$ (phase crossover frequency) is 31.7 rad/sec. This Bode diagram was obtained using MATLAB, and is contained in the M-file that is part of my *Modern*

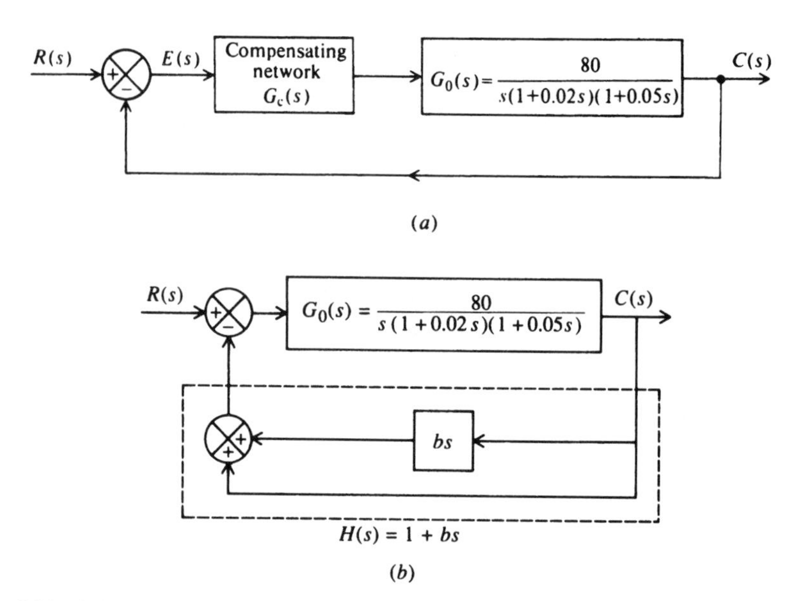

Figure 7.21 A third-order system which is to be compensated using (*a*) cascade compensation (phase-lag and phase-lead) and (*b*) rate-feedback compensation.

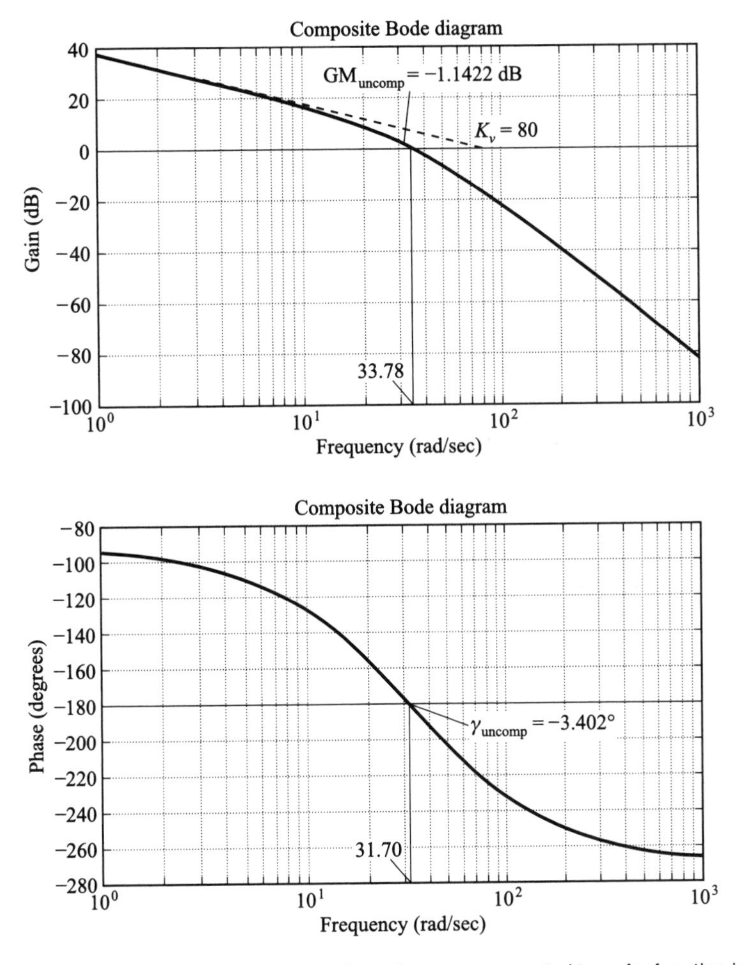

Figure 7.22 Third-order system where the uncompensated transfer function is

$$G_0(s) = \frac{80}{s(1 + 0.02s)(1 + 0.05s)}.$$

Control System Theory and Design (MCSTD) Toolbox can be retrieved free from The MathWorks, Inc. anonymous FTP server at ftp://ftp.mathworks.com/pub/books/shinners.

These results indicate that the uncompensated system is unstable. Let us next attempt to compensate this system with a phase-lag network, a phase-lead network, and the use of rate feedback in parallel with position feedback (which produces a pure zero factor). The specifications for this sytem require a minimum phase margin of 20° and a minimum gain margin of 10 dB. In addition, it is assumed that a sinusoidal disturbance at 1 rad/sec is present, and a gain of at least 35 dB is required at this frequency to nullify its effect.

1. Phase-Lag Network Compensation Let us first consider the phase-lag-network compensation case in Figure 7.21. Applying Bode's theorems in order to achieve the specified phase and gain margins, we would expect that the −20 dB decade slope in the vicinity of the new crossover frequency should not extend over too wide a range of frequencies because the relative stability that is desired is rather moderate. The phase-lag network is of the form.

$$G_c(s) = \frac{1 + Ts}{1 + \alpha Ts} \tag{7.61}$$

where $\alpha T > T$. In Section 7.2 we studied the characteristics of the phase-lag network. In particular, Figure 7.9 illustrated the Bode diagram of a general phase-lag network. Notice that this type of network is of such a nature that it attenuated all high-frequency components above $\omega = 1/T$ by a factor of $1/\alpha$. From the Bode-diagram viewpoint, this attenuating characteristic can be used for stabilization purposes by reshaping the uncompensated amplitude characteristics, so that the initial −20 dB/decade is made to cross over the 0-dB line rather than the −40-dB/decade segment. In other words, one would attempt to stabilize this system with a phase-lag network by placing the frequencies $1/\alpha T$ and $1/T$ in the range of frequencies below about 10 rad/sec. It would be desirable that the −20-dB/decade segment start at least around 10 rad/sec cross over the 0-dB line before 20 rad/sec where the amplitude–log-frequency characteristic changes to a slope of −40 dB/decade. In addition, we would not want $1/\alpha T$ to occur at less than 1 rad/sec, because an open-loop gain of 35 dB has been specified at $\omega = 1$ rad/sec. The final phase of the solution is by means of iteration. However, the procedure converges quite rapidly. Usually, two or three iterations should prove sufficient. For the requirements specified, a phase-lag network given by

$$G_c(s) = \frac{1 + 0.143s}{1 + s} \tag{7.62}$$

results in a gain crossover frequency of 11.371 rad/sec, a phase margin of 20.978° and a gain margin of 10.988 dB as shown in Figure 7.23. The frequency where the phase is −180° (phase crossover frequency) occurs at 24.22 rad/sec.

2. Phase-Lead Network Compensation. Let us next consider the phase-lead-network compensation case. The lead network has the form

$$G_c(s) = \frac{1}{\alpha}\left(\frac{1 + \alpha Ts}{1 + Ts}\right), \tag{7.63}$$

where $\alpha T > T$. In Section 7.2 we studied the characteristics of the phase-lead network. We assume that any low-frequency attenuation, which is due to the value of $1/\alpha$, is made up for by increasing the gain of the feedback control system by a like amount. The Bode diagram of a general phase-lead network was shown in Figure 7.11. From the Bode-diagram viewpoint, this type of characteristic can be used for stabilization purposes by reshaping the uncompensated amplitude characteristics so that a −40-dB/decade slope is made to cross over the 0-dB line along a synthesized −20-dB/decade slope rather than the −40-dB/decade segment. The range of frequencies where one can place the frequencies $1/\alpha T$ and $1/T$ is quite limited in this

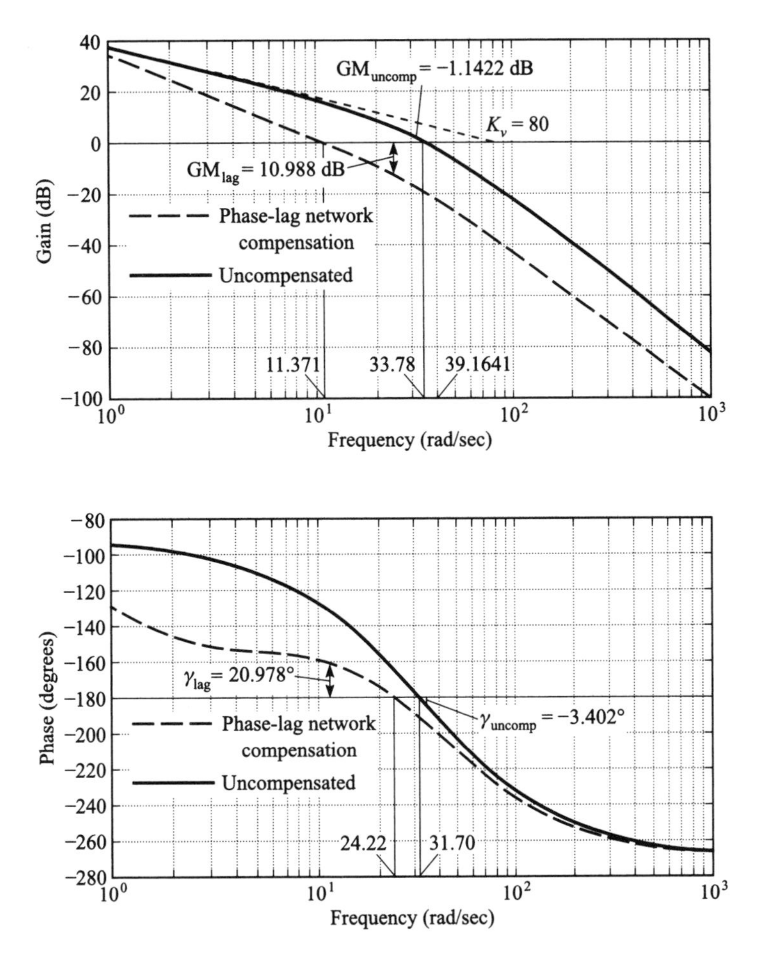

Figure 7.23 Compensation of a third-order system with a phase-lag network of $G_c(s) = (1 + 0.143s)/(1 + s)$ where the uncompensated transfer function is

$$G_0(s) = \frac{80}{s(1 + 0.02s)(1 + 0.05s)}.$$

particular problem. The value of $1/\alpha T$ can be placed between 20 and 33.78 rad/sec. The closer it is to 20 rad/sec, the greater its stabilizing effect. The further away the break $1/T$ is from 33.78 rad/sec, the greater will be the stabilizing effect of the phase-lead network. It can be seen that this particular solution does not modify the open-loop characteristics in the vicinity of 1 rad/sec and the accuracy specification of 35 dB at 1 rad/sec will easily be achieved.

We can obtain the pole and zero terms in Eq. (7.63) using a first-cut design procedure by using Eqs. (7.7) and (7.8) for the phase-lead network and the following reasoning. Equation (7.7) states the frequency where the maximum phase lead occurs, and Eq. (7.8) states the maximum value of the phase lead. In view of the analysis of the previous paragraph, it seems reasonable to place the frequency where maximum phase lead occurs around 47 rad/sec, and to make $\phi_{\max}$ equal to the phase margin desired (a conservative choice of 28° will be selected) plus the magnitude of the phase margin obtained of the uncompensated system (because it is negative) at the gain crossover frequency of 33.78 rad/sec (3.402°) plus the additional phase shift that the composite phase shift curve incurs in going from the uncompensated crossover frequency of 33.78 rad/sec to ω_{max} of 47 rad/sec (about 19°) as follows:

$$\omega_{max} = \frac{1}{T\sqrt{\alpha}} = 47 \text{ rad/sec},$$

$$\phi_{max} = \sin^{-1}\frac{(\alpha - 1)}{(\alpha + 1)} = (28^\circ + 3.4^\circ + 19^\circ).$$

Solving these two equations simultaneously, we obtain the phase-lead network

$$G_c(s) = \frac{1 + 0.059s}{1 + 0.00767s}.$$

Because this solution has "blinders" on and does not consider the proximity of other poles and zeros, it usually has to be fine-tuned. Therefore, the following phase-lead network was selected and is illustrated in Figure 7.24:

$$G_c(s) = \frac{1 + 0.04s}{1 + 0.005s}. \tag{7.64}$$

It results in a phase margin of 28.431° and a gain margin of 10.655 dB. This meets the specification requirements. Notice that the resulting phase-lead network has a low-frequency attenuation of 0.005/0.04 which must be made up by increasing the gain by a factor of 0.04/0.005 = 8. Its gain crossover frequency is 47.27 rad/sec, and the phase crossover frequency occurs at 94.06 rad/sec.

If we have a choice of using the phase-lag or phase-lead network, the phase-lag network solution would be preferable because it meets the required specifications with a narrower bandwidth than the phase-lead network case ($\omega_c = 11.371$ rad/sec for the former; $\omega_c = 47.27$ rad/sec for the latter). A feedback control system having a narrower bandwidth will reject a greater amount of noise than one having a wider bandwidth, as well as requiring less power and cost. In addition, the phase-lead network has the disadvantage of requiring a greater amount of amplification within the control system than the phase-lag network. These and other considerations are discussed further in Section 7.10, where tradeoffs of using different kinds of cascade networks and minor-loop feedback compensation methods are compared.

3. Rate Feedback in Parallel with Position Feedback Compensation. Finally, we wish to compensate the system using rate-feedback compensation in parallel with position feedback as shown in Figure 7.21*b*. The addition of rate feedback provides an $H(s)$ term of

$$H(s) = (1 + bs)$$

which represents a pure zero term that provides a phase lead of

$$\tan^{-1} b\omega = \phi_{max}.$$

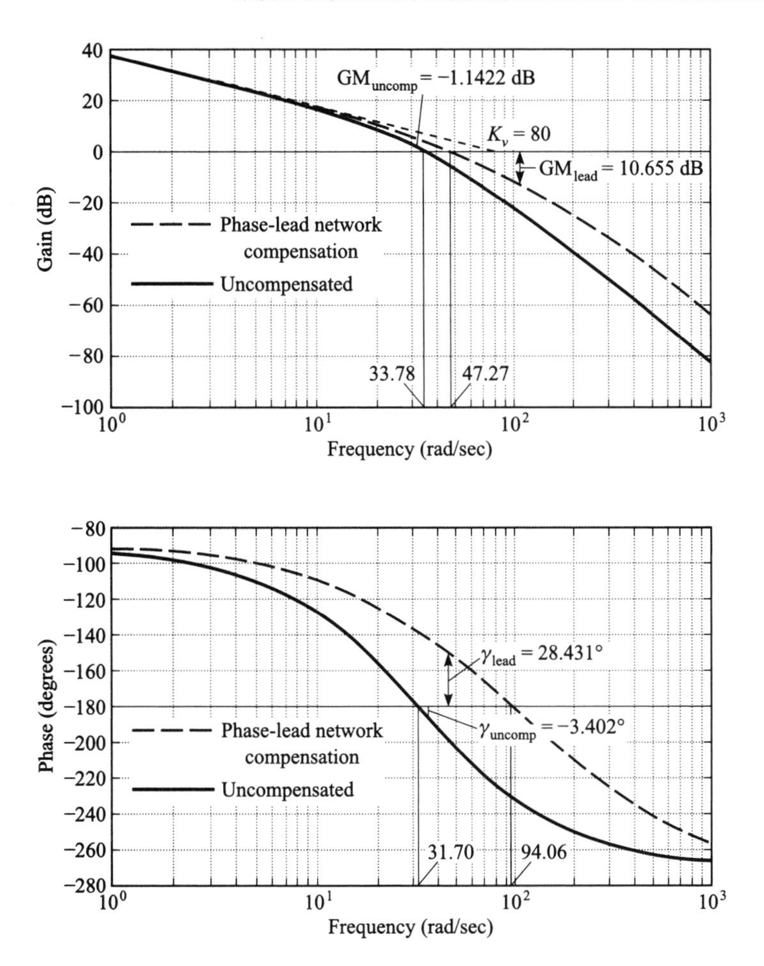

Figure 7.24 Compensation of a third-order system with a phase-lead network $G_c(s) = (1 + 0.04s)/(1 + 0.005s)$ where the uncompensated transfer function is

$$G_0(s) = \frac{80}{s(1 + 0.02s)(1 + 0.05s)}.$$

Remember that a pure zero term is an ideal compensator. Using a similar procedure as for the design of the phase-lead network, let us assume that we desire an approximate 50° phase lead at a frequency of 50 rad/sec. Therefore,

$$\begin{aligned} b(50) &= \tan 50^\circ \\ b &= 0.0238. \end{aligned}$$

The use of rate-feedback compensation is illustrated in Figure 7.25 and indicates a phase margin of 31.9728° at the gain crossover frequency of 39.1641 rad/sec. Its gain margin is infinity, because the phase never is −180° except at $\omega =$ infinity.

Figure 7.26 illustrates the Bode diagram of the uncompensated system, and the Bode diagram of the compensated systems with a phase-lag network, phase-lead

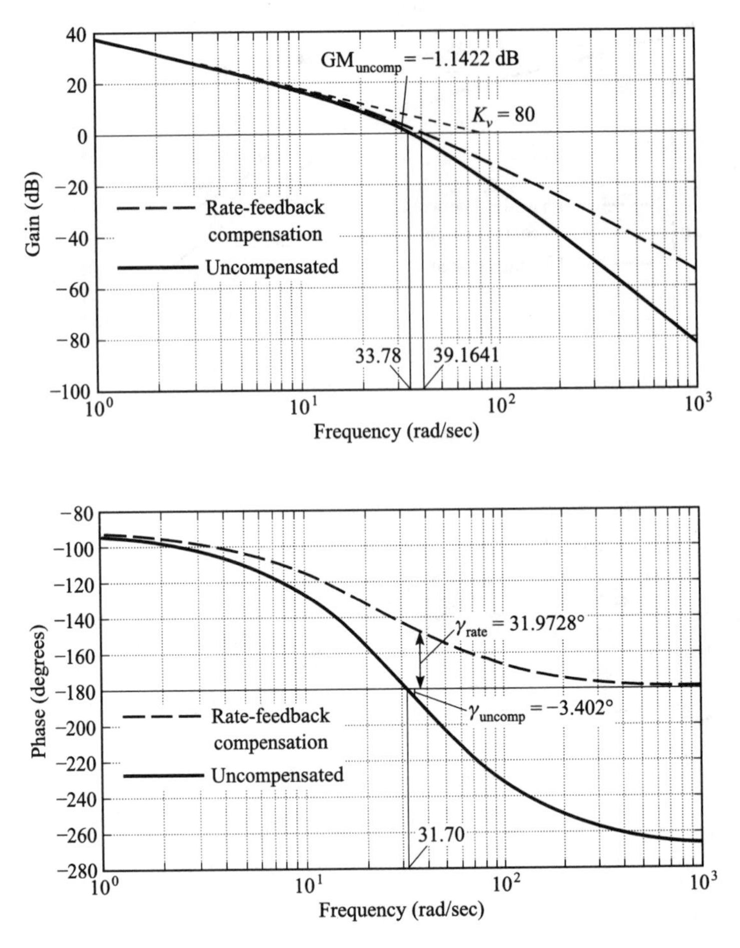

Figure 7.25 Compensation of third-order system with rate feedback ($b = 0.0238$) in parallel with position feedback where the uncompensated transfer function is

$$G_0(s) = \frac{80}{s(1 + 0.02s)(1 + 0.05s)}$$

network, and with rate feedback in parallel with position feedback all superimposed on one Bode diagram.

C. Obtaining Steady-State Error Coefficients from the Bode Diagram

The steady-state error coefficients can be determined from the Bode diagram. The definition and importance of these error coefficients have been discussed in Section 5.4. For a system having one pure integration, the velocity constant K_v can be obtained by extending the initial −20-dB/decade slope until it intersects the 0-dB line. The frequency at which it intersects this line is equal to the velocity constant. We recall from our discussion in Section 5.4 that K_v was obtained by letting s approach zero when utilizing the final-value theorem [see Eq. (5.18)]. Therefore, the pole and/or zero terms having the forms $(1 + Ts)$ or $[(Ts)^2 + 2\zeta Ts + 1]$ all

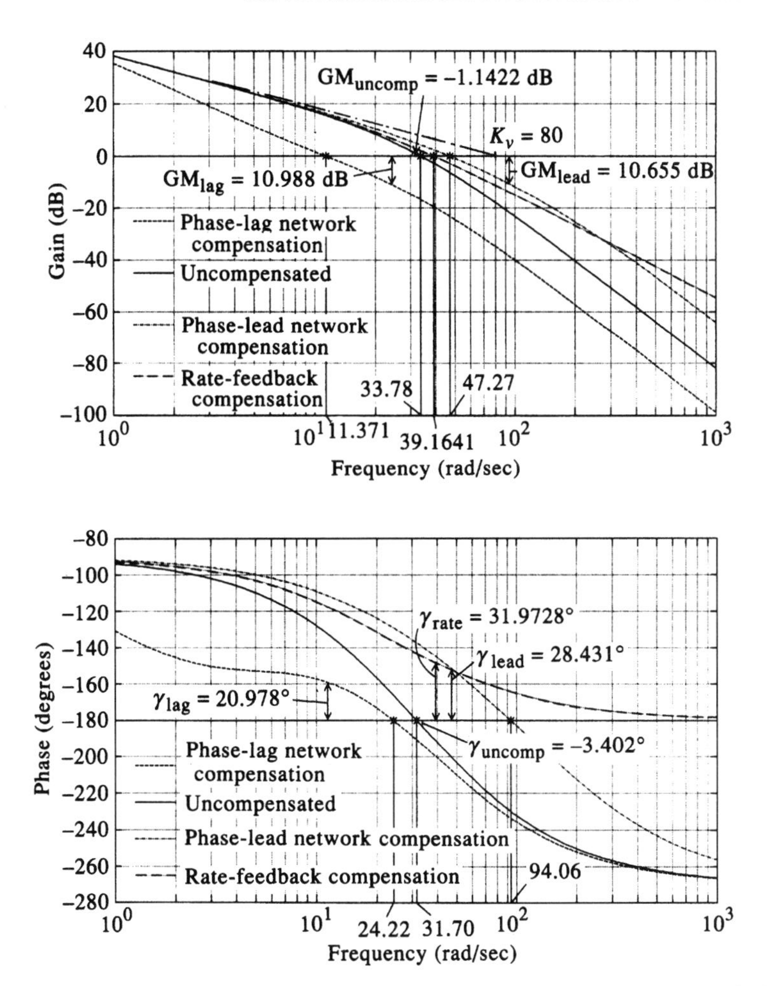

Figure 7.26 Compensation of a third-order system where the uncompensated transfer function is

$$G_0(s) = \frac{80}{s(1+0.02s)(1+0.05s)}.$$

approach unity. This permits one to obtain K_v directly by considering only the initial slope of the Bode diagram. For the Bode diagram shown in Figures 7.22–7.26, the value K_v obtained graphically is 80. As a check, using the definition given by Eq. (5.18), we obtain

$$K_v = \lim_{s \to 0} sG(s)H(s). \tag{7.65}$$

Because

$$G(s) = \frac{80}{s(1+0.02s)(1+0.05s)}, \tag{7.66}$$

$$H(s) = 1, \tag{7.67}$$

then,

$$K_v = \lim_{s \to 0} \frac{s(80)}{s(1 + 0.02s)(1 + 0.05s)} = 80. \tag{7.68}$$

It is also interesting to note that the velocity constant is the same for the uncompensated and compensated systems.

For a system which has a double pole at the origin, the acceleration constant K_a can be obtained in a similar manner. The initial −40-dB/decade slope is extended until it intersects the 0-dB line. The square of the frequency at which it intersects this line is equal to the acceleration constant.

D. Example of Compensating a System Containing a Disturbance Input

The next system we consider is illustrated in Figure 7.27. This consists of a third-order system which has an unwanted external input $U(s)$. The open-loop transfer function for the uncompensated system, $G_0(s)$, is given by

$$G_0(s) = \frac{2.2}{(1 + 0.1s)(1 + 0.4s)(1 + 1.2s)}. \tag{7.69}$$

It is desired that the steady-state error resulting from an unwanted, external step input signal at $U(s)$, should not exceed 0.1 unit. The compensation device, $G_c(s)$, is to contain amplification which will meet this accuracy requirement together with a phase-lead or phase-lag network which will provide a minimum phase margin of 20° and a minimum gain margin of 6 dB.

The value of gain K required for $G_c(s)$ will be computed first. For this calculation, $U(s)$ is assumed to be the input and $E(s)$ is assumed to be the output. The transfer function between these two points is given by

$$\frac{E(s)}{U(s)} = \frac{2.2(1 + T_2 s)}{(1 + T_2 s)(1 + 0.1s)(1 + 0.4s)(1 + 1.2s) + 2.2K(1 + T_1 s)}. \tag{7.70}$$

Setting $U(s) = 1/s$, we obtain the expression for the Laplace transform of the error, $E(s)$, as

$$E(s) = \frac{2.2(1 + T_2 s)}{s[(1 + T_2 s)(1 + 0.1s)(1 + 0.4s)(1 + 1.2s) + 2.2K(1 + T_1 s)]}. \tag{7.71}$$

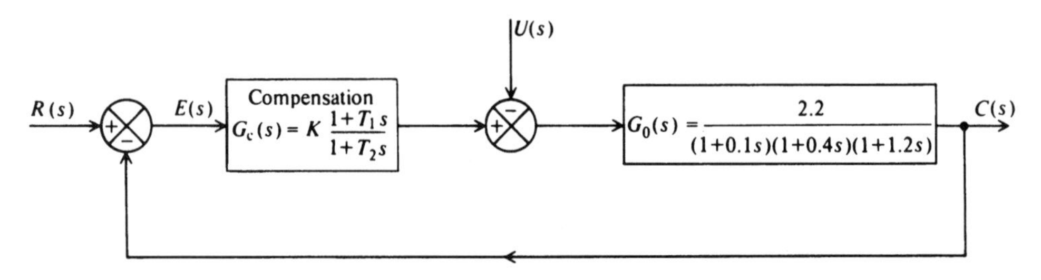

Figure 7.27 A third-order system having an unwanted external input.

The required value of K can be obtained by applying the final-value theorem to $E(s)$ and setting the result equal to 0.1 unit. We find

$$K = 9.55. \tag{7.72}$$

Let us next determine the compensating network required to achieve a phase margin of 20° and a gain margin of 6 dB. The transfer function of the uncompensated system, with $K = 9.55$, is given by

$$KG_0(s) = \frac{21}{(1+0.1s)(1+0.4s)(1+1.2s)}. \tag{7.73}$$

Its Bode diagram which is drawn in Figure 7.28, indicates a phase margin of 0.7167° (gain crossover frequency is 5.8668 rad/sec) and a gain margin of 0.267 dB (phase

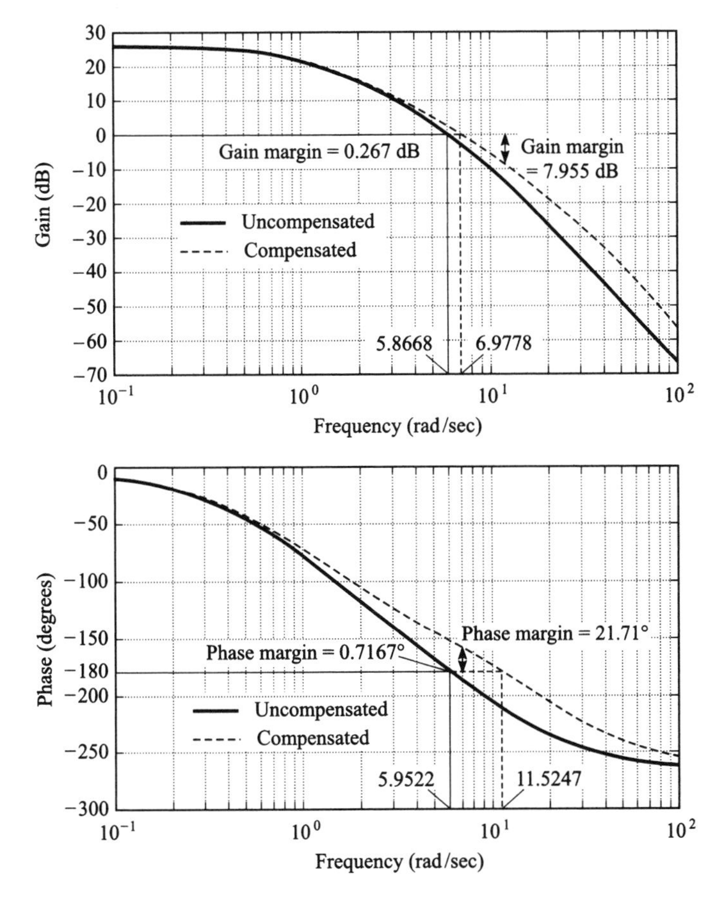

Figure 7.28 Compensation of a third-order system where

$$KG_0(s) = \frac{21}{(1+0.1s)(1+0.4s)(1+1.2s)} \text{ and } G_c(s) = \frac{1+0.167s}{1+0.05s}.$$

Solid line = uncompensated; dashed line = compensated

crossover frequency is 5.9522 rad/sec). We wish to increase the phase margin by about 20°. This Bode diagram was obtained using MATLAB and is contained in the M-file that is part of my MCSTD Toolbox and can be retrieved free from The MathWorks, Inc. anonymous FTP server at ftp://ftp.mathworks.com/pub/books/shinners.

As in the previous problem analyzed in this section, we will estimate the values of ω_{max} and ϕ_{max} for Eqs. (7.7) and (7.8), respectively, to obtain a first-cut design. We must be careful when using this approach, because we actually want to have this phase shift at the gain crossover frequency in order to achieve the desired phase margin. However, specifying α, T, and ω_{max} does not ensure that ϕ_{max} will be at the gain crossover frequency. After obtaining this first-cut design, we then fine-tune the results to obtain the specification requirements as in the previous example. For this design and an anlysis of the Bode diagram of the uncompensated system shown in Figure 7.28 a value of $\omega_{max} = 8$ rad/sec was selected. The value of ϕ_{max} selected was a conservative value of 34° which includes the difference between the minimum phase margin desired (20°) and the actual phase margin of the uncompensated system (0.7167°) plus the difference in phase shift between the gain crossover frequency of the uncompensated system (5.8668 rad/sec) and 8 rad/sec plus an extra margin to be conservative. The resulting two equations to be solved simultaneously were

$$\omega_{max} = 1/(T\sqrt{\alpha}) = 8,$$

$$\phi_{max} = \sin^{-1}\frac{\alpha - 1}{\alpha + 1} = 34^\circ.$$

The result is a phase-lead network whose transfer function is

$$\frac{1 + 0.2355s}{1 + 0.0663s}.$$

The first-cut design was adjusted a little to better fit the overall Bode diagram and the final phase-lead network selected is as follows:

$$\frac{1 + 0.167s}{1 + 0.05s}. \tag{7.74}$$

This network causes a gain crossover frequency of 6.9778 rad/sec where the phase margin is 21.71°, and a gain margin of 7.955 dB (the phase crossover frequency is at 11.5247 rad/sec). This satisfies the specification. Remember that this network will require an increase in the amplifier gain by 3.34 to make up for the attentuation of 0.05/0.167 = 0.2994 that it provides.

E. Example for Compensating a Two-Loop System Using Rate Feedback in Cascade with a Filter

The concluding problem we consider using the Bode-diagram approach consists of designing the feedback control system illustrated in Figure 7.29*a*. For this particular system, we desire that the steady-state error resulting from a velocity input of 110

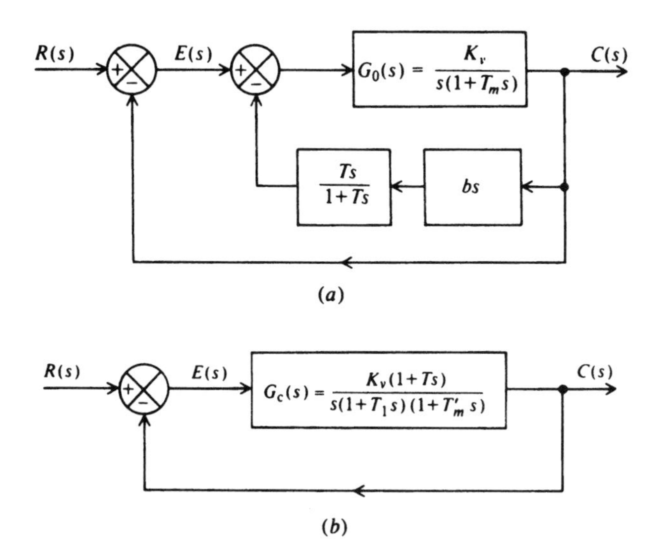

Figure 7.29 (*a*) Use of rate feedback in cascade with a high-pass filter in order to compensate a feedback control system. (*b*) Equivalent block diagram for the system shown in (*a*).

rad/sec be equal to 0.25 rad. The uncompensated open-loop transfer function $G_0(s)$ is given by

$$G_0(s) = K_v/(s(1 + T_m s)), \tag{7.75}$$

where K_v = velocity constant and T_m = motor time constant = 0.025 sec. This transfer function consists of an amplifier, positioning motor, gear train, and load. In order to achieve the required accuracy, K_v must equal

$$K_v = \frac{\omega}{\text{error}} = \frac{110 \text{ rad/sec}}{0.25 \text{ rad}} = 440/\text{sec}. \tag{7.76}$$

We want a phase margin of approximately 55° for this system. This will be achieved by means of minor-loop rate-feedback compensation which is cascaded with a simple *RC* high-pass filter (phase-lead network) in order not to increase the steady-state response error for velocity inputs. This was discussed in Section 7.3 (see Figure 7.15).

The open-loop frequency response we must plot on the Bode diagram is that obtained with the minor-rate loop closed and the outer position loop opened. Therefore, we are interested in obtaining the equivalent transfer function between $E(s)$ and $C(s)$. This is easily found to be

$$\frac{C(s)}{E(s)} = \frac{K_v(1 + Ts)}{s[(1 + T_m s)(1 + Ts) + K_v bTs]}. \tag{7.77}$$

Expanding the denominator of Eq. (7.77), we obtain the expression

$$s[T_m Ts^2 + (T_m + T + K_v bT)s + 1]. \tag{7.78}$$

This expression can be put into the form

$$s[(1 + T'_m s)(1 + T_1 s)]. \tag{7.79}$$

by defining the time constants T'_m and T_1 as

$$T'_m = \frac{T_m}{T_1} T \tag{7.80}$$

and

$$T_1 = -T'_m + T_m + (1 + K_v b)T. \tag{7.81}$$

Therefore, we may redraw Figure 7.29*a* as shown in Figure 7.29*b*. For any set of values for T_m, T, K_v, and b, we can derive T'_m and T_1 by solving the simultaneous equations (7.80) and (7.81). Another approach is to choose T and T_1 from the Bode diagram which meets the specified phase margin and solve for the required rate-feedback constant b.

The procedure we follow when compensating this system is to draw the Bode diagram for the uncompensated system in accordance with Eq. (7.75) and then fit the characteristics of the compensated system in accordance with

$$G_c(s) = \frac{C(s)}{E(s)} = \frac{K_v(1 + Ts)}{s(1 + T'_m s)(1 + T_1 s)} \tag{7.82}$$

until a phase margin of 55° is achieved. The compensated characteristic will then determine T and T_1, from which T'_m and the rate-feedback constant b can be determined. Figure 7.30 illustrates the Bode diagram of the uncompensated and compensated systems. The phase margin of the uncompensated system is 17.1°. Values of

$$T_1 = 0.33, \tag{7.83}$$

$$T = 0.033, \tag{7.84}$$

and

$$T'_m = \frac{T_m}{T_1} T = 0.0025 \quad \text{[from Eq. (7.80)]} \tag{7.85}$$

result in a phase margin of 55.36°. From Eq. (7.81), the corresponding value of rate-feedback constant b is 0.0186. Figure 7.30 was obtained using MATLAB and is contained in the M-file that is part of my MCSTD Toolbox.

The type of compensation just illustrated is used quite frequently in practice. In order to really understand what is actually happening, it is important to examine the Bode diagram of Figure 7.30 closely. The net effect of the minor-loop rate feedback has been to move the equivalent motor break frequency from $1/T_m$ to $1/T'_m$ by the ratio given in Eq. (7.80). This technique is used quite frequently to compensate power servos. The net effect of the phase-lead network, in the minor-loop feedback path, is to appear as a phase lag, for the equivalent open-loop characteristics of Figure 7.29*b*. This can be easily understood because we effectively see the reciprocal of the feedback element when looking into a closed-loop system which has an open-

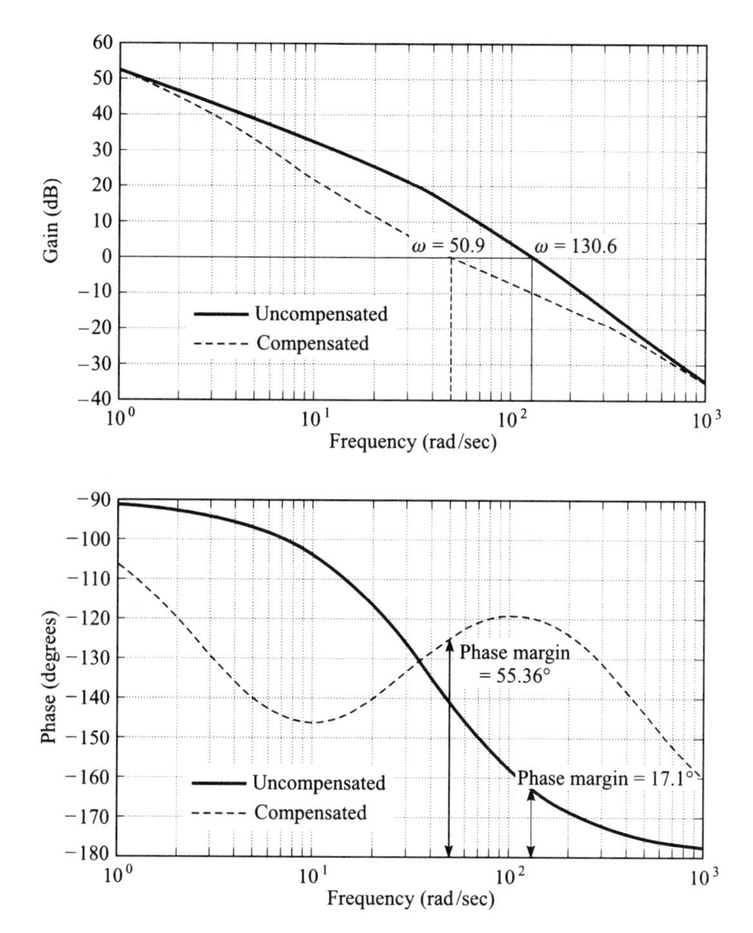

Figure 7.30 Compensation of the system shown in Figure 7.29*b*.

$$G_0(s) = \frac{440}{s(1+0.025s)} \text{ and } G_c(s) = \frac{440(1+0.033s)}{s(1+0.33s)(1+0.0025s)}.$$

loop gain much greater than unity [see Eq. (2.122)]. This is a very important fact that can be utilized to approximate the Bode diagram in preliminary designs. This point is now expanded upon in the following section.

The complete case study for the design of an angular control system for a robot's joint is presented in Section 7.4 of Chapter 7 of the accompanying volume. In this problem, the Bode diagram is used for analyzing and designing the control system.

7.7. APPROXIMATE METHODS FOR PRELIMINARY COMPENSATION AND DESIGN USING THE BODE DIAGRAM

Having presented detailed compensation methods, let us next focus our attention on approximate methods for obtaining a first cut at compensation utilizing the Bode diagram [8–10]. Although the procedures presented in this section are approximate, they are generally adequate for the preliminary stage of design. Before the system design is completed, however, the exact magnitude and phase curves should be drawn as indicated previously. The practice of utilizing approximate methods for preliminary design is generally employed as a convenience in obtaining significant time constants, gains, and phase characteristics required for a design.

A. Approximate Closed-Loop Response from the Bode Diagram

In order to develop the concept of this approach, let us consider the feedback system illustrated in Figure 7.31. The closed-loop transfer of this system is given by

$$\frac{C(s)}{R(s)} = \frac{G(s)}{1 + G(s)H(s)}. \tag{7.86}$$

Let us modify this equation into the following more convenient form:

$$\frac{C(s)}{R(s)} = \frac{1}{H(s)}\left[\frac{G(s)H(s)}{1 + G(s)H(s)}\right]. \tag{7.87}$$

As shown previously, the magnitude (in decibels) of $G(s)H(s)$ can be approximated by straight lines of constant slope when plotted against frequency on a log scale. Therefore, it appears reasonable that the term $1 + G(s)H(s)$ in Eq. (7.87) may also be dealt with in a similar approximate manner as follows:

$$1 + G(s)H(s) \approx 1 \qquad \text{for } |G(s)H(s)| < 1, \tag{7.88}$$

$$1 + G(s)H(s) \approx G(s)H(s) \qquad \text{for } |G(s)H(s)| > 1. \tag{7.89}$$

Substituting these approximations into Eq. (7.87), we find that

$$\frac{C(s)}{R(s)} \approx G(s) \qquad \text{for } |G(s)H(s)| < 1, \tag{7.90}$$

$$\frac{C(s)}{R(s)} \approx \frac{1}{H(s)} \qquad \text{for } |G(s)H(s)| > 1. \tag{7.91}$$

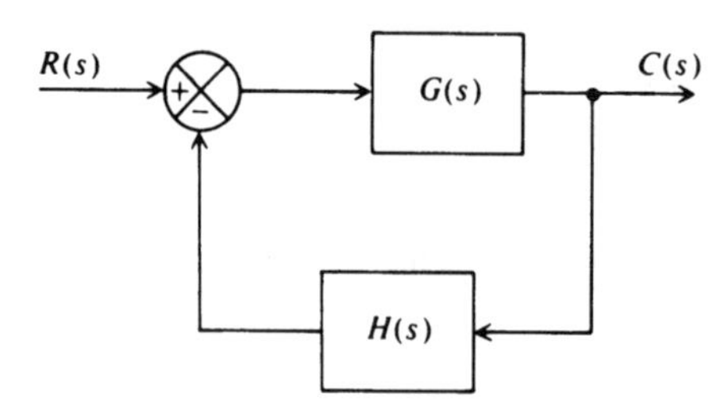

Figure 7.31 A feedback control system.

The approximations of Eqs. (7.90) and (7.91) are very useful as a first cut in the preliminary design of a control system. However, it is important to point out that they are approximate relationships, and are subject to error particularly at those frequencies where $G(s)H(s) = 1$. However, the amount by which the approximations are in error can be calculated, and this correction can then be applied to correct the approximate value. The resultant corrected value is then an exact solution.

The technique of applying the approximation of Eqs. (7.90) and (7.91) can best be illustrated through an example. Consider the feedback system of Figure 7.31, where

$$G(s) = \frac{K}{1 + Ts} \tag{7.92}$$

and

$$H(s) = 1. \tag{7.93}$$

This example will illustrate how feedback, around an element containing a time constant, reduces the time constant of that element. Substituting Eqs. (7.92) and (7.93) into Eq. (7.86), the resultant system transfer function is given by

$$\frac{C(s)}{R(s)} = \frac{K/(1 + Ts)}{1 + K/(1 + Ts)} = \frac{K}{1 + K}\frac{1}{1 + Ts/(1 + K)}. \tag{7.94}$$

For purposes of illustration, it is assumed that

$$K = 10, \tag{7.95}$$

$$T = 1 \tag{7.96}$$

in this example. For these values, Figure 7.32 illustrates the straight-line approximations for $|C(s)/R(s)|$, $|G(s)|$, and $|G(s)H(s)|$. The dashed curve representing $G(s)$ [and $G(s)H(s)$ as $H(s) = 1$] has a gain of 10 (20 dB) for frequencies lower than $\omega = 1/T = 1$ rad/sec. At frequencies higher than $\omega = 1$ rad/sec, $G(s)$ and $G(s)H(s)$ have an attentuation of 20 dB/decade. At $\omega = \omega_n = K/T = \frac{10}{1} = 10$, $G(s)$ and $G(s)H(s)$ approximately equal 0 dB. Equation (7.91) indicates that for frequencies lower than $\omega_n = 10$, the system transfer function $C(s)/R(s)$ equals $1/H(s) = 1$ and is shown by the solid line at 0 dB. For frequencies greater than ω_n, Eq. (7.90) indicates that $C(s)/R(s)$ is equal to $G(s)$ as shown. Figure 7.32 illustrates very clearly that the use of unity-gain feedback around a single time-constant element results in a reduction in its time constant. This admirable characteristic is achieved at the expense of a loss of gain.

It is important to recognize that the results obtained for the closed-loop transfer function are approximate. For example, Eq. (7.94) indicates that the gain is actually $K/(1 + K) = 10/(1 + 10) = 0.909$ instead of 1 for $\omega < \omega_n$, and the closed-loop time constant is $T/(1 + K) = 1/(1 + 10) = 0.0909$ instead of $T/K = \frac{1}{10} = 0.1$. Note that these errors decrease as the gain K increases. Generally, K is quite large and the error between the approximate and exact curves is very small.

The form of Eq. (7.87) is very well suited for determining the value of $C(s)/R(s)$ when $|G(s)H(s)| > 1$, with the bracketed term representing the difference between the

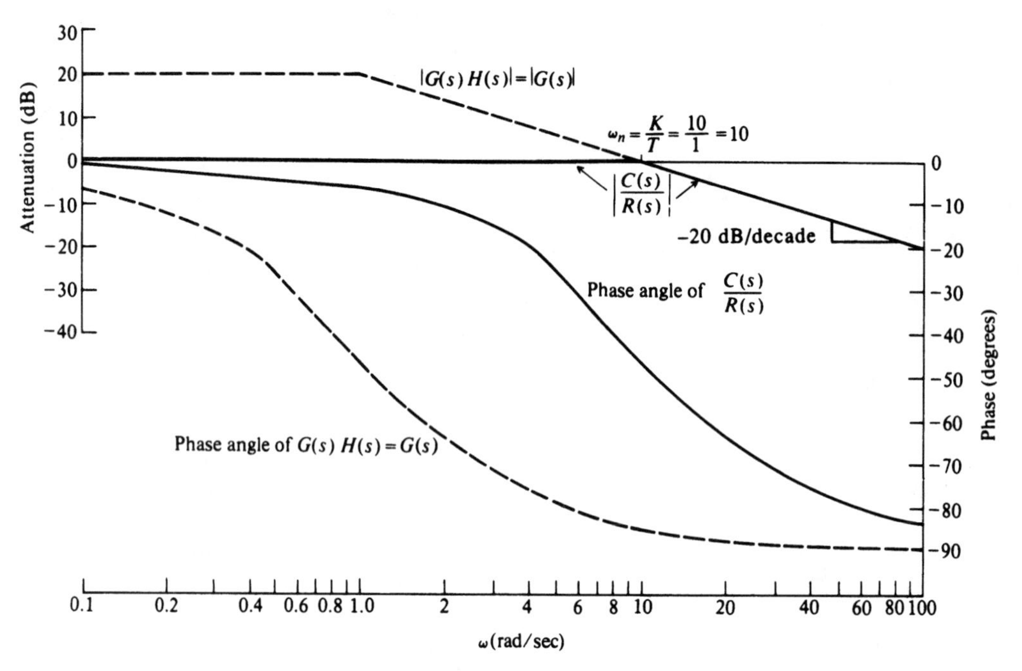

Figure 7.32 Open- and closed-loop gain and phase characteristis for the system of Figure 7.31 where $G(s) = 10/(1+s)$ and $H(s) = 1$.

approximate and exact curves. In order to determine a general analytic expression for finding the difference between the approximate and exact curves when $|G(s)H(s)| < 1$, let us reconsider Eq. (7.86) and rewrite it as follows:

$$\frac{C(s)}{R(s)} = G(s)\frac{1}{1+G(s)H(s)}. \tag{7.97}$$

Therefore,

$$\frac{C(s)}{R(s)} = G(s)\left[\frac{1/[G(s)H(s)}{1+1/[G(s)H(s)]}\right]. \tag{7.98}$$

For those frequencies where $|G(s)H(s)| < 1$, the bracketed term of Eq. (7.98) represents the error between the approximate solution

$$\frac{C(s)}{R(s)} \approx G(s) \qquad \text{for } |G(s)H(s)| < 1 \tag{7.99}$$

and the exact solution. Another way of looking at this is that the bracketed term represents a correction factor which can be used to correct the results of the approximation given by Eq. (7.90).

B. The Straight-Line Phase-Shift Approximation

For preliminary design purposes, it is usually sufficient to obtain quantitative information regarding phase shift without resorting to the exact but tedious method of calculating the phase. We have found it convenient to represent the amplitude characteristics by a straight-line approximation, and can utilize a similar technique for the phase-shift function. In order to introduce the method, let us consider the phase shift due to the transfer function representing a zero factor given by

$$G(j\omega) = 1 + j\left(\frac{\omega}{\omega_1}\right). \tag{7.100}$$

Figure 7.33*a* illustrates the exact and straight-line approximation and Figure 7.33*b* illustrates the exact phase shift and its straight-line approximation. The straight-line approximate phase shift has been constructed by drawing a straight line tangent to the actual curve at $\omega = \omega_1$.

In order to construct these straight-line phase-shift approximations by inspection, the dependence of ω_A and ω_B on ω_1 must be known. This can be obtained by first considering the slope of the actual curve at $\omega = \omega_1$ as follows:

$$\phi = \tan^{-1}\frac{\omega}{\omega_1}. \tag{7.101}$$

The slope at $\omega = \omega_1$ can be obtained by differentiating this expression:

$$\left.\frac{d\phi}{d(\ln\,\omega)}\right]_{\omega=\omega_1} = \left.\frac{d\phi}{d\omega}\,\frac{d\omega}{d(\ln\,\omega)}\right]_{\omega=\omega_1} = \left.\frac{\omega/\omega_1}{1+(\omega/\omega_1)^2}\right]_{\omega=\omega_1} = \frac{1}{2}. \tag{7.102}$$

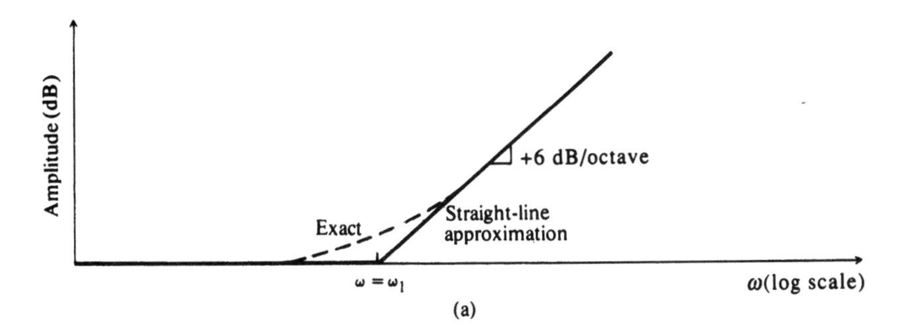

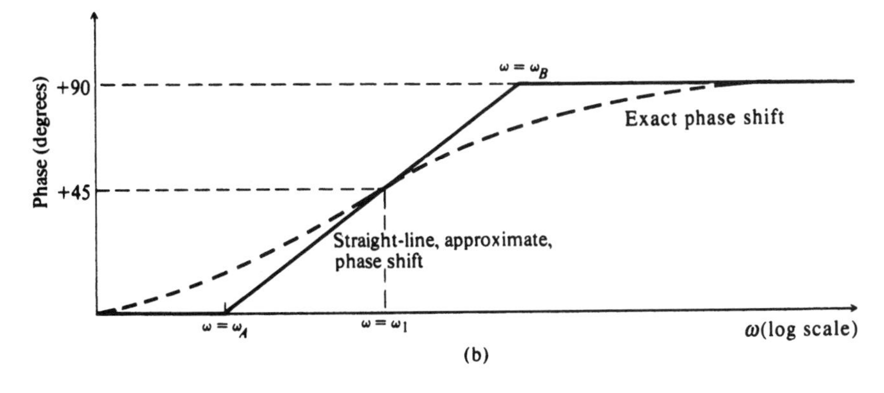

Figure 7.33 Straight-line, approximate, phase-shift methods for a zero factor term.

Knowing the slope at $\omega = \omega_1$, the intercepts at ω_A and ω_B can be obtained from

$$\frac{\pi/4}{\ln \omega_1 - \ln \omega_A} = \frac{1}{2}, \tag{7.103}$$

which gives

$$\frac{\pi}{4} = \frac{1}{2} \ln \frac{\omega_1}{\omega_A}. \tag{7.104}$$

Therefore,

$$\frac{\omega_1}{\omega_A} = \frac{\omega_B}{\omega_1} = e^{\pi/2} = 4.81. \tag{7.105}$$

It is interesting to note that if a number other than e were chosen for the logarithmic base, the slope in Eq. (7.105) would change, but the frequency ratio given by Eq. (7.105) would remain the same.

Based on this result, complicated approximate straight-line phase-shift curves can be obtained. Figure 7.34 illustrates the application of this approach for a transfer function given by

$$G(s) \frac{(1 + 10s)(1 + 0.01s)^2}{(1 + s)^2(1 + 0.001s)} \tag{7.106}$$

and compares the approximate straight-line phase-shift curve with the exact phase-shift curve. Observe that the errors obtained by using the approximate phase curve are larger than the corresponding errors of the amplitude approximation. However, the use of this approximation greatly aids the control-system engineer in obtaining a first cut preliminary design.

The example just presented contained roots that were all real. What happens if the transfer function contains underdamped quadratic factors? In order to analyze this situation, let us consider the following normalized quadratic phase-lag factor ($\zeta < 1$):

$$G(j\omega) = \frac{1}{(j\omega/\omega_n)^2 + 2\zeta(j\omega/\omega_n) + 1}. \tag{7.107}$$

It was shown in Section 6.7 that the amplitude and phase-shift terms are given respectively by [see Eq. (6.92)]:

$$-20 \log_{10} \left[\left(\frac{2\zeta\omega}{\omega_n} \right)^2 + \left(1 - \frac{\omega^2}{\omega_n^2} \right)^2 \right]^{1/2}, \tag{7.108}$$

$$\phi = -\tan^{-1} \frac{2\zeta\omega_n\omega}{\omega_n^2 - \omega^2}. \tag{7.109}$$

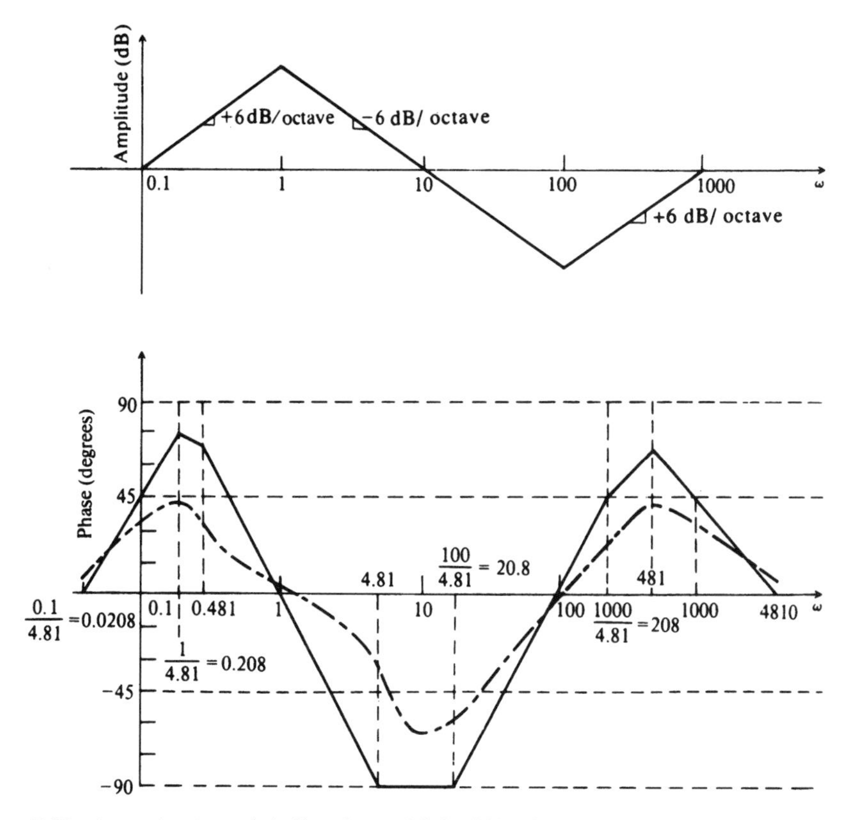

Figure 7.34 Approximate straight-line phase-shift (solid line) and exact phase-shift (short dash–long dash line) characteristics of

$$G(s) = \frac{(1 + 10s)(1 + 0.01s)^2}{(1 + s)^2(1 + 0.001s)}.$$

Let us focus attention on the phase-shift term. Differentiation of Eq. (7.109) results in the slope at $\omega = \omega_n$ being given by

$$\text{slope} = \frac{d\phi}{d(\ln \omega)} = \frac{1}{\zeta}. \tag{7.110}$$

As before, the two intercepts ω_A and ω_B can be found as follows:

$$\frac{\pi/2}{\ln \omega_n - \ln \omega_A} = \frac{1}{\zeta}. \tag{7.111}$$

Therefore,

$$\frac{\omega_n}{\omega_A} = \frac{\omega_B}{\omega_n} = e^{(\pi/2)\zeta} = 4.81^{\zeta}. \tag{7.112}$$

Exact phase characteristics corresponding to Eq. (7.107) and its straight-line approximation, based on the relationship of Eq. (7.112), are illustrated in Figure 7.35. In

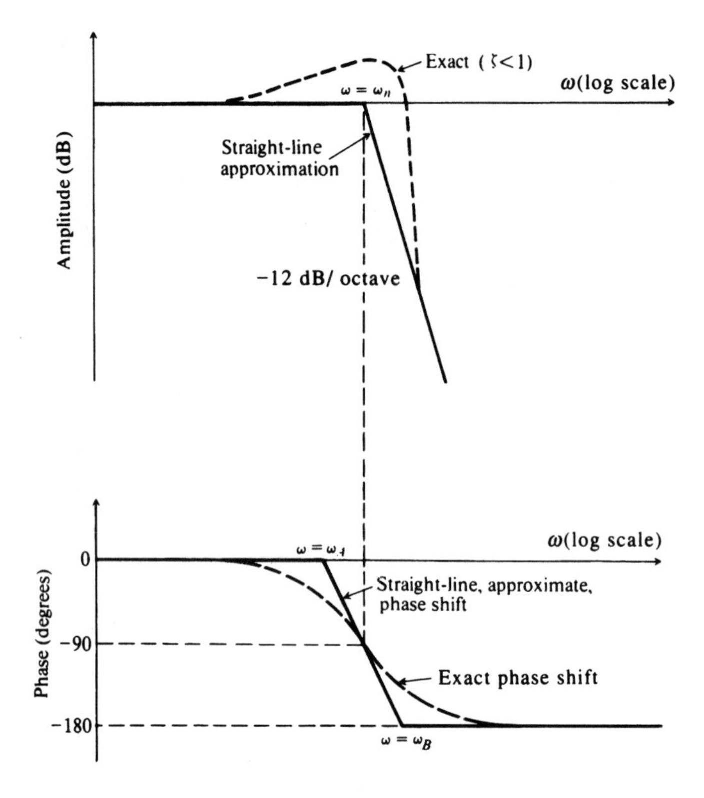

Figure 7.35 Straight-line approximate phase-shift method for a quadratic lag factor ($\zeta < 1$).

addition, the exact amplitude characteristics of Eq. (7.108) and its straight-line approximation are also illustrated. Observe from Eq. (7.112) that as ζ approaches zero, the two frequency ratios decrease and approach unity in the limits. This certainly agrees with the exact phase characteristics of a quadratic phase lag as illustrated in Figure 6.28. On the other hand, when $\zeta = 1$, the roots become real and Eq. (7.112) reduces to Eq. (7.105).

The reader is again reminded that the straight-line phase-shift approximation is not exact. It should only be used in order to obtain a first cut for preliminary design work. It shouldalso be noted that the errors of the straight-line phase-shift approximation are generally larger than the corresponding errors of the amplitude approximation. For this reason, the phase-shift approximation is not used for final design. In the final design, the actual phase-shift characteristics must be employed as previously illustrated.

7.8. COMPENSATION AND DESIGN USING THE NICHOLS CHART

The Nichols-chart method has been developed in Section 6.10. We demonstrated in that section how one could obtain the closed-loop frequency response of a feedback control system by superimposing the open-loop gain-phase characteristics onto the Nichols chart. Specifically, we obtained the closed-loop frequency response of the system shown in Figure 6.49. The intersections of the open-loop gain–phase characteristics and the Nichols closed-loop gain characteristics were shown in Figure 6.51*a*. The resulting closed-loop frequency response was illustrated in Figure 6.51*b*. It indicated a maximum value of peaking M_p of 6.758 dB (2.2) and the frequency at which it occurred, ω_p, was 8.989 rad/sec. We indicated in Sections 6.10 and 6.12 that a value of $M_p = 6.758$ dB (2.2) does not represent a good design. This section demonstrates how the control engineer may use the Nichols chart in order to achieve a specified performance.

Let us assume for this problem that an acceptable value of M_p is 1.4 (2.92 dB). This may be achieved by adding a phase-lag or phase-lead network in cascade with the forward-loop transfer function $G(s)$. A phase-lag network is used for this problem although a solution can be found as easily using a phase-lead network. We shall demonstrate that for an M_p of 2.92 dB the object is to modify the gain–phase characteristics on the Nichols chart so that it is just tangent to the 2.92 dB locus and does not enter it. By restricting the gain-phase characteristics to areas external to the $M = 2.92$ dB locus, we will have limited M_p to 2.92 dB, because the interior of this locus represents values of M greater than 2.92 dB.

Studying the characteristics of Figure 6.51*a*, we see that relatively large magnitudes of $G(j\omega)$ exist for $\omega < 3$ rad/sec. Therefore, it is not desirable to shift these magnitudes inside the $M_p = 2.92$ dB curve. In addition, it is desirable to attenuate $G(j\omega)$ by a factor of about 3 in the range of frequencies of $\omega = 5 - 12$ rad/sec. A phase-lag network, $(1 + Ts)/(1 + \alpha Ts)$, whose factor α equals 3 will achieve this if $\omega_{\max} = 1/(T\sqrt{\alpha})$ is chosen at about 1 rad/sec. Solving for αT and T, we get $\alpha T = 1.74$ and $T = 0.58$.

In order to obtain the gain-phase characteristics of the open-loop system, the Bode diagram is first drawn as indicated in Figure 7.36. Then for each value of ω the magnitude and phase of the open-loop compensated characteristics are plotted into a Nichols chart as shown in Figure 7.37. Because the open-loop gain-phase characteristics are just tangent to the constant-magnitude locus corresponding to $M = 2.976$ dB (1.4), we have achieved our goal. Notice that we have shifted $\omega = 1$ rad/sec by about $-35°$, but this does not increase M_p. Figures 7.36 and 7.37 were obtained using MATLAB, and are contained in the M-files that are part of my MCSTD Toolbox. It is important to note that the use of MATLAB and the MCSTD Toolbox permits one to obtain the Nichols chart directly without having to first obtain the Bode diagram.

7.9. COMPENSATION AND DESIGN USING THE ROOT-LOCUS METHOD

The root-locus technique has been developed previously in Sections 6.14–6.16, and digital computer programs for obtaining the root locus were presented in Section 6.17 (MATLAB) and 6.18. It is a very helpful tool that the control engineer can use in order to study the variation of gain, system parameters, and effect of compensation. We demonstrated in Section 6.14 the migration of poles in the complex plane as the gain of the system was varied from zero to infinity. We obtained the root locus for several feedback systems including the following:

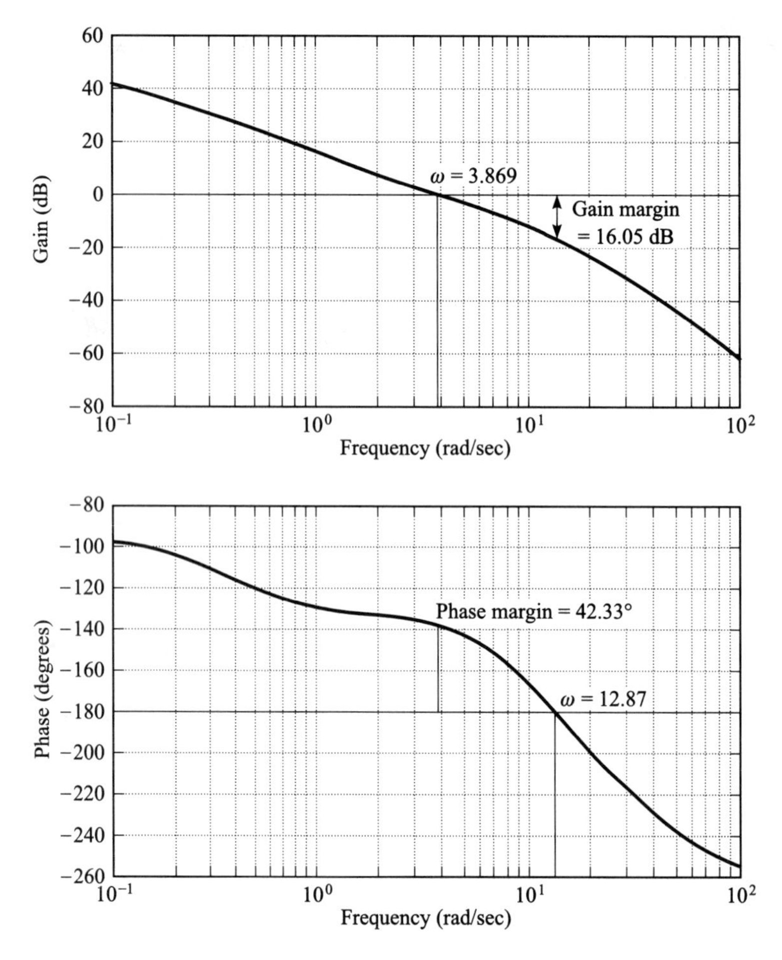

Figure 7.36 Bode diagram for the compensated system of Figure 6.49, where

$$G_c(s)G(s)H(s) = \frac{11.7(1 + 0.58s)}{s(1 + 0.05s)(1 + 0.1s)(1 + 1.74s)}.$$

$$G(s)H(s) = \frac{K}{(s + 1)(s - 1)(s + 4)^2}. \tag{7.113}$$

This was illustrated in Figure 6.60. An analysis of the root locus for this system indicated that it was always unstable, because at least one of the roots of the characteristic equation always occurred in the right-half-plane. This section demonstrates how this system may be compensated by means of a phase-lead network, and/or rate feedback in parallel with position feedback. This problem is followed by considering phase-lag-network compensation for the system illustrated in Figure 6.57. In addition, we shall demonstrate how the control engineer may determine the transient response of the compensated systems in order to meet certain specifications.

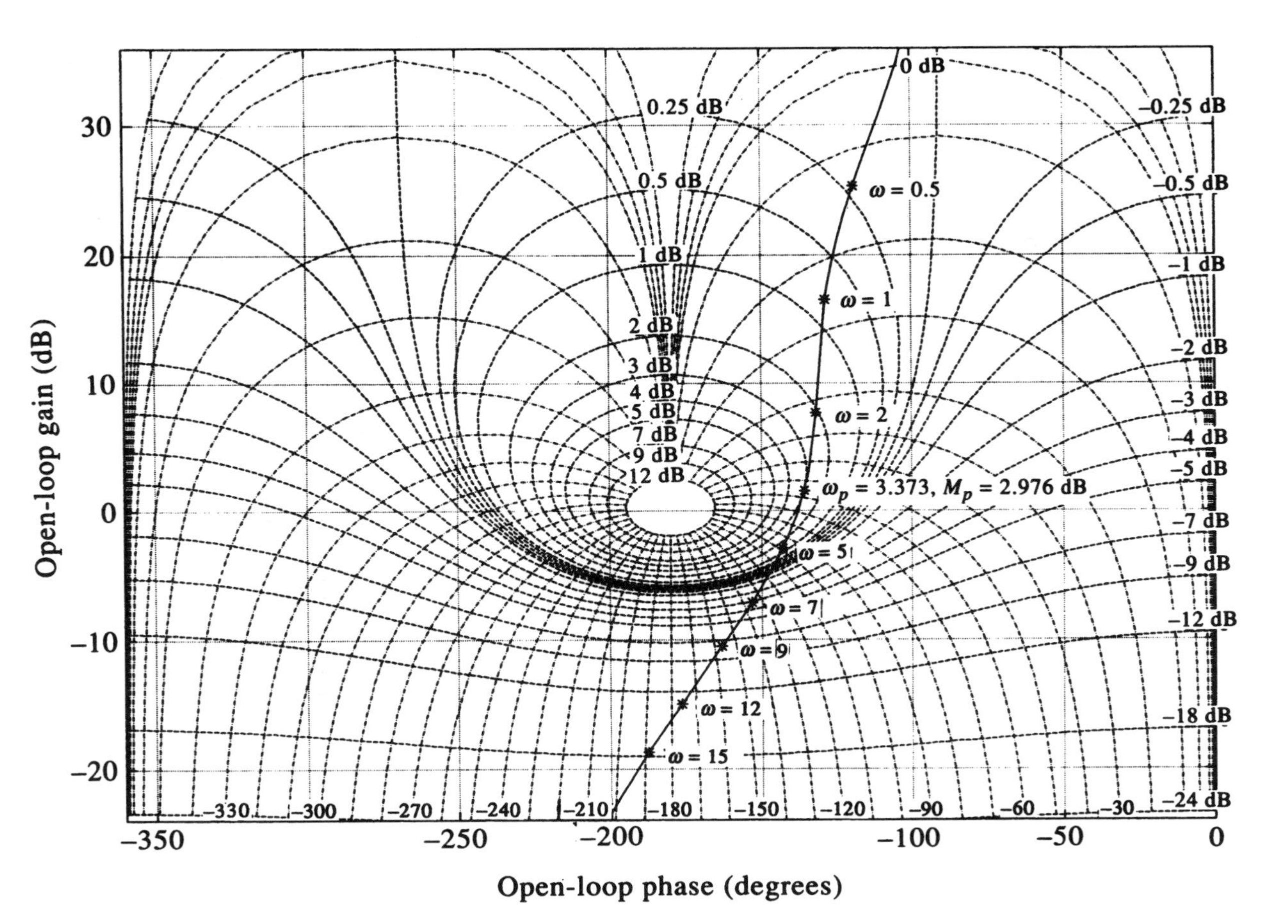

Figure 7.37 Compensation of the system shown in Figure 6.49 for $M = 2.976$ dB (1.4).

A. Example of Phase-Lead Network and Rate-Feedback Compensation

Let us attempt to stabilize the system of Figure 6.60 by means of a phase-lead network in cascade with the forward-loop transfer function. The form of its transfer function is given by

$$G_c(s) = \frac{s+\alpha}{s+\beta}. \tag{7.114}$$

We assume that the effect of the pole introduced by the phase-lead network has a negligible effect compared with its zero. Therefore, we assume that the transfer function of the cascaded phase-lead network can be approximated by the zero term only, which is also equivalent to the pure zero obtained with rate feedback in parallel with position feedback (see Figure 7.14*a*):

$$G_c(s) \approx s+\alpha. \tag{7.115}$$

The resulting value of $G_c(s)G(s)H(s)$, which is to be examined on the root locus, is given by

$$G_c(s)G(s)H(s) = \frac{K(s+\alpha)}{(s+1)(s-1)(s+4)^2}. \tag{7.116}$$

We wish to investigate the effect on stability of a variation in α as follows:

$$\begin{array}{llll} \text{Case A}: & \alpha = 0.5 & \text{Case D}: & \alpha = 4 \\ \text{Case B}: & \alpha = 1 & \text{Case E}: & \alpha = 6. \\ \text{Case C}: & \alpha = 2 & & \end{array}$$

The resulting root loci for all these cases are presented next. It is important to emphasize at this point that, although we make use of most of the anlaytic tools developed in Section 6.14, we do not use all of them, This omission is due to the fact that some of the analytic techniques developed are too complex to use for higher-order systems. For example, it is very tedious to determine the value of the gain K along the root loci utilizing the relationship given by Eq. (6.133). Fortunately, this can be obtained much more easily with the digital computer program presented in Section 6.18 or by using MATLAB (Section 6.17). The approach taken in presenting the resulting root loci for this problem is to outline the results of the 12 rules developed previously in Section 6.14, and use the digital computer programs of Section 6.18 or by using MATLAB (Section 6.17) wherever it is helpful. In addition, the values of gain obtained using the digital computer program or MATLAB which are pertinent for an intelligent evaluation of the problem will be indicated. It is my feeling that this dual approach is the best procedure to use when explaining the root-locus method.

Case A. $\alpha = 0.5$. See Figure 7.38 for the root-locus sketch.

Rule 1. There are four separate loci because the characteristic equation,

$$1 + G_c(s)G(s)H(s) = 0,$$

is a fourth-order equation.

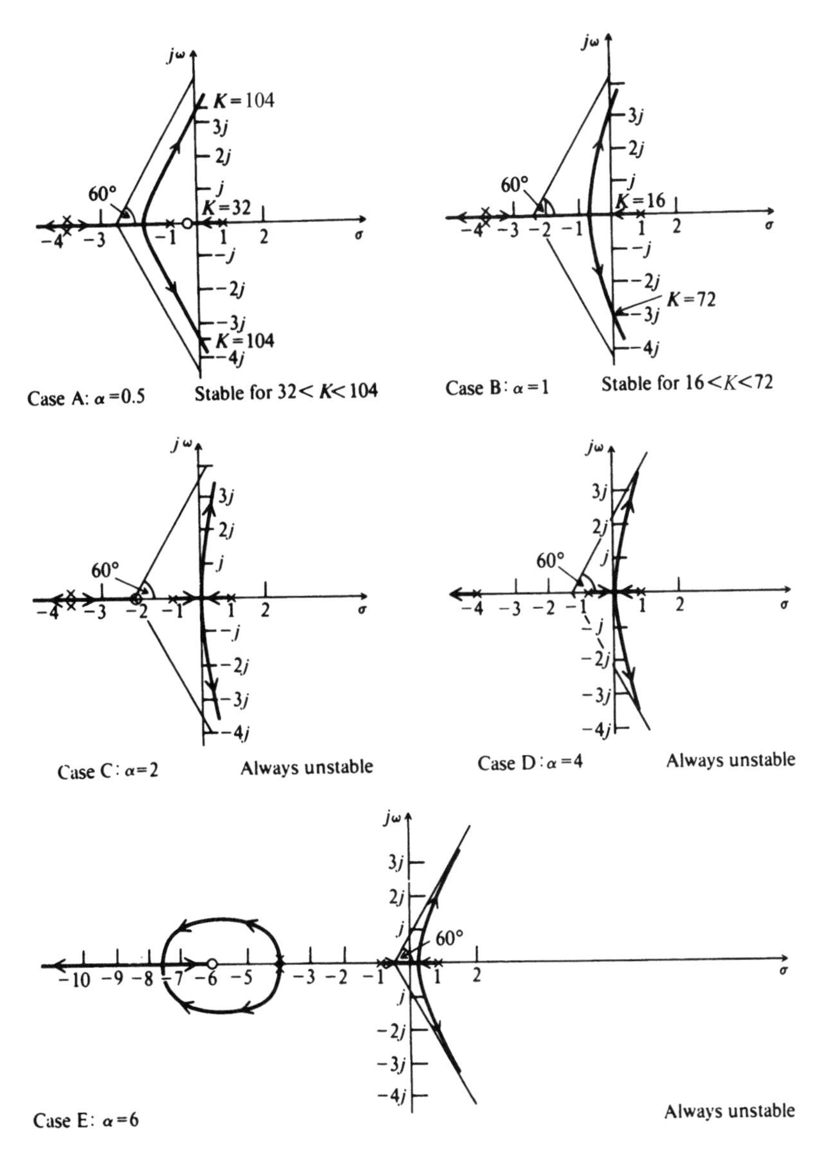

Figure 7.38 Compensation of the root locus shown in Figure 6.60 using a pure zero term where $G_c(s) = (s + \alpha)$.

Rule 2. The root locus starts ($K = 0$) from the poles located at $1, -1$, and a double pole at -4. One branch terminates ($K = \infty$) at the zero located at $\alpha = -0.5$ and three branches terminate at zeros located at infinity.

Rule 3. Complex portions of the root locus occur in complex-conjugate pairs.

Rule 4. The portions of the real axis between 1 to $-\alpha$, -1 to -4, and -4 to ∞ are part of the root locus.

Rule 5. The branches approach infinity as K becomes large at angles given by

$$\alpha_0 = \pm\frac{\pi}{3} = \pm 60^\circ,$$

and

$$\alpha_1 = \pm\frac{3\pi}{3} = \pm 180^\circ.$$

Rule 6. The intersection of the asymptotic lines and the real axis occur at

$$s_r = \frac{-8 - (-0.5)}{4 - 1} = -2.5.$$

Rule 7. Using MATLAB, we found the point of breakaway from the real axis to occur at -1.7549.

Rule 8. Using MATLAB, we found the intersection of the root locus and the imaginary axis to occur at $s = \pm j3.3$, where the gain is 104, and at the origin, where the gain is 32.

Rule 9. This rule does not apply here.

Rule 10. This rule shows that as certain of the loci turn to the right, others turn to the left to ensure that the sum of the roots is a constant.

Rule 11. This rule does not apply here.

Rule 12. The root locus does not cross itself.

The resulting root locus indicates that the system is stable when $32 < K < 104$.

Case B. $\alpha = 1$. See Figure 7.38 for the root-locus sketch. A zero at $s = -1$ cancels the pole at -1. The resulting root-locus sketch indicates that this system is stable when $16 < K < 72$. (The point of breakaway occurs at -0.667.)

Case C. $\alpha = 2$. See Figure 7.38 for the root-locus sketch. The resulting root-locus sketch indicates that the system is always unstable, because at least one of the roots of the characteristic equation always occurs in the right half-plane except for the condition where two poles exist at the origin. (The point of breakaway occurs at the origin.)

Case D. $\alpha = 4$. See Figure 7.38 for the root-locus sketch. A zero at $\alpha = 4$ cancels one of the poles located at -4. the resulting root-locus sketch indicates that the system is always unstable since at least one of the roots of the characteristic equation always occurs in the right half-plane. (The point of breakaway occurs at 0.1196.)

Case E. $\alpha = 6$. See Figure 7.38 for the root-locus sketch. The resulting root-locus sketch indicates that the system is always unstable, because at least one of the roots of the characteristic equation always occurs in the right half-plane. (The points of breakway occur at 0.1156 and −4, and a break-in occurs at −7.0611.)

Conclusions of Cases A through E

Interpretation of Figure 7.38 is quite interesting and revealing. It indicates that the exact location of the zero is very important from a stability viewpoint. Cases A and B were the only configurations which had regions of stability. As a matter of fact, the closer the zero lies to the imaginary axis, the greater is its stabilizing effect. This point is very important. Because Case A resulted in larger values of gain, it would result in a more accurate system and is, therefore, preferred to Case B.

The reader should not get the impression that α could be made extremely small (e.g. 0.0001) to satisfy this guideline because you will be paying a penalty in the size of the capacitors and resistors used in a passive phase-lead network. Remember that the form of the zero is $(s+\alpha)$, which can be modified to $\alpha[(1/\alpha)s+1]$, where $1/\alpha$ is the time constant of the zero term. Therefore, very small values of α mean that the values of the components (resistors and capacitors) would be too large which is undesirable from a practical viewpoint. If the zero term $(s+\alpha)$ is obtained using rate feedback in cascade with position feedback, as in Figure 7.14*a*, then you would have to make $1/\alpha = b$ and very small values of α would mean extremely large values of b. This would require an amplifier in cascade with the rate-feedback sensor (e.g., tachometer or rate gyro), which is undesirable due to the added cost and possible noise problems.

B. Determination of the Transient Response

The transient response of the system can also be obtained by reasoning along these lines: The transient performance is often dominated by the pair of complex conjugate poles located closest to the origin. This occurs when the other poles are far to the left of the dominant poles, or the other poles are near a zero. The resulting transient components due to these other poles are small under these conditions and diminish rapidly. For this case, the poles closest to the origin are conventionally referred to as the dominant poles. The relative dominance of the closed-loop poles is found from the ratio of the real parts of the complex-conjugate poles. As a general guideline, reasonable dominance exists if the ratios of the real parts are at least five, and there are no zeros nearby. For such conditions, the closed-loop complex-conjugate poles closest to the origin will dominate the transient response. From the discussion of Section 4.2, the expression associated with these complex poles can be given by the following expression [see Eq. (4.3)]:

$$\frac{C(s)}{R(s)} = \frac{\omega_n^2}{s^2 + 2\zeta\omega_n s + \omega_n^2} \tag{7.117}$$

where $\zeta =$ damping ratio, and $\omega_n =$ undamped natural frequency. We found in Section 4.2 that the transient response to a unit step input, for $\zeta < 1$, is given by the following expression [see Eq. (4.24)]:

$$c(t) = 1 - \frac{e^{-\zeta\omega_n t}}{\sqrt{1-\zeta^2}} \sin\left(\omega_n \sqrt{1-\zeta^2} t + \alpha\right), \qquad t \geqslant 0 \tag{7.118}$$

where

$$\alpha = \cos^{-1}(\zeta).$$

Figure 4.3 illustrated the complex-plane location of these dominant poles. The values derived for the time to the first peak [see Eq. (4.29)] and maximum percent overshoot [see Eq. (4.33)] are specifically for a second-order system whose closed-loop transfer function is given by Eq. (7.117). These quantities change if other closed-loop poles and zeros exist in addition to the dominant complex pair. However, if the ratios of the real parts of the various complex-conjugate poles are greater than five and there are no zeros nearby, then the approximation gives reasonable results. The damping ratio determined in this case using the pair of complex-conjugate roots closest to the imaginary axis is defined as the *relative damping ratio* of the control system.

Expressions for time to the first peak and percent overshoot, which consider other poles and zeros and give more accurate results, can be derived [6]. These expressions assume that

(a) Other poles are far to the left of the dominant poles, so that the amplitude of transients due to these other poles is small.
(b) Poles which are not far to the left of the dominant poles are near a zero so that the transient amplitude due to such poles is small.

The expressions, for unity-feedback systems, are given by

$$\left[t_p\right]_{\text{modified}} = \frac{1}{\sqrt{1-\zeta^2}\omega_n}\left[\frac{\pi}{2} - \sum \phi_z + \sum \phi_p\right] \tag{7.119}$$

where

$\sum \phi_z$ = sum of the angles from the zeros of C/R to one of the dominant poles,
$\sum \phi_p$ = sum of the angles from the zeros of C/R to one of the dominant poles,

and the maximum percent overshoot

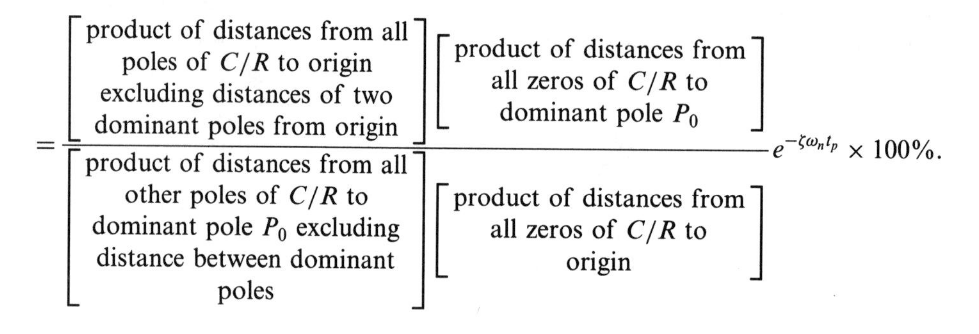

$$= \frac{\begin{bmatrix}\text{product of distances from all} \\ \text{poles of } C/R \text{ to origin} \\ \text{excluding distances of two} \\ \text{dominant poles from origin}\end{bmatrix}\begin{bmatrix}\text{product of distances from} \\ \text{all zeros of } C/R \text{ to} \\ \text{dominant pole } P_0\end{bmatrix}}{\begin{bmatrix}\text{product of distances from all} \\ \text{other poles of } C/R \text{ to} \\ \text{dominant pole } P_0 \text{ excluding} \\ \text{distance between dominant} \\ \text{poles}\end{bmatrix}\begin{bmatrix}\text{product of distances from} \\ \text{all zeros of } C/R \text{ to} \\ \text{origin}\end{bmatrix}} e^{-\zeta\omega_n t_p} \times 100\%.$$

The expression for maximum percent overshoot can be stated symbolically as

$$\text{max.\% overshoot} = \left[\left(\frac{P_1}{|P_1 - P_0|}\right)\left(\frac{P_2}{|P_2 - P_0|}\right)\left(\frac{P_3}{|P_3 - P_0|}\right)\cdots\right] \times \left[\left(\frac{|Z_1 - P_0|}{Z_1}\right)\left(\frac{|Z_2 - P_0|}{Z_2}\right)\left(\frac{|Z_3 - P_0|}{Z_3}\right)\cdots\right] e^{-\zeta\omega_n t_p} \times 100\%, \tag{7.120}$$

where the first set of brackets represents the product of the ratios of the values of s at which poles occur to their absolute distances from the dominant pole. The second set of brackets represents the product of the ratios of the absolute distances of the zeros from the dominant pole and the values of s at which the zeros occur. Let us next apply these expressions in the following design problem.

C. Example of Phase-Lag-Network Compensation and Overall System Performance

The concluding design problem we consider using the root locus consists of employing cascaded phase-lag compensation in order to improve the steady-state performance of a feedback control system. The object is to increase its gain while maintaining a good dynamic response. Specifically, we consider the system whose root locus was illustrated in Figure 6.57. For this system

$$G(s)H(s) = \frac{K}{s(s+4)(s+5)}. \tag{7.121}$$

The root locus of Figure 6.57 indicated that the system was stable when $0 < K < 180$. Let us assume that a damping factor of 0.707 achieves a desirable dynamic response for this system. In addition, we must maintain a velocity constant K_v of 30 in order to meet specified accuracy requirements. Analyzing this problem, by means of the root locus, we can find the value of K which will give the required damping factor. For example, the redrawn version of Figure 6.57 shown in Figure 7.39 indicates that a $K = 21.59$ will result in a damping ratio of 0.707 ($\alpha = \cos^{-1} 0.707 = 45°$). This value of gain does not maintain the required velocity constant of 30. The actual value of K_v resulting from $K = 21.59$ is

$$K_v = \lim_{s\to 0} sG(s)H(s) = \lim_{s\to 0} \frac{s(21.59)}{s(s+4)(s+5)} = 1.08/\text{sec}. \tag{7.122}$$

It is therefore clear that we cannot just increase the gain K to a value that produces the required velocity constant, because this would decrease the damping factor and adversely affect the transient response or cause the system to become unstable. Using the root locus for a solution, we show how these two conflicting factors can be resolved.

In order to achieve the specification requirements, we must increase the gain, but at the same time maintain the dominant complex-conjugate roots of the root locus where the value of $K = 21.59$ is shown in Figure 7.39 so that the damping ratio of 0.707 is maintained. We can accomplish this with a phase-lage network and an increase in the system gain.

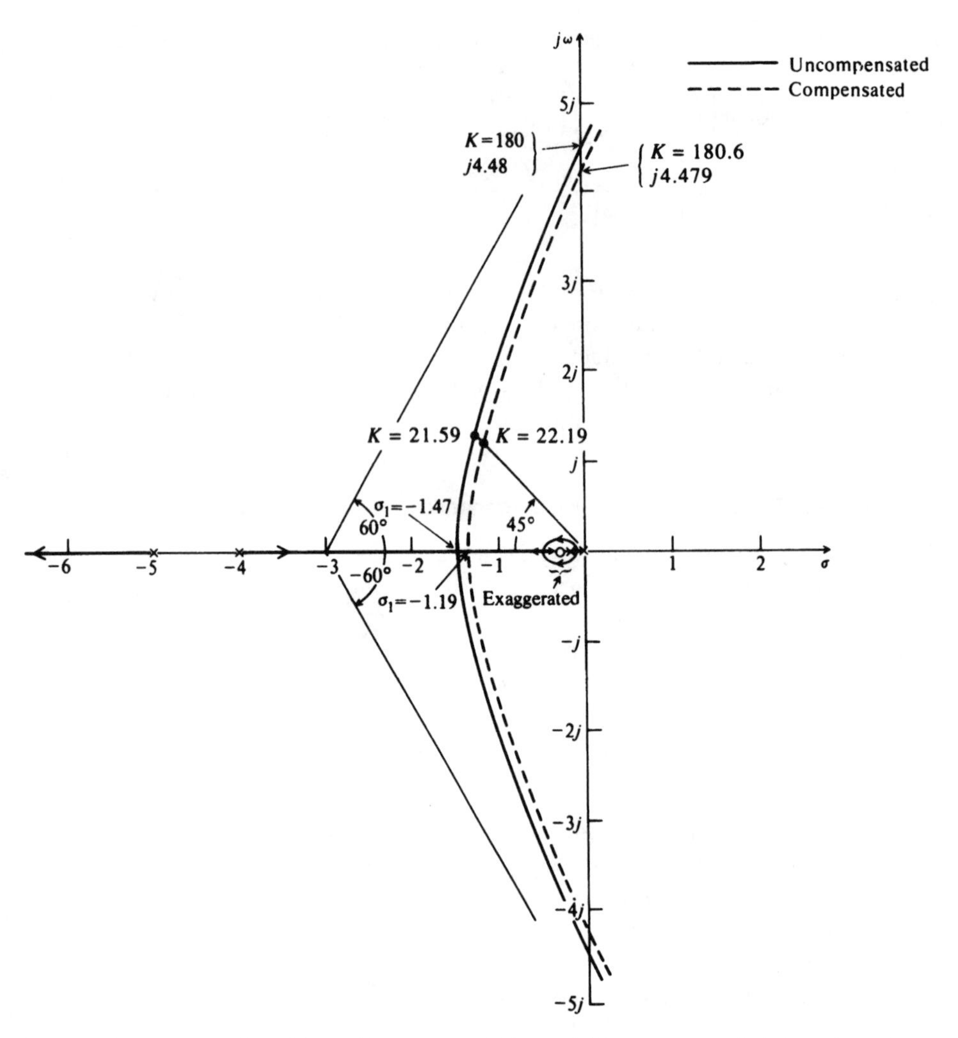

Figure 7.39 Compensation of the root locus shown in Figure 6.57 using a cascaded phase-lag network and an increase in gain.

Let us assume that the combined transfer function representing the increase in gain (n) and the phase-lag network is given by

$$G_c(s) = \frac{s + n\alpha}{s + \alpha}, \tag{7.123}$$

The form of Eq. (7.123) indicates that this compensator provides a low-frequency gain of n in addition to the phase lag, where n is also the ratio of the break frequencies. The open-loop transfer function of the compensated system is given by

$$G_c(s)G(s)H(s) = \frac{K}{s(s+4)(s+5)}\left(\frac{s + n\alpha}{s + \alpha}\right). \tag{7.124}$$

In general, the distances of $n\alpha$ and α from the origin in the s-plane are chosen to be small compared with the distances of the other zeros and poles of the uncompensated open-loop transfer function, so that the added pole and zero of the compensator will not contribute significant phase lag in the vicinity of the gain crossover frequency. This result is quite clear from a study of the Bode diagram, which we shall shortly do at the conclusion of this design (see Figure 7.40). Certainly we do not wish to add the phase-lag contribution in the vicinity of the gain crossover frequency. Therefore, the combination of pole and zero will be quite close together on the root locus and very close to the origin. The combination is usually called a dipole.

In order to complete the design, α is chosen close to the origin at 0.01 and n is chosen, using the following derivation, to achieve a $K_v = 30$:

$$K_v = 30 = \lim_{s \to 0} sG_c(s)G(s)H(s). \tag{7.125}$$

Substituting Eq. (7.124) into Eq. (7.125), we obtain

$$30 = \lim_{s \to 0} s\left(\frac{K}{s(s+4)(s+5)}\right)\left(\frac{s+n\alpha}{s+\alpha}\right),$$

or

$$30 = \frac{Kn}{(4)(5)}.$$

Because we desire that $K = 21.59$ from a transient viewpoint, we must have

$$n = \frac{30(4)(5)}{21.59} = 27.79. \tag{7.126}$$

The completed root locus for the compensated system whose open-loop transfer function is given by

$$G_c(s)G(s)H(s) = \frac{K}{s(s+4)(s+5)}\left(\frac{s+0.2779}{s+0.01}\right) \tag{7.127}$$

must now be determined. Because the dipole is added near the origin, the original root locus is not changed significantly, because the two poles and the zero near the origin tend to merge into a single pole.

Let us next determine the new resulting root locus, which is shown as the dashed curve in Figure 7.39, and analyze the effect of the dipole on it. Specifically, we wish to know whether the new root locus will indeed have $K_v = 30$. In addition, we would like to determine the transient response of the compensated system. Each of the 12 rules for constructing the root locus will be considered.

Rule 1. There are four separate branches, because the characteristic equation,

$$1 + G(s)H(s) = 0,$$

is a fourth-order equation.

Rule 2. The root locus starts ($K = 0$) from the poles located at the origin, -0.01, -4, and -5. One branch terminates ($K = \infty$) at the zero located at -0.2779 and the other three branches terminate at zeros which are located at infinity.

Rule 3. Complex portions of the root locus occur in complex-conjugate pairs.

Rule 4. The portions of the real axis between the origin and -0.01, -0.2779 and -4, and -5 to $-\infty$ are parts of the root locus.

Rule 5. The four branches approach infinity as K becomes large at angles given by

$$\alpha_0 = \pm\frac{\pi}{4-1} = \pm 60^\circ,$$
$$\alpha_1 = \pm\frac{3\pi}{4-1} = \pm 180^\circ.$$

Rule 6. The intersection of the aymptotic lines and the real axis occurs at

$$s_r = \frac{-9.01 - (-0.2779)}{4-1} = \frac{-8.73}{3} = -2.91.$$

Rule 7. The point of breakaway from the real axis can be computed from the following equation:

$$(\beta_1 - \beta_2 - \beta_3 - \beta_4 - \beta_5) = (2n+1)\pi,$$

where

$\beta_1 =$ angle from the zero at -0.2779 to the point s_1 that is located a small distance δ off the positive real axis,
$\beta_2 =$ angle from the pole at the origin to the point s_1,
$\beta_3 =$ angle from the pole at -0.01 to the point s_1,
$\beta_4 =$ angle from the pole at -4 to the point s_1,
$\beta_5 =$ angle from the pole at -5 to the point s_1.

The equation of the transition of the root locus from the real axis to a point s_1 which is a small distance δ off the axis is given by

$$\left[\left(\pi - \frac{\delta}{\sigma_1 - 0.2779}\right) - \left(\pi - \frac{\delta}{\sigma_1}\right) - \left(\pi - \frac{\delta}{\sigma_1 - 0.01}\right) - \left(\frac{\delta}{4-\sigma_1}\right) - \left(\frac{\delta}{5-\sigma_1}\right)\right] = -\pi.$$

Solving, we obtain $\sigma_1 = -1.19$, which compares with a value of -1.47 obtained from Eq. (6.151) for the uncompensated system.

Rule 8. The intersection of the root locus and the imaginary axis can be determined by applying the Routh–Hurwitz stability criterion to the characteristic equation

$$s(s+0.01)(s+4)(s+5)+K(s+0.2779)=0,$$

which becomes

$$s^4+9.01s^3+20.09s^2+(K+0.2)s+0.2779K=0.$$

The resulting Routh–Hurwitz array is given by

$$\begin{array}{c|ccc} s^4 & 1 & 20.09 & 0.2779K \\ s^3 & 9.01 & K+0.2 & \\ s^2 & 20.068-0.1111K & 0.2779K & \\ s & \dfrac{-0.111K^2+17.55K+4}{20.068-0.111K} & & \\ s^0 & 0.2779K & & \end{array}$$

An interesting situation occurs in this Routh–Hurwitz array, because the first terms of the third and fourth rows can go to zero for certain values of gain, K. When the equation

$$20.068-0.1111K=0$$

is satisfied, then a possible solution is

$$K_{\max}=180.6.$$

When the equation

$$\frac{-0.111K^2+17.55K+4}{20.068-0.111K}=0$$

is satisfied, then a possible solution is

$$K_{\max}=158.25.$$

Therefore, in order to find which is (or are) valid, let us substitute $s=j\omega$ into the characteristic equation and find out where the root locus crosses the imaginary axis. The results can be separated into a real and imaginary part and written in the following form:

$$\omega^4-20.09\omega^2+0.2779K+j\omega[-9.01\omega^2+K+0.2]=0. \tag{7.128}$$

For K to be real and imaginary parts of this equation must each equal zero. Therefore, from the imaginary part:

$$j\omega[-9.01\omega^2 + K + 0.2] = 0,$$

$$\omega = \pm\sqrt{\frac{K + 0.2}{9.01}}.$$

Now, to find the value of K which corresponds to this value of crossing of the imaginary axis, let us substitute this value of ω into the real part of Eq. (7.128). The result is the following equation:

$$K^2 - 180.33K - 36.16 = 0.$$

Therefore, we find that

$$K = 180.6$$

is the only possible real value of gain when the root locus crosses the imaginary axis. The analysis indicates that the maximum value of gain, before the system becomes unstable, is 180.6. In Eq. (6.155), we found that the uncompensated system had a maximum allowable gain of 180. Therefore, this result indicates that the dipole has an effect on the maximum allowed gain. The corresponding value of s occuring at the crossing of the imaginary axis is found to be $\pm j4.479$ by substituting $K_{\max}$ into the equation for ω. This compares with a value of $s = j4.48$ obtained previously for the uncompensated case [see Eq. (6.156)].

Rule 9. This rule does not apply to this problem.

Rule 10. This rule shows that as certain of the loci turn to the right, others turn to the left to ensure that the sum of the roots is a constant.

Rule 11. This rule is quite important to us in this case, because we want to determine the value of gain when $\zeta = 0.707$ (the intersection of a line making an angle of $+45°$ with the negative real axis and the dashed curve). For the uncompensated case, we found that $K = 21.59$. The new value can be obtained from the following expression:

$$\left|\frac{K(s + 0.2779)}{s(s + 4)(s + 5)(s + 0.01)}\right| = 1.$$

Measuring the distance from the various poles and zeros to the point of interest, we obtain

$$K = 22.19.$$

Therefore, the gain has increased slightly from 21.59 to 22.19. This will result in a slight increase of K_v from 30 to 30.8.

Rule 12. The root locus does not cross itself.

To conclude this problem, we can calculate the value of the time to the first peak and the maximum percent overshoot from Eqs. (7.119) and (7.120), respectively. These results can then be compared with the values obtained from Eqs. (4.29) and (4.33), which assumes that the transient response is completely controlled only by the pair of complex-conjugate poles located closest to the origin. The time to the first peak of the compensated system can be calculated from Eq. (7.119). In order to use this equation, the location of the other two roots must be determined. Using the computer programs (see Sections 6.17 and 6.18) for determining the root locus, these were found to be located at −6.45 and −0.26. Therefore,

$$t_p = \frac{1}{\left(\sqrt{1-\zeta^2\omega_n}\right)}\left(\frac{\pi}{2} - \sum\phi_z + \sum\phi_p\right)$$

where

$$\begin{aligned}
\zeta &= 0.707, \\
\omega_n &= 1.63 \text{ (distance from the origin of the complex plane to the two} \\
&\qquad \text{dominant poles located at } 1.15 \pm j1.15\text{—dashed curve of Figure 7.39)}, \\
\sum\phi_z &= 141^\circ = 2.47 \text{ rad}, \\
\sum\phi_p &= 136^\circ + 90^\circ + 14^\circ = 240^\circ = 4.19 \text{ rad}.
\end{aligned}$$

Substituting these values into the equation for t_p, we obtain

$$t_p = \frac{1}{(\sqrt{1-0.707^2 1.63})}(1.57 - 2.47 + 4.19) = 2.86 \text{ sec}.$$

Therefore, the time to the first peak is 2.86 sec. If one simply assumes that the transient response is governed by the pair of complex-conjugate poles located at $1.15 \pm j1.15$ and uses Eq. (4.29) to determine the time to the first peak, then the following is obtained:

$$t_p = \frac{\pi}{\omega_n\sqrt{1-\zeta^2}} \quad \text{[from Eq. (4.29)]},$$

$$t_p = \frac{\pi}{1.63\sqrt{1-0.707^2}} = \frac{3.14}{1.15} = 2.73 \text{ sec}.$$

Therefore, we see the slightly improved accuracy obtained using Eq. (7.119).

A similar analysis of the uncompensated system, utilizing Eq. (7.119), results in a time to the first peak of 2.81 sec. For this case, the two complex-conjugate poles are located at $-1.2 \pm j1.2$. The third root can be determined analytically from rule 10:

$$\begin{aligned}
\sum p_i &= \sum r_j, \\
-4 - 5 &= -1.2 + j1.2 - 1.2 - j1.2 + r, \\
r &= -6.6.
\end{aligned}$$

Therefore,

$$t_p = \frac{1}{\sqrt{1-\zeta^2\omega_n}}\left(\frac{\pi}{2} - \sum \phi_z + \sum \phi_p\right),$$

where

$$\begin{aligned}\zeta &= 0.707,\\ \omega_n &= 1.7 \text{ (distance from the origin of the complex plane to the two dominant poles—solid curve of Figure 7.39)},\\ \sum \phi_z &= 0,\\ \sum \phi_p &= 90^\circ + 13^\circ = 103^\circ = 1.8 \text{ rad.}\end{aligned}$$

Hence,

$$t_p = \frac{1}{\sqrt{1-0.707^2(1.7)}}(1.57 + 1.8) = 2.81 \text{ sec.}$$

Observe that the dipole compensation increases the time to the first peak slightly.

The maximum percent overshoot of the compensated system can be obtained from Eq. (7.120). To use this equation, we have to determine the location of the other two roots. As mentioned previously, these were found to be at -6.45 and -0.26. Therefore,

$$\begin{aligned}\text{maximum percent overshoot} &= \frac{P_1P_2}{(|P_1 - P_0|)(|P_2 - P_0|)}\frac{|Z_1 - P_0|}{Z_1}e^{-\zeta\omega_n t_p} \times 100\%\\ &= \frac{(0.26)(6.45)}{(1.45)(4.9)}\frac{(1.46)}{(0.2779)}e^{-0.707(1.63)2.86} \times 100\%\\ &= 4.52\%.\end{aligned}$$

Therefore, the resulting maximum percent overshoot is only 4.52%. If one simply assumes that the transient response is governed by the pair of complex-conjugate poles located at $1.15 \pm j1.15$ and uses Eq. (4.33) to determine the maximum percent overshoot, then the following is obtained:

$$\begin{aligned}\text{maximum percent overshoot} &= \left(e^{-\zeta\pi/\sqrt{1-\zeta^2}}\right) \times 100\%\\ &= \left(e^{-0.707\pi/\sqrt{1-0.707^2}}\right) \times 100\% = 4.3\%.\end{aligned}$$

Therefore, we see the slightly increased accuracy obtained using Eq. (7.120).

A similar analysis of the uncompensated system utilizing Eq. (7.120) results in a maximum percent overshoot of 4.48%; for this case, the third root is located at -6.6 as illustrated before:

$$\begin{aligned}\text{maximum percent overshoot} &= \frac{P_1}{|P_1 - P_0|}e^{-\zeta\omega_n t_p} \times 100\%\\ &= \frac{6.6}{5}e^{-0.707(1.7)(2.81)} \times 100\% = 4.48\%.\end{aligned}$$

Observe that the dipole compensation increases the maximum percent overshoot slightly.

The results of the transient analysis indicate that the effect of the dipole is to increase the time to the first peak slightly and to increase the maximum overshoot slightly. Of most importance, the dipole increases the velocity constant greatly, from 1.08 to 30.8.

One should note that the dominant pole concept is very useful here, the results coming quite close to the more accurate calculation. In fact, due to the ever-present uncertainty of the parameters of the actual system, it is rarely, if ever, necessary to carry out the detailed calculations indicated.

D. Comparison of the Root-Locus Method with Bode Diagrams

In our analysis of the system whose transfer function was given by Eq. (7.121) and illustrated in Figure 7.39 using the root-locus method, it was pointed out that this same system's analysis using the Bode diagram would be very interesting. It is always interesting to compare the analysis of a system using more than one of the techniques shown as was discussed in Section 6.20, where 12 commonly used transfer functions were compared from the viewpoints of the Nyquist-diagram, Bode-diagram, Nichols-chart, and root-locus methods. In this problem, it is especially useful because the root-locus solution may be difficult to comprehend initially. What did we really do when we kept the dominant complex-conjugate roots almost fixed, increased the gain and added a phase-lag network? It is very clear if we look at the uncompensated and compensated systems from the viewpoint of the Bode diagram. For purposes of the Bode-diagram analysis, the value of K in Eq. (7.121) is set equal to 21.59, and this equation is modified to the following easier form for a Bode diagram:

$$G(s)H(s) = \frac{1.08}{s(0.25s+1)(0.2s+1)}. \tag{7.129}$$

The Bode diagram of the uncompensated system is shown in Figure 7.40. This Bode diagram was obtained using MATLAB (see Section 6.8) and is contained in the M-file part of my MCSTD Toolbox disk and can be retrieved free from The MathWorks, Inc. anonymous FTP server at ftp://ftp.mathworks.com/pub/books/shinners. Notice from this diagram that this system has a gain crossover frequency of 1.025 rad/sec, a phase margin of 64.04°, and a gain margin of 18.42 dB (phase crossover frequency occurs at 4.472 rad/sec). Extension of the initial −20-dB/decade slope intersects the 0-dB line at 1.08 rad/sec indicating a K_v of 1.08 which agrees with Eq. (7.122). (The actual curve intersects the 0-dB line at 1.025 rad/sec.) We now wish to plot the Bode diagram of the compensated system whose transfer function is given by Eq. (7.127). We let $K = 22.19$ and modify Eq. (7.127) into the following equation which is easier to use for the Bode diagram:

$$G_c(s)G(s)H(s) = \frac{30.8(3.60s+1)}{s(0.25s+1)(0.2s+1)(100s+1)}. \tag{7.130}$$

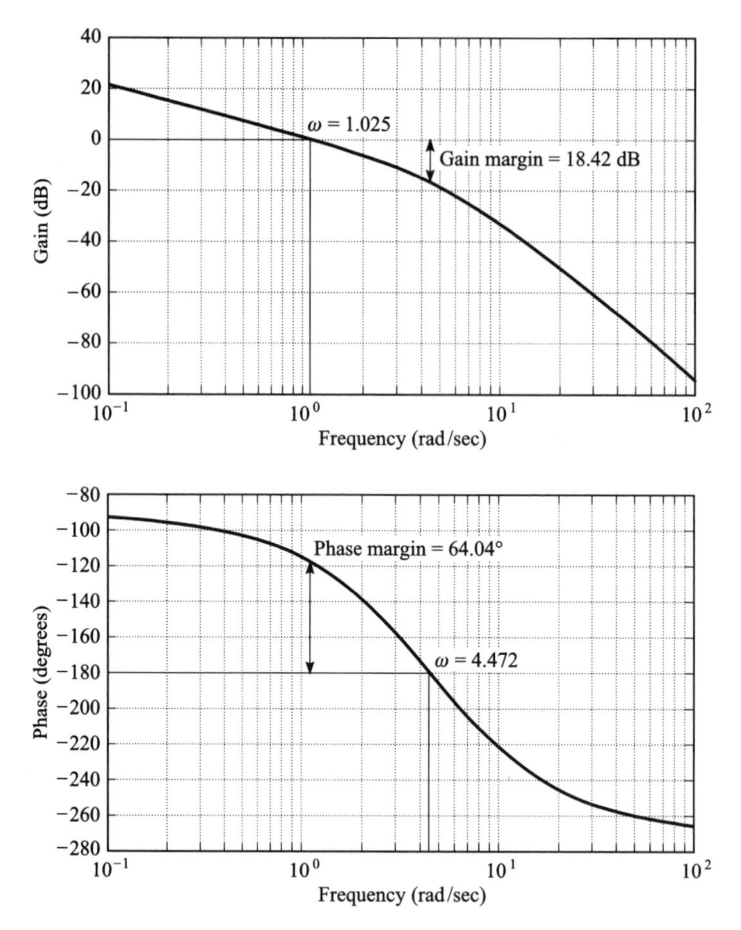

Figure 7.40 Bode diagram for the uncompensated system whose root locus is shown in Figure 7.39.

The resulting Bode diagram for the compensated system, also obtained using MATLAB, whose transfer function is given by Eq. (7.130) is shown in Figure 7.41. It has a gain crossover frequency of 1.138 rad/sec which is close to that of the uncompensated system whose Bode diagram is shown in Figure 7.40 (1.025 rad/sec), because we hardly moved the dominant complex-conjugate poles (see Figure 7.39) in going from the uncompensated to the compensated system. The phase margin of the compensated system is 49.21° (compared to 64.04° of the uncompensated system), and the gain margin of the compensated system is 16.67 dB (compared to 18.42 dB of the uncompensated system). Therefore, the uncompensated system and the compensated system have almost the same relative stability, and about the same gain crossover frequency. From the root-locus viewpoint, this means that the damping factor has hardly changed in going from the uncompensated to compensated systems. (Figure 7.39 shows that the damping factor remains constant at 0.707). However, the velocity constant of the compensated system is 30.8 [see Figure 7.41] compared to the velocity constant of the uncompensated system which is 1.08 [see Eq. (7.122)]. In conclusion, the comparison of the stabilization of this problem using the root locus (Figure 7.39) and the Bode diagram (Figures

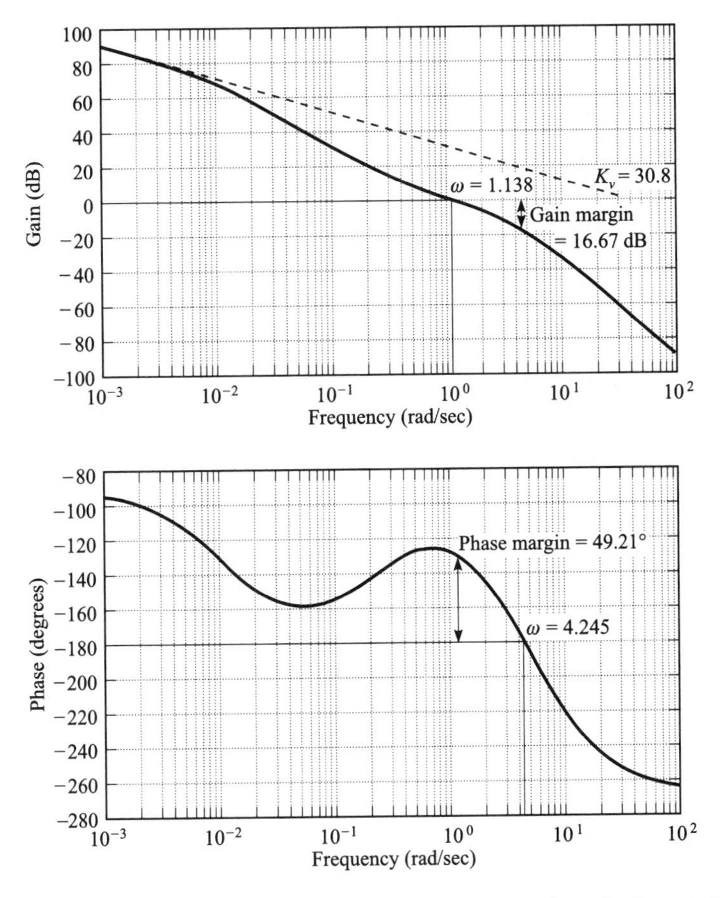

Figure 7.41 Bode diagram for the compensated system whose root locus is shown in Figure 7.39.

7.40 and 7.41) is very interesting as the information shown in the root-locus and Bode diagrams complement each other.

7.10. TRADEOFFS OF USING VARIOUS CASCADE-COMPENSATION METHODS AND MINOR-LOOP FEEDBACK

This chapter has presented the use of phase-lag, phase-lead, and phase-lag–lead cascade compensation networks, and minor-loop feedback, compensation (such as rate feedback). The question arises as to which compensation method should be used: cascade compensation or minor-loop feedback compensation. If the answer turns out to be cascade compensation, the question arises as to whether we should use phase-lag, phase-lead, or phase-lag–lead networks? This section addresses these reasonable questions by discussing the tradeoffs of these different methods of compensation, and guidelines are provided for the practical design of control systems.

I cannot provide a simple answer as to what compensation approach is the "best" method to use. Table 7.2 [11] compares the advantages and disadvantages of phase-

Table 7.2. Comparison of Compensation Methods [15]

	Phase-lag network	Phase-lead network	Phase-lag–lead network	Minor-loop rate feedback
Characteristics	a. Decreases system bandwidth	a. Increases system bandwidth	a. Increases low-frequency gain	a. Increases system bandwidth
	b. Increases low-frequency gain	b. Increases high-frequency gain	b. Increases high-frequency gain	b. Increases high-frequency gain
	c. Decreases ω_n	c. Increases ω_n	c. Increases ω_n	c. Increases ω_n
	d. Increases ζ	d. Increases ζ	d. Increases ζ	d. Increases ζ
Performance advantages	a. Improves stability	a. Improves stability	a. Improves stability	a. Improves stability
	b. Decreases high-frequency noise	b. Reduces settling time T_s	b. Reduces steady-state error	b. Reduces settling time T_s
	c. Reduces steady-state error		c. Reduces settling time T_s	c. Requires small volume
				d. Relatively independent of environment (altitude and temperature)
				e. Permits the isolation of undesirable dynamics in one portion of a control system from the complete system

Table 7.2. (continued)

	Phase-lag network	Phase-lead network	Phase-lag–lead network	Minor-loop rate feedback
Performance disadvantages	a. Increases settling time T_s b. Usually requires larger values of resistors and capacitors to achieve desired network time constants	a. Requires additional amplifier gain b. Increases high-frequency noise c. May require large values of resistors and capacitors to achieve desired network time constants	a. Increases high-frequency noise b. Usually requires large values of resistors and capacitors to achieve desired network time constants	a. Increases the steady-state error b. Increases high-frequency noise but less than resulting from a phase-lead network
Relative Cost	Least expensive	Moderately expensive	Relatively inexpensive	Usually the most expensive
Applicability	a. Very applicable when it is desired to increase the steady-state constants b. Cannot be used when the uncompensated phase at low frequencies does not equal the desired phase margin	a. Very useful when a fast transient response is desired	a. Very applicable when it is desired to increase the steady-state constants b. Very useful when a fast transient response is desired	a. Very useful when a fast transient response is desired b. Very useful for hostile environments requiring small space c. Useful for isolating undesirable dynamics

Reprinted from *Control Engineering*, May 1978.

lag, phase-lead, and phase-lag–lead networks, and minor-loop rate feedback for compensation. They are compared on the basis of characteristics, performance, applicability, and relative cost. Each of these four approaches will improve system stability if used properly, as shown in Chapters 6 and 7. Therefore, each system must be considered separately in order to determine the most effective, optimimum, or "best" approach.

If the system specification requires the need for a low-noise ("quiet") control system having a small steady-state error, then the phase-lag–network approach is the most desirable from the viewpoints of performance and cost. The phase-lead network is desirable for systems requiring a very fast response, and if the resulting increase in high-frequency noise is tolerable. The phase-lag–lead network has essentially the advantages in performance of both the phase-lag and phase-lead networks. It can result in control systems which are very fast, on a relative basis, and they can result in very small steady-state errors.

Minor-loop feedback compensation using rate-feedback from tachometers and rate gyros has the advantage that it results in a pure zero, when connected in parallel with the position feedback, and represents ideal compensation. Rate-compensating devices have the additional advantage of only requiring a small volume and they are relatively independent of the environmental conditions in which they are operating such as temperature, humidity, and altitude. This is advantageous for control systems operating in space vehicles, airplanes, and ships. For example, if a phase-lead network were used in a hostile environment, the possible use of very large resistors and capacitors could result in very large time-constant variations causing a stable control system to behave as a conditionally stable control system.

In trading off the advantages and disadvantages of minor-loop feedback compensation, it should be emphasized that although they result in systems having the highest performance, this approach is usually also the most expensive. However, in some cases, it may be the only and "best" way to go.

In the practical world of control-system applications, each case is not so straightforward that the control-system engineer can merely go to Table 7.2 and choose an approach clearly. Conflicting system requirements will usually direct the control-system engineer to more than one possible solution, and neither one of them may be the "best" method. In the final design, the choice of the compensating approach will usually be a compromise based on the tradeoffs considered.

The control-system engineer will usually select the approach based on his or her experience, subjective personal preferences, and the availability of components. As was discussed in the case of choosing the "best" method for stability analysis in Section 6.22, all of the compensation methods should be considered before one is selected. Then the "best" or most effective technique can be selected by the control-system engineer for the final design.

7.11. ILLUSTRATIVE PROBLEMS AND SOLUTIONS

This section provides a set of illustrative problems and their solutions to supplement the material presented in Chapter 7.

I7.1. A closed-loop unity–feedback control system used to position a load has a forward transfer function given by

$$G(s) = \frac{10}{2s+1}.$$

The time constant of the open-loop process is 2 sec. Determine the time constant of the closed-loop control system.

SOLUTION:

$$\frac{C(s)}{R(s)} = \frac{G(s)}{1 + G(s)H(s)} = \frac{\dfrac{10}{2s+1}}{1 + \dfrac{10}{2s+1}} = \frac{10}{2s+11}.$$

Therefore,

$$\frac{C(s)}{R(s)} = \frac{10}{11\left(\dfrac{2}{11}s + 1\right)} = \frac{10}{11(0.1818s + 1)}.$$

So, the time constant of the closed-loop control system is 0.1818 sec.

I7.2. We wish to analyze the performance of a unity-feedback second-order control system whose forward transfer function represents a process, $G_p(s)$, given by:

$$G_p(s) = \frac{500}{s(s+10)}.$$

(a) Determine the gain crossover frequency, ω_c, analytically without using the graphical Bode plot.

(b) Determine the phase margin and gain margin of this control system analytically without using the graphical Bode plot.

(c) A phase-lead network compensation, $G_c(s)$, given by

$$G_c(s) = \frac{(1 + aTs)}{(1 + Ts)}$$

is to be added in series with the process' transfer function, $G_p(s)$. Determine the values of a and T in order that the zero factor of $G_c(s)$ cancels the pole of $G_p(s)$ at $s = -10$, and the damping ratio of the control system is unity.

(d) Determine the gain crossover frequency, ω_c, analytically without using the graphical Bode plot for the compensated control system.

(e) Determine the phase margin and gain margin of this control system analytically without using the graphical Bode plot.

SOLUTION: (a)

$$\left|\frac{500}{j\omega(j\omega+10)}\right| = 1,$$

$$\left|\frac{500}{-\omega^2+10j\omega}\right| = 1.$$

Solving, we find that the gain crossover frequency, ω_c, equals 21.27 radians per second.

(b)

$$\Phi(\omega) = -90^\circ - \tan^{-1}\tfrac{\omega_c}{10} = -90^\circ - \tan^{-1} 2.127 = -155.7^\circ.$$

Therefore, the phase margin, γ, equals $180^\circ + \phi(\omega) = 24.3^\circ$.

The gain margin equals infinity because the phase of $\phi(\omega)$ is always less than -180°.

(c)

$$G(s) = G_c(s)G_p(s) = \frac{500(1+aTs)}{s(s+10)(1+Ts)} = \frac{500aT(s+1/aT)}{Ts(s+10)(s+1/T)}.$$

Therefore, we make

$$\frac{1}{aT} = 10.$$

The characteristic equation of the resulting control system is determined as follows:

$$\frac{C(s)}{R(s)} = \frac{\dfrac{500a}{s(s+\frac{1}{T})}}{1+\dfrac{500a}{s(s+\frac{1}{T})}} = \frac{500a}{s^2+\frac{1}{T}s+500a}.$$

Therefore, comparing the denominator of this equation with Eq. (4.3):

$$\omega_n^2 = 500a$$

$$2\zeta\omega_n = \frac{1}{T}$$

where $\zeta = 1$. Solving these two equations simultaneously and the previous result that $1/aT = 10$, we find that $a = 20$ and $T = 0.005$. Therefore,

$$G_c(s) = \frac{1+0.1s}{1+0.005s},$$

$$G(s) = G_c(s)G_0(s) = \frac{10,000}{s(s+200)}.$$

(d)

$$\frac{10,000}{j\omega(j\omega + 200)} = 1,$$

$$\frac{10,000}{-\omega^2 + 200j\omega} = 1.$$

Solving, we find that the gain crossover frequency of the compensated control system, ω_c, equals 48.6 rad/sec.

(e)

$$\phi(\omega) = -90^\circ - \tan^{-1}\frac{\omega_c}{200} = -90^\circ - \tan^{-1} 0.243 = -90^\circ - 13.66^\circ = -103.66^\circ.$$

Therefore, the phase margin, γ, equals $180^\circ + \phi(\omega) = 76.34^\circ$. The gain margin equals infinity because the phase of $\phi(\omega)$ is always less than 180°.

I7.3. Due to the inherent instability of a control system containing only a double integration, we wish to add a phase-lead network to the following control system:

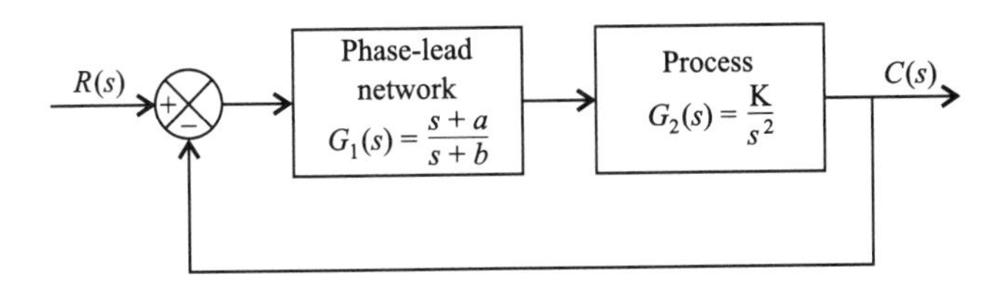

Figure I7.3i

Due to accuracy considerations, we desire the gain of the control system, K, to be 229.6. Due to component considerations, we wish the pole of the phase-lead network, b, to be located at 27. Using the root-locus method, determine the value of the zero of the phase-lead network, a, if the dominant poles of this control system are located at $-5.288 + 5.288j$ and $-5.288 - 5.288j$?

SOLUTION: The root locus for this problem is given by the plot in Figure I7.3ii.

From Eq. (6.133), we know that the absolute value of the open-loop transfer function, $G_1(s)G_2(s)$ must equal one. Therefore,

$$\left|\frac{K(s+a)}{s^2(s+27)}\right| = 1.$$

This equation can be written as follows for $K = 229.6$ and at $s = -5.288 + 5.288j$:

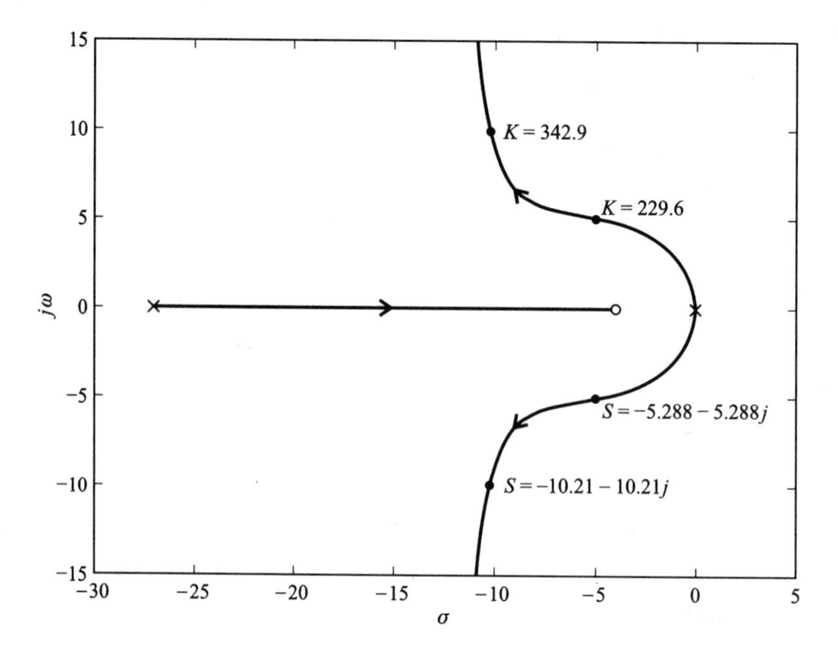

Figure I.73ii

$$\frac{229.6\sqrt{5.288^2 + (a - 5.288)^2}}{\sqrt{(5.288^2 + 5.288^2)^2}\sqrt{5.288^2 + (27 - 5.288)^2}} = 1$$

Therefore, $a = -4$.

PROBLEMS

7.1. Determine the circuit structure, the values of resistance and capitance, the gains of any amplifiers required, and the complex-plane plot for first-order networks having the following characteristics:

(a) Phase lead of 60° at $\omega = 4$ rad/sec, a minimum input impedance of 50,000 Ω, and an attentuation of 10 dB at dc.

(b) Phase lag of 60° at $\omega = 4$ rad/sec, a minimum input impedance of 50,000 Ω, and a high-frequency attenuation of 10 dB.

(c) A phase-lag-lead network having an attenuation of 10 dB for a frequency range of $\omega = 1$ to $\omega = 10$ rad/sec and an input impedance of 50,000 Ω.

In all cases, limit the maximum values of resistance to 1 MΩ and capitance to 10 μF. Furthermore, assume that the loads on the networks have essentially infinite impedance.

7.2. The system illustrated in Figure P7.2 consists of a unity-feedback loop containing a minor-rate-feedback loop.

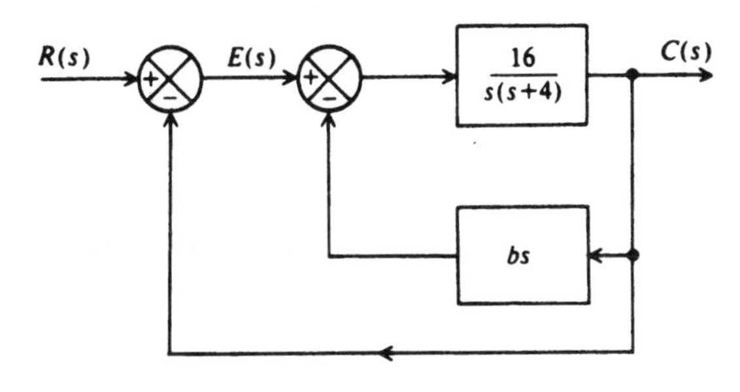

Figure P7.2

(a) Without any rate feedback ($b = 0$), determine the damping ratio, undamped natural frequency, peak overshoot of the system to a unit step input, and the steady-state error resulting from a unit ramp input.

(b) Determine the rate-feedback constant b which will increase the equivalent damping ratio of the system to 0.8.

(c) With rate feedback and a damping ratio of 0.8, determine the maximum percent overshoot of the system to a unit step input and the steady-state error resulting from a unit ramp input.

(d) Illustrate how the resulting steady-state error of the system with rate feedback to ramp inputs can be reduced to the same level if rate feedback were not used, and still maintain a damping factor of 0.8.

7.3. Repeat Problem 7.2 for the forward transfer function of the system given by $20/s(1+s)$.

7.4. Figure P7.4 illustrates the block diagram of a roll control system used to limit the roll rate excursions of a missile by providing sufficient dynamic reaction to disturbing moments [12]. The disturbance moments result from changes in bank angle and steering control deflections. The basic limitation which determines the effectiveness of the roll-control system is the response of the aileron servo.

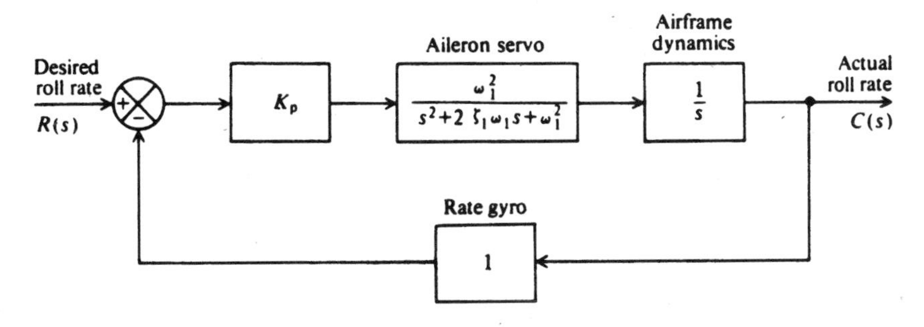

Figure P7.4

(a) Determine the transfer function, $C(s)/R(s)$, of the system illustrated in Figure P7.4.

(b) Because the transient response is governed by a pair of dominant complex-conjugate poles, specify the requirements of the aileron servo parameters in order that the equivalent damping ratio of the system is approximately 0.5, and the equivalent undamped natural frequency of the system is approximately 4 rad/sec.

7.5. A unity-feedback system has a forward transfer function given by

$$G(s) = \frac{28(1 + 0.05s)}{s(1 + s)}.$$

It is desired to compensate this system so that the resulting damping ratio is unity (critically damped).

(a) Using the classical approach, determine the time constant of a cascaded phase-lead network, containing a zero factor only, that can achieve this.

(b) Using the classical approach, determine the rate-feedback constant of a minor rate-feedback loop which can achieve critical damping.

7.6. Repeat Problem 7.5 for the forward transfer function of the system given by

$$\frac{100(1 + 0.1s)}{s(1 + 10s)}.$$

7.7. A phase-lag–lead network, whose general transfer function is given by Eq. (7.9), and whose gain and phase characteristics are illustrated in Figure 7.13, provides a phase lag at low frequencies and a phase lead at high frequencies. For the following phase-lag–lead network,

$$G(s) = K\frac{(1 + T_1 S)(1 + T_2 s)}{(\alpha + T_1 s)\left(\dfrac{1}{\alpha} + T_2 s\right)}.$$

Find the frequency ω_0 where the phase angle of $G(j\omega)$ becomes zero. For frequencies less than ω_0, this network acts as a phase-lag network; for frequencies greater than ω_0, this network acts as a phase-lead network.

7.8. It is desired that the system considered in Problem 6.32 have a minimum phase margin of 45° and a minimum gain margin of 20 dB. Cascade compensation is to be employed.

(a) Specify the time constant of a phase-lead network (or networks) that can achieve this.

(b) Repeat part (a) for a phase-lag network.

7.9. It is desired that the system considered in Problem 6.24 have a phase margin of 60° and a gain margin of 60 dB. Cascade compensation is to be employed.

Determine the time constant of a phase-lag network (or networks) that can achieve this.

7.10. The temperature-control loop of a nuclear power plant is illustrated in Figure P7.10. The transfer function of the nuclear reactor can be adequately represented by

$$G_R(s) = \frac{e^{-0.2s}}{0.4s + 1}.$$

A time delay (or transportation lag) is included in this transfer function to account for the time required to transport the fluid from the reactor to the measurement point. A proportional-plus-integral (PI) controller is used for compensation.

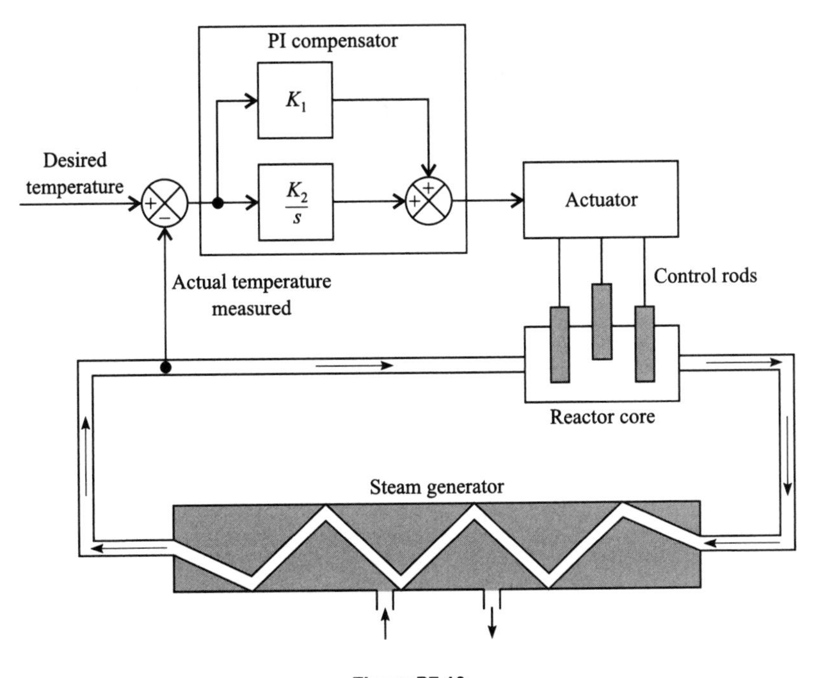

Figure P7.10

Using the Bode diagram, determine the values of K_1 and K_2 in order to achieve a phase margin of 30° and a gain margin of 4 dB.

7.11. It is desired that the system considered in Problem 6.28 have a phase margin of 45° at the crossover frequency. Determine the stabilizing element required to achieve this.

7.12. It is desired that the system considered in Problem 6.33 have a phase margin of 45° at the crossover frequency. Determine the stabilizing element required to achieve this.

7.13. The H.S. Denison, shown in Figure P7.13*a*, the first large hydrofoil seacraft built and operated in the United States [13] was designed and built by the Grumman Aerospace Corporation for the Maritime Administration of the U.S. Department of Commerce. The 80-ton hydrofoil is capable of operating at speeds of 60 knots in seas containing waves nine feet high. A simplified schematic of the automatic control system of the H.S. Denison is illustrated in Figure P7.13*b*. It consists of transducers for sensing craft motions and a

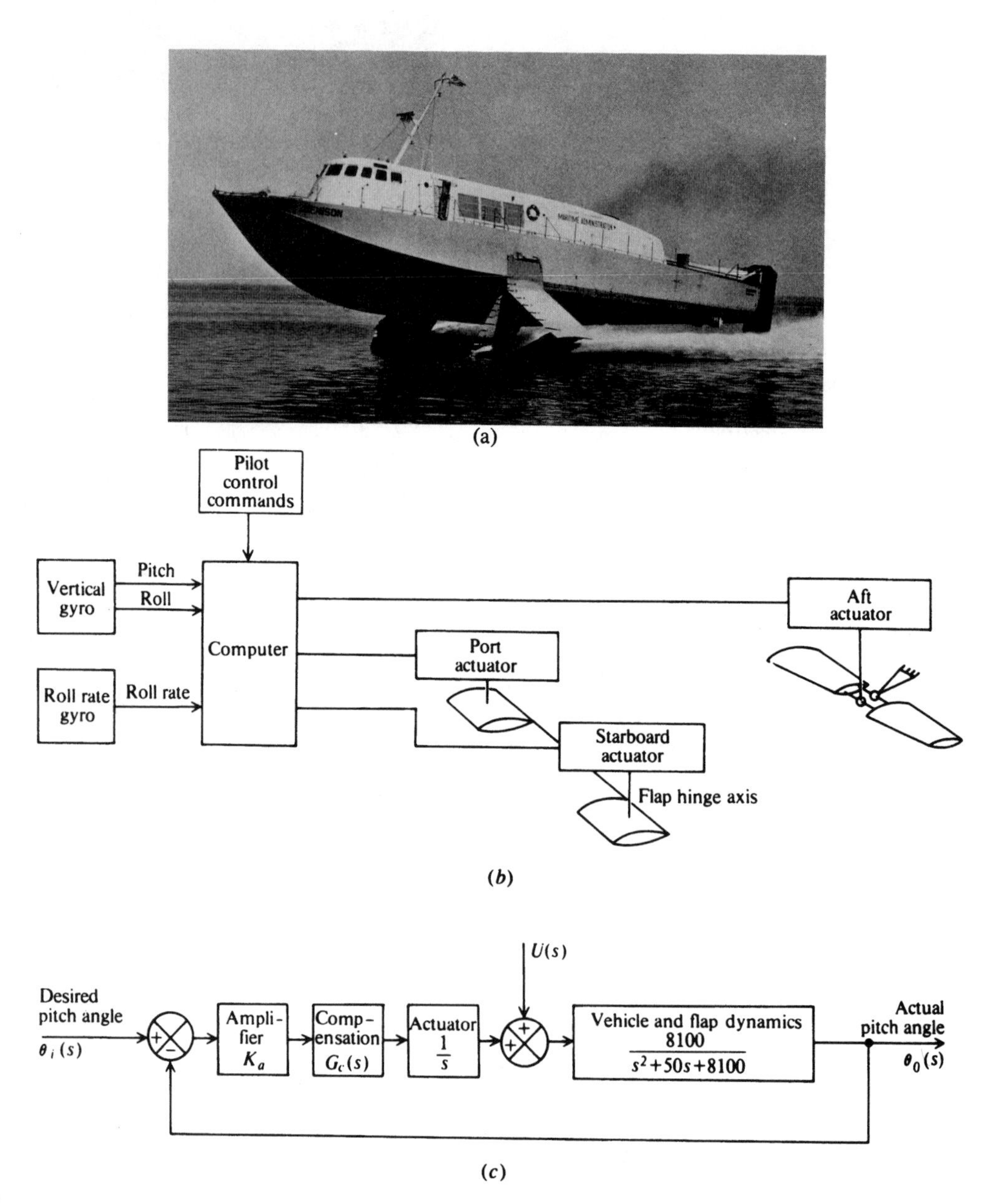

Figure P7.13 (*a*) Photograph of H. S. Denison. (Courtesy of Grumman Aerospace Corporation) (*b*) The automatic control system. (*c*) Block diagram of the pitch-control system.

computer for transmitting commands to the electrohydraulic actuators [13]. Heave rate is fed symmetrically to the forward flaps; roll and roll rate are fed differentially to the forward flaps; pitch rate is fed to the stern foil. The stabilization control system maintains level flight by means of two main surface piercing foils located ahead of the center of gravity and an all-movable submerged foil aft. An equivalent block diagram of the pitch-control system is illustrated in Figure 7.13*c*. It is desired that the craft maintain a constant level of travel despite a wave disturbance $U(s)$ whose energy is concentrated at 1 rad/sec. Assume that the specifications require that the pitch loop maintain a gain of 40 dB at 1 rad/sec in order to minimize the wave disturbance, and a crossover of 10 rad/sec for adequate response time. In addition, it is desired to have a phase margin of at least 50° at the gain crossover frequency of 10 rad/sec and a gain margin of 12 dB. Select the amplifier gain K_a and compensation network $G_c(s)$ in order to achieve these requirements.

7.14. It is desired that the system considered in Problem 6.34 have a phase margin of 45° at the gain crossover frequency. To achieve this, one or more phase-lead networks are introduced into the controller. Determine the compensation required and the resulting new crossover frequency to meet this specification.

7.15. A unity-feedback control system has a forward transfer function given by

$$G_0(s) = \frac{9}{s^2}.$$

(a) Determine the gain crossover frequency and the phase margin of this control system.

(b) A phase-lead network, as defined in Eq. (7.6), is to be added in cascade with $G_0(s)$ so that the phase margin is 70°. Determine the phase-lead network which will achieve this phase margin.

7.16. A remotely piloted aircraft (RPA) for reconnaissance purposes over heavily defended terrain is to be controlled from a ground station. Use of the RPA will eliminate loss of human life if the aircraft is destroyed due to enemy action. The conceptual diagram of the RPA system is illustrated in Figure P7.16*a*.

An equivalent block diagram of the pitch attitude axis of the RPA system is illustrated in Figure P7.16*b*.

The transportation lag T_1 represents the delay caused by the man-in-the-loop at the ground station and the time it takes to transmit the signal from the ground station to the RPA. The transportation lag T_2 represents the time it takes for the return signal to be received by the ground station from the RPA. Assume that $T_1 = 0.3$ sec and that $T_2 = 0.05$ sec.

(a) Determine analytically the gain crossover frequency needed to achieve a phase margin of 50°.

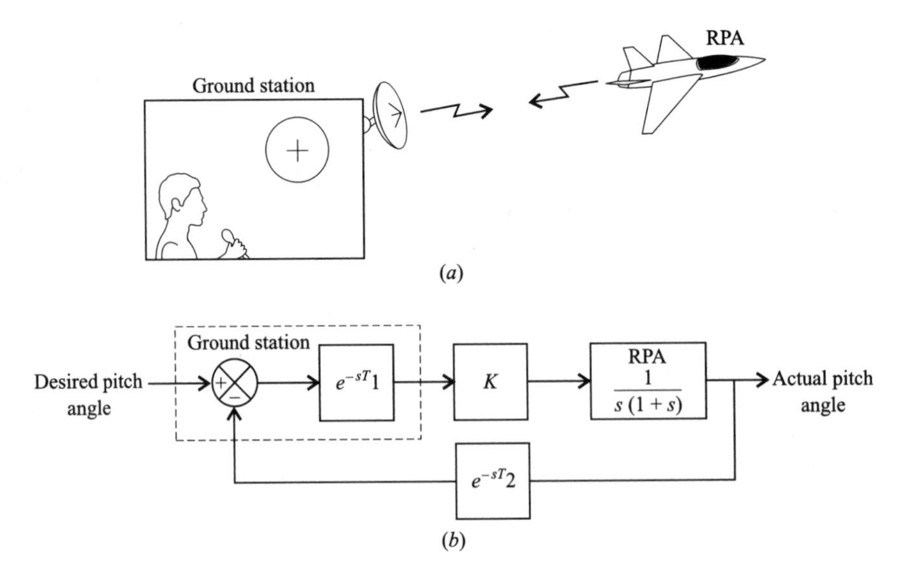

Figure P7.16

(b) Without drawing a Bode diagram, determine the value of K needed to obtain the gain crossover frequency obtained in part (a).

7.17. Space vehicles, such as the space shuttle, using wings to maneuver while reentering the Earth's atmosphere present an interesting control problem. Figure P7.17*a* illustrates a conceptual design of such a system and Figure

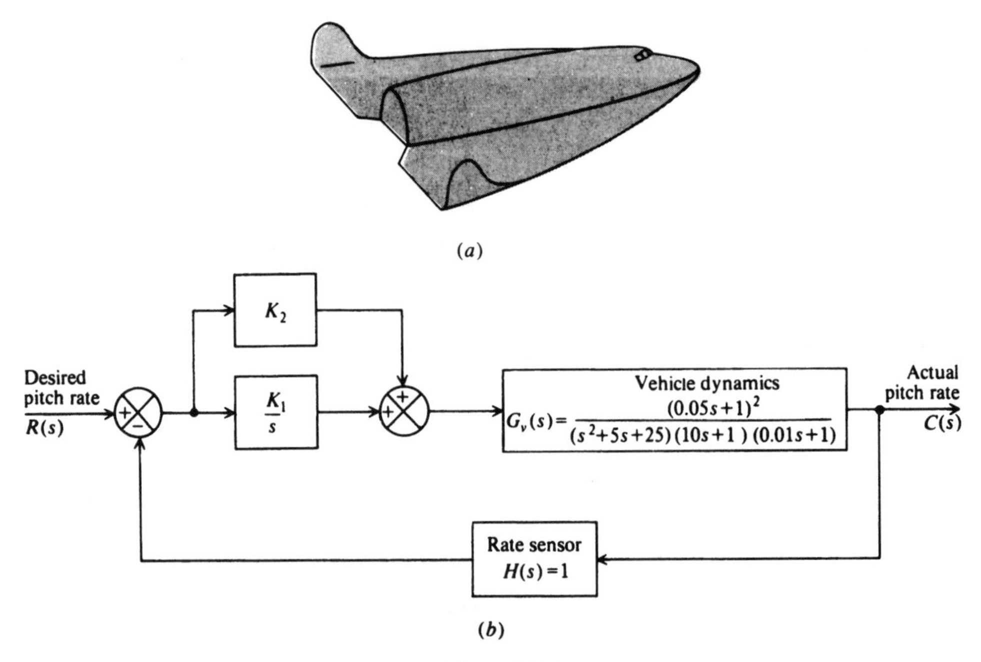

Figure P7.17

P7.17*b* indicates the block diagram of the pitch-rate control of such a system [14].

(a) Draw the Bode diagram of this system with $K_1 = 1$ and $K_2 = 0$. What are the resulting gain and phase margins?

(b) Select the values of K_1 and K_2 which will result in a gain crossover frequency of 1 rad/sec, a phase margin of at least 30° and a gain margin of at least 16 dB.

7.18. It is desired that the system considered in Problem 6.31 have a phase margin of 55° at the crossover frequency, and a gain margin of 25 dB. In order to achieve this, two phase-lead networks are introduced into the controller. This results in the controller having a transfer function given by

$$G_1(s) = K\frac{1 + T_1 s}{1 + T_2 s}\frac{1 + T_3 s}{1 + T_4 s},$$

where K is the value of gain found in part (b) of Problem 6.31. Determine T_1, T_2, T_3, T_4, and the resulting new crossover frequency to meet this specification.

7.19. The design of the Lunar Excursion Module (LEM) shown in Figure P7.19*a*, was an extremely interesting problem [15]. The control, guidance, and navigation for the LEM are provided by an all-digital system from the sensors to the gas-jet propulsion units. For purposes of this analysis, the vehicle dynamics can be approximated by a double integration, as indicated in Figure P7.19*b*, which illustrates one axis of the attitude-control system. In addition, the torque $T(s)$ is assumed to be proportional to the control signal $U(s)$. Assume that $J = 0.25$ and

$$T(s) = 2U(s).$$

Utilizing the Bode diagram for solution, determine a lead-compensation network $G_c(s)$ which will result in a crossover frequency of 5.1 rad/sec and a phase margin of 60°.

7.20. It is desired to add cascade compensation to the system considered in Problem 6.36 in order that the peak overshoot to a step input be approximately 12%.

(a) Using the Nichols chart, design a phase-lead network which can achieve this.

(b) With the compensation network chosen in part (a), determine the closed-loop amplitude and phase-frequency response.

7.21. It is desired to add cascade compensation to the system considered in Problem 6.42 in order that $M_p = 0.75$ while the same steady-state error is maintained.

(a) Design a phase-lead network to achieve this.

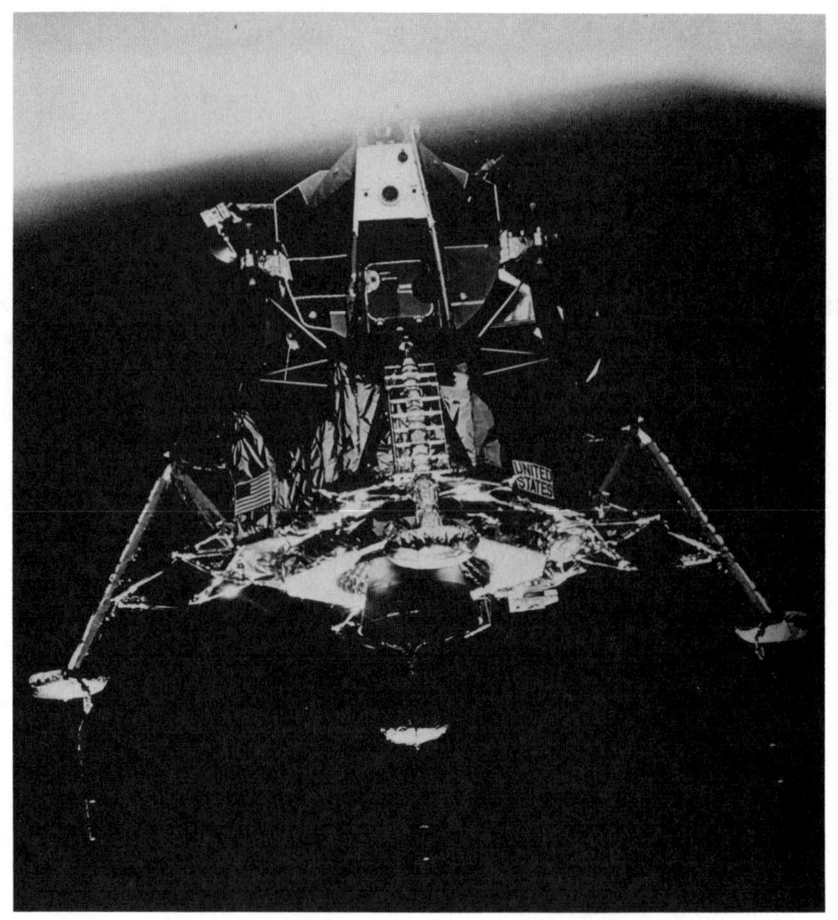

(a)

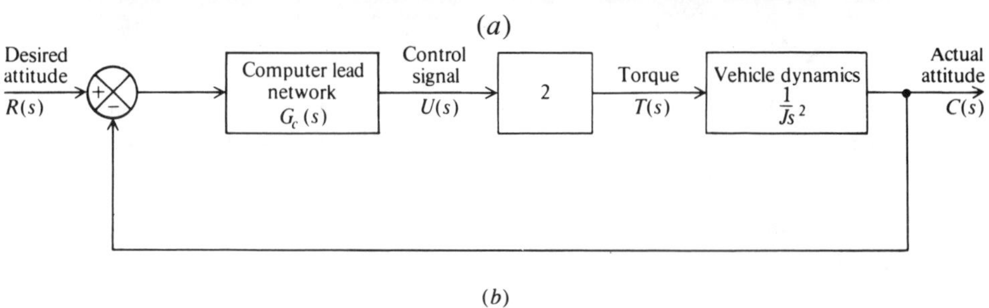

(b)

Figure P7.19

(b) With the compensation network chosen in part (a), determine the closed-loop amplitude and phase frequency response.

7.22. It is desired that the system considered in Problem 7.21 have a peak overshoot of approximately 15% to a step input.

(a) Utilizing a minor rate-feedback loop, specify the tachometer constant which can achieve this.

(b) What will be the resulting system steady-state error to a unit ramp input with the minor rate-feedback loop added?

(c) Utilizing a simple high-pass *RC* filter in cascade with the tachometer, determine the time constant of the network and the tachometer constant which will result in a 15% overshoot to a step input.

(d) What will be the steady-state error to a unit ramp when the high-pass filter is cascaded with the tachometer?

7.23. It is desired that the system considered in Problem 6.45 have a damping ratio of 0.75 for the dominant complex roots. Using the root-locus method, determine the increase in gain and phase-lag network $(s + n\alpha)/(s + \alpha)$ which can achieve this. Assume $K_v = 15$.

7.24. A unity-feedback control system has a forward transfer function given by:

$$G(s) = \frac{100}{s(1 + 0.1s)(1 + 0.01s)}$$

$$H(s) = 1.$$

(a) Draw the Bode diagram of this control system and determine the resulting phase margin and gain margin.

(b) To achieve an acceptable transient response for this system, the phase margin should be approximately 30° and the gain margin 20 dB. In addition, a sinusoidal disturbance is present at 0.1 rad/sec, and a gain of 60 dB is required at this frequency to nullify its effect. Design a passive compensation network which will achieve the desired transient response and accuracy, and also minimize the noise susceptibility of the system.

7.25. The block diagram of one axis of a robotic positioning system that uses rate feedback for compensation is illustrated in Figure P7.25.

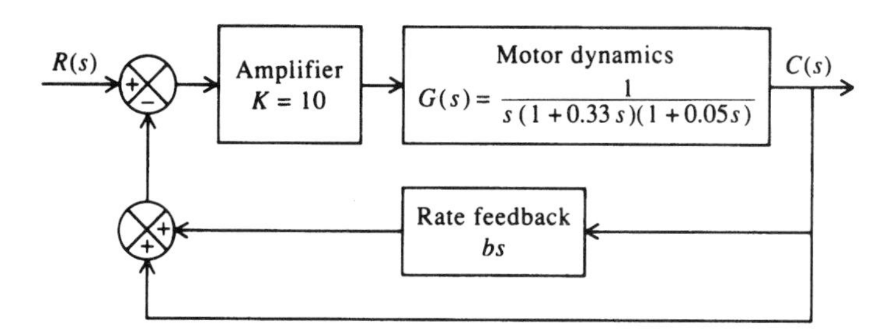

Figure P7.25

(a) Without any rate feedback ($b = 0$), determine the gain crossover frequency, phase margin and gain margin of the uncompensated system using the Bode-diagram method.

(b) For proper operation of the robot, a minimum phase margin of 65° and a minimum gain margin of 90 dB are desired. Using the Bode diagram, determine the amount of rate-feedback constant b which will achieve these requirements.

7.26. The system shown in Figure P7.26 contains a proportional plus integral (PI) controller. The PI controller contains a zero term at $s = -K$ and a pole at $s = 0$. Therefore, the PI controller has infinite gain at zero frequency, and it behaves as a phase-lag network. This feature improves the steady-state characteristics.

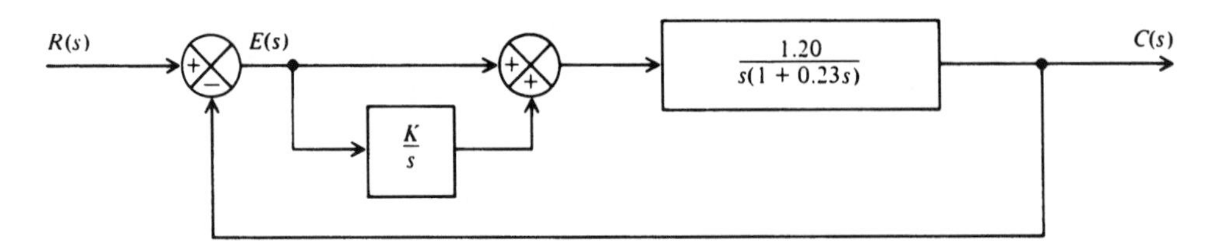

Figure P7.26

(a) Determine the steady-state error of this system to a unit step input.

(b) Determine the steady-state error of this system to a unit ramp input.

(c) Determine the range of gain K for which this system is stable.

7.27. A negative-feedback system containing unity feedback has a forward transfer function given by

$$G(s) = \frac{K(s + A)}{s(s + 2)(s + 4)}.$$

The zero factor $(s + A)$ in the numerator of this transfer is to be used to compensate the system. Utilizing the root locus, analyze the effects on system stability of the following values of A:

(a) $A = 1$, (b) $A = 2$, (c) $A = 3$,
(d) $A = 4$, (e) $A = 6$.

What conclusions can you draw from your results on the best value of A for compensating this system? What happens if A is greater than 6?

7.28. A unity-feedback control system has a forward transfer function given by:

$$G_0(s) = \frac{1}{(s + 1)(s + 8)}.$$

We wish to use a proportional plus integral (PI) controller in cascade with $G_0(s)$ to compensate this control system. The transfer function of this compensator is given by:

$$G_c(s) = 1 + \frac{K}{s}.$$

Select K so that the pair of dominant closed-loop poles for this control system is located at $s = -0.5 \pm 0.5j$.

7.29. A unity-feedback control system has the following forward transfer function:

$$G(s) = \frac{K(s+\alpha)}{s^2(s+2)}.$$

Determine the values of α so that the root locus will have zero breakaway points, not including the one at s equal zero.

7.30. The forward transfer function of a unity-feedback control system is given by the following:

$$G(s) = \frac{K}{s^2(s+2)(s+4)}$$

(a) Construct the root-locus digram for $0 \leq K \leq \infty$, and show all pertinent values on the root locus. What conclusions can you reach from this root-locus diagram?

(b) The system will be stabilized by mean of a rate-feedback element (e.g., tachometer) added in parallel to the unity feedback so that the total feedback transfer function becomes:

$$H(s) = 1 + s.$$

Construct the root locus of this compensated control system for $0 \leq K \leq \infty$, and show all pertinent values on the root locus.

7.31. The transfer functions of a negative feedback system are given by the following:

$$G(s) = \frac{K}{s(s^2+6s+10)},$$

$$H(s) = 1.$$

(a) Sketch the root locus.

(b) Determine $C(s)/R(s)$, with the denominator in factored form, if a damping ratio of 0.5 is required for the dominant roots.

7.32. Determine the increase in gain and phase-lag network compensation required to stabilize a unity-feedback system whose forward transfer function is given by

$$G(s) = \frac{K}{s(s+3)(s+4)}.$$

The requirement for the system damping ratio is 0.707, and the velocity constant is 100/sec.

7.33. A unity-feedback control system has a forward transfer function given by

$$G(s) = \frac{K(s+1)}{s(s+3)(s+6)}.$$

It is desired that the system have a velocity constant of 15 and a damping ratio of 0.75. Using the root-locus method, determine the increase in gain and phase-lag network $(s + n\alpha)/(s + \alpha)$ which can achieve this, assuming that the transient response is governed by a pair of dominant complex-conjugate poles.

7.34. A unity-feedback control system has a forward transfer function given by

$$G(s) = \frac{K(s+1)}{s(s^2+6s+9)}.$$

It is desired that the system have a velocity constant of 150 and a damping ratio of 0.75. Using the root-locus method, determine the increase in gain and phase-lag network, represented by $(s + na)/(s + a)$, which can achieve this. Assume that the transient response is governed by a pair of dominant complex-conjugate poles. Show all pertinent points on the root locus before and after compensation.

7.35. The block diagram of a positioning system is shown in Figure P7.35.

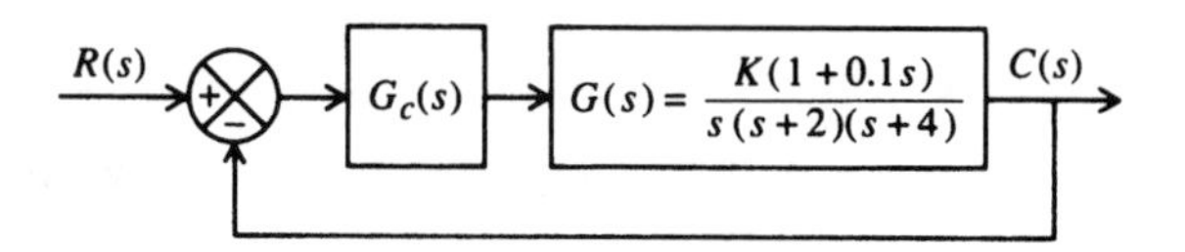

Figure P7.35

(a) Without any compensation, $G_c(s) = 1$, draw the root locus of the uncompensated system. On this diagram determine and show the following clearly:

- Point (*s*) of breakaway from the real axis
- Crossing (*s*) of the imaginary axis
- K_{max}
- All asymptotes.

(b) It is desired that the system have a velocity constant of 4 and a damping ratio of 0.707. Using the root-locus method, determine the compensation $G_c(s)$ which can achieve this. Assume that the transient response is governed by a pair of dominant complex-conjugate poles. Show all pertinent changes and points on the root locus after compensation. (An exact recalculation of the point (*s*) of breakaway, the crossing of the imaginary axis, and K_{max} are not necessary.)

7.36. In order to obtain a control system which has an infinite velocity constant, a control-system engineer designs the control system of Figure P7.36, which he or she recognizes will need some compensation network, $G_c(s)$.

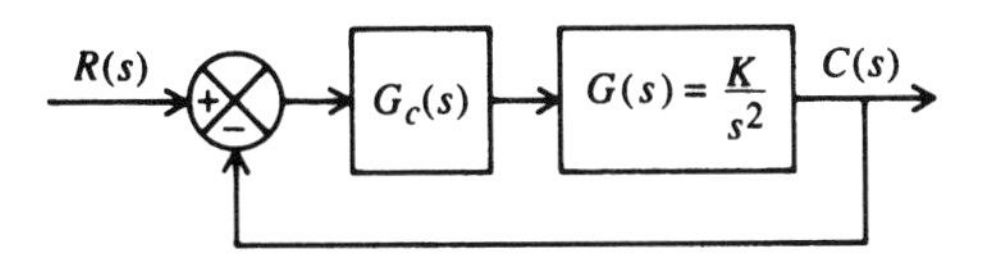

Figure P7.36

(a) Without any compensation, $G_c(s) = 1$, draw the root locus of this control system. From a practical viewpoint, is it stable?

(b) The control-system engineer desires to use a passive compensation network for $G_c(s)$, either a phase-lag or a phase-lead network. Assume that the resistors and capacitors that are available can provide a zero at -1 and a pole at -2, *or* a pole at -1 and a zero at -2. As the control-system engineer, which would you select and why? Draw the root locus for your stabilized control system (with either a phase-lag or a phase-lead network). Show all pertinent points on your root locus.

7.37. The signal-flow graph of the temperature-control loop for a xylene chemical process [16] is shown in Figure P7.37. The temperature of the process, $C(s)$, is related to the heat supplied to the process by the quadratic transfer function $G_p(s)$. Temperature is measured by a sensor having a pole at $s = -0.3$, and the output of the sensor in the form of air pressure is compared with the desired value of temperature as indicated by the reference pressure $R(s)$. The pressure difference (a measure of temperature error) actuates a pneumatic controller which provides as its output a pneumatic actuating signal applied to a steam valve. The valve, in turn, controls the flow of heat to the xylene column in order to minimize the error.

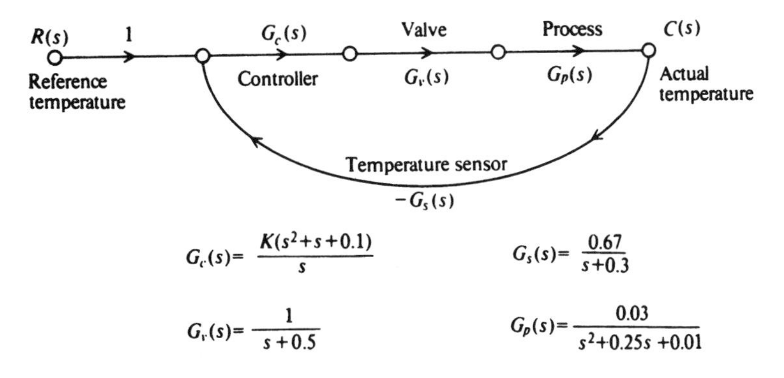

Figure P7.37

(a) Draw the root locus for this system.

(b) Determine the required gain K for a damping ratio of 0.5.

7.38. It is desired that the system considered in Problem 6.46 (b) have a damping ratio of 0.75 for the dominant complex roots. Using the root-locus method, determine the value of K which can achieve this.

7.39. A turbine speed-control system is illustrated in Figure P7.39. Assume that the transfer function of the control valve, turbine, and speed converter are as follows:

$$G_1(s) = \frac{1}{s + 0.1},$$

$$G_2(s) = \frac{0.5}{s^2 + 3s + 2},$$

$$H(s) = 1.$$

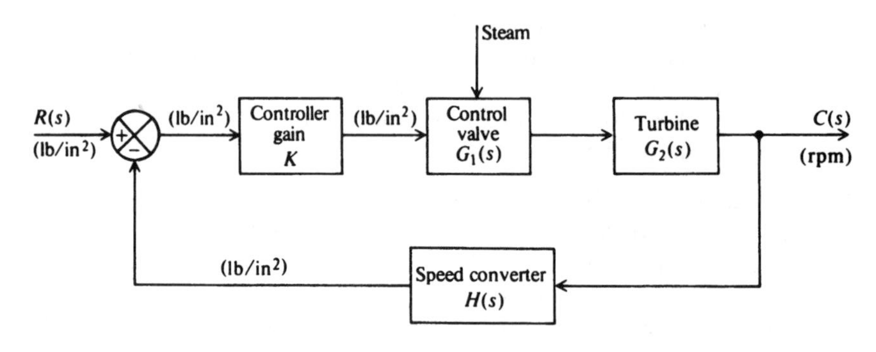

Figure P7.39

Assuming that the transient response is governed by a pair of dominant complex-conjugate poles, determine the value of the controller gain K in order that the system has a damping ratio of 0.5.

7.40. It is desired that the system considered in Problem 6.46 (c) have a damping ratio of 0.75 for the dominant complex roots. Using the root-locus method:

(a) Determine the value of K which can achieve this.

(b) Determine the values of ω_p and M_p for the value of K found in part (s) using the Nichols chart.

7.41. Determine the gain needed for the system considered in Problem 6.46 (d) which can achieve a damping ratio of 0.75.

7.42. Unlike fixed-wing aircraft, which possess a moderate degree of inherent stability, the helicopter is very unstable and requires the use of feedback loops for stabilization. A typical control system involves the use of an inner automatic stabilization loop and an outer loop which is controlled by the pilot, who inserts commands into it based on attitude errors displayed to the pilot. Figure P7.42 illustrates the pitch-control system used on the S-55 helicopter [17]. When the pilot is not utilizing the control stick, the switch S_1 is open,

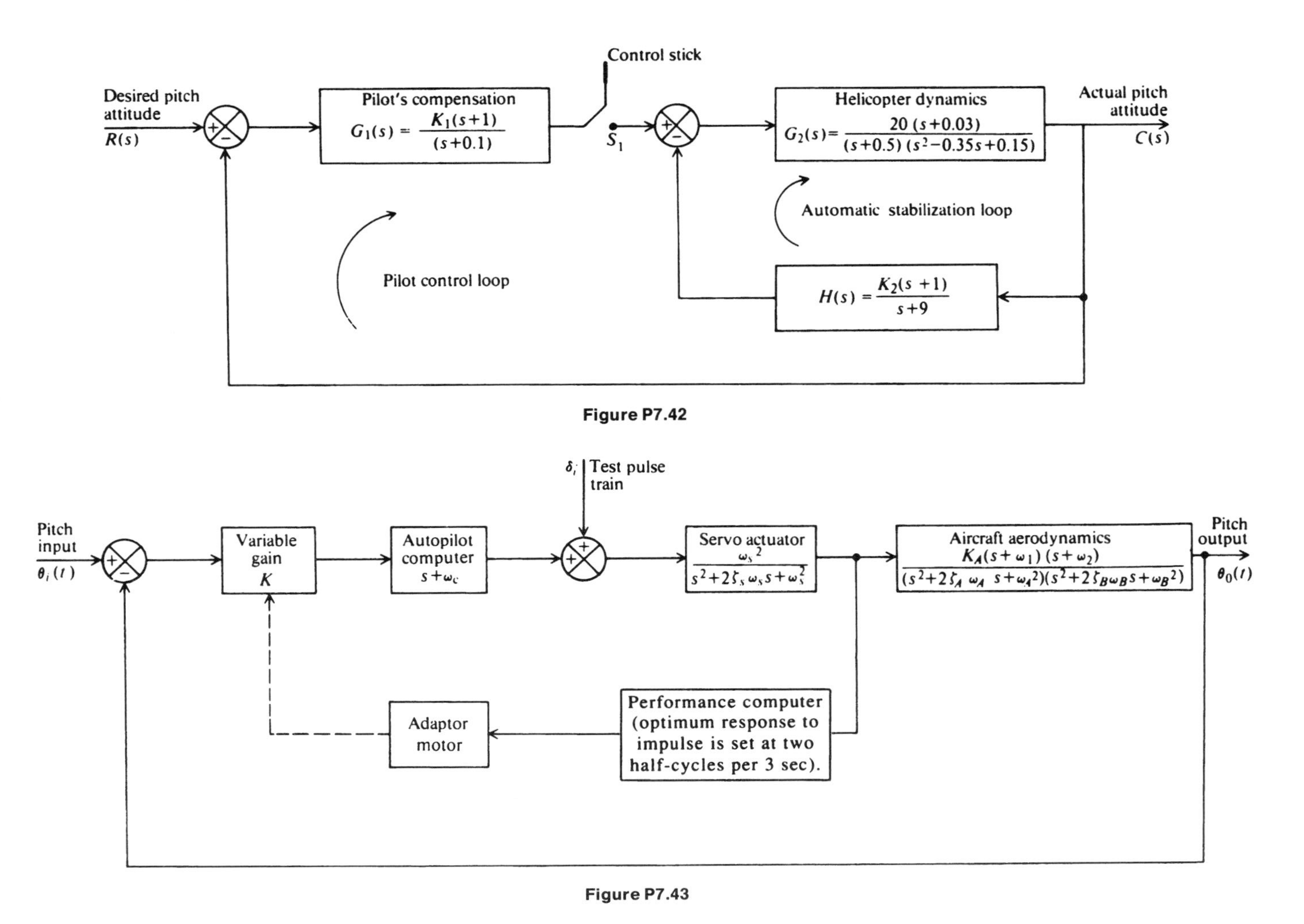

Figure P7.42

Figure P7.43

which disengages the pilot control loop. The model of the pilot's transfer function, $G_1(s)$, includes a gain factor, an anticipation time constant of 1 sec, and an error-smoothing time constant of 10 sec [17].

(a) With the pilot control loop open, plot the root locus for the automatic stabilization loop and determine the gain K_2 which results in a damping ratio of 0.5 for the dominant complex roots.

(b) Draw the root locus of the pilot control loop with K_2 set at the value determined in part (a). Determine the value of the pilot's gain compensation factor K_1 in order that the pilot control loop have a damping ratio of 0.5.

7.43. Many modern control systems are designed to be adaptive, in order that they can achieve a desired response in the presence of extreme changes in the system parameters and major external disturbances. Adaptive control systems are usually characterized by devices that automatically measure the dynamics of the controlled system and by other devices that automatically adjust the characteristics of the controlled elements based on a comparison of the measurements with some optimum figure of merit. Figure P7.43 illustrates an adaptive pitch flight control system [18]. It attempts to measure the exact location of a pair of dominant, variable, servo actuator poles that move in the complex plane, as a function of the flight conditions. The adaptive feature overcomes this problem by adjusting the gain in order to keep the location of these sensitive poles fixed in the complex plane. A test impulse train is injected into the system when the error is small. The performance computer determines the transient response of the system and compares it with an optimum desired response that is set at 2 half cycles of a transient response over a 3-sec interval of time. The performance computer is designed so that a count less than 2 over a 3-sec period will cause the adaptor motor to increase K, and a count greater than 2 over a 3-sec period will cause the adaptor motor to decrease K. Assume that the poles and zeros of the system are located in the complex plane as follows:

Table P7.43. Helicopter Control System Parameters

	Poles	Zeros	Gain
Aircraft aerodynamics	$P_a = -4 \pm j2$ $P_B = -1 \pm j1$	$\omega_1 = -3$ $\omega_2 = -2$	$K_A = 0.02$
Servo actuator	$P_C = -3 \pm j6$	—	—
Autopilot computer	—	$\omega_c = -6$	—

(a) Draw the root locus of this system.

(b) Determine the variable gain K that will result in a damping ratio of 0.3, assuming that the transient response is governed by a pair of dominant complex-conjugate poles.

7.44. Repeat Problem 7.33 for

$$G(s) = \frac{K(s+1)}{s(s^2+6s+9)}.$$

REFERENCES

1. W. R. Ahrendt, *Servomechanism Practice*. McGraw-Hill, New York, 1954.
2. J. G. Truxal, *Automatic Feedback Control System Synthesis*. McGraw-Hill, New York, 1955.
3. R. E. Kalman, "On the general theory of control systems." In *Proceedings of the First International Congress of Automatic Control*, Moscow, 1960.
4. R. E. Kalman, Y. C. Ho, and K. S. Navendra, "Controllability of linear dynamical systems," *Contrib. Diff. Equat.*, **1**, 189–213 (1961).
5. H. Lauer, R. N., Lesnick, and L. E. Matson, *Servomechanism Fundamentals*. McGraw-Hill, New York, 1960.
6. J. G. Truxal, ed. *Control Engineer's Handbook*. McGraw-Hill, New York, 1955.
7. H. W. Bode, *Network Analysis and Feedback Amplifier Design*. Van Nostrand, New York, 1945.
8. L. A. Gould, *Chemical Process Control: Theory and Applications*. Addison-Wesley, Reading, MA, 1969.
9. H. Chestnut and R. W. Mayer, *Servomechanisms and Regulating System Design*, 2nd ed., Vol. 1, Wiley, New York, 1959.
10. J. L. Bower and P. M. Schultheiss, *Introduction to the Design of Servomechanisms*. Wiley, New York, 1958.
11. S. M. Shinners, "How to approach the stability analysis and compensation of control systems." *Control Eng.* **25**(5), 62–67 (1978).
12. W. K. Waymeyer and R. W. Sporing, "Closed loop adaptation applied to missile control," In *Proceedings of the 1962 Joint Automatic Control Conference*. New York, p. 18-33.
13. R. M. Rose, *The Rough Water Performance of the H. S. Denison*, Paper No. 64–197. Am. Inst. Aeronaut. Astronaut., Washington, DC, 1984.
14. R. P. Kotfile and S. S. Oseder, "Stabilization and control of maneuvering reentry vehicle." *Sperry Eng. Rev.* **18**, 2–10 (1965).
15. F. Doennebrink and J. Russel, "LEM stabilization and control system." *AIAA/ION Guidance and Control Conference, 1965*, pp. 430–441.
16. W. A. Lynch and J. G. Truxal, *Principles of Electronic Instrumentation*. McGraw-Hill, New York, 1962.
17. L. Kaufman, "Helicopter control stick steering," *Sperry Eng. Rev.* **11**, 41–48 (1958).
18. F. C. Gregory, ed. *Proceedings of the Self-Adaptive Flight Control Systems Symposium*, WADC Techn. Rep. No. 59–49, ASTIA Doc. No. AD 209389, 1959.
19. S. M. Shinners, *Techniques of System Engineering*. McGraw-Hill, New York, 1967.

8

MODERN CONTROL-SYSTEM DESIGN USING STATE-SPACE, POLE PLACEMENT, ACKERMANN'S FORMULA, ESTIMATION, ROBUST CONTROL, AND H^{∞} TECHNIQUES

8.1. INTRODUCTION

State-space analysis was introduced in Chapter 2, and has been used in parallel with the classical frequency-domain analyses techniques presented in Chapters 3 through 7. It was shown that the state-space approach is applicable to a wider class of problems such as multiple-input/multiple-output (MIMO) control systems. Chapter 7 applied the frequency-domain approaches such as the Bode diagram, and the root locus to linear control-system design.

In the design of a control system, the question arises as to where to place the closed-loop roots. In Section 7.9 which presented the root-locus method, we could specify where to place the dominant-pair of complex-conjugate roots in order to obtain a desired transient response. However, we could not do so with great certainty because we were never sure what effect the higher-order poles would have on the second-order approximation.

The control-system design engineer desires to have design methods available which would enable the design to proceed by specifying all of the closed-loop poles of higher-order control systems. Unfortunately, the frequency-domain design methods presented in Chapter 7 do not permit the control-system engineer to specify all poles in control systems which are higher than two because they do not provide a sufficient number of unknowns for solving uniquely for the specified closed-loop poles. This problem is overcome using state-space methods which provide additional adjustable parameters, and methods for determining these parameters.

This chapter presents a modern control-system design method using state-space techniques known as *pole placement* or *pole assignment*. This design technique is similar to what we did in Section 7.9 where we placed two dominant complex-conjugate poles of the closed-loop transfer function in desired locations in order to obtain desirable transient responses. However, in this chapter, we will show how pole placement allows the control-system engineer to place all of the poles of the closed-loop transfer function in desirable locations. Ackermann's formula is also presented for designs using pole placement for application in those control systems that require feedback from state variables which are not phase variables (where each subsequent state variable is defined as the derivative of the previous state variable). A practical problem arises with the pole placement method involving cost and the availability of determining (measuring) all of the system variables needed for obtaining a solution. In many pracical control systems, all of the system state variables may not be available due to cost considerations, environmental considerations (e.g., nuclear power plant control systems), and the availability of transducers to measure certain states. For these cases, it is necessary for the control-system engineer to estimate the state variables that cannot be measured from the state variables that can be measured. Therefore, in addition to pole placement, this chapter also presents the very important subjects of controllability, observability, and estimation.

This chapter on modern control-system design also presents the design of robust control systems. Robust control systems are concerned with determining a stabilizing controller that achieves feedback performance in terms of stability and accuracy requirements, but the controller must achieve performance that is robust (insensitive) to plant uncertainty, parameter variation, and external disturbances. The design of two-degrees-of-freedom compensation control systems exhibiting desirable robustness to plant uncertainty, parameter variation, and external disturbances is presented.

This chapter concludes with an introduction to H^∞ control concepts which is a new technique that emerged in the 1980s that combines both the frequency- and time-domain approaches to provide a unified design approach. The H^∞ approach has dominated the trend of control-system development in the 1980s and 1990s. The H^∞ control-system design approach expands on the concept of robustness presented in this chapter, sensitivity (presented in Chapter 5), together with the frequency and state-variable domain techniques presented in this book. The H^∞ approach is applied to determine the optimum sensitivity for control systems.

8.2. POLE-PLACEMENT DESIGN USING LINEAR-STATE-VARIABLE FEEDBACK

Having presented methods for designing linear control systems using classical techniques, let us now look at the problem of specifying pole placement from the viewpoint of state-variable feedback [1]. In order to do this, let us first look at the basic feedback problem illustrated in Figure 8.1. This figure illustrates the concept of feeding back the states of the process in addition to that of the output. Because a linear process can be characterized by the phase-variable canonical equations

$$\dot{\mathbf{x}}(t) = \mathbf{P}\mathbf{x}(t) + \mathbf{b}u(t), \tag{8.1}$$

$$c(t) = \mathbf{L}\mathbf{x}(t), \tag{8.2}$$

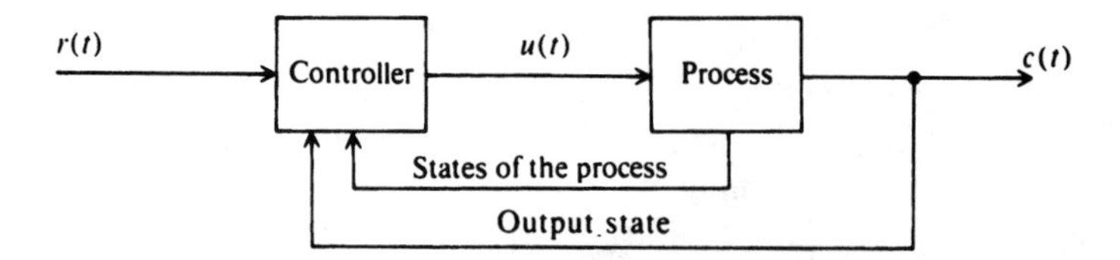

Figure 8.1 General feedback system problem illustrating feedback of the output state and the states of the process.

let us consider the configuration of Figure 8.2. It is important to observe from this figure that the control signal is generated from a knowledge of the reference input $r(t)$ and the state variables $\mathbf{x}(t)$. Note that $r(t)$, $u(t)$, and $c(t)$ represent scalars.

In general, the control input u can be represented as

$$u(t) = f(\mathbf{x}(t), r(t)).$$

Rather than considering the controller in such a broad sense, let us consider the specific condition of linear state-variable feedback where the controller weights the sum of the state variables in a linear manner. In addition, it is assumed that the controller provides a linear gain K which multiplies the difference between the reference input and the linear weighted sum of state variables fed back. Therefore, $u(t)$ can be represented as

$$u(t) = K[r(t) - (h_1 x_1(t) + h_2 x_2(t) + h_3 x_3(t) \cdots + h_n x_n(t))], \tag{8.3}$$

where h_i is defined as the ith feedback coefficient. In matrix form, $u(t)$ can be represented as

$$u(t) = K[r(t) - \mathbf{h}\mathbf{x}(t)], \tag{8.4}$$

where

$$\mathbf{h} = [h_1 \; h_2 \; h_3 \cdots h_n], \tag{8.5}$$

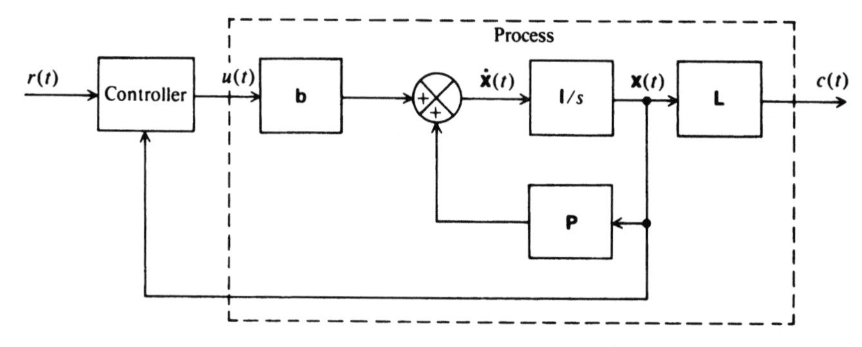

Figure 8.2 General feedback system with state-variable feedback.

$$\mathbf{x}(t) = \begin{bmatrix} x_1(t) \\ x_2(t) \\ x_3(t) \\ \vdots \\ x_n(t) \end{bmatrix}. \tag{8.6}$$

Figure 8.3 presents a matrix representation of the concept of linear-state-variable feedback, and Figure 8.4 is an example of a physical representation of a typical system as implied by Figure 8.3. In the following discussion, it is assumed that all state variables are directly available for measurement and control. In practice, this is not always possible, and techniques for modifying and extending the design procedure presented, to the case where all the state variables are not available, are also discussed in Sections 8.6 and 8.7.

How does linear feedback of the state variables affect the behavior of the process given by Eqs. (8.1) and (8.2)? This can easily be determined by substituting Eq. (8.4) into Eq. (8.1):

$$\dot{\mathbf{x}}(t) = \mathbf{P}\mathbf{x}(t) + \mathbf{b}[K(r(t) - \mathbf{h}\mathbf{x}(t)]. \tag{8.7}$$

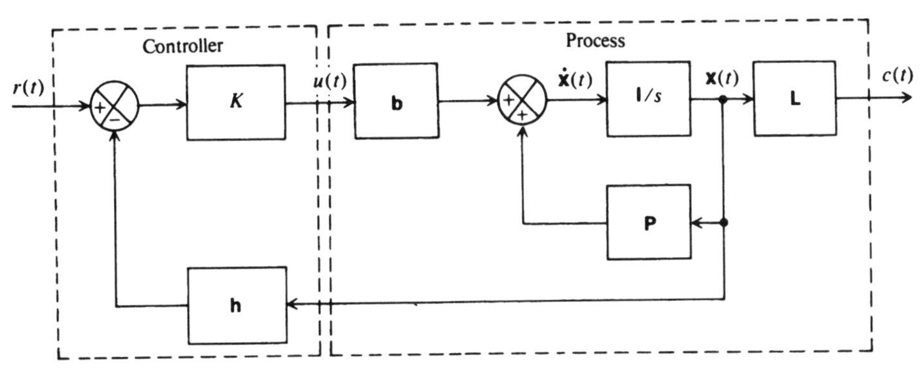

Figure 8.3 Linear-state-variable feedback representation.

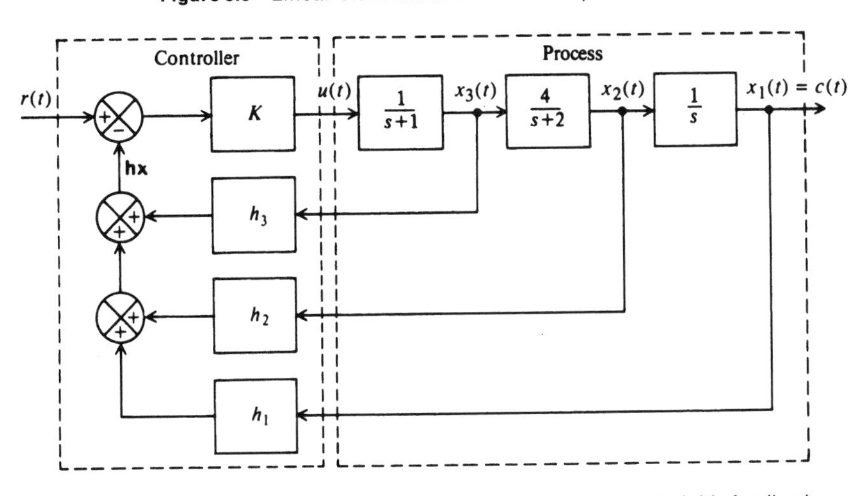

Figure 8.4 Example of pole placement design using linear-state-variable feedback.

Simplifying Eq. (8.7), and incorporating Eq. (8.2), we obtain the closed-loop equation

$$\dot{\mathbf{x}}(t) = \mathbf{P}_h\mathbf{x}(t) + K\mathbf{b}r(t), \tag{8.8}$$

$$c(t) = \mathbf{L}\mathbf{x}, \tag{8.9}$$

where

$$\mathbf{P}_h = \mathbf{P} - K\mathbf{b}\mathbf{h} \tag{8.10}$$

is the closed-loop-system matrix. Comparing Eq. (8.1) and (8.2) with (8.8) and (8.9), we observe that they are identical except that the $\mathbf{P}$ matrix has been replaced by $\mathbf{P}_h$ and $u(t)$ becomes $Kr(t)$.

How can we relate the closed-loop-system matrix $\mathbf{P}_h$ to the closed-loop transfer function, $C(s)/R(s)$? This can be accomplished by taking the Laplace transform of Eqs. (8.8) and (8.9). Because the results will be used to find a transfer function, all initial conditions are assumed to be zero:

$$s\mathbf{X}(s) = \mathbf{P}_h\mathbf{X}(s) + K\mathbf{b}R(s), \tag{8.11}$$

$$C(s) = \mathbf{L}\mathbf{X}(s). \tag{8.12}$$

Solving for $\mathbf{X}(s)$ from Eq. (8.11), we get

$$\mathbf{X}(s) = K[s\mathbf{I} - \mathbf{P}_h]^{-1}\mathbf{b}R(s). \tag{8.13}$$

The inverse matrix $[s\mathbf{I} - \mathbf{P}_h]^{-1}$ is defined as the closed-loop resolvent matrix, $\mathbf{\Phi}_h(s)$, where

$$\mathbf{\Phi}_h(s) = [s\mathbf{I} - \mathbf{P}_h]^{-1}. \tag{8.14}$$

Therefore, Eq. (8.13) may be rewritten as

$$\mathbf{X}(s) = K\mathbf{\Phi}_h(s)\mathbf{b}R(s). \tag{8.15}$$

Substituting Eq. (8.15) into Eq. (8.12), we obtain a relation between $C(s)$ and $R(s)$:

$$C(s) = K\mathbf{L}\mathbf{\Phi}_h(s)\mathbf{b}R(s). \tag{8.16}$$

Therefore, the closed-loop transfer function in terms of the closed-loop resolvent matrix is given by

$$\frac{C(s)}{R(s)} = K\mathbf{L}\mathbf{\Phi}_h(s)\mathbf{b} \tag{8.17}$$

In addition, the characteristic equation in terms of the closed-loop-system matrix can also easily be determined, simply by substituting the numerator and denominator portions of the inverse matrix, $\mathbf{\Phi}_h(s)$:

$$\frac{C(s)}{R(s)} = \frac{K\mathbf{L}[\text{adj}(s\mathbf{I} - \mathbf{P}_h]\,\mathbf{b}}{\det(s\mathbf{I} - \mathbf{P}_h)}. \tag{8.18}$$

The corresponding characteristic equation of the closed-loop system in terms of the closed-loop system matrix is given by

$$\det(s\mathbf{I} - \mathbf{P}_h) = 0. \tag{8.19}$$

Because we are concerned with synthesizing control systems in terms of pole placement design using linear-state-variable feedback concepts, we would like to force the system illustrated in Figure 8.3 into the generalized form illustrated in Figure 8.5, and study its properties. Let us first consider the derivation of $H(s)$. From Figure 8.5 we observe that

$$H(s) = \frac{\mathbf{hX}(s)}{C(s)}. \tag{8.20}$$

Substituting Eq. (8.12) into Eq. (8.20), we obtain

$$H(s) = \frac{\mathbf{hX}(s)}{\mathbf{LX}(s)}. \tag{8.21}$$

After substitution of Eq. (8.15) for $\mathbf{X}(s)$, Eq. (8.21) becomes

$$H(s) = \frac{\mathbf{h\Phi}_h(s)\mathbf{b}}{\mathbf{L\Phi}_h(s)\mathbf{b}}. \tag{8.22}$$

The term $G(s)$ can also be derived in terms of $\mathbf{\Phi}_h(s)$. The closed-loop transfer function of the system illustrated in Figure 8.5 is given by

$$\frac{C(s)}{R(s)} = \frac{KG(s)}{1 + KG(s)H(s)}. \tag{8.23}$$

Substituting Eqs. (8.17) and (8.22) into Eq. (8.23), we obtain the expression

$$G(s) = \frac{\mathbf{L\Phi}_h(s)\mathbf{b}}{1 - K\mathbf{\Phi}_h(s)\mathbf{b}}. \tag{8.24}$$

Combining Eqs. (8.22) and (8.24), the open-loop transfer function is found to be given by

$$KG(s)H(s) = \frac{K\mathbf{h\Phi}_h(s)\mathbf{b}}{1 - K\mathbf{H\Phi}_h(s)\mathbf{b}}. \tag{8.25}$$

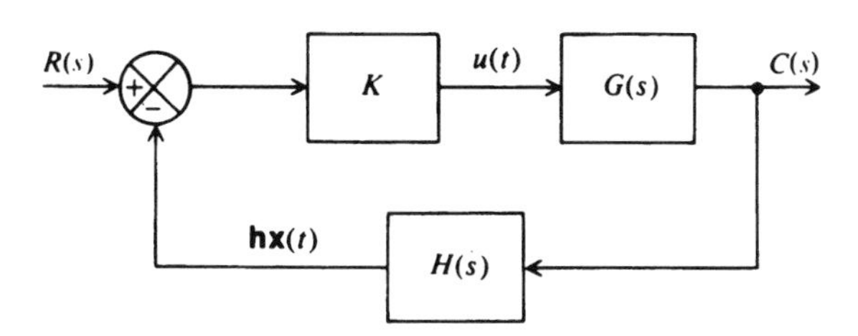

Figure 8.5 An equivalent model of Figure 8.3.

Let us compare Eqs. (8.17), (8.22)–(8.24), and (8.25) in order to draw conclusions regarding $G(s)$, $H(s)$, the open-loop transfer function $KG(s)H(s)$, and the closed-loop transfer function $C(s)/R(s)$. These characteristics will be important for designing systems using pole placement techniques with linear-state-variable feedback techniques in the following section. Based on these five equations, we can state the following properties:

1. The poles of $KG(s)H(s)$ are the poles of $G(s)$.
2. The zeros of $C(s)/R(s)$ are the zeros of $G(s)$.
3. The pole-zero excess of $C(s)/R(s)$ must be equal to the pole-zero excess of $G(s)$.

With these properties as a basis, we consider in the following section the design of control systems from the viewpoint of linear-state-variable feedback.

8.3. CONTROLLER DESIGN USING POLE PLACEMENT AND LINEAR-STATE-VARIABLE FEEDBACK TECHNIQUES

The preceding section has indicated several important relationships between open-loop and closed-loop transfer functions. This is very important in the design of control systems for the case where the closed-loop transfer function is specified and it is desired to determine the open-loop transfer function. A typical problem might specify the desired velocity constant; then use is made of Eq. (5.35) in Section 5.4 which gave the velocity constant in terms of the closed-loop poles and zeros. The problem is to determine the resulting linear-state-variable feedback system.

Let us illustrate the procedure by considering the following problem. It is desired that the closed-loop characteristics of a unity-feedback control system be given by the following parameters:

$$\omega_n = 50 \text{ rad/sec}, \qquad K_v = 35/\text{ sec}, \qquad \zeta = 0.707.$$

What form of closed-loop transfer function will satisfy these requirements? Let us first try a simple quadratic control system having a pair of complex-conjugate poles. From Eq. (5.37), such a system has a velocity constant given by

$$K_v = \frac{\omega_n}{2\zeta} = \frac{50}{2(0.707)} = 35.7/\text{ sec}.$$

Therefore, a simple quadratic control system having a pair of complex-conjugate poles will satisfy these specifications. From Eq. (4.18),

$$\cos\alpha = \zeta. \tag{8.26}$$

For a damping ratio of 0.707, $\alpha = 45^\circ$ and the relations among the complex-conjugate poles, ω_n and ζ are illustrated in Figure 8.6. Therefore, the closed-loop control system is given by

$$\frac{C(s)}{R(s)} = \frac{\omega_n^2}{s^2 + 2\zeta\omega_n s + \omega_n^2}. \tag{8.27}$$

Barnes & Noble Bookseller
235 Daniel Webster Hwy
Nashua, NH 03060
(603) 888 0533
[illegible]-12-01 S02701 R007

[illegible]

Barnes & Noble Booksellers
[illegible] 24 hours a day www.bn.com
[illegible]

Booksellers since 1873

We gladly accept returns of new and unread books and unopened music from bn.com with a bn.com receipt for store credit at the bn.com price.

Full refund issued for new and unread books and unopened music within 30 days with a receipt from any Barnes & Noble store.
Store Credit issued for new and unread books and unopened music after 30 days or without a sales receipt. Credit issued at lowest sale price.
We gladly accept returns of new and unread books and unopened music from bn.com with a bn.com receipt for store credit at the bn.com price.

Full refund issued for new and unread books and unopened music within 30 days with a receipt from any Barnes & Noble store.
Store Credit issued for new and unread books and unopened music after 30 days or without a sales receipt. Credit issued at lowest sale price.
We gladly accept returns of new and unread books and unopened music from bn.com with a bn.com receipt for store credit at the bn.com price.

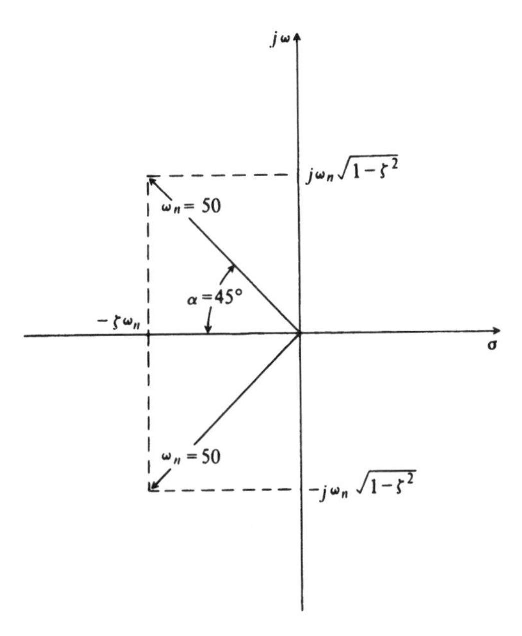

Figure 8.6 Closed-loop poles.

By substituting $\zeta = 0.707$ and $\omega_n = 50$ into Eq. (8.27), we obtain the following desired closed-loop transfer function:

$$\frac{C(s)}{R(s)} = \frac{2500}{s^2 + 70.7s + 2500}. \tag{8.28}$$

Let us assume that the open-loop process that is being controlled is illustrated in Figure 8.7. The corresponding state-variable representation is readily found to be

$$\dot{\mathbf{x}}(t) = \begin{bmatrix} 0 & 1 \\ 0 & -70 \end{bmatrix} \mathbf{x}(t) + \begin{bmatrix} 0 \\ 1 \end{bmatrix} u(t), \tag{8.29}$$

$$c(t) = [1 \quad 0]\,\mathbf{x}(t) \tag{8.30}$$

where

$$\mathbf{x}(t) = \begin{bmatrix} x_1(t) \\ x_2(t) \end{bmatrix} = \begin{bmatrix} c(t) \\ \dot{c}(t) \end{bmatrix}.$$

The resulting linear-state-variable feedback representation is illustrated in Figure 8.8. This feedback represenation can be simplified by the configuration illustrated

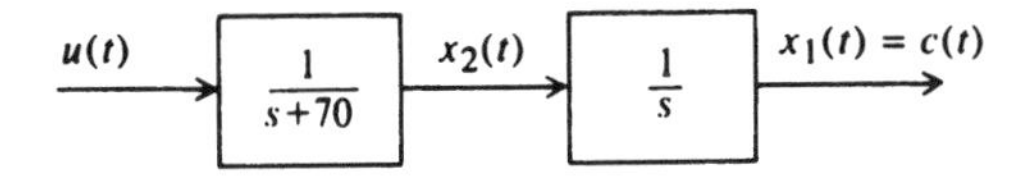

Figure 8.7 Open-loop process to be controlled.

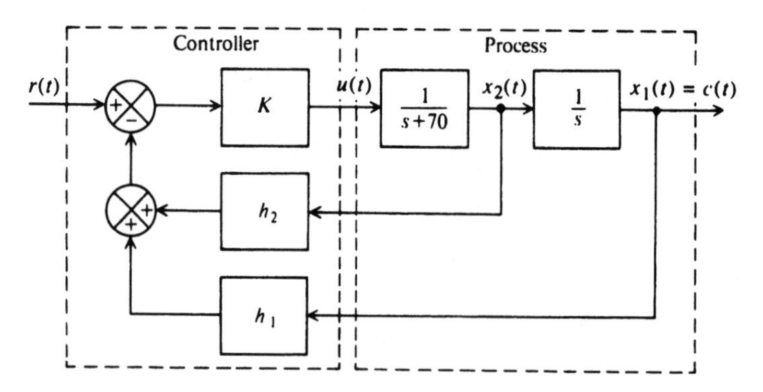

Figure 8.8 State-variable feedback representation of system.

in Figure 8.9. The resulting closed-loop transfer function is given by

$$\frac{C(s)}{R(s)} = \frac{K/[s(s+70)]}{1+(h_1+h_2 s)K/[s(s+70)]}, \tag{8.31}$$

which can be reduced to the following expression:

$$\frac{C(s)}{R(s)} = \frac{K}{s^2+(70+h_2 K)s+Kh_1}. \tag{8.32}$$

The values of $K_1 h_1$ and h_2 can be found from Eqs. (8.28) and (8.32). The following set of simultaneous equations result:

$$K = 2500, \tag{8.33}$$

$$70 + h_2 K = 70.7, \tag{8.34}$$

$$Kh_1 = 2500. \tag{8.35}$$

We have three equations and three unknowns. Solving, we find that $h_1 = 1$, $K = 2500$, and $h_2 = 2.8 \times 10^{-4}$. The final step is to draw the root locus and examine the relative stability, and the sensitivity as a function of slight gain variations. For this simple system, the final step is not necessary.

Although this simple example has been solved using block diagrams and transfer functions, it could also have been solved using the matrix-algebra approach. To illustrate this, let us pick up this problem from Eq. (8.28) which is the desirable closed-loop transfer function. We want to determine the closed-loop transfer function

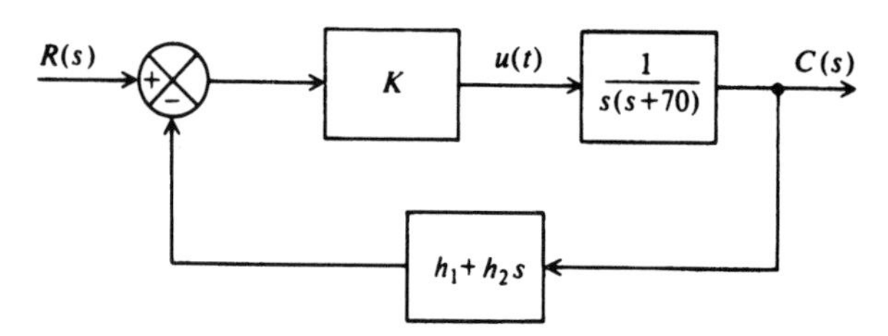

Figure 8.9 Equivalent configuration of Figure 8.8.

for the linear-state-variable-feedback control system using Eq. (8.18) and knowledge of the **P** and **B** matrices from Eq. (8.29), and the matrix **L** from Eq. (8.30) as follows:

$$\frac{C(s)}{R(s)} = \frac{K\mathbf{L}[\text{adj}\,(s\mathbf{I} - \mathbf{P} + K\mathbf{bh})]\mathbf{b}}{\det\,(s\mathbf{I} + \mathbf{P} + K\mathbf{bh})} = \frac{K[1 \quad 0]\,\text{adj}\begin{bmatrix} s & -1 \\ Kh_1 & s + 70 + Kh_2 \end{bmatrix}\begin{bmatrix} 0 \\ 1 \end{bmatrix}}{s^2 + (70 + Kh_2)s + Kh_1}. \tag{8.36}$$

Simplifying Eq. (8.36), we obtain the following:

$$\frac{C(s)}{R(s)} = \frac{K[1 \quad 0]\begin{bmatrix} s + 70 + Kh_2 & 1 \\ -Kh_1 & s \end{bmatrix}\begin{bmatrix} 0 \\ 1 \end{bmatrix}}{s^2 + (70 + Kh_2)s + Kh_1}. \tag{8.37}$$

Simplifying Eq. (8.37) results in the following expression for the closed-loop transfer function of the system:

$$\frac{C(s)}{R(s)} = \frac{K}{s^2 + (70 + Kh_2)s + Kh_1}. \tag{8.38}$$

Equation (8.38) is identical to the closed-loop transfer function we obtained in Eq. (8.32) which was obtained from the block diagram shown in Figure 8.8. Therefore, we repeat the process of setting like terms equal to each other from Eq. (8.38) and Eq. (8.28). The resulting three simultaneous equations of (8.33) through (8.35) will be identical, and the resulting parameters of $K = 1$, $h_1 = 1$, and $h_2 = 2.8 \times 10^{-4}$ will be the same as found before.

With this fundamental example as a basis, the general pole placement design procedure can be formulated as follows:

1. Determine the desired closed-loop transfer function based on the discussion of Section 5.4.
2. Determine the representation of the process to be controlled.
3. Represent the closed-loop system in terms of an equivalent linear-state-variable-feedback configuration.
4. Determine the closed-loop transfer function $C(s)/R(s)$ from the equivalent model in terms of K and **h**.
5. Equate the $C(s)/R(s)$ expressions from Steps 1 and 4 and determine K and **h**.*
6. Plot the resulting root locus of $KG(s)H(s)$ and evaluate the relative stability and sensitivity as a function of gain variations.

Let us apply this pole placement procedure next to the following more complex example. The problem concerns the control of a process in a unity-feedback closed-loop system whose transfer function is given by

$$G(s) = \frac{1}{s(s + 1)(s + 10)}. \tag{8.39}$$

*This assumes that all of the states are measurable. Sections 8.7 and 8.8 discuss the approach to be taken when all the states are not measurable.

It is assumed that the transient response of the system is governed by a pair of dominant complex-conjugate poles, and that the following parameters are desired:

$$K_v = 0.93,$$
$$\zeta = 0.707,$$
$$\omega_n = 1 \text{ rad/sec}.$$

What should the closed-loop transfer function be? From Eq. (5.37), a pair of complex-conjugate poles in the denominator would only have a velocity constant given by

$$K_v = \frac{\omega_n}{2\zeta} = 0.707. \tag{8.40}$$

Therefore, a simple pair of complex-conjugate poles is inadequate to meet the velocity constant requirement of 0.93. By examining Eq. (5.35), we conclude that a zero Z must be added to the closed-loop transfer function. How many poles should the closed-loop system have? Because

$$(N_{P_c} - N_{Z_c})_{C/R} = (N_{P_0} - N_{Z_0})_G, \tag{8.41}$$

where

$$\begin{aligned} N_{P_c} &= \text{number of closed-loop poles} = ? \\ N_{Z_c} &= \text{number of closed-loop zeros} = 1, \\ N_{P_0} &= \text{number of open-loop poles} = 3, \\ N_{Z_0} &= \text{number of open-loop zeros} = 0 \end{aligned}$$

therefore,

$$(N_{P_c} - 1) = (3 - 0), \tag{8.42}$$

and

$$N_{P_c} = 4. \tag{8.43}$$

Since the resulting unity-feedback, closed-loop transfer function has to have one zero and four poles, it has the following general form:

$$\frac{C(s)}{R(s)} = \frac{\omega_n^2 P_3 P_4}{Z} \frac{(s+Z)}{(s^2 + 2\zeta\omega_n s + \omega_n^2)(s+P_3)(s+P_4)}. \tag{8.44}$$

The value of the zero Z can be found from Eq. (5.35) as follows:

$$\frac{1}{K_v} = \frac{2\zeta}{\omega_n} + \frac{1}{P_3} + \frac{1}{P_4} - \frac{1}{Z}. \tag{8.45}$$

Due to external overall system factors in which this feedback system is to operate, it is assumed that the poles at P_3 and P_4 are specified to occur at 9 and 16, respectively. Therefore,

$$\frac{1}{0.93} = \frac{2(0.707)}{1} + \frac{1}{9} + \frac{1}{16} - \frac{1}{Z}, \tag{8.46}$$

so that $Z = 2$, and the desired closed-loop transfer function is given by

$$\frac{C(s)}{R(s)} = \frac{72(s+2)}{(s^2 + 1.414s + 1)(s+9)(s+16)} \tag{8.47}$$

or

$$\frac{C(s)}{R(s)} = \frac{72(s+2)}{s^4 + 26.4s^3 + 180.4s^2 + 299s + 144}. \tag{8.48}$$

Because the zeros of $G(s)$ must be the same as that of $C(s)/R(s)$, we must also add the factor $(s+2)$ to the numerator of $G(s)$. Then, to satisfy Eq. (8.41), we must add a pole factor $(s+\alpha)$ to the denominator of $G(s)$. The resulting compensating network to be added to $G(s)$ is given by

$$G_c(s) = \frac{s+2}{s+\alpha}, \tag{8.49}$$

where α is a pole of the open-loop transfer function which is to be determined. The resulting linear-state-variable feedback system is illustrated in Figure 8.10. The problem remaining is to select the values of K, α, and $\mathbf{h}$.

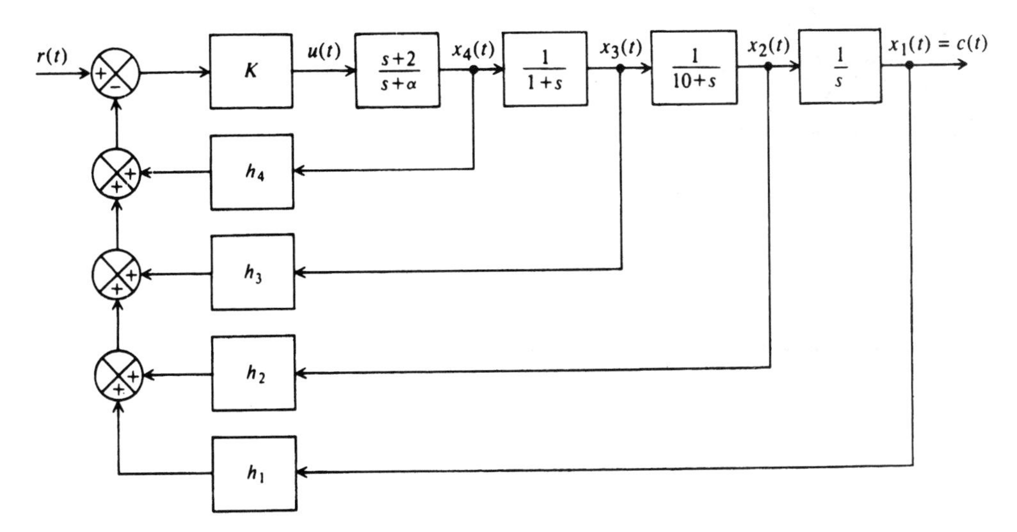

Figure 8.10 State-variable feedback representation of system.

An equivalent block diagram of this system is illustrated in Figure 8.11. The resulting closed-loop transfer function from this equivalent model is given by

$$\frac{C(s)}{R(s)} = \frac{K(s+2)}{\begin{aligned}&\{(Kh_4+1)s^4 + [K(h_3+13h_4)+911+\alpha)]s^3 \\ &+[K(h_2+12h_3+32h_4)+(10+11\alpha)]s^2 \\ &+[K(h_1+2h_2+20h_3+20h_4)+10\alpha]s+2Kh_1\}\end{aligned}} \tag{8.50}$$

Equating the two forms of $C(s)/R(s)$ given by Eqs. (8.48) and (8.50), the following set of equations is obtained:

$$\begin{aligned} K &= 72, \\ Kh_4 + 1 &= 1, \\ K(h_3 + 13h_4) + (11+\alpha) &= 26.4, \\ K(h_2 + 12h_3 + 32h_4) + (10 + 11\alpha) &= 180.4, \\ K(h_1 + 2h_2 + 20h_3 + 20h_4) + 10\alpha &= 229, \\ 2Kh_1 &= 144. \end{aligned}$$

Notice that we have six simultaneous equations with six unknowns (K, h_1, h_2, h_3, h_4, and α). Solving these equations, we obtain the following expressions:

$$\left.\begin{matrix} K = 72, & h_1 = 1, & h_2 = 0.0121 \\ h_3 = 0.0017, & h_4 = 0, & \alpha = 15.28 \end{matrix}\right\} \tag{8.51}$$

From Eq. (8.49), the resulting compensation network, $G_c(s)$, is given by

$$G_c(s) = \frac{s+2}{s+15.28} \tag{8.52}$$

which is a phase-lead network.

It is important to emphasize that α could have turned out to be negative for a different set of specifications This would be undesirable, because it would result in a zero in the right-half plane; this system would be unstable. In other cases, the system might be conditionally stable.

Our results of this pole placement example can be evaluated most conveniently on a root-locus plot. To obtain the root locus of the compensated system, the open-loop

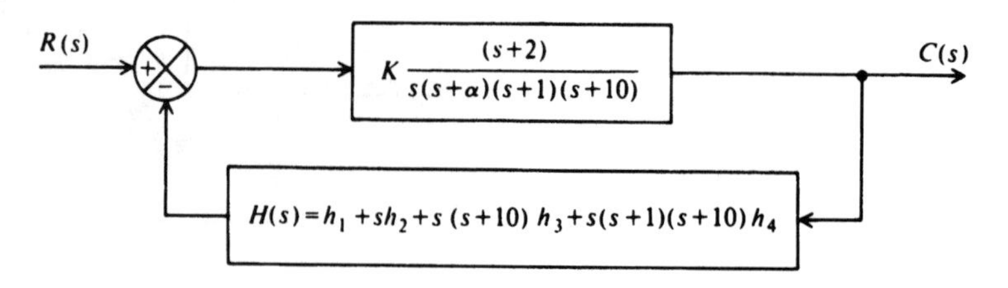

Figure 8.11 Equivalent block diagram for system illustrated in Figure 8.10.

transfer function will be obtained. For the values of the parameters found in Eq. (8.51), $H(s)$ results in the following:

$$H(s) = 0.0017[s + (8.5 + j22.4)][s + (8.5 - j22.4)]. \tag{8.53}$$

Combining Eqs. (8.39), (8.52), and (8.53) we obtain the following transfer function for the open-loop system:

$$KG_c(s)G(s)H(s) = 0.0017\frac{K(s+2)[s + (8.5 + j22.4)]\,[s + (8.5 - j22.4)]}{s(s + 15.28)(s + 10)(s + 1)} \tag{8.54}$$

The resulting root locus is plotted in Figure 8.12. Observe that the resulting root locus is stable for all values of K from zero to infinity. The locations of the dominant complex-conjugate roots for $K = 72$ are indicated.

It is important to emphasize again that the discussion of linear-state-variable feedback in this and the preceding section has assumed that all of the state variables are accessible. This is not always the case. This is analyzed further in Sections 8.6 and 8.7.

8.4. CONTROLLABILITY

The presentation of linear-state-variable feedback in Sections 8.2 and 8.3 assumed that all of the states are observable and measurable, and available to accept control signals (controllable). The concepts of controllability [2–4], and observability also play a very important role in optimal control theory (presented in the accompanying volume). Before designing a control system, we must determine whether it is

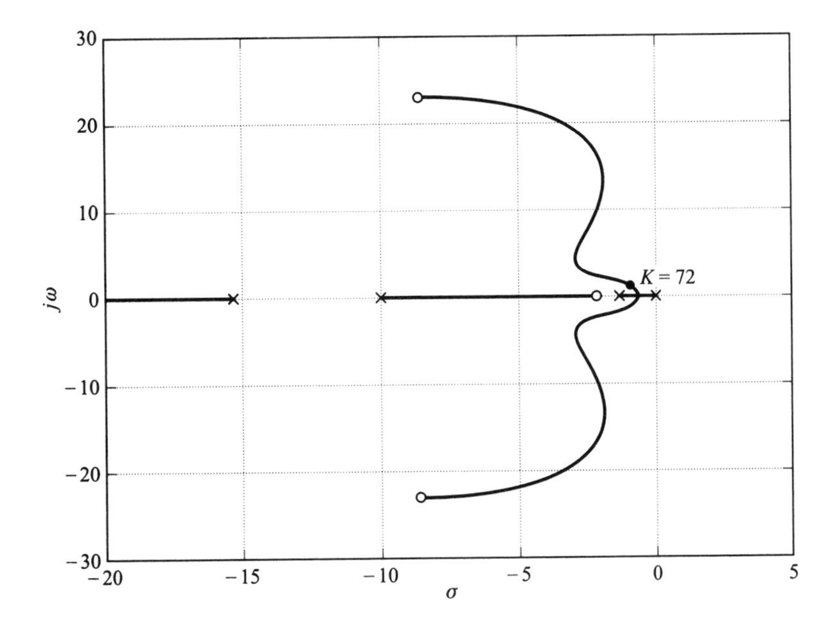

Figure 8.12 Root locus for the system of Figure 8.11 with the parameters of Eq. (8.51).

controllable and its states are observable, since the conditions on controllability and observability often govern the existence of a solution to an optimal control system. Kalman [2,3] first introduced the concepts of controllability and observability in 1960. These concepts are basic in modern optimal control theory. This section develops mathematical tests to determine controllability, and the following section presents mathematical tests for determining observability.

In order to introduce the concept of controllability, let us consider the simple open-loop system illustrated in Figure 8.13. A system is completely controllable if there exists a control which transfers every initial state at $t = t_0$ to any final state at $t = T$ for all t_0 and T. Qualitatively, this means that the system $G(s)$ is controllable if every state variable of G can be affected by the input signal $\mathbf{u}(t)$. However, if one (or several) of the state variables is (or are) not affected by $\mathbf{u}(t)$, then this (or these) state variable(s) cannot be controlled in a finite amount of time by $\mathbf{u}(t)$ and the system is not completely controllable.

A. Controllability by Inspection

As an example of a system which is not completely controllable, let us consider the signal-flow diagram illustrated in Figure 8.14. This system contains four states, only two of which are affected by $u(t)$. This input only affects the states $x_1(t)$ and $x_2(t)$. It has no effect on $x_3(t)$ and $x_4(t)$. Therefore, $x_3(t)$ and $x_4(t)$ are uncontrollable. This means that it is impossible for $u(t)$ to change $x_3(t)$ from initial state $x_3(0)$ to final state $x_3(T)$ in a finite time interval T and the system is not completely controllable.

B. The Controllability Matrix

Let us now consider this problem more precisely and establish a mathematical criterion for determining whether a system is controllable. We limit our discussion to linear constant systems. Assume that the system is described by

$$\dot{\mathbf{x}}(t) = \mathbf{P}\mathbf{x}(t) + \mathbf{B}\mathbf{u}(t), \tag{8.55}$$

$$\mathbf{c}(t) = \mathbf{L}\mathbf{x}(t). \tag{8.56}$$

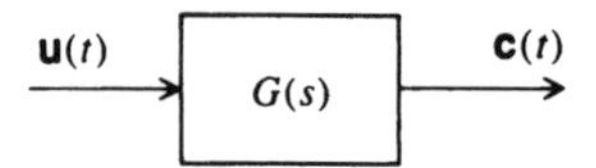

Figure 8.13 Open-loop system containing several inputs and outputs.

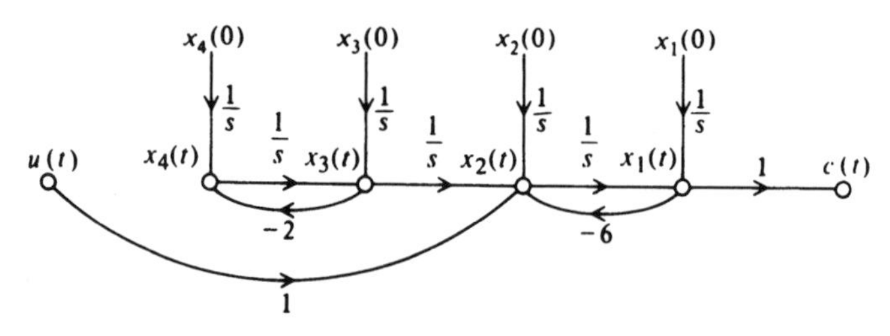

Figure 8.14 Signal-flow graph of a system that is not completely controllable.

The solution of Eq. (8.55) can be expressed as Eq. (2.265):

$$\mathbf{x}(t) = \mathbf{\Phi}(t - t_0)\mathbf{x}(t_0) + \int_{t_0}^{t} \mathbf{\Phi}(t - \tau)\mathbf{Bu}(\tau)d\tau, \quad \text{where } t \geqslant t_0. \tag{8.57}$$

Let us assume that the desired final state of our system at $t = t_f$ is zero:

$$\mathbf{x}(t_f) = \mathbf{0}. \tag{8.58}$$

Using Eqs. (2.259) and (2.261), we can write Eq. (8.57) as

$$\mathbf{x}(t_0) = -\int_{t_0}^{t_f} \mathbf{\Phi}(t_0 - \tau)\mathbf{Bu}(\tau)d\tau. \tag{8.59}$$

The state-transition matrix can be expressed from Eq. (2.322) as

$$\mathbf{\Phi}(t) = e^{\mathbf{P}t} = \sum_{n=0}^{m-1} \alpha_n(t)\mathbf{P}^n. \tag{8.60}$$

Here, $\mathbf{x}$ is an $m \times 1$ vector, $\mathbf{P}$ is an $m \times m$ matrix, $\mathbf{u}$ is an $r \times 1$ vector, $\mathbf{B}$ is an $m \times r$ matrix, and $\alpha_n(t)$ is a scalar function of t. (This form results from application of the Cayley–Hamilton theorem). Substituting Eq (8.60) into Eq. (8.59), we obtain the following expression for $\mathbf{x}(t_0)$:

$$\mathbf{x}(t_0) = -\int_{t_0}^{t_f} \sum_{n=0}^{m-1} \alpha_n(t_0 - \tau)\mathbf{P}^n\mathbf{Bu}(\tau)d\tau. \tag{8.61}$$

Because the matrices $\mathbf{P}$ and $\mathbf{B}$ are not functions of τ, we can rewrite Eq. (8.61) as

$$\mathbf{x}(t_0) = -\sum_{n=0}^{m-1} \mathbf{P}^n\mathbf{B} \int_{t_0}^{t_f} \alpha_n(t_0 - \tau)\mathbf{u}(\tau)d\tau. \tag{8.62}$$

Equation (8.62) can be rewritten as

$$\mathbf{x}(t_0) = -[\mathbf{B} \quad \mathbf{PB} \quad \mathbf{P}^2\mathbf{B} \quad \mathbf{P}^3\mathbf{B} \cdots \mathbf{P}^{m-1}\mathbf{B}] \begin{bmatrix} \mathbf{A}_0 \\ \mathbf{A}_1 \\ \mathbf{A}_2 \\ \vdots \\ \mathbf{A}_{m-1} \end{bmatrix}, \tag{8.63}$$

where

$$\mathbf{A}_n = \int_{t_0}^{t_f} \alpha_n(t_0 - \tau)\mathbf{u}(\tau)d\tau. \tag{8.64}$$

If we define

$$\mathbf{D} = [\mathbf{B} \quad \mathbf{PB} \quad \mathbf{P}^2\mathbf{B} \quad \mathbf{P}^3\mathbf{B} \cdots \mathbf{P}^{m-1}\mathbf{B}] \tag{8.65}$$

$$\mathbf{A} = [\mathbf{A}_0 \quad \mathbf{A}_1 \quad \mathbf{A}_2 \cdots \mathbf{A}_{m-1}]^T, \tag{8.66}$$

where $\mathbf{D}$, the controllability matrix, is an $m \times mr$ matrix and $\mathbf{A}$ is an $mr \times 1$ vector, then Eq. (8.63) becomes

$$\mathbf{x}(t_0) = -\mathbf{DA}. \tag{8.67}$$

For a given initial state $\mathbf{x}(t_0)$, the input $\mathbf{u}$ can be found to drive the state to $\mathbf{x}(t_f) = \mathbf{0}$ for a finite time interval $t_f - t_0$ if Eq. (8.67) has a solution. A unique solution occurs only if there is a set of m linearly independent column vectors in the matrix $\mathbf{D}$. If $\mathbf{u}$ is a scalar, then $\mathbf{D}$ is an $m \times m$ square matrix, and Eq. (8.67) represents a set of m linearly independent equations which have a solution if $\mathbf{D}$ is nonsingular or the determinant of $\mathbf{D}$ is not zero. The controllability criterion thus states that the system of Eqs. (8.55) and (8.56) is completely controllable if $\mathbf{D}$ [see Eq. (8.65)] contains m linearly independent column vectors or, if $\mathbf{u}$ is a scalar, $\mathbf{D}$ is nonsingular.

In order to illustrate this mathematical controllability concept, consider a second-order system where

$$\mathbf{P} = \begin{bmatrix} -4 & 1 \\ 0 & -2 \end{bmatrix}$$

and

$$\mathbf{B} = \begin{bmatrix} 1 \\ 0 \end{bmatrix}.$$

Then, from Eq. (8.65),

$$\mathbf{D} = [\mathbf{B} \quad \mathbf{PB}] = \begin{bmatrix} 1 & -4 \\ 0 & 0 \end{bmatrix}. \tag{8.68}$$

The resulting matrix $\mathbf{D}$ is singular (its determinant is zero) and the system is therefore not completely controllable.

As a second example, consider a second-order system where

$$\mathbf{P} = \begin{bmatrix} -3 & 2 \\ 4 & 1 \end{bmatrix}$$

and

$$\mathbf{B} = \begin{bmatrix} 0 \\ 1 \end{bmatrix}.$$

Then, from Eq. (8.65)

$$\mathbf{D} = [\mathbf{B} \quad \mathbf{PB}] = \begin{bmatrix} 0 & 2 \\ 1 & 1 \end{bmatrix}.$$

The resulting matrix $\mathbf{D}$ is nonsingular and the system, therefore, is completely controllable.

8.5. OBSERVABILITY

In order to introduce the concept of observability [2–4], let us again consider the simple open-loop system illustrated in Figure 8.13. A system is completely observable if, given the control and the output over the interval $t_0 \leqslant t \leqslant T$, one can determine the initial state $\mathbf{x}(t_0)$. Qualitatively, the system G is observable if every state variable of G affects some of the outputs in $\mathbf{c}$. It is very often desirable to determine information regarding the system states based on measurements of $\mathbf{c}$. However, if we cannot observe one or more of the states from the measurements of $\mathbf{c}$, then the system is not completely observable. We had assumed the systems were observable in our discussion of pole placement using linear-state variable feedback in Sections 8.2 and 8.3.

A. Observability by Inspection

An an example of a system which is not completely observable, let us consider the signal-flow diagram illustrated in Figure 8.15. This system contains four states, only two of which are observable. The states $x_3(t)$ and $x_4(t)$ are not connected to the output $c(t)$ in any manner. Therefore, $x_3(t)$ and $x_4(t)$ are not observable and the system is not completely observable.

B. The Observability Matrix

Let us now consider this problem more precisely and establish a mathematical criterion for determining whether a system is completely observable. Again, we limit our discussion to linear constant systems of the form

$$\dot{\mathbf{x}}(t) = \mathbf{P}\mathbf{x}(t) + \mathbf{B}\mathbf{u}(t), \tag{8.69}$$

$$\mathbf{c}(t) = \mathbf{L}\mathbf{x}(t) \tag{8.70}$$

where $\mathbf{x}$ is an $m \times 1$ vector, $\mathbf{P}$ is an $m \times m$ matrix, $\mathbf{u}$ is an $r \times 1$ vector, $\mathbf{B}$ is an $m \times r$ matrix, $\mathbf{c}$ is a $p \times 1$ vector, and $\mathbf{L}$ is a $p \times m$ matrix. The solution of Eq. (8.69) is given by [see Eq. (2.265)]

$$\mathbf{x}(t) = \mathbf{\Phi}(t - t_0)\mathbf{x}(t_0) + \int_{t_0}^{t} \mathbf{\Phi}(t - \tau)\mathbf{B}\mathbf{u}(\tau)d\tau. \tag{8.71}$$

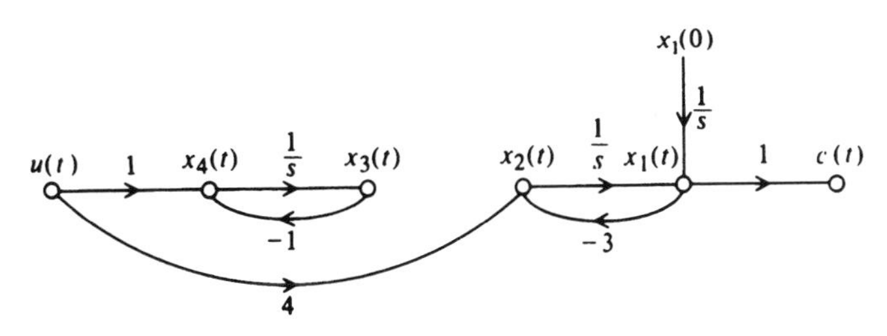

Figure 8.15 Signal-flow graph of a system that is not completely observable.

It will now be shown that observability depends on the matrices **P** and **L**. Substituting Eq. (8.71) into Eq. (8.70), we obtain

$$\mathbf{c}(t) = \mathbf{L\Phi}(t - t_0)\mathbf{x}(t_0) + \mathbf{L}\int_{t_0}^{t} \mathbf{\Phi}(t - \tau)\mathbf{Bu}(\tau)d\tau, \text{ where } t \geqslant t_0. \tag{8.72}$$

From the definition of observability we can see that the observability of $\mathbf{x}(t_0)$ depends on the term $\mathbf{L\Phi}(t - t_0)\mathbf{x}(t_0)$. Therefore, the output $\mathbf{c}(t)$ when $\mathbf{u} = \mathbf{0}$ is given by

$$\mathbf{c}(t) = \mathbf{L\Phi}(t - t_0)\mathbf{x}(t_0). \tag{8.73}$$

Substituting Eq. (8.60) into Eq. (8.73), we obtain the following expression for the output $\mathbf{c}(t)$:

$$\mathbf{c}(t) = \sum_{n=0}^{m-1} \alpha_n(t - t_0)\mathbf{LP}^n\mathbf{x}(t_0). \tag{8.74}$$

Equation (8.74) indicates that if the output $\mathbf{c}(t)$ is known over the time interval $t_0 \leqslant t \leqslant T$, then $\mathbf{x}(t_0)$ is uniquely determined from this equation if $\mathbf{x}(t_0)$ is a linear combination of $(\mathbf{L}_j\mathbf{P}^n)^T$ for $n = 0, 1, 2, \ldots, m - 1$, and $j = 1, 2, 3, \ldots, r$. The matrix $\mathbf{L}_j$ is the $1 \times m$ matrix formed by the elements of the jth row of **L**. Because $(\mathbf{L}_j\mathbf{P}^n)^T = (\mathbf{P}^T)^n\mathbf{L}_j^T$,we let **U** be the $m \times mr$ matrix defined by

$$\mathbf{U} = [\mathbf{L}^T \quad \mathbf{P}^T\mathbf{L}^T \quad (\mathbf{P}^T)^2\mathbf{L}^T \quad (\mathbf{P}^T)^3\mathbf{L}^T \cdots (\mathbf{P}^T)^{m-1}\mathbf{L}^T]. \tag{8.75}$$

The observability criterion states that the system is completely observable if there is a set of m linearly independent column vectors in the observability matrix **U**.

In order to illustrate the observability concept mathematically, consider a second-order system where

$$\mathbf{P} = \begin{bmatrix} -4 & 0 \\ 0 & -2 \end{bmatrix}$$

and

$$\mathbf{L} = [1 \quad 0].$$

Therefore,

$$\mathbf{L}^T = \begin{bmatrix} 1 \\ 0 \end{bmatrix}$$

and

$$\mathbf{P}^T\mathbf{L}^T = \begin{bmatrix} -4 & 0 \\ 0 & -2 \end{bmatrix}\begin{bmatrix} 1 \\ 0 \end{bmatrix} = \begin{bmatrix} -4 \\ 0 \end{bmatrix}.$$

Substituting these values into Eq. (8.75), we obtain

$$\mathbf{U} = [\mathbf{L}^T \quad \mathbf{P}^T\mathbf{L}^T] = \begin{bmatrix} 1 & -4 \\ 0 & 0 \end{bmatrix}.$$

Because **U** is singular, the sytem is not completely observable.

As a second example, consider the second-order system where

$$\mathbf{P} = \begin{bmatrix} 4 & -2 \\ 4 & -2 \end{bmatrix}$$

and

$$\mathbf{L} = [1 \quad 1].$$

Therefore,

$$\mathbf{L}^T = \begin{bmatrix} 1 \\ 1 \end{bmatrix}$$

and

$$\mathbf{P}^T\mathbf{L}^T = \begin{bmatrix} 4 & 4 \\ -2 & -2 \end{bmatrix}\begin{bmatrix} 1 \\ 1 \end{bmatrix} = \begin{bmatrix} 8 \\ -4 \end{bmatrix}.$$

Substituting these values into Eq. (8.75), we obtain

$$\mathbf{U} = [\mathbf{L}^T \quad \mathbf{P}^T\mathbf{L}^T] = \begin{bmatrix} 1 & 8 \\ 1 & -4 \end{bmatrix}.$$

Because **U** has two independent columns, the system is completely observable.

8.6. ACKERMANN'S FORMULA FOR DESIGN USING POLE PLACEMENT [5–7]

In addition to the method of matching the coefficients of the desired characteristic equation with the coefficients of det $(s\mathbf{I} - \mathbf{P}_h)$ as given by Eq. (8.19), Ackermann has developed a competing method. The pole placement method using the matching of coefficients of the desired characteristic equation with the coefficients of Eq. (8.19) is very useful for control systems which are represented in phase-variable form, where *phase variable* refers to systems where each subsequent state variable is defined as the derivative of the previous state variable. Some control systems require feedback from state variables which are not phase variables. Such high-order control systems can lead to very complex calculations for the feedback gains. Ackermann's method simplifies this problem by transforming the control system to phase variables, determining the feedback gains, and transforming the designed control system back to its original state-variable representation.

Let us represent a control system which is not represented in phase-variable form by the following:

$$\dot{\mathbf{y}}(t) = \mathbf{P}\mathbf{y}(t) + \mathbf{B}u(t), \tag{8.76}$$

$$c(t) = \mathbf{L}\mathbf{y}(t). \tag{8.77}$$

We will assume that the controllability matrix [see Eq. (8.65)] can be represented by

$$\mathbf{D}_y = [\mathbf{B} \ldots \mathbf{PB} \ldots \mathbf{P}^2\mathbf{B} \ldots \mathbf{P}^{m-1}\mathbf{B}]. \tag{8.78}$$

Subscript y is used to designate the original, non-phase-variable, controllability matrix. We will next assume that the control system can be transformed into the phase-variable representation using the following transformation:

$$\mathbf{y}(t) = \mathbf{A}\mathbf{x}(t). \tag{8.79}$$

Substituting Eq. (8.79) into Eq. (8.76) and Eq. (8.77), we obtain:

$$\dot{\mathbf{x}}(t) = \mathbf{A}^{-1}\mathbf{PA}\mathbf{x}(t) + \mathbf{A}^{-1}\mathbf{B}u(t), \tag{8.80}$$

$$c(t) = \mathbf{LA}\mathbf{x}(t). \tag{8.81}$$

Using the transformation of Eq. (8.79), the controllability matrix for the transformed system defined by Eqs. (8.80) and (8.81) is:

$$\mathbf{D}_x = [\mathbf{A}^{-1}\mathbf{B} \ldots (\mathbf{A}^{-1}\mathbf{PA})(\mathbf{A}^{-1}\mathbf{B}) \ldots (\mathbf{A}^{-1}\mathbf{PA})^2(\mathbf{A}^{-1}\mathbf{B}) \ldots (\mathbf{A}^{-1}\mathbf{PA})^{m-1}(\mathbf{A}^{-1}\mathbf{B})]. \tag{8.82}$$

The subscript x is used to designate the phase-variable form of the transformed control system. Equation (8.82) can be simplified to

$$\mathbf{D}_x = \mathbf{A}^{-1}[\mathbf{B} \ldots \mathbf{PB} \ldots \mathbf{P}^2\mathbf{B} \ldots \mathbf{P}^{m-1}\mathbf{B}]. \tag{8.83}$$

Therefore, substituting Eq. (8.78) into Eq. (8.83), we find that

$$\mathbf{D}_x = \mathbf{A}^{-1}\mathbf{D}_y. \tag{8.84}$$

The transformation matrix $\mathbf{A}$ can be found from

$$\mathbf{A} = \mathbf{D}_y\mathbf{D}_x^{-1}. \tag{8.85}$$

Therefore, the transformation matrix $\mathbf{A}$ can be determined from the controllability matrices defined by Eq. (8.65) and (8.83). Once the control system is transformed to the phase-variable form, the feedback gains $\mathbf{h}$ can be determined as described in Section 8.2.

Returning to Eq. (8.4) to represent $u(t)$ for the control system in phase-variable form,

$$u(t) = K[r(t) - \mathbf{hx}(t)] \tag{8.86}$$

the following is obtained by substituting Eq. (8.86) into Eq. (8.80):

$$\dot{\mathbf{x}}(t) = \mathbf{A}^{-1}\mathbf{PA}x(t) + \mathbf{A}^{-1}\mathbf{B}(K\,[r(t) - \mathbf{hx}(t)]). \tag{8.87}$$

Equation (8.87) can be simplified to the following phase-variable state equation:

$$\dot{\mathbf{x}}(t) = (\mathbf{A}^{-1}\mathbf{PA} - K\mathbf{A}^{-1}\mathbf{Bh})\mathbf{x}(t) + \mathbf{A}^{-1}\mathbf{B}Kr(t). \tag{8.88}$$

The output equation remains as shown in Eq. (8.81):

$$c(t) = \mathbf{LAx}(t). \tag{8.89}$$

Because Eqs. (8.88) and (8.89) are in phase-variable form, the rules for pole placement developed in Section 8.2 for phase-variable systems are valid for this representation. We now have to transform Eqs. (8.88) and (8.89) back to the original state and output equation representation by using the transformation provided by Eq. (8.79):

$$\mathbf{x}(t) = \mathbf{A}^{-1}\mathbf{y}(t). \tag{8.90}$$

Substituting Eq. (8.90) into Eq. (8.88), we obtain the following:

$$\mathbf{A}^{-1}\dot{\mathbf{y}}(t) = (\mathbf{A}^{-1}\mathbf{PA} - K\mathbf{A}^{-1}\mathbf{Bh})\mathbf{A}^{-1}\mathbf{y}(t) + \mathbf{A}^{-1}\mathbf{BKr}(t). \tag{8.91}$$

Therefore, Eq. (8.91) reduces to the following state equation:

$$\dot{\mathbf{y}}(t) = \mathbf{P}y(t) - K\mathbf{BhA}^{-1}\mathbf{y}(t) + \mathbf{B}Kr(t). \tag{8.92}$$

The output equation is found by substituting Eq. (8.79) in Eq. (8.89):

$$\mathbf{c}(t) = \mathbf{LAA}^{-1}\mathbf{y}(t) = \mathbf{Ly}(t). \tag{8.93}$$

By comparing Eq. (8.92) with Eqs. (8.8) and (8.10), we find that the state-variable feedback constants for the original system are

$$\mathbf{h}_y = \mathbf{h}_x\mathbf{A}^{-1}. \tag{8.94}$$

A. Example Applying Ackermann's Formula for Design using Pole Placement

We wish to apply the pole placement concepts using Ackermann's design formula to a system that uses linear-state-variable feedback. The system specifications require an overshoot of 4.33% and a settling time of 6 sec. The process transfer function is

$$\frac{C(s)}{U(s)} = \frac{(s+3)}{(s+2)(s+1)(s+4)} = \frac{(s+3)}{s^3 + 7s^2 + 14s + 8} \tag{8.95}$$

and the control system's signal-flow graph is given in Figure 8.16.

The state equation for the process illustrated in Figure 8.16 is:

$$\dot{\mathbf{y}}(t) = \mathbf{P}_y \mathbf{y}(t) + \mathbf{B}_y u(t) = \begin{bmatrix} \dot{y}_1(t) \\ \dot{y}_2(t) \\ \dot{y}_3(t) \end{bmatrix} = \begin{bmatrix} -4 & 1 & 0 \\ 0 & -1 & 1 \\ 0 & 0 & -2 \end{bmatrix} \begin{bmatrix} y_1(t) \\ y_2(t) \\ y_3(t) \end{bmatrix} + \begin{bmatrix} 0 \\ 0 \\ 1 \end{bmatrix} u(t). \tag{8.96}$$

Since we are going to need the controllability matrix $\mathbf{D}_y$ to convert this original system to phase-variable form [see Eq. (8.85)], let us compute this controllability matrix for this original system:

$$\mathbf{D}_y = [\mathbf{B}_y \quad \mathbf{P}_y \mathbf{B}_y \quad \mathbf{P}_y^2 \mathbf{B}_y] = \begin{bmatrix} 0 & 0 & 1 \\ 0 & 1 & -3 \\ 1 & -2 & 4 \end{bmatrix}. \tag{8.97}$$

The value of the determinant of $\mathbf{D}_y$ is -1, it is nonsingular, and the original control system is controllable as expected from inspection of Figure 8.16.

The next step is to transform the original system to the phase-variable form. This can easily be obtained from the transfer function $C(s)/U(s)$, which can be determined from Figure 8.16:

$$\frac{C(s)}{U(s)} = \frac{\dfrac{3}{s^3} + \dfrac{1}{s^2}}{1 + \left[\dfrac{2}{s} + \dfrac{1}{s} + \dfrac{4}{s}\right] + \left[\left(\dfrac{2}{s}\right)\left(\dfrac{1}{s}\right) + \left(\dfrac{2}{s}\right)\left(\dfrac{4}{s}\right) + \left(\dfrac{1}{s}\right)\left(\dfrac{4}{s}\right)\right] + \left[\left(\dfrac{2}{s}\right)\left(\dfrac{1}{s}\right)\left(\dfrac{4}{s}\right)\right]}. \tag{8.98}$$

Eq. (8.98) can be reduced to the following [as stated in Eq. (8.95)]:

$$\frac{C(s)}{U(s)} = \frac{(s+3)}{s^3 + 7s^2 + 14s + 8}. \tag{8.99}$$

We can determine the phase-variable state equations which represent the transfer function given by Eq. (8.99). Defining $x_1(t) = c(t)$, $x_2(t) = \dot{c}(t)$, and $x_3(t) = \ddot{c}(t)$, we obtain the following:

$$\dot{x}_1(t) = x_2(t) \tag{8.100}$$

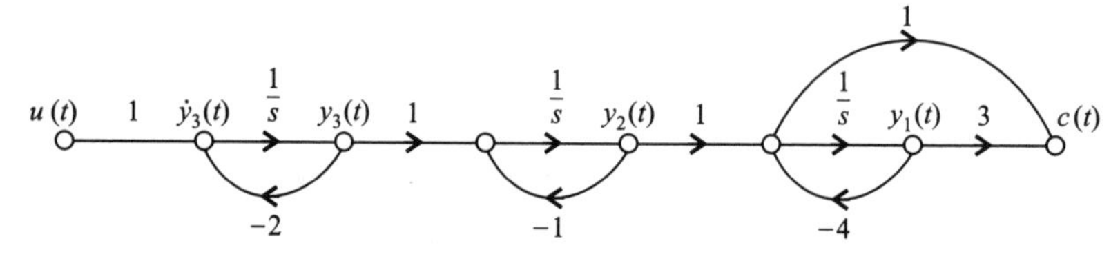

Figure 8.16 Signal-flow graph representation of process whose transfer function is defined in Eq. (8.95).

$$\dot{x}_2(t) = x_3(t) \tag{8.101}$$

$$\dot{x}_3(t) = -8x_1(t) - 14x_2(t) - 7x_3(t) + 3u(t) + \dot{u}(t) \tag{8.102}$$

Therefore, the state and output equations in matrix vector form of the phase-variable control system are given by:

$$\dot{\mathbf{x}}(t) = \mathbf{P}_x\mathbf{x}(t) + \mathbf{B}_x u(t) = \begin{bmatrix} 0 & 1 & 0 \\ 0 & 0 & 1 \\ -8 & -14 & -7 \end{bmatrix} \begin{bmatrix} x_1(t) \\ x_2(t) \\ x_3(t) \end{bmatrix} + \begin{bmatrix} 0 \\ 0 \\ 1 \end{bmatrix} (3u(t) + \dot{u}(t)), \tag{8.103}$$

$$c(t) = [3 \quad 1 \quad 0] \begin{bmatrix} x_1(t) \\ x_2(t) \\ x_3(t) \end{bmatrix} \tag{8.104}$$

The resulting controllability matrix for the phase-variable form can be obtained from Eqs. (8.65) and (8.103) and is given by:

$$\mathbf{D}_x = [\mathbf{B}_x \ \mathbf{P}_x\mathbf{B}_x \ \mathbf{P}_x^2\mathbf{B}_x] = \begin{bmatrix} 0 & 0 & 1 \\ 0 & 1 & -7 \\ 1 & -7 & 35 \end{bmatrix}. \tag{8.105}$$

Note that the determinant of $\mathbf{D}_x$ is -1 indicating that the determinant is nonsingular, and the phase-variable form is also controllable, as expected. The transformation matrix, defined in Eq. (8.85), can be determined from Eqs. (8.97) and (8.105) as follows:

$$\mathbf{A} = \mathbf{D}_y\mathbf{D}_x^{-1} = \begin{bmatrix} 0 & 0 & 1 \\ 0 & 1 & -3 \\ 1 & -2 & 4 \end{bmatrix} \begin{bmatrix} 14 & 7 & 1 \\ 7 & 1 & 0 \\ 1 & 0 & 0 \end{bmatrix} = \begin{bmatrix} 1 & 0 & 0 \\ 4 & 1 & 0 \\ 4 & 5 & 1 \end{bmatrix} \tag{8.106}$$

The next step in the procedure is to design the controller (see Figure 8.3) using the phase-variable representation, after which we will transform the design back to the original system by using Eq. (8.106). The 4.33 percent overshoot specified can be obtained from a second-order control system having a damping ratio of 0.707 [see Eq. (4.33)]. The settling time of 6 sec can be obtained from a second-order control system having a damping ratio of 0.707 and a $\omega_n = 0.943$ [see Eq. (5.41)]. Therefore, the following second-order control system can meet the design specifications:

$$\frac{C(s)}{R(s)} = \frac{\omega_n^2}{s^2 + 2\zeta\omega_n s + \omega_n^2} = \frac{0.889}{s^2 + 1.333s + 0.889}. \tag{8.107}$$

Since it was shown in Section 8.2 that zeros of closed-loop systems are zeros of open-loop systems, then we can select the third pole at $s = -3$ which will also cancel the zero at $s = -3$. Therefore, the characteristic equation of the desired closed-loop system is

$$(s^2 + 1.333s + 0.889)(s + 3) = s^3 + 4.333s^2 + 4.888s + 2.667 = 0. \tag{8.108}$$

The state and output equations for the phase-variable form with linear-state-variable-feedback can be obtained from Eqs. (8.8), (8.9), and (8.10) as follows:

$$\dot{\mathbf{x}}(t) = [\mathbf{P}_x - K\mathbf{b}_x\mathbf{h}_x]\mathbf{x}(t) + Kr(t)\mathbf{b}_x =$$

$$\begin{bmatrix} 0 & 1 & 0 \\ 0 & 0 & 1 \\ -8 & -14 & -7 \end{bmatrix} - \begin{bmatrix} 0 & 0 & 0 \\ 0 & 0 & 0 \\ Kh_1 & Kh_2 & Kh_3 \end{bmatrix} \begin{bmatrix} x_1(t) \\ x_2(t) \\ x_3(t) \end{bmatrix} + Ku(t) \begin{bmatrix} 0 \\ 0 \\ 1 \end{bmatrix}, \tag{8.109}$$

$$c(t) = [3 \quad 1 \quad 0]\mathbf{x}(t) \tag{8.110}$$

Eq. (8.109) reduces to the following:

$$\dot{\mathbf{x}}(t) = \begin{bmatrix} 0 & 1 & 0 \\ 0 & 0 & 1 \\ -8 - Kh_1 & -14 - Kh_2 & -7 - Kh_3 \end{bmatrix} \mathbf{x}(t) + \begin{bmatrix} 0 \\ 0 \\ Ku(t) \end{bmatrix}. \tag{8.111}$$

The resulting characteristic equation can be obtained from Eqs. (8.10) and (8.19):

$$\det(s\mathbf{I} - \mathbf{P}_h) = \det(s\mathbf{I} - \mathbf{P} + K\mathbf{bh}) = 0. \tag{8.112}$$

Substituting $-(\mathbf{P}_x - K\mathbf{b}_x\mathbf{h}_x)$ from Eq. (8.111) into Eq. (8.112), we obtain the following:

$$\begin{vmatrix} s & -1 & 0 \\ 0 & s & -1 \\ 8 + Kh_1 & 14 + Kh_2 & s + 7 + Kh_3 \end{vmatrix} = 0. \tag{8.113}$$

The resulting characteristic equation is given by:

$$s^3 + (7 + Kh_3)s^2 + (14 + Kh_2)s + (8 + Kh_1) = 0. \tag{8.114}$$

Comparing Eq. (8.114) and Eq. (8.108), we obtain the following three equations:

$$7 + Kh_3 = 4.333; \quad Kh_3 = -2.667, \qquad 8.115)$$

$$14 + KH_2 = 4.888; \quad Kh_2 = -9.112, \tag{8.116}$$

$$8 + Kh_1 = 2.667; \quad Kh_1 = 5.333. \tag{8.117}$$

Therefore,

$$K\mathbf{h}_x = [-5.333 \quad -9.112 \quad -2.667] \tag{8.118}$$

We will now transform $K\mathbf{h}_x$ as shown in Eq. (8.118) back to the original system using Eq. (8.94) and Eq. (8.106) as follows:

$$\mathbf{h}_y = \mathbf{h}_x\mathbf{A}^{-1} =$$

$$[-5.333 \quad -9.112 \quad -2.667] \begin{bmatrix} 1 & 0 & 0 \\ 4 & 1 & 0 \\ 4 & 5 & 1 \end{bmatrix}^{-1} = [-11.557 \quad 4.223 \quad -2.667]. \tag{8.119}$$

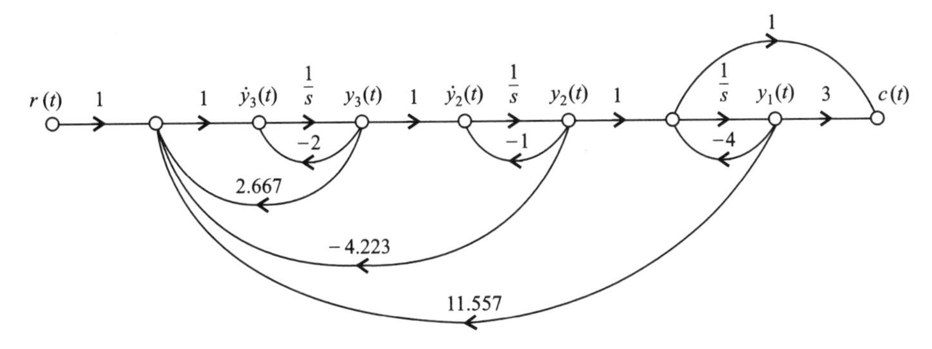

Figure 8.17 Resulting closed-loop control system with linear-state-variable-feedback designed using Ackermann's Formula.

The resulting closed-loop control system with linear-state-variable-feedback is shown in Figure 8.17.

8.7. ESTIMATOR DESIGN IN CONJUNCTION WITH THE POLE PLACEMENT APPROACH USING LINEAR-STATE-VARIABLE FEEDBACK

In the discussion of Sections 8.2 and 8.3 on linear-state-variable feedback, it was assumed that all of the states are observable and measurable, and available to accept control signals (controllable). As Sections 8.4 and 8.5 on controllability and observability have shown, some states of a feedback control system may not always be controllable and/or observable. In some systems, the system may be observable, but all of the states may not be measured due to physical limitations (e.g., chemical process control systems), or it may be due to cost restrictions that limit the use of costly sensors needed to measure all of the states. It is assumed in this section that the system is observable (no part of the system is disconnected physically from the output), but measurements are being made only on some of the states, and we wish to estimate all of the states.

Let us focus attention on the process portion of the system illustrated in Figure 8.3. We wish to use the closed-loop estimator system shown in Figure 8.18 for determining an estimate of the state vector $\mathbf{x}(t)$ and output $c(t)$ [8]. This estimator feeds back the difference between the measured output $c(t)$ and the estimated output $\hat{c}(t)$ that is obtained from a model of the process. Therefore,

$$\dot{\hat{\mathbf{x}}}(t) = \mathbf{P}\hat{\mathbf{x}}(t) + \mathbf{b}u(t) + \mathbf{M}(c(t) - \mathbf{L}\hat{\mathbf{x}}(t)), \tag{8.120}$$

where $\mathbf{M}$ defines the gain factors m_i, which are selected to obtain desirable error characteristics of the state vector $\mathbf{x}$, and $\hat{\mathbf{x}}(t)$ represents the estimate of the state $\mathbf{x}(t)$:

$$\mathbf{M} = [m_1, m_2, \ldots, m_n]^T. \tag{8.121}$$

The error in the state estimate, $\tilde{\mathbf{x}}(t)$, can be derived from

$$\tilde{\mathbf{x}}(t) = \mathbf{x}(t) - \hat{\mathbf{x}}(t). \tag{8.122}$$

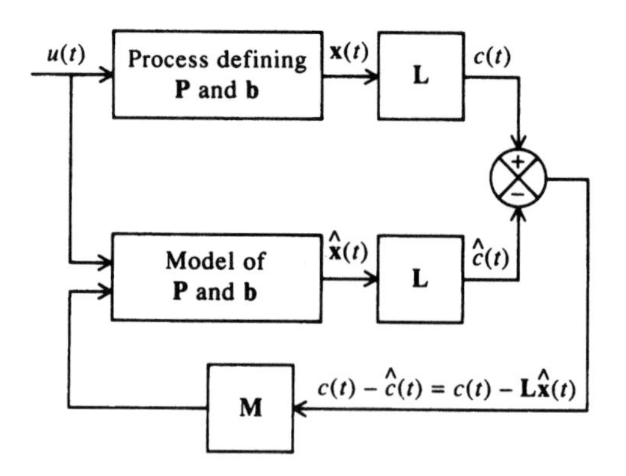

Figure 8.18 An estimator system.

The derivative of the $\tilde{\mathbf{x}}(t)$ can be obtained by subtracting $\dot{\hat{\mathbf{x}}}(t)$ [given by Eq. (8.120)] from $\dot{\mathbf{x}}(t)$ given by the system dynamics:

$$\dot{\mathbf{x}}(t) = \mathbf{P}\mathbf{x}(t) + \mathbf{b}u(t) \tag{8.123}$$

Subtracting Eq. (8.120) from Eq. (8.123), we obtain the following:

$$\dot{\mathbf{x}}(t) = \mathbf{P}\mathbf{x}(t) + \mathbf{b}u(t), \tag{8.124}$$

$$\dot{\hat{\mathbf{x}}}(t) = \mathbf{P}\hat{\mathbf{x}}(t) + \mathbf{b}u(t) + \mathbf{M}(c(t) - \mathbf{L}\hat{\mathbf{x}}(t)) \tag{8.125}$$

$$\dot{\mathbf{x}}(t) - \dot{\hat{\mathbf{x}}}(t) = \mathbf{P}(\mathbf{x}(t) - \hat{\mathbf{x}}(t)) - \mathbf{M}(c(t) - \mathbf{L}\hat{\mathbf{x}}(t)). \tag{8.126}$$

Substituting Eq. (8.2),

$$c(t) = \mathbf{L}\mathbf{x}(t) \tag{8.127}$$

into Eq. (8.126), we obtain

$$\dot{\mathbf{x}}(t) - \dot{\hat{\mathbf{x}}}(t) = \mathbf{P}(\mathbf{x}(t) - \hat{\mathbf{x}}(t)) - \mathbf{M}\mathbf{L}(\mathbf{x}(t) - \hat{\mathbf{x}}(t)). \tag{8.128}$$

Using the definition of error in the state estimate as given by (8.122), Eq. (8.128) reduces to

$$\dot{\tilde{\mathbf{x}}}(t) = \mathbf{P}\tilde{\mathbf{x}}(t) - \mathbf{M}\mathbf{L}\tilde{\mathbf{x}}(t)$$

or

$$\dot{\tilde{\mathbf{x}}}(t) = (\mathbf{P} - \mathbf{M}\mathbf{L})\tilde{\mathbf{x}}(t). \tag{8.129}$$

We have shown in Section 6.2 on the state-variable determination of the characteristic equation that a state equation, as given by Eq. (6.15),

$$\dot{\mathbf{x}}(t) = \mathbf{A}\mathbf{x}(t) \tag{8.130}$$

has a characteristic equation given by Eq. (6.22). Similarly, the characteristic equation of Eq. (8.129) is given by

$$|s\mathbf{I} - (\mathbf{P} - \mathbf{ML})| = 0. \tag{8.131}$$

The objective of the control-system engineer is to select $\mathbf{P} - \mathbf{ML}$ so that it has stable roots in order for $\tilde{\mathbf{x}}(t)$ to decay to zero. It is also desirable to have the root location produce a fast transient response so that the estimation error decays very fast to zero. Notice that the estimation error $\tilde{\mathbf{x}}(t)$ will converge to zero independent of the forcing function $u(t)$. Therefore, stability is determined from the homogeneous solution to the sysem (with $u(t) = 0$), as opposed to its particular solution (with $u(t)$ finite).

The design procedure for determining $\mathbf{M}$ is to specify the desired location of the estimator roots (e.g., $\alpha_1, \alpha_2, \ldots, \alpha_n$) from which the desired estimator characteristic equation can be specified:

$$(s + \alpha_1)(s + \alpha_2) \cdots (s + \alpha_n) = 0. \tag{8.132}$$

We can then solve for $\mathbf{M}$ by comparing the coefficients in Eqs (8.131) and (8.132).

Let us consider the design of $\mathbf{M}$ for a simple second-order system whose differential equation is given by

$$\ddot{c}(t) + \dot{c}(t) + c(t) = u(t).$$

Defining its two states as

$$\begin{aligned} x_1(t) &= c(t) \\ x_2(t) &= \dot{c}(t), \end{aligned}$$

we obtain its state equations as

$$\begin{aligned} \dot{x}_1(t) &= x_2(t) \\ \dot{x}_2(t) &= -x_1(t) - x_2(t) + u(t), \end{aligned}$$

and its output equation as

$$c(t) = [1 \quad 0]\begin{bmatrix} x_1(t) \\ x_2(t) \end{bmatrix}. \tag{8.133}$$

Therefore, for this system, we obtain $\mathbf{P}$, $\mathbf{b}$, and $\mathbf{L}$ to be as follows:

$$\begin{aligned} \mathbf{P} &= \begin{bmatrix} 0 & 1 \\ -1 & -1 \end{bmatrix}, \quad \mathbf{b} = \begin{bmatrix} 0 \\ 1 \end{bmatrix}, \\ \mathbf{L} &= [1 \quad 0]. \end{aligned} \tag{8.134}$$

Substituting Eq. (8.134) into Eq. (8.131), we obtain the following:

$$\left| \begin{bmatrix} s & 0 \\ 0 & s \end{bmatrix} - \begin{bmatrix} 0 & 1 \\ -1 & -1 \end{bmatrix} + \begin{bmatrix} m_1 \\ m_2 \end{bmatrix} [1 \quad 0] \right| = 0, \tag{8.135}$$

$$\left| \begin{bmatrix} s & 0 \\ 0 & s \end{bmatrix} - \begin{bmatrix} 0 & 1 \\ -1 & -1 \end{bmatrix} + \begin{bmatrix} m_1 & 0 \\ m_2 & 0 \end{bmatrix} \right| = 0, \tag{8.136}$$

$$\begin{vmatrix} s+m_1 & -1 \\ 1+m_2 & s+1 \end{vmatrix} = 0. \tag{8.137}$$

The resulting characteristic equation is given by

$$s^2 + (m_1 + 1)s + m_1 + m_2 + 1 = 0. \tag{8.138}$$

Where do we desire to place the two second-order roots? We could use the ITAE criterion, presented in Section 5.8, to make that determination. However, here the primary goal is to design the estimator to be very fast compared to that of the controller. Therefore, let us assume that the controller is critically damped and the controller's second-order characteristic equation has a pair of real roots located at $s = \beta_1 = \beta_2 = 2$. Let us assume that we wish the estimator to be critically damped and have the two second-order roots of the estimator located at $\alpha_1 = \alpha_2 = 20$ in Eq. (8.132), which will ensure a very fast response compared to that of the controller. Therefore, Eq. (8.132) for this example becomes

$$(s+20)^2 = 0,$$
$$s^2 + 40s + 400 = 0. \tag{8.139}$$

Comparing the coefficients of Eqs. (8.138) and (8.139), we obtain the following two sumultaneous equations to solve:

$$m_1 + 1 = 40, \tag{8.140}$$
$$m_1 + m_2 + 1 = 400. \tag{8.141}$$

Solving Eqs. (8.140) and (8.141), we obtain:

$$\mathbf{M} = \begin{bmatrix} m_1 \\ m_2 \end{bmatrix} = \begin{bmatrix} 39 \\ 360 \end{bmatrix}.$$

Designing a combined compensator of a controller and estimator is illustrated in Section 8.8.

8.8. COMBINED COMPENSATOR DESIGN INCLUDING A CONTROLLER AND AN ESTIMATOR FOR A REGULATOR SYSTEM

This section considers the combined compensator design of a controller and estimator for a regulator in which the reference input equals zero, and for the case where the reference input is finite. The block diagram of this regulator is shown in Figure 8.19, which combines the concepts illustrated in Figure 8.3 for the controller and

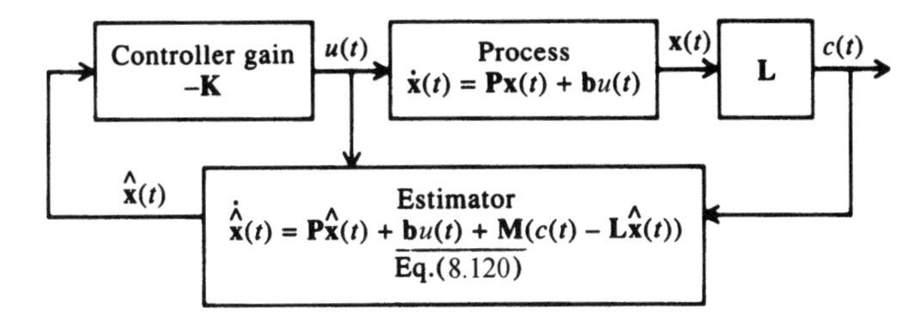

Figure 8.19 Regulator system (where the reference input $r = 0$) containing combined controller and estimator.

Figure 8.18 for the estimator. We wish to determine in this system the effect of using the estimated state vector $\hat{\mathbf{x}}(t)$, instead of $\mathbf{x}(t)$ on the system dynamics [8].

Let us consider the effect of driving the controller with $\hat{\mathbf{x}}(t)$ instead of $\mathbf{x}(t)$. From the process dynamics shown in Figure 8.3,

$$\dot{\mathbf{x}}(t) = \mathbf{P}\mathbf{x}(t) + \mathbf{b}u(t) \tag{8.142}$$

From Figure 8.19, we also know that

$$u(t) = -\mathbf{K}\hat{\mathbf{x}}(t). \tag{8.143}$$

Therefore, substituting Eq. (8.143) into Eq. (8.142), we obtain

$$\dot{\mathbf{x}}(t) = \mathbf{P}\mathbf{x}(t) - \mathbf{b}\mathbf{K}\hat{\mathbf{x}}(t). \tag{8.144}$$

In terms of the state estimation error $\tilde{\mathbf{x}}(t)$ defined in Eq. (8.122), Eq. (8.144) becomes

$$\dot{\mathbf{x}}(t) = \mathbf{P}\mathbf{x}(t) - \mathbf{b}\mathbf{K}(\mathbf{x}(t) - \tilde{\mathbf{x}}(t)). \tag{8.145}$$

Combining Eqs. (8.145) and (8.129), we obtain an overall equation for the state vector $\mathbf{x}$ and its error $\tilde{\mathbf{x}}$ as follows:

$$\begin{bmatrix} \dot{\mathbf{x}}(t) \\ \dot{\tilde{\mathbf{x}}}(t) \end{bmatrix} = \begin{bmatrix} \mathbf{P} - \mathbf{bK} & \mathbf{bK} \\ 0 & \mathbf{P} - \mathbf{ML} \end{bmatrix} \begin{bmatrix} \mathbf{x}(t) \\ \tilde{\mathbf{x}}(t) \end{bmatrix}. \tag{8.146}$$

The characteristic equation of this closed-loop combined controller and estimator system can be obtained in a manner similar to that obtained for the estimator alone [see Eq. (8.131)]:

$$\begin{vmatrix} s\mathbf{I} - \mathbf{P} + \mathbf{bK} & \mathbf{bK} \\ 0 & s\mathbf{I} - \mathbf{P} + \mathbf{ML} \end{vmatrix} = 0. \tag{8.147}$$

We can write this equation as

$$\det[s\mathbf{I} - \mathbf{P} + \mathbf{bK}]\det[s\mathbf{I} - \mathbf{P} + \mathbf{ML}] = 0 \tag{8.148}$$

The first determinant specifies the characteristic equation of the controller, and the second determinant specified the characteristic equation of the estimator [which is identical to Eq. (8.131)]. As we did for the case of the estimator alone in Section 8.7, Eq. (8.132), we can now define the combined desirable location of estimator roots

$$(s+\alpha_1)(s+\alpha_2)\cdots(s+\alpha_n)$$

and controller roots

$$(s+\beta_1)(s+\beta_2)\cdots(s+\beta_n)$$

and specify the combined estimator and controller's characteristic equation as

$$[(s+\beta_1)(s+\beta_2)\cdots(s+\beta_n)][(s+\alpha_1)(s+\alpha_2)\cdots(s+\alpha_n)]=0. \tag{8.149}$$

We can now simultaneously determine the controller gain **K** and estimator coefficients **M** by setting Eqs. (8.148) and (8.149) equal to each other:

$$\det[s\mathbf{I}-\mathbf{P}+\mathbf{bK}]\det[s\mathbf{I}-\mathbf{P}+\mathbf{ML}]=[(s+\beta_1)\cdots(s+\beta_n)] \times[(s+\alpha_1)\cdots(s+\alpha_n)]. \tag{8.150}$$

Therefore, we can observe from Eq. (8.150) that the roots of the combined controller and estimator is the sum of the controller and estimator roots found independently [8]. The primary concept to recognize in the combined compensator design for the controller and estimator is to make the estimator respond much faster than the controller, as we do not want the system's transient response limited by that of the estimator (which we can control by proper design).

It is useful to compare the modern pole placement method using the state-variable feedback method with the conventional transfer-function method for the design of the compensator as we have done in Section 8.2, when we found the open-loop transfer function $KG(s)H(s)$ in terms of $\mathbf{h}$, $\Phi_h(s)$ and $\mathbf{b}$ in Eq. (8.25). We wish to find the transfer function of the combined controller and estimator, $U(s)/C(s)$. Let us reconsider the estimator equation, Eq. (8.120), and incorporate it in the control-law equation, Eq. (8.143), because the controller is part of the compensator:

$$\dot{\hat{\mathbf{x}}}(t)=\mathbf{P}\hat{\mathbf{x}}(t)-\mathbf{bK}\hat{\mathbf{x}}(t)+\mathbf{M}(c(t)-\mathbf{L}\hat{\mathbf{x}}(t)). \tag{8.151}$$

Simplifying, we obtain

$$\dot{\hat{\mathbf{x}}}(t)=(\mathbf{P}-\mathbf{bK}-\mathbf{ML})\hat{\mathbf{x}}(t)+\mathbf{M}c(t). \tag{8.152}$$

Let us compare Eq. (8.152) with the state equation of the process:

$$\dot{\mathbf{x}}(t)=\mathbf{Px}(t)+\mathbf{b}u(t), \tag{8.153}$$

whose characteristic equation we know is given by

$$|s\mathbf{I}-\mathbf{P}|=0, \tag{8.154}$$

using the same reasoning we used in obtaining Eqs. (6.22), (8.19), and (8.131). By comparing Eqs. (8.152) and (8.153), we obtain the characteristic equation of the compensator as follows:

$$|s\mathbf{I} - (\mathbf{P} - \mathbf{bK} - \mathbf{ML})| = 0. \tag{8.155}$$

The resulting compensator may not result in a stable system because the roots of Eq. (8.155) have not been specified in advance. This is similar to our results in Section 8.3, where we found that design using linear-state-variable feeback did not ensure a stable system.

Before finding the transfer function representing the compensator, $U(s)/C(s)$, we first find the transfer function of the process from Eq. (8.153):

$$s\mathbf{X}(s) - \mathbf{x}(0) = \mathbf{PX}(s) + \mathbf{b}U(s). \tag{8.156}$$

Simplifying,

$$(s\mathbf{I} - \mathbf{P})\mathbf{X}(s) = \mathbf{b}U(s) + \mathbf{x}(0). \tag{8.157}$$

Solving for $\mathbf{X}(s)$, we obtain

$$\mathbf{X}(s) = [s\mathbf{I} - \mathbf{P}]^{-1}\mathbf{b}U(s) + [s\mathbf{I} - \mathbf{P}]^{-1}\mathbf{x}(0). \tag{8.158}$$

Since

$$c(t) = \mathbf{Lx}(t)$$

and its Laplace transform is

$$C(s) = \mathbf{LX}(s) \tag{8.159}$$

we can combine Eqs. (8.158) and (8.159) to relate the output $C(s)$ and input $U(s)$ of the process.

$$C(s) = \mathbf{L}[s\mathbf{I} - \mathbf{P}]^{-1}\mathbf{b}U(s) + \mathbf{L}[s\mathbf{I} - \mathbf{P}]^{-1}\mathbf{x}(0). \tag{8.160}$$

In order to find the transfer function of the process $C(s)/U(s)$, we assume that the initial condition, $\mathbf{x}(0)$, equals zero. Therefore,

$$\frac{C(s)}{U(s)} = \mathbf{L}[s\mathbf{I} - \mathbf{P}]^{-1}\mathbf{b}. \tag{8.161}$$

By analogy, we compare Eqs. (8.153) [and its resulting transfer function given by Eq. (8.161)] with Eq. (8.153), and conclude that the transfer function of the compensator defined by Eq. (8.152) is given by

$$\frac{U(s)}{C(s)} = -\mathbf{K}[s\mathbf{I} - \mathbf{P} + \mathbf{bK} + \mathbf{ML}]^{-1}\mathbf{M}. \tag{8.162}$$

When we determine the transfer function $U(s)/C(s)$ using this procedure, we will find that the resulting transfer function will result in a phase-lead network, a phase-lag network, or a phase-lag-lead network.

To illustrate this approach for obtaining the transfer function of the compensator, consider a process whose transfer function is given by

$$C(s)/U(s) = G(s) = 1/s^2. \tag{8.163}$$

Therefore,

$$\ddot{c}(t) = u(t).$$

Defining the state variables of this second-order system as

$$x_1(t) = c(t),$$
$$x_2(t) = \dot{c}(t),$$

we obtain the state equation to be

$$\begin{bmatrix} \dot{x}_1(t) \\ \dot{x}_2(t) \end{bmatrix} = \underbrace{\begin{bmatrix} 0 & 1 \\ 0 & 0 \end{bmatrix}}_{\mathbf{P}} \begin{bmatrix} x_1(t) \\ x_2(t) \end{bmatrix} + \underbrace{\begin{bmatrix} 0 \\ 1 \end{bmatrix}}_{\mathbf{b}} u(t), \tag{8.164}$$

and the output equation is

$$c(t) = \underbrace{[1 \quad 0]}_{\mathbf{L}} \begin{bmatrix} x_1(t) \\ x_2(t) \end{bmatrix}. \tag{8.165}$$

Let us assume that the design specification for the controller is

$$\omega_n = \sqrt{3},$$
$$\zeta = 0.58.$$

Therefore, the complex-conjugate roots of the controller are located at $-1 \pm j1.414$ and $\alpha_c(s)$ for the controller is given by

$$\alpha_c(s) = (s + 1 - j1.414)(s + 1 + j1.414) = s^2 + 2s + 3 \tag{8.166}$$

We can determine the controller gain $\mathbf{K}$ by equating like powers of s from Eq. (8.166) and that part of Eq. (8.148) concerned with the controller:

$$|s\mathbf{I} - \mathbf{P} + \mathbf{bK}| = 0. \tag{8.167}$$

Substituting for $\mathbf{P}$ and $\mathbf{b}$ from Eq. (8.164) into Eq. (8.167), we obtain the following:

$$\left| \begin{bmatrix} s & 0 \\ 0 & s \end{bmatrix} - \begin{bmatrix} 0 & 1 \\ 0 & 0 \end{bmatrix} + \begin{bmatrix} 0 \\ 1 \end{bmatrix} [K_1 \quad K_2] \right| = 0. \tag{8.168}$$

This simplifies to

$$\begin{vmatrix} s & -1 \\ K_1 & s + K_2 \end{vmatrix} = 0, \tag{8.169}$$

from which we obtain the characteristic equation of the controller:

$$s^2 + K_2 s + K_1 = 0. \tag{8.170}$$

Comparing like coefficients in Eqs. (8.166) and (8.170), we find the controller gains to be $K_1 = 3$ and $K_2 = 2$:

$$\mathbf{K} = [3 \quad 2]. \tag{8.171}$$

As discussed previously, we desire the estimator to have a much faster response than the controller. Therefore, let us assume the design specification of the estimator to be

$$\omega_n = 17,$$
$$\zeta = 0.5.$$

Therefore, the complex-conjugate roots of the estimator are located at $-8.5 \pm j14.7$, and $\alpha_e(s)$ for the estimator is given by

$$\alpha_e(s) = (s + 8.5 + j14.7)(s + 8.5 - j14.7) = s^2 + 17s + 288.3. \tag{8.172}$$

The resulting estimator feedback gain matrix is found from Eq. (8.131) [for the estimator portion of Eq. (8.148)] as follows:

$$|s\mathbf{I} - (\mathbf{P} - \mathbf{ML})| = 0. \tag{8.173}$$

Substituting the matrix values into Eq. (8.173), we obtain

$$\left| \begin{bmatrix} s & 0 \\ 0 & s \end{bmatrix} - \begin{bmatrix} 0 & 1 \\ 0 & 0 \end{bmatrix} + \begin{bmatrix} m_1 \\ m_2 \end{bmatrix} [1 \quad 0] \right| = 0, \tag{8.174}$$

which reduces to:

$$\begin{vmatrix} s + m_1 & -1 \\ m_2 & s \end{vmatrix} = 0.$$

The resulting characteristic equation in terms of m_1 and m_2 is given by

$$s^2 + m_1 s + m_2 = 0. \tag{8.175}$$

Setting like coefficients in Eqs. (8.172) and (8.175) equal to each other, we obtain $m_1 = 17$ and $m_2 = 288.3$:

$$\mathbf{M} = \begin{bmatrix} 17 \\ 288.3 \end{bmatrix}. \tag{8.176}$$

The resulting compensator transfer function is obtained by substituting parameters obtained from Eqs. (8.164), (8.165), (8.171), and (8.176) into Eq. (8.162) as follows:

$$\frac{U(s)}{C(s)} = -[3 \quad 2] \left[\begin{bmatrix} s & 0 \\ 0 & s \end{bmatrix} - \begin{bmatrix} 0 & 1 \\ 0 & 0 \end{bmatrix} + \begin{bmatrix} 0 \\ 1 \end{bmatrix} [3 \quad 2] + \begin{bmatrix} 17 \\ 288.3 \end{bmatrix} [1 \quad 0] \right]^{-1} \begin{bmatrix} 17 \\ 288.3 \end{bmatrix}. \tag{8.177}$$

This equation reduces to

$$\frac{U(s)}{C(s)} = -[3 \quad 2]\frac{\begin{bmatrix} s+2 & 1 \\ -291.3 & s+17 \end{bmatrix}\begin{bmatrix} 17 \\ 288.3 \end{bmatrix}}{s^2 + 19s + 325.3}.$$

Upon further simplification, we obtain the following:

$$\frac{U(s)}{C(s)} = -\frac{\begin{bmatrix} 3s - 576.6 & 2s + 37 \end{bmatrix}\begin{bmatrix} 17 \\ 288.3 \end{bmatrix}}{(s + 9.5 + j15.3)(s + 9.5 - j15.3)}. \tag{8.178}$$

The resulting transfer function of the compensator is given by the following equation:

$$G_{\text{comp}} = \frac{U(s)}{C(s)} = -\frac{627.6(s + 1.38)}{(s + 9.5 + j15.3)(s + 9.5 - j15.3)}. \tag{8.179}$$

Analysis of Eq. (8.179) shows that the compensator has the form of a phase-lead network with a zero at -1.38 and with two complex-conjugate poles as opposed to a simple pole as defined by the conventional phase-lead network of Eq. (7.6).

We can analyze the resulting system using the conventional root-locus and Bode-diagram methods. To obtain the open-loop transfer function for analyses, we combine Eqs. (8.163) and (8.179):

$$G(s)G_{\text{comp}}(s) = -\frac{627.6(s + 1.38)}{s^2(s + 9.5 + j15.3)(s + 9.5 - j15.3)}. \tag{8.180}$$

Replacing the specific gain of -627.6 with the variable gain K, the root locus can be evaluated from

$$G(s)G_{\text{comp}}(s) = \frac{K(s + 1.38)}{s^2(s + 9.5 + j15.3)(s + 9.5 - j15.3)}, \tag{8.181}$$

which is shown in Figure 8.20. Observe that the root locus goes through the roots chosen in Eqs. (8.166) and (8.172) when $K = 627.6$. These roots are shown in Figure 8.20 by solid dots. This figure was obtained using MATLAB, and is contained in the M-file that is part of my MCSTD Toolbox which can be retrieved free from The MathWorks, Inc. anonymous FTP server at ftp://ftp.mathworks.com/pub/books/shinners.

To draw the Bode diagram, we consider the modified form of Eq. (8.180):

$$G(s)G_{\text{comp}}(s) = \frac{866.1(0.725s + 1)}{s^2(s^2 + 19s + 325.3)}. \tag{8.182}$$

The resulting Bode diagram is drawn from the following simplification to Eq. (8.182):

$$G(s)G_{\text{comp}}(s) = \frac{2.66(0.725s + 1)}{s^2} \cdot \frac{325.3}{s^2 + 19s + 325.3}. \tag{8.183}$$

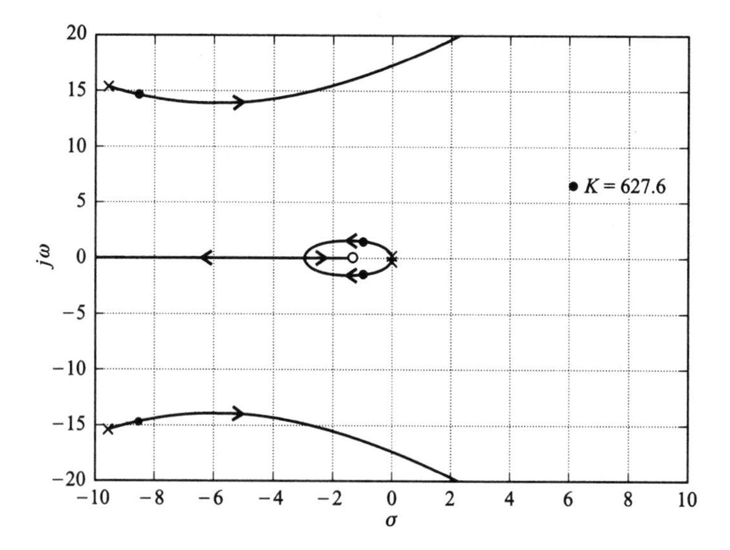

Figure 8.20 Root locus for compensated system of Figure 8.19 where $G(s)G_{\text{comp}}(s) = (K(s+1.38))/(s^2(s+9.5+j15.3)(s+9.5-j15.3))$.

The quadratic poles in the denominator have an undamped natural frequency ω_n and a damping ratio ζ given by

$$\omega_n = \sqrt{325.3} = 18.04,$$

$$\zeta = \frac{19}{2(18.04)} = 0.527.$$

The resulting Bode diagram is shown in Figure 8.21. This figure was also obtained using MATLAB and is contained in the M-file that is part of my MCSTD Toolbox.

We conclude that the uncompensated transfer function

$$G(s) = 1/s^2$$

has its phase margin increased from 0 to 51.07 degrees at its gain crossover frequency of 2.28 rad/sec., and its gain margin is increased from minus infinity to 19.13 dB (phase crossover frequency occurs at $\omega = 17.32$ rad/se) when we use the compensator of Eq. (8.179). Notice that the crossover frequency of 2.28 rad/sec is approximately consistent with the controller's closed-loop roots of $\omega_n = \sqrt{3} = 1.732$ rad/sec and $\zeta = 0.85$. This is a reasonable result, as the slower roots of the controller are more dominant than the faster estimator roots, on the system response.

A complete case study for the design of a combined controller and estimator for a regulator, using the techniques presented in Section 8.7 and 8.8, is presented in Chapter 12.

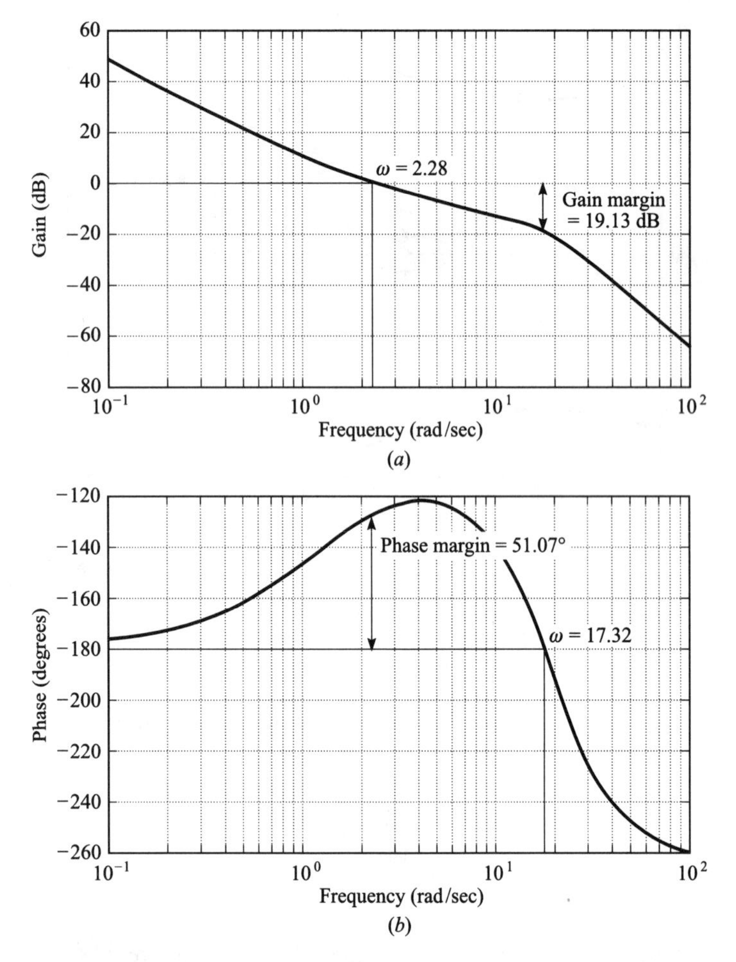

Figure 8.21 Bode diagram for compensated system of Figure 8.19 where $G(s)G_{\text{comp}}(s) = (((2.66(0.725s + 1))/s^2)(325.3/(s^2 + 19s + 325.3))$.

8.9. EXTENSION OF COMBINED COMPENSATOR DESIGN INCLUDING A CONTROLLER AND AN ESTIMATOR FOR SYSTEMS CONTAINING A REFERENCE INPUT

How can we extend the concepts developed in the previous section for regulator design, shown in Figure 8.19 (where the reference input $r(t) = 0$), to the more general problem where the reference input exists? Several methods exist which can be used to design such a system [8,9]. This section will consider the configuration illustrated in Figure 8.22. The design goal of the approach to be presented is to have the state-estimation error $\tilde{\mathbf{x}}(t)$ be independent of the reference input $r(t)$ (e.g., $\tilde{\mathbf{x}}$ should be uncontrollable from $r(t)$). This is a very important consideration, as we do not want the state-estimation error to be dependent on the type and level of the input.

Let us reconsider the controller equation (8.143),

$$u(t) = -\mathbf{K}\hat{\mathbf{x}}(t) \tag{8.184}$$

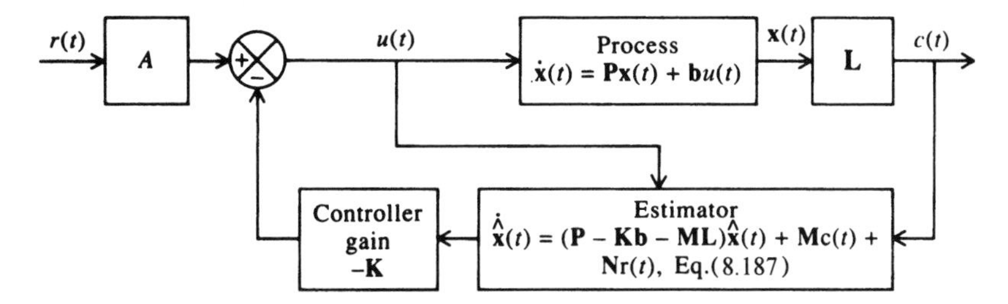

Figure 8.22 Addition of the reference input to the system shown in Figure 8.19 containing a combined controller and estimator.

and the estimator equation (8.152)

$$\dot{\hat{\mathbf{x}}}(t) = (\mathbf{P} - \mathbf{Kb} - \mathbf{ML})\hat{\mathbf{x}}(t) + \mathbf{M}c(t). \tag{8.185}$$

The reference input $r(t)$ will be introduced to these equations by adding a term $Ar(t)$ to the controller equation (8.184), and a term $\mathbf{N}r(t)$ to the estimator equation (8.185) (where $\mathbf{N}$ is a vector). Therefore, Eqs. (8.184) and (8.185) become

$$u(t) = -\mathbf{K}\hat{x}(t) + Ar(t), \tag{8.186}$$

$$\dot{\hat{\mathbf{x}}}(t) = (\mathbf{P} - \mathbf{Kb} - \mathbf{ML})\hat{\mathbf{x}}(t) + \mathbf{M}c(t) + \mathbf{N}r(t). \tag{8.187}$$

What kind of system do Eqs. (8.186) and (8.187) infer? Does it result in the configuration of Figure 8.22? To answer this question, let us first substitute

$$c(t) = \mathbf{Lx}(t) \tag{8.188}$$

into Eq. (8.187) and, thereby, eliminate the output $c(t)$ from Eq. (8.187):

$$\dot{\hat{\mathbf{x}}}(t) = (\mathbf{P} - \mathbf{Kb} - \mathbf{ML})\hat{\mathbf{x}}(t) + \mathbf{MLx}(t) + \mathbf{N}r(t). \tag{8.189}$$

To find the estimation error (which we want to be independent of $r(t)$), let us difference $\dot{\hat{\mathbf{x}}}$ [from Eq. (8.189)] and the state equation

$$\dot{\mathbf{x}}(t) = \mathbf{Px}(t) + \mathbf{b}u(t). \tag{8.190}$$

We will first substitute Eq. (8.186) into Eq. (8.190):

$$\dot{\mathbf{x}}(t) = \mathbf{Px}(t) + \mathbf{b}(-\mathbf{K}\hat{\mathbf{x}}(t) + Ar(t)). \tag{8.191}$$

Substracting Eq. (8.189) from Eq. (8.191), we obtain the following for the derivative of the estimation error:

$$\begin{aligned}\dot{\tilde{\mathbf{x}}}(t) =& \dot{\mathbf{x}}(t) - \dot{\hat{\mathbf{x}}}(t) = \mathbf{Px}(t) + \mathbf{b}(-\mathbf{K}\hat{\mathbf{x}}(t) + Ar(t)) \\ &- (\mathbf{P} - \mathbf{Kb} - \mathbf{ML})\hat{\mathbf{x}}(t) - \mathbf{MLx}(t) - \mathbf{N}r(t).\end{aligned}$$

Simplifying, we obtain the following equation:

$$\dot{\tilde{\mathbf{x}}}(t) = (\mathbf{P} - \mathbf{ML})\tilde{\mathbf{x}}(t) + (\mathbf{b}A - \mathbf{N})r(t). \tag{8.192}$$

In order to eliminate $r(t)$ from Eq. (8.192), it is necessary that

$$\mathbf{b}A = \mathbf{N}. \tag{8.193}$$

Therefore, the design criterion of the control-system engineer is to invoke Eq. (8.193). Substituting Eqs. (8.186) and (8.193) into Eq. (8.187), we obtain the following:

$$\dot{\hat{\mathbf{x}}}(t) = \mathbf{P}\hat{\mathbf{x}}(t) + \mathbf{b}(u(t) - Ar(t)) - \mathbf{ML}\hat{\mathbf{x}}(t) + \mathbf{Mc}(t) + \mathbf{b}Ar(t). \tag{8.194}$$

Simplifying, we find that

$$\dot{\hat{\mathbf{x}}}(t) = (\mathbf{P} - \mathbf{ML})\hat{\mathbf{x}}(t) + \mathbf{Mc}(\mathbf{t}) + \mathbf{b}u(\mathbf{t}) \tag{8.195}$$

or

$$\dot{\hat{\mathbf{x}}}(t) = \mathbf{P}\hat{\mathbf{x}}(t) + \mathbf{b}u(t) + \mathbf{M}(c(t) - \mathbf{L}\hat{\mathbf{x}}(t)). \tag{8.196}$$

Notice that Eq. (8.196) is the same estimator equation defined in Eq. (8.120). It is important to emphasize that this occurs only when $\mathbf{b}A = \mathbf{N}$ as defined in Eq. (8.193). Therefore we conclude that the introduction of the reference input by adding the term $Ar(t)$ in the controller equation (8.186) and a term $\mathbf{N}r(t)$ to Eq. (8.187) results in the configuration shown in Figure 8.22.

Complete design examples for the design of the controller, estimator, and compensator, with their associated root-locus and Bode-diagram analyses of the resulting design are found in Chapter 7 of the accompanying volume. In Section 7.5, the state-variable design for the controller and full-order estimator for a space vehicle is presented. In Problem 7.6, the state-variable design for the controller and full-order estimator of a chemical process control system is analyzed.

8.10. ROBUST CONTROL SYSTEMS [10–14]

Robust control is concerned with determining a stabilizing controller that achieves feedback performance in terms of stability and accuracy requirements, but the control must achieve the performance that is robust (insensitive) to plant uncertainty, parameter variation, and external disturbances. We know from the previous discussion in this book that feedback reduces the effects of external disturbances (Section 2.17) and parameter variations (Section 5.3). However, this is only achieved with relatively high loop gain which limits stability. Robust control is basically the same problem that was addressed in the 1930s by Black, Bode, and Nyquist. Modern robust control revolves around the feedback configurations illustrated in Figures 7.5, 7.6, and 7.7 which illustrate two-degrees-of-freedom compensation systems.

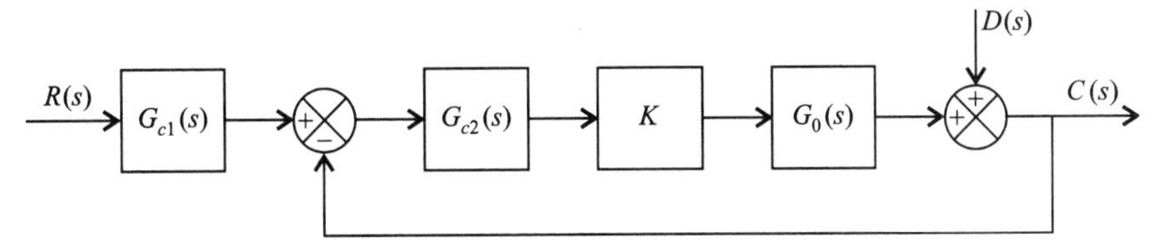

Figure 8.23 Control system with a two-degrees-of-freedom series controller $G_{c1}(s)$ and a forward-loop-controller $G_{c2}(s)$.

Let us consider the control system illustrated in Figure 8.23 which contains a disturbance $D(s)$, and it contains a two-degrees-of-freedom series controller $G_{c1}(s)$ and a forward-loop controller $G_{c2}(s)$. In this control system's operation, the amplifier gain, K, can vary.

For this control system, the overall transfer function, $C(s)/R(s)$, is given by

$$H(s) = \frac{C(s)}{R(s)} = \frac{G_{c1}(s)G_{c2}(s)KG_0(s)}{1 + KG_{c2}(s)G_0(s)}. \tag{8.197}$$

The transfer function relating the disturbance $D(s)$ to the output $C(s)$ is given by

$$N(s) = \frac{C(s)}{D(s)} = \frac{1}{1 + KG_{c2}(s)G(s)}. \tag{8.198}$$

The design approach in robust control systems is to choose the controller $G_{c1}(s)$ so that the desired closed-loop transfer function $H(s)$ is obtained, and to choose the controller $G_{c2}(s)$ so that the output, $C(s)$, is insensitive to the disturbance $D(s)$ over the frequency range in which $D(s)$ is dominant.

The sensitivity of $H(s)$ due to variations of K is given by

$$S_K^{H(s)}(s) = \frac{\dfrac{dH(s)}{H(s)}}{\dfrac{dK}{K}} = \frac{K}{H(s)}\frac{dH(s)}{dK}. \tag{8.199}$$

For the control system of Figure 8.23,

$$\frac{dH}{dK} = \frac{G_{c1}(s)G_{c2}(s)G_0(s)}{[1 + KG_{c2}(s)G_0(s)]^2}. \tag{8.200}$$

Substituting Eqs. (8.197) and (8.200) into Eq. (8.199), we obtain the following:

$$S_K^{H(s)}(s) = \frac{1}{1 + KG_{c2}(s)G_0(s)}. \tag{8.201}$$

It is important to recognize that for this control system, both $C(s)/D(s)$ given by Eq. (8.198) and the sensitivity of $H(s)$ with respect to K given by Eq. (8.201) are identical. This is a very important result which implies that we can use the same control-system techniques to suppress the affect of the disturbance $D(s)$ and robustness (insensitivity) with respect to variations of K.

A. Design Example Illustrating Robustness and Disturbance Rejection

We will next analyze how the two-degrees-of-freedom control system illustrated in Figure 8.23 can achieve a high-gain which will satisfy the robustness and performance requirements while, at the same time, minimizing the affects of the disturbance. We will analyze the system of Figure 8.23 with the following transfer function which represents a fourth-order process $KG_0(s)$:

$$KG_0(s) = \frac{142(7+s)}{s(1+s)(5+s)(20+s)}. \tag{8.202}$$

We will assume at this point that $G_{c1}(s) = G_{c2}(s) = 1$. Our problem is to investigate the affect of the variation of K. The transfer function of $KG_{c2}(s)G_0(s)$ is given by:

$$KG_{c2}(s)G_0(s) = \frac{142(7+s)}{s(1+s)(5+s)(20+s)}. \tag{8.203}$$

We want to consider the variation of the gain from 142 to double that amount (i.e., 248) and to one-half that amount (i.e., 71).

Because this control system acts as a low-pass filter, the sensitivity of $H(s)$ with respect to K is poor. The bandwidth of this control system with $K = 142$ is only 3.2 rad/sec, while the sensitivity of $H(s)$ with respect to K is expected to be greater than one at frequencies greater than 3.2 rad/sec. Figure 8.24 illustrates the unit step response of the system when $K = 142$ (the nominal value), $K = 248$, and $K = 71$. Table 8.1 lists the characteristics of the unit step transient responses and the characteristic equation roots of this control system which were obtained using

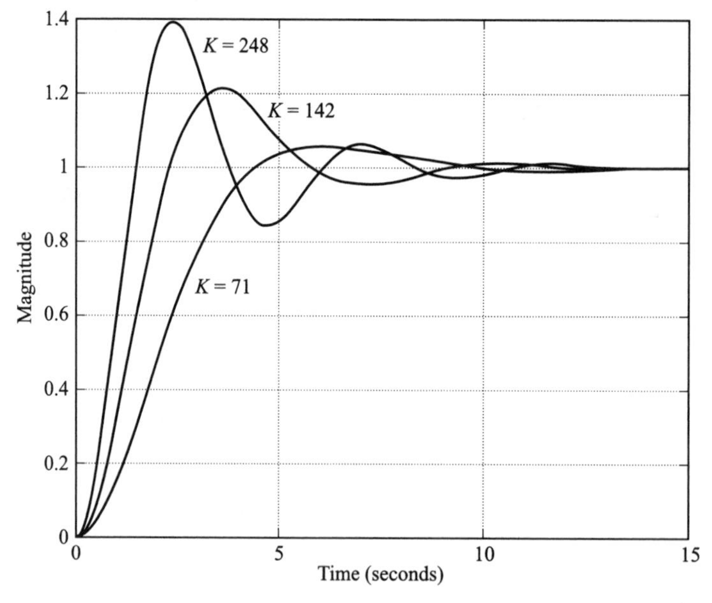

Figure 8.24 Unit step response for system of Figure 8.23 with $KG_{c2}(s)G_0(s)$ given by Eq. (8.203) and $G_{c1}(s) = 1$.

Table 8.1. Characteristics of the Control System Illustrated in Figure 8.23 where $KG_{c2}(s)G_0(s)$ is given by Eq. (8.203), and $G_{c1}(s) = 1$

K	Damping ratio	Roots of characteristic equation
248	0.277	$-5.1640,\ -20.0649,\ -0.3856 \pm j1.3348$
142	0.444	$-5.0878,\ -20.0325,\ -0.4398 \pm j0.8875$
71	0.666	$-5.0456,\ -20.01629,\ -0.4691 \pm j0.5245$

MATLAB. Observe that variations of K from its nominal value of 142 result in considerable variation in the damping ratio and the transient responses of this control system. Figure 8.25 illustrates the root loci and the location of the closed-loop, complex-conjugate, roots for the three cases being analyzed.

The design approach for this robust controller, $G_{c2}(s)$, is to place two zeros at (or near) the desired complex, conjugate-loop, poles at $-0.4398 \pm j0.8875$ for the nominal gain case of $K = 142$. Therefore,

$$G_{c2}(s) = \frac{(s + 0.4398 + j0.8875)(s + 0.4398 - j0.8875)}{0.98} \tag{8.204}$$

or

$$G_{c2}(s) = \frac{s^2 + 0.88s + 0.98}{0.98}. \tag{8.205}$$

We will approxinate $G_{c2}(s)$ as follows:

$$G_{c2}(s) = s^2 + 0.88s + 1. \tag{8.206}$$

Therefore, the forward-path transfer function of this control system with $KG_0(s)$ given by Eq. (8.202) and $G_{c2}(s)$ given by Eq. (8.206) is:

$$KG_{c2}(s)G_0(s) = \frac{K(7 + s)(s^2 + 0.88s + 1)}{s(1 + s)(5 + s)(20 + s)}. \tag{8.207}$$

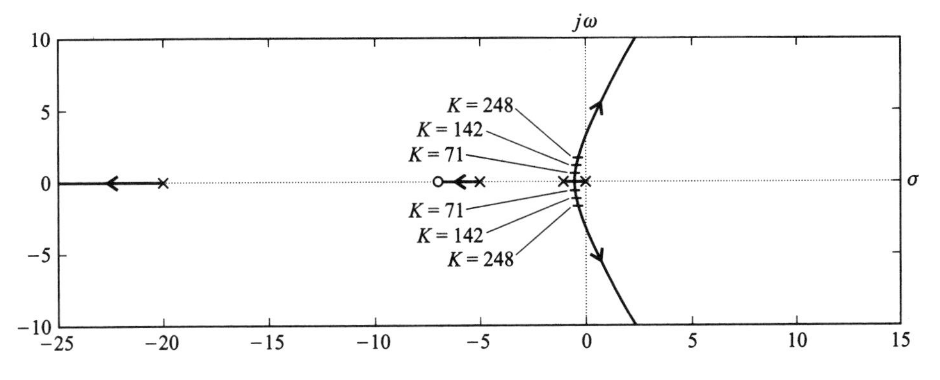

Figure 8.25 Root-locus plot for system of Figure 8.23 with $KG_{c2}(s)G_0(s)$ given by Eq. (8.203) and $G_{c1}(s) = 1$.

Table 8.2 lists the damping ratio and the characteristic equation roots of this control system, obtained using MATLAB, with the forward-loop transfer function given by Eq. (8.207). Observe that the range of the damping ratios are much closer (0.548 to 0.808) than they were before the addition of the robust controller $G_{c2}(s)$ and shown in Table 8.1 (where the damping previously varied from 0.277 to 0.666).

Figure 8.26 illustrates the root loci with the robust controller $G_{c2}(s)$ added, and the location of the closed-loop, complex-conjugate, roots for the three cases. Observe from this root locus that by locating the two zeros of the forward-loop controller $G_{c2}(s)$ near the desired characteristic equation complex- conjugate roots for $K = 142$, the sensitivity of this control system becomes much better.

It was shown in Section 8.2 during the discussion on the concept of liner-state-variable feedback that for the system illustrated in Figure 8.3, the zeros of $C(s)/R(s)$ are the zeros of $G(s)$ based on comparing Eqs. (8.17) and (8.24). Therefore, in the system we are currently analyzing in Figure 8.23, the zeros of the forward-path transfer function $KG_{c2}(s)G_0(s)$ are identical to the zeros of the closed-loop transfer function. Therefore, the closed-loop zeros of $G_{c2}(s)$ in Eq. (8.206) come close to canceling the effect of the complex-conjugate, closed-loop poles. Therefore, it is necessary to also add the series controller $G_{c1}(s)$, as illustrated in Figure 8.23, so that $G_{c1}(s)$ contains poles to cancel the zeros of $s^2 + 0.88 + 1$ of the closed-loop transfer function. Therefore, the transfer function of the forward-loop controller, $G_{c1}(s)$, is given by:

$$G_{c1}(s) = \frac{1}{s^2 + 0.88s + 1}. \tag{8.208}$$

Table 8.2. Characteristics of the Control System Illustrated in Figure 8.23 with the Forward-Loop Controller $G_{c2}(s)$ added and where $KG_{c2}(s)G_0(s)$ is given by Eq. (8.207).

K	Damping ratio	Roots of characteristic equation
248	0.548	$-6.3715, -47.3367, -0.4459 \pm j0.6814$
142	0.644	$-6.0358, -33.3566, -0.4538 \pm j0.5392$
71	0.808	$-5.6914, -26.5282, -0.4652 \pm j0.3387$

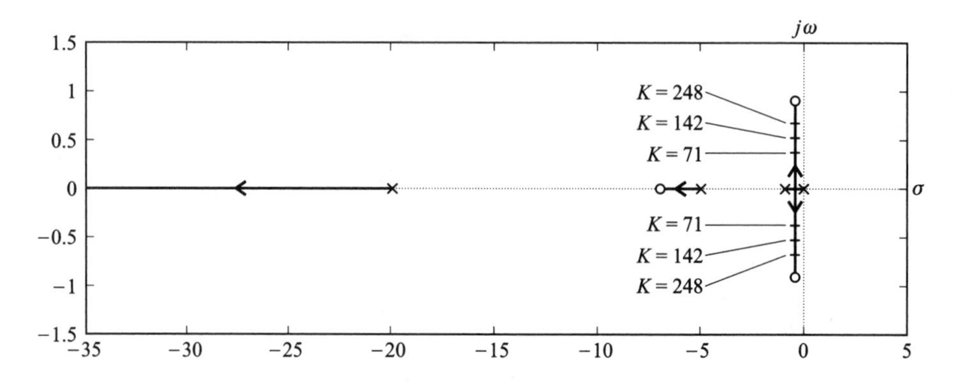

Figure 8.26 Root-locus plot for system shown in Figure 8.23 with $KG_{c2}(s)G_0(s)$ given by Eq. (8.207).

The unit step response of this control system with the forward-path transfer function of the control system given by Eq. (8.207), with $K = 71$, 142, and 248, and with the forward-loop controller transfer function given by Eq. (8.208) is illustrated in Figure 8.27. Comparing these unit step responses with those in Figure 8.24, we conclude that this control system has been made to be much less sensitive to variations in K. For example, the maximum percent overshoot of the transient responses for the original system illustrated in Figure 8.24 ranged from 5.6% (for $K = 71$) to 39% (for $K = 288$). However, the control system designed to be robust has a transient response as illustrated in Figure 8.27 which shows that the maximum percent overshoot varies from 1.4% (for $K = 71$) to only 13% (for $K = 288$). In addition, as we pointed out before in comparing Eqs. (8.198) and (8.201), which are identical, the robustness with respect to variations in K will also provide disturbance suppression using the same control-system techniques.

Since the disturbance suppression attributes are a function of frequency, let us examine the frequency characteristics of the $C(s)/D(s)$ transfer function. Substituting Eq. (8.203) into Eq. (8.198), we obtain the following transfer function for $C(s)/D(s)$ for the case where the forward-loop controller $G_{c2}(s)$ has a gain equal to one:

$$\frac{C(s)}{D(s)} = \frac{s^4 + 26s^3 + 125s^2 + 100s}{s^4 + 26s^3 + 125s^2 + 242s + 994}. \tag{8.209}$$

Substituting Eq. (8.207) into Eq. (8.198), we obtain the following transfer function for $C(s)/D(s)$ for the case where the forward-loop controller $G_{c2}(s)$ has the transfer function given by Eq. (8.206)

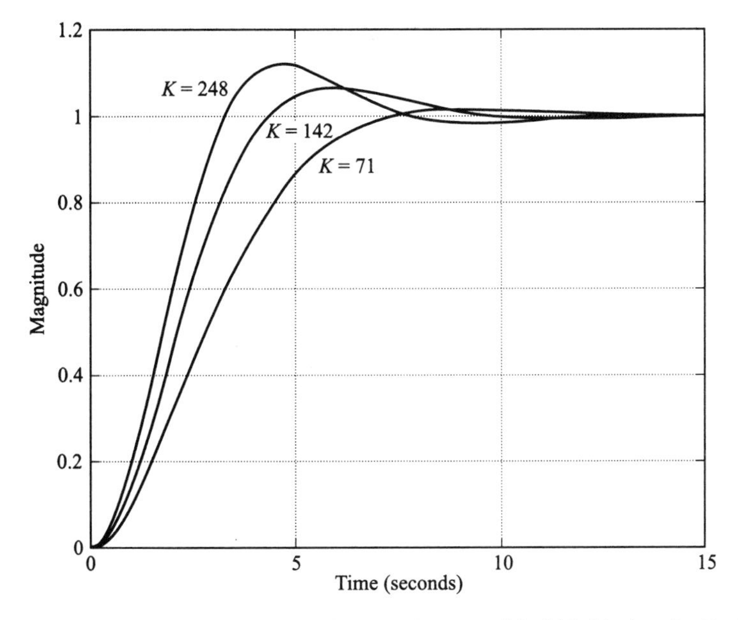

Figure 8.27 Unit step response for system of Figure 8.23 with $KG_{c2}(s)G_0(s)$ given by Eq. (8.207) and $G_{c1}(s)$ given by Eq. (8.208).

$$\frac{C(s)}{D(s)} = \frac{s^4 + 26s^3 + 125s^2 + 100s}{s^4 + 168s^3 + 1244s^2 + 1117s + 994}. \tag{8.210}$$

Figure 8.28 is a plot of the frequency response of the disturbance suppression transfer fucntion $C(s)/D(s)$ as defined in Eqs. (8.209) and (8.210). It shows that the disturbance suppression of the control system at low frequencies is approximately the same with and without $G_{c2}(s)$ in the control system. However, the addition of $G_{c2}(s)$, as defined by Eq. (8.206), greatly improves the disturbance suppression attributes of the control system at high frequencies. This is consistent with our expectations.

Robust control-system design principles are being applied to many, modern, practical control systems. For example, the reader is referred to Reference 15 which presents the design for a robust control system for preventing car skidding. A complete case study for the design of a robust control system for controlling the flaps of a hydrofoil is presented in Section 6.7 of the accompanying volume.

8.11. AN INTRODUCTION TO H^{∞} CONTROL CONCEPTS [16,17]

In the presentation of this book, the classical frequency-domain approach and the modern state-variable time-domain approach have been presented in parallel. It has

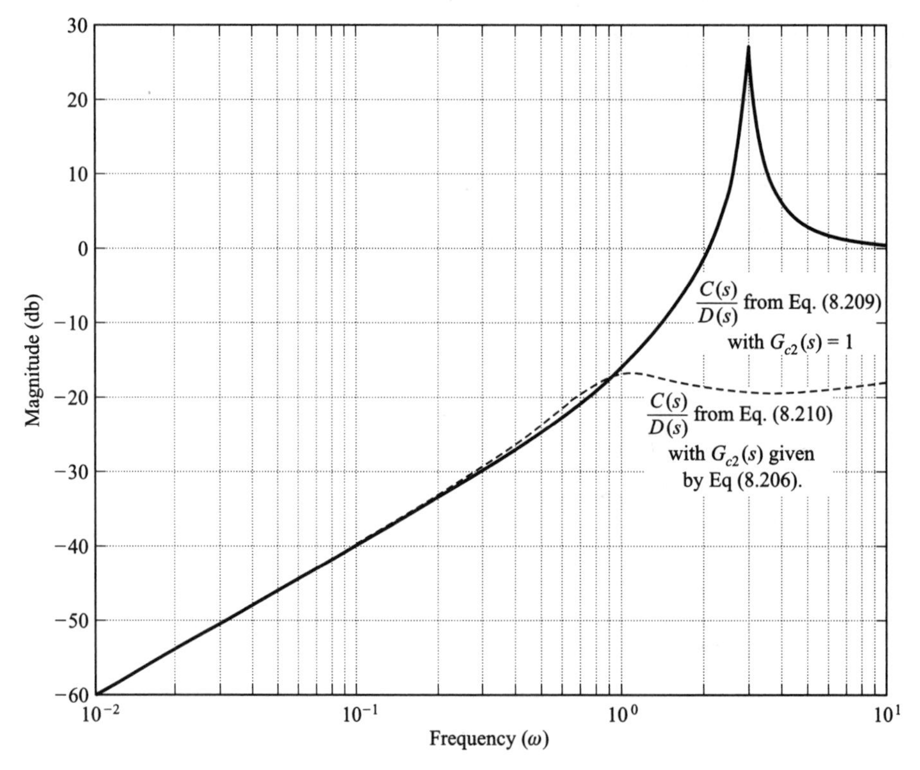

Figure 8.28 Frequency response of the disturbance suppression transfer functions defined in Eqs. (8.209) and (8.210).

been shown that they complement each other. Prior to the 1960s, the frequency-domain approach predominated. With the advent of the space race, the availability of practical digital computers, modern optimal control theory, and the state-variable approach in the early 1960s, the pendulum swung to the time-domain approach. The 1960s and 1970s saw an abundant amount of work performed on applying modern optimal control theory, which is presented in Chapter 11 in this book. In the early 1980s, a new technique has emerged known as H^{∞} control theory which combines both the fequency- and time-domain approaches to provide a unified answer. Zames is given credit for its introduction with his paper in the *IEEE Transactions on Automatic Control* [16]. The H^{∞} approach has dominated the trend of control-system development in the 1980s and 1990s. A complete treatment of the subject of H^{∞} is complex, and beyond the scope of this book. However, we can expand the concepts of robustness (introduced in Section 8.10) and sensitivity (introduced in Chapter 5), together with the frequency and state-variable domain techniques presented in this book to introduce the basic concepts of H^{∞} control theory and apply it to some simple problems. This is the objective of this and the following sections.

A. Sensitivity of Control Systems Containing Disturbances and Measurement Noise

Let us extend our understanding of sensitivity developed in Chapter 5 to the control system illustrated in Figure 8.29 which contains a disturbance $U(s)$ and feedback sensor measurement noise $N(s)$. Although noise is a stochastic process, we will assume in this analysis that it can be represented as a deterministic process. This initial analysis will focus on the single-input single-output (SISO) system shown in Figure 8.29 which contains a disturbance $U(s)$ and feedback sensor measurement noise $N(s)$. We will then discuss the extension of these results to multiple-input multiple-output (MIMO) systems.

Using Mason's theorem, we can write the following relationship for the output $C(s)$ by inspection of Figure 8.29:

$$C(s) = \frac{G(s)}{1 + G(s)}[R(s) - N(s)] + \frac{1}{1 + G(s)}U(s). \tag{8.211}$$

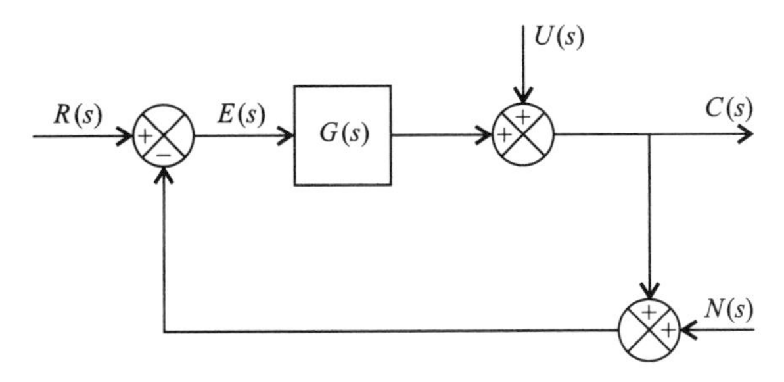

Figure 8.29 Control system containing a disturbance $U(s)$ and sensor measurement observation noise $N(s)$.

We can also use Mason's theorem to write the following relationship for the control-system error $E(s)$ by inspection of Figure 8.29:

$$E(s) = \frac{1}{1 + G(s)}[R(s) - U(s) - N(s)]. \tag{8.212}$$

From the analysis in Section 5.3 on sensitivity, we know that the *sensitivity* of the transfer function $C(s)/R(s)$ to changes in $G(s)$ for the control system shown in Figure 8.29 is given by the following expression:

$$S_{G(s)}^{C(s)/R(s)} = \frac{1}{1 + G(s)}. \tag{8.213}$$

It is interesting to observe that this sensitivity is also the transfer function from $U(s)$ to $-E(s)$:

$$-\frac{U(s)}{E(s)} = S_{G(s)}^{C(s)/R(s)} = \frac{1}{1 + G(s)}. \tag{8.214}$$

In the MIMO case, the sensitivity function in Eq. (8.213) is modified so that $G(s)$ becomes the matrix of transfer functions between the inputs and the outputs, and the value of 1 is replaced with the identity matrix.

The transfer function $C(s)/R(s)$ of the control system shown in Figure 8.29 is given by the following:

$$\frac{C(s)}{R(s)} = \frac{G(s)}{1 + G(s)}. \tag{8.215}$$

This expression is also known as the *complementary sensitivity function* which is defined as follows:

$$T(s) = \frac{G(s)}{1 + G(s)}. \tag{8.216}$$

It is very important to recognize that the sum of the sensitivity function given by Eq. (8.213), and the complementary sensitivity function given by Eq. (8.216) equals one:

$$S(s) + T(s) = 1. \tag{8.217}$$

We can express the error function $E(s)$ in Eq. (8.212) in terms of the sensitivity function. Let us assume that the feedback sensor measurement noise is zero. Therefore, substituting Eq. (8.213) into Eq. (8.212), we obtain the following:

$$E(s) = S(s)[R(s) - U(s)]. \tag{8.218}$$

Equation (8.218) is a very important relationship because it states that we should make the sensitivity function S small to make the control system error $E(s)$ small. In addition, we know that since Eq. (8.213) represents the sensitivity function as well as the transfer function between $U(s)$ and $-E(s)$ [see Eq. (8.214)], we want the sensitivity function $S(s)$ to be as small as possible for both disturbance rejection and for making the control-system error small.

The conclusion of this analysis is that we want to make $S(s)$ as small as possible. However, is this possible over the entire range of frequencies? Let us analyze Eq. (8.213) to determine if this is feasible. Since $G(s)$ approaches zero as s approaches infinity in practical systems, then

$$\lim_{s\to\infty} S(s) = 1. \tag{8.219}$$

The result of Eq. (8.213) is that we can only make the sensitivity function $S(s)$ small over low and mid-range frequencies, but not at high frequencies.

Ideally, what would we like to make the complementary sensitivity function? Analyzing Eq. (8.217), we would like to make the complementary sensitivity function $T(s)$ equal to one because that would then result in the sensitivity function $S(s)$ being equal to zero. However, we know from Eq. (8.216) that $T(s)$ approaches zero as $G(s)$ approaches infinity. Therefore, we can design the complementary sensitivity function $T(s)$ to approximate one at low and mid-range frequencies, but not at high frequencies. Figure 8.30 illustrates the representative frequency responses of the sensitivity function $S(s)$ and the complementary sensitivity function $T(s)$.

The transfer function relating the sensor measurement noise $N(s)$ to the output $C(s)$ can be obtained from Eq. (8.211) by setting $R(s) = U(s) = 0$:

$$\frac{C(s)}{N(s)} = -\frac{G(s)}{1+G(s)}. \tag{8.220}$$

Observe that Eq. (8.220) is the negative of the complementary sensitivity function $T(s)$ defined in Eq. (8.216). Therefore,

$$T(s) = -\frac{C(s)}{N(s)} = \frac{G(s)}{1+G(s)}. \tag{8.221}$$

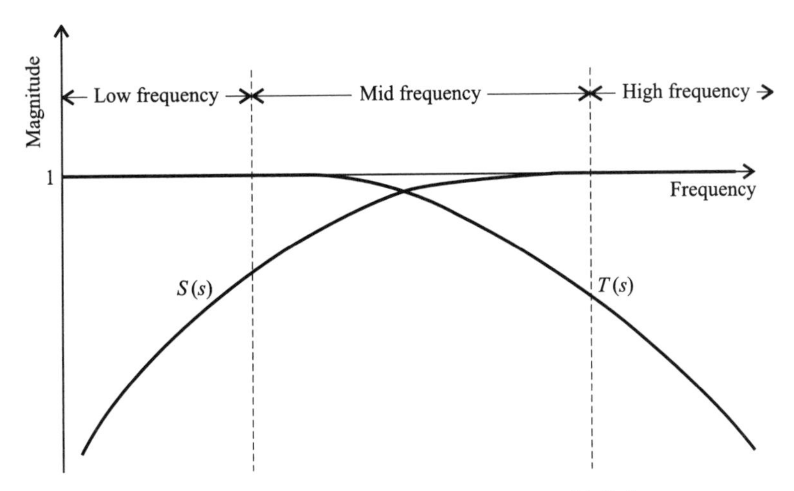

Figure 8.30 Representative sensitivity and complementary sensitivity frequency responses.

This results in a problem to the control-system engineer because we want the transfer function $C(s)/N(s)$ to be as small as possible, but the desirable $T(s)$ is large at low and mid-range frequencies as illustrated in Figure 8.30. Equation (8.217) showed that we would want to make $T(s) = 1$ ideally in order to drive the sensitivity function $S(s) = 0$. Therefore, the control-system engineer must make a tradeoff here between allowable feedback sensor measurement noise and the complementary sensitivity function (which also affects the sensitivity function [see Eq. (8.217)]. The basic tradeoff is to determine allowable noise and permissible sensitivity.

B. Desirable Control-System Transfer-Function Characteristics

From Figure 8.29 and Eq. (8.211), we can state the relationship between the reference input $R(s)$, the disturbance $U(s)$, and the feedback sensor measurement noise $N(s)$ with the output $C(s)$ as follows:

$$C(s) = H_r(s)R(s) + H_u(s)U(s) + H_n(s)N(s) \tag{8.222}$$

where

$$H_r(s) = \frac{C(s)}{R(s)} = \frac{G(s)}{1 + G(s)}, \tag{8.223}$$

$$H_u(s) = \frac{C(s)}{U(s)} = \frac{1}{1 + G(s)}, \tag{8.224}$$

$$H_n(s) = \frac{C(s)}{N(s)} = -\frac{G(s)}{1 + G(s)}. \tag{8.225}$$

In terms of the error function $E(s)$, as defined in Eq. (8.212), we define $E(s)$ as follows:

$$E(s) = [1 - H_r(s)]R(s) - H_u(s)U(s) + H_n(s)N(s). \tag{8.226}$$

It is interesting to examine the effect of the feedback sensor noise on the control system. To do this, let us assume that all the other external inputs are zero:

$$R(s) = U(s) = 0. \tag{8.227}$$

Let us assume that $n(t)$ has a Fourier transform although, in practice, noise usually does not have a Fourier transform. Using Parseval's theorem, we can determine the integral-squared error (ISE) which was defined in Chapter 5 by Eq. (5.45):

$$\int_0^{\infty} e^2(t)\,dt = \frac{1}{2\pi}\int_{-\infty}^{\infty} |H_n(j\omega)|^2 |N(j\omega)|^2\,d\omega. \tag{8.228}$$

By focusing on the square of the error, the ISE penalizes both positive and negative values of the error. The function $|N(j\omega)|^2$ is defined as the energy density spectrum of ω.

Let us try to define the desirable frequency characteristics from a practical viewpoint for $H_r(j\omega)$, $H_u(j\omega)$, and $H_n(j\omega)$. The reference input $R(s)$ and the disturbance $U(s)$ are usually low-frequency signals. Therefore, it is desired to have $H_r(j\omega) \approx 1$ and $H_u(j\omega) \approx 0$ at low frequencies. In practice, $H_r(j\omega) \approx 0$ and $H_u(j\omega)$ and $H_n(j\omega) \approx 1$ at high frequencies. In practice, we try to make $H_r(j\omega) \approx 1$, and $H_u(j\omega)$ and $H_n(j\omega) \approx 0$ over the frequency range from 0 to the gain crossover frequency ω_c. These are ideals and, in practice, we can only approximate these. For example, Figure 6.51*b* shows the closed-loop frequency response of Figure 6.51*a*. This closed-loop frequency response corresponds to $H_r(s)$ for the system shown in Figure 6.51*a*.

In practice, $G(s)$ is much greater than one at low frequencies. Therefore, Eq. (8.223) reduces to

$$H_r(s) = \frac{C(s)}{R(s)} = \frac{G(s)}{1 + G(s)} \approx 1 \tag{8.229}$$

and Eqs. (8.224) and (8.225) reduce to

$$H_u(s) = \frac{C(s)}{U(s)} = \frac{1}{1 + G(s)} \approx \frac{1}{G(s)}, \tag{8.230a}$$

$$H_n(s) = \frac{C(s)}{N(s)} = -\frac{G(s)}{1 + G(s)} \approx -1 \tag{8.230b}$$

at low frequencies. Therefore, a high loop gain is very desirable for frequencies in the passband (defined as frequencies between 0 and ω_c) because it then approximates $H_r(j\omega) \approx 1$ and $H_u(j\omega)$ is very small. However, the sensor measurement observation noise remains a problem because $H_n(s)$ is approximately -1. The control engineer has two choices regarding $N(s)$: (a) Design the sensor so that its measurement noise is very small; (b) tradeoff how high the loop gain is designed so that it minimizes the effect of the disturbance input $U(s)$, but the loop gain should not be too high so that the effect of the noise is tolerable.

C. Extension of Sensitivity and Complementary Sensitivity Concepts to Multivariable Control Systems

Many modern control systems contain several inputs and outputs. Therefore, it is important to extend our understanding of sensitivity and complementary sensitivity to multiple-input multiple-output (MIMO) control systems.

Figure 8.31 illustrates the block diagram of a MIMO control system where $\mathbf{r}(t)$ and $\mathbf{c}(t)$ are vectors, and $\mathbf{D}(s)$ and $\mathbf{G}(s)$ are matrices. The dimensions of $\mathbf{D}(s)$ are $n \times i$, and the dimensions of $\mathbf{G}(s)$ are $i \times n$, where $n = \dim[\mathbf{C}(s)]$ and $i = \dim[\mathbf{m}(s)]$. By analogy to Eq. (8.213) for the SISO control system, the sensitivity function for the MIMO control system is given by:

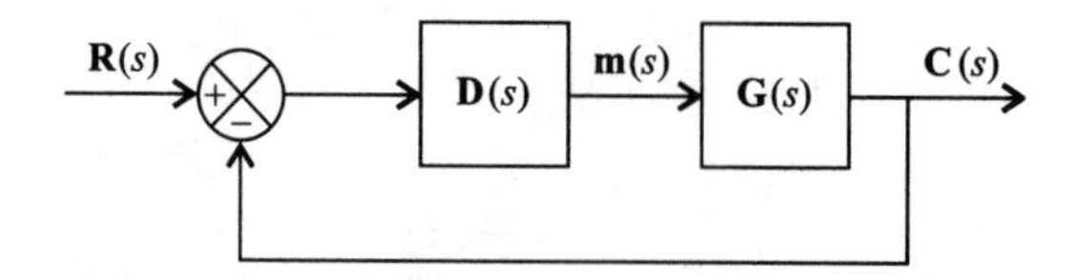

Figure 8.31 Block diagram of a MIMO control system.

$$\mathbf{S}_{\mathbf{G}(s)\mathbf{D}(s)}^{\mathbf{C}(s)/\mathbf{R}(s)} = [\mathbf{I} + \mathbf{D}(s)\mathbf{G}(s)]^{-1}. \tag{8.231}$$

By analogy to Eq. (8.216) for the SISO control system, the complementary sensitivity function for the MIMO control system is given by

$$\mathbf{T}(s) = [\mathbf{I} + \mathbf{D}(s)\mathbf{G}(s)]^{-1}\mathbf{D}(s)\mathbf{G}(s). \tag{8.232}$$

The sensitivity and complementary sensitivity functions are both $n \times n$ matrices. By analogy to Eq. (8.217) for the SISO case, the sum of the sensitivity and complementary functions for the MIMO case is given by

$$\mathbf{S}(s) + \mathbf{T}(s) = \mathbf{I} \tag{8.233}$$

where **I** represents the identity matrix.

D. Example for Finding S (*s*) and T(*s*) for a MIMO Control System

The open-loop transfer function matrix for a two-input, two-output control system to be analyzed is given by

$$\mathbf{D}(s)\mathbf{G}(s) = \begin{bmatrix} \dfrac{1}{s} & \dfrac{1}{s+1} \\ 1 & \dfrac{2}{s(s+2)} \end{bmatrix}. \tag{8.234}$$

Let us determine the sensitivity and the complementary sensitivity functions.

The sensitivity function for this MIMO control system is obtained from Eq. (8.231) as follows:

$$\mathbf{S}(s) = [\mathbf{I} + \mathbf{D}(s)\mathbf{G}(s)]^{-1} = \begin{bmatrix} \dfrac{1}{s} + 1 & \dfrac{1}{s+1} \\ 1 & \dfrac{2}{s(s+2)} + 1 \end{bmatrix}^{-1}. \tag{8.235}$$

Therefore,

$$\mathbf{S}(s) = \frac{1}{-s^3 - 2s^2 + 2s + 2} \begin{bmatrix} 2s(s+1) & -s^2(s+2) \\ -s^2(s+1)(s+2) & s(s+1)(s+2) \end{bmatrix}. \tag{8.236}$$

The complementary sensitivity function $\mathbf{T}(s)$ can be found from

$$\mathbf{T}(s) = \mathbf{I} - \mathbf{S}(s) = \begin{bmatrix} 1 & 0 \\ 0 & 1 \end{bmatrix} - \frac{1}{-s^3 - 2s^2 + 2s + 2} \begin{bmatrix} 2s(s+1) & -s^2(s+2) \\ -s^2(s+1)(s+2) & s(s+1)(s+2) \end{bmatrix}. \tag{8.237}$$

Simplifying this expression, we obtain

$$\mathbf{T}(s) = \frac{1}{-s^3 - 2s^2 + 2s + 2} \begin{bmatrix} -s^3 - 4s^2 + 2 & s^2(s+2) \\ s^2(s+1)(s+2) & -2s^3 - 5s^2 + 2 \end{bmatrix}. \tag{8.238}$$

8.12. FOUNDATIONS OF H^∞ CONTROL THEORY

Based on the preceding results, the concepts of modern H^∞ control theory will be presented. The very basic problem that H^∞ as presented by Zames in Reference 16 focuses on is sensitivity reduction of feedback control systems as an optimization problem, and it is separated from the problem of stabilization. The technique is concerned with the effects of feedback on uncertainty, where the uncertainty may be in the form of an additive disturbance $U(s)$ as illustrated in Figure 8.29. H^∞ control theory approaches the problem from the point of view of classical sensitivity theory, which has been presented in Chapter 5 and subsections 8.11A–8.11D, with the difference that feedback will not only reduce but also optimize sensitivity in an appropriate sense.

H^∞ control theory is a complex subject. The purpose of presenting it in this book, which is designed for the undergraduate student and the practicing engineer, is to introduce it and motivate the reader with an interest in this field to review some of the recent research papers which have been written on this subject [18–20]. In its basic form, H^∞ control theory attempts to minimize the supremum function over the entire frequency range

$$u = \sup_{\omega} |S(j\omega)W(j\omega)| \tag{8.239}$$

where $S(j\omega)$ is the sensitivity function and $W(j\omega)$ is a weighting function. We can view the magnitude of the product $S(j\omega)W(j\omega)$ as the magnitude of the weighted sensitivity. The weighting function emphasizes that low sensitivity is more important at low frequencies than higher frequencies. Therefore, by emphasizing the minimization of the magnitude of the product, the result is that the weighting function is greatest at those frequencies where the sensitivity is the smallest—namely at low frequencies.

The general solution to Eq. (8.239), obtained from function analysis, has the following general form [21]:

$$S(s)W(s) = kB(s) \tag{8.240}$$

where k is a constant and $B(s)$ is known as the Blaschke product. For the problem being considered in this book where the process $G(s)$ is stable (e.g., all of its poles are in the left-half of the s-plane and are of minimum phase), then $B(s) = 1$ and the constant k is any desirable, small, real number. Therefore, Eq. (8.240) states that we want to make the weighting function large for those low frequencies where we want to make the sensitivity small. Conversely, we want to make the weighting function small for those high frequencies where the sensitivity is large.

Let us examine in detail the result when $G(s)$ has all its poles in the left-half of the s-plane, and $B(s) = 1$. Therefore,

$$|S(j\omega)| = \frac{|k|}{|W(j\omega)|} \tag{8.241}$$

which implies that the shape of the magnitude of the sensitivity curve is the inverse of the weighting function. Therefore, the shape of the optimum $|S(j\omega)|$ is independent of the process $G(s)$, and only depends on the magnitudes of the constant k and the weighting function $W(j\omega)$.

A. Application of the Theory to an Example

Let us try to pull all the concepts presented in this section together by giving an example. We will consider a simple SISO control system, and determine its sensitivity function, s, complementary sensitivity function, T, and the weighting function, W. We will also assume that the constant k in Eq. (8.240) is equal to one.

For the illustrative example, let us consider a unity-feedback control system whose forward transfer function $G(s)$ is given by

$$G(s) = \frac{2}{s(s+1)}. \tag{8.242}$$

Substituting Eq. (8.242) into Eqs. (8.213), (8.216), and (8.241), we can calculate the sensitivity S, the complementary sensitivity T, and the weighting function W, respectively. The result is shown in Figure 8.32. The result agrees with the theoretical results expected from the presentation of this section. The sensitivity S is very small at low frequencies, and approaches one at very high frequencies. The weighting function W is the inverse of the sensitivity function, and is very large at low frequencies and also approaches one at very high frequencies. The complementary sensitivity function equals one minus the sensitivity function (see Eq. (8.217)), and equals one at very low frequencies and is very small at very high frequencies.

B. The H^∞ Process for the General SISO Case

Let us assume that in the general SISO case, the proces $G(s)$ has poles in the right half-plane. For this general case, the H^∞ process would proceed in the following step-by-step manner:

1. The weighting function $W(s)$ would be selected which would have the following characteristics:

- It must have no zeros in the right half-plane.

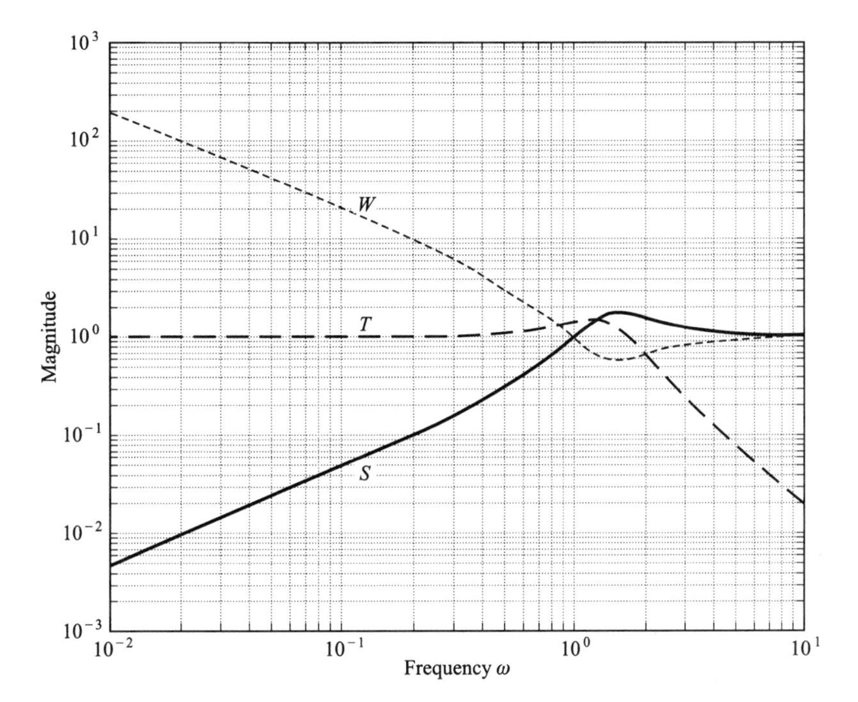

Figure 8.32 *S*, *T* and *W* for a unity-feedback system with $G(s) = 2/s(s+1)$.

- It must contain any poles that $G(s)$ has on the $j\omega$ axis.
- It must not contain $j\omega$ poles at zeros of $G(s)$.

2. Determine the Blaschke product $B(s)$ from knowledge of the poles of $G(s)$. If $G(s)$ has no poles in the right half-plane as illustrated in the preceding example in subsection 8.11A, $B(s) = 1$. Let us next consider the general case where $G(s)$ has poles in the right half-plane. Therefore, from Eq. (8.240),

$$S(p_n) = k\frac{B(p_n)}{W(p_n)} = 0, \quad n = 1, 2, 3, \ldots, N \tag{8.243}$$

where p_n are poles of $G(s)$ in the right half-plane.

3. $S(p_n)$ has to satisfy the following interpolation conditions for simple poles and zeros of $G(s)$ in the right half-plane:

- S equals zero at right-half poles, and one at right-half zeros.
- T is one at right-half poles and zero at right-half zeros.

This can be shown as follows. Let us consider the control system in Figure 8.33. The sensitivity of $C(s)/R(s)$ to $D(s)G(s)$ is given by

$$S_{D(s)G(s)}^{C(s)/R(s)} = \frac{1}{1 + D(s)G(s)}. \tag{8.244}$$

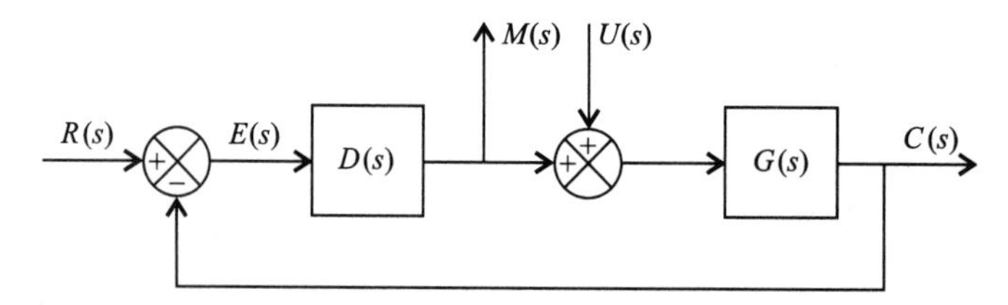

Figure 8.33 General SISO control system containing a reference input, $R(s)$, and a disturbance signal, $U(s)$.

The complementary sensitivity function is given by

$$T(s) = 1 - S(s) = \frac{D(s)G(s)}{1 + D(s)G(s)}. \tag{8.245}$$

The transfer functions between the inputs $R(s)$ and $U(s)$, and the outputs $M(s)$ and $C(s)$ are given as follows:

$$M(s) = \frac{D(s)}{1 + D(s)G(s)} R(s) - \frac{D(s)G(s)}{1 + D(s)G(s)} U(s). \tag{8.246}$$

In terms of $T(s)$ given by Eq. (8.245), Eq. (8.246) can be written as

$$M(s) = \frac{T(s)}{G(s)} R(s) - T(s)U(s). \tag{8.247}$$

Therefore,

$$C(s) = G(s)[M(s) + U(s)] = G(s)\left[\frac{D(s)}{1 + D(s)G(s)} R(s) - \frac{D(s)G(s)}{1 + D(s)G(s)} U(s) + U(s)\right]. \tag{8.248}$$

Equation (8.248) can be reduced to the following:

$$C(s) = \frac{D(s)G(s)}{1 + D(s)G(s)} R(s) + \frac{G(s)}{1 + D(s)G(s)} U(s) \tag{8.249}$$

In terms of $T(s)$ and $S(s)$, Eq. (8.249) can be rewritten as follows:

$$C(s) = T(s)R(s) + G(s)S(s)U(s) \tag{8.250}$$

Analysis of Eq. (8.247) reveals the following:

- If $T(s)/G(s)$ is to be stable, then $T(s)$ must cancel the right-half zero(s) of $G(s)$.

Analysis of Eq. (8.250) reveals the following:

- If $G(s)S(s)$ is to be stable, then $S(s)$ must cancel the pole(s) in the right-half plane of $G(s)$.

These two very important *interpolation conditions* imply that for simple zeros and poles which are in the right half-plane, $S(s)$ equals one at the right-half-plane zeros, and $S(s)$ equals zero at the right-half-plane zeros, and $S(s)$ equals zero at the right-half plane poles. Similarly, $T(s)$ equals zero at the right-half-plane zeros, and $T(s)$ equals one at the right-half-plane poles.

4. Form the Blaschke product. Returning to Eq. (8.243), and recognizing that $S(p_n)$ must satisfy these interpolation conditions, what should $B(p_n)$ and $W(p_n)$ be? Let us assume that $G(s)$ has poles in the right half of the s-plane, and let p_n, $n = 1, 2, \ldots, N$, be poles of $G(s)$. Therefore, Eq. (8.243) must be satisfied for the interpolation conditions. Because the weighting function $W(s)$ has no poles in the right half-plane, it cannot equal infinity. Therefore, it is necessary to have the Blaschke product $B(s)$ contain the zeros of $S(p_n) : B(p_n) = 0$, $n = 1, 2, 3 \ldots, N$. This means that the Blaschke product accounts for these zeros as follows:

$$B(s) = B_p(s)B'(s) \tag{8.251}$$

where

$$B_p(s) = \left[\frac{(-s + p_1)}{(s + p_1)}\right] \cdots \left[\frac{(-s + p_N)}{(s + p_N)}\right]. \tag{8.252}$$

We can find $B'(s)$ from the interpolation conditions previously shown. For example, let us assume that these zeros in the right half-plane which force $S(s)$ to equal one at these zeros are $z_1, z_2, z_3, \ldots, z_M$. From Eqs. (8.243) and (8.251),

$$S(z_m)W(z_m) = W(z_m) = kB'(z_m)B_p(z_m), \quad m = 1, 2, 3, \ldots, M. \tag{8.253}$$

This equation can be simplified to

$$kB'(z_m) = \frac{W(z_m)}{B_p(z_m)}, \quad m = 1, 2, 3, \ldots, M. \tag{8.254}$$

Equation (8.254) represents M equations which must be solved. It can be shown that if $M = 1$, then $B'(s) = 1$. If M is greater than one, then

$$B'(s) = \left(\frac{-s + c_1}{s + c_1}\right)\left(\frac{-s + c_2}{s + c_2}\right) \cdots \left(\frac{-s + c_{M-1}}{s + c_{M-1}}\right) \tag{8.255}$$

where c_i are complex values with the positive real parts chosen, together with the value of k, to satisfy Eq. (8.254).

5. Solve for $S(s)$. Having solved $B'(s)$, the optimal sensitivity $S(s)$ is then calculated from the following equation which is obtained by substituting Eq. (8.251) into Eq. (8.240):

$$S(s) = \frac{kB'(s)B_p(s)}{W(s)}. \tag{8.256}$$

The resulting $S(s)$ is stable because $B'(s)$ and $B_p(s)$ are stable, and since $W(s)$ has no poles in the right half-plane. Because $|B(j\omega)| = 1$

$$|S(j\omega)| = \frac{|k|}{|W(j\omega)|}. \tag{8.257}$$

Therefore, this equation shows that the magnitude of the sensitivity curve is the inverse of the magnitude of the weighting function curve. It is also interesting to observe that the optimal $|S(j\omega)|$ is independent explicitly of $G(s)$. However, $G(s)$ does effect the value of k.

C. An Example of the SISO Case Containing a Pole in the Right Half-Plane

Consider the control system illustrated in Figure 8.33 where

$$D(s)G(s) = \frac{(-s+2)}{(-s+3)(s+4)} \tag{8.258}$$

and the weighting function $W(s)$, shown in Figure 8.34, is given by

$$W(s) = \frac{(0.2Ts+1)}{(2Ts+1)}. \tag{8.259}$$

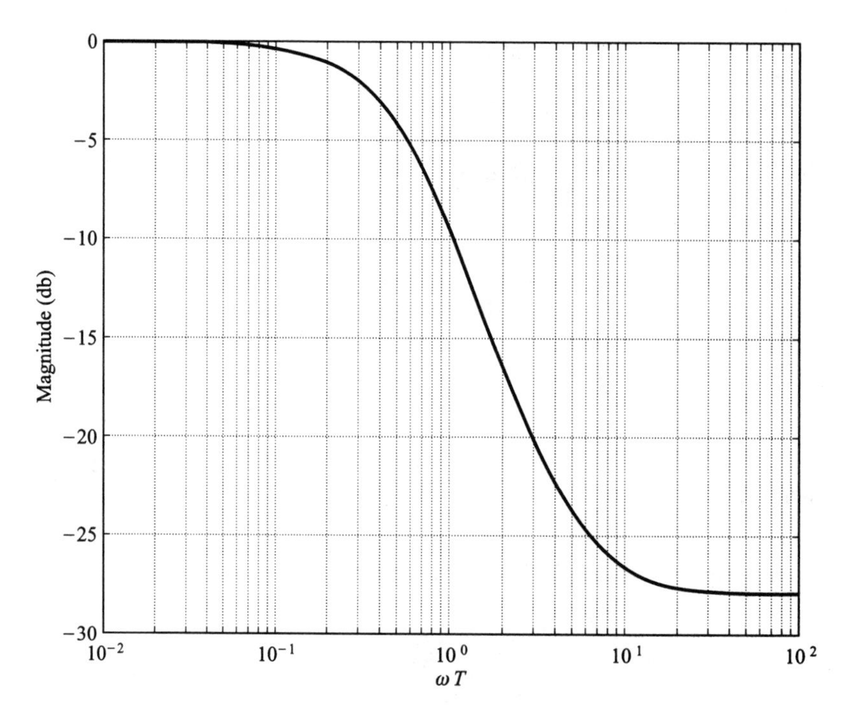

Figure 8.34 Weighting function $W(s)$.

This weighting function gives more weight to low frequencies (up to the bandwidth 1/2T) than high frequencies. From Eq. (8.252), the right-half pole of $G(s)$ at $s = 3$ results in $B_p(s)$ given by

$$B_p(s) = \frac{-s+3}{s+3}. \tag{8.260}$$

Because $D(s)G(s)$ has only one zero at $s = 2$ in the right half-plane of the s-plane, then $B'(s) = 1$. Therefore, from Eq. (8.254), we can find k as follows:

$$k = \frac{W(2)}{B_p(2)} = \frac{(0.4T+1)\,(5)}{(4T+1)\,(1)} = 5\frac{(0.4T+1)}{(4T+1)}. \tag{8.261}$$

Substituting k from Eq. (8.261), $B_p(s)$ from Eq. (8.260), and $W(s)$ from Eq. (8.259) into Eq. (8.256) we can solve for the optimal sensitivity $S(s)$:

$$S(s) = 5\frac{(0.4T+1)}{(4T+1)}\frac{(-s+3)}{(s+3)}\frac{(2Ts+1)}{(0.2Ts+1)}. \tag{8.262}$$

We can obtain the absolute value of the optimal sensitivity function from Eq. (8.257) as follows:

$$|S(j\omega)| = 5\frac{(0.4T+1)}{(4T+1)}\sqrt{\frac{(4\omega^2T^2+1)}{(0.04\omega^2T^2+1)}}. \tag{8.263}$$

Figure 8.35 illustrates the absolute value of the sensitivity function obtained from Eq. (8.263) for values of $T = 0.2$, 2, and 20. It is concluded from this curve that the sensitivity is smaller (better) at low frequencies if the bandwidth is not too high (or T is not too small).

8.13. LINEAR ALGEBRAIC ASPECTS OF CONTROL-SYSTEM DESIGN COMPUTATIONS [21–24]

The fundamental problems of linear control-system theory were reconsidered at the end of the 1960s and the beginning of the 1970s. The algebraic characteristics of the control-system problems were reestablished and have resulted in a better understanding of the foundations of linear control-system theory.

This section focuses on the interplay between some recent results and methodologies in numerical linear algebra and their application to problems arising in control systems. Let us reconsider the phase-variable canonical form representation of a control system:

$$\dot{\mathbf{x}}(t) = \mathbf{P}\mathbf{x}(t) + \mathbf{B}\mathbf{u}(t), \tag{8.264}$$

$$\mathbf{c}(t) = \mathbf{L}\mathbf{x}(t). \tag{8.265}$$

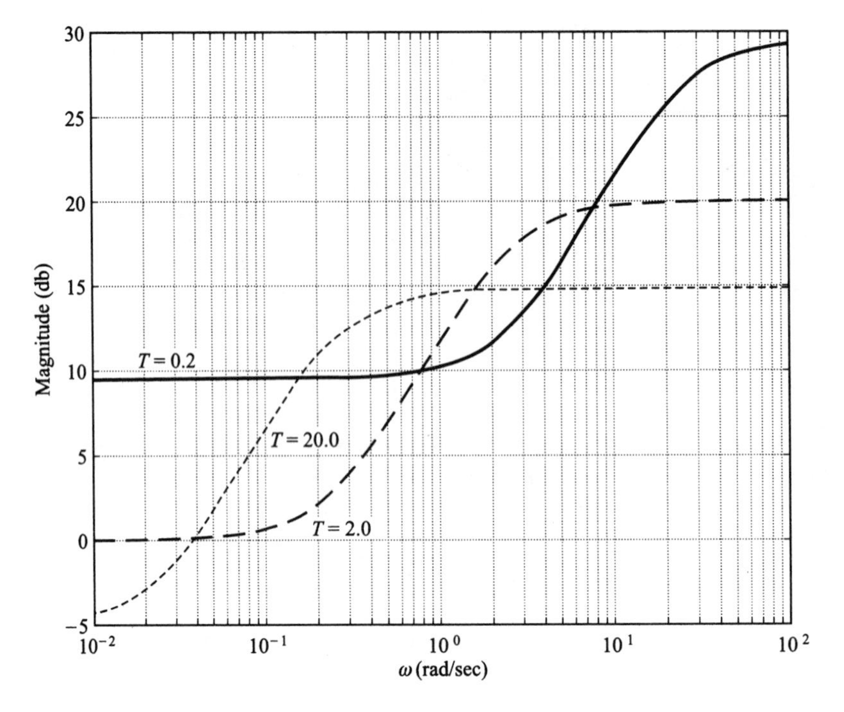

Figure 8.35 Optimal sensitivity frequency characteristic as a function of T.

We can represent this canonical form representation in terms of its frequency response by using Eqs. (8.10) and (8.18):

$$\frac{C(s)}{R(s)} = \frac{K\mathbf{L}[(s\mathbf{I} - \mathbf{P} + K\mathbf{bh})]\mathbf{b}}{\det(s\mathbf{I} - \mathbf{P} + K\mathbf{bh})}. \tag{8.266}$$

The problem of computing $C(s)/R(s)$, or $C(j\omega/R(j\omega)$, from knowledge of K, $\mathbf{P}$, $\mathbf{b}$, $\mathbf{L}$, and $\mathbf{h}$, efficiently is a practical problem of great interest. Reference 23 presents an efficient and generally applicable algorithm to solve this problem. The approach recommended performs an initial reduction of $\mathbf{P}$ to upper Hessenberg form $\mathbf{H}$, rather than solving the linear equation (with dense, unstructured $\mathbf{P}$) $(j\omega\mathbf{I} - \mathbf{P})\mathbf{X} = \mathbf{b}$, which would require $O(n^3)$ operations for each successive value of ω. The orthogonal matrices used to effect the Hessenberg form of $\mathbf{P}$ are incorporated into $\mathbf{b}$ and $\mathbf{L}$ giving $\tilde{\mathbf{b}}$ and $\tilde{\mathbf{L}}$. Therefore, as ω varies, the coefficient matrix in the linear equation $(j\omega\mathbf{I} - \mathbf{H})\mathbf{X} = \tilde{\mathbf{b}}$ remains in upper Hessenberg form. The advantage is that $\mathbf{X}$ can be determined in $O(n^2)$ operations rather than $O(n^3)$ as before, which is a very significant saving. In addition, this methodology is numerically very stable and has the advantage of being independent of the eigenstructure (possibly ill-conditioned) of $\mathbf{P}$.

This methodology can be extended to state-variable models in implicit form. For example, let us replace Eq. (8.264) by

$$\mathbf{E}\dot{\mathbf{x}}(t) = \mathbf{P}\mathbf{x}(t) + \mathbf{b}u(t) \tag{8.267}$$

Therefore, Eq. (8.266) can be replaced with the following:

$$\frac{C(s)}{R(s)} = \frac{K\mathbf{L}[adj(s\mathbf{E} - \mathbf{P} + K\mathbf{bh})]\mathbf{b}}{\det(s\mathbf{I} - \mathbf{P} + K\mathbf{bh})}. \tag{8.268}$$

We can use the initial triangular/Hessenberg reduction to again reduce the problem to one of updating the diagonal of a Hessenberg matrix and, therefore, an $O(n^2)$ linear equation problem. [24].

8.14. ILLUSTRATIVE PROBLEMS AND SOLUTIONS

This section provides a set of illustrative problems and their solutions to supplement the material presented in Chapter 8.

I8.1. The state and output equations of a second-order control system are given by the following:

$$\frac{dx_1(t)}{dt} = -3x_1 + 4t(t),$$
$$\frac{dx_2(t)}{dt} = -x_2(t) + u(t),$$
$$c(t) = x_1(t)$$

where $x_1(t)$ and $x_2(t)$ represent the system states, $c(t)$ is the system's output, and $u(t)$ represents its input.

(a) Determine whether the system is controllable.

(b) Determine whether the system is observable.

SOLUTION: (a) From Eq. (8.65) controllability can be determined for this second-order system from:

$$\mathbf{D} = [\mathbf{B} \quad \mathbf{PB}].$$

The phase variable canonical form of the state and output equations can be written as:

$$\begin{bmatrix} \dot{x}_1(t) \\ \dot{x}_2(t) \end{bmatrix} = \begin{bmatrix} -3 & 0 \\ 0 & -1 \end{bmatrix} \begin{bmatrix} x_1(t) \\ x_2(t) \end{bmatrix} + \begin{bmatrix} 4 \\ 1 \end{bmatrix} u(t),$$

$$c(t) = [1 \quad 0] \begin{bmatrix} x_1(t) \\ x_2(t) \end{bmatrix}$$

Therefore the companion matrix, **P** is given by

$$\mathbf{P} = \begin{bmatrix} -3 & 0 \\ 0 & -1 \end{bmatrix}$$

and the input vector, **B**, is given by

$$\mathbf{B} = \begin{bmatrix} 4 \\ 1 \end{bmatrix}$$

and the output matrix is given by:

$$\mathbf{L} = [1 \quad 0]$$

Therefore, the matrix **D** is given by:

$$\mathbf{D} = \begin{bmatrix} 4 & -12 \\ 1 & -1 \end{bmatrix}$$

and the system is controllable.

(b) Observability can be determined for this second-order control system from Eq. (8.75) where

$$\mathbf{U} = [\mathbf{L}^T \quad \mathbf{P}^T\mathbf{L}^T]$$

Therefore, the matrix U is given by:

$$\mathbf{U} = \begin{bmatrix} 1 & -3 \\ 0 & 0 \end{bmatrix}.$$

So, the system is unobservable because **U** is singular.

I8.2. Synthesize a system using linear-state-variable feedback that has a closed-loop transfer function given by

$$\frac{C(s)}{R(s)} = \frac{10(s+6)}{s^3 + 4s^2 + 10s + 20}.$$

Assume that the proces to be controlled has a transfer function given by

$$G(s) = \frac{2}{s(0.5s+1)}.$$

In your solution, show the following:

(a) Synthesis of the linear-state-variable-feedback system.

(b) Identification of any compensation network needed to satisfy the synthesis. What kind of network is it?

(c) Check of the stability of the resulting system synthesized using the root-locus method.

SOLUTION: (a)

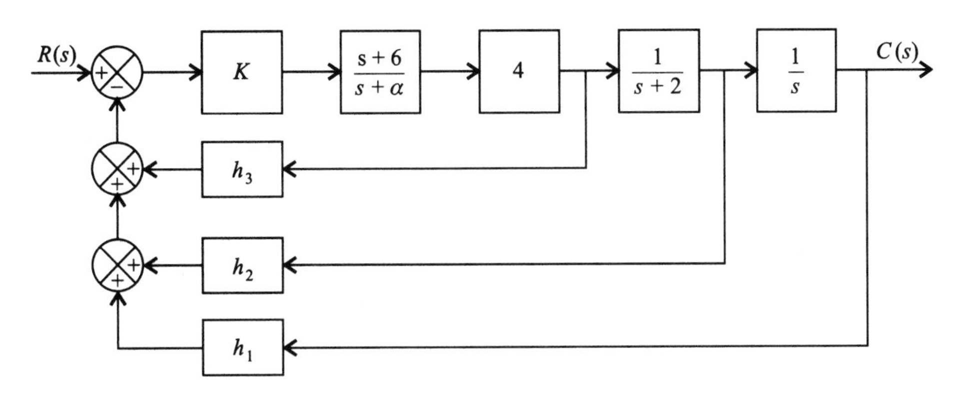

Figure I8.2(i)

$$\frac{C(s)}{R(s)} = \frac{\dfrac{4K(s+6)}{s(s+2)(s+\alpha)}}{1 + \dfrac{4K(s+6)h_3}{(s+\alpha)} + \dfrac{4K(s+6)h_2}{(s+\alpha)(s+2)} + \dfrac{4K(s+6)h_1}{s(s+\alpha)(s+2)}}.$$

This equation can be reduced to the following:

$$\frac{C(s)}{R(s)} =$$

$$\frac{4K(s+6)}{(1+4kh_3)s^3 + (\alpha + 2 + 32Kh_3 + 4Kh_2)s^2 + (2\alpha + 48Kh_3 + 24Kh_2 + 4Kh_1)s + 24Kh_1}.$$

Comparing the coefficients of this equation for the synthesized linear-state-variable-feedback system and that of the desired closed-loop transfer function given by

$$\frac{C(s)}{R(s)} = \frac{10(s+60)}{s^3 + 4s^2 + 10s + 20}$$

we obtain the following five simultaneous equations to be solved:

$$\begin{aligned} 10 &= 4K, \\ 1 + 4Kh_3 &= 1, \\ \alpha + 2 + 32Kh_3 + 4KH_2 &= 4, \\ 2\alpha + 48Kh_3 + 24Kh_2 + 4Kh_1 &= 10, \\ 24Kh_1 &= 20. \end{aligned}$$

Therefore, we obtain the following results:

$$K = 2.5;\ \ h_1 = 0.33;\ h_2 = 0.0675;\ \ h_3 = 0;\ \alpha = 1.325.$$

(b) The compensation network is given by

$$\frac{(s+6)}{(s+1.325)}$$

which is a phase-lag network.

(c) The root locus is drawn from the following transfer function:

$$G(s)H(s) = \frac{4(s+6)}{s(s+2)(s+\alpha)}[h_1 + sh_2] = \frac{K(s+6)(s+4.9)}{s(s+1.325)(s+2)}$$

The following root-locus diagram shows that this control system is stable for $0 < K < \infty$:

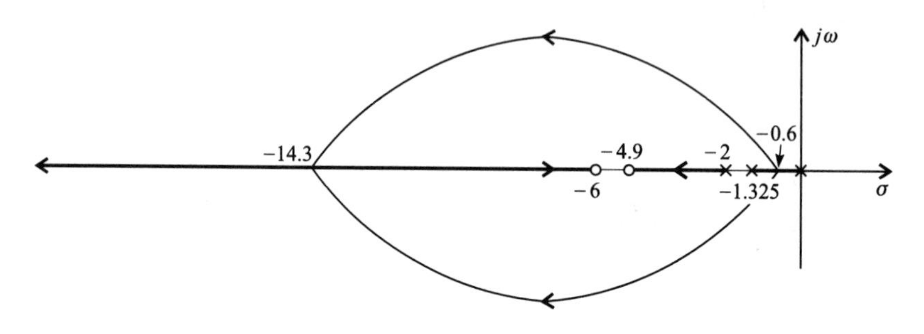

Figure I8.2(ii)

I8.3. Design a third-order controller whose three roots are located at the following locations in the s-plane: -1; -1; -12. The transfer function of the system to be controlled is given by

$$\frac{C(s)}{U(s)} = G(s) = \frac{1}{s(s+1)(s+2)}.$$

Determine the controller gain matrix, **K**.

SOLUTION: The desired location of the controller roots are located at:

$$\alpha_c(s) = (s+1)^2(s+12) = s^3 + 14s^2 + 25s + 12 \qquad \text{(I8.3-1)}$$

We need to determine the companion matrix, **P**, and the input gain vector, **b**, so that we can solve for the controller gain matrix, **K** from Eq. (8.167). The states of this third-order control system are defined as follows:

Let $x_1 = c(t)$; $x_2 = \dot{c}(t)$; $x_3 = \ddot{c}(t)$. Therefore, the state equations are given by:

$$\dot{x}_1(t) = x_2(t),$$
$$\dot{x}_2(t) = x_3(t),$$
$$\dot{x}_3(t) = -8x_1(t) - 2x_2(t) - 3x_3(t) + u(t).$$

The state equations in their vector and matrix format are given by:

$$\begin{bmatrix} \dot{x}_1(t) \\ \dot{x}_2(t) \\ \dot{x}_3(t) \end{bmatrix} = \begin{bmatrix} 0 & 1 & 0 \\ 0 & 0 & 1 \\ 0 & -2 & -3 \end{bmatrix} \begin{bmatrix} x_1(t) \\ x_2(t) \\ x_3(t) \end{bmatrix} + \begin{bmatrix} 0 \\ 0 \\ 1 \end{bmatrix} u(t).$$

Therefore, the companion matrix, **P**, is given by:

$$\mathbf{P} = \begin{bmatrix} 0 & 1 & 0 \\ 0 & 0 & 1 \\ 0 & -2 & -3 \end{bmatrix}.$$

The controller gain **K** can be determined from Eq. (8.167) as follows:

$$|s\mathbf{I} - \mathbf{P} + \mathbf{bK}| = 0.$$

Substituting into Eq. (8.167), we obtain the following:

$$\left| \begin{bmatrix} s & 0 & 0 \\ 0 & s & 0 \\ 0 & 0 & s \end{bmatrix} - \begin{bmatrix} 0 & 1 & 0 \\ 0 & 0 & 1 \\ 0 & -2 & -3 \end{bmatrix} + \begin{bmatrix} 0 \\ 0 \\ 1 \end{bmatrix} [K_1 \quad K_2 \quad K_3] \right| = 0.$$

This can be simpified to

$$\begin{vmatrix} s & -1 & 0 \\ 0 & s & -1 \\ K_1 & 2 + K_2 & s + 3 + K_3 \end{vmatrix} = 0$$

which can be reduced to the following:

$$s^3 + (3 + K_3)s^2 + (2 + K_2)s + K_1 = 0. \tag{I8.3-2}$$

Setting coefficients of Eqs. (I8.3-1) and (I8.3-2) equal to each other, we obtain the following:

$$\begin{aligned} 14 &= 3 + K_3, \\ 25 &= 2 + K_2, \\ 12 &= K_1. \end{aligned}$$

Therefore, we solve these three equations and find that:

$$\begin{aligned} K_1 &= 12, \\ K_2 &= 23, \\ K_3 &= 11. \end{aligned}$$

Therefore, the controller gain matrix, **K**, is given by:

$$\mathbf{K} = [12 \quad 23 \quad 11].$$

I8.4. Design a third-order estimator whose three roots are located at the following locations in the s-plane: -3.5; -3.5; -15. The transfer function of the process is given by

$$G(s)H(s) = \frac{1}{s(s+3)(s+4)}$$

Determine the estimator gain vector **M**.

SOLUTION: The desired location of the estimator roots is at:

$$\alpha_e(s) = (s+3.5)^2(s+15) = s^3 + 22s^2 + 117.25s + 183.75. \qquad \text{(I8.4-1)}$$

We need to determine the companion matrix, **P**, the input gain vector, **b**, and the output matrix, **L**, so that we can determine the estimator gain factors in the **M** matrix from Eq. (8.131). The states of this third-order control system are defined as follows:

Let $x_1(t) = c(t)$; $x_2(t) = \dot{c}(t)$; $x_3(t) = \ddot{c}(t)$. Therefore, the state equations are given by:

$$\begin{aligned}
\dot{x}_1(t) &= x_2(t),\\
\dot{x}_2(t) &= x_3(t),\\
\dot{x}_3(t) &= -12x_2(t) - 7x_3(t) + r(t).
\end{aligned}$$

The state equations in their vector and matrix form are given by:

$$\begin{bmatrix} \dot{x}_1(t) \\ \dot{x}_2(t) \\ \dot{x}_3(t) \end{bmatrix} = \begin{bmatrix} 0 & 1 & 0 \\ 0 & 0 & 1 \\ 0 & -12 & -7 \end{bmatrix} \begin{bmatrix} x_1(t) \\ x_2(t) \\ x_3(t) \end{bmatrix} + \begin{bmatrix} 0 \\ 0 \\ 1 \end{bmatrix} r(t).$$

Therefore, the companion matrix, **P** and the input vector, **b**, are

$$\mathbf{P} = \begin{bmatrix} 0 & 1 & 0 \\ 0 & 0 & 1 \\ 0 & -12 & -7 \end{bmatrix}; \quad \mathbf{b} = \begin{bmatrix} 0 \\ 0 \\ 1 \end{bmatrix}.$$

The output equation is given by:

$$c(t) = [1 \quad 0 \quad 0] \begin{bmatrix} x_1(t) \\ x_2(t) \\ x_3(t) \end{bmatrix}.$$

Therefore, the output matrix is given by

$$\mathbf{L} = [1 \quad 0 \quad 0].$$

The **M** matrix can be determined from Eq. (8.131) as follows:

$$|s\mathbf{I} - (\mathbf{P} - \mathbf{ML})| = 0$$

Substituting into Eq. (8.131), we obtain the following:

$$\left| \begin{bmatrix} s & 0 & 0 \\ 0 & s & 0 \\ 0 & 0 & s \end{bmatrix} - \begin{bmatrix} 0 & 0 & 1 \\ 0 & 0 & 1 \\ 0 & -12 & -7 \end{bmatrix} + \begin{bmatrix} m_1 \\ m_2 \\ m_3 \end{bmatrix} [1 \quad 0 \quad 0] \right| = 0.$$

This can be simplified to

$$\begin{vmatrix} s + m_1 & -1 & 0 \\ m_2 & s & -1 \\ m_3 & 12 & s+7 \end{vmatrix} = 0$$

which can be reduced to the following:

$$s^3 + (7 + m_1)s^2 + (12 + 7m_1 + m_2)s + (12m_1 + 7m_2 + m_3) = 0. \qquad \text{(I8.4-2)}$$

Setting like coefficients of Eqs. (I8.4-1) and (I8.4-2), equal to each other, we obtain the following:

$$\begin{aligned} 22 &= 7 + m_1, \\ 117.25 &= 12 + 7m_1 + m_2, \\ 183.75 &= 12m_1 + 7m_2 + m_3. \end{aligned}$$

Therefore, we solve these equations and find that:

$$\begin{aligned} m_1 &= 15, \\ m_2 &= 0.25, \\ m_3 &= 2. \end{aligned}$$

Therefore, the **M** vector is given by:

$$\mathbf{M} = \begin{bmatrix} 15 \\ 0.25 \\ 2 \end{bmatrix}.$$

I8.5. Repeat the problem solved in Section 8.6 using Ackermann's Formula for a process transfer function given by

$$\frac{C(s)}{U(s)} = \frac{(s+3)}{(s+4)(s+2)(s+8)}$$

and the control system's signal flow graph is given by:

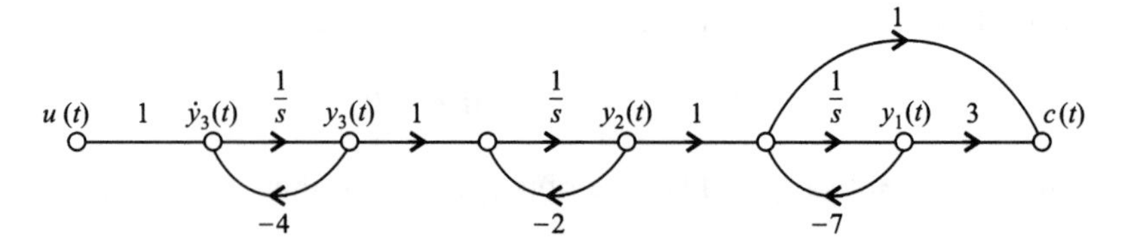

Figure I8.5(i)

(a) Determine the state equations for the process.

(b) Determine the controllability matrix for this original system.

(c) Transform the original system to the phase variable form, and determine the state and output equations.

(d) Determine the controllability matrix for the phase variable form.

(e) Determine the transformation matrix **A**.

(f) Design a contoller assuming that the dominant pair of complex-conjugage roots has a damping ratio of 0.707, and it results in a settling time of 4 sec. Select the third pole at the location of the zero of the process to be controlled. What is the resulting characteristic equation of the desired closed-loop control system?

(g) Determine the state and ouput equations for the phase-variable form with linear-state-variable-feedback.

(h) What is the resulting characteristic equation for the set of equations found in part (g)?

(i) Determine the linear-state-variable-feedback constants for the control system in phase-variable form.

(j) Transform the linear-state-variable-feedback constants back to the original system using the transformation matrix **A**.

(k) Draw the resulting closed-loop control system with linear-state-variable-feedback.

SOLUTION: (a)

$$\dot{\mathbf{y}}(t) = \mathbf{P}_y y(t) + \mathbf{B}_y u(t) = \begin{bmatrix} \dot{y}_1(t) \\ \dot{y}_2(t) \\ \dot{y}_3(t) \end{bmatrix} = \begin{bmatrix} -7 & 1 & 0 \\ 0 & -1 & 1 \\ 0 & 0 & -2 \end{bmatrix} + \begin{bmatrix} 0 \\ 0 \\ 1 \end{bmatrix} u(t)$$

(b)

$$\mathbf{D}_y = [\mathbf{B}_y \quad \mathbf{P}_y\mathbf{B}_y \quad \mathbf{P}_y^2\mathbf{B}_y] = \begin{bmatrix} 0 & 0 & 1 \\ 0 & 1 & -3 \\ 1 & -2 & 4 \end{bmatrix}$$

The value of the determinant of $\mathbf{D}_y$ is -1, it's nonsingular, and the original control system is controllable as expected from inspection of the signal-flow graph.

(c) From the given signal-flow graph:

$$\frac{C(s)}{U(s)} = \frac{\dfrac{3}{s^3} + \dfrac{1}{s^2}}{1 + \left[\dfrac{2}{s} + \dfrac{1}{s} + \dfrac{7}{s}\right] + \left[\left(\dfrac{2}{s}\right)\left(\dfrac{1}{s}\right) + \left(\dfrac{2}{s}\right)\left(\dfrac{7}{s}\right) + \left(\dfrac{1}{s}\right)\left(\dfrac{7}{s}\right)\right] + \left[\left(\dfrac{2}{s}\right)\left(\dfrac{1}{s}\right)\left(\dfrac{7}{s}\right)\right]}$$

This transfer function can be reduced to the following:

$$\frac{C(s)}{U(s)} = \frac{(s+3)}{s^3 + 10s^2 + 23s + 14}$$

Defining $x_1(t) = c(t)$, $x_2(t) = \dot{c}(t)$, and $x_3(t) = \ddot{c}(t)$, we obtain

$$\dot{x}_1(t) = x_2(t)$$
$$\dot{x}_2(t) = x_3(t)$$
$$\dot{x}_3(t) = 14x_1(t) - 23x_2(t) - 10x_3(t) + 3u(t) + \dot{u}(t)$$

Therefore, the state and output equations in matrix vector form of the phase variable control system are given by:

$$\dot{\mathbf{x}}(t) = \mathbf{P}_x\mathbf{x}(t) + \mathbf{B}_x u(t) = \begin{bmatrix} 0 & 1 & 0 \\ 0 & 0 & 1 \\ -14 & -23 & -10 \end{bmatrix} \begin{bmatrix} x_1(t) \\ x_2(t) \\ x_3(t) \end{bmatrix} + \begin{bmatrix} 0 \\ 0 \\ 1 \end{bmatrix} (3u(t) + \dot{u}(t))$$

$$c(t) = [3 \quad 1 \quad 0] \begin{bmatrix} x_1(t) \\ x_2(t) \\ x_3(t) \end{bmatrix}$$

(d) The resulting controllability matrix for the phase variable form can be obtained from

$$\mathbf{D}_x = [\mathbf{B}_x \quad \mathbf{P}_x\mathbf{B}_x \quad \mathbf{P}_x^2\mathbf{B}_x] = \begin{bmatrix} 0 & 0 & 1 \\ 0 & 1 & -10 \\ 1 & -10 & 77 \end{bmatrix}$$

Note that the determinant of $\mathbf{D}_x$ is -1 indicating that the determinant is nonsingular, and the phase variable form is also controllable, as expected.

(e) The transformation matrix, defined in Eq. 8.85 is

$$\mathbf{A} = \mathbf{D}_y\mathbf{D}_x^{-1} = \begin{bmatrix} 0 & 0 & 1 \\ 0 & 1 & -3 \\ 1 & -2 & 4 \end{bmatrix} \begin{bmatrix} 23 & 10 & 1 \\ 10 & 1 & 0 \\ 1 & 0 & 0 \end{bmatrix} = \begin{bmatrix} 1 & 0 & 0 \\ 7 & 1 & 0 \\ 7 & 8 & 1 \end{bmatrix}$$

(f) A second-order control system having a damping ratio of 0.707 with a settling time of 4 seconds results in a $\omega_n = 1.414$ (see Eq. 5.41). Therefore, the following second-order control system can meet the design specifications:

$$\frac{C(s)}{R(s)} = \frac{\omega_n^2}{s^2 + 2\zeta\omega_n s + \omega_n^2} = \frac{2}{s^2 + 2s + 2}$$

With third pole selected at $s = 3$, which also cancels the zero at $s = -3$, the characteristic equation of the desired closed-loop system is given by:

$$(s^2 + 2s + 2)(s + 3) = s^3 + 5s^2 + 8s + 6 = 0$$

(g) The state and output equations for the phase-variable form with linear-state-variable-feedback can be obtained from Eqs. 8.8 and 8.10 as follows:

$$\dot{\mathbf{x}}(t) = [\mathbf{P}_x - K\mathbf{b}_x\mathbf{h}_x]\mathbf{x}(t) + Kr(t)\mathbf{b}_x = \left[\begin{bmatrix} 0 & 1 & 0 \\ 0 & 0 & 1 \\ -14 & -23 & -10 \end{bmatrix} - \begin{bmatrix} 0 & 0 & 0 \\ 0 & 0 & 0 \\ Kh_1 & Kh_2 & Kh_3 \end{bmatrix}\right] \times \begin{bmatrix} x_1(t) \\ x_2(t) \\ x_3(t) \end{bmatrix} + Kr(t)\begin{bmatrix} 0 \\ 0 \\ 1 \end{bmatrix}$$

$$c(t) = [3 \quad 1 \quad 0]\mathbf{x}(t)$$

The state equation reduces to the following:

$$\dot{\mathbf{x}}(t) = \begin{bmatrix} 0 & 1 & 0 \\ 0 & 0 & 1 \\ -14 - Kh_1 & -23 - Kh_2 & -10 - Kh_3 \end{bmatrix}\mathbf{x}(t) + \begin{bmatrix} 0 \\ 0 \\ Kr(t) \end{bmatrix}$$

(h) The resulting characteristic equation can be obtained from Eqs. 8.10 and 8.19:

$$\det(s\mathbf{I} - \mathbf{P}_h) = (s\mathbf{I} - \mathbf{P} + K\mathbf{bh}) = 0$$

Therefore,

$$\begin{vmatrix} s & -1 & 0 \\ 0 & s & -1 \\ 14 + Kh_1 & 23 + Kh_2 & s + 10 + Kh_3 \end{vmatrix} = 0$$

The resulting characteristic equation is given by:

$$s^3 + (10 + Kh_3)s^2 + (23 + Kh_2)s + (14 + Kh_1) = 0$$

(i) Comparing this characteristic equation (see part h) with the characteristic equation of the desired closed-loop system (see part f), results in the following:

$$10 + Kh_3 = 5; \quad Kh_3 = -5$$
$$23 + Kh_2 = 8; \quad Kh_2 = -15$$
$$14 + Kh_1 = 6; \quad Kh_1 = -8$$

Therefore,

$$K\mathbf{h}_x = [-8 \quad -15 \quad -5]$$

(j) We will now transform $K\mathbf{h}_x$ back to the original system using Eq. 8.94 as follows:

$$\mathbf{h}_y = \mathbf{h}_x\mathbf{A}^{-1} = [-8 \quad -15 \quad -5]\begin{bmatrix} 1 & 0 & 0 \\ -7 & 1 & 0 \\ 49 & -8 & 1 \end{bmatrix}^{-1} = [-148 \quad 25 \quad -5]$$

(k) The resulting closed-loop control system with linear-state-variable-feedback is the following:

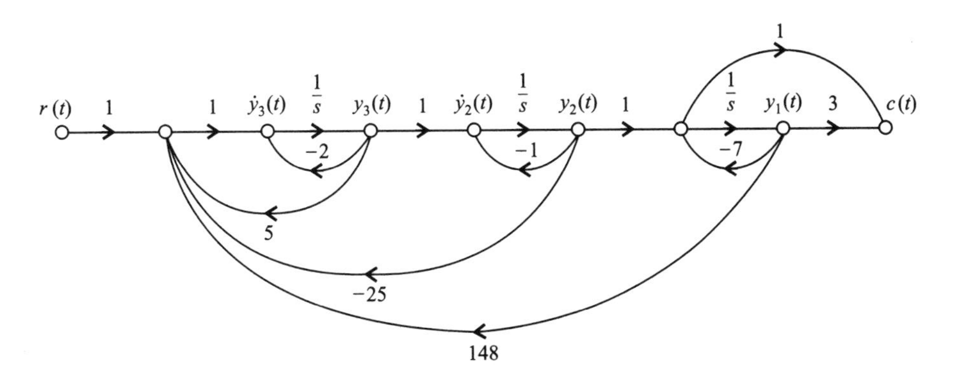

Figure I8.5(ii) Resulting closed-loop control system wth linear-state-variable-feedback designed using Ackermann's formula.

I8.6. Design a combined compensator of a regulator containing a controller and estimator for the system illustrated in Figure I8.6(i).

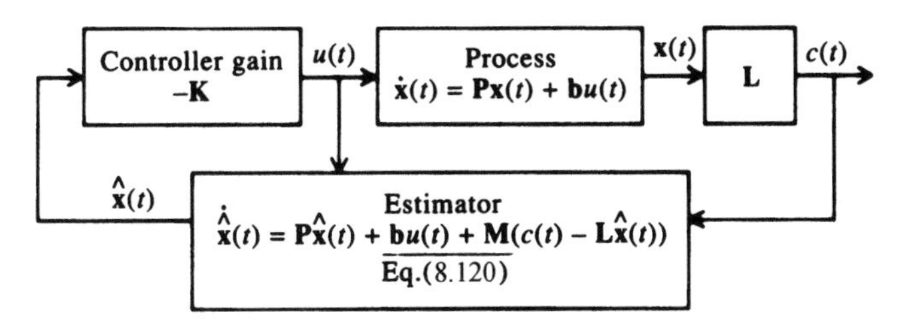

Figure I8.6(i)

Assume that the transfer function of the process is given by

$$\frac{C(s)}{U(s)} = G(s) = \frac{4}{(s+2)^2}.$$

Assume that the design specification of the controller is that it is critically damped with $\omega_n = 3$ rad/sec, and that the estimator is also critically damped with $\omega_n = 30$ rad/sec.

(a) Find the controller's gain coefficients' matrix.

(b) Find the estimator's coefficients' vector.

(c) Determine the transfer function of the compensator for the combined controller and estimator, $U(s)/C(s)$.

(d) *Sketch* the root locus of the compensated system. From your sketch, is the system unstable, stable, or conditionally stable?

SOLUTION: **(a)** The desired location of the controller roots is at:

$$\alpha_c(s) = (s+3)^2 = s^2 + 6s + 9 \tag{I8.6-1}$$

We need to determine the companion matrix, **P**, and the input gain vector, **b**, so that we can solve for the controller gain matrix, **K**, from Eq. (8.167). The states of this second-order control system are defined as follows:

Let $x_1(t) = c(t)$; $x_2(t) = \dot{c}(t)$. Therefore, the state equations are given by:

$$\dot{x}_1(t) = x_2(t)$$
$$\dot{x}_2(t) = -4x_1(t) - 4x_2(t) + 4u(t).$$

The state and output equations in their vector and matrix format are given by:

$$\begin{bmatrix} \dot{x}_1(t) \\ \dot{x}_2(t) \end{bmatrix} = \begin{bmatrix} 0 & 1 \\ -4 & -4 \end{bmatrix} \begin{bmatrix} x_1(t) \\ x_2(t) \end{bmatrix} + \begin{bmatrix} 0 \\ 4 \end{bmatrix} u(t)$$

$$c(t) = [1 \quad 0] \begin{bmatrix} x_1(t) \\ x_2(t) \end{bmatrix}.$$

Therefore, the companion matrix, **P**, the input vector, **b**, and the output matrix, **L** are given by:

$$\mathbf{P} = \begin{bmatrix} 0 & 1 \\ -4 & -4 \end{bmatrix}; \quad \mathbf{b} = \begin{bmatrix} 0 \\ 4 \end{bmatrix}; \quad \mathbf{L} = [1 \quad 0].$$

The controller gain **K** can be determined from Eq. (8.167) as follows:

$$|s\mathbf{I} - \mathbf{P} + \mathbf{bK}| = 0.$$

Substituting into Eq. (8.167), we obtain the following:

$$\left|\begin{bmatrix} s & 0 \\ 0 & s \end{bmatrix} - \begin{bmatrix} 0 & 1 \\ -4 & -4 \end{bmatrix} + \begin{bmatrix} 0 \\ 4 \end{bmatrix}[K_1 \quad K_2]\right| = 0.$$

This equation can be simplified to

$$\begin{vmatrix} s & -1 \\ 4 + 4K_1 & s + 4 + 4K_2 \end{vmatrix} = 0.$$

which can be reduced to the following:

$$s^2 + 4(1 + K_2)s + 4(1 + K_1) = 0. \qquad \text{(I8.6-2)}$$

Setting like coefficients of Eqs. (I8.6-1) and (I8.6-2) equal to each other, we obtain the following:

$$6 = 4(1 + K_2),$$
$$9 = 4(1 + K_1).$$

Therefore, we solve these two equations and find that:

$$K_1 = 1.25$$
$$K_2 = 0.5$$

Therefore, the controller gain matrix, **K**, is given by:

$$\mathbf{K} = [1.25 \quad 0.5]$$

(b) The desired location of the estimator roots is at:

$$\alpha_e(s) = (s + 30)^2 = s^2 + 60s + 900. \qquad \text{(I8.6-3)}$$

The **M** matrix can be determined from Eq. (8.131) as follows:

$$|\mathbf{sI} - (\mathbf{P} - \mathbf{ML})| = 0$$

Substituting into Eq. (8.131), we obtain the following:

$$\left|\begin{bmatrix} s & 0 \\ 0 & s \end{bmatrix} - \begin{bmatrix} 0 & 1 \\ -4 & -4 \end{bmatrix} + \begin{bmatrix} m_1 \\ m_2 \end{bmatrix}[1 \quad 0]\right| = 0.$$

This can be simplified to

$$\begin{vmatrix} s + m_1 & -1 \\ 4 + m_2 & s + 4 \end{vmatrix} = 0$$

which can be reduced to the following:

$$s^2 + (m_1 + 4)s + 4m_1 + 4 + m_2 = 0 \tag{I8.6-4}$$

Setting like coefficients of Eqs. (I8.6-3) and (I8.6-4) equal to each other, we obtain the following:

$$60 = m_1 + 4; \quad m_1 = 56$$

$$900 = 4m_1 + 4 + 4m_2; \quad m_2 = 672.$$

Therefore, the **M** vector is given by:

$$\mathbf{M} = \begin{bmatrix} 56 \\ 672 \end{bmatrix}$$

(c) The transfer function of the compensator, $U(s)/C(s)$ can be determined from Eq. (8.162):

$$\frac{U(s)}{C(s)} = -\mathbf{K}[s\mathbf{I} - \mathbf{P} + \mathbf{bK} + \mathbf{ML}]^{-1}\mathbf{M}$$

Substituting values for **K**, **P**, **B**, and **L** found in part (a), and **M** found in part (b), we obtain the following:

$$\frac{U(s)}{C(s)} = -[1.25 \quad 0.5]\left[\begin{bmatrix} s & 0 \\ 0 & s \end{bmatrix} - \begin{bmatrix} 0 & 1 \\ -4 & -4 \end{bmatrix} + \begin{bmatrix} 0 \\ 4 \end{bmatrix}[1.25 \quad 0.5] + \begin{bmatrix} 56 \\ 672 \end{bmatrix}[1 \quad 0]\right]^{-1}\begin{bmatrix} 56 \\ 672 \end{bmatrix}$$

which reduces to

$$\frac{U(s)}{C(s)} = -[1.25 \quad 0.5]\begin{bmatrix} s+56 & -1 \\ 681 & s+6 \end{bmatrix}^{-1}\begin{bmatrix} 56 \\ 672 \end{bmatrix}.$$

Further reduction reduces this equation to the following:

$$\frac{U(s)}{C(s)} = \frac{[1.25 \quad 0.5]\begin{bmatrix} s+6 & 1 \\ -681 & s+56 \end{bmatrix}\begin{bmatrix} 56 \\ 672 \end{bmatrix}}{s^2 + 62s + 1017}.$$

This equation reduces the following expression for the transfer function of the compensator for the combined controller and estimator, $U(s)/C(s)$:

$$\frac{U(s)}{C(s)} = -\frac{406(s + 2.48)}{s^2 + 62s + 1017}.$$

This is a phase-lead network having the zero at $s = -2.48$.

(d) The root locus for the compensated system is plotted from the following expression:

$$G(s)G_{\text{comp}}(s) = G(s)\frac{U(s)}{C(s)} = -\frac{1}{(s+2)^2}\frac{406(s+2.48)}{(s+31+j7.48)(s+31-j7.48)}.$$

The root locus shown in Figure I8.6(ii) shows that this is a conditionally stable system that is stable for $< K < 61, 102$. Observe that the root locus goes through the roots chosen for the controller roots ($s = -3, -3$) and estimator roots ($s = -30, -30$).

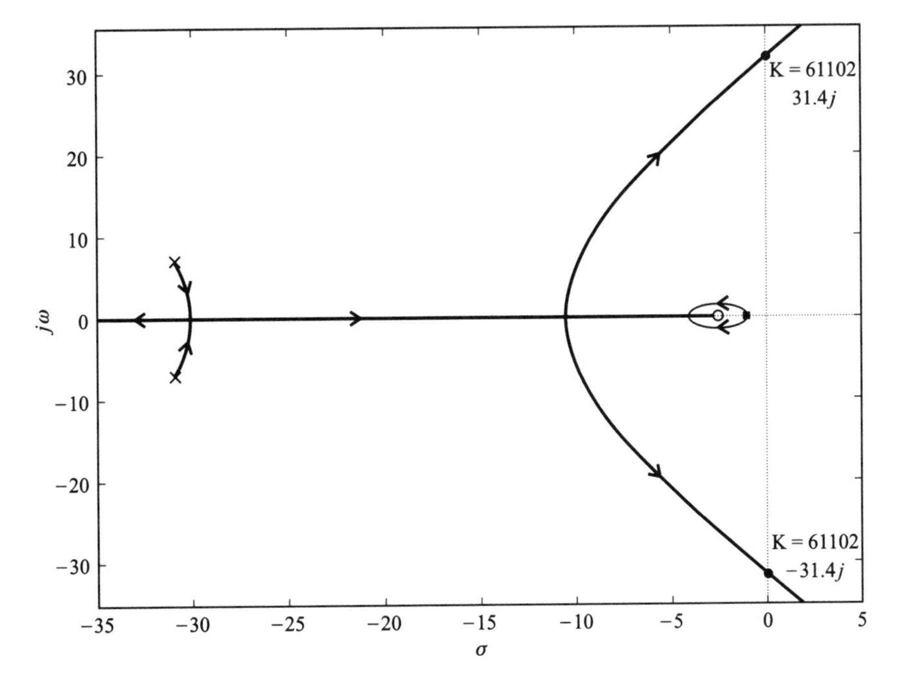

Figure I8.6(ii)

I8.7. The system of Problem I8.6 is modified to respond to a reference input, $r(t)$, as shown in the following block diagram:

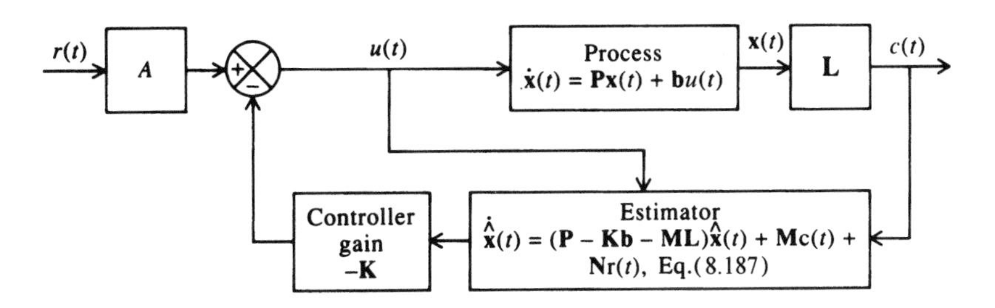

Figure I8.7

Assume that the process, combined compensator containing a controller and an estimator, and the specifications for the controller and estimator are exactly the same as in Problem I8.6. Assuming that $A = 4$, determine the requirement on **M** so that the state-estimation error is independent of the reference input, $r(t)$.

SOLUTION: We can determine the requirement on **N** so that the state-estimation error is independent of the reference input, $r(t)$, from Eq. (8.193):

$$\mathbf{b}A = \mathbf{N}.$$

From Problem I8.6, we know that the input vector, **b**, is given by

$$\mathbf{b} = \begin{bmatrix} 0 \\ 4 \end{bmatrix}.$$

Substituting **b** and $A = 4$ into Eq. (8.193), we obtain the following:

$$\begin{bmatrix} 0 \\ 4 \end{bmatrix} 4 = \begin{bmatrix} n_1 \\ n_2 \end{bmatrix}$$

Therefore, **N** is given by

$$\mathbf{N} = \begin{bmatrix} 0 \\ 16 \end{bmatrix}.$$

I8.8. We wish to design a robust control system containing a two-degrees-of-freedom series controller $G_{cl}(s)$ and a forward-loop-controller $G_{c2}(s)$ which is illustrated as follows:

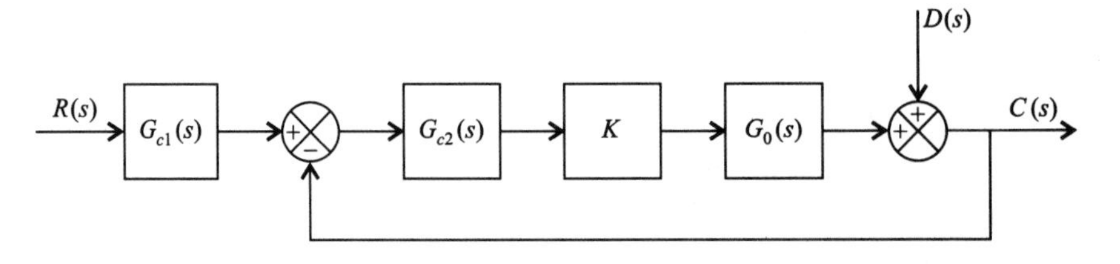

Figure I8.8(i)

The transfer function $KG_0(s)$ is

$$KG_0(s) = \frac{150(s+12)}{s(s+2)(s+8)(s+25)}. \tag{I8.8-1}$$

The gain of this transfer function can vary significantly, and the value of 150 is only the nominal amount. During its operation, the gain has been known to go as low as 75 and as high as 300. The objective of the control-system engineer is to design the robust control system shown so that the effect of this wide gain variation is minimized on the control system's transient response.

(a) Determine the control system's transient response to a unit step input for $K = 75$, 150, and 300, assuming that $G_{c1}(s) = G_{c2}(s) = 1$.

(b) Draw the root locus for the conditions defined in part (a).

(c) From the root locus drawn in part (b), determine the location of the roots for the closed loop system for gains of 75, 150, and 300, and the damping and overshoot, for the dominant set of complex-conjugate roots, for the three sets of gains being considered.

(d) Design the robust controller $G_{c2}(s)$.

(e) Determine the transfer function $KG_{c2}(s)G_0(s)$.

(f) Plot the root locus for the transfer function determined in part (e), and determine the damping, overshoot, and the roots of the characteristic equations for gains of 75, 150, and 300.

(g) Determine the design of the forward-loop controller $G_{c1}(s)$ in this two-degrees-of-freedom control system.

(h) With $G_{c1}(s)$, as determined in part (g) added to this control system, determine the transient response of this control system to a unit step input for $K = 75$, 150, and 300.

(i) What conclusions can you reach from the resulting transient responses found in parts (a) and (h)?

SOLUTION: **(a)** Figure I8.8(ii) illustrates the unit step responses for this control system for $K = 75$, 150, and 300 which was obtained using MATLAB. Observe from this figure that the peak overshoot varies from 27.3% for $K = 75$, to 69.3% for $K = 150$, and 76.3% for $K = 300$.

(b) The root locus obtained from MATLAB is shown in Figure I8.8(iii).

(c) Table I8.8(i) lists the damping, overshoot, and the characteristic equation roots for the dominant set of complex-conjugate roots of this control system.

Observe from this table the considerable variation in the damping ratio and the maximum percent overshoot (based on the dominant complex-conjugate roots) for the three cases being analyzed.

(d) The robust controller has the following pair of complex-conjugate zeros to cancel the poles of the system where $K = 150$:

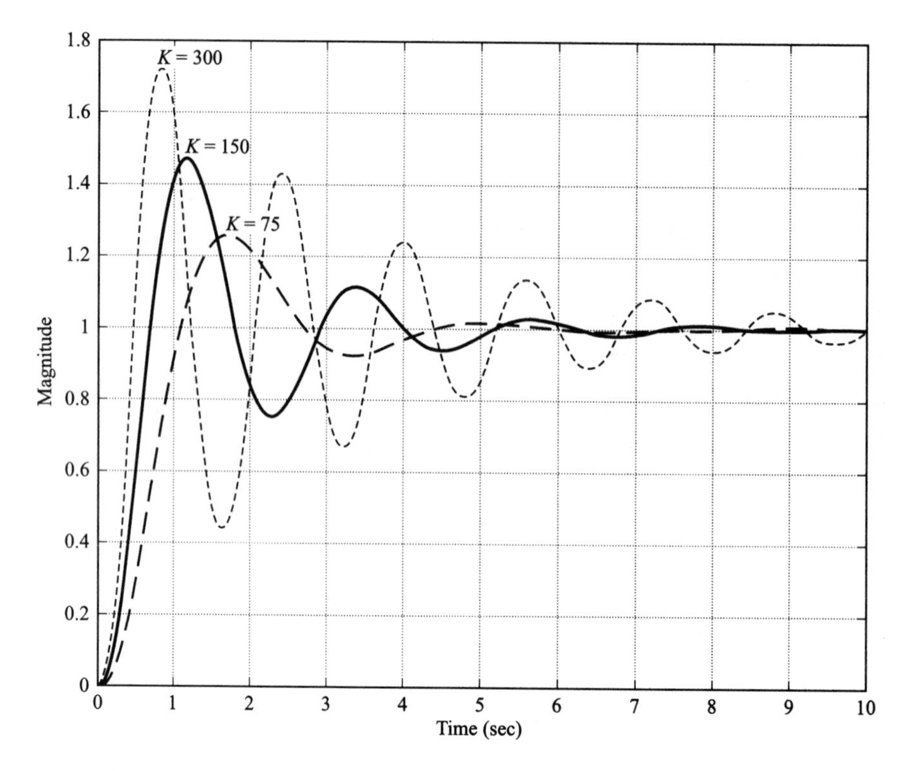

Figure I8.8(ii) Unit step response of $KG(s) = K(s+12)/(s+2)(s+8)(s+25)$.

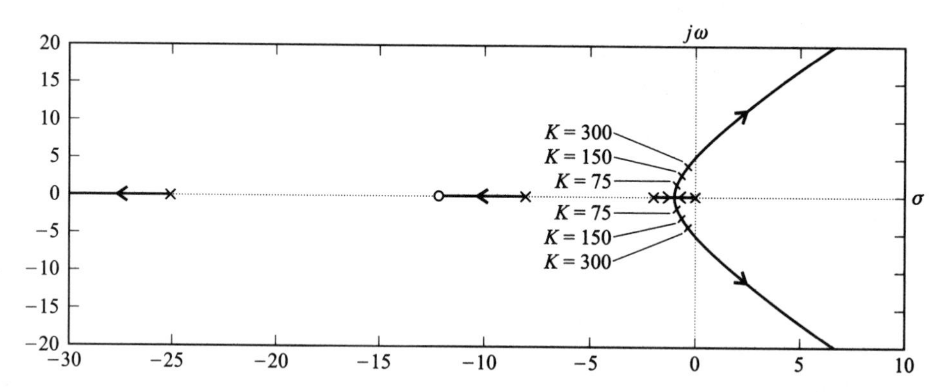

Figure I8.8(iii) Root-locus plot for system shown in Figure I8.8(i) for $KG_0(s)$ as defined in Eq. (I8.8-1) and $G_{c1}(s) = G_{c2}(s) = 1$.

Table I8.8(i)

K	Damping ratio	Max % overshoot	Roots of characteristic equation
300	0.08577	76.3	$-8.9274,\ -25.3890,\ -0.3418 \pm j3.9707$
150	0.21528	69.3	$-8.5592,\ -25.1969,\ -0.6219 \pm j2.8213$
75	0.38176	27.3	$-8.3154,\ -25.0991,\ -0.7928 \pm j1.9193$

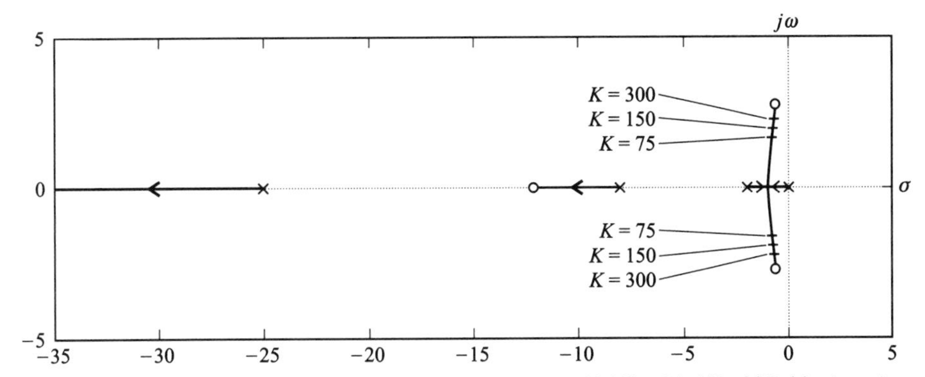

Figure I8.8(iv) Root-locus plot for system shown in Figure I8.8(i) with $KG_{c2}(s)G_0(s)$ given by Eq. (I8.8-3).

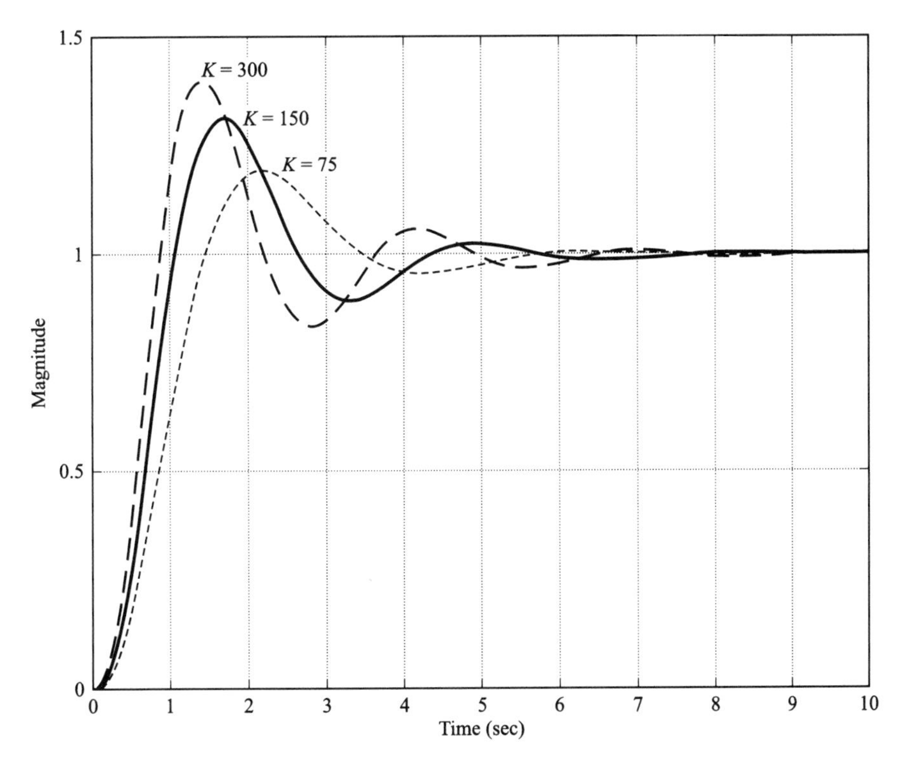

Figure I8.8(v) Unit step response of $KG(s) = K(s+12)/s(s+2)(s+8)(s+25)$ with robust control.

Table I8.8(ii)

K	Damping ratio	Max. % overshoot	Roots of characteristic equation
300	0.27822	40.2%	$-10.9999, -58.6295, -0.6573 \pm j2.2694$
150	0.34267	31.7%	$-10.3725, -41.1929, -0.7033 \pm j1.9282$
75	0.4575	19.9%	$-9.6604, -32.7831, -0.7713 \pm j1.4990$

$$G_{c2}(s) = \frac{s^2 + 1.2439s + 8.3463}{8.3463} = \frac{(s + 0.6219 + 2.8263j)(s + 0.6219 - 2.8263j)}{8.3463} \tag{I8.8–2}$$

(e)

$$KG_{c2}(s)G_0(s) = K\frac{(s^2 + 1.2439s + 8.3463)}{8.3463} \frac{(s + 12)}{s(s + 2)(s + 8)(s + 25)}. \tag{I8.8–3}$$

(f) The root locus, which was obtained using MATLAB, is illustrated in Figure I8.8(iv). From this root locus, the damping ratio, maximum percent overshoot, and roots of the characteristic equation are shown in Table I8.8(ii):

(g) The design of the forward-loop controller $G_{c1}(s)$ is the reciprocal of the robust controller $G_{c2}(s)$:

$$G_{c1}(s) = \frac{8.3463}{s^2 + 1.2439s + 8.3463}. \tag{I8.8–4}$$

(h) The resulting transient response of this control system to a unit step input for $K = 75$, 150, and 300 is illustrated in Figure I8.8(v).

(i) The resulting transient response illustrated in Figure I8.8(v) shows that the maximum percent overshoots with the robust control design have been greatly reduced from those illustrated in Figure I8.8(ii). The new robust control-system design truly exhibits admirable robust control features. Table I8.8(iii) compares the transient response with and without robust control:

Table I8.8(iii)

K	Maximum percent overshoot without robust control	Maximum percent overshoot with robust control
300	76.3%	40.2%
150	69.3%	31.7%
75	27.3%	19.9%

In addition to the reduction in the maximum percent overshoots, observe that the range of the maximum percent overshoots is reduced from 76.3%/27.3% = 2.79 without robust control to 40.2/19.9% = 2.02. That is a very significant reduction which robust control makes possible. Therefore, the robust system is much less sensitive to variations in K. In addition, the robustness with respect to variations in K will also provide disturbance suppression to $D(s)$.

I8.9. The optimal sensitivity function for the control system illustrated in Figure I8.9(i) is to be determined in the H^∞ sense.

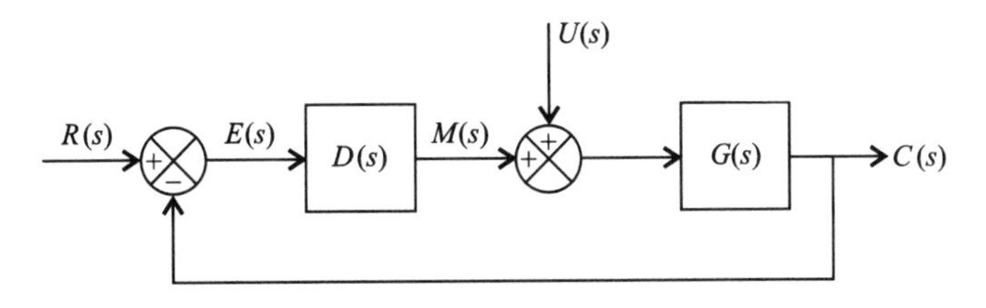

Figure I8.9(i)

The transfer function of the process $G(s)$ is given by

$$G(s) = \frac{(-s+4)}{(-s+5)(s+6)}.$$

The weighting function W(s) is given by

$$W(s) = \frac{(0.3Ts+1)}{(3Ts+1)}.$$

(a) Determine the value of $B_p(s)$.

(b) Determine the value of the constant k.

(c) Determine the value of the optimal sensitivity function $S(s)$.

(d) Determine the absolute value of the optimal sensitivity function and plot the results for $T = 0.2$, 2, and 20 sec.

(e) What conclusions can you reach from these curves?

SOLUTION: (a) From Eq. (8.252), the right-half pole of $G(s)$ at $s = 5$ results in

$$B_p(s) = \frac{-s+5}{s+5}.$$

(b) Because $G(s)$ has only one zero in the right half-plane of the s-plane, then $B'(s) = 1$. Therefore, from Eq. (8.254), we can find k as follows:

$$k = \frac{W(4)}{B_p(4)} = \frac{[0.3(4)T+1]}{[3(4)T+1]}\frac{9}{1} = 9\frac{(1.2T+1)}{(12T+1)}.$$

(c) Substituting into Eq. (8.256) for k, $B'(s)$, $B_p(s)$, and $W(s)$, we obtain the following:

$$S(s) = 9\frac{(1.2T+1)}{(12T+1)}\frac{(-s+5)}{(s+5)}\frac{(3Ts+1)}{(0.3Ts+1)}.$$

(d) The absolute value of the optimal sensitivity function can be obtained from Eq. (8.257) as follows:

$$|S(j\omega)| = 9\frac{(1.2T+1)}{(12T+1)}\sqrt{\frac{(9\omega^2T^2+1)}{(0.09\omega^2T^2+1)}}.$$

The absolute value of the sensitivity function is plotted in Figure I8.9(ii) for the following values of T: 0.2, 2, and 20.

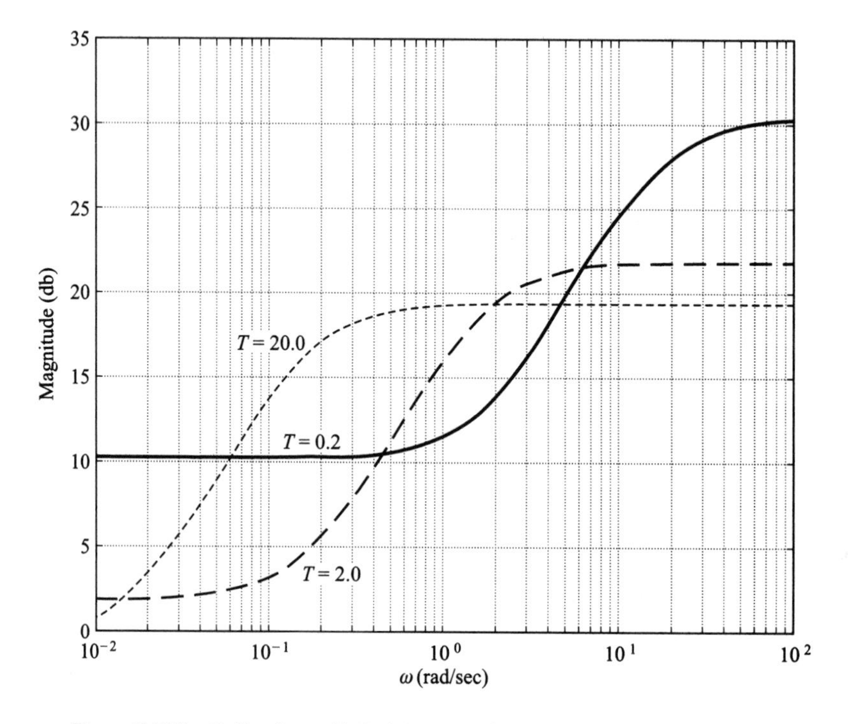

Figure I8.9(ii) Optimal sensitivity frequency characteristic as a function of T.

(e) It is concluded from these curves that the sensitivity is smaller (better) at low frequencies if the bandwidth is not too high (or T is not too small).

PROBLEMS

8.1. The control system illustrated in Figure P8.1(i) contains linear-state-variable-feedback elements h_1 and h_2.

(a) Determine the gain K and the linear-state-variable-feedback constants h_1 and h_2 so that the resulting control system represents a zero steady-state

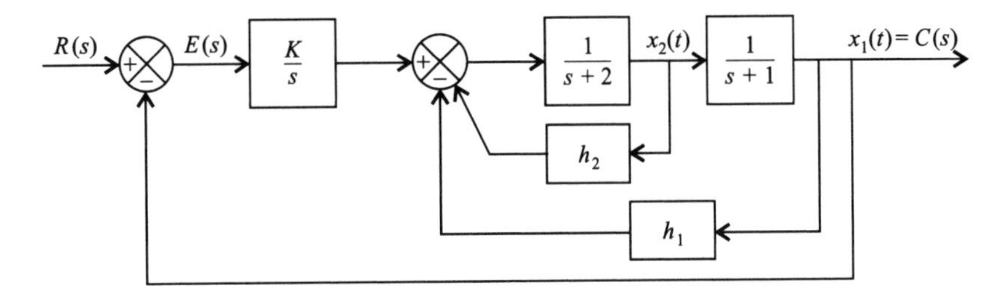

Figure P8.1(i)

step error system and its characteristic equation contsins roots at $-2+j$, $-2-j$, and -8.

(b) The same performance can be obtained as in part (a) if we implement a series controller, $G_c(s)$, instead of using linear-state-variable-feedback as illustrated in the configuration shown in Figure P8.1(ii).

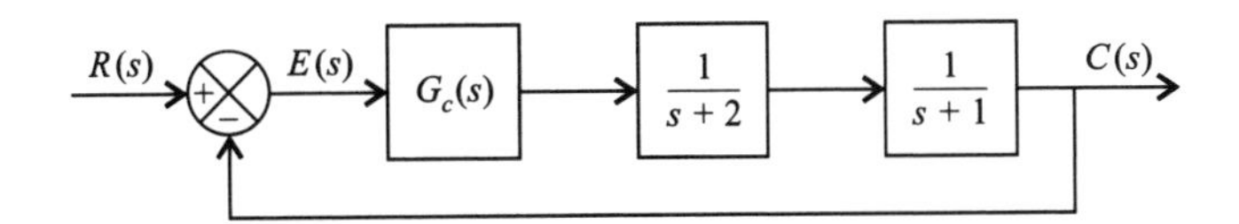

Figure P8.1(ii)

Determine the transfer function of $G_c(s)$ in terms of K, h_1, and h_2 obtained in part (a) and the other system parameters provided.

8.2. A control system containing a controller and a process are illustrated in the block diagram in Figure P8.2(i).

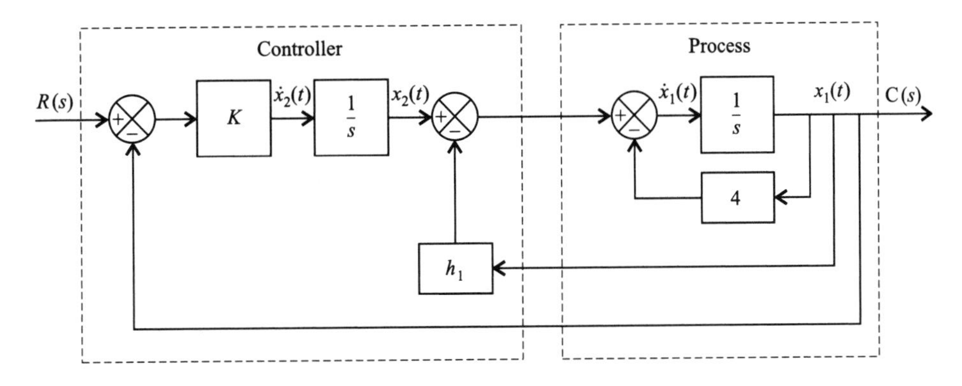

Figure P8.2(i)

(a) Determine the state equations of this control system.

(b) Determine the characteristic equation from knowledge of **P**.

(c) Determine the constant h_1 and the gain K if the roots of the characteristic equation are at -6 and -8.

(d) Instead of using the controller configuration, the control-system engineer wishes to design a "proportional plus integral controller" whose transfer function, $G_c(s)$, is given by:

$$G_c(s) = K_p + \frac{K_I}{s}$$

and is shown in figure P8.2(ii).

Determine the values of K_p and K_I so that the roots of the characteristic equation are also at -6 and -8.

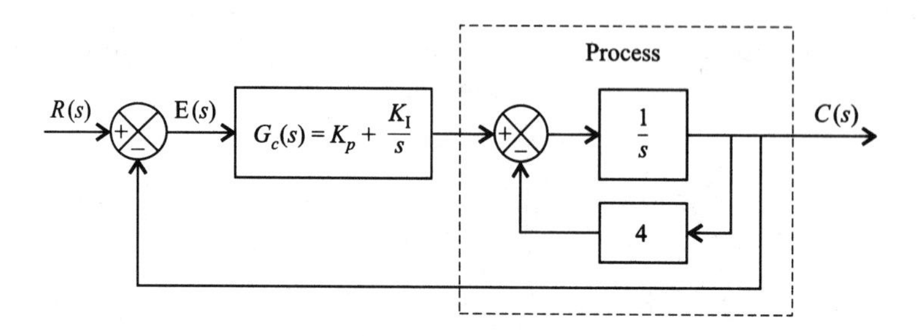

Figure P8.2(ii)

(e) Which design would you select, the controller configuration of part (a) or the proportional plus integral controller of part (d).

8.3. The controllability and observability of the control system shown in Figure P8.3 is to be determined.

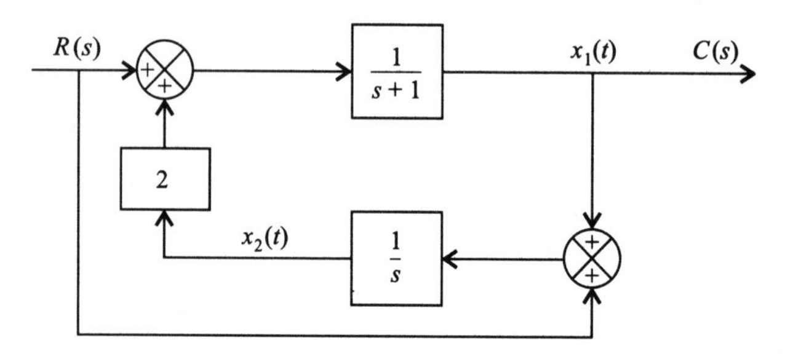

Figure P8.3

(a) Find the state and output equations of this control system.

(b) Determine the **D** matrix. Is the system controllable?

(c) Determine the **U** matrix. Is the system observable?

8.4. Synthesize a system utilizing linear-state-variable feedback that has closed-loop poles existing at $-9, 0$ and $-16, 0$ and can satisfy the following specifications:

$$K_v = 1, \quad \zeta = 0.707, \quad \omega_n = 1.$$

It is assumed that the process to be controlled has a transfer function given by $G(s) = 20/[s(s+1)(s+10)]$, and the transient response is governed by a pair of dominant complex-conjugate poles.

8.5. Repeat Problem 8.4 for the following specifications:

$$\frac{C(s)}{R(s)} = \frac{72(s+2)}{(s^2 + 1.414s + 1)(s+9)(s+16)},$$

$$G(s) = \frac{10}{s(1+5s)(1+0.5s)}.$$

8.6. Repeat Problem 8.4 for the following specifications:

$$\frac{C(s)}{R(s)} = \frac{8}{s^3 + 6s^2 + 10s + 8},$$

$$G(s) = \frac{1}{s(s+1)}.$$

8.7. It is desired to synthesize a system using linear-state-variable feedback which has a closed-loop pole at $-4, 0$ and can satisfy the following specifications:

$$K_v = 1, \quad \zeta = 0.707, \quad \omega_n = 1.$$

It is assumed that the process to be controlled has a transfer function given by

$$G(s) = \frac{20}{s(s+1)},$$

and the transient response is governed by a pair of dominant complex-conjugate poles.

(a) Can the system specifications of velocity constant be met with a simple second-order system?

(b) What must be added to the closed-loop system transfer function?

(c) What must be added to the open-loop system transfer function?

(d) Synthesize the block-diagram configuration of the linear-state-variable feedback configuration for the resulting system.

(e) Determine all unknown network constants and linear-state-variable feedback coefficients.

(f) Draw the root locus, and determine whether the resulting synthesized system is stable. If it shows that the system is unstable, what might be done to stabilize the system?

8.8. It is desired to synthesize a system using linear-state-variable feedback which can satisfy the following specifications:

$$K_v = 100/\text{sec}, \quad \omega_n = 100, \quad \zeta = 0.5.$$

The process to be controlled has a transfer function given by

$$G(s) = \frac{1}{s(2s+1)}$$

and assume that the transient response is governed by a pair of dominant complex-conjugate poles.

(a) Can the system specifications be met with a simple second-order system?

(b) Synthesize the block-diagram configuration of the linear-state-variable feedback for the resulting system.

(c) Determine all unknown constants (if any) and all linear-state-variable feedback coefficients.

(d) Determine whether the resulting synthesized system is stable using the root-locus method for solution.

8.9. Synthesize a system using linear-state-variable feedback that has a closed-loop transfer function given by

$$\frac{C(s)}{R(s)} = \frac{10(s+4)}{s^3 + 6s^2 + 22s + 40}.$$

It is assumed that the process to be controlled has a transfer function given by

$$G(s) = \frac{4}{s(0.5s+1)}$$

In your solution, show the following:

(a) Synthesis of the linear-state-variable feedback system.

(b) Identification of any compensation network needed to satisfy the synthesis.

(c) Check of the stability of the resulting system synthesized using the root-locus method.

8.10. A control system is defined by the following:

$$\mathbf{P} = \begin{bmatrix} -1 & 0 & 0 \\ 1 & -1 & 0 \\ 0 & 0 & 1 \end{bmatrix}, \quad \mathbf{L} = [0 \quad 1 \quad 1], \quad \mathbf{B} = \begin{bmatrix} 0 \\ 1 \\ 0 \end{bmatrix}$$

(a) Determine whether the system is controllable.

(b) Determine whether the system is observable.

8.11. The state and output equations of a second-order system are given by the following:

$$\begin{aligned} \dot{x}_1(t) &= -8x_1(t) + 4u(t), \\ \dot{x}_2(t) &= -2x_2(t) + u(t), \\ c(t) &= x_1(t) \end{aligned}$$

where $x_1(t)$ and $x_2(t)$ represent the system state, $c(t)$ is its output, and $u(t)$ is its input.

(a) Determine whether the system is controllable.

(b) Determine whether the system is observable.

8.12. A control system is defined by the following:

$$\mathbf{P} = \begin{bmatrix} -1 & 0 & 0 \\ 0 & -2 & 1 \\ 0 & 0 & -2 \end{bmatrix}, \quad \mathbf{L} = [1 \quad 0 \quad 0], \quad \mathbf{B} = \begin{bmatrix} 1 \\ 0 \\ 1 \end{bmatrix}.$$

(a) Determine whether the system is controllable.

(b) Determine whether the system is observable.

8.13. Design a second-order controller which is to be critically damped, and whose two second-order roots are located at -7 in the s-plane. The transfer function of the system to be controlled is given by

$$\frac{C(s)}{U(s)} = G(s) = \frac{1}{(s+1)^2}.$$

Find the vector **K**.

8.14. Repeat Problem 8.13 if the controller is critically damped, but its two second-order roots are now located at -4 instead of -7.

8.15. Design a second-order estimator for a system which is to have a damping ratio of 0.5, and whose estimator has two second-order roots located at -30 in the s-plane. Find the vector **M**. Assume that $\omega_n = 1$ rad/sec.

8.16. Repeat Problem 8.15 if the two estimator roots are located at -15 instead of -30.

8.17. Design a second-order estimator which satisfies the ITAE criterion for a zero steady-state step error system. Assume that $\omega_n = 25$ rad/sec.

8.18. We wish to design a regulator system (where the reference input equals zero) containing a combined controller and estimator as illustrated in Figure 8.19. The process' transfer function is given by the following expression:

$$\frac{C(s)}{U(s)} = G(s) = \frac{1}{s(s+1)}.$$

(a) Determine the state and output equations of this process.

(b) We wish to design the controller whose design specifications are for $\omega_n = 2$ rad/sec and for $\zeta = 1$. Determine the controller gain matrix, **K**.

(c) We wish to design the estimator to have a much faster response than the controller. Therefore, we will design the estimator to have $\omega_n = 20$ rad/sec and $\zeta = 1$. Determine the estimator gain matrix **M**.

(d) Determine the resulting compensator's transfer function, $U(s)/C(s)$.

(e) We wish to extend the design of this sytem to the case of a system containing a reference input, $r(t)$, as shown in Figure 8.22. Determine the vector **N** so that the estimator error is independent of $r(t)$. Assume that $A = 10$.

8.19. Repeat the problem solved in Section 8.6 using Ackermann's Formula for a process transfer function given by

$$\frac{C(s)}{U(s)} = \frac{(s+2)}{(s+1)(s+4)(s+8)}$$

and the control system's signal-flow graph is shown in Figure P8.19.

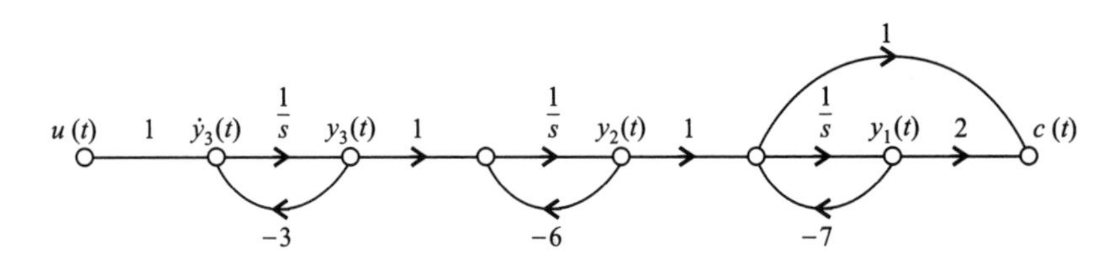

Figure P8.19

(a) Determine the state equations for the process.

(b) Determine the controllablity matrix for this original system. Is the system controllable?

(c) Transform the original system to the phase-variable form, and determine the state and output equations.

(d) Determine the controllability matrix for the phase variable form.

(e) Determine the transformation matrix **A**.

(f) Design a controller assuming that the dominant pair of complex-conjugate roots has a damping ratio of 0.707, and it results in a settling time of 4 sec. Select the third pole at the location of the zero of the process to be controlled. What is the resulting characteristic equation of the desired closed-loop control system?

(g) Determine the state and output equations for the phase-variable form with linear-state-variable feedback.

(h) What is the resulting characteristic equation for the set of equations found in part (g)?

(i) Determine the linear-state-variable feedback constants for the control system in phase-variable form.

(j) Transform the linear-state-variable feedback constants back to the original system using the transformation matrix **A**.

(k) Draw the resulting closed-loop control system with linear-state-variable feedback.

8.20. Determine the closed-loop transfer function of the resulting closed-loop control system developed using Ackermann's Formula and shown in Figure 8.17, and compare it with the desired closed-loop transfer function given by Eq. (8.107). Do they agree?

8.21. The design of the positioning system of a tracking radar is a very interesting control-system problem. Figure P8.21(*a*) illustrates a tracking radar which has wind torque disturbances acting on it, and the equivalent block diagram for one axis of the tracking radar's positioning loop is illustrated in Figure P8.21(*b*). The block diagram illustrated represents a two-degrees-of-freedom robust control system which has the dual capability of minimizing the effects of the wind torque disturbance $D(s)$ while also being robust (insensitive) to variations in the gain K. The transfer function for the forward-loop transfer function for the tracking radar $G_0(s)$ is given by

$$G_0(s) = \frac{(1 + 0.4s)}{s(1 + s)(1 + 0.15s)}.$$

(a) Determine the transient response of this control system to a unit step input at $R(s)$ assuming that $G_{c1}(s) = G_{c2}(s) = 1$ and K has a nominal value of 40, but can also vary as low as 10 and as high as 160.

(b) Draw the root locus for the conditions in part (a). Determine the location of the closed roots for $K = 10$, 40, and 160, and determine the damping ratio for the dominant complex-conjugate roots for these three cases.

(c) Design $G_{c2}(s)$ so that it contains the zeros which cancel the complex-conjugate roots for the case of nominal gain 40.

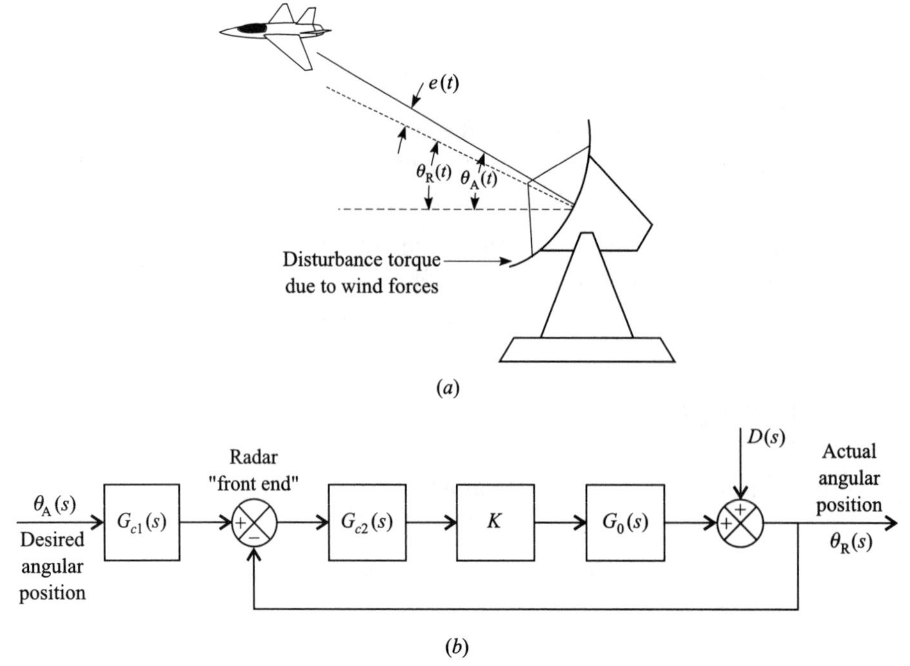

Figure P8.21 A tracking radar conceptually illustrating disturbance torques due to wind forces (*a*), and the equivalent block diagram of the tracking radar's positioning loop (*b*).

(d) Draw the root locus for the conditions of part (c). Determine the location of the closed-loop roots for $K = 10$, 40, and 160, and determine the damping ratio for the dominant complex-conjugate roots for these three cases.

(e) Design the robust controller $G_{c1}(s)$, and plot the resulting transient response of this control system containing $G_{c1}(s)$ and $G_{c2}(s)$, which was designed in part (c).

(f) How do the transient responses of part (e) compare to the transient responses of part (a)? Discuss the robustness of your resulting design to variations in the gain K.

8.22. The optimum sensitivity for the control system illustrated in Figure P8.22 is to be determined in the H^{∞} sense. The proces $G(s)$ has a transfer function given by

$$G(s) = \frac{(-s + 3)}{(-s + 4)(s + 5)}.$$

The weighting function, $W(s)$, is given by

$$W(s) = \frac{(0.5Ts + 1)}{(5Ts + 1)}$$

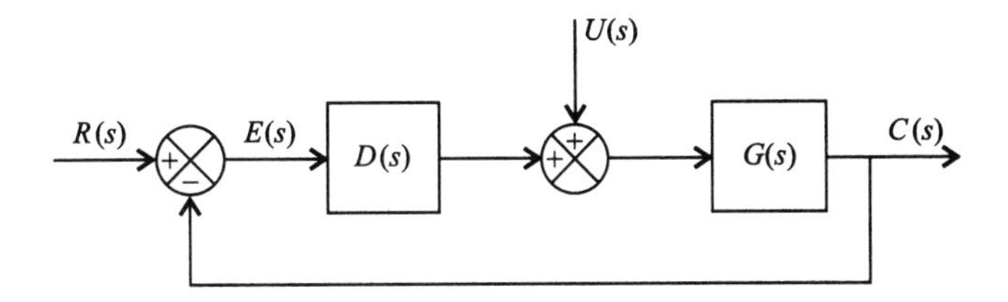

Figure P8.22

(a) Determine $B_p(s)$.

(b) Determine $B'(s)$.

(c) Determine k.

(d) Determine $S(s)$.

(e) Determine the magnitude of the optimum value of the sensitivity function, $|S(j\omega)|$ as a function of T.

(f) From your results for $|S(j\omega)|$ in part (e), plot the magnitude of the optimum sensitivity function as a function of ω for $T = 0.2$, 2 and 20.

REFERENCES

1. J. L. Melsa and D. G. Schultz, *Linear Control Systems*. McGraw-Hill, New York, 1969.
2. R. E. Kalman, '"On the general theory of control systems." In *Proceedings of the First International Congress of Automatic Control*, Moscow, 1960.
3. R. E. Kalman, Y. C. Ho, and K. S. Navendra, "Controllability of linear dynamical systems." *Contrib. Differ. Equat.*, 189–213 (1961).
4. E. G. Gilbert, "Controllability and observability in multivariable control systems." *J. Control*, Ser. (A) **1**, Soc. Ind. Appl. Math., 128–151 (1963).
5. J. E. Ackermann, "Der Entwurf linearer regelungs Systems in Zustandstraum." *Regelungstech, Prozess-Datenverarb*, **7**, 297–300 (1972).
6. K. Ogatta, *Modern Control Engineering*, 3rd ed., Prentice-Hall, Englewood Cliffs, NJ, 1997.
7. N. S. Nise, *Conrrol Systems Engineering*, 2nd ed., Addison–Wesley, Reading, MA, 1995.
8. G. F. Franklin, J. D. Powell, and A. Emami-Naeini, *Feedback Control of Dynamic Systems*, 2nd ed. Addison–Wesley, Reading, MA, 1991.
9. A. Emami-Naeini and G. F. Franklin, "Zero assignment in the multivariable robust servomechanism," *Proc IEEE Conf. Dec. Control., 1982*, pp. 891–893.
10. Modern CACSD Using the Robust Control Toolbox by R. Y. Chiang and M. G. Safonou, *Proc. Conf. on Aerospace and Computational Control*, Oxnard, CA, August 28–30, 1989.
11. P. Dorato, *Robust Control*, IEEE Press, New York, 1987.
12. P. Dorato, "Case Studies in Robust Control Design." *IEEE Proceedings of the Decision and Control Conference*, December 1990, pp. 2030–3031.
13. B. C. Kuo, *Automatic Control Systems*. 7th ed., Prentice-Hall, Englewood Cliffs, NJ, 1995.

14. P. A. Ioannou and J. Sun, *Robust Adaptive Control*. Prentice-Hall Professional Technical Reference, Des Moines, IA, 1995.
15. J. Ackermann, "Robust Control Prevents Car Skidding." *IEEE Control Systems*, **17**, pp. 23–31 (1997).
16. G. Zames, "Feedback and Optimal Sensitivity; Model Reference Transformations, Multiplicative Seminorms, and Approximate Inverses," *IEEE Transactions on Automatic Control*, **AC-26**, 301–320 (1981).
17. P. R. Belanger, *Control Engineering*. Saunders College Publishing, Harcourt Brace and Company, Fort Worth, 1995.
18. Z. Gajic and M.Lelic, *Modern Control System Engineering*. Prentice-Hall, 1996.
19. B. A. Francis, J. W. Helton, and G. Zames, "H^{∞}-Optimal Feedback Controllers for Linear Multivariable Systems. *IEEE Trans. Automatic Control*, **AC-29**, 888–900 (1984).
20. J. C. Doyle, K. Glover, P. O. Khargonekhav, and B. A. Francis,"State Space Solutions to Standard H^2 and H^{∞} Control Problems." *IEEE Trans. Automatic Control*, **34**, 831–847 (1989).
21. A. J. Laub, "Numerical Linear Algebra Aspects of Control Design Computations." *IEEE Trans. Automatic Control*, **AC-30**, 97–108 (1985).
22. K. J. Åström and B. Wittenmark, *Computer-Controlled Systems, Theory and Design*. Prentice-Hall, Englewood Cliffs, NJ, 1990.
23. A. J. Laub, "Efficient multivariable frequency response calculations." *IEEE Trans. Automatic Control*, **AC-26**, 407–408 (1981).
24. C. B. Moler and G. W. Stewart, "An algorithm for generalized matrix eigenvalue problems." *SIAM J. Numer. Anal.*, **10**, 241–256 (1973).

APPENDIX A

LAPLACE-TRANSFORM TABLE

Laplace transform, $F(s)$	Time function, $f(t), t \geqslant 0$
1	$\delta(t_0)$, unit impulse at $t = t_0$
$\frac{1}{s}$	$U(t)$, unit step function
$\frac{1}{s^2}$	t
$\frac{2}{s^3}$	t^2
$\frac{n!}{s^{n+1}}$	t^n
$\frac{1}{s+a}$	e^{-at}
$\frac{1}{(s+a)(s+b)}$	$\frac{e^{-at} - e^{-bT}}{b-a}$
$\frac{1}{(s+a)^n}$	$\frac{1}{(n-1)!} t^{n-1} e^{-at}$
$\frac{s+\alpha}{(s+a)(s+b)}$	$\frac{1}{(b-a)}[(\alpha - e)e^{-at} - (\alpha - b)e^{-bt}]$
$\frac{1}{(s+a)(s+b)(s+c)}$	$\frac{e^{-at}}{(b-a)+(c-a)} + \frac{e^{-bt}}{(c-b)(a-b)} + \frac{e^{-ct}}{(a-c)(b-c)}$
$\frac{\omega}{s^2+\omega^2}$	$\sin \omega t$
$\frac{s}{s^2+\omega^2}$	$\cos \omega t$
$\frac{\omega}{(s+\omega)^2}$	$\omega t e^{-\omega t}$
$\frac{1}{(1+Ts)^n}$	$\frac{t^{n-1} e^{-t/T}}{T^n (n-1)!}$

Laplace transform, $F(s)$	Time function, $f(t), t \geqslant 0$
$\dfrac{1}{s(1+Ts)}$	$1 - e^{-t/T}$
$\dfrac{1}{s(1+Ts)^2}$	$1 - \dfrac{t+T}{T} e^{-t/T}$
$\dfrac{\omega}{(s+a)^2+\omega^2}$	$e^{-at} \sin \omega t$
$\dfrac{(s+a)}{(s+a)^2+\omega^2}$	$e^{-at} \cos \omega t$
$\dfrac{\omega^2}{s(s^2+2\zeta\omega_n s+\omega_n^2)}$	$1 - \dfrac{e^{-\zeta\omega_n t}}{\sqrt{1-\zeta^2}} \sin(\omega_n \sqrt{1-\zeta^2}\, t + \alpha)$, where $\cos \alpha = \zeta$
$\dfrac{\omega_n^2}{s^2(s^2+2\zeta\omega_n s+\omega_n^2)}$	$t - \dfrac{2\zeta}{\omega_n} + \dfrac{1}{\omega_n\sqrt{1-\zeta^2}} e^{-\zeta\omega_n t} \sin(\omega_n\sqrt{1-\zeta^2}\, t - \theta)$, where $\theta = 2\tan^{-1} \dfrac{\sqrt{1-\zeta^2}}{-\zeta}$
$\dfrac{s}{(1+Ts)(s^2+\omega_n^2)}$	$\dfrac{-1}{(1+T^2\omega_n^2)} e^{-t/T} + \dfrac{1}{\sqrt{1+T^2\omega_n^2}} \cos(\omega_n t - \theta)$, where $\theta = \tan^{-1} \omega_n T$
$\dfrac{s}{(s^2+\omega_n^2)^2}$	$\dfrac{1}{2\omega_n} t \sin \omega_n t$
$\dfrac{1}{(s+b)\left[(s+a)^2+\omega^2\right]}$	$\dfrac{e^{-bt}}{(b-a)^2+\omega^2} + \dfrac{e^{-at}\sin(\omega t-\theta)}{\omega\left[(b-a)^2+\omega^2\right]^{1/2}}$, where $\theta = \tan^{-1} \dfrac{\omega}{b-a}$
$\dfrac{2abs}{\left[s^2+(a+b)^2\right]\left[s^2+(a-b)^2\right]}$	$\sin at \sin bt$
$\dfrac{1+as+bs^2}{s^2(1+T_1 s)(1+T_2 s)}$	$t + (a - T_1 - T_2) + \dfrac{b - aT_1 + T_1^2}{T_1 - T_2} e^{-t/T_1} - \dfrac{b - aT_2 + T_2^2}{T_1 - T_2} e^{-t/T_2}$

APPENDIX B

PROOF OF THE NYQUIST STABILITY CRITERION

The Nyquist stability criterion can be derived from Cauchy's residue theorem, which states that

$$\frac{1}{2\pi j}\int_C g(s)\mathrm{ds} = \sum \text{residues of } g(s) \text{ at the poles enclosed by the closed contour } C. \tag{B1}$$

Let us replace $g(s)$ by $f'(s)/f(s)$, where $f(s)$ is a function of s which is single valued on and within the closed contour C and analytic on C. Observe that the singularities of $f'(s)/f(s)$ occur only at the zeros and poles of $f(s)$. The residue may be found at each singularity with multiplicity of the order of zeros and poles taken into account. The residues in the zeros of $f(s)$ are positive and the residues in the poles of $f(s)$ are negative. Therefore, if $f(s)$ is not equal to zero along C, and if there are not at most a finite number of singular points that are all poles within the contour C, then

$$\frac{1}{2\pi j}\int_C \frac{f'(s)}{f(s)}ds = Z - P, \tag{B2}$$

where $Z =$ number of zeros of $f(s)$ within C, with due regrd for their multiplicity of order, and $P =$ number of poles of $f(s)$ within C, with due regard for their multiplicity of order. The left-hand side of Eq. (B2) may be written as

$$\frac{1}{2\pi j}\int_C \frac{f'(s)}{f(s)}ds = \frac{1}{2\pi j}\int_C d[\ln f(s)]. \tag{B3}$$

In general, $f(s)$ will have both real and imaginary parts along the contour C. Therefore, its logarithm can be written as

$$\ln f(s) = \ln |f(s)| + j \,\underline{/f(s)}. \tag{B4}$$

If we assume that $f(s)$ is not zero anywhere on the contour C, the integration of Eq. (B3) results in the expression

$$\frac{1}{2\pi j}\int_C d[\ln f(s)] = \frac{1}{2\pi j}[\ln|f(s)| + j\,\angle f(s)]_{s_1}^{s_2}, \tag{B5}$$

where, s_1 and s_2 denote the arbitrary beginning and end of the closed contour C as it is followed. Because $|f(s)|$ returns to its initial value in completing the closed curve,

$$\frac{1}{2\pi j}\int_C d[\ln f(s)] = \frac{1}{2\pi}\left[\angle f(s_2) - \angle f(s_1)\right]. \tag{B6}$$

Therefore, Eq. (B6) can be rewritten as

$$\frac{1}{2\pi j}\int_C \frac{f'(s)}{f(s)}ds = \frac{1}{2\pi} \times \text{net change in angle of } f(s) \text{ as } s \text{ is varied over the contour } C. \tag{B7}$$

By equating Eqs. (B2) and (B7), we obtain

$$Z - P = \frac{1}{2\pi} \times \text{net change in angle of } f(s) \text{ as } s \text{ is varied over the contour } C. \tag{B8}$$

Equation (B8) states that the excess of zeros over poles of $f(s)$ within the contour C equals $1/2\pi$ times the net change in angle (i.e., equals the number of net encirclements of the origin) of $f(s)$ as s is varied over the contour C. Let N be this number. Then Eq. (B8) can be written as

$$Z - P = N, \tag{B9}$$

where the contour C is traversed in a clockwise direction, and where an encirclement is defined as being positive if it also is in a clockwise direction. The number Z must be zero for the system to be stable.

This relationship, which is known as Cauchy's principle of the argument, is the basis of Nyquist's stability criterion. To make use of this principle in applying Nyquist's stability criterion, let us consider the feedback control system of Figure B1. Let

$$f(s) = 1 + G(s)H(s)$$

and examine the number of times $f(s)$ encircles the origin as s traverses the contour of Figure B2. (It is assumed that there are no poles on the imaginary axis, except at

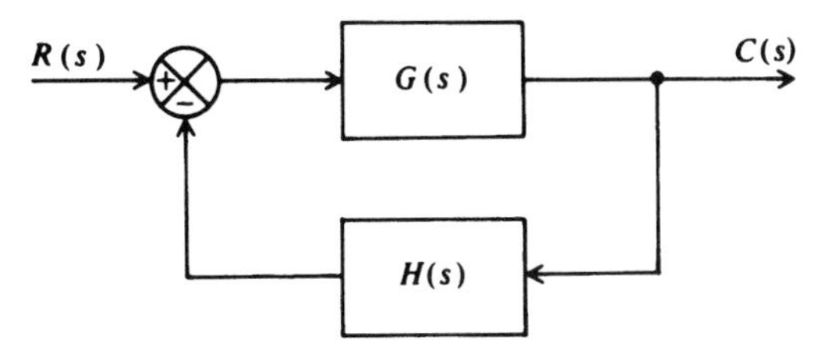

Figure B1

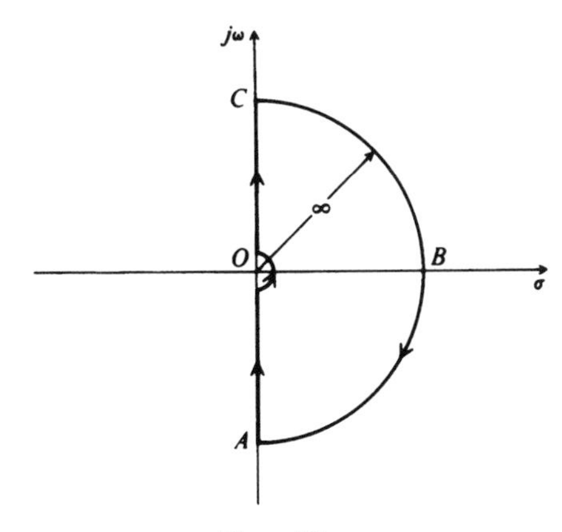

Figure B2

the origin.) Observe that the origin of $f(s)$ corresponds to $G(s)H(s) = -1$. Therefore, if $G(s)H(s)$ is plotted for the contour of Figure B2, the number of times that $G(s)H(s)$ encircles the point $-1 + j0$ equals the number of zeros minus the number of poles of $1 + G(s)H(s)$ for s in the right half-plane.

Let us examine the contour of Figure B2 in detail. Because

$$\lim_{s\to\infty} |G(s)H(s)| = 0 \tag{B10}$$

for all physical systems, the points along the infinite arc ABC contract to the origin for $\omega = 0$. Along the small semicircle about the origin, let

$$s = re^{j\theta} = \sigma + j\omega. \tag{B11}$$

Because $G(s)H(s)$ is usually a rational function with a denominator of higher power than that of the numerator, then

$$G(s)H(s) = \lim_{r\to 0} G(re^{j\theta})H(re^{j\theta}) \tag{B12}$$

will map into a segment of an infinite circle. Along the imaginary axis, $s = j\omega$, so that we are concerned with $G(j\omega)H(j\omega)$. Because $G(-j\omega)H(-j\omega)$ is a conjugate of $G(j\omega)H(j\omega)$, the two functions are symmetric about the real axis and

$G(-j\omega)H(-j\omega)$ is a reflection of $G(j\omega)H(j\omega)$ about the real axis. Therefore, it is only necessary to plot $G(j\omega)H(j\omega)$ from $\omega = 0$ to ∞.

Therefore, N can be determined by plotting $G(j\omega)H(j\omega)$ and observing the number of encirclements of the $-1 + j0$ point; P can be determined by inspection of the $G(s)H(s)$ expression; and Z can be found by using Eq. (B9).

ANSWERS TO SELECTED PROBLEMS

CHAPTER 1

1.1. An electrical signal proportional to the difference between the desired heading (gyroscopic setting) and original heading is amplified by the power amplifier. The amplified signal drives the motor which turns the rudder until the desired heading and actual heading of the ship are in agreement, and the corresponding electrical signal that is proportional to the difference between these two headings is zero. This is an open-loop control system as there is no feedback signal.

1.2. The rudder's positioning can be made into a closed-loop control system by fastening a resistor to the rudder in a similar manner to the resistor that is fixed to the ship's frame. An electrical feedback signal can then be obtained of the actual position, which can be appropriately compared with that of the desired, or reference, position.

1.3. If the reference temperature of the thermostat is changed, the reference input to the control system changes, and an electrical error signal resuls. The electric hot-water heater will then change the temperature of the water until the difference between the reference input and actual temperatures is zero.

1.4. A change in the ambient temperature surrounding the tank manifests itself as a disturbance input within the heating control system. The explanation of the system's resulting control action is similar to that discussed in the book for a disturbance occurring in an automatic speed-control system (see Figure 1.10).

1.6. The speed of an internal combustion engine can be varied by adjusting the spark setting and fuel-air mixture. By utilizing a tachometer to feed back an electrical signal proportional to the internal combusion engine's speed and comparing it with a reference voltage that is proportional to the desired speed, the spark setting and fuel-air mixture can be theoretically adjusted to control the speed of the engine.

1.7. **(a)** The basic system would require that the elevator's actuating signal be proportional to posiiton, velocity, and acceleration. Theoretically, this could be obtained by feeding back electrical signals proportional to position, velocity, and acceleration, and comparing these signals with reference signals representing the desired values. In practice, this can be simplified by placing integrators properly within the control system in order that only electrical signals proportional to position, and perhaps velocity, need be sensed.

(b) Specifications must be placed on the velocity and acceleration capabilities in order that they be limited to safe values from the passenger's viewpoint.

1.8. Assuming that safety brakes are not utilized in the feedback control system devised, the man entering the elevator acts as a disturbance force in the feedback control system. The functioning of the control system's resulting action is similar to that discussed in Chapter 1 for a disturbance torque occuring in an automatic speed-control system (see Figure 1.10).

1.17. (Part 4) See Figure A1.17.

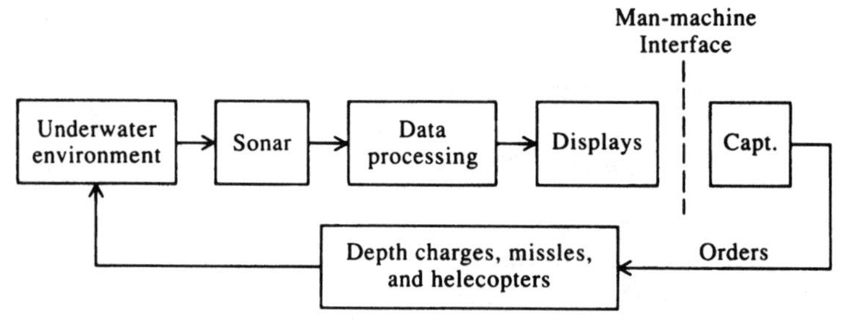

Figure A1.17

1.19. See Figure A1.19.

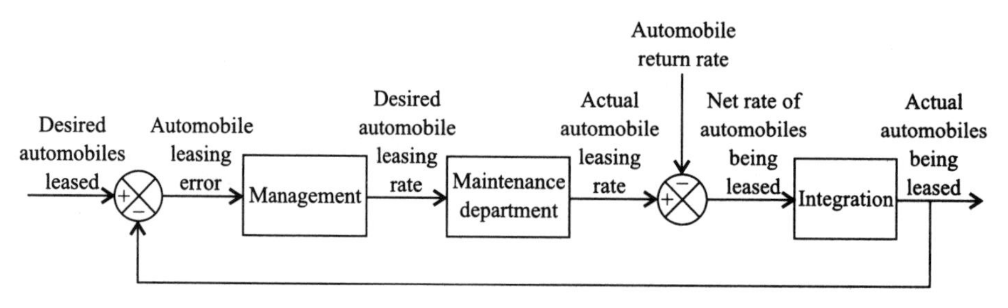

Figure A1.19

CHAPTER 2

2.2. $v(t) = 2.5(1 - e^{-6t})U(t)$.

2.5. **(a)** $Y(s) = \dfrac{16s^2 + 60s + 4}{s(2s+3)(s+1)}$,

(b) $Y(s) = \dfrac{1.33}{s} - \dfrac{66.7}{2s+3} + \dfrac{40}{s+1}$,

(c) $y(t) = (1.33 - 33.3e^{-1.5t} + 40e^{-t})U(t)$.

2.8. The final-value theorem cannot be used for this function to determine its final value, because $F(s)$ is not analytic in the right half-plane (pole exists at $s = 2$).

2.9. **(a)** $f_A(t) = (-5te^{-2t} - 10e^{-et} - 0.0315e^{-10.48t} + 10.0315e^{-1.52t})U(t)$,

(b) $f_B(t) = (-\frac{5}{6}e^{-4t} + \frac{3}{2}e^{4t} - \frac{2}{3}e^{-t})U(t)$,

(c) $f_C(t) = (0.5 - 0.25e^{-7.47t} - 0.250e^{-0.53t})U(t)$.

2.13. Output of first system: $-2e^{-2t}U(t)$.

Output of second system: $(4e^{-2t} - 8te^{-2t})U(t)$.

2.18.

$$\frac{C(s)}{R(s)} = \frac{\begin{matrix} G_1(s)G_2(s)G_3(s)G_5(s)G_6(s) \\ +G_1(s)G_7(s)G_5(s)G_6(s)[1 + G_3(s)H_3(s)] \\ +G_1(s)G_2(s)G_3(s)G_8(s)[1 + G_5(s)H_5(s)] \end{matrix}}{\begin{matrix} 1 + G_3(s)H_3(s) + G_5(s)H_5(s) \\ +G_2(s)G_3(s)G_4(s)G_5(s)G_6(s)H_2(s) \\ +G_7(s)G_5(s)G_6(s)H_2(s) \\ +G_2(s)G_3(s)G_8(s)H_2(s) \\ +G_3(s)H_3(s)G_5(s)H_5(s) \\ +G_3(s)H_3(s)G_7(s)G_5(s)G_6(s)H_2(s) \\ +G_5(s)H_5(s)G_2(s)G_3(s)G_8(s)H_2(s) \end{matrix}}$$

2.20.

(a)

$$\frac{C(s)}{R(s)} = \frac{G_0(s)[1 + G_4(s)H_2(s)] + G_1(s)G_2(s)G_3(s)G_4(s)G_5(s)}{1 + G_1(s)G_2(s)H_1(s) + G_4(s)H_2(s) + G_2(s)G_3(s)G_4(s)G_5(s)H_3(s) + G_1(s)G_2(s)H_1(s)G_4(s)H_2(s)}$$

(b)

$$\frac{E(s)}{R(s)} = \frac{1 + G_4(s)H_2(s) + G_2(s)G_3(s)G_4(s)G_5(s)H_3(s)}{1 + G_1(s)G_2(s)H_1(s) + G_4(s)H_2(s) + G_2(s)G_3(s)G_4(s)G_5(s)H_3(s) + G_1(s)G_2(s)H_1(s)G_4(s)H_2(s)}$$

2.21.

(a)

$$\frac{C_1(s)}{R_2(s)} = \frac{G_5(s)G_9(s)G_3(s)G_4(s)}{\begin{aligned}1 &+ G_2(s)G_3(s)G_4(s)H_1(s)\\ &+ G_6(s)G_7(s)G_8(s)H_2(s)\\ &+ G_9(s)G_{10}(s)G_7(s)G_8(s)H_2(s)\\ &+ G_2(s)G_3(s)G_4(s)H_1(s)G_6(s)G_7(s)G_8(s)H_2(s)\end{aligned}}$$

$$\frac{C_2(s)}{R_2(s)} = \frac{G_5(s)G_6(s)G_7(s)G_8(s)[1 + G_2(s)G_3(s)G_4(s)H_1(s)] + G_5(s)G_9(s)G_{10}(s)G_7(s)G_8(s)}{\begin{aligned}1 &+ G_2(s)G_3(s)G_4(s)H_1(s)\\ &+ G_6(s)G_7(s)G_8(s)H_2(s)\\ &+ G_9(s)G_{10}(s)G_7(s)G_8(s)H_2(s)\\ &+ G_2(s)G_3(s)G_4(s)H_1(s)G_6(s)G_7(s)G_8(s)H_2(s)\end{aligned}}$$

(b)

$$\frac{C_1(s)}{R_1(s)} = \frac{G_1(s)G_2(s)G_3(s)G_4(s)[1 + G_6(s)G_7(s)G_8(s)H_2(s)]}{\begin{aligned}1 + G_2(s)G_3(s)G_4(s)H_1(s) + G_6(s)G_7(s)G_8(s)H_2(s) + G_9(s)G_{10}(s)G_7(s)G_8(s)H_2(s) + \ldots\\ \ldots + [G_2(s)G_3(s)G_4(s)H_1(s)]\,[G_6(s)G_7(s)G_8(s)H_2(s)]\end{aligned}}$$

$$\frac{C_2(s)}{R_1(s)} = \frac{G_1(s)G_2(s)G_{10}(s)G_7(s)G_8(s)}{\begin{aligned}1 + G_2(s)G_3(s)G_4(s)H_1(s) + G_6(s)G_7(s)G_8(s)H_2(s) + G_9(s)G_{10}(s)G_7(s)\ldots\\ \ldots G_8(s)H_2(s) + [G_2(s)G_3(s)G_4(s)H_1(s)]\,[G_6(s)G_7(s)G_8(s)H_2(s)]\end{aligned}}$$

2.25.

$$\frac{C(s)}{R(s)} = \frac{G_1(s)G_2(s)G_3(s)G_4(s)}{\begin{aligned}[1 &+ G_2(s)G_3(s)H_3(s) + G_3(s)G_4(s)H_4(s)\\ &+ G_1(s)G_2(s)G_3(s)H_2(s)\\ &- G_1(s)G_2(s)G_3(s)G_4(s)H_1(s)].\end{aligned}}$$

2.26.

$$\frac{E(s)}{R(s)} = \frac{1 + G_2(s)G_3(s)H_3(s) + G_3(s)G_4(s)H_4(s)}{\begin{aligned}[1 &+ G_2(s)G_3(s)H_3(s) + G_3(s)G_4(s)H_4(s)\\ &+ G_1(s)G_2(s)G_3(s)H_2(s)\\ &- G_1(s)G_2(s)G_3(s)G_4(s)H_1(s)].\end{aligned}}$$

2.28

$$\frac{E_5(s)}{E_2(s)} = \frac{G_2(s)[1 + G_1(s)H_1(s) + G_3(s)H_3(s) + G_4(s)H_4(s) + G_1(s)H_1(s)G_3(s)H_3(s) + G_1(s)H_1(s)G_4(s)H_4(s)+}{\Delta}\ldots$$

$$\ldots\frac{+G_3(s)H_3(s)G_4(s)H_4(s) + G_1(s)H_1(s)G_3(s)H_3(s)G_4(s)H_4(s)]}{\Delta}.$$

where (all parameters in Δ are a function of s):

$$\begin{aligned}\Delta = 1 &- (-G_1H_1 - G_2H_2 - G_3H_3 - G_4H_4 - G_2G_3H_5 - G_1G_2G_3G_4H_0)\\ &+((G_1, H_1)(G_2H_2) + (G_1H_1)(G_3H_3) + (G_1H_1)(G_4H_4) + (G_2H_2)(G_3H_3)\\ &+(G_2H_2)(G_4H_4) + (G_3H_3)(G_4H_4) + (G_1H_1)(G_2G_3H_5)\\ &+(G_4H_4)(G_2G_3H_5)) - (-G_1H_1)(-G_2H_2)(-G_3H_3) + (-G_2H_2)(-G_3H_3)(-G_4H_4)\\ &+(-G_1H_1)(-G_3H_3)(-G_4H_4) + (-G_1H_1)(-G_2H_2)(-G_4H_4)\\ &+(-G_1H_1)(-G_2G_3H_5)(-G_4H_4)) + ((G_1H_1)(G_2H_2)(G_3H_3)(G_4H_4))\end{aligned}$$

2.35.

(a)

$$\frac{C(s)}{D(s)} = \frac{1 - G_d(s)G_1(s)G_2(s)}{1 + G_1(s)G_2(s)}.$$

(b)

$$G_d(s) = \frac{1}{G_1(s)G_2(s)}.$$

2.39. One possible solution to this problem is shown in Figure A2.39.

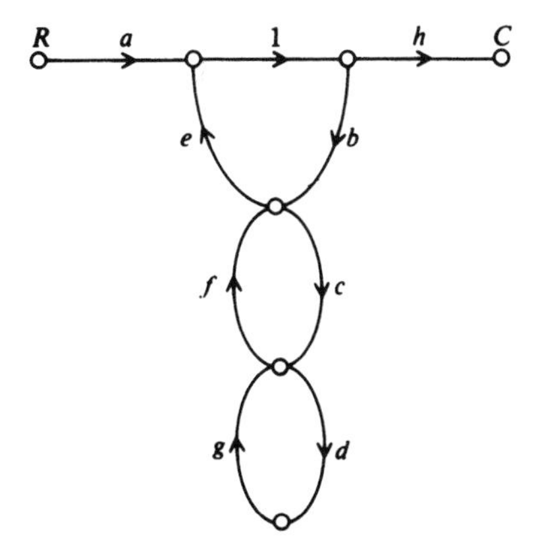

Figure A2.39

2.42. (a) Defining $x_1(t) = c(t)$ and $x_2(t) = \dot{c}(t)$, the plant dynamics become

$$\dot{x}_1(t) = x_2(t), \quad \dot{x}_2(t) = 2x_2(t) - x_1(t),$$

or in the vector/matrix form

$$\dot{\mathbf{x}}(t) = \mathbf{P}\mathbf{x}(t), \quad \mathbf{c}(t) = \mathbf{L}\mathbf{x}(t),$$

where

$$\mathbf{x}(t) = \begin{bmatrix} x_1t) \\ x_2(t) \end{bmatrix}, \quad \dot{\mathbf{x}}(t) = \begin{bmatrix} \dot{x}_1(t) \\ \dot{x}_2(t) \end{bmatrix}, \quad \mathbf{P} = \begin{bmatrix} 0 & 1 \\ -1 & -2 \end{bmatrix},$$
$$\mathbf{L} = [1 \quad 0].$$

(b) With $x_1(t) = c(t)$ and $x_2(t) = \dot{c}(t)$, the plant dynamics become

$$\dot{x}_1(t) = x_2(t), \quad \dot{x}_2(t) = -2x_2(t) - x_1(t) + A,$$

or in vector/matrix form

$$\dot{\mathbf{x}}(t) = \mathbf{P}\mathbf{x}(t) + \mathbf{B}\mathbf{u}(t), \quad \mathbf{c}(t) = \mathbf{L}\mathbf{x}(t),$$

where

$$\mathbf{x}(t) = \begin{bmatrix} x_1(t) \\ x_2(t) \end{bmatrix}, \quad \dot{\mathbf{x}}(t) = \begin{bmatrix} \dot{x}_1(t) \\ \dot{x}_2(t) \end{bmatrix}, \quad \mathbf{P} = \begin{bmatrix} 0 & 1 \\ -1 & -2 \end{bmatrix},$$
$$\mathbf{B} = \begin{bmatrix} 0 & 0 \\ 0 & 1 \end{bmatrix}, \quad \mathbf{u}(t) = \begin{bmatrix} 0 \\ A \end{bmatrix},$$
$$\mathbf{L} = [1 \quad 0].$$

(c) With $x_1(t) = c(t), x_2(t) = \dot{c}(t)$, and $x_3(t) = \ddot{c}(t)$, the plant dynamics becomes

$$\begin{aligned}\dot{x}_1(t) &= x_2(t),\\ \dot{x}_2(t) &= x_3(t),\\ \dot{x}_3(t) &= -2x_1(t) - 2x_2(t) - 3x_3(t),\end{aligned}$$

or, in vector/matrix form,

$$\dot{\mathbf{x}}(t) = \mathbf{P}\mathbf{x}(t), \quad \mathbf{c}(t) = \mathbf{L}\mathbf{x}(t),$$

where

$$\mathbf{x}(t) = \begin{bmatrix} x_1(t) \\ x_2(t) \\ x_3(t) \end{bmatrix}, \quad \dot{\mathbf{x}}(t) = \begin{bmatrix} \dot{x}_1(t) \\ \dot{x}_2(t) \\ \dot{x}_3(t) \end{bmatrix}, \quad \mathbf{P} = \begin{bmatrix} 0 & 1 & 0 \\ 0 & 0 & 1 \\ -2 & -2 & -3 \end{bmatrix},$$
$$\mathbf{L} = [1 \quad 0 \quad 0].$$

(d) With $x_1(t) = c(t)$, $x_2(t) = \dot{c}(t)$, and $x_3(t) = \ddot{c}(t)$, the plant dynamics become

$$\begin{aligned}\dot{x}_1(t) &= x_2(t),\\ \dot{x}_2(t) &= x_3(t),\\ \dot{x}_3(t) &= -2x_1(t) - 2x_2(t) - 3x_3(t) + A,\end{aligned}$$

or, in vector/matrix form,

$$\dot{\mathbf{x}}(t) = \mathbf{P}\mathbf{x}(t) + \mathbf{B}\mathbf{u}(t), \quad \mathbf{c}(t) = \mathbf{L}\mathbf{x}(t),$$

where

$$\mathbf{x}(t) = \begin{bmatrix} x_1(t) \\ x_2(t) \\ x_3(t) \end{bmatrix}, \quad \dot{\mathbf{x}}(t) = \begin{bmatrix} \dot{x}_1(t) \\ \dot{x}_2(t) \\ \dot{x}_3(t) \end{bmatrix}, \quad \mathbf{P} = \begin{bmatrix} 0 & 1 & 0 \\ 0 & 0 & 1 \\ -2 & -2 & -3 \end{bmatrix},$$

$$\mathbf{B} = \begin{bmatrix} 0 & 0 & 0 \\ 0 & 0 & 0 \\ 0 & 0 & 1 \end{bmatrix}, \quad \mathbf{u}(t) = \begin{bmatrix} 0 \\ 0 \\ A \end{bmatrix}, \quad \mathbf{L} = [1 \quad 0 \quad 0].$$

2.43. With $x_1(t) = \theta_1(t)$, $x_2(t) = \dot{\theta}_1(t)$, $x_3(t) = \theta_2(t)$, $x_4(t) = \dot{\theta}_2(t)$, $x_5(t) = \theta_3(t)$, and $x_6(t) = \dot{\theta}_3(t)$, the plant dynamics become

$$\begin{aligned} \dot{x}_1(t) &= x_2(t), \\ \dot{x}_2(t) &= -\omega_0 x_6(t) + L_1/I, \\ \dot{x}_3(t) &= x_4(t), \\ \dot{x}_4(t) &= L_2/I, \\ \dot{x}_5(t) &= x_6(t), \\ \dot{x}_6(t) &= \omega_0 x_2(t) + L_3/I, \end{aligned}$$

or in vector/matrix form

$$\dot{\mathbf{x}}(t) = \mathbf{P}\mathbf{x}(t) + \mathbf{B}\mathbf{u}(t),$$

where

$$\mathbf{P} = \begin{bmatrix} 0 & 1 & 0 & 0 & 0 & 0 \\ 0 & 0 & 0 & 0 & 0 & -\omega_0 \\ 0 & 0 & 0 & 1 & 0 & 0 \\ 0 & 0 & 0 & 0 & 0 & 0 \\ 0 & 0 & 0 & 0 & 0 & 1 \\ 0 & \omega_0 & 0 & 0 & 0 & 0 \end{bmatrix},$$

$$\mathbf{B} = \begin{bmatrix} 0 & 0 & 0 & 0 & 0 & 0 \\ 0 & 1/I & 0 & 0 & 0 & 0 \\ 0 & 0 & 0 & 0 & 0 & 0 \\ 0 & 0 & 0 & 1/I & 0 & 0 \\ 0 & 0 & 0 & 0 & 0 & 0 \\ 0 & 0 & 0 & 0 & 0 & 1/I \end{bmatrix}, \quad \mathbf{u}(t) = \begin{bmatrix} 0 \\ L_1 \\ 0 \\ L_2 \\ 0 \\ L_3 \end{bmatrix}.$$

2.45. With $x_1(t) = v(t)$ and $x_2(t) = \dot{v}(t)$, the plant dynamics become

$$\begin{aligned} \dot{x}_1(t) &= x_2(t), \\ \dot{x}_2(t) &= -x_1(t) + u(1 - x_1^2(t))x_2(t). \end{aligned}$$

2.47. The differential equation of the system is given by

$$\ddot{c}(t) + 4\dot{c}(t) + c(t) = r(t).$$

With $x_1(t) = c(t)$ and $x_2(t) = \dot{c}(t)$, the plant dynamics become

$$\dot{x}_1(t) = x_2(t), \quad \dot{x}_2(t) = -x_1(t) - 4x_2(t) + r(t),$$

or in vector/matrix form

$$\dot{\mathbf{x}}(t) = \mathbf{P}\mathbf{x}(t) + \mathbf{B}r(t),$$

where

$$\mathbf{P} = \begin{bmatrix} 0 & 1 \\ -1 & -4 \end{bmatrix}, \quad \mathbf{B} = \begin{bmatrix} 0 \\ 1 \end{bmatrix}, \quad \mathbf{x}(t) = \begin{bmatrix} x_1(t) \\ x_2(t) \end{bmatrix}, \quad \dot{\mathbf{x}}(t) = \begin{bmatrix} \dot{x}_1(t) \\ \dot{x}_2(t) \end{bmatrix}.$$

2.61. With $x_1(t) = c(t)$, $x_2(t) = \dot{c}(t)$, and $x_3(t) = \ddot{c}(t)$, the state equations become

$$\begin{aligned} \dot{x}_1(t) &= x_2(t), \\ \dot{x}_2(t) &= x_3(t), \\ \dot{x}_3(t) &= -400x_1(t) - 820x_2(t) - 412x_3(t) + 400r(t), \end{aligned}$$

or in vector/matrix form

$$\dot{\mathbf{x}}(t) = \mathbf{P}\mathbf{x} + \mathbf{B}r(t),$$

where

$$P = \begin{bmatrix} 0 & 1 & 0 \\ 0 & 0 & 1 \\ -400 & -820 & -412 \end{bmatrix}, \quad \mathbf{B} = \begin{bmatrix} 0 \\ 0 \\ 400 \end{bmatrix},$$

$$\mathbf{x}(t) = \begin{bmatrix} x_1(t) \\ x_2(t) \\ x_3(t) \end{bmatrix}, \quad \dot{x}(t) = \begin{bmatrix} \dot{x}_1(t) \\ \dot{x}_2(t) \\ \dot{x}_3(t) \end{bmatrix}.$$

The output equation is given by

$$c(t) = \mathbf{L}\mathbf{x}(t),$$

where

$$\mathbf{L} = [1 \quad 0 \quad 0].$$

2.64. $\Phi(t) =$

$$\begin{bmatrix} 1 & 7+0.00305e^{-6.86t}-7.3e^{-0.14t} & 1+0.0217e^{-6.86t}-1.065e^{-0.14t} \\ 0 & -0.0209e^{-6.86t}+1.02e^{-0.14t} & -0.149e^{-6.86t}+0.149e^{-0.14t} \\ 0 & 0.149e^{-6.86t}-0.149e^{-0.14t} & 1.02e^{-6.86t}-0.0208e^{-0.14t} \end{bmatrix} U(t).$$

2.75 (a) $$\Phi(t) = \begin{bmatrix} 0.2+0.8e^{-5t} & 0.4-0.4e^{-5t} \\ 0.4-0.4e^{-5t} & 0.8+0.2e^{-5t} \end{bmatrix} U(t),$$

(b) $x_1(t) = 44,000 + 156,000e^{-5t},$

$x_2(t) = 88,000 - 78,000e^{-5t},$

(c) $t = 0.334.$

2.77. (a) $$\Phi(t) = \begin{bmatrix} -2e^{-3t}+3e^{-2t} & -e^{-3t}+e^{-2t} \\ 6e^{-3t}-6e^{-2t} & 3e^{-3t}-2e^{-2t} \end{bmatrix} U(t),$$

(b) $c(t) = -40e^{-3t} + 50e^{-2t}, \quad t \geq 0$

$\dot{c}(t) = 120e^{-3t} - 100e^{-2t}, \quad t \geq 0.$

2.78. (a) $$\mathbf{\Phi}(t) = \begin{bmatrix} e^t & -2+2e^t \\ 0 & 1 \end{bmatrix}, t \geq 0$$

(b) $$\mathbf{x}(t) = \begin{bmatrix} -3+3e^t-2t \\ t \end{bmatrix}, t \geq 0$$

2.79. $$\mathbf{\Phi}(t) = \begin{bmatrix} e^{3t} & -1.333+1.333e^{3t} \\ 0 & U(t) \end{bmatrix}, t \geq 0$$

(b) $$\mathbf{x}(t) = \begin{bmatrix} -0.778+0.778e^{3t}-1.333t \\ t \end{bmatrix}, t \geq 0$$

CHAPTER 3

3.1. (a) $$f(t) = \frac{B_1 M_2}{K_2+K_3}\frac{d^3y(t)}{dt^3} + \left[M_2 + \frac{M_2K_1}{K_2+K_2}\right]\frac{d^2y(t)}{dt^2} + B_1\frac{dy(t)}{dt} + K_1y(t)$$

(b) $$\frac{Y(s)}{F(s)} = \frac{1}{\dfrac{B_1M_2}{K_2+K_3}s^3 + \left[M_2 + \dfrac{M_2K_1}{K_2+K_3}\right]s^2 + B_1s + K_1}.$$

3.2. (b)

$$\frac{Y(s)}{F(s)} = \frac{(B_1 s + K_1)(B_2 s + K_2)(B_3 s + K_3)}{[ABCD - AB(B_3 s + K_3)^2 - AD(B_2 s + K_2)^2 - CD(B_1 s + K_1)^2 + (B_1 s + K_1)^2(B_3 s + K_3)^2]},$$

where

$$\begin{aligned} A &= M_1 s^2 + B_1 s + K_1, \\ B &= M_2 s^2 + (B_1 + B_2)s + (K_1 + K_2), \\ C &= M_3 s^2 + (B_2 + B_3)s + (K_2 + K_3), \\ D &= M_4 s^2 + B_3 + K_3. \end{aligned}$$

3.3. (b)

$$\frac{\theta(s)}{T(s)} = \frac{(K_2 + s)^2 - X(s)Y(s)}{W(s)\big[(K_2 + B_2 s)^2 - X(s)Y(s)\big] + Y(s)(K_1 + s)^2},$$

where

$$\begin{aligned} W(s) &= J_1 s^2 + B_1 s + K_1, \\ X(s) &= J_2 s^2 + (B_1 + B_2)s + (K_1 + K_2), \\ Y(s) &= J_3 s^2 + (B_2 + B_3)s + (K_2 + K_3). \end{aligned}$$

3.4. (a)

$$\frac{d^2\theta_0(t)}{dt^2} + \frac{B}{J}\frac{d\theta_0(t)}{dt} + \frac{K}{J}\theta_0(t) = \frac{K'}{J}\frac{d\theta_i(t)}{dt},$$

(b)

$$\frac{\theta_0(s)}{\theta_i(s)} = \frac{(K'/J)s}{s^2 + (B/J)s + K/J}.$$

3.5.

$$\frac{\theta_0(s)}{E_f(s)} = \frac{9.43}{s(1.11s + 1)(0.227s + 1)(0.028s + 1)}.$$

3.7. Defining the following terms

$$\begin{aligned} T_f &= L_f/R_f, & \gamma'' &= B(R_g + R_m + R)R_f/K_T K_e, \\ T_a'' &= (L_g + L_m)/(R_g + R_m + R), & \gamma''' &= BK_g R/K_T K_e, \\ T_0 &= JK_g R/K_T K_e, & T_m'' &= J(R_g + R_m + R)R_f/K_T K_e. \end{aligned}$$

(a)

$$\begin{aligned} & T_m'' T_a'' T_f \frac{d^4\theta_0(t)}{dt^4} + [T_m''(T_a'' + T_f) + \gamma'' T_a'' T_f]\frac{d^3\theta_0(t)}{dt^3} \\ & + [T_m'' + \gamma''(T_a'' + T_f) + T_0 + L_f]\frac{d^2\theta_0(t)}{dt^2} \\ & + [\gamma'' + \gamma''' + R_f]\frac{d\theta_0(t)}{dt} = \frac{K_g}{K_e} e_f(t) \end{aligned}$$

(b)
$$\frac{\theta_0(s)}{E_f(s)} = \frac{K_g/K_e}{\begin{array}{l}\{s[T_m''T_a''T_f s^3 + [T_m''(T_a'' + T_f) + \gamma'' T_a'' T_f]s^2 \\ +[T_m'' + \gamma''(T_a'' + T_f) + T_0 + L_f]s + \gamma'' + \gamma''' + R_f]\}\end{array}}$$

3.8. (b)
$$\frac{e_b(s)}{E_a(s)} = \frac{\dfrac{R_2}{R_1 + R_2}\left[-T_a T_m s^2 - (T_m' + \gamma T_a)s + \dfrac{(R_1 + R_2)}{R_2} - \gamma'\right]}{T_a T_m s^2 + (T_n'' + \gamma T_a)s + (\gamma'' + 1)}$$

where

$$T_a = L_{CF}/R_{CF}, \quad T_m = JR_{CF}/K_e K_T, \quad \gamma = \frac{R_{CF}B}{K_e K_T},$$
$$T_m' = \frac{J[R_{CF} - (R_1/R_2)R_{AC}]}{K_e K_T}, \quad T_m'' = \frac{J[R_{CF} + R_{AC}]}{K_e K_T},$$
$$\gamma' = \frac{B[R_{CF} - (R_1/R_2)R_{AC}]}{K_e K_T}, \quad \gamma'' = \frac{B(R_{CF} + R_{AC})}{K_e K_T}.$$

3.11. $\dfrac{C(s)}{R(s)} =$

$$\frac{(1/V_m)(\partial Q/\partial r)}{[Vs/(K_B V_m^2)](Ms^2 + Bs + K) + (1/V_m^2)(L - \partial Q/\partial P)(Ms^2 + Bs + K) + s}.$$

3.13.
$$\frac{\theta_0(s)}{E_c(s)} = \frac{2.29}{s(0.0102s + 1)}.$$

3.14.
$$\frac{\theta_0(s)}{E_c(s)} = \frac{K_m'}{s(T_m's + 1)},$$

where

$$K_m' = \frac{K_e'}{B - K_n'}, \quad K_e' = NK_e,$$
$$K_n' = N^2 K_n$$
$$T_m' = \frac{J_{\text{total}}}{B - K_n'}, \quad J_{\text{total}} = J_{\text{motor}}N^2 + J_2.$$

3.15.
$$\frac{\theta_0(s)}{E_c(s)} = \frac{0.0636}{s(0.0102s + 1)}.$$

3.21. (a) MR^2.

(b)
$$T_D(t) = J_m \frac{d^2\theta_0(t)}{dt^2} + (M_p R^2)\frac{d^2\theta_0(t)}{dt^2} + B_m \frac{d\theta_0(t)}{dt}.$$

(c)
$$\frac{\theta_0(s)}{T_D(s)} = \frac{1}{(J_m + M_p R^2)s^2 + B_m s}.$$

(d)
$$X(t) = R\theta_0(t).$$

(e)
$$\frac{X(s)}{\theta_0(s)} = R.$$

(f)
$$\frac{X(s)}{T_D(s)} = \frac{R}{s[(J_m + M_p R^2)s + B_m]}.$$

3.22.
$$a_{LA} = 1.1111 a_{LB}.$$

Therefore, we conclude that System A has the higher initial load acceleration.

3.24.
$$\frac{\theta_0(s)}{E_c(s)} = \frac{0.365}{s(0.0543s + 1)}.$$

3.27.
$$\frac{\theta_0(s)}{E_c(s)} = \frac{0.606}{s(0.0572s + 1)}.$$

CHAPTER 4

4.1. (a) $\omega_n = 47.3$ rad/sec. $\zeta = 1.028$.

(b) Percent overshoot = 0, time to peak = ∞.

(c) The error as a function of time can be plotted from the following expression:

$$e(t) = 2.62e^{-37.2t} - 1.63e^{-60.2t}, \quad t \geqslant 0.$$

4.6 (a)
$$\frac{C(s)}{R(s)} = \frac{K_a q}{s^2 + \sqrt{q}K_r s + qK_a}.$$

(b) $\omega_n = 8$ rad/sec, $\zeta = 0.5$.

(c) Maximum percent overshoot = 16.4%, time to peak = 0.452 sec.

4.11. $K_a = 2.79$.

4.13. (a) $\omega_n = 4.47$ rad/sec.

(b) Damping ratio = 0.37.

(c) Maximum percent overshoot = 29%.

(d) $t_p = 0.76$. sec.

4.14. (a) $\omega = 4.47$ rad/sec.

(b) Damping ratio = 0.56.

(c) Maximum percent overshoot = 12%.

(d) $t_p = 0.85$ sec.

4.16.

$$K_m = 1.4591,$$

$$T_m = 0.05457 \text{ sec}.$$

4.19. **(c)** $c(t) = 1 + 0.1314e^{-1.9835t} - 1.1314e^{-1.9838t}$

CHAPTER 5

5.1. **(a)** $S_{K_1}^T = 1.$

(b) $S_{K_2}^T = -\dfrac{1}{10^{-3}s^2 + 10^{-3}s + 1}.$

(c) $S_G^T = \dfrac{s^2 + s}{s^2 + s + 1000}.$

5.2. **(a)**

$$S_{G_1}^T = \frac{s^3 + 182s^2 + 8200s}{s^3 + 182s^2 + 8200s + 2000}.$$

(b)

$$S_{G_2}^T = \frac{s^3 + 102s^2 + 200s}{s^3 + 182s^2 + 8200s + 2000}.$$

(c)

$$S_{H_2}^T = \frac{-(80s^2 + 8000s)}{s^3 + 182s^2 + 8200s + 2000}.$$

(d)

$$S_{K_3}^T = 1.$$

(e)

In the vicinity of $\omega = 1 : S_{G_2}^T, S_H^T, S_{G1}^T, S_{K_3}^T.$

5.6.

$$S_{T(s)}^{H(s)} = -\frac{Ts^2}{Ts^2 + s + K}.$$

5.17. **(a)** 4.

(b) Infinity.

5.20. **(a)**

$$e_{ss} = -\frac{1}{\lim\limits_{s \to 0} G(s) + b}.$$

(b)

$$G(s) = \frac{1}{s}.$$

5.24.

$$A = \frac{2\zeta}{\omega_n}.$$

5.26. **(a)** 1.

(b) Infinity.

(c) Add a second pure integration to $G(s)$.

5.27. **(a)** Zero.

(b) 2.4.

(c) Infinity.

5.30. **(a)** On the assumption that $J_L \gg J_m N^2$,

$$G(s) = \frac{C(s)}{E(s)} = \frac{2.1}{s(0.11s + 1)}.$$

(b) $\omega_n = 4.37$ rad/sec; $\zeta = 10.3$.

(c) 0%; ∞ sec.

(d) Zero.

(e) 0.476.

(f) Infinity.

5.38. $K_1 = 0.112, \quad \beta_1 = 0.77, \quad \beta_2 = 4.36.$

5.44. **(a)** $e_{ss} = 0$

(b) $e_{ss} = 0.1$

(c) $K = 20$

5.45. $K_m = 0.255$

CHAPTER 6

6.3. **(a)** $s^3 + s^2 + Ks + 4 = 0.$

(b) Range of K where the system is stable:

$$K > 4.$$

6.8. **(a)** Stable.

(b) Unstable with two poles in right half-plane.

(c) Stable.

(d) Stable.

6.13. $C + 0.3 - A > 0, \quad T - B + D > 0.$

6.14. **(a)** Unstable with two poles in right half-plane.

(b) Unstable with two poles in right half-plane.

(c) Unstable with two poles in right half-plane.

(d) Unstable with two poles in right half-plane.

6.18. System is unstable with two roots in the right half-plane.

6.19. (a) Unstable with two poles in right half-plane.

(b) Unstable with two poles in right half-plane.

(c) Stable.

6.25. (a) Pertinent characteristics of the Bode diagam are as follows:

1. Crossover frequency = 5.72 rad/sec.
2. Phase margin = $-10.53°$.
3. Gain margin = -4.44 dB.

(b) Pertinent characteristics of the Bode diagram are as follows:

1. Crossover frequency = 0.7216 rad/sec.
2. Phase margin = $-17.76°$.

(c) Pertinent characteristics of the Bode diagrams are as follows:

1. Crossover frequency = 2.935 rad/sec.
2. Phase margin = $-24.73°$.
3. Gain margin = $-\infty$.

(d) Pertinent characteristics of the Bode diagram are as follows:

1. Crossover frequency = 0.2792 rad/sec.
2. Phase margin = $16.24°$.
3. Gain margin = ∞.

6.27.

$$G(s)H(s) = \frac{100}{s(1+0.1s)}.$$

6.30. (a) Phase margin = $43.21°$, gain margin = 15.56 dB.

(b) Phase margin = $43.21°$, gain margin = 15.56 dB.

(d) $K_p = \infty,\ K_v = 2,\ K_a = 0.$

(e) 2.5 rad.

6.32. (b) Crossover frequency = 3.757 rad/sec, phase margin = $-8.891°$; gain margin = -3.098 dB, system is unstable.

(c) $K_p = \infty,\ K_v = 10,\ K_a = -0.$

(d) 4 rad.

6.33. (b) $K_a = 45.$

(c) Crossover frequency = 1.498 rad/sec, phase margin = $-8.522°$, gain margin = $-\infty$; system is unstable.

6.34. (a) Crossover frequency = 4.483 rad/sec, phase margin = 32.44°, gain margin = 8.657 dB.

(b) $K = 9.9$.

(c) Crossover frequency = 13.61 rad/sec, phase margin = $-30.62°$, gain margin = -11.26 dB.

6.35. (c) $\omega_p = 3.742$ rad/sec, $M_p = 6.30$ dB.

(d) No.

6.36. (c) $\omega_p = 2.472$ rad/sec, $M_p = 1.304$ dB.

(d) No.

6.37. (a) The Bode diagram indicates a crossover frequency of 23.51 rad/sec and a phase margin of 20.61°.

(c) $\omega_p = 23.46$ rad/sec, $M_p = 8.929$ dB.

(d) No.

6.42. (a) $K = 1.91$.

(b) $\omega_p = 2.472$ rad/sec, $M_p = 1.304$ dB.

6.43. (b) $M_p = 5.283$ dB, $\omega_p = 8.165$ rad/sec.

6.44. (b) System is unstable.

6.45. (a) Pertinent characteristics of the root are as follows:

1. Root locus occurs along the real axis between the origin and -1, and between -10 and $-\infty$.
2. The asymptotes intersect the real axis at -3.67 with angles of $\pm 60°$.
3. The point of breakaway of the root locus from the real axis occurs at -0.49.

(b) $K = 11$.

6.46. (a) The root locus lies along the imaginary axis.

(b) Pertinent characteristics of the root locus are as follows:

1. Root locus occurs along the real axis at the origin and between -1 and -10.
2. The asymptotes intersect the real axis at -4.5 with angles of $\pm 90°$.
3. The points of breakaway of the root locus from the real axis occur at 0, -2.5, and -4.
4. The root locus indicates that the system is stable.

(c) Pertinent characteristics of the root locus are as follows:

1. Root locus occurs along the real axis between the origin and -1, and at -4.
2. The asymptotes intersect the real axis at -3.5 with angles of $\pm 90°$.
3. The root locus indicates that the system is stable for all values of gain.

(d) Pertinent characteristics of the root locus are as follows:

1. Root locus occurs along the real axis at the origin, -0.1, and between -4 and -5.
2. The asymptotes interesect the real axis at -4.4 with angles of $\pm 90°$.
3. The points of breakaway of the root locus from the real axis occur at 0, -0.1, and -4.5.
4. The root locus indicates that the system is stable.

6.47. Pertinent characteristics of the root locus are as follows:

1. Root locus occurs along the real axis at the origin, between -0.01 and -0.1, and between -1.67 and $-\infty$.
2. The point of breakaway of the root locus from the real axis occurs at 0 and -3.4.
3. The root locus crosses the imaginary axis at $\pm j0.386$.
4. The root locus indicates that the system is stable for $0.14 < K < \infty$.

6.68. (a) Pertinent characteristics of the root locus are as follows:

1. Root locus occurs along the real axis between the origin and -2, and between -4 and -7.
2. Asymptotes intersect the real axis at 0.5 with angles of $\pm 90°$.
3. The point of breakaway of the root locus from the real axis occurs at -0.92.

(b) $K_{\max} = 4.8$ at $\pm 7.4833j$.

6.71. Pertinent characteristics of the root locus are as follows:

1. Root locus occurs along the real axis between the origin and -3.125, and from -8 to -100.
2. Asymptotes intersect the real axis at -47.56 with angles of $\pm 90°$.
3. The points of breakaway of the root locus from the real axis occur at -1.74 and at -46.22; a point of break-in occurs at -15.58.
4. The root locus indicates that the system is stable.

6.72. $K = 1.414$

6.73. **(a)** Asymptotes are $\pm 60^\circ$ which intersect the real axis at -2. Root locus exists on the real axis from -2 to minus infinity.

(b) $K_{\max} = 64$, and the root locus intersects the imaginary axis at $\pm j3.46$.

CHAPTER 7

7.1. **(a)** The electrical network illustrated in Table 2.4, items 3, with

$$R_1 = 690\,\mathrm{k}\Omega, \quad R_2 = 51\,\mathrm{k}\Omega, \quad C_1 = 1.38\mu\mathrm{F}$$

and cascaded with an amplifier whose gain is 4.6 will satisfy the specifications

(b) The electrical network illustrated in Table 2.4, item 4, with

$$R_1 = 690\,\mathrm{k}\Omega, \quad R_2 = 51\,\mathrm{k}\Omega, \quad C_2 = 1.38\mu\mathrm{F}$$

and cascaded with an amplifier whose gain is 4.6 will satisfy the specifications.

(c) The electrical network illustrated in Table 2.4, item 5, with

$$R_1 = 595\,\mathrm{k}\Omega, \quad R_2 = 25\,\mathrm{k}\Omega, \quad C_1 = 1.68\,\mu\mathrm{F} \quad C_2 = 4\,\mu\mathrm{F}$$

will satisfy the specifications.

7.2. **(a)** $\zeta = 0.5$; $\omega_n = 4$ rad/sec; maximum percent overshoot $=$ 16.4%; steady-state error $=$ 0.25.

(b) $b = 0.15$.

(c) Maximum percent overshoot $=$ 1.5%; steady-state error $=$ 0.4.

(d) Add a high-pass filter in cascade with the tachometer as illustrated in Figure 7.15.

7.5. **(a)** $G_c(s) = 1 + 0.382s$.

(b) $b = 0.382$.

7.7. $\omega_0 = \dfrac{1}{\sqrt{T_1 T_2}}$.

7.8. **(a)** $G_c(s) = \dfrac{0.5s + 1}{0.014s + 1}\,\dfrac{0.05s + 1}{0.01s + 1}$;

the attenuation of 1/178.6 due to the two lead networks is assumed to be compensated for by increasing the gain of the amplifier by 178.6. The result is a phase margin of 46.9° at a gain crossover frequency of 6.4 rad/sec, and a gain margin of 32.8 dB at a phase crossover frequency of 68 rad/sec.

(b) $\dfrac{12.5s + 1}{250s + 1}$.

7.11. A phase-lead network whose transfer function is

$$\frac{1}{15}\frac{1+s}{1+0.067s}$$

The result is a phase margin of 61.8° at a gain crossover frequency of 0.49 rad/sec, and a gain margin of 22.4 dB at a phase crossover frequency of 3.08 rad/sec, assuming that the attenuation of the phase-lead network of $\frac{1}{15}$ is compensated for by boosting the gain of the amplifier by a factor of 15. Therefore, this phase-lead network achieves the design specification of a minimum phase margin of 60°.

7.12. Same answer as for Problem 7.11.

7.14. Two cascaded phase-lead networks are required to meet these specifications. Their transfer function, together with the gain found in Problem 6.34, is given by

$$G_1(s) = 9.9\frac{1+0.2s}{1+0.01s}\frac{1+0.1s}{1+0.005s}.$$

The constant attenuation of 400 due to the characteristics of these two phase-lead networks must be made up by an amplification increase of 400. The resulting phase margin is 46.09° at a gain crossover frequency of 55.58 rad/sec.

7.15.

(a) $\omega_c = 3$ rad/sec.

(b) $$G_c(s) = \frac{1}{40}\frac{(1+0.904s)}{(1+0.0226s)}.$$

It is assumed that the attenuation of 1/40 is compensated for by increasing the gain of the amplifier. This phase-lead network provides a phase margin of 71.86° at a new gain crossover frequency of 8.078 radians/second.

7.20. (a) $G_c(s) = \dfrac{1+0.416s}{1+0.384s}$

The constant attenuation of 1.08 due to the characteristics of this phase-lead network must be made up by an amplification increase of 1.08.

(b) For part (a), $\omega_p = 2.375$ rad/sec, $M_p = 1.154$ dB.

7.21. (a) $G_c(s) = \dfrac{0.5s+1}{0.001s+1}\dfrac{0.05s+1}{0.001s+1};$

The attenuation due to the phase-lead network is assumed to be compensated for by increasing the gain of the amplifiers.

(b) For the phase-lead network, $M_p = 0.7513$ dB, $\omega_p = 3.864$ rad/sec.

7.23. The phase-lag network is given by

$$\frac{s+0.0357}{s+0.001}.$$

7.27. **(a)** $b = 0.25$.

(b) $e_{ss} = 0.3024$.

(c) The transfer function of the rate feedback path is given by

$$\frac{5s}{1+5s}0.25s.$$

(d) $e_{ss} = 0.0524$.

7.28. $K = 4.22$.

7.38. $K = 2.226$

7.40. **(a)** $K = 129.3$.

(b) $M_p = 4.24$ dB; $\omega_p = 10.67$ rad/sec.

7.41 $K = 25.57$.

CHAPTER 8

8.2.

(a) $x_1(t) = -4x_1(t) - h_1x_1(t) + x_2(t)$

$x_2(t) = -Kx_1(t) + Kr(t)$

(b) $s^2 + 4s + h_1s + K = 0$.

(c) $h_1 = 10$,

$K = 48$.

(d) $K_I = 64$,

$K_p = 10$.

(e) The proportional plus integral controller configuration of part (d) produces a zero term which would aid in the stability compensation of this control system. The controller configuration of part (a) does not have this feature, and it would be more difficult to stabilize.

8.6. The synthesized system is illustrated in Figure A8.6, where

$$K = 8, \quad h_1 = 1, \quad h_2 = \frac{5}{8}, \quad h_3 = \frac{5-\alpha}{8}.$$

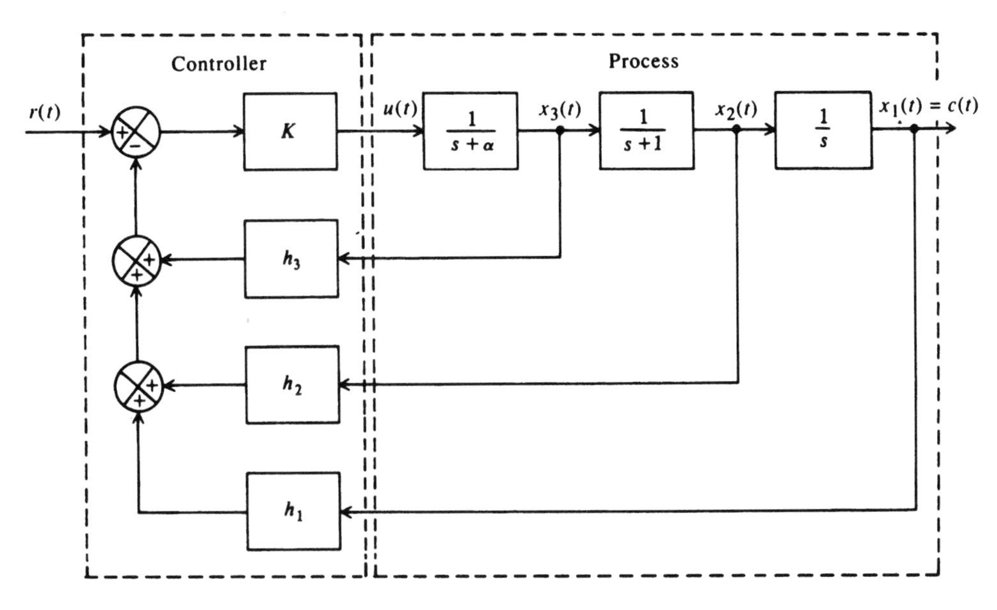

Figure A8.6

A root locus analysis indicates that a good choice of α is approximately 2.5. Therefore, h_3 becomes $\frac{5}{16}$.

8.8. **(a)** Yes.

(b) see Figure A8.8.

(c) $K = 20,000$,

$h_1 = 1$,

$h_2 = 0.00995$.

(d) Root locus shows that the system is stable.

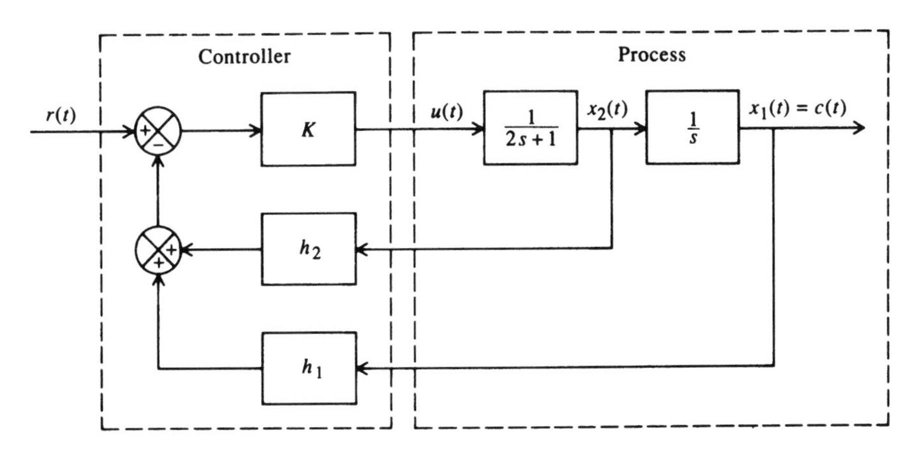

Figure A8.8

8.10. **(a)** $\mathbf{D} = \begin{bmatrix} 0 & 0 & 0 \\ 1 & -1 & 1 \\ 0 & 0 & 0 \end{bmatrix}$;

System is not controllable.

(b) $\mathbf{U} = \begin{bmatrix} 0 & 1 & -2 \\ 1 & -1 & 1 \\ 1 & 1 & 1 \end{bmatrix}$;

System is observable.

8.11. **(a)** $\mathbf{D} = \begin{bmatrix} 4 & -32 \\ 1 & -2 \end{bmatrix}$;

System is controllable.

(b) $\mathbf{U} = \begin{bmatrix} 1 & -8 \\ 0 & 0 \end{bmatrix}$;

System is unobservable.

8.13. $\mathbf{K} = \begin{bmatrix} K_1 \\ K_2 \end{bmatrix} = \begin{bmatrix} 12 \\ 48 \end{bmatrix}.$

8.15. $\mathbf{M} = \begin{bmatrix} m_1 \\ m_2 \end{bmatrix} = \begin{bmatrix} 59 \\ 840 \end{bmatrix}.$

8.17. $\mathbf{M} = \begin{bmatrix} m_1 \\ m_2 \end{bmatrix} = \begin{bmatrix} 34 \\ 590 \end{bmatrix}.$

8.20. $H(s) = \dfrac{Y(s)}{U(s)} = \dfrac{1}{s^2 + 1.333s + 0.889}$

INDEX